Signaling and Communication in Plants

Volume 2

Series Editor

František Baluška
Department of Plant Cell Biology, IZMB, University of Bonn, Kirschallee 1,
D-53115 Bonn, Germany

More information about this series at http://www.springer.com/series/8094

P. Vidhyasekaran

Plant Hormone Signaling Systems in Plant Innate Immunity

Springer

P. Vidhyasekaran
Plant Pathology
Tamil Nadu Agricultural University
Coimbatore, Tamil Nadu, India

ISSN 1867-9048 ISSN 1867-9056 (electronic)
ISBN 978-94-017-9284-4 ISBN 978-94-017-9285-1 (eBook)
DOI 10.1007/978-94-017-9285-1
Springer Dordrecht Heidelberg New York London

Library of Congress Control Number: 2014950145

Printed on acid-free paper

Springer is part of Springer Science+Business Media (www.springer.com)

Contents

1 Introduction 1

1.1 Plant Innate Immunity 2

1.2 Salicylic Acid Signaling 2

1.3 Jasmonate Signaling 4

1.4 Ethylene Signaling 5

1.5 Abscisic Acid Signaling 6

1.6 Auxin Signaling 7

1.7 Cytokinins 8

1.8 Gibberellins 8

1.9 Brassinosteroids 9

1.10 Plant Hormone Signaling Network 10

1.11 Can Molecular Manipulation of Plant Hormone Signaling Network Help the Plant to Win the War Against Pathogens? 12

References 13

2 Salicylic Acid Signaling in Plant Innate Immunity 27

2.1 Salicylic Acid as an Endogenous Immune Signal in Plants 29

2.2 Biosynthesis of Salicylic Acid in Plants 29

2.2.1 Phenylalanine Pathway 29

2.2.2 Isochorismate Pathway 31

2.2.3 Role of Regulatory Proteins (EDS1, EDS4, PAD4, EDS5, SID2) in Salicylic Acid Biosynthesis 31

2.2.4 An RNA-Binding Protein (RBP) May Be Involved in SA Biosynthesis Pathway 34

2.2.5 GH3.5 Is Involved in Salicylic Acid Biosynthesis 34

2.2.6 Role of CDR1 Gene in SA Biosynthesis 35

2.2.7 Role of FMO1 Gene in SA Biosynthesis Pathway 36

2.2.8 Cytokinin May Be Involved in Activation of Salicylic Acid Biosynthesis 36

2.2.9 Some Transcription Factors May Be Involved in Accumulation of Salicylic Acid 36

2.3 Upstream of Salicylic Acid Signaling System 39
2.3.1 G-Proteins Trigger Salicylic Acid Biosynthesis in SA Signaling System 39
2.3.2 Calcium Signaling May Act Upstream of Salicylic Acid Accumulation 40
2.3.3 MAP Kinases May Act Upstream of Salicylic Acid Accumulation 43
2.3.4 Reactive Oxygen Species May Act Upstream of Salicylic Acid Accumulation 44
2.3.5 Nitric Oxide May Act Upstream of Salicylic Acid Accumulation 45
2.4 Downstream Events in Salicylic Acid Signaling 46
2.4.1 Generation of Salicylic Acid Conjugates 46
2.4.2 ROS Signaling System May Act Downstream of SA Accumulation 47
2.4.3 NO May Act Downstream of SA Accumulation 48
2.4.4 MAPK Signaling Cascade May Act Downstream in SA Signaling System 48
2.5 SA Signaling Induces Increased Expression of Transcription Factors to Activate SA-Responsive Defense-Related Genes 49
2.5.1 SA Induces WRKY Transcription Factors 49
2.5.2 SA Induces ERF Transcription Factors 51
2.6 NPR1 Is Master Regulator of SA Signaling 52
2.6.1 NPR1 Acts Downstream of SA Signal 52
2.6.2 SA Controls Nuclear Translocation of NPR1 52
2.6.3 SA Modulates Proteasome-Mediated Degradation of NPR1 53
2.6.4 NPR1 Interacting Proteins 55
2.6.5 SA-Dependent NPR1-Activated Transcription Factors 55
2.6.6 SA-Induced Expression of PR Genes, Independent of NPR1 57
2.7 Role of SUMO in SA Signaling System 57
2.8 SA Induces Transcription of Various Defense Genes 58
2.9 Role of SA Signaling in Stomatal Closure-Related Immune Responses Against Bacterial Pathogens 58
2.10 SA Induces Resistance Against Viruses by Modulating AOX-Mediated Alternative Respiratory Pathway 59
2.11 SA Triggers Small RNA-Directed RNA Silencing System 60
2.12 Enhancement of Small RNA-Directed RNA Silencing by Salicylate Signaling System 62
2.13 Interplay Between SA-Induced AOX-Mediated Redox Signaling and SA-Induced Small RNA-Directed RNA Silencing 62
2.14 Salicylic Acid Signaling Is Involved in Induction of Systemic Acquired Resistance 64

2.15 Mobile Long-Distance Signals for Induction of Systemic Acquired Resistance 65
2.15.1 Search for Long-Distance Mobile Signal 65
2.15.2 Methyl Salicylate May Be a Mobile Signal 65
2.15.3 DIR1 and Glycerol-3-Phosphate-Dependent Factor Mobile Signal Complex 68
2.15.4 Azelaic Acid May Be a Mobile Signal 70
2.15.5 Dehydroabietinal as a Mobile Signal 70
2.15.6 Pipecolic Acid as an SAR Long-Distance Signal 71
2.16 Role of Mediator Complex in SA-Mediated Systemic Acquired Resistance 73
2.17 Salicylic Acid Triggers Priming and Induces Systemic Acquired Resistance 75
2.17.1 What Is SA-Triggered Priming? 75
2.17.2 Accumulation of Dormant MAPKs May Be Involved in SA-Triggered Priming 76
2.17.3 Histone Modifications May Be Involved in Gene Priming in SA-Induced SAR 77
2.17.4 NPR1 May Be Involved in Chromatin Modification-Induced Priming 78
2.17.5 Histone Replacement May Be Instrumental for Priming of SA-Responsive Loci 79
2.18 Next-Generation Systemic Acquired Resistance 80
2.19 Crosstalk Between Salicylate and Jasmonate Signaling Systems 81
2.19.1 Antagonism Between SA and JA Signaling Systems 81
2.19.2 SA May Block JA Biosynthesis 81
2.19.3 SA May Suppress JA-Responsive Gene Expression 82
2.19.4 NPR1 in the Cytosol Modulates Crosstalk Between SA and JA Signaling Systems 83
2.19.5 Role of Glutaredoxin and TGA Transcription Factors in the SA–JA Crosstalk 85
2.19.6 Role of MAP Kinase 4 (MPK4) in SA and JA Crosstalk 86
2.19.7 SA May Suppress JA Signaling by Targeting GCC-Box Motifs in JA-Responsive Promoters 87
2.19.8 JA May Inhibit SA Signaling 88
2.19.9 Synergism Between SA and JA Signaling Pathways 88
2.20 Crosstalk Between SA and ET Signaling Systems 89
2.21 Crosstalk Between SA and ABA Signaling Systems 89
2.22 Crosstalk Between SA and Auxin Signaling Systems 89
2.23 Negative Regulation of Salicylate-Mediated Immunity by Brassinosteroid Signaling 90
2.24 SA Signaling System May Induce Resistance Against a Wide Range of Pathogens 90

2.24.1 SA Signaling System Is Involved in Conferring Fungal and Oomycete Disease Resistance 90
2.24.2 SA Signaling System Is Involved in Conferring Bacterial Disease Resistance 91
2.24.3 SA Signaling System Is Involved in Conferring Virus Disease Resistance 93
2.25 Pathogens May Suppress SA Signaling System to Cause Disease.. 94
2.25.1 Pathogens May Secrete Effectors to Suppress SA Signaling System 94
2.25.2 Pathogen Produces Toxin and Suppresses SA Signaling System to Promote Disease Development 94
2.25.3 Pathogen Manipulates the Antagonistic Effect Between SA and JA Signaling Systems to Promote Disease Development 96
References 96

3 Jasmonate Signaling System in Plant Innate Immunity 123
3.1 Jasmonate Signaling System Is a Key Component in PAMP-Triggered Innate Immunity 124
3.2 Biosynthesis of Jasmonates 125
3.3 Jasmonate Biosynthesis Intermediate OPDA in Defense Signaling 126
3.4 JA Metabolites Involved in Defense Signaling 126
3.4.1 Methyl Jasmonate 126
3.4.2 Jasmonoyl-Isoleucine 128
3.5 Upstream of JA Biosynthesis 129
3.5.1 PAMP Triggers Enhanced Biosynthesis and Accumulation of JA 129
3.5.2 G-Proteins in the Induction of JA Biosynthesis 130
3.5.3 G-Proteins-Activated Polyamine Synthesis in Triggering JA Biosynthesis 131
3.5.4 Calcium Signature Triggers JA Biosynthesis 134
3.5.5 Role of ROS in JA Biosynthesis Pathway 135
3.5.6 Role of NO in JA Biosynthesis Pathway 135
3.5.7 Mitogen-Activated Protein Kinases Functioning Upstream in JA Biosynthesis Pathway 137
3.5.8 Systemin Triggers JA Biosynthesis in Tomato 138
3.6 Jasmonate Receptor Complex in JA Signal Perception 139
3.6.1 COI1, an F-Box Protein, Is a Jasmonate Receptor 139
3.6.2 COI1–JAZ Receptor Complex 140
3.6.3 InsP5 Potentiates JA Perception by COI1–JAZ1 Complex 140
3.6.4 JA-Ile Promotes Physical Interaction Between JAZ1 and COI1 140

3.7 JA Signaling Pathway 141
3.7.1 JAZ Proteins Suppress JA Signaling 141
3.7.2 Role of COI1 Protein in the Degradation of JAZ Proteins by E3 Ubiquitin Ligase 142
3.7.3 Role of JA-Ile in the JAZ Degradation by 26S Proteasome 143
3.7.4 MYC2, MYC3, and MYC4 Transcription Factors Regulate JA-Responsive Gene Expression 144
3.8 Mediator Complex Regulates Transcription of JA-Responsive Genes by Interacting with Transcription Factors 147
3.9 MAP Kinases May Regulate the Downstream Events in JA Signaling Pathway 150
3.10 Histone Acetylation May Regulate JA-Mediated Signaling Systems 152
3.11 JA-Induced Pep1 Peptide Amplifies JA Downstream Signaling to Induce JA-Responsive Genes 153
3.12 Transcription Factors Acting Downstream of JA in Defense Signaling System 154
3.13 JA Signaling System-Activated Defense Genes 156
3.14 JA Signaling System Triggers Immune Responses Against Necrotrophic Pathogens 157
3.15 JA and Ethylene Signaling Pathways May Operate Concomitantly in Plant Innate Immune System 158
3.15.1 Cooperative Function of JA and ET Signaling Pathways in Plant Innate Immunity 158
3.15.2 ERF Transcription Factors May Concurrently Modulate JA and ET Signaling Pathways in Plant Immune System 159
3.15.3 Role of Ethylene Transcription Factors EIN3 and EIL1 in JA/ET Signaling Synergy 160
3.15.4 Ethylene Signaling System May Protect JA Signaling System Against Its SA-Mediated Suppression 161
3.16 JA Signaling May Suppress SA Signaling System 162
3.17 Suppression of JA Signaling by SA Signaling System 163
3.17.1 SA Suppresses Biosynthesis of JA 163
3.17.2 SA Suppresses JA Signaling System by Targeting GCC-Box Motifs in JA-Responsive Promoters 164
3.17.3 Role of WRKY62 Transcription Factor in the Suppression of JA Signaling by SA 165
3.17.4 Role of WRKY70 and MYB Transcription Factors in the Suppression of JA Signaling by SA 166
3.17.5 WRKY50 and WRKY51 Transcription Factors May Modulate JA Signaling Suppression by SA 166
3.17.6 Role of TGA Transcription Factors in the Suppression of JA Signaling by SA 167

3.18 Interplay Between JA and Abscisic Acid Signaling Systems in Plant Immune Responses 168
3.19 Crosstalk Between JA Signaling and Small RNA Signaling Systems 169
3.20 JA Signaling in Induced Systemic Immunity 171
3.20.1 JA Signaling Plays Major Role in Induced Systemic Resistance 171
3.20.2 Mobile Signal Involved in Induced Systemic Resistance 173
3.20.3 Priming in Induced Systemic resistance 173
References 174

4 Ethylene Signaling System in Plant Innate Immunity 195
4.1 Ethylene Signaling Is an Important Component in Plant Innate Immunity 196
4.2 Ethylene Biosynthesis in Plants 197
4.2.1 Enzymes Involved in Ethylene Biosynthesis 197
4.2.2 Pathogen Infection Triggers Enhanced Expression of Ethylene Biosynthesis Genes 199
4.2.3 PAMPs/HAMPs Induce Expression of ET Biosynthesis Genes and Trigger ET Biosynthesis 199
4.2.4 G-Proteins May Trigger Ethylene Biosynthesis Pathway 200
4.2.5 Role of Ca^{2+} Influx-Mediated Ca^{2+} Signature in Ethylene Biosynthesis 201
4.2.6 Role of Calcium-Dependent Protein Kinase (CDPK) in Induction of Ethylene Biosynthesis 202
4.2.7 Reactive Oxygen Species May Trigger Transcription of Ethylene Biosynthesis Genes 202
4.2.8 Nitric Oxide May Trigger Activation of Ethylene Biosynthesis Enzymes 204
4.2.9 MAP Kinase Cascades May Induce Biosynthesis of Ethylene 204
4.2.10 Role of Ubiquitin–Proteasome in Ethylene Biosynthesis 208
4.3 Ethylene Signal Transduction Downstream of Ethylene Biosynthesis 208
4.3.1 Ethylene Signal Perception by Membrane-Bound Receptor Complex 208
4.3.2 Ethylene Receptors Physically Interact with CTR1 and Transmit the Ethylene Signal 211
4.3.3 EIN2 Acts as the Central Regulator of Ethylene Signaling 212
4.3.4 Regulation of the Interaction of EIN2 and Ethylene Receptors by Protein Phosphorylation 213

4.3.5 EIN3/EIL Family of Proteins Functioning Downstream of EIN2 in Ethylene Signaling Pathway 215
4.3.6 ETR1–RTE1-Mediated CTR1-Independent Ethylene Signaling Pathway 218
4.4 ERF Transcription Factors Functioning Downstream in Ethylene Signaling System 219
4.5 ROS and NO Signaling Systems Activate Transcription of Ethylene-Responsive Genes 221
4.6 MAPK Cascade May Regulate Ethylene Signaling System........... 222
4.7 Ethylene Signaling Triggers Transcription of Plant Pattern Recognition Receptors (PRRs) in PAMP–PRR Signaling System 223
4.8 Ethylene Triggers Ca^{2+} Influx in Downstream Ethylene Signaling System 224
4.9 Ethylene and Jasmonate Signaling Interdependency in Triggering Plant Immune Responses 225
4.10 Ethylene Induces Transcription of Defense-Related Genes............ 228
4.11 Ethylene Signaling System Modulates Plant Immune Signaling System Triggering Resistance or Susceptibility Against Different Pathogens 229
References.. 231

5 Abscisic Acid Signaling System in Plant Innate Immunity 245
5.1 Abscisic Acid as a Multifaceted Plant Hormone Signal Triggering or Suppressing Plant Defense Responses 246
5.2 ABA Biosynthesis in Innate Immune Responses 251
5.2.1 Pathogen/PAMP Triggers Biosynthesis and Accumulation of ABA 251
5.2.2 ABA Biosynthesis Pathway 251
5.2.3 G-Proteins May Be Involved in ABA Biosynthesis 253
5.3 ABA Perception and Signal Transduction 254
5.3.1 ABA Signaling Pathway 254
5.3.2 ABA Receptors 256
5.3.3 PYR/PYL/RCAR Negatively Regulates PP2C 258
5.3.4 ABA-Bound PYR/PYL/RCAR Can Shift ABA Signaling Status to “Active” State.................................... 259
5.3.5 ABA-Induced PP2C Phosphatase Inhibition Leads to SnRK2 Protein Kinase-Activated Phosphorylation of ABA-Responsive Genes... 259
5.3.6 Phosphatases in ABA Signaling Network......................... 260
5.3.7 Role of SnRK2 Protein Kinase in ABA Signaling............ 261
5.3.8 Phospholipase D in ABA Signaling Pathway..................... 264
5.4 ABA Signaling Events Downstream of PYR/PYL/RCAR-PP2C–SNRK2 Signaling Complex.. 265
5.4.1 Role of G-Proteins in ABA Downstream Signaling.......... 265

5.4.2 Role of Ca^{2+} Signaling System in ABA Downstream Signaling ... 268
5.4.3 ABA Activates ROS Signaling System Downstream of ABA Signaling System ... 270
5.4.4 Nitric Oxide (NO) Acts Downstream of H_2O_2 in ABA Signaling System ... 271
5.4.5 MAP Kinases Function Downstream of ABA Signaling System ... 271
5.4.6 ABA Regulates the Expression of Several Transcription Factors ... 272
5.5 Systemic Movement of ABA and Intercellular ABA Signaling Pathway ... 273
5.5.1 AtABCG25 Is Involved in the Intercellular Transport of ABA in ABA Signaling Pathway ... 273
5.5.2 AtABCG40 Is Involved in Intercellular Transport of ABA in ABA Signaling Pathway ... 274
5.6 Interplay Between ABA and JA Signaling Systems ... 274
5.6.1 ABA Signaling and JA Signaling Pathways May Be Interconnected ... 274
5.6.2 ABA and JA May Act Cooperatively in the Induction of Defense Genes ... 276
5.6.3 ABA May Suppress JA-Activated Defense Responses ... 276
5.6.4 Role of Mediator Subunit MED25 in ABA and JA Signaling Interplay ... 277
5.7 Interplay Between ABA and SA Signaling Systems ... 278
5.7.1 ABA May Suppress SA Biosynthesis ... 278
5.7.2 Suppression of SA Signaling System by ABA ... 279
5.7.3 Reciprocal Antagonistic Interaction Between ABA and SA Signaling Systems ... 280
5.7.4 Synergistic Interaction Between ABA and SA Signaling Systems ... 281
5.8 Interplay Between ABA and Ethylene Signaling Systems ... 282
5.8.1 ABA Activates Ethylene Biosynthesis and Ethylene Signaling Pathway ... 282
5.8.2 Ethylene Signaling Triggers ABA Biosynthesis ... 283
5.8.3 Synergistic and Antagonistic Interaction Between ABA and Ethylene Signaling Systems ... 285
5.9 ABA Signaling System May Trigger Defense Responses Against Pathogens ... 285
5.9.1 ABA Signaling Is Involved in Conferring Resistance Against a Wide Range of Pathogens ... 285
5.9.2 ABA Signaling System Triggers Callose Deposition and Confers Disease Resistance ... 286
5.9.3 ABA Signaling Cascade May Trigger Stomatal Closure Immune Responses ... 287

5.9.4 ABA Signaling May Modulate Other Hormone Signaling Systems and Trigger Defense Responses 288
5.10 ABA Signaling System May Confer Susceptibility Against Pathogens 289
5.10.1 ABA Induces Susceptibility Against Fungal and Bacterial Pathogens 289
5.10.2 ABA May Suppress Plant Immune Responses and Induce Susceptibility 289
5.10.3 ABA May Modulate JA, SA, and ET Signaling Pathways and Confer Susceptibility Against Pathogens 290
5.11 Pathogens May Suppress Host Defense Mechanisms by Activating ABA Signaling System to Cause Disease 291
5.12 Pathogens May Hijack ABA Signaling Pathway to Cause Disease 291
5.13 Pathogen Produces Toxins/Effectors and Suppresses ABA-Dependent Defenses 292
References 292

6 Auxin Signaling System in Plant Innate Immunity 311
6.1 Auxin as a Signaling Molecule 312
6.2 Auxin Biosynthesis 312
6.3 Auxin Signaling Pathway 314
6.3.1 Auxin-Binding Proteins/Receptors 314
6.3.2 Auxin–IAA Proteins 314
6.3.3 Auxin Response Factor (ARF) Proteins 315
6.3.4 Auxin-Inducible Gene Expression 316
6.3.5 Ubiquitin–Proteasome System in Auxin Signaling Pathway 316
6.3.6 Auxin Homeostasis 319
6.3.7 Auxin Transport 320
6.4 Pathogen Infection Elevates Auxin Biosynthesis in Plants 321
6.5 Antagonism Between Auxin Signaling and PAMPs/Elicitors-Triggered Signaling Systems 322
6.6 Antagonism Between Auxin Signaling and HAMP/Endogenous Elicitor-Triggered Signaling Systems 324
6.7 Interplay Between Auxin Signaling and Mitogen-Activated Protein Kinase Mediated Signaling Systems 326
6.8 Nitric Oxide Modulates Auxin Signaling 327
6.9 Interaction Between Auxin and Salicylic Acid (SA) Signaling Systems 327
6.9.1 Repression of Auxin Signaling Pathway by Salicylic Acid 327
6.9.2 Auxin Signaling Compromises the Induction of SA Signaling 328

6.9.3 Auxin Response Gene (GH3) Modulates SA Signaling 329
6.10 Role of Auxin Signaling in Systemic Acquired Resistance (SAR) 330
6.11 Interactions Between Auxin and Jasmonate Signaling Systems 331
6.12 Interaction Between Auxin and Ethylene Signaling Systems 333
6.13 Interaction Between Small RNAs and Auxin Signaling Systems 335
6.14 Auxin Signaling May Promote Susceptibility 336
6.14.1 Enhanced Auxin Levels Promote Susceptibility 336
6.14.2 Role of Auxin Receptors in Promoting Disease Susceptibility 337
6.14.3 Role of Aux/IAA Proteins in Promoting Susceptibility 338
6.14.4 Role of the Auxin-Responsive GH3 Genes in Promoting Disease Susceptibility 339
6.14.5 Conjugated Auxin Promotes Plant Disease Susceptibility 341
6.14.6 Role of Auxin Transport System in Promoting Disease Susceptibility 342
6.15 Auxin Signaling May Promote Plant Disease Resistance 342
6.15.1 Overexpression of Auxin-Responsive Genes Promote Disease Resistance 342
6.15.2 Auxin Response Factors Modulate Plant Defense Responses 343
6.15.3 Exogenous Application of Auxin Induces Plant Disease Resistance 344
References 344

7 Cytokinin Signaling System in Plant Immunity 359
7.1 Cytokinin Signaling in Plant Immune System 360
7.2 Cytokinin Biosynthesis 360
7.3 Cytokinin Degradation 362
7.4 Cytokinin Signal Perception and Transduction 362
7.4.1 Cytokinin Receptors 362
7.4.2 Cytokinin Phosphorelay Signaling System 363
7.5 Cytokinin-Responsive Genes 366
7.6 Cytokinins May Be Involved in Triggering Defense Responses 367
7.6.1 Cytokinins Confer Resistance Against Pathogens 367
7.6.2 Cytokinin Augments Plant Immune Responses by Enhancing Callose Deposition 368
7.6.3 Cytokinin May Trigger Accumulation of Antimicrobial Phytoalexins to Confer Disease Resistance 368
7.6.4 Cytokinins Induce Priming of Plant Cells for Activation of Defense-Related Genes 368
7.6.5 Cytokinin May Modulate SA Signaling System to Trigger Immune Responses 369

7.6.6 Cytokinins May Induce Resistance Independently of SA Signaling System 369
7.6.7 Cytokinins May Modulate Redox Signaling to Trigger Immune Responses 370
7.7 Cytokinins May Induce Susceptibility 370
7.8 Interplay Between Cytokinin and SA Signaling Pathways in Plant Immune System 371
7.8.1 Cytokinin May Enhance SA Biosynthesis 371
7.8.2 Type-B ARR Interacts with TGA3 of SA Signaling Pathway to Trigger Immune Responses 372
7.8.3 Type-A ARRs Negatively Regulate SA-Dependent Immune Responses 373
7.8.4 Cytokinin Synergistically Acts with SA to Trigger Immune Responses 374
7.9 Interaction Between Cytokinin and Abscisic Acid Signaling Systems 374
7.10 Interplay Between Cytokinin and Auxin Signaling Systems in Plant Immunity 376
References 377

8 Gibberellin Signaling in Plant Innate Immunity 383
8.1 Role of Gibberellins in Plant Immune Responses 383
8.2 Biosynthesis of Gibberellins 384
8.3 GA Signaling Pathway 384
8.3.1 GA Signal Receptors 384
8.3.2 DELLA Proteins, Repressors of GA Signaling 386
8.3.3 Suppression of the Repressive Activity of DELLAs by Proteasome-Dependent Degradation of DELLAs 386
8.4 GA Triggers Susceptibility or Resistance Against Different Pathogens 388
8.4.1 GA Triggers Resistance Against Pathogens 388
8.4.2 GAs May Negatively Regulate Plant Defense Responses and Induce Susceptibility 388
8.5 Interplay of GA Signaling System with SA Signaling System in Modulating Plant Immune System 389
8.6 Interplay of GA and JA Signaling Systems in Modulating Plant Immune System 391
8.6.1 Antagonistic Interaction Between GA and JA Signaling Systems 391
8.6.2 JA Induces Enhanced Expression of DELLA Genes Involved in GA Signaling 391
8.6.3 DELLAs Modulate JA Responses by Degrading JAZ Proteins and/or Sequestering JAZs into Inactive Complexes 392

8.6.4 GA Attenuates the Expression of JA-Responsive Genes 393
8.7 Interplay Between GA and Brassinosteroids Signaling Systems in Plant Immune Responses 394
8.8 Interplay Between GA and Auxin Signaling Systems 395
8.9 GA May Be Involved in Triggering Systemic Acquired Resistance (SAR) 395
8.10 Pathogen May Suppress GA Signaling Pathway to Cause Disease 396
References 396

9 Brassinosteroid Signaling in Plant Immune System 403
9.1 Brassinosteroids Modulate Plant Immune Responses 404
9.2 Biosynthesis of Brassinosteroids 404
9.2.1 Pathogen Triggers Brassinosteroid (BR) Biosynthesis 404
9.2.2 BR Biosynthesis via Mevalonate Pathway 405
9.2.3 Early and Late C-6 Oxidation Pathways Involved in BR Biosynthesis 407
9.2.4 C-22 Oxidation Branch in Brassinosteroid Biosynthetic Pathway 408
9.2.5 Homeostasis of Brassinosteroids 408
9.3 Brassinosteroid Signaling System 411
9.3.1 BRI1 as a Brassinosteroid Receptor 411
9.3.2 BRL1 as an Additional BR Receptor 412
9.3.3 BAK1 Acts as a Co-receptor in BR Signal Reception 413
9.3.4 Autophosphorylation of BRI1 and BAK1 414
9.3.5 BKI1, a Negative Regulator of BRI1 Signaling 415
9.3.6 BR Signaling Events Downstream of BR Signal Perception 415
9.3.7 Brassinosteroid-Signaling Kinases (BSKs) 418
9.3.8 BRI1 SUPPRESSOR 1 (BSU1) Phosphatase 419
9.3.9 MSBP1 Negatively Regulates Brassinosteroid Signaling 420
9.3.10 CDG1 in BR Signal Transduction 420
9.3.11 BIN2 Negatively Regulates BR Signaling 421
9.3.12 Protein Phosphatase 2A (PP2A) 422
9.3.13 BZR1 and BES1 (BZR2) Transcription Factors 422
9.3.14 Function of 14-3-3 Proteins in Regulation of Activities of BZR Transcription Factors 424
9.4 Pathogen Modulates Brassinosteroid Signaling System in Infected Plants 425
9.5 BR Signaling Triggers Plant Disease Resistance 427
9.6 BAK1 in the BR Signaling Pathway Triggers Plant Disease Resistance 427

9.7 BR Signaling Machinery Negatively Regulates Plant Immune Responses and Induces Susceptibility 428
9.8 Brassinosteroid Signaling Negatively Regulates Salicylate-Mediated Immunity 428
9.9 BR Signaling Negatively Regulates Gibberellic Acid (GA)-Mediated Plant Immune Responses 429
9.10 Interplay Between BR and PAMP–PRR Signaling Systems 430
9.10.1 PAMP–PRR Signaling Complex 430
9.10.2 Crosstalk Between BR Biosynthesis Pathway and PAMP–PRR Signaling 431
9.10.3 Overexpression of BRI1 Antagonizes BAK1-Mediated PAMP–PRR Signaling 431
9.10.4 BR-Responsive Transcription Regulator BZR1 May Suppress PAMP–PRR Signaling System 433
9.10.5 Antagonistic Regulation of PAMP–Triggered Immunity by the bHLH transcription Factor HBI1 433
9.10.6 Activation of BRI1 Pathway Leads to Inhibition of PAMP-Triggered Immunity 433
9.10.7 Inhibition of PRR-Mediated Immune Signaling by BR Perception 435
9.11 Pathogen Hijacks Brassinosteroid Signaling Machinery to Cause Disease 435
9.12 Crosstalk Between BR and Other Hormone Signaling Systems 435
References 436

Index 445

Chapter 1
Introduction

Abstract The plant hormones salicylic acid (SA), jasmonates (JA), ethylene (ET), abscisic acid (ABA), auxin (AUX), cytokinin (CK), gibberellin (GA), and brassinosteroid (BR) play an important role in intercellular and systemic signaling systems triggering expression of various defense-responsive genes. The SA–JA–ET signaling systems are considered as the backbone of the plant immune signaling system, while ABA, auxin, cytokinin, GA, and BR are involved in modulating plant immune responses by regulating host defense responses triggered by the SA–JA–ET signaling systems. SA signaling is required for the manifestation of systemic acquired resistance (SAR). Methyl salicylate, dehydroabietinal, pipecolic acid, azelaic acid, a lipid transfer protein (DIR1), a lipid-derived molecule (glycerol 3-phosphate), and a glycerol-3-phosphate-dependent factor have been reported as mobile signaling components in SA-induced SAR. Mediators MED16 and MED15 are involved in triggering SA-mediated SAR. SAR is associated with priming of defense responses, and the priming results in a faster and stronger induction of defense responses after pathogen attack. Histone modifications are systemically set during a priming event. The priming can be inherited epigenetically, and descendants of primed plants exhibit next-generation systemic acquired resistance. DNA methylation plays an important role in the transgenerational SAR. JA signaling triggers systemic immunity called "induced systemic resistance (ISR)." JA-Ile may be the mobile signal involved in the induction of ISR. The ISR involves priming of JA-dependent responses. MED25, MED16, and MED8 subunits of the Mediator complex interact with several transcription factors known to function in the control of JA-associated gene expression. Ethylene may act as a two-faceted player in the plant immune response network, triggering resistance or susceptibility against different pathogens. ABA signal perception and signal transduction pathway includes PYR/PYL/RCAR (an abscisic acid receptor), type 2C protein phosphatase (PP2C, a negative regulator), and SNF1-related protein kinase (SnRK2, a positive regulator). Auxin binds to TIR1/AFB nuclear receptors, which are F-box subunits of SCF ubiquitin ligase complex. The auxin signal is then modulated by the Aux/IAA repressors and the auxin response factor (ARF) transcription factors. Auxin signaling is also involved in triggering SAR. Auxin signaling increases SA levels, which trigger SAR. Cytokinins regulate the host defense responses either positively or negatively depending on the concentrations of cytokinins available at the infection site. Key components in the GA signaling pathway include the GA receptor GID1, the DELLA proteins, and the F-box proteins. GA regulates plant immune responses by

P. Vidhyasekaran, *Plant Hormone Signaling Systems in Plant Innate Immunity*, Signaling and Communication in Plants 2, DOI 10.1007/978-94-017-9285-1_1

modulating JA and SA signaling systems. Plant hormones act in concert. Plant hormone signaling pathways are not simple linear and isolated cascades, but can crosstalk with each other. Architecture of plant immune signaling networks may not be static and may vary with the invading pathogen genotype. Crosstalk between different hormone signaling pathways in the signaling network helps the plant to choose the effective defense strategy to follow, depending on the type of pathogen it is encountering. The crosstalk also allows the pathogens to manipulate plants to their own benefit by shutting down the specific hormone signaling pathway involved in triggering defense responses and hijacking the signaling pathway involved in induction of susceptibility.

1.1 Plant Innate Immunity

Plants are endowed with innate immune system, which has a high potential to detect and fight against viral, bacterial, oomycete, and fungal pathogens and protect the crop plants against a wide range of diseases (Vidhyasekaran 2004, 2007, 2014; Lacombe et al. 2010; Segonzac and Zipfel 2011; Alkan et al. 2012; Dubery et al. 2012; Denancé et al. 2013; Kim and Hwang 2014). The plant innate immune system is a sleeping system in unstressed healthy plants (Vidhyasekaran 2014). Specific signals are needed to activate the "sleeping" immune system. Pathogen-associated molecular patterns (PAMPs) of invading pathogens have been found to be potential signals to activate the plant innate immunity. These PAMP signals are perceived by the plant pattern recognition receptors (PRRs), and the PAMP–PRR signaling complex activates the plant immune system (Iriti and Faoro 2009; Nicaise et al. 2009; Petutschnig et al. 2010; Shinya et al. 2010; Schulze et al. 2010; Segonzac and Zipfel 2011; Yang et al. 2013). The plant immune system uses several second messengers to encode information generated by the PAMPs and deliver the information downstream of PRRs to proteins which decode/interpret signals and initiate defense gene expression (Mersmann et al. 2010; Boudsocq et al. 2010; Hwang and Hwang 2011). Plant hormones such as salicylic acid (SA), jasmonates (JA), ethylene (ET), abscisic acid (ABA), auxin (AUX), cytokinin (CK), gibberellin (GA), and brassinosteroid (BR) have been reported to play an important role in intercellular and systemic signaling systems triggering expression of various defense-responsive genes (Denancé et al. 2013; Yang et al. 2013; Alazem et al. 2014; Kim and Hwang 2014).

1.2 Salicylic Acid Signaling

Salicylic acid (SA) is an important endogenous immune signal in the induction of disease resistance responses in plants (Anand et al. 2008; Fung et al. 2008; Garcion et al. 2008; Mukherjee et al. 2010; Dempsey et al. 2011; Liu et al. 2011a, b; Argueso et al. 2012; Fu et al. 2012; Denancé et al. 2013; Gimenez-Ibanez and Solano 2013; Yang et al. 2013). Infection of plants by necrotizing pathogens, which induce the accumulation of SA, or treatment of plants with synthetic compounds, which are able to trigger SA signaling, induces systemic acquired resistance (SAR). SAR is a heightened state of defense against a broad spectrum of pathogens activated throughout a plant following a local infection (Liu et al. 2011a, b). SA signaling is required for the manifestation of SAR (Du et al. 2012; Návarová et al. 2012; Shah and Zeier 2013).

Methyl salicylate, dehydroabietinal, pipecolic acid, and azelaic acid are the systemic signal molecules involved in the activation of SAR (Jung et al. 2009; Manosalva et al. 2010; Chaturvedi et al. 2012; Návarová et al. 2012; Shah and Zeier 2013). A lipid transfer protein (DIR1), a lipid-derived molecule (glycerol 3-phosphate), and a glycerol-3-phosphate-dependent factor have been reported as lipid-based mobile signaling components in SA-induced SAR (Kachroo et al. 2001, 2004; Chaturvedi et al. 2008; Jung et al. 2009; Chanda et al. 2011). The mediators MED16 and MED15/NRB4 have been shown to be involved in triggering SA-mediated SAR. Mediator is a multiprotein complex that functions as a transcriptional coactivator (Conaway and Conaway 2011a, b; Kidd et al. 2011a; Canet et al. 2012; An and Mou 2013). MED16 may regulate the function of NPR1 in inducing SAR (Zhang et al. 2012).

SAR is associated with priming of defense (Kohler et al. 2002; Jung et al. 2009; Luna et al. 2012; Slaughter et al. 2012), and the priming results in a faster and stronger induction of defense mechanisms after pathogen attack (Conrath 2011). Jaskiewicz et al. (2011) showed that histone modifications are systemically set during a priming event. These modifications might create a memory of the primary infection that is associated with an amplified reaction to the second stress condition. The priming can be inherited epigenetically from disease-exposed plants, and descendants of primed plants exhibit next-generation systemic acquired resistance (Luna et al. 2012; Slaughter et al. 2012). The transgenerational SAR was found to be sustained over one stress-free generation, indicating an epigenetic basis of the phenomenon (Luna et al. 2012). DNA methylation may also play an important role in the transgenerational SAR (Luna et al. 2012).

NPR1 functions as a transcriptional co-activator in a TGA2–NPR1 complex after SA treatment. Binding of SA causes a conformational change in NPR1 that is accompanied by the release of the C-terminal transactivation domain from the N-terminal autoinhibitory BTB/POZ domain (Wu et al. 2012). Pathogen/PAMP exposure induces SA accumulation (Durrant and Dong 2004), and the induced SA controls the nuclear translocation of NPR1 through cellular redox changes (Spoel and Dong 2012). The SA-induced changes in gene expression have been found to have a link to chromatin remodeling, such as histone modifications and histone replacement. The recruitment of chromatin-modifying complexes to SA-responsive loci controls their basal and SA-induced expression (March-Diaz et al. 2008; van den Burg and Takken 2009, 2010; Jaskiewicz et al. 2011). SA signaling triggers transcription of a multitude of defense-related genes in plants (Krinke et al. 2007). Small RNA-directed RNA silencing is a potent immune surveillance system targeting foreign nucleic acids of invading pathogens (Ding and Voinnet 2007; Jaubert et al. 2011). SA signaling may enhance the efficiency of RNA silencing pathway in triggering immune responses against viruses by activating RdRP/RDR (Diaz-Pendon et al. 2007).

1.3 Jasmonate Signaling

Several metabolites of jasmonates (JA) have been reported to act as signal molecules in triggering plant immunity. Among them, (+)-7-iso-jasmonoyl-L-isoleucine (JA-Ile) is the major bioactive form of the hormone JA (Kombrink 2012; Wasternack and Hause 2013). JA signaling systems modulate plant immune responses and confer resistance or susceptibility against different pathogens (Méndez-Bravo et al. 2011; Moffat et al. 2012). JA receptor is a three-molecule co-receptor complex, consisting of COI1, JAZ, and inositol pentakisphosphate, all of which are indispensable for high-affinity hormone binding (Sheard et al. 2010). The JA receptor JAZ proteins are suppressors of jasmonate signaling (Chini et al. 2007; Howe 2010; Wasternack and Hause 2013). Repression of JA response genes involves binding of JAZ to NINJA, which contains an EAR motif that recruits the corepressor TPL, which may silence gene expression. COI1 is involved in the degradation of the repressors of the JA signaling pathway through SCF^{COI1}-dependent ubiquitin (Thines et al. 2007). In response to stress cues that activate JA-Ile synthesis, high levels of JA-Ile promote SCF^{COI1}-mediated ubiquitination and subsequent degradation of JAZs by the 26S proteasome. JAZ degradation relieves TPL-mediated repression of gene expression (Howe 2010; Wasternack and Hause 2013). Acetylation of the core histones in nucleosomes plays an important role in gene regulation (Wu et al. 2008), and histone deacetylation has been shown to be involved in COI1-mediated activation of JA-inducible transcription factors (Wang et al. 2008b). MED25, MED16, and MED8 subunits of the Mediator complex interact with several transcription factors (TFs) known to function in the control of JA-associated gene expression (Kidd et al. 2009; Cevik et al. 2012; Iñigo et al. 2012; Wathugala et al. 2012; Zhang et al. 2012). MYC2 is a master regulator of the JA signaling pathway. Several ERF, bHLH, WRKY, MYB, NAC, and bZIP transcription factors have been shown to act downstream of JA signaling system triggering expression of JA-responsive defense genes (Nurmberg et al. 2007; Wasternack 2007; Zander et al. 2010; Cheng et al. 2011; Le Hénanff et al. 2013).

JA signaling triggers systemic immunity conferring resistance against a wide range of pathogens, and the systemic immunity triggered by JA is called "induced systemic resistance (ISR)" as against "systemic acquired resistance (SAR)" induced by SA. ISR is dependent mostly on jasmonic acid (Kravchuk et al. 2011; Niu et al. 2011; Weller et al. 2012; Zamioudis and Peterse 2012; Bakker et al. 2013; Martinez-Medina et al. 2013). ISR is triggered mostly by biocontrol agents (BCAs) and necrotrophic fungi. Several chemicals and PAMPs/MAMPs have also been reported to trigger ISR (Kravchuk et al. 2011). JA-Ile may be the mobile signal involved in the induction of ISR (Sato et al. 2011; Matsuura et al. 2012). JA-Ile may be synthesized de novo and transported into systemic tissues (Matsuura et al. 2012). The ISR has been shown to be mainly based on priming JA-dependent responses (Martinez-Medina et al. 2013). Treatment with *Trichoderma hamatum* T382 primes the plant (ISR-prime), resulting in an accelerated activation of defense responses against *B. cinerea* during ISR-boost in *Arabidopsis thaliana* (Mathys et al. 2012). Hexanoic acid-treated plants infected with the necrotrophic pathogen *Botrytis cinerea* showed priming in the expression of the JA-responsive genes *PDF1.2*, *PR-4*, and *VSP1* in *Arabidopsis* (Kravchuk et al. 2011).

1.4 Ethylene Signaling

Ethylene (ET) signaling system is an important component in plant innate immune system (Berr et al. 2010; Nie et al. 2011; Shakeel et al. 2013). Pathogen invasion or PAMP application results in enhanced expression of ET biosynthetic genes leading to enhanced ET biosynthesis (Qutob et al. 2006; Denoux et al. 2008; Mur et al. 2008; Zhu et al. 2011a; Nambeesan et al. 2012; Vidhyasekaran 2014). Ethylene signal transduction is initiated by the binding of ethylene to its membrane-bound receptors ETR1, ERS1, ETR2, EIN4, and ERS2 (Grefen et al. 2008; Zhao and Guo 2011; Shakeel et al. 2013). The five-member family of ethylene receptors act as negative regulators in the ethylene signaling pathway (Qu and Schaller 2004; Gao and Schaller 2009). CTR1 functions as a key mediator of ethylene signal transduction, acting just downstream of the receptors. It negatively regulates ET signaling (Guo et al. 2004; Yoo et al. 2008). All the five ethylene receptors physically interact with CTR1 (Gao et al. 2003). EIN2 is a positive regulator of the ethylene signaling pathway and acts downstream of CTR1 (Ju et al. 2012; Qiao et al. 2012; Shakeel et al. 2013). Upon ethylene binding to the ethylene receptors, the ethylene receptors inactivate CTR1, potentially through propagation of conformational changes in the receptor–CTR1 protein complex. EIN2 becomes dephosphorylated and this results in proteolytic cleavage and release of C-terminal domain of EIN2 (Ju and Chang 2012; Ju et al. 2012; Qiao et al. 2012; Wen et al. 2012; Shakeel et al. 2013). The C-terminal domain of EIN2 translocates to the nucleus. In the nucleus, EIN2 either directly or indirectly activates the transcription factors EIN3 and EIN3-like1 (EIL1) to initiate the transcriptional response to ethylene (Qiao et al. 2012; Wen et al. 2012; Shakeel et al. 2013). EIN3 binds to the promoter sequence of the ethylene-inducible transcription factor ERF1 (Solano et al. 1998). ERFs contain a single DNA-binding domain. ERFs have been shown to bind specifically to the GCC-box that is found in several promoters of the pathogenesis-related (PR) genes as well as ethylene- and jasmonate-inducible defense genes (Yamamoto et al. 1999; Gutterson and Reuber 2004). ERF2 might play a major role in the elicitor-induced GCC-box-mediated transcription of defense genes (Yamamoto et al. 1999). The transcription factor ERF1 has been shown to induce transcription of several defense-related *PR* genes (Zhou et al. 2005; Johansson et al. 2006; Jung et al. 2007; Oñate-Sánchez et al. 2007).

Ethylene may act as a two-faceted player in the plant immune response network, triggering resistance or susceptibility against different pathogens (De Vleesschauwer et al. 2008, 2010; Gaige et al. 2010; Akagi et al. 2011; Son et al. 2012). ET has also been reported to be involved in systemic immunity (Zhu et al. 2011a). Ethylene has been shown to be an integral part of PAMP-triggered immunity. Ethylene perception and signaling are crucial for the PRR *FLS2* gene transcription (Boutrot et al. 2010). FLS2 promoter revealed the presence of nine potential EIN3/EIL-binding sites (Boutrot et al. 2010), suggesting that EIN3 may bind to the promoter of the *FLS2* gene to influence its transcription. Plants mutated in the key ethylene signaling protein EIN2 are impaired in all the PRR FLS2-mediated responses. The EIN3 and EIN3-like transcription factors, which depend on EIN2 activity for their accumulation, directly controlled the transcription of the PRR *FLS2* gene transcription (Boutrot et al. 2010).

1.5 Abscisic Acid Signaling

The phytohormone abscisic acid (ABA) plays a multifaceted role in plant immunity (Cao et al. 2011; Alazem et al. 2014). ABA induces defense responses (Asselbergh et al. 2008; Fan et al. 2009; Garcia-Andrade et al. 2011) or suppresses the immune responses depending upon the type of plant–pathogen interactions (Feng et al. 2012; Sánchez-Vallet et al. 2012; Yazawa et al. 2012). Pathogen/pathogen-associated molecular patterns (PAMPs) induce increase in ABA accumulation triggering disease resistance (Whenham et al. 1986) or susceptibility (Mohr and Cahill 2003; Koga et al. 2004; Schmidt et al. 2008). ABA signal perception and signal transduction pathway includes PYR/PYL/RCAR (an abscisic acid receptor), type 2C protein phosphatase (PP2C, a negative regulator), and SNF1-related protein kinase (SnRK2, a positive regulator) (Umezawa et al. 2010, 2013). In the presence of accumulated ABA, the PYR/PYL/RCAR receptor proteins disrupt the interaction between the SnRK2s and PP2Cs, thus preventing the PP2C-mediated dephosphorylation of the SnRK2s and resulting in the activation of the SnRK2 kinases (Fujii et al. 2009; Miyazono et al. 2009; Umezawa et al. 2009; Vlad et al. 2009; Raghavendra et al. 2010). The SnRK2 kinases phosphorylate and activate downstream transcription factors, which initiate transcription at ABA-responsive promoter elements (Sheard and Zheng 2009). The *cis*-regulatory elements responsible for the ABA regulation of gene expression share a conserved motif, ACGTGGC, which is known as ABA-responsive element (ABRE). ABRE appears in the promoters of many defense genes (Adie et al. 2007). The ABA-induced gene regulation is mediated by a subfamily of basic leucine zipper class transcription factors referred to as ABRE-binding factors (ABFs, also referred to as AREBs) (Choi et al. 2005; Furihata et al. 2006). The bZIP-type transcription factors AREBs/ABFs bind ABRE and transactivate downstream gene expression in *Arabidopsis* (Furihata et al. 2006). MYC2, MYB2, BOS1, and WRKY transcription factors are also involved in ABA-inducible gene expression (Abe et al. 2003; Mengiste et al. 2003; Anderson et al. 2004; Xie et al. 2005). ABA signaling system is involved in the induction of callose (β-1,3-glucan) deposition (Yazawa et al. 2012), which is involved in conferring disease resistance. Stomatal closure is an innate immune response involved in bacterial disease resistance (Hettenhausen et al. 2012). ABA induces stomatal closure immune response (Saito et al. 2008; Hossain et al. 2011; Munemasa et al. 2011; Hubbard et al. 2012; Sugiyama et al. 2012; Uraji et al. 2012). ABA regulates plant immune responses mostly by modulating other plant hormone signaling systems involved in the immune signaling systems. Antagonistic interaction between ABA and JA signaling pathways modulates defense gene expression and disease resistance (Anderson et al. 2004; Chen et al. 2012). ABA negatively regulates SA-dependent immune responses (Sánchez-Vallet et al. 2012). ABA signaling system suppresses SA signaling system and vice versa (Alazem et al. 2014). ABA signaling may also act synergistically with SA signaling in triggering plant immune responses (Seo and Park 2010). Ethylene signaling pathway triggers ABA biosynthesis pathway (Wasilewska et al. 2008; Hauser et al. 2011; Liu et al. 2012). ABA-induced resistance against the brown spot pathogen *Cochliobolus miyabeanus* in rice involves repression of ethylene signaling (De Vleesschauwer et al. 2010) Synergistic interaction between ABA and ethylene signaling systems has also been reported (Sánchez-Vallet et al. 2012).

1.6 Auxin Signaling

The plant hormones SA, JA, and ET signaling systems are considered as the backbone of the plant immune signaling system. In contrast, the plant hormone auxin is considered as a simple signaling molecule involved in modulation of those hormone signaling pathways activating or suppressing the plant defense responses (Hayashi 2012; Peer et al. 2013; Sauer et al. 2013; Tatsuki et al. 2013). Auxin binds to TIR1/AFB nuclear receptors, which are F-box subunits of SCF ubiquitin ligase complex (Dharmasiri et al. 2005; Mockaitis and Estelle 2008; Parry et al. 2009; Calderon-Villalobos et al. 2010). The auxin signal is then modulated by the Aux/IAA repressors and the auxin response factor (ARF) transcription factors (Hayashi 2012). The specificity of the auxin-regulated gene expression is regulated by the expression of these regulatory proteins (Hayashi 2012). Auxin signaling appears to be mostly involved in disease susceptibility rather than in disease resistance (O'Donnell et al. 2003; Kidd et al. 2011b). Elevated levels of endogenous plant IAA have been observed during pathogen infection, and the susceptible reaction seems to be associated with rapidly increased endogenous biosynthesis of IAA (Fu et al. 2011). Auxins may induce susceptibility by inducing the formation of the conjugated forms of auxin through the action of GH3 proteins (Fu and Wang 2011; González-Lamothe et al. 2012). Fungal and bacterial pathogens hijack the host auxin metabolism in *Arabidopsis thaliana*, leading to the accumulation of a conjugated form of the hormone, indole-3-acetic acid (IAA)–Asp, to promote disease development. IAA–Asp increases pathogen progression in the plant by regulating the transcription of virulence genes (González-Lamothe et al. 2012). Auxin may suppress the induction of SA signaling and induce susceptibility (Robert-Seilaniantz et al. 2011).

Auxin also induces resistance against some pathogens, probably modulating auxin homeostasis, polar auxin transport, and expression of ARF transcription factors (Robert-Seilaniantz et al. 2011; Mah et al. 2012). Auxin signaling has been reported to be involved in triggering systemic acquired resistance (SAR). Auxin signaling increases SA levels, which trigger SAR (Rock and Sun 2005). Auxin is highly mobile and is involved in SA-triggered SAR (Truman et al. 2010). Auxin signaling induces disease resistance by triggering accumulation of SA, which positively regulates defense responses, and by suppressing both JA and ABA signaling systems, which negatively regulate defense responses (Truman et al. 2010). Auxin triggers biosynthesis of ethylene (Tatsuki et al. 2013). Application of ethylene biosynthetic precursor ACC triggers an increase in the rate of IAA biosynthesis (Swarup et al. 2007), and ACC treatments also increase IAA transport (Negi et al. 2008). Ethylene has been shown to induce susceptibility (De Vleesschauwer et al. 2010; Pantelides et al. 2013) or resistance (De Vleesschauwer et al. 2008; Gaige et al. 2010; Zhu et al. 2011a; Nambeesan et al. 2012) against various pathogens.

1.7 Cytokinins

Cytokinins act as both long-range and local signals (Hwang and Sakakibara 2006) and play an important role in modulation of plant innate immunity (Choi et al. 2010; Grosskinsky et al. 2011; Naseem et al. 2012; Pieterse et al. 2012). Cytokinin signaling system may regulate positively or negatively the plant defense responses (Choi et al. 2010; Argueso et al. 2012). Cytokinins may regulate the host defense responses either positively or negatively depending on the concentrations of cytokinins available at the infection site (Babosha 2009; Argueso et al. 2012). Cytokinin may modulate SA signaling system to trigger immune responses (De Vleesschauwer et al. 2010; Pantelides et al. 2013). ABA treatment decreases the expression of several genes involved in cytokinin biosynthesis and degradation (Tsai et al. 2012) Cytokinin antagonistically impacts the signaling of auxin (Stepanova and Alonso 2011). Synergism between auxin and cytokinin signaling has also been reported (Hwang et al. 2012). Elevated plant auxin levels enhance susceptibility by repressing the defense-related *PR1* gene expression (Kazan and Manners 2009), while elevated cytokinin levels mediate resistance and induction of *PR1* (Choi et al. 2011; Naseem et al. 2012).

1.8 Gibberellins

Gibberellins (GA) are important plant growth hormones involved in plant innate immunity. GA modulates plant defense responses, mostly by regulating SA–JA–ET signaling systems. Key components in the GA signaling pathway include the GA receptor GID1, the DELLA proteins, and the F-box proteins (Hauvermale et al. 2012; Davière and Achard 2013). Upon GA binding, the GA-dependent conformational change causes the GID1 N-terminal helical lid domain to behave like "molecular glue" to form the GA–GID1–DELLA complex. DELLAs repress GA-dependent defense responses, whereas GA relieves their repressive activity (Achard and Genschik 2009). GA lifts DELLA repression by targeting DELLA for destruction via the ubiquitin–proteasome pathway. GA regulates the plant innate immune responses either positively or negatively. It induces susceptibility or resistance against different fungal and bacterial pathogens (Navarro et al. 2008; Yang et al. 2008; De Vleesschauwer et al. 2012; Qin et al. 2013). GA modulates plant disease resistance or susceptibility by inducing the degradation of DELLA proteins (Hauvermale et al. 2012). GA regulates plant immune responses by modulating JA and SA signaling systems (Navarro et al. 2006; Yang et al. 2008; Qi et al. 2014). It also enhances SAR against pathogens (Xia et al. 2010). GA expedites SA accumulation (Navarro et al. 2008; Alonso-Ramirez et al. 2009) and promotes resistance against pathogens by degrading DELLA proteins (Navarro et al. 2008). Loss-of-function mutants in DELLAs, the suppressors of GA signaling up-regulate the SA-mediated defense and down-regulate JA/ET-mediated defense in *Arabidopsis* (Robert-Seilaniantz et al. 2007). GA interacts antagonistically with JA signaling (Yang et al. 2013), while GA attenuates the JA-induced expression of a number of JA-responsive genes (Cao et al. 2006; Hou et al. 2008, 2010). Both JA and GA signaling systems modulate plant immune responses. JA interferes with gibberellin signaling cascade. JA delays GA-mediated DELLA protein degradation. The JAZ protein JAZ9 inhibits RGA (a DELLA protein) interaction with the transcription factor PIF3 (phytochrome-interacting factor 3) (Yang et al. 2012). JA signaling has been shown to activate expression of DELLA genes involved in GA signaling pathway. JA-induced *RGL3* expression works via the COI1/MYC2-dependent signaling pathway. JA-mediated induction of RGL3 expression was abolished in the *coi1-1* mutant, indicating that *RGL3* is downstream of COI1 (Wild et al. 2012). Brassinosteroids (BR) negatively regulate innate immune responses induced by GAs. BR and GA cause cross-inhibitory effects on the reciprocal hormone biosynthesis pathways to interact in a mutually antagonistic manner (De Vleesschauwer et al. 2012). Pathogen triggers overexpression of DELLA proteins to suppress GA-regulated defense responses (De Vleesschauwer et al. 2012).

1.9 Brassinosteroids

Brassinosteroids (BRs) are growth-promoting steroidal hormones in plants, and they are also involved in plant innate immunity (Nakashita et al. 2003; Bajguz and Hayat 2009; Divi et al. 2010; Jaillais et al. 2011; Albrecht et al. 2012; Owens et al. 2012; Wang 2012; Vriet et al. 2012). BR signals are perceived by the plasma membrane receptor BRI1 and co-receptor BAK1. Several positive (BSK1, BSU1, PP2A, CDG1) and negative (BKI1, BIN2, MSBP1, and 14-3-3) regulators of BR signaling control the activities of BZR1 and BES1 family of transcription factors, which regulate the expression of hundreds to thousands of genes for various BR responses (Wang et al. 2008a; Kim et al. 2009; Li et al. 2010). BRs either positively (Khripach et al. 2000; Nakashita et al. 2003) or negatively (Albrecht et al. 2012; Belkhadir et al. 2012; De Vleesschauwer et al. 2012; Nahar et al. 2013) regulate plant innate immunity. BAK1, a key component in BR signaling pathway (Schwessinger et al. 2011), is involved in triggering plant disease resistance by modulating JA signaling system (Yang et al. 2011). BR may also induce susceptibility to pathogens (De Vleesschauwer et al. 2012). BR negates disease resistance conferred by the SA synthetic analog benzothiadiazole, suggesting negative crosstalk between BR and SA signaling pathways. BR-mediated suppression of SA defenses occurs downstream of SA biosynthesis, but upstream of NPR1 and OsWRKY45 in the SA signaling pathway (De Vleesschauwer et al. 2012). BR triggers the expression of GA repressor proteins and suppresses GA-induced defense responses (De Vleesschauwer et al. 2012). Crosstalk between PAMP–PRR signaling and BR synthesis pathway has been reported (Albrecht et al. 2012). Increasing the endogenous pool of bioactive BR antagonizes flg22-induced responses (Belkhadir et al. 2012). BRI1–BAK signaling modulates PAMP–PRR signaling pathway. BAK1 is a common co-receptor for the PRRs activated by various PAMPs. Similarly BAK1 is a co-receptor for the BR receptor, BRI1. Signaling downstream of BAK1 differs between BRI1 and FLS2 (PRR for flg22) pathways (Lu et al. 2010). Pathogen infection results in elevation of BR signal processing (De Vleesschauwer et al. 2012). Pathogens may exploit BRs as virulence factors and hijack the plant BR machinery to cause disease (De Vleesschauwer et al. 2012).

1.10 Plant Hormone Signaling Network

Plant hormones activate different signaling pathways inducing distinctly different defense genes (Liu et al. 2007; Spoel et al. 2007; Mitsuhara et al. 2008; van Verk et al. 2008; Cevik et al. 2012). These signaling pathways are not simple linear and isolated cascades, but can crosstalk with each other (Tsuda et al. 2009; Verhage et al. 2010; Yang et al. 2013). Architecture of plant immune signaling networks may not be static and may vary with the pathogen genotype invasion. Both antagonism and synergism between SA and JA signaling systems have been widely reported in plants (Robert-Seilaniantz et al. 2011; El Rahman et al. 2012; Pieterse et al. 2012; Thaler et al. 2012; Zander et al. 2012; Zheng et al. 2012; Gimenez-Ibanez and Solano 2013; Van der Does et al. 2013). Crosstalk between JA and ET (Melotto et al. 2008; Pré et al. 2008; Bari and Jones 2009; Grant and Jones 2009; Pauwels and Goossens 2011; Robert-Seilaniantz et al. 2011; Zhu et al. 2011b), SA and ET (Leon-Reyes et al. 2010), JA and GA (Yang et al. 2013), SA and auxin (Robert-Seilaniantz et al. 2011), and SA and ABA (Xu et al. 2013) has been reported. The DELLA protein RGL3 in the GA pathway represses the SA pathway (Wild et al. 2012). In contrast, the DELLA protein enhances the expression of JA-dependent expression

(Wild et al. 2012). The interplay between SA, JA, and GA signaling pathways has been reported in *Arabidopsis* (Wild et al. 2012). Synergism between ABA and SA signaling systems has also been reported in *Arabidopsis* (Chen et al. 2013). Mutual interactions between stress-specific hormones such as SA and JA/ET are regarded as the central backbone of the immunity (Pieterse et al. 2012). However, the growth-promoting hormones (auxin, cytokinins, gibberellic acid, and abscisic acid) either inhibit or potentiate this balance in mediating the protection or susceptibility of the plant against the invading pathogen (Pieterse et al. 2012; Naseem et al. 2012; Naseem and Dandekar 2012).

Plant hormones act in concert (Naseem et al. 2012; Naseem and Dandekar 2012). Crosstalk between defense signaling pathways may provide the plant with a powerful regulatory potential, which helps the plant to "decide" which defensive strategy to follow, depending on the type of attacker it is encountering (De Vos et al. 2005). Plants modulate the relative abundance of SA, JA, and ET levels; modify the expression of defense-related genes; and coordinate complex interactions between defense signaling pathways to activate an effective defense response against attack by various types of pathogens (Bari and Jones 2009). Crosstalk between the hormone signaling systems fine-tunes the defense responses in the plant immune system (Grant and Jones 2009; El Rahman et al. 2012; Pieterse et al. 2009, 2012). Argueso et al. (2012) showed that cytokinin up-regulates plant immunity via an elevation of SA-dependent defense responses and SA in turn feedback inhibits cytokinin signaling. The cross-talk between cytokinin and SA signaling networks may help plants to fine-tune defense responses against pathogens (Argueso et al. 2012). DELLAs, the repressors of GA signaling, promote susceptibility to virulent biotrophs and resistance to necrotrophs, partly by altering the relative strength of salicylic acid and jasmonic acid (Navarro et al. 2008).

Crosstalk between defense signaling pathways may also allow pathogens to manipulate plants to their own benefit by shutting down induced defense through influences on the signaling network. Infection with biotrophic *Pseudomonas syringae*, which induces SA-mediated defense, renders plants more susceptible to the necrotrophic pathogen *Alternaria brassicicola* by suppressing JA signaling pathway (Spoel et al. 2007). *Botrytis cinerea* manipulates the antagonistic effects between immune pathways to promote disease development in tomato (El Oirdi et al. 2011). The rice bacterial blight pathogen *Xanthomonas oryzae* pv. *oryzae* has been shown to hijack the rice ABA machinery to cause disease (Xu et al. 2013). This immune-suppressive effect of ABA may be due to suppression of SA-mediated defenses that normally serve to limit pathogen growth (Xu et al. 2013).

Concentration of the plant hormones in the signaling network may also alter the immune responses. Treatment of rice plants with increasing concentrations of gibberellic acid (GA) enhanced resistance to *Pythium graminicola* in a concentration-dependent manner. Conversely, depletion of endogenous GA levels using the GA biosynthesis inhibitor uniconazole promoted disease susceptibility (De Vleesschauwer et al. 2012). Enhanced biosynthesis of ethylene induced by ACC treatment decreases SA- and JA-associated defense signaling (Shen et al. 2011). A transient synergistic enhancement in the expression of genes associated with either JA or SA signaling was observed when both jasmonic acid and salicylic acid were applied at low concentrations. However, antagonism was observed at more prolonged treatment times or at higher concentrations. Similar results were also observed when adding the jasmonate precursor α-linolenic acid with salicylic acid (Mur et al. 2006). These results suggest that the outcomes of JA–SA interactions depend on the relative concentration of JA and SA.

Activation of multiple hormone signaling pathways may induce resistance against a wide range of pathogens (Hénanff et al. 2013). JA signaling system triggers resistance against necrotrophic pathogens (McGrath et al. 2005; Zheng et al. 2006; Méndez-Bravo et al. 2011; El Rahman et al. 2012), while SA signaling is involved in triggering resistance against biotrophic and hemibiotrophic pathogens (Thaler et al. 2004; Nie 2006; De Vos et al. 2006; Spoel et al. 2007; Jelenska et al. 2007; El Oirdi et al. 2011). By contrast, the SA–JA–ET–ABA signaling network triggers expression of several defense genes and confers resistance to both necrotrophic and biotrophic pathogens (Hénanff et al. 2013). *Bacillus cereus* induces systemic resistance against pathogens by simultaneously activating SA-, JA-, and ET-dependent signaling pathways (Niu et al. 2011).

1.11 Can Molecular Manipulation of Plant Hormone Signaling Network Help the Plant to Win the War Against Pathogens?

Crosstalk between the different hormone signaling systems fine-tunes the defense responses against biotrophic and necrotrophic fungal, oomycete, bacterial, and viral pathogens (Pieterse et al. 2009; Méndez-Bravo et al. 2011; El Rahman et al. 2012; Pieterse et al. 2012; Xu et al. 2013). Crosstalk between different hormone signaling pathways in the hormone signaling network helps the plant to choose the effective defense strategy to follow depending on the type of pathogen it is encountering (De Vos et al. 2005; Niu et al. 2011; Nambeesan et al. 2012; Wang et al. 2012; Hénanff et al. 2013). The crosstalk also allows the pathogens to manipulate plants to their own benefit by shutting down the specific hormone signaling pathway involved in triggering defense responses and hijacking the signaling pathway involved in induction of susceptibility (de Torres-Zabala et al. 2007; Katsir et al. 2008; Xu et al. 2013). Can we manipulate specific signaling system to activate defense responses and suppress the action of pathogens in hijacking the signaling pathway triggering susceptibility? It has been reported that concentration of the plant hormones in the signaling network can alter the strong and fast expression of specific hormones (Mur et al. 2006; Shen et al. 2011; De Vleesschauwer et al. 2012). Manipulation of the signaling network may be a potential strategy to enhance activation and improvement of plant immunity for crop disease management. This book describes the molecular basis of plant hormone-induced immune responses in plants to develop technologies for effective management of crop diseases. Enhancing disease resistance through altered regulation of plant immunity systems would be durable and publicly acceptable (Yamamizo et al. 2006; Shao et al. 2008; Gust et al. 2010; Lacombe et al. 2010; Vidhyasekaran 2014).

References

Abe H, Urao T, Seki M, Shinozaki K, Yamaguchi-Shinozaki Y (2003) Arabidopsis AtMYC2 (bHLH) and AtMYB2 (MYB) function as transcriptional activators in abscisic acid signaling. Plant Cell 15:63–78

Achard P, Genschik P (2009) Releasing the brakes of plant growth: how GAs shut down DELLA proteins. J Exp Bot 60:1085–1092

Adie BA, Perez-Perez J, Godoy M, Sanchez-Serrano JJ, Schmelz EA, Solano R (2007) ABA is an essential signal for plant resistance to pathogens affecting JA biosynthesis and activation of defenses in *Arabidopsis*. Plant Cell 19:1665–1681

Akagi A, Dandekar AM, Stotz HU (2011) Resistance of *Malus domestica* fruit to *Botrytis cinerea* depends on endogenous ethylene biosynthesis. Phytopathology 101:1311–1321

Alazem M, Lin K-Y, Lin N-S (2014) The abscisic acid pathway has multifaceted effects on the accumulation of *Bamboo mosaic virus*. Mol Plant Microbe Interact 27:177–189

Albrecht C, Boutrot F, Segonzac C, Schwessinger B, Gimenez-Ibanez S, Chinchilla D, Rathjen JP, de Vries SC, Zipfel C (2012) Brassinosteroids inhibit pathogen-associated molecular pattern-triggered immune signaling independent of the receptor kinase BAK1. Proc Natl Acad Sci U S A 109:303–308

Alkan N, Fluhr R, Prusky D (2012) Ammonium secretion during *Colletotrichum coccodes* infection modulates salicylic and jasmonic acid pathways of ripe and unripe tomato fruit. Mol Plant Microbe Interact 25:85–96

Alonso-Ramirez A, Rodriguez D, Reyes D, Jiménez JA, Nicolás G, López-Climent M, Gómez-Cadenas A, Nicolás C (2009) Evidence for a role of gibberellins in salicylic acid-modulated early plant responses to abiotic stress in *Arabidopsis* seeds. Plant Physiol 150:1335–1344

An C, Mou Z (2013) The function of the Mediator complex in plant immunity. Plant Signal Behav 8(3):e23182

Anand A, Uppalapati SR, Ryu C-M, Allen SN, Kang L, Tang Y, Mysore KS (2008) Salicylic acid and systemic acquired resistance play a role in attenuating crown gall disease caused by *Agrobacterium tumefaciens*. Plant Physiol 146:703–715

Anderson JP, Badruzsaufari E, Schenk PM, Manners JM, Desmond OJ, Ehlert C, Maclean DJ, Ebert PR, Kazan K (2004) Antagonistic interaction between abscisic acid and jasmonate-ethylene signaling pathways modulates defense gene expression and disease resistance in *Arabidopsis*. Plant Cell 16:3460–3479

Argueso CT, Ferreira FJ, Epple P, To JPC, Hutchison CE, Schaller GE, Dangl JL, Kieber JJ (2012) Two-component elements mediate interactions between cytokinin and salicylic acid in plant immunity. PLoS Genet 8(1):e1002448

Asselbergh B, Achuo AE, Höfte M, Van Gijsegem F (2008) Abscisic acid deficiency leads to rapid activation of tomato defence responses upon infection with *Erwinia chrysanthemi*. Mol Plant Pathol 9:11–24

Babosha AV (2009) Regulation of resistance and susceptibility in wheat powdery mildew pathosystem with exogenous cytokinins. J Plant Physiol 166:1892–1903

Bajguz A, Hayat S (2009) Effects of brassinosteroids on the plant responses to environmental stresses. Plant Physiol Biochem 47:1–8

Bakker PAHM, Doornbos RF, Zamioudis C, Berendsen RL, Pieterse CMJ (2013) Induced systemic resistance and the rhizosphere microbiome. Plant Pathol J 29:136–143

Bari R, Jones JDG (2009) Role of plant hormones in plant defence responses. Plant Mol Biol 69:473–488

Belkhadir Y, Jaillais Y, Epple P, Balsemäo-Pires E, Dangl JL, Chory J (2012) Brassinosteroids modulate the efficiency of plant immune responses to microbe-associated molecular patterns. Proc Natl Acad Sci U S A 109:297–302

Berr A, McCallum EJ, Alioua A, Heintz D, Heitz T, Shen WH (2010) Arabidopsis histone methyltransferase SET DOMAIN GROUP8 mediates induction of the jasmonate/ethylene pathway genes in plant defense response to necrotrophic fungi. Plant Physiol 154:1403–1414

Boudsocq M, Willmann MR, McCormack M, Lee H, Shan L, He P, Bush J, Cheng SH, Sheen J (2010) Differential innate immune signalling via Ca^{2+} sensor protein kinases. Nature 464:418–422

Boutrot F, Segonzac C, Chang KN, Qiao H, Ecker JR, Zipfel C (2010) Direct transcriptional control of the *Arabidopsis* immune receptor FLS2 by the ethylene-dependent transcription factors EIN3 and EIL1. Proc Natl Acad Sci U S A 107:14502–14507

Calderon-Villalobos LI, Tan X, Zheng N, Estelle M (2010) Auxin perception – structural insights. Cold Spring Harb Perspect Biol 2010:a005546

Canet JV, Dobón A, Tornero P (2012) Non-recognition-of BTH4, an *Arabidopsis* mediator subunit homolog, is necessary for development and response to salicylic acid. Plant Cell 24:4220–4235

Cao Y, Wu Y, Zheng Z, Song F (2006) Overexpression of the rice EREBP-like gene *OsBIERF3* enhances disease resistance and salt tolerance in transgenic tobacco. Physiol Mol Plant Pathol 67:202–211

Cao FY, Yoshioka K, Desveaux D (2011) The roles of ABA in plant-pathogen interactions. J Plant Res 124:489–499

Cevik V, Kidd BN, Zhang P, Hill C, Kiddle S, Denby KJ, Holub EB, Cahill DM, Manners JM, Schenk PM, Beynon J, Kazan K (2012) MEDIATOR25 acts as an integrative hub for the regulation of jasmonate-responsive gene expression in Arabidopsis. Plant Physiol 160:541–555

Chanda B, Xia Y, Mandal MK, Yu K, Sekine K-T, Gao Q-M, Selote D, Hu Y, Stromberg A, Navarre D, Kachroo A, Kachroo P (2011) Glycerol-3-phosphate is a critical mobile inducer of systemic immunity in plants. Nat Genet 43:421–427

Chaturvedi R, Krothapalli K, Makandar R, Nandi A, Sparks AA, Roth MR, Welti R, Shah J (2008) Plastid omega3-fatty acid desaturase-dependent accumulation of a systemic acquired resistance inducing activity in petiole exudates of *Arabidopsis thaliana* is independent of jasmonic acid. Plant J 54:106–117

Chaturvedi R, Venables B, Petros RA, Nalam V, Li M, Wang X, Takemoto LJ, Shah J (2012) An abietane diterpenoid is a potent activator of systemic acquired resistance. Plant J 71:161–172

Chen R, Jiang H, Li L, Zhai Q, Qi L, Zhou W, Liu X, Li H, Zheng W, Sun J, Li C (2012) The Arabidopsis mediator subunit MED25 differentially regulates jasmonate and abscisic acid signaling through interacting with MYC2 and ABI5 transcription factors. Plant Cell 24:2898–2916

Chen L, Zhang L, Li D, Wang F, Yu D (2013) WRKY8 transcription factor functions in the TMV-cg defense response by mediating both abscisic acid and ethylene signaling in Arabidopsis. Proc Natl Acad Sci U S A 110(21):E1963–1971. doi:10.1073/pnas.1221347110

Cheng Z, Sun L, Qi T, Zhang B, Peng W, Liu Y, Xie D (2011) The bHLH transcription factor MYC3 interacts with the jasmonate ZIM-domain proteins to mediate jasmonate response in Arabidopsis. Mol Plant 4:279–288

Chini A, Fonseca S, Fernández G, Adie B, Chico JM, Lorenzo O, Garcia-Casado G, López-Vidriero I, Lozano FM, Ponce MR, Micol JL, Solano R (2007) The JAZ family of repressors is the missing link in jasmonate signaling. Nature 448:666–671

Choi H-L, Park H-J, Park JJ, Kim S, Im M-Y, Seo H-H, Kim Y-W, Hwang I, Kim SY (2005) Arabidopsis calcium-dependent protein kinase AtCPK32 interacts with ABF4, a transcriptional regulator of abscisic acid-responsive gene expression, and modulates its activity. Plant Physiol 139:1750–1761

Choi J, Huh SU, Kojima M, Sakakibara H, Paek K-H, Hwang I (2010) The cytokinin-activated transcription factor ARR2 promotes plant immunity via TGA3/NPR1-dependent salicylic acid signaling in *Arabidopsis*. Dev Cell 19:284–295

Choi J, Choi D, Lee R, Ryu CM, Hwang I (2011) Cytokinins and plant immunity: old foes or new friends. Trends Plant Sci 16:388–394

Conaway RC, Conaway JW (2011a) Function and regulation of the Mediator complex. Curr Opin Genet Dev 21:225–230
Conaway RC, Conaway JW (2011b) Origin and activity of the Mediator complex. Semin Cell Dev Biol 22:729–734
Conrath U (2011) Molecular aspects of defence priming. Trends Plant Sci 16:524–531
Daviȇre J-M, Achard P (2013) Development at a glance. Gibberellin signaling in plants. Development 140:1147–1151
de Torres-Zabala M, Truman W, Bennett MH, Lafforgue G, Mansfield JW, Rodriguez Egea P, Bogre L, Grant M (2007) *Pseudomonas syringae* pv. *tomato* hijacks the *Arabidopsis* abscisic acid signalling pathway to cause disease. EMBO J 26:1434–1443
De Vleesschauwer D, Djavaheri M, Bakker PAHM, Höfte M (2008) *Pseudomonas fluorescens* WCS374r-induced systemic resistance in rice against *Magnaporthe oryzae* is based on pseudobactin-mediated priming for a salicylic acid-repressible multifaceted defense response. Plant Physiol 148:1996–2012
De Vleesschauwer D, Yang Y, Vera Cruz C, Höfte M (2010) Abscisic acid-induced resistance against the brown spot pathogen *Cochliobolus miyabeanus* in rice involves MAP kinase-mediated repression of ethylene signaling. Plant Physiol 152:2036–2052
De Vleesschauwer D, Van Buyten E, Satoh K, Balidion J, Mauleon R, Choi I-R, Vera-Cruz C, Kikuchi S, Höfte M (2012) Brassinosteroids antagonize gibberellin- and salicylate-mediated root immunity in rice. Plant Physiol 158:1833–1846
De Vos M, Van Oosten VR, Van Poecke RMP, Van Pelt JA, Pozo MJ, Mueller MJ, Buchala AJ, Métraux J-P, Van Loon LC, Dicke M, Pieterse CMJ (2005) Signal signature and transcriptome changes of *Arabidopsis* during pathogen and insect attack. Mol Plant Microbe Interact 18:923–937
De Vos M, Van Zaanen W, Koornneef A, Korzelius JP, Dicke M, Van Loon LC, Pieterse CMJ (2006) Herbivore-induced resistance against microbial pathogens in *Arabidopsis*. Plant Physiol 142:352–363
Dempsey DA, Vlot AC, Wildermuth MC, Klessig DF (2011) Salicylic acid biosynthesis and metabolism. Arabidopsis Book 9:e0156. doi:10.1199/tab.0156
Denancé N, Sánchez-Vallet A, Goffner D, Molina A (2013) Disease resistance or growth: the role of plant hormones in balancing immune responses and fitness costs. Front Plant Sci 4:Article 155. 1–12
Denoux C, Galletti R, Mammarella N, Gopalan S, Werck D, De Lorenzo G, Ferrari S, Ausubel FM, Dewdney J (2008) Activation of defense response pathways by OGs and Flg22 elicitors in Arabidopsis seedlings. Mol Plant 1:423–445
Dharmasiri N, Dharmasiri S, Estelle M (2005) The F-box protein TIR1 is an auxin receptor. Nature 435:441–445
Diaz-Pendon JA, Li F, Li WX, Ding SW (2007) Suppression of antiviral silencing by cucumber mosaic virus 2b protein in *Arabidopsis* is associated with drastically reduced accumulation of three classes of viral small interfering RNAs. Plant Cell 19:2053–2063
Ding S-W, Voinnet O (2007) Antiviral immunity directed by small RNAs. Cell 130:413–426
Divi UK, Rahman T, Krishna P (2010) Brassinosteroid-mediated stress tolerance in *Arabidopsis* shows interactions with abscisic acid, ethylene and salicylic acid pathways. BMC Plant Biol 10:151
Du Q, Zhu W, Zhao Z, Qian X, Xu Y (2012) Novel benzo-1,2,3-thiadiazole-7-carboxylate derivatives as plant activators and the development of their agricultural applications. J Agric Food Chem 60:346–353
Dubery IA, Sanabria NM, Huang JC (2012) Nonself perception in plant innate immunity. Adv Exp Med Biol 738:79–107
Durrant WE, Dong X (2004) Systemic acquired resistance. Annu Rev Phytopathol 42:185–209
El Oirdi M, El Rahman TA, Rigano L, El Hadrami A, Rodriguez MC, Daay F, Vojnov A, Bouarab K (2011) *Botrytis cinerea* manipulates the antagonistic effects between immune pathways to promote disease development in tomato. Plant Cell 23:2405–2421
El Rahman TA, El Oirdi M, Gonzalez-Lamothe R, Bouarab K (2012) Necrotrophic pathogens use salicylic acid signaling pathway to promote disease development in tomato. Mol Plant Microbe Interact 25:1584–1593

Fan J, Hill L, Crooks C, Doerner P, Lamb C (2009) Abscisic acid has a key role in modulating diverse plant-pathogen interactions. Plant Physiol 150:1750–1761

Feng DX, Tasset C, Hanemian M, Barlet X, Hu J, Trémousaygue D, Deslandes L, Marco Y (2012) Biological control of bacterial wilt in *Arabidopsis thaliana* involves abscisic acid signalling. New Phytol 194:1035–1045

Fu J, Wang S (2011) Insights into auxin signaling in plant-pathogen interactions. Front Plant Sci 2:Article 74. 7

Fu J, Liu H, Li Y, Yu H, Li X, Xiao J, Wang S (2011) Manipulating broad-spectrum disease resistance by suppressing pathogen-induced auxin accumulation in rice. Plant Physiol 155:589–602

Fu ZQ, Yang S, Saleh A, Wang W, Ruble J, Oka N, Mohan R, Spoel SH, Tada Y, Zheng N, Dong X (2012) NPR3 and NPR4 are receptors for the immune signal salicylic acid in plants. Nature 486:228–232

Fujii H, Chinnusamy V, Rodrigues A, Rubio S, Antoni R, Park SY, Cutler SR, Sheen J, Rodriguez PL, Zhu JK (2009) *In vitro* reconstitution of an abscisic acid signalling pathway. Nature 462:660–664

Fung RWM, Gonzalo M, Fekete C, Kovacs LG, He Y, Marsh E, McIntyre LM, Schchtman DP, Qiu W (2008) Powdery mildew induces defense-oriented reprogramming of the transcriptome in a susceptible but not in a resistant grapevine. Plant Physiol 146:236–249

Furihata T, Manuyama K, Fujita Y, Umezawa T, Yoshida R, Shinozaki K, Yamaguchi-Shinozaki K (2006) Abscisic acid-dependent multisite phosphorylation regulates the activity of a transcription activator AREB1. Proc Natl Acad Sci U S A 103:1988–1993

Gaige AR, Ayella A, Shuai B (2010) Methyl jasmonate and ethylene induce partial resistance in *Medicago truncatula* against the charcoal rot pathogen *Macrophomina phaseolina*. Physiol Mol Plant Pathol 74:412–416

Gao Z, Schaller GE (2009) The role of receptor interactions in regulating ethylene signal transduction. Plant Signal Behav 4:1152–1153

Gao Z, Chen YF, Randlett MD, Zhao XC, Findell JL, Kieber JJ, Schaller GE (2003) Localization of the Raf-like kinase CTR1 to the endoplasmic reticulum of Arabidopsis through participation in ethylene receptor signaling complexes. J Biol Chem 278:34725–34732

Garcia-Andrade J, Ramirez V, Flors V, Vera P (2011) Arabidopsis *ocp3* mutant reveals a mechanism linking ABA and JA to pathogen-induced callose deposition. Plant J 67:783–794

Garcion C, Lohmann A, Lamodiere E, Catinot J, Buchala A, Doermann P, Métraux J-P (2008) Characterization and biological function of the *ISOCHORISMATE SYNTHASE2* gene of the Arabidopsis. Plant Physiol 147:1279–1287

Gimenez-Ibanez S, Solano R (2013) Nuclear jasmonate and salicylate signaling and crosstalk in defense against pathogens. Front Plant Sci 4:Article 72. 1–11

González-Lamothe R, El Oirdi M, Brisson N, Bouarab K (2012) The conjugated auxin indole-3-acetic acid-aspartic acid promotes plant disease development. Plant Cell 24:762–777

Grant MR, Jones JDG (2009) Hormone (Dis)harmony moulds plant health and disease. Science 324:750–752

Grefen C, Städele K, Ruzicka K, Obrdlik P, Harter K, Horák J (2008) Subcellular localization and in vivo interactions of the *Arabidopsis thaliana* ethylene receptor family members. Mol Plant 1:308–320

Grosskinsky DK, Naseem M, Abdelmohsen UR, Plickert N, Engelke T, Griebel T, Zeier J, Novák O, Strnad M, Pfeifhofer H, Graaff EVD, Simon U, Roitsch T (2011) Cytokinins mediate resistance against *Pseudomonas syringae* in tobacco through increased antimicrobial phytoalexin synthesis independent of salicylic acid signaling. Plant Physiol 157:815–830

Guo Z, Chen XJ, Wu XL, Ling JQ, Xu P (2004) Overexpression of the AP2/EREBP transcription factor OPBP1 enhances disease resistance and salt tolerance in tobacco. Plant Mol Biol 55:607–618

Gust AA, Brunner F, Nürnberger T (2010) Biotechnological concepts for improving plant innate immunity. Curr Opin Biotechnol 21:204–210

Gutterson N, Reuber TL (2004) Regulation of disease resistance pathways by AP2/ERF transcription factors. Curr Opin Plant Biol 7:465–471

Hauser SY, Waadt R, Schroeder JJL (2011) Evolution of abscisic acid synthesis and signalling mechanism. Curr Biol 21:R346–R355

Hauvermale AL, Ariizumi T, Steber CM (2012) Gibberellin signaling: a theme and variations on DELLA repression. Plant Physiol 160:83–92

Hayashi K-I (2012) The interaction and integration of auxin signaling components. Plant Cell Physiol 53:965–975

Hénanff GL, Profizi C, Courteaux B, Rabenoelina F, Gérard C, Clément C, Baillieul F, Cordelier S, Dhondt-Cordelier S (2013) Grapevine NAC1 transcript factor as a convergent node in developmental processes, abiotic stresses, and necrotrophic/biotrophic pathogen tolerance. J Exp Bot. doi:10.1093/jxb/ert277

Hettenhausen C, Baldwin IT, Wu J (2012) Silencing *MPK4* in *Nicotiana attenuata* enhances photosynthesis and seed production but compromises abscisic acid-induced stomatal closure and guard cell-mediated resistance to *Pseudomonas syringae* pv. *tomato* DC3000. Plant Physiol 158:759–776

Hossain MA, Munemasa S, Uraji M, Nakamura Y, Mori IC, Murata Y (2011) Involvement of endogenous abscisic acid in methyl jasmonate-induced stomatal closure in *Arabidopsis*. Plant Physiol 156:430–438

Hou X, Hu WW, Shen L, Lee IY, Tao Z, Han JH, Yu H (2008) Global identification of DELLA target genes during *Arabidopsis* flower development. Plant Physiol 147:1126–1142

Hou X, Lee LYC, Xia K, Yan Y, Yu H (2010) DELLAs modulate jasmonate signaling via competitive binding of JAZs. Dev Cell 19:884–894

Howe GA (2010) Ubiquitin ligase-coupled receptors extend their reach to jasmonate. Plant Physiol 154:471–474

Hubbard KE, Siegel RS, Valerio G, Brandt B, Schroeder JI (2012) Abscisic acid and CO_2 signalling via calcium sensitivity in guard cells, new CDPK mutant phenotypes and a method for improved resolution of stomatal stimulus–response analyses. Ann Bot 109:5–17

Hwang IS, Hwang BK (2011) The pepper mannose-binding lectin gene *CaMBL1* is required to regulate cell death and defense responses to microbial pathogens. Plant Physiol 155:447–463

Hwang I, Sakakibara H (2006) Cytokinin biosynthesis and perception. Physiol Plant 126:528–538

Hwang I, Sheen J, Muller B (2012) Cytokinin signaling networks. Annu Rev Plant Biol 63:353–380

Iñigo S, Alvarez MJ, Strasser B, Califano A, Cerdán PD (2012) PFT1, the MED25 subunit of the plant Mediator complex, promotes flowering through CONSTANS dependent and independent mechanisms in Arabidopsis. Plant J 69:601–612

Iriti M, Faoro F (2009) Chitosan as a MAMP, searching for a PRR. Plant Signal Behav 4:66–68

Jaillais Y, Hothorn M, Belkhadir Y, Dabi T, Nimchuk ZL, Meyerowitz EM, Chory J (2011) Tyrosine phosphorylation controls brassinosteroid receptor activation by triggering membrane release of its kinase inhibitor. Genes Dev 25:232–237

Jaskiewicz M, Conrath U, Peterhänsel C (2011) Chromatin modification acts as a memory for systemic acquired resistance in the plant stress response. EMBO Rep 12:50–55

Jaubert M, Bhattacharjee S, Mello AF, Perry KL, Moffett P (2011) ARGONAUTE2 mediates RNA silencing anti-viral defenses against *Potato virus X* in Arabidopsis. Plant Physiol 156:1556–1564

Jelenska J, Yao N, Vinatzer BA, Wright CM, Brodsky JL, Greenberg JT (2007) A J domain virulence effector of *Pseudomonas syringae* remodels host chloroplasts and suppresses defenses. Curr Biol 17:499–508

Johansson A, Staal J, Dixelius C (2006) Early responses in the *Arabidopsis- Verticillium longisporum* pathosystem are dependent on *NDR1*, JA- and ET-associated signals via cytosolic NPR1 and *RFO1*. Mol Plant Microbe Interact 19:958–969

Ju C, Chang C (2012) Advances in ethylene signaling: protein complexes at the endoplasmic reticulum (ER) membrane. AoB Plants 2012:pls031. doi:10.1093/aobpla/pls031

Ju C, Yoon GM, Shemansky JM, Lin DY, Ying ZI, Chang J, Gamett WM, Kessenbrock M, Groth G, Tucker ML, Cooper B, Kieber JJ, Chang C (2012) CTR1 phosphorylates the central regulator EIN2 to control ethylene hormone signaling from the ER membrane to the nucleus in *Arabidopsis*. Proc Natl Acad Sci U S A 109:19486–19491

Jung C, Lyou SH, Yeu SY, Kim MA, Rhee S, Kim M, Lee JS, Choi YD, Cheong J-J (2007) Microarray-based screening of jasmonate-responsive genes in *Arabidopsis thaliana*. Plant Cell Rep 26:1053–1063

Jung HW, Tschaplinski TJ, Wang L, Glazebrook J, Greenberg JT (2009) Priming in systemic plant immunity. Science 324:89–91

Kachroo P, Shanklin J, Shah J, Whittle EJ, Klessig DF (2001) A fatty acid desaturase modulates the activation of defense signaling pathways in plants. Proc Natl Acad Sci U S A 98:9448–9453

Kachroo A, Venugopal SC, Lapchyk L, Falcone D, Hildebrand D, Kachroo P (2004) Oleic acid levels regulated by glycerolipid metabolism modulate defense gene expression in Arabidopsis. Proc Natl Acad Sci U S A 101:5152–5157

Katsir L, Schilmiller A, Staswick P, He SY, Howe GA (2008) COI1 is a critical component of a receptor for jasmonate and the bacterial virulence factor coronatine. Proc Natl Acad Sci U S A 105:7100–7105

Kazan K, Manners JM (2009) Linking development to defense: Auxin in plant-pathogen interactions. Trends Plant Sci 14:373–382

Khripach V, Zhabinskii V, Groot AD (2000) Twenty years of brassinosteroids: steroidal plant hormones warrant better crops for XX1 century. Ann Bot 86:441–447

Kidd BN, Edgar CI, Kumar KK, Aitken EA, Schenk PM, Manners JM, Kazan K (2009) The Mediator complex subunit PFT1 is a key regulator of jasmonate-dependent defense in *Arabidopsis*. Plant Cell 21:2237–2252

Kidd BN, Cahill DM, Manners JM, Schenk PM, Kazan K (2011a) Diverse roles of the Mediator complex in plants. Semin Cell Dev Biol 22:741–748

Kidd BN, Kadoo NY, Dombrecht B, Tekeoglu M, Gardiner DM, Thatcher LF, Aitken EA, Schenk PM, Manners JM, Kazan K (2011b) Auxin signaling and transport promote susceptibility to the root-infecting fungal pathogen Fusarium oxysporum in Arabidopsis. Mol Plant Microbe Interact 24:733–748

Kim DS, Hwang BK (2014) An important role of the pepper phenylalanine ammonia-lyase gene (PAL1) in salicylic acid-dependent signaling of the defence response to microbial pathogens. J Exp Bot. doi:10:1093/jxb/eru109

Kim T-W, Guan S, Sun Y, Deng Z, Tang W, Shang JX, Sun Y, Burlingame AL, Wang ZY (2009) Brassinosteroid signal transduction from cell-surface receptor kinases to nuclear transcription factors. Nat Cell Biol 11:1254–1260

Koga H, Dohi K, Mori M (2004) Abscisic acid and low temperatures suppress the whole plant-specific resistance reaction of rice plants to the infection of *Magnaporthe grisea*. Physiol Mol Plant Pathol 65:3–9

Kohler A, Schwindling S, Conrath U (2002) Benzothiadiazole-induced priming for potentiated responses to pathogen infection, wounding, and infiltration of water into leaves requires the *NPR1/NIM1* gene in *Arabidopsis*. Plant Physiol 128:1046–1056

Kombrink E (2012) Chemical and genetic exploration of jasmonate biosynthesis and signaling paths. Planta 236:1351–1366

Kravchuk Z, Vicedo B, Flors V, Camañes G, González-Bosch C, Garcia-Agustin P (2011) Priming for JA-dependent defenses using hexanoic acid is an effective mechanism to protect Arabidopsis against *B. cinerea*. J Plant Physiol 186:359–366

Krinke O, Ruelland E, Valentová O, Vergnolle C, Renou J-P, Taconnat L, Flemr M, Burketova L, Zachowski A (2007) Phosphatidylinositol 4-kinase activation is an early response to salicylic acid in *Arabidopsis* suspension cells. Plant Physiol 144:1347–1359

Lacombe S, Rougon-Cardoso A, Sherwood E, Peeters N, Dahlbeck D, van Esse HP, Smoker M, Rallapalli G, Thomma BPHJ, Stakawicz B, Jones JDG, Zipfel C (2010) Interfamily transfer of a plant pattern-recognition receptor confers broad-spectrum bacterial resistance. Nat Biotechnol 28:365–369

Le Hénanff G, Profizi C, Courteaux B, Rabenoelina F, Gérard C, Clément C, Baillieul F, Cordelier S, Dhondt-Cordelier S (2013) Grapevine NAC1 transcription factor as a convergent node in developmental processes, abiotic stresses, and necrotrophic/biotrophic pathogen tolerance. J Exp Bot. doi:10.1093/jxb/ert277

Leon-Reyes A, Du Y, Koornneef A, Proietti S, Körbes AP, Memelink J, Pieterse CMJ, Ritsema T (2010) Ethylene signaling renders the jasmonate response of *Arabidopsis* insensitive to future suppression by salicylic acid. Mol Plant Microbe Interact 23:187–197

Li Y, Zhang Q, Zhang J, Wu L, Qi Y, Zhou J-M (2010) Identification of microRNAs involved in pathogen-associated molecular pattern-triggered plant innate immunity. Plant Physiol 152:2222–2231

Liu PP, Montgomery TA, Fahlgren N, Kasschau KD, Nonogaki H, Carrington JC (2007) Repression of AUXIN RESPONSE FACTOR10 by microRNA160 is critical for seed germination and post-germination stages. Plant J 52:133–146

Liu P-P, von Dahl CC, Klessig DF (2011a) The extent to which methyl salicylate is required for signaling systemic acquired resistance is dependent on exposure to light after infection. Plant Physiol 157:2216–2226

Liu P-P, von Dahl CC, Park S-W, Klessig DF (2011b) Interconnection between methyl salicylate and lipid-based long-distance signaling during the development of systemic acquired resistance in *Arabidopsis* and tobacco. Plant Physiol 155:1762–1768

Liu D, Chen X, Liu J, Ye J, Guo Z (2012) The rice ERF transcription factor *OsERF922* negatively regulates resistance to *Magnaporthe oryzae* and salt tolerance. J Exp Bot 63:3899–3911

Lu D, Wu S, Gao X, Zhang Y, Shan L, He P (2010) A receptor-like cytoplasmic kinase, BIK1, associates with a flagellin receptor complex to initiate plant innate immunity. Proc Natl Acad Sci U S A 107:495–501

Luna E, Bruce TJA, Roberts MR, Flors V, Ton J (2012) Next generation systemic acquired resistance. Plant Physiol 158:844–853

Mah KM, Uppalapati SR, Tang Y, Allen S, Shuai B (2012) Gene expression profiling of *Macrophomina phaseolina* infected *Medicago truncatula* roots reveals a role for auxin in plant tolerance against the charcoal rot pathogen. Physiol Mol Plant Pathol 79:21–30

Manosalva PM, Park S-W, Forouhar F, Tong L, Fry WE, Klessig DF (2010) *Methyl Esterase 1* (*StMES1*) is required for systemic acquired resistance in potato. Mol Plant Microbe Interact 23:1151–1163

March-Diaz R, Garcia-Domînguez M, Lozano-Juste J, León J, Florencio FJ, Reyes JC (2008) Histone H2A.Z and homologues of components of the SWR1 complex are required to control immunity in *Arabidopsis*. Plant J 53:475–487

Martinez-Medina A, Fernández I, Sánchez-Guzmán MJ, Jung SC, Pascual JA, Pozo MJ (2013) Deciphering the hormonal signalling network behind the systemic resistance induced by *Trichoderma harzianum* in tomato. Front Plant Sci 4:206

Mathys J, De Cremer K, Timmermans P, Van Kerckhove S, Lievens B, Vanhaecke M, Cammue BP, De Coninck B (2012) Genome-wide characterization of ISR induced in *Arabidopsis thaliana* by *Trichoderma hamatum* T382 against *Botrytis cinerea* infection. Front Plant Sci 3:108. doi:10.3389/fpls.2012.00108

Matsuura H, Takeishi S, Kiatoka N, Sato C, Sueda K, Masuta C, Nabeta K (2012) Transportation of de novo synthesized jasmonoyl isoleucine in tomato. Phytochemistry 83:25–33

McGrath KC, Dombrecht B, Manners JM, Schenk PM, Edgar CI, Maclean DJ, Scheible W-R, Udvardi MK, Kazan K (2005) Repressor- and activator-type ethylene response factors functioning in jasmonate signaling and disease resistance identified via a genome-wide screen of Arabidopsis transcription factor gene expression. Plant Physiol 139:949–959

Melotto M, Underwood W, He SY (2008) Role of stomata in plant innate immunity and foliar bacterial diseases. Annu Rev Phytopathol 46:101–122

Méndez-Bravo A, Calderón-Vázquez C, Ibarra-Laclette E, Raya-González J, Ramirez-Chávez E, Molina-Torres J, Guevara-Garcia AA, López-Bucio J, Herrera-Estrella L (2011) Alkamides activate jasmonic acid biosynthesis and signaling pathways and confer resistance to *Botrytis cinerea* in *Arabidopsis thaliana*. PLoS One 6(11):e27251

Mengiste T, Chen X, Salmeron J, Dietrich R (2003) The *Botrytis Susceptible 1* gene encodes an R2R3MYB transcription factor protein that is required for biotic and abiotic stress responses in Arabidopsis. Plant Cell 15:2551–2565

Mersmann S, Bourdais G, Rietz S, Robatzek S (2010) Ethylene signaling regulates accumulation of the FLS2 receptor and is required for the oxidative burst contributing to plant immunity. Plant Physiol 154:391–400

Mitsuhara I, Iwai T, Seo S, Yanagawa Y, Kawahigasi H, Hirose S, Ohkawa Y, Ohashi Y (2008) Characteristic expression of twelve *PR1* family genes in response to pathogen infection, wounding, and defense-related signal compounds. Mol Genet Genomics 279:415–427

Miyazono K, Miyakawa T, Sawano Y, Kubota K, Kang HJ, Asano A, Miyauchi Y, Takahashi M, Zhi Y, Fujita Y, Yoshida T, Kodaira KS, Yamaguchi-Shinozaki K, Tanokura M (2009) Structural basis of abscisic acid signalling. Nature 462:609–614

Mockaitis K, Estelle M (2008) Auxin receptors and plant development: a new signaling paradigm. Annu Rev Cell Dev Biol 24:55–80

Moffat CS, Ingle RA, Wathugala DL, Saunders NJ, Knight H, Knight MR (2012) ERF5 and ERF6 play redundant roles as positive regulators of JA/Et-mediated defense against *Botrytis cinerea* in Arabidopsis. PLoS One 7(4):e35995. doi:10.137/journal.pone.0035995

Mohr PG, Cahill DM (2003) Abscisic acid influences the susceptibility of *Arabidopsis thaliana* to *Pseudomonas syringae* pv. *tomato* and *Peronospora parasitica*. Funct Plant Biol 30:461–469

Mukherjee M, Larrimore KE, Ahmed NJ, Bedick TS, Barghouthi NT, Traw MB, Barth C (2010) Ascorbic acid deficiency in *Arabidopsis* induces constitutive priming that is dependent on hydrogen peroxide, salicylic acid, and the *NPR1* gene. Mol Plant Microbe Interact 23:340–351

Munemasa S, Mori IC, Murata Y (2011) Methyl jasmonate signaling and signal crosstalk between methyl jasmonate and abscisic acid in guard cells. Plant Signal Behav 6:939–941

Mur LAJ, Kenton P, Atzorn R, Miersch O, Wasternack C (2006) The outcomes of concentration-specific interactions between salicylate and jasmonate signaling include synergy, antagonism, and oxidative stress leading to cell death. Plant Physiol 140:249–262

Mur LAJ, Laarhoven LJJ, Harren FJM, Hall MA, Smith AK (2008) Nitric oxide interacts with salicylate to regulate biphasic ethylene production during the hypersensitive response. Plant Physiol 148:1537–1546

Nahar K, Kyndt T, Hause B, Höfte M, Gheysen G (2013) Brassinosteroids suppress rice defense against root-knot nematodes through antagonism with jasmonate pathway. Mol Plant Microbe Interact 26:106–115

Nakashita H, Yasuda M, Nitta T, Asami T, Fujioka S, Arai Y, Sekimata K, Takatsuto S, Yamaguchi I, Yoshida S (2003) Brassinosteroid functions in a broad range of disease resistance in tobacco and rice. Plant J 33:887–898

Nambeesan S, AbuQamar S, Laluk K, Mattoo AK, Mickelbart MV, Ferruzzi MG, Mengiste T, Handa AK (2012) Polyamines attenuate ethylene-mediated defense responses to abrogate resistance to *Botrytis cinerea* in tomato. Plant Physiol 158:1034–1045

Naseem M, Dandekar T (2012) The role of auxin-cytokinin antagonism in plant-pathogen interactions. PLoS Pathog 8(11):e1003026. doi:10.1371/journal ppat 1003026

Naseem M, Philippi N, Hussain A, Wangorsch G, Ahmed N, Dandekar T (2012) Integrated systems view on networking by hormones in *Arabidopsis* immunity reveals multiple crosstalk for cytokinin. Plant Cell 24:1793–1814

Návarová H, Bernsdorff F, Döring A-C, Zeier J (2012) Pipecolic acid, an endogenous mediator of defense amplification and priming, is a critical regulator of inducible plant immunity. Plant Cell 24:5123–5141

Navarro L, Dunoyer P, Jay F, Arnold B, Dhamasiri N, Estelle M, Voinnet O, Jones JDG (2006) A plant miRNA contributes to antibacterial resistance by repressing auxin signaling. Science 312:436–439

Navarro L, Bari R, Achard P, Lison P, Nemri A, Harberd NP, Jones JDG (2008) DELLAs control plant immune responses by modulating the balance of jasmonic acid and salicylic acid signaling. Curr Biol 18:650–655

Negi S, Ivanchenko MG, Muday GK (2008) Ethylene regulates lateral root formation and auxin transport in *Arabidopsis thaliana*. Plant J 55:175–187

Nicaise V, Roux M, Zipfel C (2009) Recent advances in PAMP-triggered immunity against bacteria: Pattern recognition receptors watch over and raise the alarm. Plant Physiol 150:1638–1647

Nie X (2006) Salicylic acid suppresses *Potato virus Y* isolate N:O-induced symptoms in tobacco plants. Phytopathology 96:255–263

Nie H, Wu Y, Yao C, Tang D (2011) Suppression of *edr2*-mediated powdery mildew resistance, cell death and ethylene-induced senescence by mutations in *ALD1* in *Arabidopsis*. J Genet Genomics 38:137–148

Niu Y, Figueroa P, Browse J (2011) Characterization of JAZ-interacting bHLH transcription factors that regulate jasmonate responses in Arabidopsis. J Exp Bot 62:2143–2154

Nurmberg PL, Knox KA, Yun B-W, Morris PC, Shafiei R, Hudson A, Loake GJ (2007) The developmental selector *AS1* is an evolutionarily conserved regulator of the plant immune response. Proc Natl Acad Sci U S A 104:18795–18800

O'Donnell PJ, Schmelz EA, Moussatche P, Lund ST, Jones JB, Klee HJ (2003) Susceptible to intolerance - a range of hormonal actions in a susceptible *Arabidopsis* pathogen response. Plant J 33:245–257

Oñate-Sánchez L, Anderson JP, Young J, Singh KB (2007) AtERF14, a member of the ERF family of transcription factors, plays a nonredundant role in plant defense. Plant Physiol 143:400–409

Owens RA, Tech KB, Shao JY, Sano T, Baker CJ (2012) Global analysis of tomato gene expression during *Potato spindle tuber viroid* infection reveals a complex array of changes affecting hormone signaling. Mol Plant Microbe Interact 25:582–598

Pantelides S, Tjamos SE, Pappa S, Kargakis M, Paplomatas EJ (2013) The ethylene receptor ETR1 is required for *Fusarium oxysporum* pathogenicity. Plant Pathol 62:1302–1309

Parry G, Calderon-Villalobos LI, Prigge M, Peret B, Dharmasiri S, Itoh H, Lechner E, Gray WM, Bennett M, Estelle M (2009) Complex regulation of the TIR1/AFB family of auxin receptors. Proc Natl Acad Sci U S A 106:22540–22545

Pauwels L, Goossens A (2011) The JAZ proteins: A crucial interface in the jasmonate signaling cascade. Plant Cell 23:3080–3100

Peer WA, Cheng Y, Murphy AS (2013) Evidence of oxidative attenuation of auxin signaling. J Exp Bot 64:2629–2639

Petutschnig EK, Jones AME, Serazetdinova L, Lipka U, Lipka V (2010) The Lysin motif receptor-like kinase (LysM-RLK) is a major chitin binding protein in *Arabidopsis thaliana* and subject to chitin-induced phosphorylation. J Biol Chem 285:28902–28911

Pieterse CMJ, Leon-Reyes A, Van der Ent S, Van Wees SCM (2009) Networking by small-molecule hormones in plant immunity. Nat Chem Biol 5:308–316

Pieterse CMJ, Leon-Reyes A, van der Does D, Verhage A, Koornneef A, van Pelt JA, van Wees SCM (2012) Networking by small-molecule hormones in plant immunity. Induced resistance against insects and diseases. IOBC-WPRS Bull 83:77–80

Pré M, Atallah M, Champion A, De Vos M, Pieterse CMJ, Memelink J (2008) The AP2/ERF domain transcription factor ORA59 integrates jasmonic acid and ethylene signals in plant defense. Plant Physiol 147:1347–1357

Qi T, Huang H, Wu D, Qi Y, Song S, Xie D (2014) *Arabidopsis* DELLA and JAZ proteins bind the WD-repeat/bHLH/MYB complex to modulate gibberellin and jasmonate signaling synergy. Plant Cell 26:1118–1133

Qiao H, Shen ZS, Huang C, Schmitz RJ, Urich MA, Briggs SP, Ecker JR (2012) Processing and subcellular trafficking of ER-tethered EIN2 control response to ethylene gas. Science 338:390–393

Qin X, Liu JH, Zhao WS, Chen XJ, Guo ZJ, Peng YL (2013) Gibberellin 20-oxidase gene *OsGA20ox3* regulates plant stature and disease development in rice. Mol Plant Microbe Interact 26:227–239

Qu X, Schaller GE (2004) Requirement of the histidine kinase domain for signal transduction by the ethylene receptor ETR1. Plant Physiol 136:2961–2970

Qutob D, Kemmerling B, Brunner F, Küfner I, Engelhardt S, Gust AA, Luberacki B, Seitz HU, Stahl D, Rauhut T, Glawischnig E, Schween G, Lacombe B, Watanabe N, Lam E, Schlichting

R, Scheel D, Nau K, Dodt G, Hubert D, Gijzen M, Nürnberger T (2006) Phytotoxicity and innate immune responses induced by NEP1-like proteins. Plant Cell 18:3721–3744

Raghavendra AS, Gonugunta VK, Christmann A, Grill L (2010) ABA perception and signaling. Trends Plant Sci 15:395–401

Robert-Seilaniantz A, Grant MR, Jones JDG (2011) Hormone crosstalk in plant disease and defense: more than just jasmonate-salicylate antagonism. Annu Rev Phytopathol 49:2621–2627

Robert-Seilaniantz A, Navarro L, Bari R, Jones JDG (2007) Pathological hormone imbalances. Curr Opin Plant Biol 10:372–379

Rock CD, Sun X (2005) Crosstalk between ABA and auxin signaling pathways in roots of *Arabidopsis thaliana* (l.) Heynh. Planta 222:98–106

Saito N, Munemasa S, Nakamura Y, Shimoishi Y, Mori IC, Murata Y (2008) Roles of RCN1, regulatory A subunit of protein phosphatase 2A, in methyl jasmonate signaling and signal crosstalk between methyl jasmonate and abscisic acid. Plant Cell Physiol 49:1396–1401

Sánchez-Vallet A, López C, Ramos B, Delgado-Cerezo M, Riviere P, Llorente F, Fernández PV, Miedes E, Estevez JM, Grant M, Molina A (2012) Disruption of abscisic acid signaling constitutively activates Arabidopsis resistance to the necrotrophic fungus *Plectosphaerella cucumerina*. Plant Physiol 160:2109–2124

Sato C, Aikawa K, Sugiyama S, Nabeta K, Masuta C, Matsuura H (2011) Distal transport of exogenously applied jasmonoyl-isoleucine with wounding stress. Plant Cell Physiol 52:509–517

Sauer M, Robert S, Kleine-Vehn J (2013) Auxin: simply complicated. J Exp Bot 64:2565–2577

Schmidt K, Pflugmacher M, Klages S, Maser A, Mock A, Stahl DJ (2008) Accumulation of the hormone abscisic acid (ABA) at the infection site of the fungus *Cercospora beticola* supports the role of ABA as a repressor of plant defence in sugar beet. Mol Plant Pathol 9:661–673

Schulze B, Mentzel T, Jehle A, Mueller K, Beeler S, Boller T, Felix G, Chinchilla D (2010) Rapid heteromerization and phosphorylation of ligand-activated plant transmembrane receptors and their associated kinase BAK1. J Biol Chem 285:9444–9451

Schwessinger B, Roux M, Kadota Y, Ntoukakis V, Sklenar J, Jones A, Zipfel C (2011) Phosphorylation-dependent differential regulation of plant growth, cell death, and innate immunity by the regulatory receptor-like kinase BAK1. PLoS Genet 7(4):e1002046

Segonzac C, Zipfel C (2011) Activation of plant pattern-recognition receptors by bacteria. Curr Opin Microbiol 1(4):54–61

Seo PJ, Park C-M (2010) MYB96-mediated abscisic acid signals induce pathogen resistance response by promoting salicylic acid biosynthesis in *Arabidopsis*. New Phytol 186:471–483

Shah J, Zeier J (2013) Long-distance communication and signal amplification in systemic acquired resistance. Front Plant Sci 4:Article 30. doi:10.3389/fpls.2013.00030

Shakeel SN, Wang X, Binder BM, Eric Schaller G (2013) Mechanisms of signal transduction by ethylene: overlapping and non-overlapping signalling roles in a receptor family. AoB Plants 5:plt010. doi:10.1093/aobpla/plt010

Shao M, Wang J, Dean RA, Lin Y, Gao X, Hu S (2008) Expression of a harpin-encoding gene in rice confers durable nonspecific resistance to Magnaporthe grisea. Plant Biotechnol J 6:73–81

Sheard LB, Zheng N (2009) Plant biology signal advance for abscisic acid. Nature 462:575–576

Sheard LB, Tan X, Mao H, Withers J, Ben-Nissan G, Hinds TR, Kobayashi Y, Hsu EF, Sharon M, Browse J, He SY, Rizo J, Howe GA, Zheng N (2010) Jasmonate perception by inositol-phosphate-potentiated COI1-JAZ co-receptor. Nature 468:400–405

Shen X, Liu H, Yuan B, Li X, Xu C, Wang S (2011) OsEDR1 negatively regulates rice bacterial resistance via activation of ethylene biosynthesis. Plant Cell Environ 34:179–191

Shinya T, Osada T, Desaki Y, Hatamoto M, Yamanaka Y, Hirano H, Takai R, Che F-S, Kaku H, Shibuya N (2010) Characterization of receptor proteins using affinity cross-linking with biotinylated ligands. Plant Cell Physiol 51:262–270

Slaughter A, Daniel X, Flors V, Luna E, Hohn B, Mauch-Mani B (2012) Descendants of primed Arabidopsis plants exhibit resistance to biotic stress. Plant Physiol 158:835–843

Solano R, Stepanova A, Chao Q, Ecker JR (1998) Nuclear events in ethylene signaling: a transcriptional cascade mediated by ETHYLENE-INSENSITIVE3 and ETHYLENE-RESPONSE FACTOR1. Genes Dev 12:3703–3714

Son GH, Wan J, Kim HJ, Nguyen XC, Chung W-S, Hong JC, Stacey G (2012) Ethylene-responsive element-binding factor 5, ERF5, is involved in chitin-induced innate immunity response. Mol Plant Microbe Interact 25:48–60

Spoel SH, Dong X (2012) How do plants achieve immunity? Defence without specialized immune cells. Nat Rev Immunol 12:89–100

Spoel SH, Johnson JS, Dong X (2007) Regulation of tradeoffs between plant defenses against pathogens with different lifestyles. Proc Natl Acad Sci U S A 104:18842–18847

Stepanova AN, Alonso JM (2011) Bypassing transcription: a shortcut in cytokinin-auxin interactions. Dev Cell 4:606–610

Sugiyama Y, Uraji M, Watanabe-Sugimoto M, Okuma E, Munemasa S, Shimoishi Y, Nakamura Y, Mori IC, Iwai S, Murata Y (2012) FIA functions as an early signal component of abscisic acid signal cascade in *Vicia faba* guard cells. J Exp Bot 63:1357–1365

Swarup R, Perry P, Hagenbeek D, Van Der Straeten D, Beemster GT, Sandberg G, Bhalerao R, Ljung K, Bennett MJ (2007) Ethylene upregulates auxin biosynthesis in Arabidopsis seedlings to enhance inhibition of root cell elongation. Plant Cell 19:2186–2196

Tatsuki M, Nakajima N, Fujii H, Shimada T, Nakano M, Hayashi K-I, Hayama H, Yoshioka H, Nakamura Y (2013) Increased levels of IAA are required for system 2 ethylene synthesis causing fruit softening in peach (*Prunus persica* L. Batsch). J Exp Bot 64:1049–1059

Thaler JS, Humphrey PT, Whiteman NK (2012) Evolution of jasmonate and salicylate signal crosstalk. Trends Plant Sci 17:260–270

Thaler JS, Owen B, Higgins VJ (2004) The role of the jasmonate response in plant susceptibility to diverse pathogens with a range of lifestyles. Plant Physiol 135:530–538

Thines B, Katsir L, Melotto M, Niu Y, Mandaokar A, Liu G, Nomura K, He SY, Howe GA, Browse J (2007) JAZ repressor proteins are targets of the SCF^{COI1} complex during jasmonate signaling. Nature 448:661

Truman WM, Bennett MH, Turnbull CGN, Grant MR (2010) *Arabidopsis* auxin mutants are compromised in systemic acquired resistance and exhibit aberrant accumulation of various indolic compounds. Plant Physiol 152:1562–1573

Tsai YC, Weir NR, Hill K, Zhang W, Kim HJ, Shiu S-H, Schaller GE, Kieber JJ (2012) Characterization of genes involved in cytokinin signaling and metabolism from rice. Plant Physiol 158:1666–1684

Tsuda K, Sato M, Stoddard T, Glazebrook J, Katagiri F (2009) Network properties of robust immunity in plants. PLoS Genet 5(12):e1000772, pp 16

Umezawa T, Sugiyama N, Mizoguchi M, Hayashi S, Myouga F, Yamaguchi-Shinozaki K, Ishihama Y, Hirayama T, Shinozaki K (2009) Type 2C protein phosphatases directly regulate abscisic acid-activated protein kinases in *Arabidopsis*. Proc Natl Acad Sci U S A 106:17588–17593

Umezawa T, Nakashima K, Miyakawa T, Kuromori T, Tanokura M, Shinozaki K, Yamaguchi-Shinozaki K (2010) Molecular basis of the core regulatory network in ABA responses: sensing, signaling and transport. Plant Cell Physiol 51:1821–1839

Umezawa T, Sugiyama N, Takahashi F, Anderson JC, Ishihama Y, Peck SC, Shinozaki K (2013) Genetics and phosphoproteomics reveal a protein phosphorylation network in the abscisic acid signaling pathway in *Arabidopsis thaliana*. Sci Signal 6(270):rs8. doi:10.1126/scisignal.2003509

Uraji M, Katagiri T, Okuma E, Te W, Hossain M-A, Masuda C, Miura A, Nakamura Y, Mori I-C, Shinozaki K, Murata Y (2012) Cooperative function of PLDδ and PLDα1in abscisic acid-induced stomatal closure in Arabidopsis. Plant Physiol 159:450–460

van den Burg HA, Takken FLW (2009) Does chromatin remodeling mark systemic acquired resistance? Trends Plant Sci 14:286–294

van den Burg HA, Takken FLW (2010) SUMO-, MAPK- and resistance protein-signaling converge at transcription complexes that regulate plant innate immunity. Plant Signal Behav 5:1507–1601

Van der Does D, Leon-Reyes A, Koornneef A, Van Verk MC, Rodenburg N, Pauwels L, Goossens A, Körbes AP, Memelink J, Ritsema T, Van Wees SCM, Pieterse CMJ (2013) Salicylic acid suppresses jasmonic acid signaling downstream of SCF^{COI1}-JAZ by targeting GCC promoter motifs via transcription factor ORA59. Plant Cell 25:744–761

van Verk MC, Pappaaioannou D, Neeleman L, Bol JF, Linthorst HJM (2008) A novel WRKY transcription factor is required for induction of *PR-1a* gene expression by salicylic acid and bacterial elicitors. Plant Physiol 140:1983–1995

Verhage A, van Wees SCM, Pieterse CMJ (2010) Plant immunity: It's the hormones talking, but what do they say? Plant Physiol 154:536–540

Vidhyasekaran P (2004) Concise encyclopedia of plant pathology. Haworth Press, Binghamton, p 619

Vidhyasekaran P (2007) Fungal pathogenesis in plants and crops: molecular biology and host defense mechanisms, 2nd edn. CRC Press, Taylor Francis Group, Boca Raton, p 510

Vidhyasekaran P (2014) PAMP signals in plant innate immunity: signal perception and transduction. Springer, Dordrecht, p 442

Vlad F, Rubio S, Rodrigues A, Sirichandra C, Belin C, Robert N, Leung J, Rodriguez PL, Lauriere C, Meriot S (2009) Protein phosphatases 2C regulate the activation of the Snf1-related kinase OST1 by abscisic acid in *Arabidopsis*. Plant Cell 21:3170–3184

Vriet C, Russinova E, Reuzeau C (2012) Boosting crop yields with plant steroids. Plant Cell 24:842–857

Wang Z-Y (2012) Brassinosteroids modulate plant innate immunity at multiple levels. Proc Natl Acad Sci U S A 109:7–8

Wang X, Kota U, He K, Blackburn K, Li J, Goshe MB, Clouse SD (2008a) Sequential transphosphorylation of the BRI1/BAK1 receptor kinase complex impacts early events in brassinosteroid signaling. Dev Cell 15:220–235

Wang Z, Cao G, Wang X, Miao J, Liu X, Chen Z, Qu L-J, Gu H (2008b) Identification and characterization of COI1-dependent transcription factor genes involved in JA-mediated response to wounding in *Arabidopsis* plants. Plant Cell Rep 27:125–135

Wang Z, Tan X, Zhang Z, Gu S, Li G, Shi H (2012) Defense to *Sclerotinia sclerotiorum* in oilseed rape is associated with the sequential activations of salicylic acid signaling and jasmonic acid signaling. Plant Sci 184:75–82

Wasilewska A, Vlad F, Sirichandra C, Redko Y, Jammes F, Valon C, Frei dit Frey N, Leung J (2008) An update on abscisic acid signaling in plants and more…. Mol Plant 1:198–217

Wasternack C (2007) Jasmonates: an update on biosynthesis, signal transduction and action in plant stress response, growth and development. Ann Bot 100:681–697

Wasternack C, Hause B (2013) Jasmonates: biosynthesis, perception, signal transduction and action in plant stress response, growth and development. An update to the 2007 review in *Annals of Botany*. Ann Bot 111:1021–1058

Wathugala DL, Henisley PA, Moffat CS, Cremelie P, Knight MR, Knight H (2012) The Mediator subunit SFR6/MED16 controls defence gene expression mediated by salicylic acid and jasmonate responsive pathways. New Phytol 195:217–230

Weller DM, Mavrodi DV, Van Pelt JA, Pieterse CMJ, Van Loon LC, Bakker PAHM (2012) Induced systemic resistance (ISR) in *Arabidopsis thaliana* against *Pseudomonas syringae* pv. tomato by 2,4-diacetylphloroglucinol-producing *Pseudomonas fluorescens*. Phytopathology 102:403–412

Wen X, Zhang C, Ji Y, Zhao Q, He W, An F, Jiang L, Guo H (2012) Activation of ethylene signaling is mediated by nuclear translocation of the cleaved EIN2 carboxyl terminus. Cell Res 22:1613–1616

Whenham RJ, Fraser RSS, Brown LP, Payne JA (1986) Tobacco mosaic virus-induced increase in abscisic acid concentration in tobacco leaves – Intracellular location in light and dark-green areas, and relationship to symptom development. Planta 168:592–598

Wild M, Davière JM, Cheminant S, Regnault T, Baumberger N, Heintz D, Baltz R, Genschik P, Achard P (2012) The Arabidopsis DELLA RGA-LIKE3 is a direct target of MYC2 and modulates jasmonate signaling responses. Plant Cell 24:3307–3319

Wu K, Zhang L, Zhou C, Yu C-W, Chalkam V (2008) HDA6 is required for jasmonate response, senescence and flowering in *Arabidopsis*. J Exp Bot 59:225–234
Wu Y, Zhang D, Chu Jee Y, Boyle P, Wang Y, Brindle Ian D, De Luca V, Despr½s C (2012) The Arabidopsis NPR1 protein is a receptor to the plant defense hormone salicylic acid. Cell Rep 1:639–647
Xia Y, Yu K, Navarre D, Seebold K, Kachroo A, Kachroo P (2010) The *glabra1* mutation affects cuticle formation and plant responses to microbes. Plant Physiol 154:833–846
Xie Z, Zhang ZL, Zou X, Huang J, Ruas P, Thompson D, Shen QJ (2005) Annotations and functional analyses of the rice WRKY gene superfamily reveal positive and negative regulators of abscisic acid signaling in aleurone cells. Plant Physiol 137:176–189
Xu J, Audenaert K, Hofte M, De Vleesschauwer D (2013) Abscisic acid promotes susceptibility to the rice leaf blight pathogen *Xanthomonas oryzae* pv. *oryzae* suppressing salicylic acid-mediated defenses. PLoS One 8:e67413
Zheng X-Y, Spivey NW, Zeng W, Liu P-P, Fu ZQ, Klessig DF, He SY, Dong X (2012) Coronatine promotes *Pseudomonas syringae* virulence by activating a signaling cascade that inhibits salicylic acid accumulation. Cell Host Microbe 11:587–596
Yamamizo C, Kuchimura K, Kobayashi A, Katou S, Kawakita K, Jones JDG, Doke N, Yoshioka H (2006) Rewiring mitogen-activated protein kinase cascade by positive feedback confers potato blight resistance. Plant Physiol 140:681–692
Yamamoto S, Suzuki K, Shinshi H (1999) Elicitor-responsive, ethylene-independent activation of GCC box-mediated transcription that is regulated by both protein phosphorylation and dephosphorylation in cultured tobacco cells. Plant J 20:571–579
Yang D-L, Li Q, Deng Y-W, Lou Y-G, Wang M-Y, Zhou G-X, Zhang Y-Y, He Z-H (2008) Altered disease development in the *eui* mutants and *Eui* overexpressors indicates that gibberellins negatively regulate rice basal disease resistance. Mol Plant 1:528–537
Yang DH, Hettenhausen C, Baldwin IT, Wu J (2011) The multifaceted function of BAK1/SERK3: plant immunity to pathogens and responses to insect herbivores. Plant Signal Behav 6:1322–1324
Yang DL, Yao J, Mei CS, Tong XH, Zeng LJ, Li Q, Xiao LT, Sun TP, Li J, Deng W, Lee CM, Thomashow MF, Yang Y, He Z, He SY (2012) Plant hormone jasmonate prioritizes defense over growth by interfering with gibberellin signaling cascade. Proc Natl Acad Sci U S A 109:E1192–E1200
Yang D-L, Yang Y, He Z (2013) Roles of plant hormones and their interplay in rice immunity. Mol Plant 6:675–685
Yazawa K, Jiang C-J, Kojima M, Sakakibara H, Takatsuji H (2012) Reduction of abscisic acid levels or inhibition of abscisic acid signaling in rice during the early phase of *Magnaporthe oryzae* infection decreases its susceptibility to the fungus. Physiol Mol Plant Pathol 78:1–7
Yoo SD, Cho Y-H, Tena G, Xiong Y, Sheen J (2008) Dual control of nuclear EIN3 by bifurcate MAPK cascades in C_2H_4 signalling. Nature 451:789–795
Zamioudis C, Peterse CMJ (2012) Modulation of host immunity by beneficial microbes. Mol Plant Microbe Interact 25:139–150
Zander M, La Camera S, Lamotte O, Metraux JP, Gatz C (2010) *Arabidopsis thaliana* class-II TGA transcription factors are essential activators of jasmonic acid/ethylene-induced defense responses. Plant J 61:200–210
Zander M, Chen S, Imkampe J, Thurow C, Gatz C (2012) Repression of the *Arabidopsis thaliana* jasmonic acid/ethylene-induced defense pathway by TGA-interacting glutaredoxins depends on their C-terminal ALWL motif. Mol Plant 5:831–840
Zhang X, Wang C, Zhang Y, Sun Y, Mou Z (2012) The Arabidopsis mediator complex subunit16 positively regulates salicylate-mediated systemic acquired resistance and jasmonate/ethylene-induced defense pathways. Plant Cell 24:4294–4309
Zhao Q, Guo HW (2011) Paradigms and paradox in the ethylene signaling pathway and interaction network. Mol Plant 4:626–634
Zheng Z, Qamar SA, Chen Z, Mengiste T (2006) Arabidopsis WRKY33 transcription factor is required for resistance to necrotrophic fungal pathogens. Plant J 48:596–605

Zhou C, Zhang L, Duan J, Miki B, Wu K (2005) Histone deacetylase19 is involved in jasmonic acid and ethylene signaling of pathogen response in *Arabidopsis*. Plant Cell 17:1196–1204

Zhu L, Li Y, Li L, Yang J, Zhang M (2011a) Ethylene is involved in leafy mustard systemic resistance to *Turnip mosaic virus* infection through the mitochondrial alternative oxidase pathway. Physiol Mol Plant Pathol 76:166–172

Zhu Z, An F, Feng Y, Li P, Xue LAM, Jiang Z, Kim JM, To TK, Li W, Zhang X, Yu Q, Dong Z, Chen WQ, Seki M, Zhou JM, Guo H (2011b) Derepression of ethylene-stabilized transcription factors (EIN3/EIL1) mediates jasmonate and ethylene signaling synergy in *Arabidopsis*. Proc Natl Acad Sci U S A 108:12539–12544

Chapter 2
Salicylic Acid Signaling in Plant Innate Immunity

Abstract Plants are endowed with innate immune system to protect against invading pathogens. The innate immune system serves as a surveillance system against possible attack by viral, bacterial, fungal, and oomycete pathogens. The innate immune system is a sleeping giant to fight against pathogens, and specific signals are needed to activate them. The pathogen's signature, pathogen-associated molecular pattern (PAMP), switches on the plant innate immune system. The PAMPs are perceived as alarm signals by plant pattern recognition receptors (PRRs), which have a "receptor" and a "signaling domain" in one molecule to perceive and transduce the PAMP signal. Several second messengers are involved in delivering the message generated by the PAMP/PRR signaling complex to plant hormone signals. Salicylic acid (SA) is the important endogenous plant hormone signal in delivering the extracellular PAMP message into the plant cell to initiate the transcription of defense genes. PAMP signaling system generates specific Ca^{2+} signature in the cytosol, which triggers SA biosynthesis. The information encoded in calcium signature is decoded by an array of calmodulins. A calmodulin-binding protein, CBP60g, has been shown to be involved in activating SA biosynthesis. Calcium signature signals transduced to calmodulin-binding protein CBP60g trigger activation of isochorismate synthase in SA biosynthesis pathway. ROS also acts upstream of SA accumulation. H_2O_2 causes an intracellular accumulation of benzoic acid (BA), and the conversion of BA to SA is catalyzed by benzoic acid 2-hydroxylase (BA2H), an inducible enzyme that is synthesized de novo in response to increased BA level. Nitric oxide (NO) activates SA biosynthesis pathway, by inducing phenylalanine ammonia lyase (PAL) which is a key enzyme in biosynthesis of salicylic acid. Several MAP kinase cascades have been shown to act upstream of SA signaling system. SA signaling induces increased expression of transcription factors to activate SA-responsive defense-related genes. NPR1 is a master regulator of the SA-mediated induction of defense genes. NPR1 directly binds SA and binding of SA occurs through $Cys^{521/529}$ via the transition metal copper. Nuclear localization of NPR1 protein is essential for its function. In the absence of pathogen challenge, NPR1 is retained in the cytoplasm. Without induction, NPR1 protein forms an oligomer and is excluded from the nucleus. Pathogen/PAMP exposure induces SA accumulation, and the induced SA controls the nuclear translocation of NPR1

P. Vidhyasekaran, *Plant Hormone Signaling Systems in Plant Innate Immunity*, Signaling and Communication in Plants 2, DOI 10.1007/978-94-017-9285-1_2

through cellular redox changes. In the absence of pathogen challenge, NPR1 is continuously cleared from the nucleus by proteasome, which restricts its co-activator activity to prevent untimely activation of defense responses. Two NPR1 paralogues, NPR3 and NPR4, have been identified as adaptor proteins of the CUL3 E3 ligase, and they target NPR1 degradation in an SA concentration-dependent manner. At increased SA concentration after infection, SA binds to NPR4, and NPR1, freed from NPR4 binding, activates transcription of defense genes. NPR1 is a cofactor of TGA transcription factors, and it enhances binding of TGA transcription factors to the promoter of *PR1* gene to activate transcription of *PR1* gene. Systemic acquired resistance (SAR) is a salicylic acid-dependent heightened state of defense against a broad spectrum of pathogens activated throughout a plant following a local infection. Methyl salicylate, methyl salicylate esterase, a lipid transfer protein (DIR1), a lipid-derived molecule (glycerol-3-phosphate)-dependent factor, azelaic acid, dehydroabietinal, and pipecolic acid have been suggested to be the systemic mobile signal molecules involved in SAR. Some Mediators have been shown to be involved in triggering SA-mediated SAR. Mediator is a multiprotein complex that functions as a transcriptional coactivator. SAR is associated with priming of defense, and the priming results in a faster and stronger induction of defense mechanisms after pathogen attack. Some dormant MAPKs have been suggested to be important components required for priming. Pipecolic acid is an endogenous mediator of defense priming. SAR involves extensive reprogramming of transcription. SA mediates changes in the expression pattern of about 1,000–2,000 genes. Such a broad effect on gene transcription may be associated with extensive chromatin remodeling. The chromatin remodeling may involve substitution of canonical histones in the octamer by histone variants, in a process known as histone replacement. Chromatin structure is important for the regulation of gene expression, and chromatin states could control cellular memory. The primed genes may be poised for enhanced activation of gene expression by the histone modification in chromatin. There may be a tight correlation between histone modification patterns and gene priming, and also there may be a histone memory for information storage in the plant stress response. NPR1 may be involved in the chromatin modification-induced priming. NPR1 plays important role in inducing high levels of chromatin modification on promoters of the transcription factor genes. Chromatin remodeling may be instrumental for priming of SA-responsive loci to enable their enhanced reactivation upon subsequent pathogen attack. The priming can be inherited epigenetically from disease-exposed plants, and descendants of primed plants exhibit next-generation systemic acquired resistance. The descendants of primed plants showed a faster and higher accumulation of transcripts of defense-related genes in salicylic acid signaling pathway and enhanced disease resistance upon challenge inoculation with virulent pathogens. The transgenerational SAR was found to be sustained over one stress-free generation, indicating an epigenetic basis of the phenomenon. DNA methylation may also play an important role in transgenerational SAR. The transgenerational SAR is transmitted by hypomethylated genes that direct priming of SA-dependent defenses in the following generations.

2.1 Salicylic Acid as an Endogenous Immune Signal in Plants

Salicylic acid (SA) is an important endogenous immune signal in the induction of disease resistance response in plants (Anand et al. 2008; Fung et al. 2008; Garcion et al. 2008; Mukherjee et al. 2010; Dempsey et al. 2011; Liu et al. 2011a, b; Argueso et al. 2012; Fu et al. 2012; Denancé et al. 2013; Gimenez-Ibanez and Solano 2013; Yang et al. 2013). An increase in endogenous concentration of SA after an infection has been reported in many plant–pathogen interactions, and this increase is correlated to the activation of defense mechanisms (Garcion and Métraux 2006; Iwai et al. 2007; Nobuta et al. 2007; Spoel et al. 2007; Anand et al. 2008; Fabro et al. 2008; Garcion et al. 2008). SA signaling system activates not only local resistance but also systemic acquired resistance (SAR) observed in distal (systemic) tissues. SAR is an SA-dependent heightened defense to a broad spectrum of pathogens that is activated throughout a plant following local infection (Liu et al. 2011a).

Infection of plants by necrotizing pathogens, which induce the accumulation of SA, or treatment of plants with synthetic compounds, which are able to trigger SA signaling, causes the induction of a unique physiological state called "priming" (Slaughter et al. 2012; Po-Wen et al. 2013). SAR is associated with priming of defense (Kohler et al. 2002; Jung et al. 2009; Luna et al. 2012), and the priming results in a faster and stronger induction of defense mechanisms after pathogen attack (Conrath 2011; Po-Wen et al. 2013). The priming can be inherited epigenetically from disease-exposed plants (Pastor et al. 2013a), and descendants of primed plants exhibit next-generation systemic acquired resistance (Slaughter et al. 2012; Luna et al. 2012). The transgenerational SA-induced SAR has also been reported (Luna et al. 2012; Pieterse 2012). These studies show that SA is an important immune signal in plants triggering local, systemic, and also transgenerational systemic disease resistance.

2.2 Biosynthesis of Salicylic Acid in Plants

2.2.1 Phenylalanine Pathway

Different pathways for synthesis of SA have been reported. SA may be synthesized via phenylalanine pathway (Ogawa et al. 2006; Garcion and Métraux 2006; Sawada et al. 2006) or isochorismate pathway (Wildermuth et al. 2001; Garcion et al. 2008). SA is synthesized via both the isochorismate pathway and phenylalanine pathway in *Arabidopsis* (Mauch-Mani and Slusarenko 1996; Wildermuth et al. 2001; Ferrari et al. 2003). However, Ogawa et al. (2006) have shown that SA is synthesized predominantly via the phenylalanine pathway in tobacco.

SA may be synthesized from *trans*-cinnamic acid in the phenylalanine pathway by two different routes. One involves side-chain decarboxylation of cinnamic acid to benzoic acid, followed by 2-hydroxylation to SA (Fig. 2.1). Alternatively,

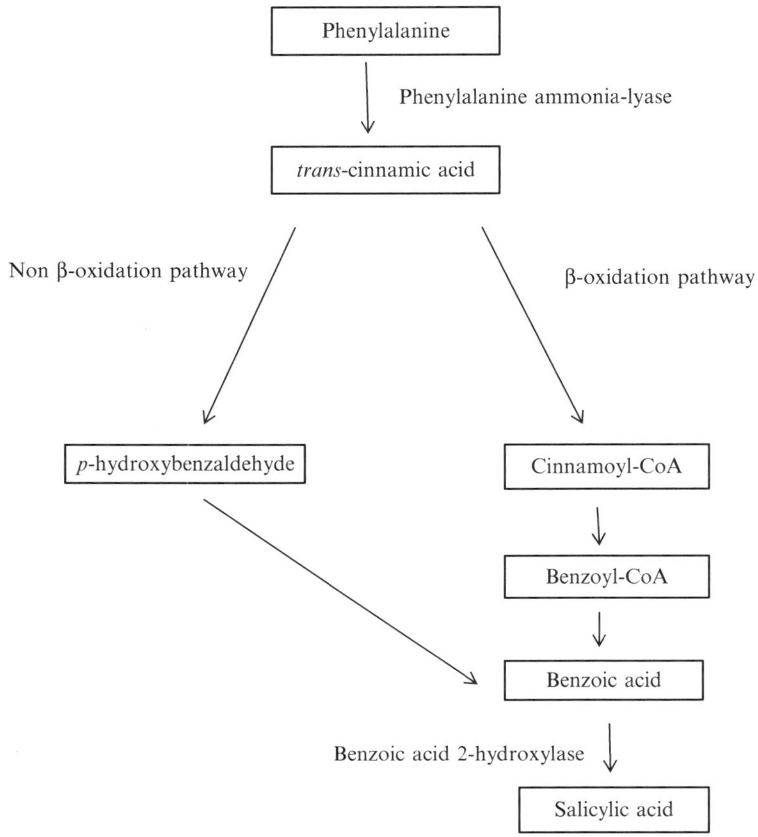

Fig. 2.1 Biosynthesis of salicylic acid in the phenylalanine pathway through benzoic acid route (Adapted from Lee et al. 1995; León et al. 1995; Ribnicky et al. 1998; Chong et al. 2001)

cinnamic acid could be first 2-hydroxylated to *o*-coumaric acid and then decarboxylated to SA (Fig. 2.2) (Lee et al. 1995). *Trans*-cinnamic acid is decarboxylated by two different mechanisms. The first mechanism may operate via β-oxidation, while the other one may be via nonoxidative chain-shortening mechanism (Lee et al. 1995; Fig. 2.1). In the β-oxidation pathway, *trans*-cinnamic acid is converted into cinnamoyl-CoA followed by benzoyl-CoA and then to benzoic acid. In the non-β-oxidation pathway, *p*-hydroxybenzaldehyde is the intermediate in the biosynthesis of benzoic acid (Ribnicky et al. 1998; Chong et al. 2001). Benzoic acid is converted to SA by the enzyme benzoic acid 2-hydroxylase (BA2H). BA2H is a soluble cytochrome P450 oxygenase (León et al. 1993, 1995; Fig. 2.1).

Biosynthesis of SA through *o*-coumaric acid has been demonstrated in tomato seedlings infected with *Agrobacterium tumefaciens* (Chadha and Brown 1974), while biosynthetic pathway via benzoic acid has been reported in tobacco leaves infected with *Tobacco mosaic virus* (Yalpani et al. 1993) and in rice leaves (Silverman et al. 1995).

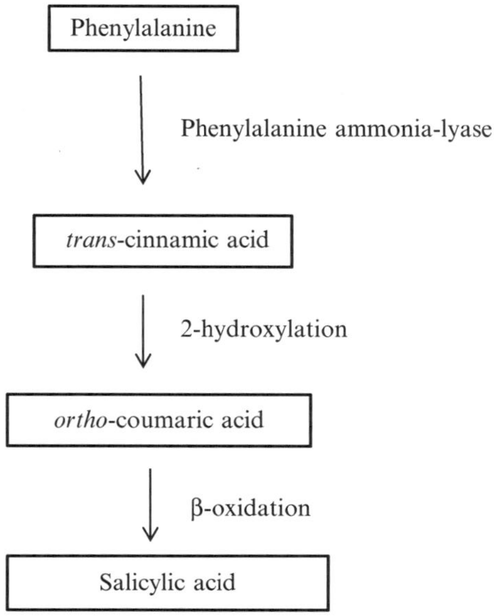

Fig. 2.2 Biosynthesis of salicylic acid in the phenylalanine pathway through *o*-coumaric acid route (Lee et al. 1995)

2.2.2 Isochorismate Pathway

SA synthesis may also proceed through the isochorismate pathway (Wildermuth et al. 2001; Fig. 2.3). In this pathway, SA synthesis proceeds from chorismate via isochorismate, and isochorismate synthase (ICS) is the enzyme involved in the conversion of chorismate to isochorismate (Gaille et al. 2002, 2003). Two ICS genes, *ICS1* and *ICS2*, have been identified in *Arabidopsis* (Garcion et al. 2008). Isochorismate pyruvate lyase (IPL) may be involved in the conversion of isochorismate to SA (Wildermuth et al. 2001; Strawn et al. 2007). However, this enzyme has not yet been reported in plants. No gene encoding proteins similar to bacterial IPL could be detected in *Arabidopsis* (Chen et al. 2009). Two *Arabidopsis* genes, *PBS3* and *EPS*, have been shown to be important in SA biosynthesis. *PBS3* encodes a member of the acyl-adenylate/thioester-forming enzyme family, and *EPS1* encodes a member of the BAHD acyltransferase superfamily. PBS3 (AVR_{ppHB} Susceptible 3) and EPS1 (enhanced pseudomonas susceptibility1) may be the regulatory molecules involved in biosynthesis of SA (Chen et al. 2009; van Verk et al. 2011; Fig. 2.3). The isochorismate pathway has been shown to be active in tomato (Uppalapati et al. 2007) and tobacco (Catinot et al. 2008).

2.2.3 Role of Regulatory Proteins (EDS1, EDS4, PAD4, EDS5, SID2) in Salicylic Acid Biosynthesis

Several regulatory proteins are involved in upstream of SA signaling and involved in SA production in *Arabidopsis* (Fig. 2.4). Mutations in *eds1* (for *enhanced disease susceptibility1*), *eds4*, or *pad4* (for *phytoalexin-deficient4*) lead to reduced SA levels

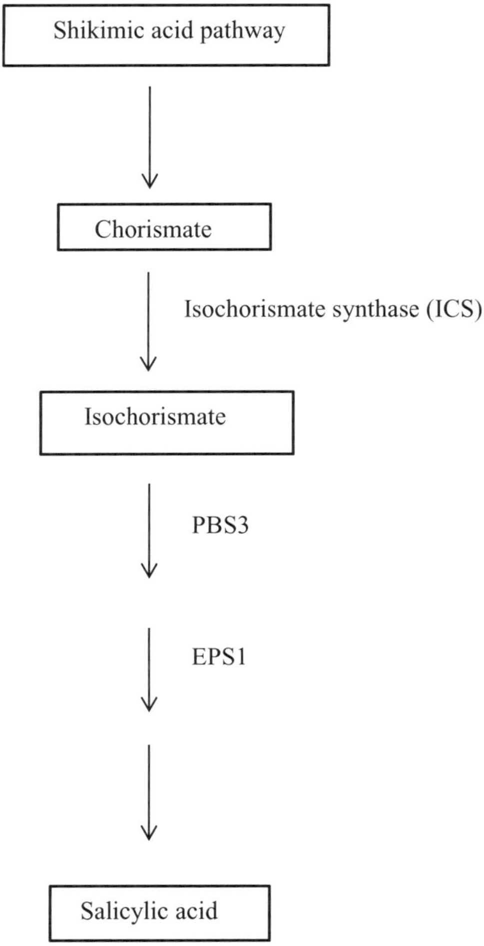

Fig. 2.3 Biosynthesis of salicylic acid in isochorismate pathway (Adapted from Wildermuth et al. 2001; Gaille et al. 2003; Chen et al. 2009)

in infected leaves (Zhou et al. 1998; Gupta et al. 2000; Feys et al. 2001). EDS1 is required for SA production, and it controls SA production to amplify defense signals (Rustérucci et al. 2001; Eulgem et al. 2004). PAD4 is a key regulator acting upstream of SA (Lippok et al. 2007). *Arabidopsis* plants carrying *pad4* mutations have a defect in accumulation of SA upon pathogen infection (Zhou et al. 1998). PAD4 is required for amplification of weak signals to a level sufficient for activation of SA signaling (Jirage et al. 1999). The PAD4 protein sequence displays similarity to triacylglycerol lipases and other esterases (Jirage et al. 1999). EDS1 heterodimerizes with PAD4, and their nuclear localization is important for subsequent steps in the immune signaling pathway (Feys et al. 2001). It is suggested that EDS1 and PAD4 transduce ROS-derived signals leading to SA production (Rustérucci et al. 2001; Mateo et al. 2004; Wiermer et al. 2005). EDS1 and PAD4 may have a fundamental role in transducing redox signals. EDS1 forms several molecularly and

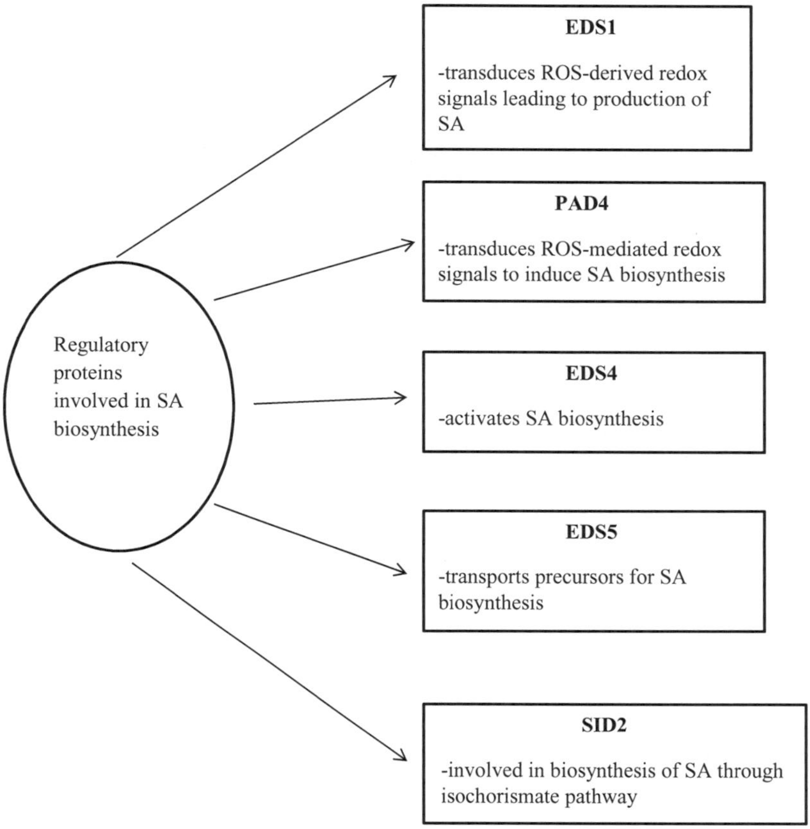

Fig. 2.4 Regulatory proteins involved in salicylic acid biosynthesis

spatially distinct complexes with PAD4 and an in vivo signaling partner, SAG101 (for Senescence-associated gene 101) (Wiermer et al. 2005; Xing and Chen 2006).

SID2 is an isochorismate synthase that is involved in the biosynthesis of SA (Wildermuth et al. 2001). The presence of *sid2* (for *SA induction deficient2*) mutant allele depresses SA synthesis in *Arabidopsis* and compromises the pathogen-induced expression of the *PR1* gene (Wildermuth et al. 2001). The results suggest that SID2 is involved in the biosynthesis of SA through isochorismate pathway. Another gene *EDS5/SID1* encodes a protein, which transports precursors for SA biosynthesis. EDS5 exhibits homology to multidrug and toxin extrusion (MATE) transporter proteins from animals (Nawrath et al. 2002). EDS5 expression requires PAD4, placing EDS5 downstream of PAD4 (Nawrath et al. 2002). EDS4 plays a role in SA signaling and in SA-induced systemic acquired resistance (Gupta et al. 2000). The *eds4* mutant plants showed reduced accumulation of SA induced by *Pseudomonas syringae* pv. *maculicola* infection (Gupta et al. 2000). These results suggest that SA biosynthesis pathway consists of a linear pathway in which EDS1, PAD4, and EDS4 activate EDS5/SID1and SID2, which produce SA (Glazebrook et al. 2003).

2.2.4 An RNA-Binding Protein (RBP) May Be Involved in SA Biosynthesis Pathway

Gene expression is controlled at both transcriptional and post-transcriptional levels. RNA-binding proteins (RBP) are involved in multiple post-transcriptional processes. After protein-coding genes are transcribed into pre-mRNA by RNA polymerase II, processing and modification steps, such as splicing, are required to produce functional mRNA that is ready for export from the nucleus to the cytoplasm (Lorkovic 2009). The cytoplasmic mRNAs can be translated or degraded. RBP can regulate all of these processes. Plant RBPs are characterized by the presence of RNA-binding domains, such as the RNA recognition motif (RRM) or the K-homology (KH) domain (Lorkovic 2009). Many RBPs have been shown to be involved in plant innate immunity (Fu et al. 2007). An RBP from *A. thaliana*, AtRBP-defense related 1 (AtRBP-DR1), is involved in triggering defense responses against *Pseudomonas syringae* pv. *tomato* DC3000 (Qi et al. 2010). Loss-of-function mutants of *AtRBP-DR1* were found to be more susceptible to the bacterial pathogen than the wild-type plants. Transgenic *Arabidopsis* plants overexpressing *AtRBP-DR1* were developed, and these transgenic plants showed high accumulation of SA (Qi et al. 2010). The loss-of-function *AtRBP-DR1* mutant plants accumulated less SA. AtRBP-DR1 overexpression lines showed higher mRNA levels of *SID2*. The *SID2* gene encodes an isochorismate synthase, which is involved in SA biosynthesis. Activation of the SA pathway by AtRBP-DR1 overexpression was dependent on SID2. The SA-related phenotype in the overexpression line was found to be fully dependent on *SID2* (Qi et al. 2010). Overexpression of *AtRBP-DR1* led to high accumulation of SA, and the plants showed enhanced resistance to *P. syringae* pv. *tomato* DC3000 (Qi et al. 2010). Collectively, these studies showed that the RNA-binding protein DR1 activates the expression of *SID2* gene and enhances isochorismate synthase activity, which triggers the SA biosynthesis (Fig. 2.5; Qi et al. 2010).

2.2.5 GH3.5 Is Involved in Salicylic Acid Biosynthesis

Another gene, *GH3.5* (*Gretchen Hagen3.5*), has been shown to be involved in SA accumulation in *Arabidopsis thaliana* (Zhang et al. 2007b). *GH3.5* is a member of the GH3 family of early auxin-responsive genes in *A. thaliana*. It encodes a protein possessing in vitro adenylation activity on SA. The *A. thaliana* mutant overexpressing GH3.5 showed higher levels of SA compared with wild-type plants inoculated with avirulent strains of *P. syringae*. GH3.5 positively modulates

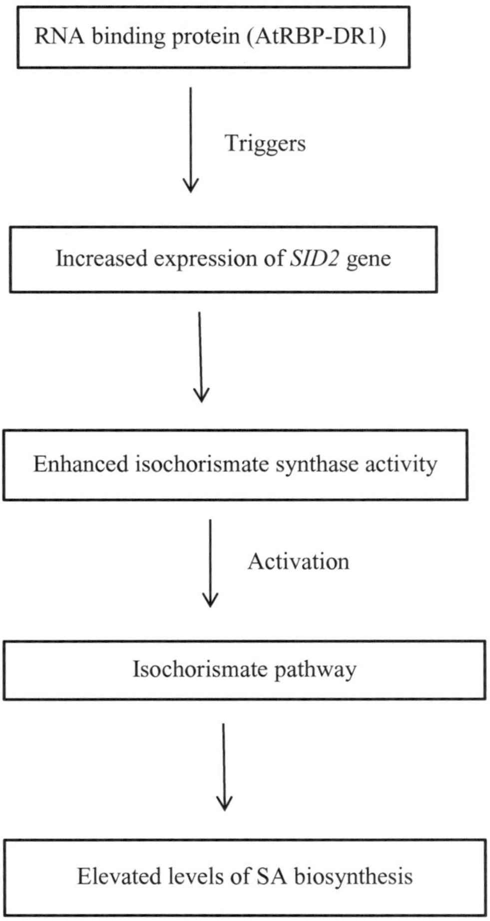

Fig. 2.5 Role of RNA binding protein in triggering salicylic acid biosynthesis (Adapted from Qi et al. 2010)

the SA pathway to enhance plant defense response through elevating SA biosynthesis, activating SA-induced genes, WRKYs, and basal defense-related genes (Zhang et al. 2007b).

2.2.6 *Role of CDR1 Gene in SA Biosynthesis*

An *Arabidopsis* gene, *CDR1* (for *constitutive disease resistance1*), has been shown to take part in reactive oxygen species-mediated SA accumulation in *Arabidopsis* (Xia et al. 2004). It induces accumulation of SA and also induces oxidative burst, which involves H_2O_2 production. The *CDR1* gene encodes an apoplastic protein that shares significant sequence similarity to aspartic proteases. CDR1 contains two active sites with the conserved motifs Asp–Thr–Gly–Ser and

Asp–Ser–Gly–Thr, respectively (Xia et al. 2004). It has been suggested that CDR1 activation might lead to the generation of an endogenous extracellular peptide elicitor. The CDR1-released peptide elicitor might rapidly induce the oxidative burst, and the generated H_2O_2 would have activated SA biosynthesis and accumulation (Xia et al. 2004).

2.2.7 Role of FMO1 Gene in SA Biosynthesis Pathway

Flavin-dependent monooxygenase1 (*FMO1*) gene has been shown to be involved in *EDS1*-mediated SA biosynthesis pathway. *FMO1* positively regulates the *EDS1* pathway (Bartsch et al. 2006). *FMO1* is transcriptionally upregulated in response to superoxide generation but not by hydrogen peroxide (Olszak et al. 2006). An intrinsic activity of EDS1 and PAD4 is to process ROS-derived signals in the defense signaling (Rustérucci et al. 2001; Mateo et al. 2004). FMO may alter the redox state of the EDS1 system (Bartsch et al. 2006) and activate the linear pathway in which EDS1, PAD4, and EDS4 activate EDS5/SID1and SID2, which produce SA (Glazebrook et al. 2003).

2.2.8 Cytokinin May Be Involved in Activation of Salicylic Acid Biosynthesis

Cytokinins are well-known plant growth hormones, and they are also involved in modulation of plant immune responses (Choi et al. 2010, 2011). Cytokinin has been shown to activate SA biosynthesis in *Arabidopsis* (Choi et al. 2010). Within 4 h of the PAMP flg22 application, the expression of *Ics1*, a key SA biosynthetic gene, was hyperactivated 2.7-fold in the presence of the cytokinin t-zeatin (Fig. 2.6; Choi et al. 2010). The increased expression of the *Ics1* gene encoding isochorismate synthase led to elevated SA production (Choi et al. 2010). The results suggest that cytokinin may play an important role in SA biosynthesis pathway.

2.2.9 Some Transcription Factors May Be Involved in Accumulation of Salicylic Acid

Some MYB (*MY*ELO*B*LAST) transcription factors have been shown to play an important role in SA biogenesis. Several *MYB* transcription factor genes are found in plants. They are characterized by the presence of a highly conserved MYB domain at their N-termini (Du et al. 2009). MYB transcription factors contain one or more MYB domains (Stracke et al. 2001). MYB proteins are classified into subfamilies depending on the number of conserved repeats of the MYB domain they

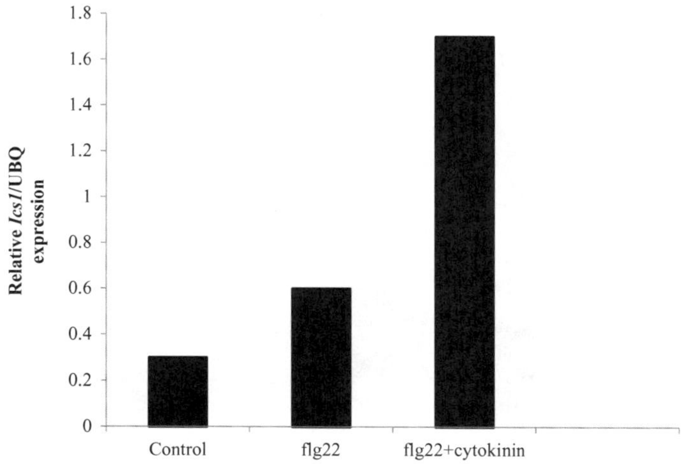

Fig. 2.6 Enhanced expression of the SA biosynthetic gene *ICS1* in cytokinin-treated *Arabidopsis* plants (Adapted from Choi et al. 2010)

contain (Mengiste et al. 2003). A common feature of MYB proteins is the presence of a functional DNA-binding domain that typically consists of one to three repeats of the MYB domain. The three MYB repeats are referred to as R1, R2, and R3. Each repeat is about 50–53 amino acids long and encodes three α-helices, with the second and third helices forming a helix-turn-helix structure which intercalates in the major groove of DNA when bound to it. MYB repeats typically contain regularly spaced tryptophan residues, which build a central tryptophan cluster in the three-dimensional helix-turn-helix fold (Du et al. 2009). MYB transcription factors bind to the *cis-regulatory* element such as MYB-boxes (Laquitaine et al. 2006).

The MYB gene *MYB96* has been shown to be involved in SA biosynthetic pathway (Seo and Park 2010). The *MYB96* gene was rapidly induced by the bacterial pathogen-associated molecular pattern (PAMP) flg22 treatment, and the transcript abundance reached the peak within 1 h. The induction of *SID2* gene, encoding the key SA biosynthesis enzyme isochorismate synthase (ICS), was initiated 2 h after the treatment and reached the peak after flg22 application. The *SID2* gene was significantly up-regulated in the activation-tagging *myb96-1d* line but slightly suppressed in the *myb96-1* mutant (Seo and Park 2010). SA is also synthesized from phenylalanine by phenylalanine ammonia lyase (PAL) activity, although its contribution to endogenous SA content is relatively lower than that of the isochorismate pathway (Lee et al. 1995). The *PAL1* gene was also induced moderately in the activation-tagging *myb96-1d* line (Seo and Park 2010). SA biosynthesis was elevated in the activation-tagging myb96-1d line, and the endogenous concentration of SA was sevenfold higher in the activation-tagging *myb96-1d* line (Fig. 2.7; Seo and Park 2010). These results suggest that PAMP triggers *MYB96* induction, and the induced *MYB96* gene positively regulates the *SID2* gene and also *PAL1* gene to a certain extent, which would result in SA accumulation.

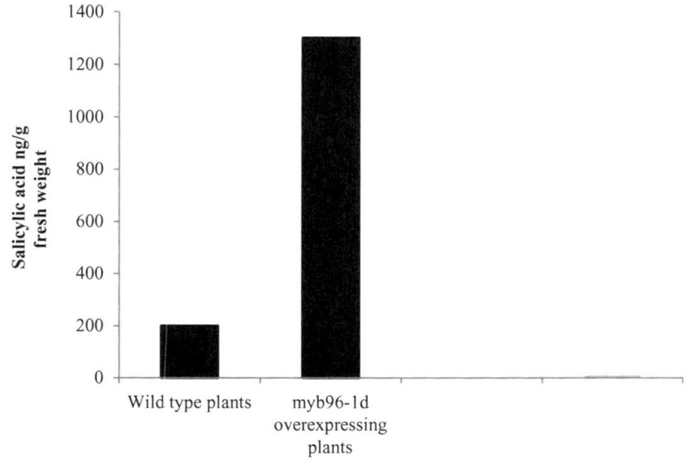

Fig. 2.7 Accumulation of salicylic acid in *Arabidopsis* plants overexpressing *myb96-1d* (Adapted from Seo and Park 2010)

Another transcription factor AtMYB30 has been shown to be involved in SA biosynthesis. AtMYB30 has been shown to be involved in a signaling cascade that modulates (directly or indirectly) SA synthesis. SA operates through the potentiation of reactive oxygen species (ROS) generation, and subsequently, the *ICS* gene encoding isochorismate synthase is positively regulated by AtMYB30, resulting in increased SA production (Raffaele et al. 2006). Alterations of *AtMYB30* expression (overexpression, depletion by antisense strategy, T-DNA insertion mutant) have been shown to modulate SA levels in *Arabidopsis* (Raffaele et al. 2006). *SID1*/*EDS5* and *SID2* which are involved in SA biosynthesis may be activated by AtMYB30 (Raffaele et al. 2006). These studies suggest that AtMYB30 is involved in a signaling cascade that modulates SA synthesis.

Two WRKY transcription factors have been shown to activate SA biosynthetic genes *ICS1* and *PBS3* resulting in accumulation of SA (van Verk et al. 2011). WRKY28 positively regulates the expression of *ICS1* involved in SA biosynthesis pathway. WRKY28 binding sites positioned -445 and -460 base pairs upstream of the transcription start site have been identified in the *ICS1* promoter (van Verk et al. 2011). WRKY46 induces enhanced expression of *PBS3* gene. Overexpression of *WRKY46* resulted in increased accumulation of *PBS3* mRNA, suggesting that WRKY46 is a transcriptional activator of *PBS3* (van Verk et al. 2011). PBS3 (GH3.12) is also involved in pathogen-induced accumulation of SA-O-β-glucoside (SAG), and it regulates SA-dependent defense responses (Nobuta et al. 2007).

Some transcription factors may also act upstream of SA signaling and induce SA biosynthesis pathway. NtWIF is a transcription factor activated upon phosphorylation by wound-induced protein kinase (WIPK) in tobacco plants. The transgenic tobacco plants overexpressing *NtWIF* exhibited 50-fold higher levels of SA (Waller et al. 2006), suggesting that the transcription factor acts upstream of SA biosynthesis.

2.3 Upstream of Salicylic Acid Signaling System

2.3.1 G-Proteins Trigger Salicylic Acid Biosynthesis in SA Signaling System

Several early signaling events have been shown to be involved in accumulation of SA in plant tissues. G-proteins (Yoda and Sano 2003; Fujiwara et al. 2006), Ca^{2+} signaling network system (Garcia-Brugger et al. 2006), MAPK signaling systems (Andreasson et al. 2005; Waller et al. 2006; Zhang et al. 2007a), ROS signaling system (León et al. 1995; Torres et al. 2006; Ahn et al. 2007), and NO signaling (Durner et al. 1998) act upstream of SA accumulation.

Guanosine triphosphate (GTP)-binding proteins (G-proteins) are the regulatory GTPases (Yalowsky et al. 2010; Zhang et al. 2011). The regulatory GTPases have the ability to bind GTP and hydrolyze it to guanosine diphosphate (GDP). GDP locks G-proteins into their inactive state, while GTP locks G-proteins into their activated state. Active or inactive states of G-proteins depend on the binding of GTP or GDP, respectively (Mucha et al. 2011; Wu et al. 2011). These GTPases act as a simple binary switch (the "off" GDP-bound and the "on" GTP-bound states) (Mucha et al. 2011; Wu et al. 2011). Upon stimulation by an upstream pathogen-associated molecular pattern (PAMP) signal, a G-protein-coupled receptor (GPCR) converts the GDP-bound inactive form of the GTPase into the GTP-bound active form through GDP/GTP replacement (Yang 2002; Oki et al. 2009; Pandey et al. 2010).

G-proteins are involved in plant innate immune signaling system initiated by PAMP signal (Gelli et al. 1997; Park et al. 2000; Zhang et al. 2011). They play an important role in SA biosynthesis. G-proteins may trigger salicylic acid biosynthesis in SA signaling system (Beffa et al. 1995). Cholera toxin from *Vibrio cholerae* is a multimeric protein consisting of A1, A2, and five B subunits. The A1 subunit catalyzes the ADP-ribosylation of heterotrimeric G-protein Gα subunit, and this process irreversibly blocks the GTPase activity of G-proteins leading to the sustained activation of the downstream signaling pathway (Beffa et al. 1995). Cholera toxin does not activate G-proteins directly; it acts to maintain the active state of G-proteins with bound GTP (Beffa et al. 1995). Transgenic tobacco plants expressing A1 subunit of cholera toxin were developed, and tissues of these transgenic plants showed accumulation of high levels of salicylic acid (Beffa et al. 1995). The results suggest that the active form of G-proteins triggers the SA biosynthesis. Sano et al. (1994) reported that expression of a small G-protein in transgenic tobacco abnormally induced salicylic acid in response to an external stimulus. Transgenic tobacco plants expressing a rice gene encoding small GTPase, *rgp1*, showed high accumulation of salicylic acid (Yoda and Sano 2003). These studies reveal that G-proteins are involved in SA biosynthesis.

2.3.2 *Calcium Signaling May Act Upstream of Salicylic Acid Accumulation*

2.3.2.1 Ca^{2+} Signature May Modulate SA Biosynthesis and Accumulation Pathway

The PAMP-activated calcium signaling is modulated by calcium signatures. Ca^{2+} signatures (single calcium transients, oscillations, or waves) are generated in the cytosol and in noncytosolic locations including the nucleus and chloroplast through the coordinated action of Ca^{2+} influx and efflux pathways (Luan et al. 2002; Lecourieux et al. 2006; McAinsh and Pittman 2009). Cytosolic Ca^{2+} signals result from a combined action of Ca^{2+} influx through channels and Ca^{2+} efflux through pumps and cotransporters (McAinsh and Pittman 2009; Ward et al. 2009; Boursiac et al. 2010). Influx occurs down the electrochemical gradient through various ion channels, such as voltage-gated channels or Ca^{2+}-permeable cyclic nucleotide-gated channels (CNGCs) or glutamate-gated ion channels (Qi et al. 2010; Moeder et al. 2011; Vatsa et al. 2011; Price et al. 2012; Vincill et al. 2012). Efflux requires energy-dependent Ca^{2+} pumps (autoinhibited Ca^{2+}-ATPases (ACAs) and ER-type Ca^{2+}-ATPases) (McAinsh and Pittman 2009; Boursiac et al. 2010). The cytoplasmic Ca^{2+} signal is shaped by the balance of activity between Ca^{2+} influx and efflux (Boursiac et al. 2010).

Disruption of the vacuolar calcium ATPases in *Arabidopsis* results in the activation of salicylic acid signaling pathway, probably by generating specific Ca^{2+} signature in the cytosol, which triggers SA biosynthesis (Boursiac et al. 2010). A double knockout mutation of the vacuolar Ca^{2+} pumps ACA1 and ACA11 in *Arabidopsis thaliana* resulted in the activation of SA signaling system triggering programmed cell death. Initiation and spread of hypersensitive response that protects plants from pathogens could also be suppressed by disrupting the production of SA in *Arabidopsis* mutants with combined *aca4/11* mutations and a *sid2* (for *salicylic acid induction-deficient2*) mutation. SID2 is an isochorismate synthase that is involved in the biosynthesis of SA (Wildermuth et al. 2001). These studies suggest that disruption of the vacuolar calcium ATPases may result in the activation of *SID2*-mediated SA biosynthesis pathway (Boursiac et al. 2010).

2.3.2.2 Role of Calmodulin-Binding Proteins in SA Biosynthesis

In plant cells, the calcium ion is a ubiquitous intracellular second messenger carrying the signals generated by perception of PAMPs by pattern recognition receptors (PRRs) of host plants (Kwaaitaal et al. 2011; Nürnberger and Küfner 2011; Ranf et al. 2011; Segonzac and Zipfel 2011; Hamada et al. 2012; Stael et al. 2012). Second messengers are molecules that are used by plants to encode information and deliver it downstream to proteins which decode/interpret signals and initiate cellular responses (e.g., changes in enzyme activity, gene expression) (Snedden and Fromm

2001). Calcium ion acts as a signal carrier (Allen et al. 2000). Calcium signaling is modulated by specific calcium signatures (Lecourieux et al. 2006). The calcium signatures are recognized by different calcium sensors to transduce calcium-mediated signals into downstream events (Wang et al. 2012; Hashimoto et al. 2012).

The Ca^{2+} signature controls diverse cellular processes via Ca^{2+} sensors (DeFalco et al. 2010). Calmodulins are the important Ca^{2+} sensor relays identified in plants (Kang et al. 2006; Takabatake et al. 2007). They function through bimolecular interactions. They undergo a conformational change induced by Ca^{2+} before interacting with and changing the activity or structure of the target protein (Lecourieux et al. 2006). Calmodulins are involved in decoding calcium signals (Lecourieux et al. 2006; Kudla et al. 2010).

The information encoded in calcium signature is decoded by an array of calmodulins (Sanders et al. 2002; Hashimoto et al. 2012). Thus, the extracellular signals are transmitted to calmodulins, and the calmodulin-binding proteins on receiving signals from the calmodulins activate enzymes and trigger transcription of specific genes (Lecourieux et al. 2006; Ma and Berkowitz 2007; Dodd et al. 2010; Reddy et al. 2011). A calmodulin-binding protein, CBP60g, has been shown to be involved in activating SA biosynthesis (Wang et al. 2009). Overexpression of CBP60g in *Arabidopsis* caused elevated SA accumulation, increased expression of the defense genes, and enhanced defense responses and enhanced resistance to *Pseudomonas syringae* (Wan et al. 2012). CBP60g has been shown to participate in SA signaling biosynthesis and accumulation (Wang et al. 2009). It has been suggested that the signal coming from CBP60g may act upstream from SA synthesis, as SA levels are reduced in *cbp60g* mutants (Wang et al. 2009).

The effect of *cbp60g* mutant in SA biosynthesis was most similar to that of *pad4* mutant, suggesting that CBP60 may act upstream of PAD4 (Wang et al. 2009). PAD4, a key regulator of SA signaling system, contributes to SA levels. The *pad4* mutant plants showed reduced accumulation of SA after PAMP treatment (Tsuda et al. 2008). It has been shown that the effect of *cbp60g* mutant in SA biosynthesis was almost similar to that of *sid2* mutant (Wang et al. 2009). It suggests that CBP60g may also act upstream of SID2, an isochorismate synthase that is involved in the biosynthesis of SA (Garcion et al. 2008; Wang et al. 2009, 2011; Truman and Glazebrook 2012). Plants carrying *cbp60g* null mutations were compromised in the induction of *SID2* and accumulation of SA (Wang et al. 2009). The calmodulin-binding protein *CBP60g* has been shown to bind to the promoter region of *SID2* (Zhang et al. 2010). A central domain of CBP60g was found to bind to an oligomer with the sequence GAAATTTTGG selected from the *SID2* promoter (Zhang et al. 2010). CBP60g was shown to bind specifically to a 10-mer oligonucleotide with the sequence GAAATTTTGG (Truman and Glazebrook 2012). CBP60g is strongly induced in response to PAMPs treatment (Wang et al. 2009). PAMP-triggered signaling is greatly affected by the loss of *CBP60g* (Wang et al. 2011). Loss of CBP60g severely impacts the plant's ability to produce SA in response to bacterial inoculation and renders the plant susceptible to infection. These results suggest that the calmodulin-binding protein CBP60g binds with *SID2* gene and promotes SA biosynthesis through activation of SID2 (Fig. 2.8).

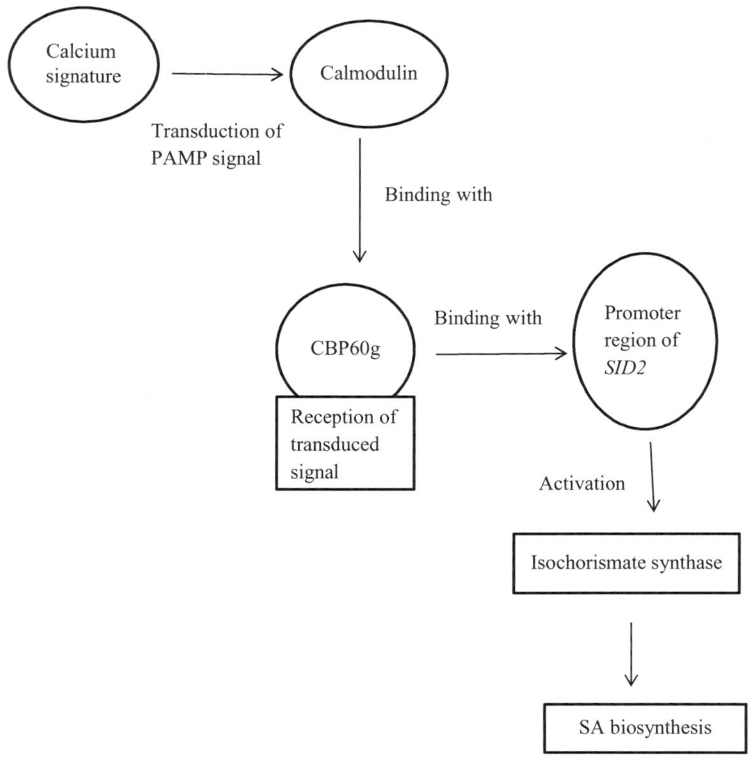

Fig. 2.8 Calcium signature signals transduced to calmodulin-binding protein CBP60g trigger activation of isochorismate synthase in SA biosynthesis pathway (Adapted from Wang et al. 2009, 2011; Truman and Glazebrook 2012; Wan et al. 2012)

One family of CaM-binding proteins, designated as the calmodulin-binding transcription activator (CAMTA) family, resembles a group of putative transcription activators identified in the human genome (Bouché et al. 2002; Galon et al. 2010). This family of proteins contains a transcription activation domain and two types of DNA-binding domains designated the CG1 domain and the transcription factor immunoglobulin domain, ankyrin repeats, and a varying number of IQ CaM-binding motifs (Bouché et al. 2002). *Arabidopsis thaliana* contains six *CAMTA* genes (*AtCAMTA1–AtCAMTA6*). CAMTAs comprise a conserved family of transcription factors (Bouché et al. 2002). The C-terminal CaM-binding domain of CAMTAs mediates interactions with calmodulin (Kudla et al. 2010). CAMTA3 directly interacts with the promoter of the *EDS1* gene, a regulator of salicylic acid levels, and represses its expression (Du et al. 2009). The results suggest that some CaM-binding proteins may negatively regulate SA biosynthesis.

2.3.3 *MAP Kinases May Act Upstream of Salicylic Acid Accumulation*

PAMPs are perceived as danger signals by PRRs, and the PAMP–PRR complex activates the plant innate immunity. Mitogen-activated protein kinase cascades are major signal transduction systems functioning downstream of PRRs upon perception of PAMP elicitor signals (Hettenhausen et al. 2012; Zhang et al. 2012c). A typical MAPK signaling module consists of three interconnected protein kinases: a MAP kinase kinase kinase (MAPKKK or MEKK [for *M*APK/*E*xtracellular signal-regulated kinase *K*inase *K*inase]), a MAP kinase kinase (MAPKK or MKK), and a MAP kinase (MAPK or MPK) (Mészáros et al. 2006; Li et al. 2012). MAPKs function at the bottom of the three-kinase cascade and are activated by MAPKKs through phosphorylation. The activity of MAPKKs is, in turn, regulated by MAPKKKs via phosphorylation. MAPKKKs receive PAMP signals from upstream receptors/sensors to activate the MAPK signaling system (Hirt 2000; Ichimura et al. 2002; Teige et al. 2004; Li et al. 2012).

Several MAP kinase cascades have been shown to act upstream of SA signaling system. MPK4, its upstream MAP kinase kinases MKK1 and MKK2, and the MAP kinase kinase kinase MEKK1 form a cascade that negatively regulates SA signaling system in *Arabidopsis* (Petersen et al. 2000; Mészáros et al. 2006; Qiu et al. 2008a; Pitzschke et al. 2009). The *mpk4* plants exhibit constitutive systemic acquired resistance, including elevated salicylic acid levels and increased resistance to virulent pathogens (Petersen et al. 2000). The *mekk1/mkk1/mkk2* double mutants also display similar elevated levels of SA (Petersen et al. 2000; Gao et al. 2008; Qiu et al. 2008b). The results suggest that the three-kinase module involving MEKK1–MKK1/2–MPK4 negatively regulates SA biosynthesis and signaling system. Two mitogen-activated protein kinases, salicylic acid-induced protein kinase (SIPK) and wound-induced protein kinase (WIPK), are involved in SA signaling system (Ren et al. 2006; Kallenbach et al. 2010; Meldau et al. 2012). WIPK and SIPK function cooperatively to control SA biosynthesis (Seo et al. 2007). Silencing of WIPK/SIPK induced an increase in SA levels (Kobayashi et al. 2010), suggesting that WIPK and SIPK negatively regulate accumulation of SA. EDR1 (ENHANCED DISEASE RESISTANCE 1) is a MAPKK Kinase (MAPKKK), which functions at the top of a MAP kinase cascade. The *edr1* (*enhanced disease resistance 1*) gene encodes a putative MAPKKK, which negatively regulates SA signaling system. All *edr1-associated* phenotypes are suppressed by mutations that reduce SA production (*pad4* and *eds1*) (Frye et al. 2001). The results suggest that EDR1 may negatively regulate PAD4- and EDS1-mediated SA biosynthesis pathway.

Some MAP kinases positively regulate SA signaling pathway in plant innate immunity. Activation of OsMPK6, an ortholog of AtMPK4, resulted in accumulation of SA and enhanced expression of SA signaling genes (Shen et al. 2010). The *MPK7* gene from cotton, *GhMPK7*, has been found to have a role in activating defense responses in plants. Overexpression of the cotton *MAPK* gene in *Nicotiana benthamiana* induced rapid and strong expression of SA pathway genes (Shi et al. 2010). The MAPKK gene detected in *Arabidopsis*, *MKK7*, positively regulates plant basal and systemic acquired resistance. MKK7 has been shown to trigger accumulation of SA, and the increases in SA levels resulted in enhanced expression of PR genes (Zhang et al. 2007a). The activation-tagged *bud1* mutant, in which the expression of *MKK7* is increased, accumulates SA, exhibits constitutive PR gene expression, and displays enhanced resistance to both the oomycete pathogen *Hyaloperonospora parasitica* and *Pseudomonas syringae* pv. *maculicola* (Zhang et al. 2007a). These results suggest that the MAPKs may act upstream of SA biosynthesis.

2.3.4 Reactive Oxygen Species May Act Upstream of Salicylic Acid Accumulation

Hydrogen peroxide (H_2O_2), which is a reactive oxygen species, can also act as a potent signaling molecule (Lehtonen et al. 2012; Petrov and van Breusegem 2012). Pathogen-associated molecular pattern (PAMP) treatment or pathogen infection triggers rapid production of H_2O_2 (Asada 2006; Sagi and Fluhr 2006; Vidhyasekaran 2007; Lehtonen et al. 2012; Petrov and van Breusegem 2012). NADPH oxidases and cell wall peroxidases are the two important groups of enzymes involved in H_2O_2 production (Bolwell et al. 1995, 1998; Suzuki et al. 2011; Daudi et al. 2012; Lehtonen et al. 2012; O'Brein et al. 2012). In rose cells H_2O_2 is produced by a plasma membrane NADPH oxidase, whereas in bean cells H_2O_2 is derived directly from cell wall peroxidases (Bolwell et al. 1998).

H_2O_2 stimulates salicylic acid (SA) biosynthesis in tobacco (León et al. 1995). SA is synthesized in tobacco leaves from benzoic acid (BA) after elicitation (Yalpani et al. 1993). Elicitors trigger the oxidative burst, which results in production of H_2O_2. H_2O_2 causes an intracellular accumulation of BA (León et al. 1993). León et al. (1995) showed that when H_2O_2 was infiltrated into tobacco leaves, the levels of BA increased by more than fivefold. The conversion of BA to SA is catalyzed by benzoic acid 2-hydroxylase (BA2H), an inducible enzyme that is synthesized de novo in response to increased BA level (León et al. 1993). Higher BA levels induce the accumulation of BA2H protein in the cells and provide more substrates for this enzyme. It has been shown that H_2O_2 stimulates BA2H activity (León et al. 1995). BA2H is a soluble Cyt P-450 monooxygenase that uses molecular oxygen for the 2-hydroxylation of benzoic acid (León et al. 1995). In vitro activation of BA2H peroxides was inhibited by a catalase inhibitor 3-amino-1,2,4-triazole (León et al. 1995). The results suggest that the increased 2-hydroxylation activity may be due to the additional oxygen arising from the H_2O_2 degraded by catalase. The catalase-mediated release of molecular oxygen from peroxide may lead to the activation of BA2H, resulting in enhanced accumulation of SA (Fig. 2.9; León et al. 1995).

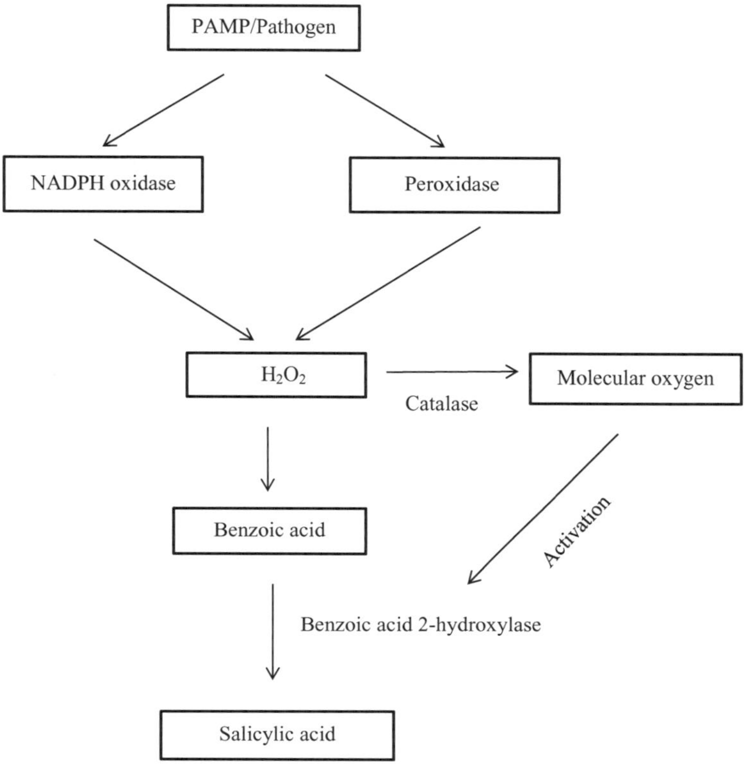

Fig. 2.9 Hydrogen peroxide-stimulated salicylic acid biosynthesis pathway (Adapted from Yalpani et al. 1993; León et al. 1993, 1995; Petrov and van Breusegem 2012)

2.3.5 Nitric Oxide May Act Upstream of Salicylic Acid Accumulation

Nitric oxide (NO) is a key mediator for rapid induction of plant immune signaling systems (Neill et al. 2008; Perchepied et al. 2010; Chun et al. 2012; Bellin et al. 2013). An oomycete PAMP/elicitor triggers a NO burst within minutes in tobacco cells (Foissner et al. 2000; Lamotte et al. 2004). A transient burst of NO has been observed in the roots of *Arabidopsis thaliana* as an early response after contact with *Verticillium longisporum* (Tischner et al. 2010). The NO burst has been reported to occur prior to activation of the SA signaling system that eventually activates the transcription of defense genes.

NO has been reported to activate SA biosynthesis pathway. NO is involved in the production of salicylic acid (Durner et al. 1998). NO donors produce SA accumulation (Durner et al. 1998). NO induces phenylalanine ammonia lyase (PAL) which is a key enzyme in the biosynthesis of salicylic acid (Neill et al. 2002). NO triggers

UDP-glucose/SA glucosyltransferase that converts SA to SA β-glucoside, a conjugated and stable form of SA (Zago et al. 2006). NO is capable of inducing expression of the pathogenesis-related protein PR1, which is known to be induced by SA signaling system (Neill et al. 2002). Expression level of SA-inducible *PR1* gene rises following administration of NO donors or expression of recombinant NO synthase in tobacco (Levine et al. 1994). NO is required for the full function of SA as an SAR (systemic acquired resistance) inducer (Song and Goodman 2001).

2.4 Downstream Events in Salicylic Acid Signaling

2.4.1 Generation of Salicylic Acid Conjugates

SA levels in healthy plants are normally low, but rapidly increase upon infection or in induced resistant plants (Malamy et al. 1992; Zhou et al. 1998; Shapiro and Gutsche 2003). Since excess amount of free SA is phytotoxic, diseased/stressed plants regulate free SA levels in part by glucosylation forming SA-O-β-glucoside (SAG) (Fig. 2.10; Enyedi et al. 1992; Malamy et al. 1992; Chong et al. 2001; Shapiro and Gutsche 2003; Song et al. 2008). The conversion of free SA to SAG is catalyzed by UDP-Glc/SA glucosyltransferases (SAGT) (Lim et al. 2002; Song 2006).

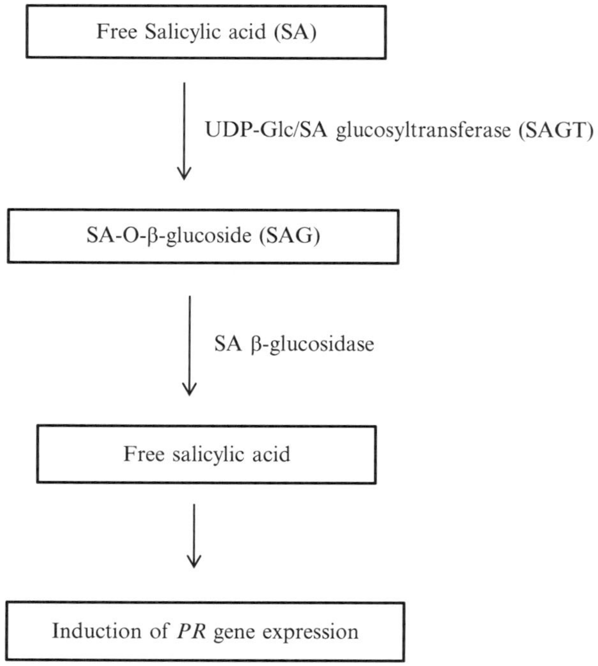

Fig. 2.10 Conversion of free SA to SAG and then to free SA to regulate free SA levels for activation of plant immune responses (Adapted from Seo et al. 1995; Dean et al. 2005; Song et al. 2008)

SA is synthesized in the chloroplast (Strawn et al. 2007), while SAG is formed in the cytoplasm and then transported into the vacuole (Dean et al. 2005). Vacuolar localization of SAG suggests that SAG is primarily a storage form of SA (Nobuta et al. 2007). SAG can be rapidly hydrolyzed to form free SA by endogenous hydrolases (Hennig et al. 1993; Shulaev et al. 1997). SAG is biologically inactive and does not induce *PR* gene expression but can be readily converted back to free active SA by an SA β-glucosidase (Seo et al. 1995).

PBS3 (for *avrPphB* susceptible) (also referred to as GH3.12) is a member of the GH3 protein family of acyl-adenylate/thioester-forming enzymes. GH3 family members act as phytohormone–amino acid synthetases (Staswick and Tiryaki 2004). The *pbs3* mutants are compromised in pathogen-induced accumulation of SAG and expression of *PR1* gene (Nobuta et al. 2007). Exogenous application of SA was sufficient to restore SAG accumulation, *PR1* gene expression, and enhanced disease resistance. The results suggest that the GH3 acyl adenylase is involved in the accumulation of SAG (Nobuta et al. 2007). It is also suggested that GH3 may act directly on SA as an amino acid synthetase to form an SA–amino acid conjugate (Nobuta et al. 2007; Zhang et al. 2007b). The amino acid conjugation may play a critical role in SA metabolism and induced defense responses (Nobuta et al. 2007).

2.4.2 ROS Signaling System May Act Downstream of SA Accumulation

Several effector proteins involved in SA-induced defense responses have been identified. The first protein identified as a salicylic acid-binding protein is the cytosolic (peroxisomal) tobacco catalase (CAT) that reversibly binds SA (Chen et al. 1993; Conrath et al. 1995). SA inhibits CAT's H_2O_2-degrading activity (Durner and Klessig 1996; Wendehenne et al. 1998). SA-mediated inhibition of CAT may generate H_2O_2, which may activate the ROS signaling system inducing expression of defense genes (Chen et al. 1993). Peroxidase has also been reported to interact directly with SA (Durner and Klessig 1995). ROS signaling system may act both upstream and downstream of SA signaling (Slaymaker et al. 2002).

A second specific high-affinity SA-binding protein, SABP2, has been identified in tobacco cytoplasm (Du and Klessig 1997). The third SA-binding protein identified in tobacco is SABP3 (Slaymaker et al. 2002). SABP3 has been detected in the soluble fraction of purified tobacco leaf chloroplasts. SABP3 binds SA with high affinity. Partial sequencing of SABP3 indicated that it is the chloroplast carbonic anhydrase (Slaymaker et al. 2002). SABP3 shows antioxidant activity, and SA may inhibit the antioxidant activity by binding with SABP3. SA's ability to inhibit antioxidant enzymes may play a role in ROS levels, which might activate a positive feedback loop that amplifies SA production and induces defense responses (Slaymaker et al. 2002).

2.4.3 NO May Act Downstream of SA Accumulation

Downstream of SA accumulation, SA has been shown to activate NO synthesis in *Arabidopsis* (Zottini et al. 2007). Many NO-regulated enzymes, including aconitase or catalase, are regulated by SA, suggesting that NO signaling system acts downstream of SA (Durner et al. 1997; Clark et al. 2000). A tight interrelationship between NO and SA in plant defense has been reported (Kumar and Klessig 2000). NO activates salicylate-induced protein kinase (SIPK) in tobacco. Studies with transgenic *NahG* tobacco revealed that SA is required in the NO-mediated induction of SIPK. SIPK may function downstream of SA in the NO signaling pathway (Kumar and Klessig 2000).

NO serves as a key redox-active signal for the activation of various SA downstream defense responses (Klessig et al. 2000). NPR1 and TGA1 are key redox-controlled regulators of SA-induced systemic acquired resistance (SAR) in plants. The translocation of NPR1 into the nucleus has been shown to be promoted by NO (Lindermayr et al. 2010).

NO acts substantially in cellular signal transduction through stimulus-coupled S-nitrosylation of cysteine residues (Benhar et al. 2008). The addition of an NO moiety to a cysteine (Cys) thiol to form an S-nitrosothiol (SNO) is termed S-nitrosylation (Malik et al. 2011). NO reacts rapidly with glutathione (GSH) to yield S-nitrosoglutathione (GSNO) (Espunya et al. 2012). GSNO acts synergistically with SA in SAR (Espunya et al. 2012).

NPR1 is an important regulator of SAR downstream of SA (Mou et al. 2003; Zhang et al. 2003). The events downstream of SA include an increase of NO (Krinke et al. 2007; Zottini et al. 2007). NO is required for the full function of NPR1 in SA-triggered SAR (Song and Goodman 2001). Nuclear localization of NPR1 protein is essential for its function (Kinkema et al. 2000; Meur et al. 2006). Without induction by SA, NPR1 protein forms an oligomer and is excluded from the nucleus. Redox changes cause monomeric NPR1 to emerge and accumulate in the nucleus and activate PR gene expression (Kinkema et al. 2000; Mou et al. 2003). NPR1 is sequestered in the cytoplasm as an oligomer through intermolecular disulfide bonds. NO-mediated S-nitrosylation of NPR1 by S-nitrosoglutathione (GSNO) at Cys156 facilitates the NPR1 oligomerization, which maintains protein homeostasis upon SA induction (Tada et al. 2008). These results suggest that NO is involved in the action of NPR1 in triggering SAR. It also has been shown that NO-induced nitrosoglutathione could act as a long-distance phloematic signal in SA-induced SAR (Durner and Klessig 1999).

2.4.4 MAPK Signaling Cascade May Act Downstream in SA Signaling System

MAP kinase signaling cascade has been widely reported to act upstream in SA biosynthesis pathway (Zhang et al. 2007a; Shen et al. 2010; Shi et al. 2010). It is also known that SA activates MAPK signaling cascade downstream of SA signaling pathway (Zhang et al. 1998; Kumar and Klessig 2000; Zhang and Liu 2001; Cheong et al. 2003; Uppalapati et al. 2004; Brodersen et al. 2006). SA strongly induced p48 and p44 MAPKs in pea (Uppalapati et al. 2004). It activated a 48-kD MAP kinase and SIPK in tobacco (Zhang and Klessig 1997; Zhang and Liu 2001). A MAPKK encoding gene *GhMKK5* from cotton has been isolated and characterized. *GhMKK5* is significantly induced by SA (Zhang et al. 2012a). The expression of SA signaling system-inducible *PR1a* and *PR5* was greatly elevated in *GhMKK5*-overexpressing *Nicotiana benthamiana* (Zhang et al. 2012a). Another SA signaling pathway gene *NPR1*, which is involved in systemic acquired resistance (SAR) response, was also significantly increased in *GhMKK5*-overexpressing plants (Zhang et al. 2012a). The results suggest that GhMKK5 may act downstream in SA signaling system.

EDR1 (ENHANCED DISEASE RESISTANCE 1) is a MAPKK Kinase (MAPKKK), which functions at the top of a MAP kinase cascade. The *NahG* transgene, which lowers endogenous SA levels, also suppresses *EDR1* expression in *Arabidopsis* (Frye et al. 2001), suggesting that SA triggers expression of *EDR1*. The *EDR1* negatively regulates SA signaling system (Frye et al. 2001).

SA activated the salicylic acid-inducible protein kinase (SIPK) in tobacco (Kumar and Klessig 2000). Studies with transgenic *NahG* tobacco revealed that SA is required for the induction of SIPK and the SIPK may function downstream of SA in the SA signaling pathway (Kumar and Klessig 2000). Another MAPK, MPK4, has been shown to act downstream of SA in the SA signaling system (Petersen et al. 2000). The *mpk4* knockout mutant shows elevated SA levels and constitutively expresses pathogenesis-related (PR) genes (Petersen et al. 2000). Expression of the bacterial NahG salicylate hydroxylase in *mpk4* plants abolishes PR gene expression, indicating the role of the MAPK in SA-mediated signaling system (Petersen et al. 2000; Brodersen et al. 2006). A substrate for MPK4 has been identified and it was designated MKS1 (for *M*AP *K*inase 4 *S*ubstrate *1*) (Andreasson et al. 2005). MPK4 interacts with the nuclear protein MKS1 that in turn interacts with two WRKY transcription factors, WRKY25 and WRKY33 (Andreasson et al. 2005). The molecular phenotypes of plants over- or under-expressing *MKS1* indicate that it mediates some effects of MPK4 on SA-mediated resistance responses. The results suggest that the MKS1 is required for SA-dependent resistance in *Arabidopsis* (Andreasson et al. 2005). The transcription factors WRKY25 and WRKY33 may function as downstream components of the MPK4-mediated signaling pathway and contribute to repression of SA-dependent disease resistance response (Andreasson et al. 2005).

2.5 SA Signaling Induces Increased Expression of Transcription Factors to Activate SA-Responsive Defense-Related Genes

2.5.1 SA Induces WRKY Transcription Factors

Several transcription factors are involved in activation of transcription of defense genes. Proteins containing WRKY zinc-finger motifs constitute a class of transcription factors. WRKY is localized to the nucleus of plant cells and recognizes DNA molecules containing the TTGAC(C/T) W-box sequence (Zheng et al. 2006). A common feature of all WRKY proteins is the WRKY domain, a highly conserved stretch of about 60 amino acids. Each WRKY domain contains a C-terminal-located novel zinc finger and the strictly conserved amino acid sequence WRKYGQK (tryptophan–arginine–lysine–tyrosine–glycine–glutamine–lysine) at its N terminus. This sequence is required for proper folding of the zinc finger and for DNA binding (Maeo et al. 2001). The WRKY domain binds specifically to various W-box elements, all containing a TGAC core sequence (Wang et al. 1998).

The promoters of a large number of defense-related genes including *PR* genes contain W-box sequences that are recognized by WRKY proteins and are necessary for the inducible expression of these defense genes (Yu et al. 2001). WRKY transcription factors are important regulators of SA-dependent defense responses (Maleck et al. 2000; Wang et al. 2006). In *Arabidopsis*, 74 WRKY transcription factors have been reported (Eulgem et al. 2000), and most of these transcription factors are involved in plant defense responses. Forty-nine of 72 WRKY genes tested were differentially regulated in *Arabidopsis* plants treated with SA (Dong et al. 2003). WRKY proteins have been shown to be strongly involved in alteration of gene expression in response to SA treatment (Eulgem et al. 1999; Dellagi et al. 2000; Asai et al. 2002; Chen and Chen 2002; Yoda et al. 2002; Kalde et al. 2003).

SA induces increased expression of WRKY6, WRKY7, WRKY11, WRKY17, WRKY18, WRKY25, WRKY26, WRKY33, WRKY38, WRKY40, WRKY46, WRKY53, WRKY54, WRKY60, WRKY62, WRKY63, WRKY64, WRKY67, WRKY70, and WRKY75 in *Arabidopsis* (Chen and Chen 2002; Dong et al. 2003; Kalde et al. 2003; Li et al. 2004; Knoth et al. 2007; Mao et al. 2007; Miao and Zentgraf 2007; Zheng et al. 2006, 2007). Expression of WRKY70 is activated by SA, and NPR1 is required for induction of the transcription factor (Li et al. 2004). Overexpression of WRKY70 induced enhanced expression of SA-responsive PR genes (Li et al. 2004). WRKY70 functions downstream of ROS and SA (Knoth et al. 2007). WRKY11 and WRKY17 function as negative regulators of WRKY70 (Journot-Catalino et al. 2006).

Expression of WRKY62 is induced by SA (Mao et al. 2007). SA-inducible WRKY53 transcription factor interacts with a JA-inducible protein, which triggers resistance against bacterial and fungal pathogens in *Arabidopsis* (Miao and Zentgraf 2007). WRKY18 is an SA-inducible transcription factor. It positively modulates defense-related gene expression (Chen and Chen 2002). WRKY18 physically interacts with structurally related *Arabidopsis* WRKY40 and WRKY60, resulting in altered DNA-binding activities (Xu et al. 2006). While constitutive expression of WRKY18 enhanced resistance to *Pseudomonas syringae*, its coexpression with WRKY40 or WRKY60 made plants more susceptible to the pathogen (Xu et al. 2006). SA may induce the activation of WRKY1 through its action on SA-inducible protein kinase (SIPK). SIPK phosphorylates and activates WRKY1, and coexpression of WRKY1 and SIPK results in more rapid induction of defense-related cell death than overexpression of either alone (Menke et al. 2005). A rice WRKY gene, OsWRKY13, activates both SA synthesis-related genes and SA-responsive genes. It suggests that OsWRKY13 activates genes activating both upstream and downstream of SA (Qiu et al. 2007). A *Vitis vinifera* transcription factor VvWRKY1 is induced by SA. It encodes a polypeptide of 151 amino acids, and it specifically interacts with the W-box in various nucleotide contexts (Marchive et al. 2007). The potato transcription factor StWRKY is also induced by SA (Dellagi et al. 2000).

A novel WRKY transcription factor, NtWRKY12, has been shown to be induced by SA. NtWRKY12 belongs to the class of transcription factors in which the WRKY sequence is followed by a GKK rather than a GQK sequence. The binding sequence of NtWRKY12 (WK-box TTTTCCAC) deviated significantly from the consensus sequence (W-box TTGAC[C/T]) shown to be recognized by WRKY factors with the GQK sequence (van Verk et al. 2008). NtWRKY12 binds with specific binding sites in the *PR1a* promoter in positions -564 (box WK1) and -859 (box WK2). NtWRKY12 and TGA1a act synergistically in *PR1a* expression induced by salicylic acid (van Verk et al. 2008).

Some of the WRKY transcription factors may act as negative regulators of SA-mediated defense gene expression. Some WRKY factors have been shown to negatively regulate pathogen-induced *PR* gene expression (Journot-Catalino et al. 2006; Wang et al. 2006; Xu et al. 2006; Zheng et al. 2007). WRKY48 acts as a negative regulator of SA-induced *PR1* gene expression in *A. thaliana* and basal resistance to *Pseudomonas syringae* (Xing et al. 2008). SA induces WRKY7 gene in *A. thaliana*. However, WRKY7 is a transcriptional repressor in *Arabidopsis*, and transgenic plants overexpressing WRKY7 showed reduced expression of defense-related genes, including *PR1* (Kim et al. 2006). SA induces transcription of WRKY33. However, SA-mediated PR gene expression is retarded by WRKY33, suggesting that WRKY33 negatively acts with SA in inducing defense gene expression (Zheng et al. 2006). WRKY25 is positively regulated by SA signaling pathway. However, its overexpression reduced expression of the SA-regulated PR1 gene. WRKY25 is a substrate of *Arabidopsis* MAP kinase 4, a repressor of SA-dependent defense responses (Zheng et al. 2007).

2.5.2 SA Induces ERF Transcription Factors

SA induces another type of transcription factors, called ethylene response factors (ERF). ERF transcriptional factors are a subfamily of the APETELA2 (AP2) transcription factor family and contain a single DNA-binding domain. The target sequence for ERF transcription factors is the GCC-box that is found in several promoters of PR genes (Grennan 2008). SA induces expression of *TaERF3* in wheat (*Triticum aestivum*) (Zhang et al. 2007b).

2.6 NPR1 Is Master Regulator of SA Signaling

2.6.1 NPR1 Acts Downstream of SA Signal

NPR1 (for *non-expresser of PR gene1*, also known as *NIM1* [*nonimmunity1*] and *SAI1* [*salicylic acid inducible1*]) gene is a master regulator of the SA-mediated induction of defense genes. NPR1 is an important regulator of responses downstream of SA (Mou et al. 2003; Zhang et al. 2003). NPR1 expression levels become elevated upon induction by SA (Cao et al. 1997). *Arabidopsis npr1/nim1* mutants are impaired in their ability to induce PR gene expression even after treatment with SA (Chern et al. 2008), suggesting the function of NPR1 downstream of SA. NPR1 contains a bipartite nuclear localization sequence and two potential protein–protein interaction domains: an ankyrin repeat domain and a BTB/POZ (for Broad complex, Tramtrack, and a Bric-a-brac/Pox virus and Zinc finger) domain (Cao et al. 1997; Ryals et al. 1997). NPR1 functions as a transcriptional co-activator in a TGA2–NPR1 complex after SA treatment. This function requires the BTB/POZ domain and the oxidation of NPR1 Cys521 and Cys529 (Rochon et al. 2006).

NPR1 directly binds SA and binding of SA occurs through $Cys^{521/529}$ via the transition metal copper. NPR1 binds both SA and copper through $^{Cys521/529}$. Removal of metals through chelation abolishes the binding of SA by NPR1, even in the presence of $Cys^{521/529}$. Binding of SA causes a conformational change in NPR1 that is accompanied by the release of the C-terminal transactivation domain from the N-terminal autoinhibitory BTB/POZ domain (Wu et al. 2012).

2.6.2 SA Controls Nuclear Translocation of NPR1

Nuclear localization of NPR1 protein is essential for its function (Kinkema et al. 2000; Meur et al. 2006). In the absence of pathogen challenge, NPR1 is retained in the cytoplasm. Without induction, NPR1 protein forms an oligomer and is excluded from the nucleus. Pathogen/PAMP exposure induces SA accumulation (Durrant and Dong 2004), and the induced SA controls the nuclear translocation of NPR1 through cellular redox changes (Spoel and Dong 2012). Redox changes cause monomeric NPR1 to emerge and accumulate in the nucleus and activate PR gene expression (Kinkema et al. 2000; Mou et al. 2003). NPR1 is sequestered in the cytoplasm as an oligomer through intermolecular disulfide bonds. After induction by SA, the disulfide bonds are reduced, releasing NPR1 monomers into the nucleus. S-Nitrosylation of NPR1 by S-nitrosoglutathione (GSNO) at Cys156 facilitates its oligomerization, which maintains protein homeostasis upon SA induction. Conversely, the SA-induced NPR1 oligomer-to-monomer reaction is catalyzed by thioredoxins (TRX). Mutants in both NPR1 Cys156 and TRX compromised NPR1-mediated disease resistance response. Thus, the regulation of NPR1 is through opposing action of GSNO and TRX. These studies suggest that SA regulates the conversion of NPR1 from an oligomeric to monomeric form, which leads to its nuclear location (Fig. 2.11; Tada et al. 2008).

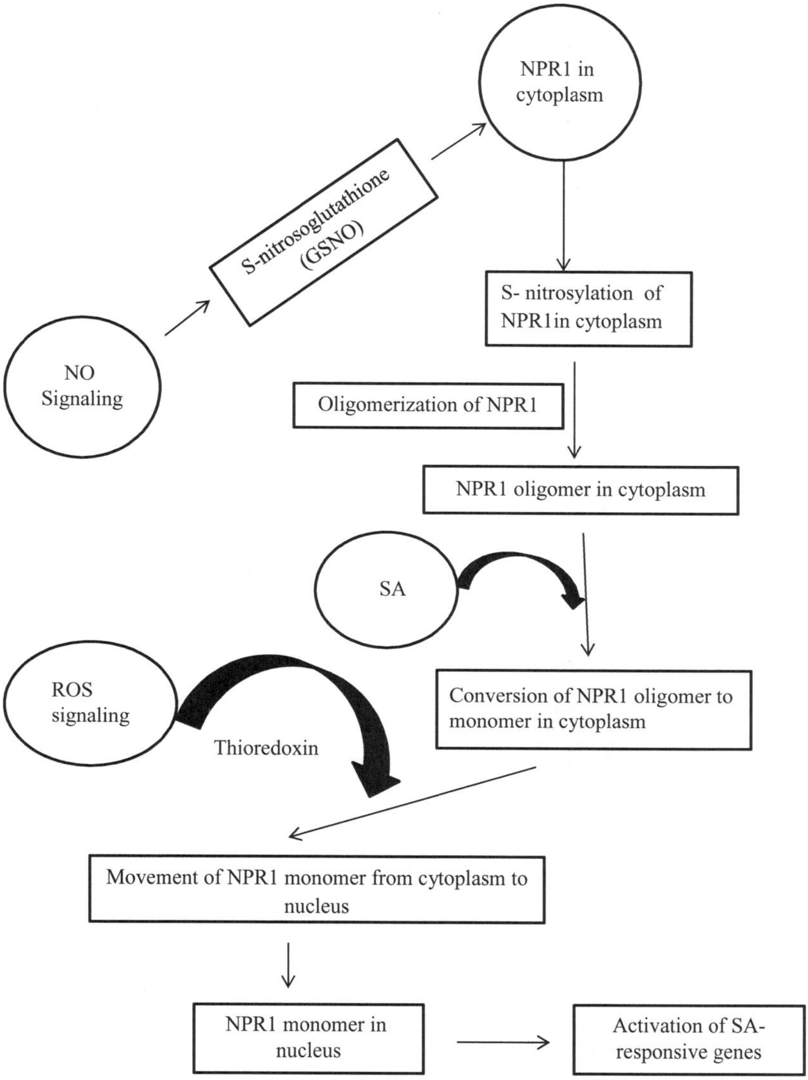

Fig. 2.11 Regulation of NPR1 function through the action of S-nitrosoglutathione and thioredoxin (Adapted from Tada et al. 2008)

2.6.3 SA Modulates Proteasome-Mediated Degradation of NPR1

In the absence of pathogen challenge, NPR1 is continuously cleared from the nucleus by proteasome, which restricts its co-activator activity to prevent untimely activation of defense responses (Spoel et al. 2009). NPR1 protein is degraded by proteasome, and the proteasome-mediated degradation of NPR1 plays an important role in basal disease resistance (Spoel et al. 2009).

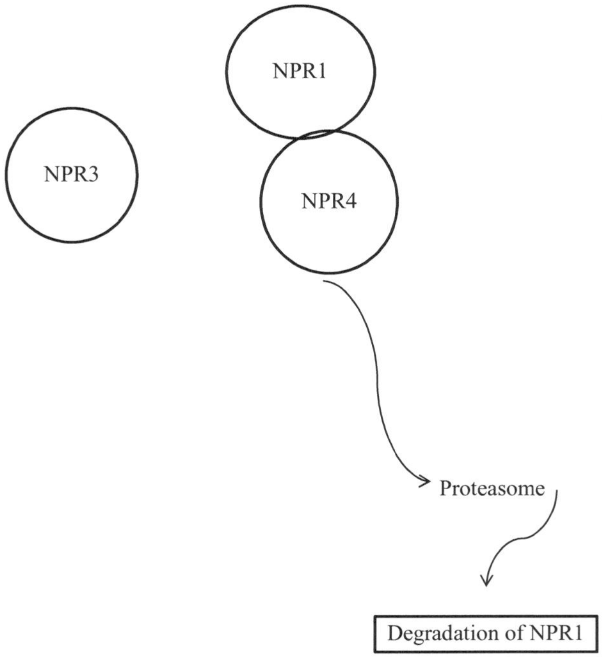

Fig. 2.12 Role of NPR4 in the degradation of NPR1 in plants not exposed to pathogen/PAMP (Adapted from Fu et al. 2012; Moreau et al. 2012)

Proteasomes are large protein complexes located in the nucleus and the cytoplasm (Peters et al. 1994). Proteasomes regulate the concentration of particular proteins. The proteins are tagged for degradation by a small protein called ubiquitin (Pickart and Eddins 2004). Ubiquitin ligase (E3) recognizes the specific protein to be ubiquitinated and catalyzes the transfer of ubiquitin to the target protein (Dreher and Callis 2007). E3 ligases play a key role in the ubiquitin–proteasome system. A Cullin 3 (CUL3) E3 ligase has been found to degrade NPR1 protein (Fu et al. 2012).

Two NPR1 paralogues, NPR3 and NPR4, have been identified as adaptor proteins of the CUL3 E3 ligase. Both NPR3and NPR4 contain the BTP domain as well as an extra protein–protein interaction domain (ankyrin repeat), which are typical for CUL3 substrate adaptors (Fu et al. 2012). The NPR3 and NPR4 adaptor proteins of CUL3 E3 ligase target NPR1 degradation in an SA concentration-dependent manner. At low SA levels found in unchallenged plants, NPR1 is unavailable to induce defense genes, since it is targeted through its binding to NPR4 for degradation in proteasomes (Fig. 2.12; Fu et al. 2012; Moreau et al. 2012). Both NPR3 and NPR4 could bind SA. However, NPR3 had lower affinity in binding SA than NPR4. Accordingly, the binding of SA to NPR3 was slower than NPR4. NPR4 has been shown to have several SA-binding sites (Fu et al. 2012). Hence, it is suggested that SA at low concentrations, NPR1 readily binds with NPR4 and not with NPR3 (Moreau et al. 2012).

At increased SA concentration after infection, SA binds to NPR4 disrupting its interaction with NPR1. NPR1, freed from NPR4 binding, may play its role in activation of defense genes (Fig. 2.13; Moreau et al. 2012).

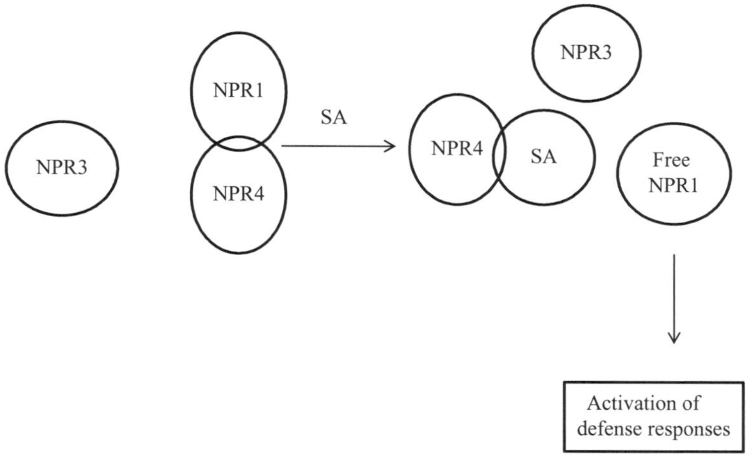

Fig. 2.13 SA binds to NPR4 and frees NPR1 from NPR4 to activate defense gene expression (Adapted from Moreau et al. 2012)

2.6.4 NPR1 Interacting Proteins

Two classes of proteins interact with NPR1/NIM1. The first class includes *Arabidopsis* NIMIN1 (for *NIM In*teracting protein1), NIMIN2, and NIMIN3; tobacco NIMIN2-like proteins (Zwicker et al. 2007); and rice NRR (for *N*egative *R*egulator of disease *R*esistance). Knockout and RNA silencing of *NIMIN1* led to enhanced *PR1* gene expression (Weigel et al. 2005). Constitutive expression of *NtNIMIN2a* led to delayed *PR1* induction, and suppression of *NtNIMIN2a* transcripts enhanced the accumulation of PR1 protein (Zwicker et al. 2007).

A rice gene *NRR* has been identified, and NRR interacts with both the *Arabidopsis* NPR1 protein and the rice NH1 (for rice *N*PR1 *h*omolog*1*) (Chern et al. 2005). Rice NRR protein shows limited similarity to the *Arabidopsis* and tobacco NIMIN2 proteins, only in the NPR1 interaction. NRR behaves similar to NIMIN2 proteins in its interaction with NPR1 (Chern et al. 2005). NRR binds to NPR1 in vivo in a protein complex to inhibit transcriptional activation of *PR* genes, and that NRR contains a transcription repression domain for active repression. Expression of NRR in *Arabidopsis* results in suppression of *PR* gene induction by SA and resistance to pathogens (Chern et al. 2008).

2.6.5 SA-Dependent NPR1-Activated Transcription Factors

The second class of NPR1 interacting proteins belongs to TGA family members of basic-region leucine zipper (bZIP) transcription factors (Zhang et al. 1999; Després et al. 2000; Chern et al. 2001; Fitzgerald et al. 2005). There are 10 TGA transcription factors in *Arabidopsis* of which seven (TGA1–TGA7) have been characterized with respect to their interaction with NPR1 (Jakoby et al. 2002). TGA2, TGA3, TGA5, TGA6, and TGA7 interact with NPR1 in planta when transiently expressed (Després et al. 2000; Kim and Delaney 2002). The other two TGA factors, TGA1 and TGA4 were found to bind NPR1 only in SA-induced leaves. Reduction of two Cys residues that are uniquely present in TGA1 and TGA4 is responsible for this SA-dependent interaction (Després et al. 2003). Wheat contains several HBP-1b/TGA factors (Mikami et al. 1994). In rice, four NPR1-interacting TGA and TGA-like factors have been identified (Chern et al. 2001). The rice TGA2, TGA2.1, binds with NPR1 (Chern et al. 2001). Rice TGA2.1 binds to oligonucleotides containing the as-1-like (for "activation sequence-1-like") element from the PR1 gene promoter and to the promoter of the rice chitinase gene, RCH10 (Chern et al. 2001).

TGA factors bind to the *as-1* (*cis*-acting) element (Johnson et al. 2003; Rochon et al. 2006). Increased binding to the *as-1* element was observed with SA-induced plant extracts. Binding of in vitro-synthesized TGA2 to the *as-1* element was enhanced in the presence of NPR1 (Després et al. 2000). Depletion of TGA2 and TGA3 from nuclear extracts resulted in reduced protein binding to the *as-1* element (Johnson et al. 2003). TGA2 and TGA3 were recruited to the PR1 promoter in vivo in an NPR1-dependent manner (Johnson et al. 2003). NPR1 may transiently interact with the DNA unbound fraction of TGA2 to promote its recruitment to an active form on cognate target promoters (Johnson et al. 2008). These results suggest that NPR1 enhances the binding of TGA factors to the promoter (Fig. 2.14). TGA-regulated gene expression is very complex. The *PR1* promoter has both positive and negative regulatory *cis*-elements. Mutation in the *LS7* element that contains a TGA-binding site resulted in complete loss of gene induction, whereas mutation in another TGA-binding element (*LS5*) augmented gene expression (Johnson et al. 2003, 2008).

A canonical DNA-binding domain is absent in NPR1. It regulates PR gene expression as a cofactor of the TGA transcription factors, which interact with NPR1

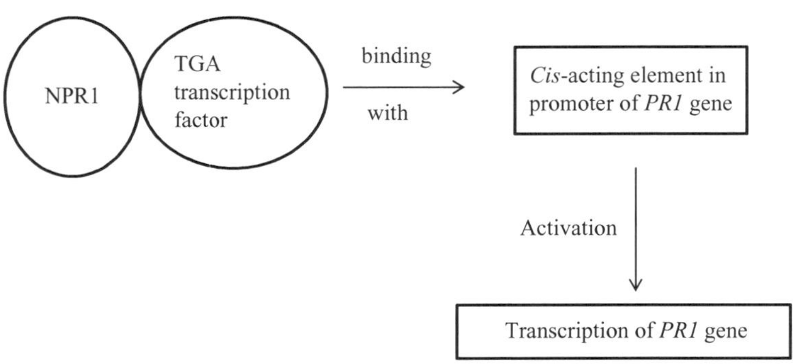

Fig. 2.14 NPR1 as a cofactor of TGA transcription factor enhances binding of TGA transcription factors to the promoter of *PR1* gene to activate transcription of *PR1* gene (Adapted from Fan and Dong 2002; Johnson et al. 2008)

(Fan and Dong 2002). Molecular studies of TGA factors strongly suggest differential effects of TGAs in regulating PR gene expression (Kesarwani et al. 2007). TGA1, TGA3, TGA4, and TGA6 have a positive role in PR gene expression. However, TGA1 and TGA4 have only moderate effects on PR gene expression. TGA2 has repressor activity on PR gene expression, and it acts as a negative regulator of PR genes (Kesarwani et al. 2007). The PR1 promoter contains both positive (LS7) and negative (LS4 and LS5) *cis*-elements. Before induction, TGA2 represses the *PR* genes through interaction with the negative elements. TGA1, TGA3, and TGA6 may bind to the positive element in the uninduced state. This does not lead to *PR* gene expression in the presence of TGA2. It is suggested that upon induction, TGA2 repression is removed and transcription is activated by NPR1 in association with the positive TGA transcription factors, including TGA1, TGA3, TGA5, and TGA6 (Kesarwani et al. 2007).

2.6.6 SA-Induced Expression of PR Genes, Independent of NPR1

In some cases, SA may trigger the induction of *PR* genes, independent of NPR1 (Desveaux et al. 2004). *PR1* is positively regulated in an SA-dependent, but NPR1 independent, manner by the "Whirly" transcription factor, WHY1 in *Arabidopsis* (Desveaux et al. 2004). In rice, *WRKY45* acts in the SA signaling pathway independently of rice NPR1 (NH1) (Shimono et al. 2007). Two defense-related genes, encoding a glutathione *S*-transferase and a cytochrome P450, were found to be regulated downstream of *WRKY45* but were not regulated by the rice NPR1 (NH1). The results suggest the apparent independence of WRKY45- and NH1-dependent pathways. In *Arabidopsis*, most SA-responsive genes are regulated downstream of NPR1, and this regulation is mediated by several WRKY transcription factors acting downstream of NPR1. In rice, WRKY45 and NH1 constitute apparently independent signaling pathways. GST and cytochrome P450 genes are regulated by WRKY45 but not by NH1. *PR1a* and *PR1b* are dependent on NH1 and are conditionally dependent also on WRKY45. Post-translational signal flow from WRKY45 would account for the dependence of these genes on both NH1 and WRKY45. The *lipoxygenase* gene does not appear to depend on either NH1 or WRKY45, so it may be regulated by another regulator (Shimono et al. 2007).

2.7 Role of SUMO in SA Signaling System

The SA-induced changes in gene expression have been found to have a link to chromatin remodeling, such as histone modifications and histone replacement. The recruitment of chromatin-modifying complexes to SA-responsive loci controls their basal and SA-induced expression (March-Diaz et al. 2008; van den Burg and Takken

2009, 2010; Jaskiewicz et al. 2011). Basal repression of these loci may require the post-translational modifier SUMO (for SMALL UBIQUITIN-LIKE MODIFIER). SUMO conjugation has been reported to control the activity, assembly, and disassembly of chromatin-modifying complexes to transcription complexes (van den Burg and Takken 2009). SUMO conjugation determines recruitment and activity of chromatin-modifying enzymes and thereby indirectly controls SA-induced gene expression (van den Burg and Takken 2010).

The *SIZ1* gene, which encodes an *Arabidopsis* SUMO E3 ligase, regulates innate immunity. Mutant *siz1* plants exhibit constitutive SAR characterized by elevated accumulation of salicylic acid and increased resistance to *Pseudomonas syringae* pv. *tomato*. Transfer of the *NahG* gene to *siz1* plants results in reversal of these phenotypes back to wild type. Analyses of the double mutants *npr1 siz1*, *pad4 siz1*, *ndr1 siz1* revealed that *SIZ1* controls SA signaling (Lee et al. 2007).

2.8 SA Induces Transcription of Various Defense Genes

SA triggers transcription of a multitude of genes in plants. More than 790 genes have been shown to be regulated by SA in *Arabidopsis* (Krinke et al. 2007), and most of them are defense response genes. SA induces *PR1* gene expression in *A. thaliana* (Takahashi et al. 2004; Edgar et al. 2006; Kim et al. 2006; Xing et al. 2008). It induced expression of *BGL* (Edgar et al. 2006) and *BG3* (Stein et al. 2006) genes encoding β-1,3-glucanase, *PR4* gene (Stein et al. 2006), and *PR5* gene (Takahashi et al. 2004; Stein et al. 2006) in *A. thaliana*. SA also triggers the activation of various genes encoding different chitinases in *Arabidopsis* (Stein et al. 2006).

SA induces *PR1a* (van Verk et al. 2008), *PR1b* (Nie 2006), and *PR2* genes expression in tobacco (Waller et al. 2006). The expression of acidic *PR1*, *PR2*, and *PR3* genes was up-regulated by SA treatment in tobacco (Niki et al. 1998). SA induces *PR1* expression in potato (Navarre and Mayo 2004). SA induces *OsPR1b* in rice and its expression is downstream of OsWRKY71 (Liu et al. 2007). Acidic *PR1* and *PR2* promoters are induced via SA-dependent signaling pathway in tomato (Hondo et al. 2007).

SA treatment induced peroxidase activity and suppressed replication/accumulation of *Potato virus Y* in tobacco (Nie 2006). SA induced callose deposition in *Arabidopsis* (DebRoy et al. 2004). It also induced transcription of the disease resistance genes *RPW8.1* and *RPW8.2* in *Arabidopsis* (Xiao et al. 2003).

2.9 Role of SA Signaling in Stomatal Closure-Related Immune Responses Against Bacterial Pathogens

SA may trigger defense responses by preventing bacterial entry into host tissue. Bacterial entry into host tissue is a critical first step in causing infection by bacterial pathogens in plants (Melotto et al. 2006; Gudesblat et al. 2009). Stomata serve as passive ports of bacterial entry during infection. The PAMP-activated innate immune response in *Arabidopsis* involves stomatal closure, which occurs within the first hour of contact with plant tissue (Melotto et al. 2006). The stomatal defense against bacterial pathogens observed in *Arabidopsis* was shown to be under the control of the defense signaling molecule SA. The regulation of stomatal defense by SA suggests that stomatal defense is an integrated part of the SA-regulated immune system (Melotto et al. 2006). The bacterial pathogen *Pseudomonas syringae* pv. *tomato* reopens the closed stomata and counteracts the PAMP-induced/SA-mediated stomatal closure, by producing virulence factors, specifically the phytotoxin coronatine. The toxin overcomes or suppresses SA-dependent defenses (Brooks et al. 2005) and probably suppresses the stomatal defense.

2.10 SA Induces Resistance Against Viruses by Modulating AOX-Mediated Alternative Respiratory Pathway

SA is an important component in the signal transduction pathway leading to systemic resistance that is effective against a broad spectrum of oomycete, fungal, bacterial, and viral pathogens. The hallmark of SA-mediated disease resistance system is the induction of pathogenesis-related (PR) proteins through the function of *NPR1* gene which acts downstream of SA production. PR proteins show antifungal, antioomycete, and antibacterial activity (Vidhyasekaran 2007), but so far no antiviral activity of PR proteins has been reported. It is suggested that SA signaling pathway inducing resistance against viruses may be different from the known resistance pathways inducing resistance against bacterial and fungal pathogens (Ji and Ding 2001).

Mitochondrial signaling processes regulate some aspects of SA-induced virus resistance (Hammerschmidt 2009; Carr et al. 2010). Reactive oxygen species (ROS) are constantly generated within mitochondria as by-products of respiratory electron transport chain activity. Perturbation in this ROS pool or in mitochondrial redox can function in the intracellular signal transduction (Maxwell et al. 1999). This form of signaling is influenced by alternative oxidase (AOX). AOX is a mitochondrial enzyme that is the sole component of the alternative respiratory pathway (Fu et al. 2010; Lee et al. 2011b).

Salicylhydroxamic acid (SHAM), an inhibitor of the mitochondrial enzyme AOX, antagonized the inhibitory effect of SA to viruses but not to fungal or bacterial pathogens (Chivasa et al. 1997). SA-mediated resistance to *Cucumber mosaic virus* (CMV) is attributed to the inhibition of systemic virus movement, which also is sensitive to SHAM (Naylor et al. 1998). Both inhibition and enhancement of AOX have been shown to inhibit the induction of virus resistance, probably by disruption of redox signaling in the mitochondrion (Gilliland et al. 2003; Murphy et al. 2004).

SA-induced resistance against viruses uses signal transduction pathways different from those regulating PR gene expression, constituting a different branch of the SA activity program, in which expression of AOX-1 is involved (Wong et al. 2002).

SA-mediated resistance against viruses follows a distinct branch of SA signaling pathway in tobacco, which is independent of PR gene expression, sensitive to SHAM, and associated with *Aox* induction (Chivasa et al. 1997; Murphy et al. 1999). An early and transient induction of *AOX-1* was observed in response to *Plum pox virus* (PPV) in wild-type plants (Alamillo et al. 2006). Induced resistance against *Turnip crinkle virus* also is SA dependent, yet NPR1, ethylene, and JA independent (Kachroo et al. 2000).

2.11 SA Triggers Small RNA-Directed RNA Silencing System

Small RNA-directed RNA silencing is a potent immune surveillance system targeting foreign nucleic acids of invading pathogens (Ding and Voinnet 2007; Jaubert et al. 2011). The RNA silencing pathway in plants presents a formidable defense against viral pathogens (Qu and Morris 2005). Double-stranded RNA (dsRNA) is the starting point of the antiviral RNA silencing system. RNA silencing is triggered by dsRNA which is commonly generated during plant virus replication (Willmann et al. 2011). In case of single-stranded RNA (ssRNA) viruses, the viral RNA-dependent RNA polymerase (RdRP) encoded by the plant copies a plus-sense ssRNA generating a dsRNA molecule (Qi et al. 2009; Garcia-Ruiz et al. 2010; Wang et al. 2010). In case of other RNA viruses, the two strands do not anneal but can fold into highly structured molecules that have dsRNA regions (Alvarado and Scholthof 2009). In case of geminiviruses, the RNAs transcribed from their circular genomes act as a source of dsRNA (Chellappan et al. 2005). Viroids form hairpin structures, which contain intervals of dsRNA (Papaefthimiou et al. 2001). The nucleic acids of the viral pathogens may be the pathogen-associated molecular patterns (PAMPs), and Ding (2010) suggested that the viral double-stranded RNA (vdsRNA) is the viral PAMP.

The plant innate immune system (defense surveillance system) detects the presence of dsRNA as aberrant RNA molecule (Wypijewski et al. 2009) and generates small RNAs. The generated small RNAs direct the antiviral machinery to cleave and destroy the invading viral genome (Alvarado and Scholthof 2009). In the RNA silencing pathway, small interfering RNAs (siRNAs) are derived from perfectly paired double-stranded RNA (dsRNA) precursors. These dsRNA precursors are derived by the action of RNA-dependent RNA polymerase (RDR or RdRP) (Katiyar-Agarwal and Jin 2010). Primary small RNAs are generated from dsRNA precursors by the action of DCL enzymes. These small RNAs can serve as primers for host RNA-dependent RNA polymerases to generate additional dsRNA targets for DCL enzymes to amplify the silencing signal (Voinnet 2008; Vaistij and Jones 2009; Garcia-Ruiz et al. 2010). The silencing signal is amplified by the action of RNA-dependent RNA polymerase (RDR/RdRP) (Voinnet 2008). The increased expression of RDR1 increases the resistance based on RNA silencing against viruses (Leibman et al. 2011).

Tobacco lines deficient in the inducible RdRP activity were obtained by expressing antisense RNA for the *NtRDRP1* gene in transgenic plants. When infected by *Tobacco mosaic virus* (TMV), these transgenic plants accumulated significantly higher levels of viral RNA and developed more severe disease symptoms than wild-type tobacco plants. After infection by a strain of *Potato virus X* that does not spread in wild-type tobacco plants, the transgenic *NtRDRP1* antisense plants accumulated virus and developed symptoms not only locally in inoculated leaves but also systemically in upper uninoculated leaves (Xie et al. 2001). These results suggest that RdRP plays an important role in plant antiviral immunity.

Several studies have indicated that the RDRs (RdRPs) are inducible by SA (Ji and Ding 2001; Gilliland et al. 2003; Yu et al. 2003; Jovel et al. 2011, Lee et al. 2011b). The activity of a tobacco RdRP was increased in SA-treated plants. Biologically active SA analogs capable of activating plant defense also induced the RdRP activity, whereas biologically inactive analogs did not (Xie et al. 2003). A tobacco *RdRP* gene, *NtRDRP1*, was isolated and found to be induced by treatment with SA or its biologically active analogs (Xie et al. 2003). *Arabidopsis* RNA-dependent RNA polymerase gene (*AtRdRP1*) encoding RdRP1 is induced by salicylic acid treatment in *Arabidopsis* (Yu et al. 2003).

Transgenic rice plants expressing the *Arabidopsis NPR1* (*AtNPR1*) gene showed a higher susceptibility to infection by the *Rice yellow mottle virus* (RYMV), which correlated with a mis-regulation of RYMV-responsive expression of the SA-regulated RNA-dependent RNA polymerase 1 gene (*RDR1*) (Quilis et al. 2008). RDR1-dependent viral siRNAs confer SA-dependent resistance against *Cucumber mosaic virus* by directing non-cell-autonomous antiviral silencing (Diaz-Pendon et al. 2007). Application of exogenous SA enhances virus resistance possibly by amplifying the activity of RDR1 (Ji and Ding 2001; Yu et al. 2003; Jovel et al. 2011). The RdRP gene from cotton, *GhRdRP*, is induced by salicylic acid (SA) and 5-chloro salicylic acid (5-CSA) (Gao et al. 2009).

The viral suppressor of RNA silencing (VSR) 2b protein detected in CMV suppressed non-cell-autonomous transgene silencing and SA-mediated virus resistance. The 2b expressed from the CMV genome drastically reduced the accumulation of 21-, 22-, and 24-nt classes of viral siRNAs produced by DCL4, DCL2, and DCL3, respectively. Disease symptoms caused by CMV-Δ2b in *Arabidopsis* mutants defective in antiviral silencing were as severe as those caused by CMV. The results suggest that the VSR 2b may have an indirect role in virus virulence. Production of CMV siRNAs without 2b interference depended largely on RDR1 inducible by SA. It has been suggested that 2b inhibits the production of RDR1-dependent viral siRNAs that confer SA-dependent virus resistance by directing non-cell-autonomous antiviral silencing (Diaz-Pendon et al. 2007). Collectively, these results suggest that SA signaling may enhance the efficiency of RNA silencing pathway in triggering immune responses against viruses by activating RdRP/RDR.

2.12 Enhancement of Small RNA-Directed RNA Silencing by Salicylate Signaling System

Both SA signaling system (Lee et al. 2011b) and RNA silencing system (Leibman et al. 2011) have been shown to be involved in plant immunity against viral pathogens. When both the systems are expressed simultaneously, resistance responses against the virus diseases appear to be enhanced (Jovel et al. 2011). *Tomato ringspot virus* (ToRSV, a *Nepovirus* sp.) systemically infects many herbaceous plants. In these host plants, viral RNA accumulates in symptomatic leaves and in young, asymptomatic leaves that emerge late in infection, suggesting systemic spread of this virus in host plants. However, systemic infection by the virus was restricted in tobacco, and viral RNA did not accumulate in asymptomatic young emerging leaves. ToRSV-derived siRNAs and SA signaling-derived *PR1a* transcripts were detected only in tissues that contained viral RNA, indicating local induction of RNA silencing and SA-dependent defense responses. Lesion size and viral systemic spread were reduced with SA pretreatment but enhanced in *NahG* transgenic plants deficient in SA accumulation (Jovel et al. 2011). It suggests that SA treatment enhances local defense responses and also it is exclusively involved in suppression of systemic infection by the virus in tobacco. Knocking down the SA-inducible RNA-dependent RDR1 exacerbated the necrotic reaction but did not affect viral systemic spread. The results suggest that SA signaling and not RNA silencing is involved in restriction of systemic infection (Jovel et al. 2011).

Plum pox virus (PPV) is able to replicate in inoculated leaves of *Nicotiana tabacum*, but is defective in systemic movement in this host. PPV produced a systemic infection in transgenic tobacco expressing the P1/HC-Pro, a viral suppressor of RNA silencing (VSR). The results indicate the involvement of RNA silencing in defense against systemic infection of PPV in *N. tabacum* (Alamillo et al. 2006). PPV was able to move to upper noninoculated leaves of tobacco plants expressing bacterial hydroxylase (NahG) that degrades SA. Replication and accumulation of PPV were higher in the locally infected leaves of plants deficient in SA or expressing the VSR. Accumulation of virus-derived small RNA was reduced in the NahG transgenic plants. Systemic infection of PPV was highly enhanced in NahG×P1/HC-Pro double-transgenic plants. Collectively, these results suggest that SA signaling system might act as an enhancer of the function of RNA silencing system in tobacco (Alamillo et al. 2006).

2.13 Interplay Between SA-Induced AOX-Mediated Redox Signaling and SA-Induced Small RNA-Directed RNA Silencing

The gene encoding RNA-dependent RNA polymerase 1 (RDR1) is involved in RNA silencing-mediated basal resistance to several viruses (Rakshandehroo et al. 2009; Leibman et al. 2011). Reducing the accumulation of tobacco *RDR1* (*NtRDR1*)

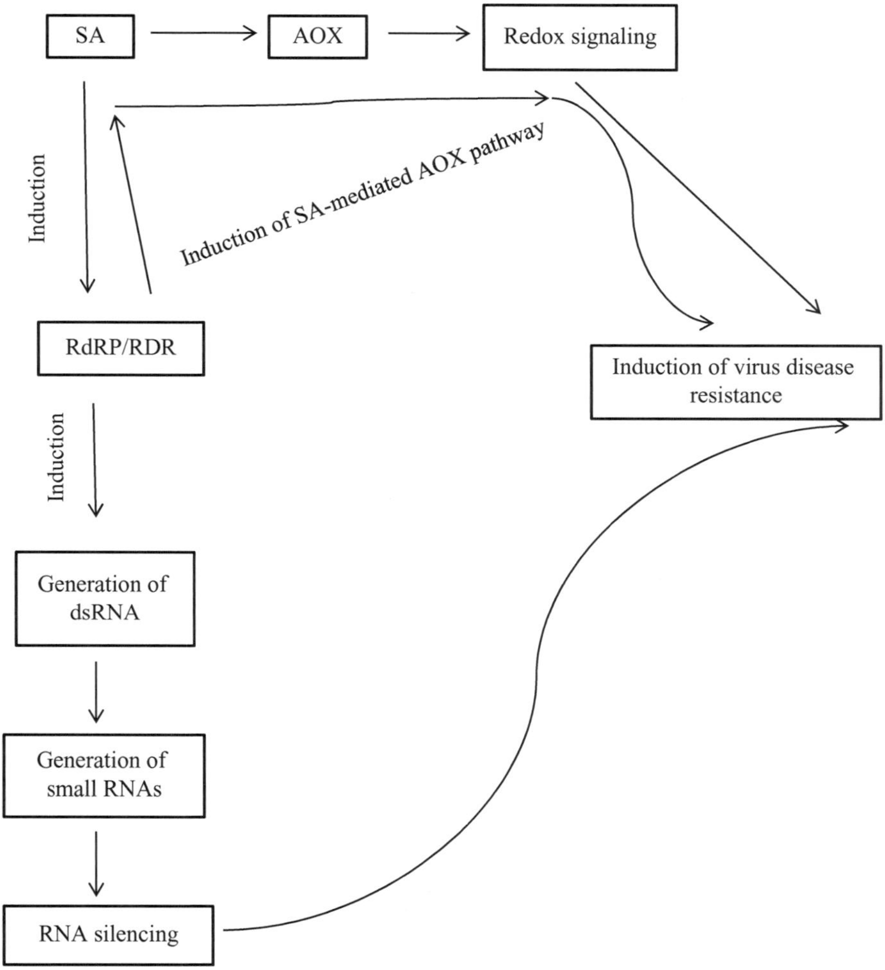

Fig. 2.15 Interplay of SA-triggered AOX-mediated ROS signaling and RNA silencing pathways in inducing virus disease resistance (Adapted from Ji and Ding 2001; Rakshandehroo et al. 2009; Jovel et al. 2011)

transcript by RNA inhibition mediated by transgenic expression of a double-stranded RNA hairpin corresponding to part of *RDR1* gene resulted in little or no induction of accumulation of *RDR1* transcripts after infection by *Potato virus Y* (PVY). Reduced accumulation of *NtRDR1* transcripts also resulted in lower or no induced expression of SA-inducible AOX gene encoding mitochondrial alternative oxidase (Rakshandehroo et al. 2009). These results suggest that RDR1 is involved in both RNA silencing and SA signaling defense pathways and also demonstrate the interplay between these two defense pathways (Fig. 2.15).

A viral suppressor of RNA silencing, 2b of CMV has been found to be an antagonist of the SA-mediated virus resistance mechanism (Ji and Ding 2001).

Systemic infection of CMV-Δ2b (CMV-Δ2b contains a U to A substitution at nucleotide 2,420 of RNA2 that converts the fourth codon of the 2b gene into a stop codon) but not CMV was completely blocked by SA treatment, indicating a direct correlation between Cmv2b expression and the suppression of SA-mediated virus resistance (Ji and Ding 2001). Expression of *cmv2b* was linked tightly to inhibition of SA inducibility of *Aox* (alternative oxidase) (Ji and Ding 2001). These results suggest that a virus-encoded suppressor of RNA silencing protein is able to interfere with the SA-mediated resistance mechanism. An increased in vivo SA accumulation, resulting from CMV infections or an exogenous application, potentiates a systemic RNA silencing antiviral defense response in tobacco plants that is sensitive to 2b VSR protein (Ji and Ding 2001).

2.14 Salicylic Acid Signaling Is Involved in Induction of Systemic Acquired Resistance

Systemic acquired resistance (SAR) is a heightened state of defense against a broad spectrum of pathogens activated throughout a plant following a local infection (Liu et al. 2011a). The induced resistance is detected in the uninoculated systemic tissue of a pathogen-infected plant (Liu et al. 2011b). Localized attack by a necrotizing pathogen induces SAR to subsequent attack by a broad range of normally virulent pathogens. Salicylic acid (SA) accumulation is required for activation of local defenses at the initial site of attack, and in the distant pathogen-free organs for the induction of SAR. SA accumulation and signaling in these distant pathogen-free organs are primed to further increase to higher levels upon challenge with a pathogen (Maldonado et al. 2002; Jung et al. 2009; Návarová et al. 2012; Shah and Zeier 2013).

SA signaling has been shown to be required for the manifestation of SAR (Nandi et al. 2004; Du et al. 2012). Isochorismate synthase activity is required for SA synthesis (Gaille et al. 2002, 2003; Garcion et al. 2008). The *Arabidopsis ics1* mutant, which is deficient in isochorismate synthase 1 activity, has been found to be SAR deficient (Wildermuth et al. 2001; Jung et al. 2009; Chaturvedi et al. 2012). Transgenic plants expressing the SA-degrading enzyme salicylate hydroxylase encoded by the *Pseudomonas putida nahG* gene were found to be deficient in expressing SAR (Vernooij et al. 1994; Lawton et al. 1995). The *FMO1* (*FLAVIN-DEPENDENT MONOOXYGENASE1*) gene is required for the systemic accumulation of SA in distant pathogen-free leaves, and the *FMO1* gene is required for the induction of SAR (Chaturvedi et al. 2012; Shah and Meier 2013). Methyl esterase 1 (StMES1) is required for SAR in potato (Manosalva et al. 2010), and MES is required for accumulation of free SA in the distal systemic tissue, the tissue that does not receive the primary (initial) infection (Park et al. 2007). *NPR1* gene is a master regulator of the SA-mediated induction of defense genes (Mou et al. 2003; Zhang et al. 2003; Chern et al. 2008). NPR1 directly binds SA (Wu et al. 2012) and activates SA signaling system. NPR1 is involved in triggering SAR (Maier et al. 2011; Hermann et al. 2013), and the *npr1* mutant of *Arabidopsis thaliana* is deficient in SAR (Durrant and Dong 2004).

2.15 Mobile Long-Distance Signals for Induction of Systemic Acquired Resistance

2.15.1 Search for Long-Distance Mobile Signal

The establishment of SAR may require translocation of a signal from the pathogen-inoculated leaf to the distal organs, where salicylic acid-dependent defenses are activated (Chaturvedi et al. 2008). Some mobile signal(s) produced at the site of primary infection may have to carry the SAR message and travel throughout the plant. Search for the mobile signals involved in SAR is going on for the past several decades (Yalpani et al. 1991; Vernooij et al. 1994; Park et al. 2007; Truman et al. 2007; Chaturvedi et al. 2008; Jung et al. 2009; Park et al. 2009; Shah 2009; Manosalva et al. 2010; Chanda et al. 2011; Liu et al. 2011a, b; Návarová et al. 2012). It was initially suggested that SA itself may be the mobile signal. Accumulation of SA in the phloem and its requirement to activate SAR led to conclusion that SA may be the mobile signal (Yalpani et al. 1991). However, grafting experiments with tobacco plants expressing the bacterial *NahG* gene, which encodes the SA-degrading enzyme SA hydroxylase, showed that SA may not be the mobile signal. *Tobacco mosaic virus* (TMV)-infected NahG rootstocks were still capable of generating the signal for induction of SAR in wild-type scions, despite their inability to accumulate SA (Vernooij et al. 1994).

Subsequently, several mobile signals have been identified (Fig. 2.16). Methyl salicylate (MeSA) was identified as a mobile signal in inducing SAR (Park et al. 2007, 2009; Vlot et al. 2008a, b; Manosalva et al. 2010). In addition to MeSA, several candidate mobile signals that are linked to lipid metabolism and translocation have been reported (Kachroo et al. 2001; 2004; Nandi et al. 2003, 2004; Chaturvedi et al. 2008; Jung et al. 2009; Chanda et al. 2011). A lipid transfer protein (DIR1), a lipid-derived molecule (glycerol 3-phosphate), and a glycerol-3-phosphate-dependent factor have been reported as lipid-based mobile signaling components in SA-induced SAR. Azelaic acid is the other lipid signal, which is involved in SAR (Jung et al. 2009). An abietane diterpenoid, dehydroabietinal (DA) has been found to be a potent activator of SAR (Chaturvedi et al. 2012). The nonprotein amino acid pipecolic acid, a common Lys catabolite, has also been identified as a systemic signal molecule (Návarová et al. 2012; Shah and Zeier 2013). Thus, several signaling molecules may be involved in long-distance signaling transport to induce SAR (Fig. 2.16).

2.15.2 Methyl Salicylate May Be a Mobile Signal

It has been suggested that methyl salicylate (MeSA) may be a mobile signal involved in SAR. MeSA serves as a long-distance phloem-mobile SAR signal in tobacco and *Arabidopsis* (Manosalva et al. 2010). MeSA is synthesized from salicylate (SA) via a reaction catalyzed by S-adenosyl-L-methionine (SAM): salicylic acid carboxyl

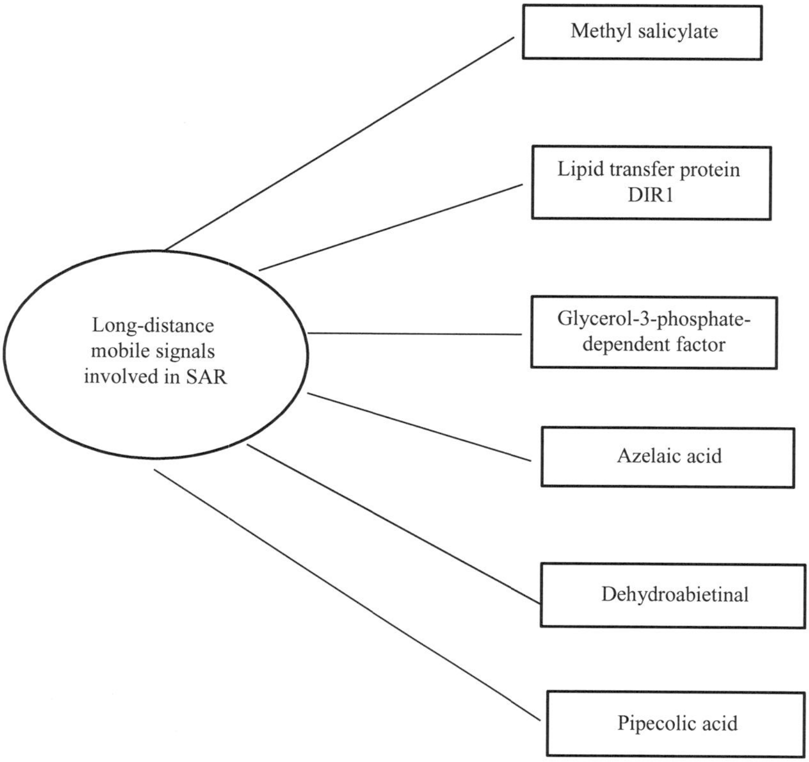

Fig. 2.16 Long-distance mobile signal complex in induction of systemic acquired resistance

methyltransferase (SAMT) with a methyl group donated by the molecule SAM (Kwon et al. 2009). SAMT genes are induced by pathogen or SA treatment (Xu et al. 2006; Koo et al. 2007). Altering expression of SAMT1 compromises SAR (Liu et al. 2010). SAR development appears to require SAMT1 in the primary infected tissue to produce MeSA (Park et al. 2007). The enhanced expression of *SAMT1* gene results in the conversion of total SA to MeSA (Kwon et al. 2009).

MeSA is biologically inactive in induction of defense response. It has been demonstrated that it is a phloem-mobile signal. MeSA moves systemically to the noninfected tissues and is associated with the induced systemic resistance (Kumar and Klessig 2008; Park et al. 2009). By contrast, SA is a highly active signal inducing expression of defense genes, but SA is transported upward only in small amounts via xylem (Rocher et al. 2006). It seems that MeSA which moves systemically through phloem may be converted to active SA form in the distal systemic tissue, the tissue that does not receive the primary (initial) infection (Park et al. 2007, 2009).

A high-affinity SA-binding protein (SABP) termed SABP2 has been identified and characterized from tobacco (Du and Klessig 1997; Kumar and Klessig 2003; Forouhar et al. 2005; Kumar et al. 2006; Tripathi et al. 2010). Orthologs of tobacco

SABP2 have been identified in *Arabidopsis thaliana* (Vlot et al. 2008a, b) and potato (Manosalva et al. 2010). SABP2 is a lipase belonging to the α/β hydrolase superfamily. It is present in extremely low abundance and specifically binds SA with high affinity (Kumar and Klessig 2003). It has strong esterase activity with methyl salicylate as the substrate (Forouhar et al. 2005). Thus, SABP2 is a methyl salicylate esterase whose function is to convert biologically inactive methyl salicylate to active SA (Kumar and Klessig 2008; Vlot et al. 2008a, b; Manosalva et al. 2010). SABP2's esterase activity is inhibited in the initially primary infected tissue by SA binding in its active site; this facilitates accumulation of methyl salicylate, which is then translocated through the phloem to systemic tissue for perception and processing by SABP2 to SA (Kumar and Klessig 2008; Park et al. 2009). It has been demonstrated that the tobacco SAMT (NtSAMT1) activity, and thus MeSA biosynthesis, is required in the primary infected leaves where the SAR signal is produced. In contrast, MeSA esterase (MSE) activity is needed in the uninoculated systemic leaves, where the SAR signal is perceived and processed (Park et al. 2007). MeSA does not induce defense responses (Seskar et al. 1998), and it must be converted to SA by an MSE for biological activity. The SABP2's MSE activity must be inhibited in the primary infected tissue (by SA binding in its active site pocket) to facilitate the accumulation of sufficient levels of MeSA to signal SAR (Park et al. 2007, 2009). A synthetic SA analog (2,2,2,2′-tetra-fluoroacetophenone) that inhibits MSE activity blocks SAR development in tobacco, potato, and *Arabidopsis* (Park et al. 2009). Collectively these studies suggest that MeSA is a critical mobile signal and on conversion of MeSA by MSE to SA, SA, and not MeSA, activates the expression of defense genes (Fig. 2.17).

Attaran et al. (2009) showed that MeSA production is not essential for induction of systemic acquired resistance (SAR) in *Arabidopsis*. MeSA production increased in *Arabidopsis* leaves inoculated with *Pseudomonas syringae*; however, most MeSA was emitted into the atmosphere, and only small amounts were retained. In several *Arabidopsis* defense mutants, the abilities to produce MeSA and to establish SAR did not coincide. T-DNA insertion lines defective in expression of a pathogen-responsive SA methyltransferase gene were completely devoid of induced MeSA production but increased systemic SA levels and developed SAR upon local *P. syringae* inoculation (Attaran et al. 2009). Further, it has been demonstrated that MeSA production induced by *P. syringae* depended on the JA pathway but the JA biosynthesis or downstream signaling was not required for SAR (Attaran et al. 2009). These results suggested that MeSA is dispensable for SAR in *Arabidopsis*.

However, Liu et al. (2011b) showed that the length of light exposure that plants receive after the primary infection determined the extent to which MeSA is required for SAR signaling. When the primary infection occurred late in the day and as a result infected plants received very little light exposure before entering the night/dark period, MeSA and its metabolizing enzymes were essential for SAR development. In contrast, when infection was in the morning followed by 3.5 h or more of exposure to light, SAR developed in the absence of MeSA. These studies suggest that length of light exposure that plants receive after the primary infection may determine the role of MeSA in SAR.

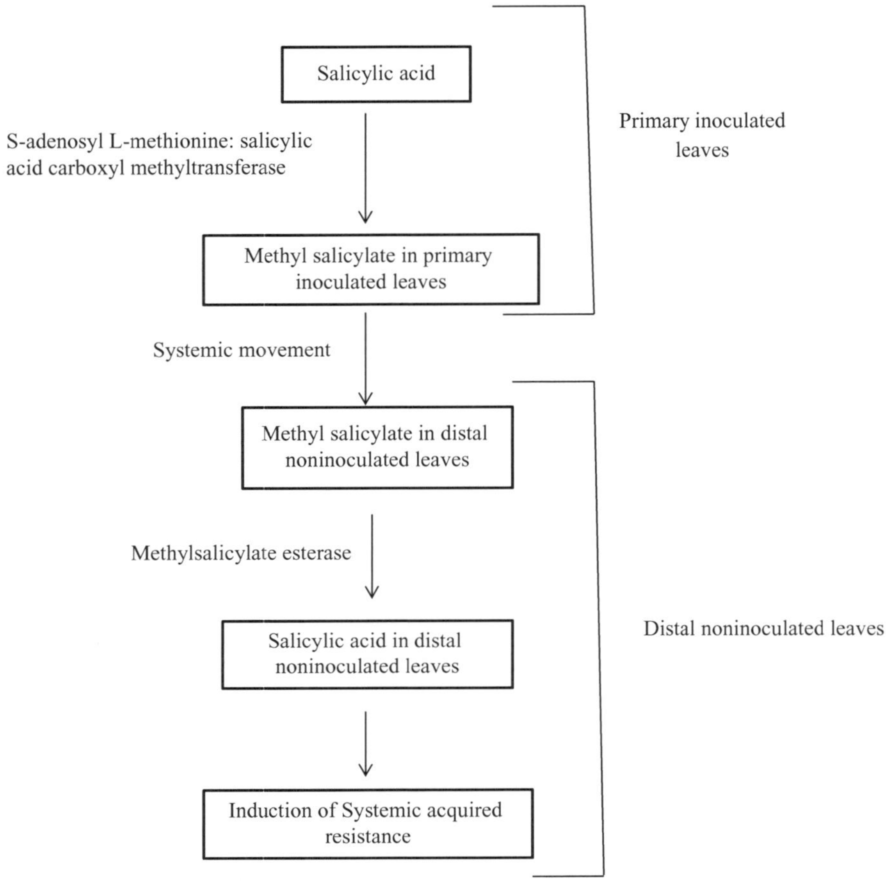

Fig. 2.17 Role of salicylic acid carboxyl methyltransferase, methyl salicylate, and methyl salicylate esterase in induction of systemic acquired resistance (Adapted from Park et al. 2007, 2009; Kwon et al. 2009)

2.15.3 DIR1 and Glycerol-3-Phosphate-Dependent Factor Mobile Signal Complex

Maldonado et al. (2002) reported another mobile signal molecule involved in SAR in *Arabidopsis*. They characterized an *Arabidopsis* mutant, *dir1-1* (*D*efective in *I*nduced *R*esistance 1-1), which is specifically compromised in SAR. These plants contained a mutation in a putative apoplastic protein with homology to family 2 lipid transfer proteins, and they are compromised for SAR but not for local resistance. The *dir1-1* mutant exhibited wild-type local resistance to avirulent and virulent *Pseudomonas syringae*, but pathogenesis-related (PR) gene expression was abolished in uninoculated distant leaves, and *dir1-1* failed to develop SAR to virulent *Pseudomonas syringae* or *Peronospora parasitica* (Maldonado et al. 2002).

The results suggest that the *dir1-1* is defective in the production or transmission from the inoculated leaf of an essential mobile signal.

Glycerol 3-phosphate (G3P) has been shown to be an inducer of SAR. Genetic mutants defective in G3P biosynthesis cannot induce SAR but can be rescued when G3P is supplied exogenously (Chanda et al. 2011). Radioactive tracer experiments showed that a glycerol-3-phosphate derivative was translocated to distal tissues and this translocation required the lipid transfer protein, DIR1. It was also observed that glycerol 3-phosphate was required for the translocation of DIR1 to distal tissues, which occurs through the symplast. The *dir1* mutant plants accumulated reduced levels of glycerol 3-phosphate (Chanda et al. 2011). These results suggest the cooperative interaction of the lipid transfer protein (DIR1) and the lipid-derived molecule (glycerol 3-phosphate) orchestrates the induction of SAR in plants (Maldonado et al. 2002).

A plastid glycerol-3-phosphate-derived factor has also been shown as a mobile signal molecule. Petiole exudates (PeXs) collected from *Arabidopsis* leaves inoculated with an avirulent (Avr) *Pseudomonas syringae* strain promoted resistance when applied to *Arabidopsis*, tomato, and wheat (Chaturvedi et al. 2008). *Arabidopsis* FATTY ACID DESATURASE7 (FAD7), SUPPRESSOR OF FATTY ACID DESATURASE DEFICIENCY1 (SFD1), and SFD2 genes have been shown to be required for accumulation of the SAR-inducing activity. In contrast to Avr PeX from wild-type plants, Avr PeXs from *fad7*, *sfd1*, and *sfd2* mutants were unable to activate SAR when applied to wild-type plants (Chaturvedi et al. 2008). The *sfd1* mutation diminished the SAR-associated accumulation of elevated levels of SA and *PR1* gene transcript in the distal leaves of plants previously inoculated with an avirulent pathogen (Nandi et al. 2004). FAD7, SFD1, and SFD2 are known to be involved in plastid glycerolipid biosynthesis (Chaturvedi et al. 2008). The *sfd1* mutants had defects in lipid composition; in particular levels of the plastid-localized 34:6-monogalactosyldiacylglycerol were lower in the *sfd1* mutant (Chaturvedi et al. 2008, 2012).

Biologically induced SAR was compromised in the *sfd1* mutant (Chaturvedi et al. 2008). *SFD1* encodes a plastid-localized dihydroxyacetone phosphate (DHAP) reductase that synthesizes glycerol 3-phosphate (G3P). SFD1's DHAP reductase activity and its localization to the plastids, suggesting that SFD1 synthesized G3P, or a product thereof, are required for the accumulation and long-distance transport of a SAR signal (Lorenc-Kukula et al. 2012; Shah and Zeier 2013). In *Arabidopsis*, G3P levels were elevated in the pathogen-inoculated and the distal pathogen-free leaves (Chanda et al. 2011). Locally applied ^{14}C-labeled G3P could not be recovered in the systemic leaves (Chanda et al. 2011). The result suggests that G3P per se is unlikely to be the systemically translocated SAR signal and rather, a G3P-dependent factor is likely involved in long-distance signaling (Shah and Zeier 2013).

SFD1- and DIR1-dependent factors may function together in long-distance signaling. G3P, when co-applied with DIR1 protein, was capable of enhancing systemic disease resistance (Chanda et al. 2011). G3P levels were lower in *dir1* mutant, suggesting that the DIR1 and the G3P-dependent factor are required for systemic translocation of each other (Chanda et al. 2011).

2.15.4 Azelaic Acid May Be a Mobile Signal

A nine-carbon dicarboxylic acid, azelaic acid, has been found to accumulate in the vascular sap of *Arabidopsis* plants infected by the bacterial pathogen *Pseudomonas syringae* (Jung et al. 2009). Levels of azelaic acid increased in plants exposed to pathogens and triggered systemic acquired resistance (Zahn 2009). Azelaic acid had several properties of a long-distance resistance-priming signal (Park et al. 2009). The accumulated azelaic acid primed plants to accumulate SA upon infection by the pathogen. Azelaic acid induced the expression of *AZI1* (for *Azelaic acid induced 1*) gene. Mutation of the *AZI1* gene resulted in the specific loss of SAR triggered by azelaic acid (Jung et al. 2009). The results suggest that the predicted secreted protein AZI1 is important for generating vascular sap that confers disease resistance. The azelaic acid induced local and systemic resistance against *P. syringae* in *Arabidopsis* (Jung et al. 2009).

Azelaic acid may be synthesized by oxidation of 9-oxononanoic acid synthesized from fatty acids by the action of 9-lipoxygenase and hydroperoxide lyase (Vicente et al. 2012). The importance of azelaic acid as the systemic molecule in inducing SAR is doubtful. It has been shown that azelaic acid is not essential for the establishment of SAR per se (Návarová et al. 2012; Zoeller et al. 2012). However, when azelaic acid is translocated, it can add to the strength of systemic immunity observed during SAR (Shah and Zeier 2013).

2.15.5 Dehydroabietinal as a Mobile Signal

Dehydroabietinal (DA), an abietane diterpenoid, has been found to be an activator of SAR (Chaturvedi et al. 2012). It is a potent inducer of SAR and it is active when applied as picomolar solutions to leaves of *Arabidopsis*, tomato, and tobacco. DA was purified from vascular sap of *Arabidopsis thaliana* leaves treated with SAR-inducing microbe. Deuterated DA when applied to *Arabidopsis* leaves was rapidly transported out of the leaf and recovered from the untreated leaves (Chaturvedi et al. 2012). Locally applied DA was translocated throughout the plant and systemically induced the accumulation of SA and *PR1* expression in the untreated leaves. DA-induced SAR was attenuated in the SA-deficient *NahG* transgenic, *ics1 ics2* double mutant, and *npr1* mutant plants. It suggests that DA functions upstream of SA accumulation and signaling. *FMO1* gene has been shown to be required for systemic SA accumulation in DA-treated plants and DA-induced SAR (Chaturvedi et al. 2012; Shah and Zeier 2013). DA in vascular sap was redistributed into a SAR-inducing "signaling DA" pool that was associated with a trypsin-sensitive high-molecular-weight fraction (Chaturvedi et al. 2012). It suggests that DA-orchestrated SAR involves a vacular sap protein(s). However, the proteins involved in this process have not yet been characterized.

2.15.6 Pipecolic Acid as an SAR Long-Distance Signal

SA accumulates locally at site of the bacterial pathogen *Pseudomonas syringae* pv. *maculicola* (*Psm*) infection and also in distal noninoculated leaves (Návarová et al. 2012). SA was not transported from inoculated to distal leaves, but it systemically accumulated in the distal leaves (Vernooij et al. 1994). Systemic SA accumulation proceeds via regulation of *ICS1* and de novo SA biosynthesis (Attaran et al. 2009). *ALD1* (*AGD2-LIKE DEFENSE RESPONSE PROTEIN1*) and *FMO1* (*FLAVIN-DEPENDENT MONOOXYGENASE1*) are also involved in systemic SA accumulation and SAR (Mishina and Zeier 2006; Návarová et al. 2012). Pathogen-induced expression of *ALD1*, *FMO1*, and *ICS1* in inoculated tissue proceeds independently of SA (Metraux 2002; Song et al. 2004a, b; Bartsch et al. 2006; Mishina and Zeier 2006).

The nonprotein amino acid pipecolic acid (Pip) has been identified as a mobile long-distance signal in the SAR induction pathway. Pip strongly accumulated in *Arabidopsis* leaves inoculated with the SAR-inducing *Pseudomonas syringae* pv. *maculicola* (*Psm*) and in leaves treated with the bacterial PAMPs flg22 and LPS (Návarová et al. 2012). It also accumulated in distal, noninoculated leaves of *Psm-treated* plants, and there was tenfold systemic increase at 48 h after local *Psm* inoculation (Návarová et al. 2012).

The SA accumulation in distal leaves may be triggered by perception of mobile long-distance signals (Shah 2009). Significantly enhanced systemic levels of Pip but not of SA were observed at 36h after *Psm* inoculation, suggesting that Pip accumulation precedes SA accumulation in systemic leaves at the onset of SAR. Systemic Pip levels start to significantly rise before marked elevations of SA are detectable in the systemic tissue (Návarová et al. 2012). Systemic accumulation of Pip was markedly reduced in the SAR-defective *fmo1* and *ics1* mutants; localized bacterial treatment still provoked significant increases of Pip in distal leaves of these mutants. By contrast, systemic levels of Pip were not enhanced in the *npr1* and *pad4* mutants (Návarová et al. 2012).

ALD1, an aminotransferase with Lys converting activity, may be involved in Pip biosynthesis. ALD1 is important for the activation of both local and systemic defenses in *Arabidopsis* (Song et al. 2004a). ALD1 has aminotransferase activity with strong substrate preference for Lys (Song et al. 2004b). Δ1-Piperideine-2-carboxylic acid and ε-amino-α-ketocaproic acid are direct reaction products of an ALD1-catalyzed Lys aminotransferase reaction. However, it is not yet known how Lys transamination products are converted to Pip (Návarová et al. 2012). *ALD1* transcript levels strongly increased in *Psm*-inoculated plants in both local and distal leaves (Návarová et al. 2012).

Pipecolate-deficient *ald1* plants fail to accumulate SA in distal leaf tissue following pathogen inoculation and are fully compromised in SAR (Jing et al. 2011; Návarová et al. 2012). The *ald1* plants regain the ability for systemic SA accumulation and SAR establishment, when Pip is exogenously applied to the whole plant prior to pathogen treatment. The results suggest that Pip accumulation is critical for systemic SA production and SAR (Návarová et al. 2012). Pip feeding of plants prior to inoculation boosts pathogen-triggered induction of SA biosynthesis and defense-related gene expression, suggesting that Pip strongly amplifies pathogen-triggered defense responses. It has been suggested that the early systemic increase of Pip at the onset of SAR functions as an initial trigger for signal amplification leading to the systemic increase in SA (Návarová et al. 2012).

FMO1 (Flavin-dependent monooxygenase1) has been identified as an important regulator of SAR. Localized *P. syringae* inoculation triggered enhanced expression of several hundred genes in the distal leaves of *Arabidopsis* plants. This massive switch in gene expression at the systemic plant level is totally lost in the *fmo1* mutant (Mishina and Zeier 2006). *FMO1* is necessary for the systemic accumulation of SA upon SAR induction (Mishina and Zeier 2006). The *fmo1* mutant fails to establish Pip-induced resistance to bacterial infection. These results suggest that FMO1 functions downstream of Pip and upstream of SA in SAR (Návarová et al. 2012). FMO1 may be involved in the oxidation of Pip in the Pip signal amplification pathway (Návarová et al. 2012). *Fmo1* is necessary for systemic resistance induction (Liu et al. 2011a). Pip enhances both its own biosynthesis and downstream signaling in SAR via amplification of pathogen-triggered *ALD1* and *FMO1* expression, indicating a positive feedback amplification loop with Pip as a central player (Návarová et al. 2012; Shah and Zeier 2013).

PAD4 (Phytoalexin-deficient4) and NPR1 are also involved in Pip-mediated resistance (Mishina and Zeier 2006; Jing et al. 2011). PAD4 promotes pathogen-induced Pip production, and it is also required for resistance promoted by Pip application (Návarová et al. 2012). PAD4 seems to exert its central defense regulatory role via transcriptional control of *ALD1*, *FMO1*, and *ICS1* (Shah and Zeier 2013). A substantial increase of Pip biosynthesis was observed 10 h after the challenge infection in SAR-induced plants, indicating a strong potentiation of Pip biosynthesis during SAR (Návarová et al. 2012).

SA content increases in the distal leaves of plants that were inoculated with pathogen on other leaves. SA content increase was fully dependent on *ALD1-mediated* Pip biosynthesis and downstream signaling gene *FMO1* (Song et al. 2004a; Mishina and Zeier 2006; Návarová et al. 2012; Shah and Zeier 2013). The results suggest that Pip triggers de novo biosynthesis of SA on distal leaves. Conversely, systemic Pip accumulation in the distal leaves strongly relies on *FMO1* and *ICS1*-mediated SA biosynthesis (Návarová et al. 2012). The results suggest that strong SAR establishment depends on effective signal amplification involving feedback mechanisms that integrate Pip and SA signaling (Fig. 2.18; Shah and Zeier 2013).

Pip also induces priming of SAR responses. Pip application to *Arabidopsis* plants significantly increased *ALD1* transcript levels and strongly primed plants for pathogen-triggered expression of *ALD1*, indicating a positive regulatory role for Pip on its own biosynthesis. Pip feeding also led to strong priming of *Psm*-triggered *FMO1* and *PR1* gene expression (Návarová et al. 2012). Pip-deficient *ald1* mutant is defective in SAR priming, and it suggests that endogenously accumulating Pip mediates priming during SAR (Návarová et al. 2012).

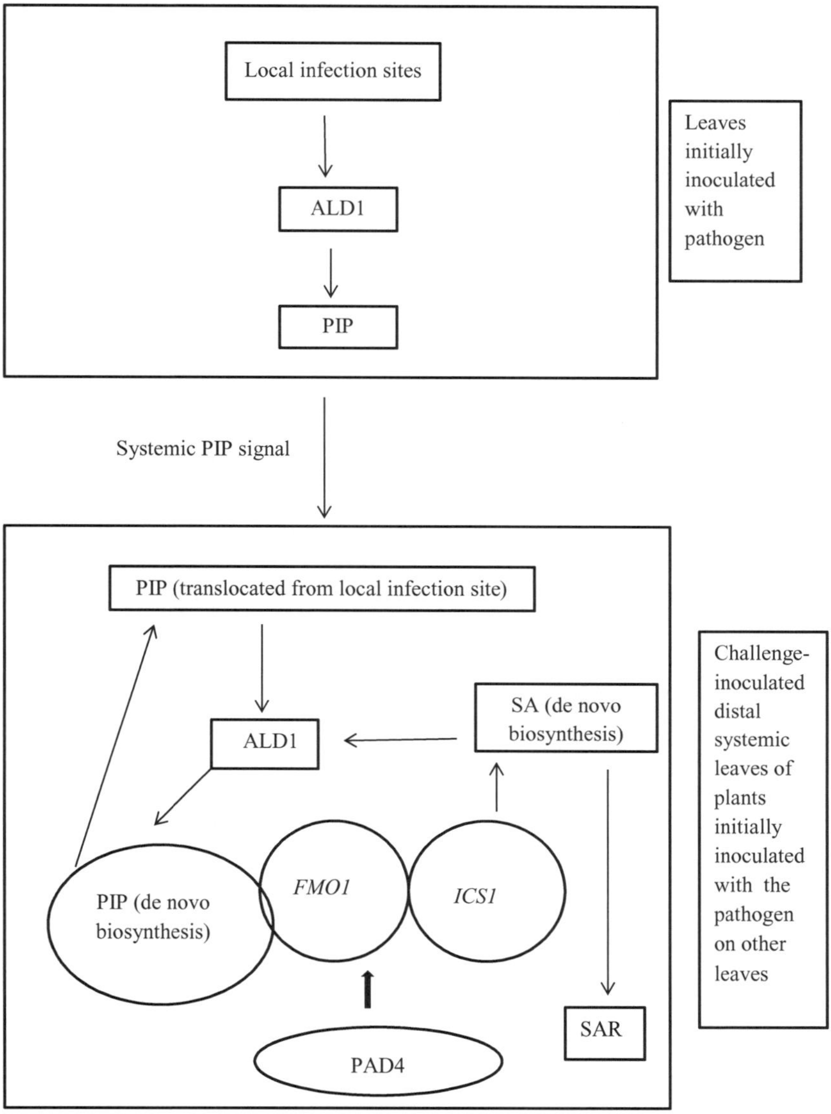

Fig. 2.18 Feedback amplification of Pip and SA in the SAR pathway (Adapted from Návarová et al. 2012; Shah and Zeier 2013)

2.16 Role of Mediator Complex in SA-Mediated Systemic Acquired Resistance

Some Mediators have been shown to be involved in triggering SA-mediated SAR. Mediator is a multiprotein complex that functions as a transcriptional coactivator (Kidd et al. 2011; An and Mou 2013). Mediator fine-tunes gene-specific and pathway-specific transcriptional reprogramming by interacting with specific activators/repressors together with general transcription factors at the promoter site (Conaway and Conaway 2011a, b; An and Mou 2013). Mediator performs both general and specific roles in regulating gene transcription. Mediator subunits may regulate a wide range of signaling pathways through selectively interacting with specific transcription factors (Cevik et al. 2012; Chen et al. 2012; Wathugala et al. 2012).

The *Arabidopsis* Mediator subunit16 (MED16) has been shown to be an essential positive regulator of SAR (Zhang et al. 2012b). Mutations in MED16 reduced NPR1 protein levels (Zhang et al. 2012b). The transcription co-activator NPR1 is the master regulator of SAR in *Arabidopsis* (Maier et al. 2011; Shah and Zeier 2013), and MED16 may regulate the function of NPR1 in inducing SAR. MED16 might regulate SA responsiveness via the modulation of NPR1 protein accumulation. Following *Pseudomonas syringae* inoculation, *med16* mutant plants locally and systemically accumulated SA to levels similar to that in wild-type plants but impaired in PR gene expression (Zhang et al. 2012b). The results suggest that MED16 may act downstream of SA signaling and upstream of NPR1 in SAR signaling pathway (Fig. 2.19; Zhang et al. 2012b). It is suggested that MED16 might relay signals from the SA pathway to the transcription machinery. *Med16* knockout lines exhibited increased susceptibility to both avirulent and virulent strains of *P. syringae* and were unable to establish SAR (Zhang et al. 2012b). The Mediator subunit SFR6/MED16 in *Arabidopsis* has been shown to be involved in conferring resistance against *P. syringae*. The *srf6* mutants were more susceptible to *P. syringae*. They exhibited weaker

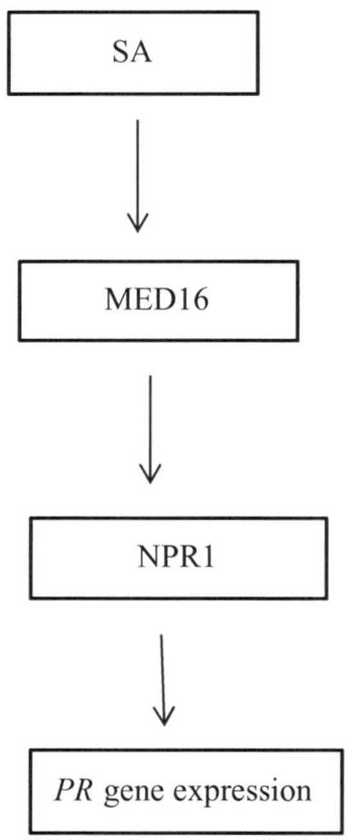

Fig. 2.19 MED16 acts downstream of SA and triggers NPR1 accumulation to trigger *PR* gene expression (Adapted from Zhang et al. 2012b)

expression of the SAR marker *PR* gene than wild-type *Arabidopsis* after direct application of SA (Wathugala et al. 2012). These results suggest that MED16 plays an important role in induction of SAR.

The Mediator NRB4 (Non-recognition-of-BTH4) has been shown to be involved in SAR in *Arabidopsis* (Canet et al. 2012). NRB4 is allelic to Mediator subunit 15 (MED15). Plants carrying weak *nrb4* alleles exhibit strong insensitivity and show attenuated SAR to *P. syringae*. Like *npr1* mutant, *nrb4* mutants fail to develop SA-induced resistance and overaccumulate SA during the *P. syringae* infection process (Canet et al. 2012). NRB4 may function downstream of NPR1 in SA signaling pathway. Overexpression of NPR1 in *med15/nrb4* mutant does not restore the response to SA. It has been shown that MED15/NRB4 is not required for NPR1 stability and subcellular localization and no protein–protein interaction exists between MED15 and NPR1 (Canet et al. 2012). The results suggest that MED15/NRB4 may act downstream of NPR1 in regulating SA response.

2.17 Salicylic Acid Triggers Priming and Induces Systemic Acquired Resistance

2.17.1 What Is SA-Triggered Priming?

Infection of plants by necrotizing pathogens, which induce the accumulation of SA, or treatment of plants with synthetic compounds, which are able to trigger SA signaling, causes the induction of a unique physiological state called "priming" (Conrath 2009, 2011; Camañes et al. 2012; Slaughter et al. 2012). When a treatment puts a plant in a state of increased alertness with no or only minimal gene induction, it is called priming (Conrath et al. 2002, 2006). After localized foliar infections by diverse pathogens or by treatment with synthetic compounds, plants develop whole-plant immunity, called SAR (Conrath et al. 2006; Beckers and Conrath 2007; Conrath 2009; Mukherjee et al. 2010). In this process, distal (systemic) leaves become primed to activate stronger defense response upon secondary infection (Beckers et al. 2009; Slaughter et al. 2012). SAR is associated with priming of defense (Kohler et al. 2002; Jung et al. 2009; Luna et al. 2012), and the priming results in a faster and stronger induction of defense mechanisms after pathogen attack (Conrath 2011). The potentiated induction of defense responses only becomes apparent after further pathogen challenge and has often been overseen when studying SAR (Kohler et al. 2002; Conrath et al. 2006; Ahmad et al. 2010). Priming of pathogen-induced gene allows the plant to react more effectively to a subsequent invader (Conrath et al. 2002, 2006; Conrath 2009). In contrast to constitutive activation of defense responses, priming does not require major metabolic changes when no pathogens are present. Therefore, it forms a low-cost defense strategy while acting against a broad spectrum of attackers (Van Hulten et al. 2006; Pieterse and Dicke 2007; Van der Ent et al. 2008).

This defense priming offers disease protection that is effective against a broad spectrum of virulent pathogens (Ahmad et al. 2010). SA or its functional analog benzo(1,2,3)thiadiazole-7-carbothioic acid S-methyl ester (BTH), an SA biosynthesis-inducing chemical probenazole, and a nonprotein amino acid β-aminobutyric acid are known to induce priming and confer resistance against viral (Mandal et al. 2008), bacterial (Flors et al. 2008), oomycete (Hamiduzzaman et al. 2005; Ziadi et al. 2008), and fungal (Hukkanen et al. 2007; Umemura et al. 2009) diseases.

2.17.2 Accumulation of Dormant MAPKs May Be Involved in SA-Triggered Priming

Although the priming phenomenon has been known for decades, the molecular basis of SA-triggered priming is poorly understood (Conrath 2011). Mitogen-activated protein kinase (MAPK) cascades are major pathways downstream of sensors/receptors that transduce extracellular stimuli into intracellular responses in plants (Liu et al. 2003; Pedley and Martin 2005; Hettenhausen et al. 2012; Zhang et al. 2012a). The MAPKs transduce extracellular stimuli into intracellular transcription factors through activation of Ca^{2+}, ROS, SA, JA, and ethylene-dependent signaling systems and enhance expression of defense-related genes in plant innate immune system (Vidhyasekaran 2014). Some dormant MAPKs have been suggested to be important components required for priming in *Arabidopsis*, and the prestress deposition of these inactive kinases may be a possible mechanism of priming during the development of systemic acquired resistance (Beckers et al. 2009). MPK3 and functionally redundant MPK6 have been found to be important components for full priming in *Arabidopsis*. The resistance-inducing avirulent strains of *P. syringae* pv. *tomato* DC3000 and *P. syringae* pv. *phaseolicola* induced SA accumulation and MPK3 expression. Both SA and the SA-related compounds benzothiadiazole (BTH) and 4-chloro-SA activate *MPK3* gene expression and induce priming and SAR. In contrast, another SA-related compound 3-hydroxybenzoic acid did not induce *MPK3* gene expression, priming, and disease resistance. This strong correlation between the ability of avirulent bacteria and various SA-related compounds to activate *MPK3* gene expression and their capacity to prime plants for augmented defense gene activation and induced resistance suggests that MPK3 plays a role in priming (Beckers et al. 2009). Similarly another MAPK gene, *MPK6*, was also found to be involved in priming process. However, the BTH-induced accumulation of *MPK6* transcript and protein was less pronounced. Both MPK3 and MPK6 accumulate in an inactive form during priming of *Arabidopsis* with BTH (Beckers et al. 2009). Both MPK3 and MPK6 displayed greater activity in *Arabidopsis* plants which are primed and subsequently challenged with the virulent *P. syringae* pv. *maculicola*. These two enzymes were more strongly activated in primed plants than in nonprimed plants. Priming of defense gene expression and induced resistance were lost or reduced in *mpk3* or *mpk6* mutants (Beckers et al. 2009). These results suggest that prestress deposition of the signaling components MPK3 and MPK6 is a critical step in priming plants for full induction of defense responses during induced resistance.

2.17.3 Histone Modifications May Be Involved in Gene Priming in SA-Induced SAR

Chromatin is the combination of DNA and proteins that make up the contents of the nucleus of a cell. Chromatin structure is important for the regulation of gene expression (Ouyang and Gill 2009; Jaskiewicz et al. 2011). The primary function of chromatin is to package DNA into a smaller volume to fit in the cell. The first level of compaction is achieved by nucleosomal packaging of DNA. Each nucleosome comprises 147 bp of DNA wrapped around a histone octamer that consists of two molecules each of histone proteins H2A, H2B, H3, and H4 (Probst et al. 2003, 2004; Shi and Fang 2011). The primary protein components of chromatin are histones that compact the DNA (Bender 2004).

The histone proteins are subject to various covalent modifications including acetylation, methylation, phosphorylation, and ADP-ribosylation (Berger 2002; Britton et al. 2011; Chen et al. 2011). Acetylation of lysines in the amino-terminal tails of histones H3 and H4 has been associated with active genes (Eberharter and Becker 2002). Lysine acetylation is an important modification involved in the regulation of gene expression (Arif et al. 2010). In histone methylation, lysine and arginine residues are methylated, and up to three methyl groups are added to each residue. Specific methylation patterns are associated with gene activation and repression. The strongest correlation between histone methylation and gene activity has been reported for trimethylation of Lys4 on histone H3 (H3K4me3) on promoters and coding sequences of active genes (Ruthenburg et al. 2007).

Jaskiewicz et al. (2011) showed that histone modifications are systemically set during a priming event. These modifications might create a memory of the primary infection that is associated with an amplified reaction to a second stress condition. It is known that chromatin states could control cellular memory (Zhang 2008). The primed genes may be poised for enhanced activation of gene expression by histone modification (Jaskiewicz et al. 2011).

In the SAR response, defense genes in the infected and distant tissue show the "priming" phenomenon; they are able to respond faster and/or to a greater extent to a subsequent challenge (Kohler et al. 2002; Conrath 2009). The promoters of many of these genes contain at least one "W-box" that provides binding sites for WRKY transcription factors (Rushton et al. 2010). Genes encoding WRKY factors are themselves transcriptionally induced by either pathogen infection or treatment with PAMPs (Dong et al. 2003).

The SA analog BTH induces priming (Kohler et al. 2002; Beckers et al. 2009). Priming of defense genes for amplified response to secondary stress can be induced by application of salicylic acid or its synthetic analogue acibenzolar-*S*-methyl. BTH treatment induced a typical priming effect on *WRKY29*, *WRKY6*, and *WRKY53* genes (Jaskiewicz et al. 2011). The expression of the *WRKY* genes was enhanced in BTH-treated plants when the plants were stressed. Water infiltration, a stress-inducing cell death after BTH treatment, resulted in strongly enhanced gene activation, compared with plants that were stressed without previous BTH treatment. On the WRKY29 promoter, trimethylation (H3K4me3) and dimethylation (H3K4me2) of H3K4 and all acetylations tested increased after BTH application although this did not induce *WRKY29* transcription. Thus, chromatin marks normally associated with active genes are set by the priming stimulus before gene activation. After previous priming, a stress stimulus enhanced some of the modifications – H3K4me2, H3K9ac, and H4K12ac – on WRKY29 (Jaskiewicz et al. 2011). Collectively, these results suggest that prestress application of BTH induced chromatin modifications on *WRKY* gene promoters that facilitate the activation of gene expression by subsequent stress (Jaskiewicz et al. 2011).

Localized *Pseudomonas syringae* pv. *maculicola* infection primed the *WRKY* gene promoters in remote leaves to an augmented response to secondary stress, and the transcriptional responses in distal leaves were similar to those observed with BTH. On the three *WRKY29*, *WRKY6*, and *WRKY53* gene promoters, clear increases in histone modifications were observed after pathogen infection. The response amplitude after perception of the systemic signals for SAR was similar to that observed after BTH treatment. Thus, pathogen exposure induces one or more systemic signals that are stored in the form of histone modifications. Enhanced trimethylation of H3K4 concomitant with gene priming was a common feature of the assayed *WRKY* promoters (Jaskiewicz et al. 2011). Collectively, these studies suggest that there may be a tight correlation between histone modification patterns and gene priming and also there may be a histone memory for information storage in the plant stress response (Jaskiewicz et al. 2011).

2.17.4 NPR1 May Be Involved in Chromatin Modification-Induced Priming

Mutants that are attenuated in pathogen defense are often also compromised in gene priming. The *npr1* mutant of *Arabidopsis thaliana* is deficient in SAR (Durrant and Dong 2004) and cannot be primed for enhanced gene expression (Kohler et al. 2002; Beckers et al. 2009). BTH induced trimethylation of H3K4 on the *WRKY29* promoter in the wild-type *Arabidopsis thaliana* plants, but not on the priming-deficient *npr1* mutant. The results were similar to *WRKY6* and *WRKY53* expression and histone modification. Neither gene augmented expression after BTH treatment and stress stimulus in the *npr1* mutant. This correlated with the impaired ability of *npr1* to induce high H3K4me3 levels on the *WRKY6* and *WRKY53* promoters in response to BTH (Jaskiewicz et al. 2011). The results suggest that NPR1 plays important role in inducing high levels of chromatin modification on promoters of WRKY genes.

2.17.5 Histone Replacement May Be Instrumental for Priming of SA-Responsive Loci

Salicylic acid is known to trigger a long-lasting systemic acquired resistance (SAR). SAR involves extensive reprogramming of transcription (Thibaud-Nissen et al. 2006; Wang et al. 2006). SA mediates changes in the expression pattern of about 1,000–2,000 genes. Such a broad effect on gene transcription may be associated with extensive chromatin remodeling, which would require the involvement of specific chromatin remodeling complexes (March-Diaz et al. 2008). The chromatin remodeling may involve substitution of canonical histones in the octamer by histone variants, in a process known as histone replacement (Kamakka and Biggins 2005). One of the histone variants, H2A.Z, has been linked to both transcriptional repression and activation. Histone H2A.Z is incorporated into chromatin through the action of a multisubunit complex termed SWR1 in yeast. *Arabidopsis* PIE1 (for PHOTOPERIOD-INDEPENDENT EARLY FLOWERING1) is homologous to the yeast SWR1 protein. The histone variant H2A.Z, but not canonical H2A histone, interacts with PIE1. N-terminal region of *PIE1* specifically interacts with *Arabidopsis* H2A.Z histone variants. A role for PIE1 and histone H2A.Z has been observed in the priming process in SAR (March-Diaz et al. 2008).

To study the role of PIE1 in SAR, *pie1* mutant plants were developed. A total of 622 genes were differentially expressed between *pie1* mutant and wild-type (WT) plants. Most of the mis-regulated genes in *pie1* plants were related to SAR response. Several genes mis-regulated in *pie1* plants were also mis-regulated in wild-type plants treated with an analogue of SA, benzo(1,2,3)thiadiazole-7-carbothioic acid (BTH), which induces SAR (March-Diaz et al. 2008). These results suggest that PIE1 plays an important role in SA-mediated SAR.

The *pie1* mutant plants showed increased expression of several genes involved in SA accumulation. The level of *EDS5* mRNA increased in *pie1* mutant plants. EDS5 is involved in triggering SA accumulation. The amount of ICS1 mRNA increased about twofold in the absence of PIE1. ICS1 (also called SID2) encodes isochorismate synthase 1, an enzyme that is required for the synthesis of SA. PIE1 may also regulate downstream of the SA signal regulating NPR1 targets (March-Diaz et al. 2008). *pie1* mutant plants also showed increased resistance to virulent *Pseudomonas syringae* pv. *tomato*.

Comparison of gene expression profiles revealed that 65 % of the genes were misregulated in *pie1* mutant plants and the majority of mis-regulated genes were related to SA-dependent immunity. Several SAR marker genes were constitutively expressed in *pie1* plants. Such variations observed at the molecular level resulted in enhanced resistance to *Pseudomonas syringae* pv. *tomato*. Collectively these results suggest that PIE1 chromatin remodeling complex maintains a repressive state of the SAR response and this response is related to priming (March-Diaz et al. 2008). When PIE1 is inactivated, SAR is induced. It may be possible that defense genes are poised for enhanced activation during SAR by replacement on gene promoters of histone H2A with its variant H2A.Z (March-Diaz et al. 2008). Chromatin remodeling may be instrumental for priming of SA-responsive loci to enable their enhanced reactivation upon subsequent pathogen attack (van den Burg and Takken 2009).

2.18 Next-Generation Systemic Acquired Resistance

The priming can be inherited epigenetically from disease-exposed plants, and descendants of primed plants exhibit next-generation systemic acquired resistance (Luna et al. 2012; Slaughter et al. 2012). Slaughter et al. (2012) compared the reactions of *Arabidopsis thaliana* plants that had been either primed with an avirulent isolate of *Pseudomonas syringae* pv. *tomato* or primed with β-aminobutyric acid (BABA). The descendants of primed plants showed a faster and higher accumulation of transcripts of defense-related genes in salicylic acid signaling pathway and enhanced disease resistance upon challenge inoculation with a virulent isolate of *P. syringae*. The progeny of the primed plants was also more resistant against the oomycete pathogen *Hyaloperonospora arabidopsidis*. When transgenerationally primed plants were subjected to an additional priming treatment, their descendants displayed an even stronger primed phenotype (Slaughter et al. 2012). The results suggest that plants can inherit a sensitization for the priming phenomenon and the primed state of plants is transferred to their progeny.

The transgenerational SAR was found to be sustained over one stress-free generation, indicating an epigenetic basis of the phenomenon (Luna et al. 2012). The progeny from *Pseudomonas syringae* pv. *tomato*-inoculated *Arabidopsis thaliana* plants (P_1) displayed reduced responsiveness of jasmonic acid (JA)-inducible genes and enhanced susceptibility to the necrotrophic fungus *Alternaria brassicicola* (Luna et al. 2012). JA signaling is known to activate defense mechanisms against necrotrophic pathogens (McGrath et al. 2005; Zheng et al. 2006; Jakob et al. 2007; Korolev et al. 2008; Pré et al. 2008), and such signaling is suppressed in the P1 progeny which shows enhanced expression of SA-inducible genes. This shift in SA and JA responsiveness was not associated with changes in corresponding hormone levels. Instead, SA-inducible promoters of *PR1*, *WRKY6*, and *WRKY53* in P1 progeny plants were enriched with acetylated histone H3 at lysine 9, a chromatin mark associated with a permissive state of transcription. In contrast, the JA-inducible promoter of PDF1.2 showed increased H3 triple methylation at lysine 21, a mark related to repressed gene transcription (Luna et al. 2012). The results suggest the importance of chromatin remodeling in the transgenerational SAR. P1 progeny from defense regulatory mutant *npr1-1* failed to develop transgenerational defense phenotypes (Luna et al. 2012). It demonstrates the important role for NPR1 in expression of transgenerational defense responses.

DNA methylation may also play important role in transgenerational SAR (Luna et al. 2012). DNA methylation is initiated by DRM (Domains-Rearranged methyltransferase) and CMT3 (Chromomethyltransferase3) that are methyltransferases (Cao et al. 2003; Henderson et al. 2010; Huang et al. 2010; Greenberg et al. 2011; Naumann et al. 2011). Neither *drm* nor *cmt3* mutants affected the maintenance of preestablished RNA-directed CpG methylation. However, *drm* mutants showed a nearly complete loss of asymmetric methylation and a partial loss of CpNpG methylation. The remaining asymmetric and CpNpG methylation was dependent on the activity of CMT3, showing that DRM and CMT3 act redundantly to maintain non-CpG methylation. These DNA methyltransferases appear to act downstream of siRNAs, since *drm1 drm2 cmt3* triple mutants show a lack of non-CpG methylation but elevated levels of siRNAs. Further it has been demonstrated that DRM activity is required for the initial establishment of DNA methylation in all sequence contexts including CpG, CpNpG, and asymmetric sites (Cao et al. 2003).

The *drm1 drm2 cmt3* triple mutant has been shown to be affected in non-CpG DNA methylation. This mutant mimicked the transgenerational SAR. The bacterial pathogen *P. syringae* pv. *tomato* induces DNA hypomethylation in *Arabidopsis* (Luna et al. 2012). Collectively these results suggest that transgenerational SAR is transmitted by hypomethylated genes that direct priming of SA-dependent defenses in the following generations.

2.19 Crosstalk Between Salicylate and Jasmonate Signaling Systems

2.19.1 Antagonism Between SA and JA Signaling Systems

Antagonism between salicylic acid (SA) and JA signaling pathways has been widely reported in plants (Farmer et al. 2003; Spoel et al. 2003; Li et al. 2004; Brodersen et al. 2006; Mao et al. 2007; Koornneef and Pieterse 2008; Zander et al. 2010; El Rahman et al. 2012; Pieterse et al. 2012; Robert-Seilaniantz et al. 2011; Zander et al. 2012; Gimenez-Ibanez and Solano 2013; Van der Does et al. 2013). Antagonistic effect between SA and JA signaling has been reported in tobacco in inducing PR protein genes. The expression of acidic *PR1*, *PR2*, and *PR3* genes was upregulated by SA treatment, and it was suppressed in the presence of JA in a dose-dependent manner, while *PR2*, *PR5*, and *PR6* genes were upregulated by JA and suppressed by SA (Niki et al. 1998). The tobacco basic PR protein gene *NtPRB1b* responds positively to JA and ET signaling systems, but negatively to SA (Niki et al. 1998). JA signaling antagonizes SA signaling system in *Arabidopsis thaliana* (Kloek et al. 2001). The *Arabidopsis* mutants, which accumulate high levels of SA, show repression of JA-mediated pathway (Kachroo et al. 2003a, b). SA promotes disease development caused by necrotrophic pathogens by suppressing JA signaling system (El Rahman et al. 2012).

2.19.2 SA May Block JA Biosynthesis

Transgenic *Arabidopsis* plants expressing NahG, which were unable to accumulate SA, produced 25-fold higher levels of JA in response to infection by *Pseudomonas syringae* pv. *tomato* DC3000 (Spoel et al. 2003). The results suggest that SA may suppress JA biosynthesis. SA blocks JA biosynthesis in tomato leaves (Peña-Cortés et al. 1993). SA produced during pathogen infection plays an important role in the suppression of JA biosynthesis. Lipoxygenase (LOX) is the key enzyme in the JA biosynthesis pathway. Pathogen-induced SA suppresses *LOX2* gene encoding LOX2 in *Arabidopsis* (Spoel et al. 2003). Compared with wild-type plants, transgenic NahG plants showed enhanced expression of *LOX2* and accordingly synthesized 25-fold higher levels of JA during pathogen infection (Spoel et al. 2003). Allene oxide synthase (AOS) is another enzyme involved in JA biosynthesis (Laudert and Weiler 1998). Accumulation of allene oxide synthase transcripts in flax leaves is inhibited by salicylic acid (Harms et al. 1998). SA has been suggested to suppress the expression of the JA biosynthetic enzymes lipoxygenase (Spoel et al. 2003) and allene oxide synthase (AOS) (Harms et al. 1998). Leon-Reyes et al. (2010b) showed that the JA biosynthesis genes *LOX2* encoding lipoxygenase, *AOS* encoding allene oxide synthase, *AOC2* encoding allene oxide cyclase, and *OPR3* encoding 12-oxo-phytodienoate reductase were all repressed by SA in *Arabidopsis*. Collectively these studies suggest that SA may suppress the biosynthesis of JA, resulting in the suppression of JA signaling system (Fig. 2.20).

2.19.3 SA May Suppress JA-Responsive Gene Expression

SA strongly antagonized the JA signaling pathway, resulting in the downregulation of a large set of JA-responsive genes, including the marker genes *PDF1.2* and *VSP2* in *Arabidopsis thaliana* (Leon-Reyes et al. 2010b). In the *Arabidopsis aos/dde2* mutant, the expression of the JA biosynthesis enzyme allene oxide synthase (AOS) was completely blocked. Mutant *aos/dde2* plants did not express the JA-responsive marker genes *PDF1.2* or *VSP2* in response to infection with the necrotrophic fungus *Alternaria brassicicola*. Bypassing JA biosynthesis by exogenous application of methyl jasmonate (MeJA) rescued this JA-responsive phenotype in *aos/dde2*. Application of SA suppressed MeJA-induced *PDF1.2* expression to the same level in the *aos/dde2* mutant as in wild-type plants (Leon-Reyes et al. 2010b). The results suggest that SA-mediated suppression of JA-responsive gene expression is targeted at a position downstream of the JA biosynthetic pathway.

SA has been shown to suppress JA signaling downstream of the COI1–JAZ receptor complex (Van der Does et al. 2013). SA is a potent suppressor of JA-inducible gene expression (Doares et al. 1995; Harms et al. 1998; Gupta et al. 2000). Salicylic acid inhibits synthesis of proteinase inhibitors in tomato leaves induced by JA (Doares et al. 1995). The expression of the JA-responsive genes *VSP* and *PDF1.2* was enhanced strongly in the transgenic SA-degrading *NahG* gene expressing *Arabidopsis* plants (Spoel et al. 2003). The results suggest that in wild-type plants JA signaling is inhibited by SA that accumulates during pathogen infection.

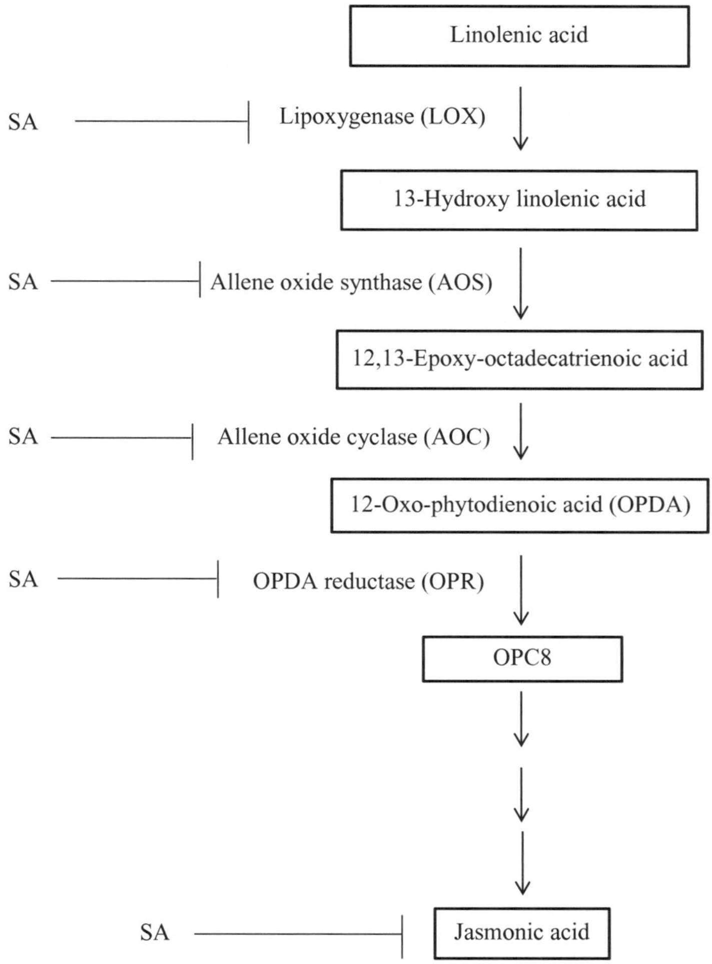

Fig. 2.20 Inhibition of JA biosynthesis enzymes by SA (Adapted from Spoel et al. 2003; Leon-Reyes et al. 2010b)

2.19.4 NPR1 in the Cytosol Modulates Crosstalk Between SA and JA Signaling Systems

NPR1 (non-expresser of PR gene1), the transcription cofactor, has been shown to be involved in the suppressive action of SA against JA signaling (Spoel et al. 2003). NPR1 forms a high-molecular-weight oligomeric complex in the cytosol. SA accumulation causes changes in cellular redox potential, and it leads to the formation of NPR1 monomers that are translocated into the nucleus. However, the NPR1 homomeric complex remains in the cytoplasm (Mou et al. 2003). SA-induced suppression of the

JA response is controlled by the function of NPR1 in the cytosol (Spoel et al. 2003; Koornneef and Pieterse 2008). NPR1 is translocated to the nucleus upon activation by SA, where it facilitates the activation of SA-responsive PR genes. In the cytosol, the remaining SA-activated NPR1 pool may be involved in the suppression of JA-responsive gene expression, either by facilitating the delivery of negative regulators of JA-responsive genes to the nucleus or by inhibiting positive regulators of JA-responsive gene expression (Spoel et al. 2003).

The NPR1 5'UTR contains several W-boxes, which are binding sites for WRKY transcription factors (Wasternack 2007). Some of the NPR1-activated transcription factors are involved in suppressing the activation of JA-responsive genes. The transcription factor WRKY62 is synergistically induced by MeJA and SA (Mao et al. 2007). MeJA inducibility of *WRKY62* is mediated by NPR1. Analysis of *wrky62* mutant and WRKY62-overexpressing plants showed that *WRKY62* suppresses JA-responsive gene expression (Mao et al. 2007). *WRKY62* may function downstream of cytosolic NPR1 and regulate SA-mediated inhibition of JA signaling (Mao et al. 2007).

The *Arabidopsis* transcription factor, WRKY70, is involved in crosstalk between SA and JA signaling pathways. WRKY70 is a common transcription factor in SA- and JA-mediated signaling pathways. It acts downstream of SA and *NPR1* (Knoth et al. 2007) and inhibits JA-responsive genes (Li et al. 2004). JA negatively regulates *WRKY70* transcript levels, whereas SA regulates *WRKY70* transcription positively (Li et al. 2004). *WRKY70* mRNA levels are enhanced in a *coi1* background, suggesting that JA represses its expression (Li et al. 2004). WRKY70 acts as a positive regulator of SA-responsive genes, while it acts as negative regulator of the expression of JA-responsive genes downstream of JA (Li et al. 2004, 2006; Journot-Catalino et al. 2006). WRKY70 activates SA-inducible *PR* genes and represses the JA-inducible gene *PDF2.1* (Fig. 2.21; Li et al. 2006, 2007). WRKY70-mediated suppression of most of the JA-responsive genes appeared to require functional NPR1 (Li et al. 2006).

Transgenic *Arabidopsis* plants overexpressing *WRKY70* have been developed. These plants showed high resistance to the biotrophic powdery mildew pathogen *Golovinomyces cichoracearum* (=*Erysiphe cichoracearum*), but showed enhanced susceptibility to the necrotrophic fungal pathogen *Alternaria brassicicola* (Li et al. 2006). It is known that SA induces resistance against biotrophic pathogens, whereas JA confers resistance against necrotrophic pathogens. The results suggest that WRKY70 positively regulates SA signaling system while it negatively regulates JA signaling system.

WRKY11 and WRKY17 of *Arabidopsis* have also been implicated in SA/JA crosstalk. In a double mutant in which *WRKY11* and *WRKY17* were knocked out, transcripts of SA-responsive genes accumulated to higher levels, whereas those of JA-responsive genes were significantly lower. Expression of *WRKY70* was up-regulated in this double mutant, suggesting that WRKY11 and WRKY17 function as negative regulators of WRKY70 (Journot-Catalino et al. 2006).

The transcription factor OsWRKY13 activates salicylic acid synthesis-related genes and SA-responsive genes, while it suppresses JA synthesis-related genes and

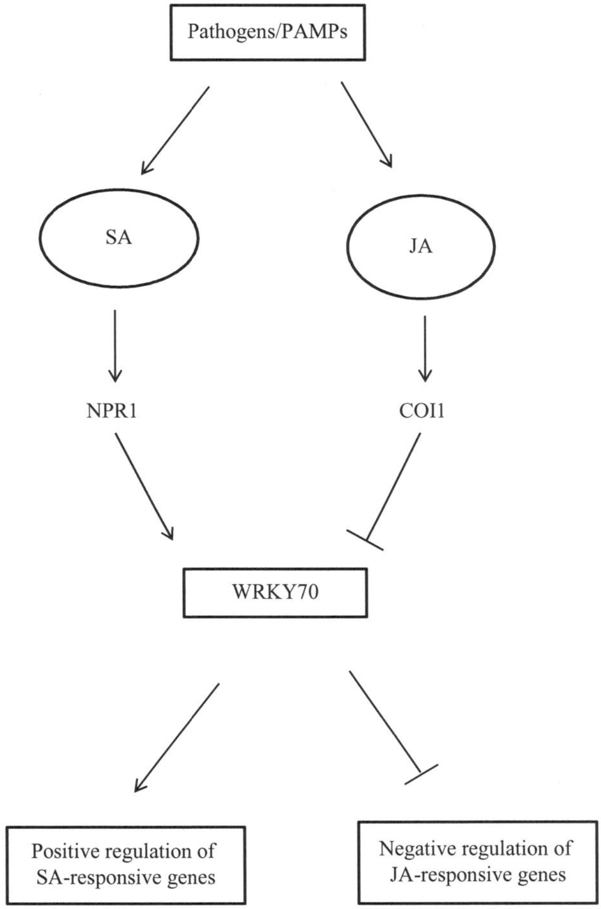

Fig. 2.21 Role of WRKY70 in crosstalk between SA and JA signaling pathways (Adapted from Li et al. 2004, 2006, 2007)

JA-responsive genes (Qiu et al. 2007). It was observed that OsWRKY13 bound to the promoters of its own and at least three other genes in SA- and JA-dependent signaling pathways (Qiu et al. 2007).

2.19.5 Role of Glutaredoxin and TGA Transcription Factors in the SA–JA Crosstalk

SA/JA crosstalk may also be redox regulated. The antagonistic effect of SA on JA-responsive gene transcription has been shown to be linked to SA-induced changes in the cellular redox potential (Koornneef and Pieterse 2008). A glutaredoxin GRX480 has been reported to be involved in SA/JA crosstalk (Ndamukong et al. 2007). Glutaredoxins mediate redox regulation of proteins by their capacity to catalyze disulfide transitions. Basic/leucine zipper-type transcription factors of the TGA family

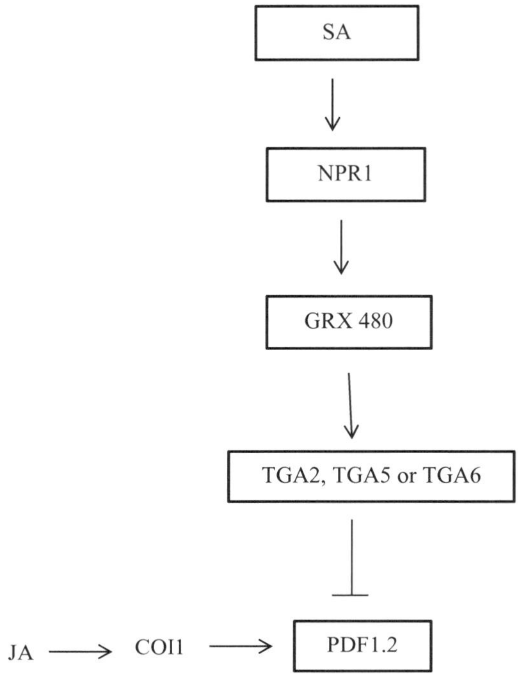

Fig. 2.22 SA-inducible glutaredoxin (GRX480) interacts with TGA factors and suppresses JA-responsive *PDF1.2* transcription (Adapted from Ndamukong et al. 2007)

interact with GRX480. Expression of GRX480 was found to be inducible by SA and dependent on NPR1. Transgenic *Arabidopsis* plants expressing *GRX480* showed near wild-type expression of SA-inducible responses, while transcription of JA-dependent defensing gene *PDF1.2* was suppressed in the transgenic plants. Suppression of JA-inducible gene depended on the presence of TGA factors. The interaction between GRX480 and TGA transcription factors was essential for the GRX480-dependent crosstalk. It is suggested that SA-activated NPR1 induces GRX480, which in turn interacts with TGA transcription factors (TGA2, TGA5, or TGA6) to suppress JA-responsive gene transcription (Fig. 2.22; Ndamukong et al. 2007).

2.19.6 Role of MAP Kinase 4 (MPK4) in SA and JA Crosstalk

A MAP kinase, MPK4 has been shown to be involved in SA and JA crosstalk. It acts as a negative regulator of SA signaling and as a positive regulator of JA signaling in *Arabidopsis* (Petersen et al. 2000). Inactivation of MPK4 in the mutant *mpk4 Arabidopsis* plants resulted in increased expression of SA-responsive genes and suppression of JA-responsive genes (Petersen et al. 2000). The target of MPK4 is its substrate MAP KINASE4 SUBSTRATE 1 (MKS1), and phosphorylation of MKS1 by MPK4 represses SA signaling. MSK1 interacts with the WRKY transcription factors WRKY25 and WRKY33, and both of them can be phosphorylated by MPK4 (Andreasson et al. 2005). Overexpression of both WRKY25 and WRKY33 results in repression of SA signaling responses, suggesting that these transcription factors, after activation by phosphorylation by the action of MPK4, suppress SA signaling system (Zheng et al. 2006, 2007). By contrast, *wrky33* mutant plants showed reduced expression of JA-responsive genes, suggesting that WRKY33 after phosphorylation by MPK4 activates JA signaling system (Zheng et al. 2006). These studies reveal that MPK4 is involved in crosstalk between SA and JA signaling systems.

EDS1 and PAD4 are defense regulators, and they affect SA/JA signal antagonism as activators of SA and as repressors of JA defenses (Wiermer et al. 2005; Brodersen et al. 2006). EDS1 and PAD4 regulate accumulation of SA in *Arabidopsis* (Bartsch et al. 2006). The *eds1* and *pad4* mutant *Arabidopsis* plants were highly susceptible to the bacterial pathogen *Pseudomonas syringae*. Constitutive overexpression of *EDS1* and *PAD4* enhanced resistance to the bacterial pathogen (Xing and Chen 2006). Enhanced resistance to the bacterial pathogen in transgenic *EDS1*- and *PAD4*-overexpressing lines was associated with a quicker and strong induction of SA-regulated defense gene *PR1* after infection with the pathogen. By contrast, *eds1* and *pad4* mutants were tolerant to the fungal pathogen *Botrytis cinerea*. Constitutive overexpression of *EDS1* and *PAD4* restored the susceptibility of the mutants to *Botrytis cinerea*. The negative role of EDS1 and PAD4 in plant response to the necrotrophic fungal pathogen *B. cinerea* correlated with their negative effects on induction of jasmonate-regulated *PDF1.2* gene expression (Xing and Chen 2006). The function of EDS1 and PAD4 in the antagonistic interaction between SA and JA depends on a MAP kinase, MPK4 in *Arabidopsis* (Brodersen et al. 2006).

Both SA-repressing and JA-activating functions of EDS1 and PAD4 appear to depend on MPK4 kinase activity in *Arabidopsis*. Mutations in *EDS1* and *PAD4* genes suppress derepression of the SA pathway and suppress the JA pathway in *mpk4* mutant plants (Brodersen et al. 2006). These results suggest that EDS1/PAD4 affect SA/JA signal antagonism as activators of SA and as repressors of JA defenses and MPK4 negatively regulates both of these functions.

2.19.7 SA May Suppress JA Signaling by Targeting GCC-Box Motifs in JA-Responsive Promoters

It has been now well demonstrated that SA suppresses JA pathway. The suppression of the JA pathway by SA may function downstream of the E3 ubiquitin ligase Skp–Cullin–F-box complex SCF^{COI1}, which targets ZIM-domain transcriptional repressor proteins (JAZs) for proteasome-mediated degradation (Chini et al. 2007; Thines et al. 2007; Pauwels and Goossens 2011; Van der Does et al. 2013). SA does not affect the stability or JA-induced degradation of JAZs. It was observed that the 1-kb promoter regions of JA-responsive genes that were suppressed by SA were significantly enriched in the JA-responsive

GCC-box motifs. SA may suppress JA signaling by targeting GCC-box motifs in JA-responsive promoters via a negative effect on the accumulation of the APETALA2/Ethylene response factor (AP2/ERF)-type transcriptional activator ORA59 (Octadecanoid-responsive Arabidopsis 59). The GCC-box motif is overrepresented in JA-responsive promoters that are suppressed by SA, and this promoter was found to be sufficient for SA-mediated suppression of JA-induced gene expression (Van der Does et al. 2013). SA reduced the accumulation of the GCC-box binding transcription factor ORA59 but not that of ERF1. Collectively these studies show that the SA pathway inhibits JA signaling by targeting GCC-box motifs in JA-responsive promoters (Gimenez-Ibanez and Solano 2013; Van der Does et al. 2013).

2.19.8 JA May Inhibit SA Signaling

It is widely reported that SA inhibits JA signaling and there are also reports indicating that JA also may inhibit SA signaling. In tobacco, MeJA induced the transcription of a glucosyltransferase, a homologue of UDP-glucose/salicylic acid glucosyltransferase in tobacco (Lee and Raskin 1999) and *Arabidopsis* (Sasaki et al. 2001). It is suggested that MeJA inhibits salicylic acid signaling through the induction of the glucosyltransferase, because the sugar conjugate of salicylic acid is inactive (Sasaki et al. 2001).

2.19.9 Synergism Between SA and JA Signaling Pathways

Synergism between SA and JA pathways has also been reported (Mur et al. 2006). A transient synergistic enhancement in the expression of genes associated with either JA (*PDF1.2* [defensin] and *Thi1.2* [thionin]) or SA (*PR1* [PR1a in tobacco]) signaling was observed when both jasmonic acid and salicylic acid were applied at low concentrations. However, antagonism was observed at more prolonged treatment times or at higher concentrations. Similar results were also observed when adding the jasmonate precursor α-linolenic acid with salicylic acid (Mur et al. 2006). These results suggest that the outcomes of JA–SA interactions depend on the relative concentration of JA and SA.

Synergism between JA/ET pathways and SA pathway in inducing resistance against *Plectosphaerella cucumerina* in *Arabidopsis* has been reported (Berrocal-Lobo et al. 2002). Components of the ET and JA pathways work in combination with SA-mediated resistance response in the defense of *Arabidopsis* against *P. syringae* and *Hyaloperonospora parasitica* (Clarke et al. 2000).

2.20 Crosstalk Between SA and ET Signaling Systems

Synergism between SA and ET signaling systems in rice has been reported (Mitsuhara et al. 2008). SA and ET work synergistically and induce resistance against *Cucumber mosaic virus* in *Arabidopsis* (Takahashi et al. 2002). A positive cooperation between ET and SA pathways was observed in *Arabidopsis* plants in response to *Plectosphaerella cucumerina* (Berrocal-Lobo et al. 2002). SA and ET/JA signaling pathways synergistically induced the expression of *PDF1.2* in *Arabidopsis* (Devadas et al. 2002). A negative crosstalk between SA and ET has also been shown in *Arabidopsis* in inducing resistance to *Pseudomonas syringae* pv. *tomato* (Berrocal-Lobo et al. 2002).

2.21 Crosstalk Between SA and ABA Signaling Systems

Abscisic acid suppresses salicylic acid-dependent signaling mechanisms (Audenaert et al. 2002). ABA suppresses SA accumulation in *Arabidopsis* infected with *Pseudomonas syringae* pv. *tomato* (Mohr and Cahill 2007). The ABA-deficient mutant of tomato had greater SA-mediated responses, suggesting the antagonistic interaction between ABA and SA signaling systems (Thaler and Bostock 2004). ABA reduced SA defense gene expression in *Arabidopsis* (Adie et al. 2007). ABA inhibits the accumulation of SA and the expression of SA-mediated defense genes (Mohr and Cahill 2007; Spoel and Dong 2008). ABA suppresses inducible innate immune responses by down-regulating SA biosynthesis and SA-mediated defenses in *Arabidopsis* (de Torres et al. 2009). ABA has been shown to promote susceptibility to the rice bacterial blight pathogen *Xanthomonas oryzae* pv. *oryzae* by suppressing SA-mediated defenses (Xu et al. 2013). In contrast, ABA signals have been shown to promote SA biosynthesis in *Arabidopsis* (Seo and Park 2010).

2.22 Crosstalk Between SA and Auxin Signaling Systems

SA has been shown to repress auxin signaling pathway and induce resistance against diseases (Wang et al. 2007). The SA analog BTH treatment repressed 21 genes related to auxin signal transduction. It also affected auxin homeostasis by lowering the levels of free auxin (Wang et al. 2007). Auxin may also suppress the induction of SA signaling. In plants overexpressing the auxin receptor *AFB1*, the accumulation of SA or the induction of SA response pathway is impaired after *P. syringae* pv. *tomato* DC3000 challenge inoculation in *Arabidopsis* (Robert-Seilaniantz et al. 2011).

2.23 Negative Regulation of Salicylate-Mediated Immunity by Brassinosteroid Signaling

Brassinosteroid (BR) signaling induced susceptibility to the rice root pathogen *Pythium graminicola* by suppressing the plant innate immunity. BR treatment induced susceptibility, while an inhibitor of BR signaling, brassinazole (BRZ), induced disease resistance activating plant immunity. These results suggest that BRs are negative regulators of innate immunity in rice plants. In contrast, SA signaling triggered immune responses against the pathogen in rice plants. Application of the brassinosteroid brassinolide (BL) negated resistance conferred by the SA synthetic analog benzothiadiazole (BTH), suggesting negative crosstalk between BR and SA signaling pathways. Exogenous application of BL decreased the pathogen-induced strong upregulation of the transcription of the SA regulatory genes *OsNPR1* and *OsWRKY45*. The temporal expression of both the SA regulatory genes was inversely correlated with that of the BR marker genes *OsBLE2*, *OsCPD1*, and *OsRAVL1* (De Vleeschauwer et al. 2012). These studies suggest that BR-mediated suppression of SA defenses occurs downstream of SA biosynthesis, but upstream of NPR1 and OsWRKY45 in the SA signaling pathway (De Vleeschauwer et al. 2012).

2.24 SA Signaling System May Induce Resistance Against a Wide Range of Pathogens

2.24.1 SA Signaling System Is Involved in Conferring Fungal and Oomycete Disease Resistance

SA signaling induces resistance against a wide range of pathogens. It is highly effective against biotrophic and hemibiotrophic fungal and oomycete pathogens and bacterial and viral pathogens. The SA analog benzothiadiazole (BTH), which activates SA signaling downstream of SA accumulation, induces resistance against the biotrophic rust fungal pathogen *Puccinia helianthi* in sunflower plants (Amzalek and Cohen 2007). SA induces resistance against the biotrophic powdery mildew pathogen *Erysiphe orontii* in *A. thaliana* (Thomma et al. 2001). It also induces resistance against the powdery mildew pathogen *Oidium* sp. in tobacco (Nakashita et al. 2002). SA signaling induces resistance against the biotrophic downy mildew oomycete pathogen *Hyaloperonospora parasitica* in *Arabidopsis* (Genger et al. 2008). Probenazole, which activates SA signaling system at SA accumulation stage as well as at NPR1 stage, triggers resistance against *H. parasitica* in *Arabidopsis* (Yoshioka et al. 2001).

SA signaling system is involved in triggering resistance against hemibiotrophic rice blast fungal pathogen *Magnaporthe oryzae* by inducing various defense genes

(Oostendorp et al. 2001). BTH enhanced resistance against the fungal pathogen *Leptosphaeria maculans* in *Brassica napus* plants. It triggered SAR, which was dependent on SA signaling system (Potlakayala et al. 2007). BTH induced resistance against the anthracnose pathogens *Colletotrichum destructivum* in cowpea seedlings (Latunde-Dada and Lucas 2001) and *Colletotrichum orbiculare* in cucumber (Cools and Ishii 2002; Deepak et al. 2006). The SA analog BTH treatment triggered accumulation of SA-inducible PR proteins and conferred resistance against the hemibiotrophic oomycete pathogen *Phytophthora palmivora* in papaya (Zhu et al. 2003).

SA signaling is not effective against the control of necrotrophic pathogens. It confers enhanced susceptibility to the necrotrophic fungal pathogen *Botrytis cinerea* (Kachroo et al. 2001, 2003a, b; Nandi et al. 2003, 2005; Genger et al. 2008). SA has been shown to promote disease development caused by *B. cinerea* through NPR1, by suppressing the JA-dependent defense genes (El Rahman et al. 2012). SA signaling pathway contributes to disease development caused by another necrotrophic pathogen *Alternaria solani* in tomato (El Rahman et al. 2012).

2.24.2 SA Signaling System Is Involved in Conferring Bacterial Disease Resistance

SA signaling system induces resistance against various bacterial pathogens. The SA analog BTH induced resistance against the fire blight pathogen *Erwinia amylovora* in apple leaf tissues by modulating redox systems (Sklodowska et al. 2010). SA-mediated defense responses promote resistance against the rice bacterial blight pathogen *Xanthomonas oryzae* pv. *oryzae* (Xu et al. 2013). The *NPR1* gene is a key regulator of SA-mediated signaling system and acts downstream of SA signal (Chern et al. 2008; Mukherjee et al. 2010). Transgenic rice plants overexpressing *NH1*, a rice homolog of *Arabidopsis NPR1* involved in SA signaling, show high levels of resistance to *X. oryzae* pv. *oryzae* (Chern et al. 2005). BTH enhanced resistance against the bacterial pathogen *Pseudomonas syringae* pv. *maculicola* in *Brassica napus* plants. It triggered SAR, which was dependent on SA signaling system (Potlakayala et al. 2007).

MYB96 transcription factor-mediated abscisic acid (ABA) signals promote SA biosynthesis, and the activated SA signaling system triggers resistance against the bacterial pathogen *Pseudomonas syringae* in *Arabidopsis* (Seo and Park 2010). SA signaling system induces resistance against *Pseudomonas syringae* pv. *tomato* in *Arabidopsis* (Thaler and Bostock 2004; Genger et al. 2008; Camañes et al. 2012). Transgenic *Arabidopsis* plants overexpressing *AtRBP-DR1*, the gene encoding RNA-binding protein, were developed, and these transgenic plants showed high accumulation of SA (Qi et al. 2010). Overexpression of *AtRBP-DR1* led to enhanced resistance to *P. syringae* pv. *tomato* DC3000, suggesting that SA induces resistance against the bacterial pathogen (Qi et al. 2010).

The *RAR1* and *SGT1b* genes are required for SA accumulation in *Arabidopsis*, and both are required in a genetically additive manner for induction of disease resistance (Zhou et al. 2008). Overexpression of *OsRAR1* and *OsSGT1* in rice significantly induced basal resistance to the bacterial pathogen *X. oryzae* pv. *oryzae* (Wang et al. 2008). Silencing of the regulator genes *RAR1* and *SGT1-2* involved in SA-mediated defense resulted in susceptibility to *Pseudomonas syringae* in soybean (Fu et al. 2009). These results suggest that SA signaling system is involved in inducing resistance against bacterial pathogens.

The genes encoding WRKY transcription factors functioning in SA signaling system also induce resistance against bacterial pathogens (Miao and Zentgraf 2007; Xing et al. 2008). SA induces the interaction of the transcription factors WRKY38 and WRKY62 with Histone Deacetylase19 (HDA19) and fine-tunes plant basal defense responses against *P. syringae* in *Arabidopsis* (Kim et al. 2008). Constitutive expression of the SA-inducible transcription factor *WRKY18* in *Arabidopsis* enhanced resistance to *P. syringae* (Xu et al. 2006). Overexpression of *OsWRKY13* activates SA synthesis-related genes and SA-responsive genes and enhances resistance to rice bacterial blight (Qiu et al. 2007).

SA signaling system may also induce susceptibility against bacterial pathogens. A rice gene *NRR* interacts with *NPR1* in the SA signaling pathway and suppresses the induction of defense genes, making the plants susceptible to pathogens (Chern et al. 2008). Overexpression of the transcription factors *WRKY38* and *WRKY62* involved in SA signaling system in *Arabidopsis* confers susceptibility to *P. syringae* (Kim et al. 2007). Another pathogen-induced, SA-dependent transcription factor WRKY7 is a negative regulator of host defense. *WRKY7*-overexpressing *Arabidopsis* plants supported more growth of *P. syringae* and developed more severe symptoms than wild-type plants (Kim et al. 2006). Overexpression of the transcription factor *WRKY48*, which is a negative regulator of SA-induced defense responses, suppresses basal resistance to *P. syringae* and makes the *Arabidopsis* plants susceptible to the pathogen (Xing et al. 2008).

CMPA (3-chloro-1-methyl-1H-pyrazole-5-carboxylic acid) protects rice from infection by the bacterial blight pathogen *X. oryzae* pv. *oryzae* (Nakashita et al. 2003; Nishioka et al. 2005). CMPA acts downstream of SA accumulation and acts in the SA signaling pathway between SA production and NPR1 activity (Yasuda 2007). It enhances resistance of tobacco to *Pseudomonas syringae* pv. *tabaci* (Yasuda et al. 2003).

Probenazole induces resistance against rice bacterial blight (*X. oryzae* pv. *oryzae*) by inducing various defense genes (Oostendorp et al. 2001). It induces the expression of S-adenosyl-L-methionine (SAM): salicylic acid carboxyl methyltransferase (SAMT) in *Atropa belladonna*. The enzyme is responsible for the conversion of SA to methyl salicylate (MeSA) (Kwon et al. 2009). MeSA is involved in triggering SA-induced systemic acquired resistance. Tobacco plants treated with probenazole showed increased levels of free and total SA (Midoh and Iwata 1996; Nakashita et al. 2002). The results suggest that probenazole intervenes in SA signaling system at SA accumulation stage as well as at NPR1 stage to trigger resistance against *X. oryzae* pv. *oryzae* (Yoshioka et al. 2001).

Activation or suppression of SA signaling system may induce resistance or susceptibility against the crown gall disease caused by *Agrobacterium tumefaciens* in *Nicotiana benthamiana* (Anand et al. 2008). SA specifically induces resistance against *A. tumefaciens*. SA decreases virulence and attachment of the bacteria to plant cells. It inhibits induction of virulence (*vir*) genes and the *repABC* operon. Plant genes involved in SA biosynthesis and signaling are important determinants for *A. tumefaciens* infectivity on plants. Silencing of *ICS* (isochorismate synthase, the enzyme involved in the biosynthesis of SA), *NPR1* (the regulator of SA signaling), and *SABP2* (SA-binding protein 2) in *N. benthamiana* enhanced *Agrobacterium* infection (Anand et al. 2008). These results suggest that induction of SA signaling induces resistance, while suppression of it induces susceptibility (Anand et al. 2008).

2.24.3 SA Signaling System Is Involved in Conferring Virus Disease Resistance

SA signaling system triggers defense responses against various virus diseases. The SA analog BTH is involved in triggering SA signaling system downstream of SA biosynthesis. It is effective in controlling various virus diseases, such as *Tobacco mosaic virus* (TMV) in tobacco (Friedrich et al. 1996; Waller et al. 2006), *Cucumber mosaic virus* (CMV) in tomato (Anfoka 2000), and *Tomato spotted wilt virus* (TSWV) in tobacco (Mandal et al. 2008). BTH was found to be effective in controlling TSWV in field-grown tomato (Momol et al. 2004).

SA signaling system confers resistance against *Cauliflower mosaic virus* in *Arabidopsis* (Love et al. 2005) Compromising early SA accumulation increases viral dispersal during lesion establishment in TMV-infected tobacco (Mur et al. 1997). Induced resistance against *Turnip crinkle virus* in *Arabidopsis* also is SA dependent (Kachroo et al. 2000).

SA appears to be a key signal molecule in activating defense responses in potato. The wild-type potato plants showed resistance to *Potato virus X* (PVX) avirulent strain ROTH1. SA-deficient transgenic potato plant lines overexpressing the bacterial enzyme salicylate hydroxylase (NahG), which degrades SA, were developed. These transgenic plants showed a decreased disease resistance response to infection by the virus strain (Sänchez et al. 2010). The results suggest that SA content determines disease resistance or susceptibility to the virus disease in potato.

SA signaling system may also be involved in inducing susceptibility to virus diseases. SA signaling system has been shown to enhance virulence of *Clover yellow vein virus* (ClYVV) in susceptible pea cultivars. The virus induced the SA-responsive chitinase gene (*SA-CHI*) in the susceptible pea cultivar. Application of SA or its analog BTH enhanced virulence of a mutant of the virus with reduced virulence. BTH enhanced disease symptoms induced by the wild-type virulent strain of ClYVV (Atsumi et al. 2009). These results suggest that the SA signaling may be associated with susceptibility in pea–*Clover yellow vein virus* interaction.

2.25 Pathogens May Suppress SA Signaling System to Cause Disease

2.25.1 Pathogens May Secrete Effectors to Suppress SA Signaling System

It is widely reported that SA signaling system triggers defense responses against a wide range of pathogens. Virulent pathogens have to overcome the immune responses induced by SA signaling system to cause disease. Pathogens produce effectors to suppress the plant immune responses. The effector proteins target basic innate immunity in plants (Metz et al. 2005; Block et al. 2008; Bartetzko et al. 2009; Boller and He 2009; de Jonge and Thomma 2009; de Jonge et al. 2010; Khang et al. 2010; Kwon 2010; Song and Yang 2010; Szczesny et al. 2010; Valent and Khang 2010; Block and Alfano 2011; Thomma et al. 2011; Wu et al. 2011).

Several *Pseudomonas syringae* strains use a specialized type III secretion system to deliver effector proteins into host cells to subvert host defense mechanisms, thereby promoting pathogenesis (Hauck et al. 2003; DebRoy et al. 2004; Jelenska et al. 2007; Göhre et al. 2008). *P. syringae* pv. *tomato* effector AvrPtoB (now called $HopAB2_{PtoDC3000}$) suppresses basal defense in *Arabidopsis* (de Torres et al. 2006). The effectors HopPtoM and AvrE of *P. syringae* pv. *tomato* suppress the SA-dependent basal immunity in *Arabidopsis* (DebRoy et al. 2004). Effectors of HopPtoM/AvrE family appear to be widespread in plant-pathogenic bacteria, and they appear to target the defense signaling pathways (DebRoy et al. 2004). The effector HopI1 (previously named HopPmaI) isolated from *P. syringae* pv. *maculicola* strain PmaES4326 suppresses SA accumulation in plant cells (Jelenska et al. 2007). HopI1 expression in *Arabidopsis* plant cells resulted in a 60 % decrease in the level of SA-inducible *PR1* gene transcript and approximately 50 % lower free and total SA levels. It suggests that HopI1 suppresses SA accumulation and SA-dependent defenses (Jelenska et al. 2007).

2.25.2 Pathogen Produces Toxin and Suppresses SA Signaling System to Promote Disease Development

The bacterial pathogen *Pseudomonas syringae* produces coronatine that mimics jasmonic acid-isoleucine and promotes opening of bacterial entry, bacterial growth in the apoplast, and systemic susceptibility (Zheng et al. 2012). Coronatine was found to activate three homologous NAC transcription factor (TF) genes through direct activity of the transcription factor MYC2. The activated NAC TFs repress the *isochorismate synthase 1* (ICS1) gene, which is involved in SA biosynthesis.

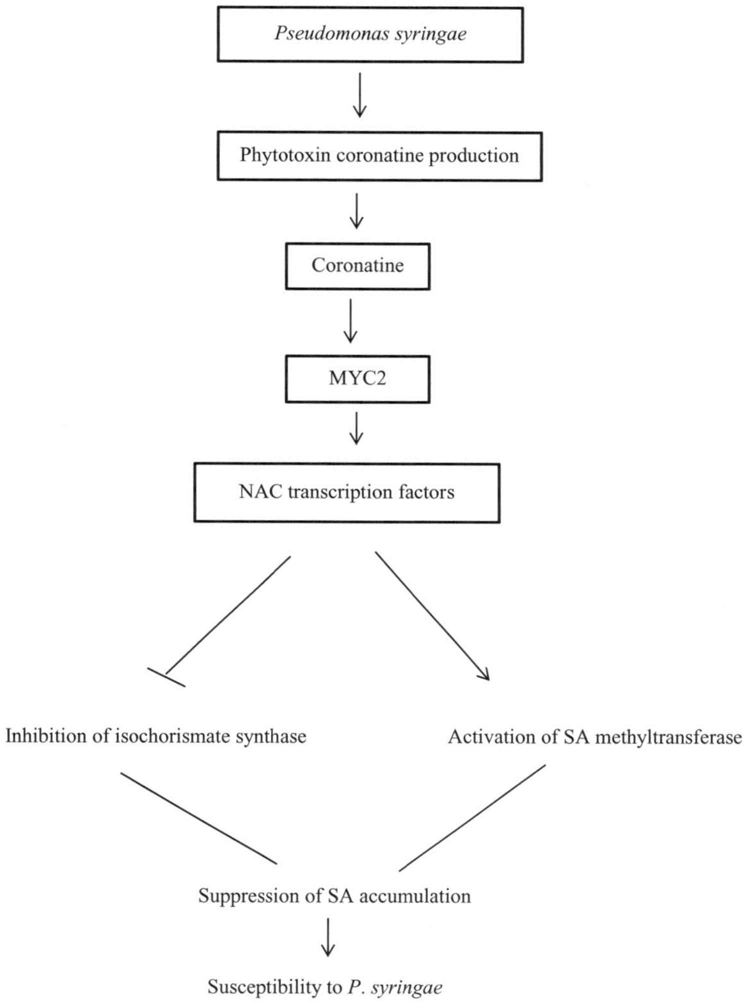

Fig. 2.23 *Pseudomonas syringae* produces phytotoxin coronatine and suppresses SA signaling system to cause disease (Adapted from Zheng et al. 2012)

These TFs also activate expression of *BSMT1* gene encoding SA methyltransferase, which converts SA to the inactive volatile methyl SA (MeSA) (Zheng et al. 2012). These results suggest that the toxin produced by the bacterial pathogen promotes systemic susceptibility by suppressing SA signaling, and these studies also suggest that SA signaling is important in conferring resistance against the bacterial pathogen (Fig. 2.23; Zheng et al. 2012).

2.25.3 Pathogen Manipulates the Antagonistic Effect Between SA and JA Signaling Systems to Promote Disease Development

SA signaling system is known to induce resistance against the bacterial pathogen *P. syringae* pv. *tomato* in *Arabidopsis* (Brooks et al. 2005; Jelenska et al. 2007). When *Arabidopsis* plants were inoculated with *P. syringae* pv. *tomato*, there was substantial increase in *WRKY33* transcript level (Dong et al. 2003; Zheng et al. 2006). WRKY33 induces JA-mediated signaling system, and the induced JA signaling system reduces the SA-mediated signaling system by its antagonistic function (Zheng et al. 2006; van Verk et al. 2008). It suggests that the pathogen suppresses the SA signaling pathway, which is needed to induce resistance against the pathogen, to cause disease.

Activation of SA-mediated pathway appears to confer resistance to *Pseudomonas syringae* pv. *tomato* DC3000. The pathogen produces the toxin coronatine, which induces JA-signaling pathway while suppressing the SA-dependent pathway (Katsir et al. 2008). It suggests that the pathogen is able to suppress the signaling pathway which may suppress its pathogenesis and diverts the signals for activation of an inefficient pathway resulting in defense response that cannot inhibit its pathogenesis (Spoel et al. 2003, 2007; Uppalapati et al. 2007). The PR1 protein is required for the inhibition of *P. syringae* pv. *tomato* (Laurie-Berry et al. 2006), and the induction of this pathogenesis-related protein is dependent on SA accumulation (van Verk et al. 2008). The toxin produced by the pathogen triggers JA signaling, which suppresses SA-mediated defenses. Thus, it appears that the bacterial pathogen utilizes the toxin to manipulate JA signaling both to suppress SA-mediated responses and to promote symptom development (Brooks et al. 2005; Laurie-Berry et al. 2006; Koornneef and Pieterse 2008).

References

Adie BA, Perez-Perez J, Godoy M, Sanchez-Serrano JJ, Schmelz EA, Solano R (2007) ABA is an essential signal for plant resistance to pathogens affecting JA biosynthesis and activation of defenses in *Arabidopsis*. Plant Cell 19:1665–1681

Ahmad S, Gordon-Weeks R, Pickett J, Ton J (2010) Natural variation in priming of basal resistance: from evolutionary origin to agricultural exploitation. Mol Plant Pathol 11:817–827

Ahn I-P, Kim S, Lee Y-H, Suh S-C (2007) Vitamin B_1-induced priming is dependent on hydrogen peroxide and the *NPR1* gene in *Arabidopsis*. Plant Physiol 143:838–848

Alamillo JM, Saénz P, Garcia JA (2006) Salicylic acid-mediated and RNA-silencing defense mechanisms cooperate in the restriction of systemic spread of plum pox virus in tobacco. Plant J 48:217–227

Allen GJ, Chu SP, Schumacher K, Shimazaki CT, Vafeados D, Kemper A, Hawke SD, Tallman G, Tsien RY, Harper JF, Chory J, Schroeder JI (2000) Alteration of stimulus-specific guard cell calcium oscillations and stomatal closing in *Arabidopsis det3* mutant. Science 289:2338–2342

Alvarado VY, Scholthof HB (2009) Plant responses against invasive nucleic acids: RNA silencing and its suppression by plant viral pathogens. Semin Cell Dev Biol 20:1032–1040

Alvarez ME, Nota F, Cambiagno DA (2010) Epigenetic control of plant immunity. Mol Plant Pathol 11:563–576

Amzalek E, Cohen Y (2007) Comparative efficacy of systemic acquired resistance-inducing compounds against rust infection in sunflower plants. Phytopathology 97:179–186

An C, Mou Z (2013) The function of the Mediator complex in plant immunity. Plant Signal Behav 8:e23182

Anand A, Uppalapati SR, Ryu C-M, Allen SN, Kang L, Tang Y, Mysore KS (2008) Salicylic acid and systemic acquired resistance play a role in attenuating crown gall disease caused by *Agrobacterium tumefaciens*. Plant Physiol 146:703–715

Andreasson E, Jenkins T, Brodersen P, Thorgrimsen S, Petersen NHT, Zhu S, Qiu J-L, Micheelsen P, Rocher A, Petersen M, Newman A, Nielsen HB, Hirt H, Somssich I, Mattsson O, Mundy J (2005) The MAP kinase substrate MKS1 is a regulator of plant defense responses. EMBO J 24:2579–2589

Anfoka GH (2000) Benzo-(1,2,3)-thiadiazole-7-carbothioic acid-*S*-methyl ester induces systemic resistance in tomato (*Lycopersicon esculentum* Mill. cv. volledung) to cucumber mosaic virus. Crop Prot 19:401–405

Argueso CT, Ferreira FJ, Epple P, To JPC, Hutchison CE, Schaller GE, Dangl JL, Kieber JJ (2012) Two-component elements mediate interactions between cytokinin and salicylic acid in plant immunity. PLoS Genet 8(1):e1002448

Arif M, Selvi BR, Kundu TK (2010) Lysine acetylation: the tale of a modification from transcription regulation to metabolism. ChemBioChem 11:1501–1504

Asada K (2006) Production and scavenging of reactive oxygen species in chloroplasts and their functions. Plant Physiol 141:391–396

Asai T, Tena G, Plotnikova J, Willmann MR, Chiu W-L, Gómez-Gómez L, Boller T, Ausubel FM, Sheen J (2002) MAP kinase signalling cascade in *Arabidopsis* innate immunity. Nature 415:977–983

Atsumi G, Kagaya U, Kitazawa H, Nakahara KS, Uyeda I (2009) Activation of the salicylic acid signaling pathway enhances *Clover yellow vein virus* virulence in susceptible pea cultivars. Mol Plant-Microbe Interact 22:166–175

Attaran E, Zeier TE, Griebel T, Zeier J (2009) Methyl salicylate production and jasmonate signaling are not essential for systemic acquired resistance in *Arabidopsis*. Plant Cell 21:954–971

Audenaert K, De Meyer GB, Höfte MM (2002) Abscisic acid determines basal susceptibility of tomato to *Botrytis cinerea* and suppresses salicylic acid- dependent signaling mechanisms. Plant Physiol 128:491–501

Azevedo C, Sadanandom A, Kitagawa K, Freialdenhoven A, Shirasu K, Schulze-Lefert P (2002) The RAR1 interactor SGT1, an essential component of *R* gene-triggered disease resistance. Science 295:2073–2076

Azevedo C, Betsuyaku S, Peart J, Takahashi A, Noel L, Sadanandom A, Casals C, Parker J, Shirasu K (2006) Role of SGT1 in resistance protein accumulation in plant immunity. EMBO J 25:2007–2016

Bartetzko V, Sonnewald S, Vogel F, Hartner K, Stadler R, Hammes UZ, Bornke F (2009) The *Xanthomonas campestris* pv. *vesicatoria* type III effector protein XopJ inhibits protein secretion: evidence for interference with cell wall-associated defense responses. Mol Plant-Microbe Interact 22:655–664

Bartsch M, Gobbato E, Bednarek P, Debey S, Schultze J, Bautor J, Parker JE (2006) Salicylic acid-independent ENHANCED DISEASE SUSCEPTIBILITY1 signaling in *Arabidopsis* immunity and cell death is regulated by the monooxygenase FMO1 and the nudix hydrolase *NUDT7*. Plant Cell 18:1038–1051

Beckers GJM, Conrath U (2007) Priming for stress resistance: from the lab to the field. Curr Opin Plant Biol 10:425–431

Beckers GJM, Jaskiewicz M, Liu Y, Underwood WR, He SY, Zhang S, Conrath U (2009) Mitogen-activated protein kinases 3 and 6 are required for full priming of stress responses in *Arabidopsis thaliana*. Plant Cell 21:944–953

Beffa R, Szell M, Meuwly P, Pay A, Vogeli-Lange R, Metraux JP, Neuhaus G, Meins F, Ferenc N (1995) Cholera toxin elevates pathogen resistance and induces pathogenesis-related gene expression in tobacco. EMBO J 14:5753–5761

Bellin D, Asai DM, Yoshioka H (2013) Nitric oxide as a mediator for defense responses. Mol Plant Microbe Interact 26:271–277

Bender J (2004) Chromatin-based silencing mechanisms. Curr Opin Plant Biol 7:521–526

Benhar M, Forrester MT, Hess DT, Stamler JS (2008) Regulated protein denitrosylation by cytosolic and mitochondrial thioredoxins. Science 320:1050–1054

Berger SL (2002) Histone modifications in transcriptional regulation. Curr Opin Genet Dev 2:142–148

Berrocal-Lobo M, Molina A, Solano R (2002) Constitutive expression of ETHYLENE-RESPONSE-FACTOR1 in *Arabidopsis* confers resistance to several necrotrophic fungi. Plant J 29:23–32

Block A, Alfano JR (2011) Plant targets for *Pseudomonas syringae* type III effectors: virulence targets or guarded decoys? Curr Opin Microbiol 14:39–46

Block A, Li G, Fu ZQ, Alfano JR (2008) Phytopathogen type III effector weaponry and their plant targets. Curr Opin Plant Biol 11:396–403

Boller T, He SY (2009) Innate immunity in plants: an arms race between pattern recognition receptors in plants and effectors in microbial pathogens. Science 324:742–744

Bolwell GP, Butt VS, Davies DR, Zimmerlin A (1995) The origin of the oxidative burst in plants. Free Radical Res 23:517–532

Bolwell GP, Davies DR, Gerrish C, Auh C-K, Murphy TM (1998) Comparative biochemistry of the oxidative burst produced by rose and French bean cells reveals two distinct mechanisms. Plant Physiol 116:1379–1385

Bouché N, Scharlat A, Snedden WA, Bouchez D, Fromm H (2002) A novel family of calmodulin-binding transcription activators in multicellular organisms. J Biol Chem 277:21851–21861

Boursiac Y, Lee SM, Romanowsky S, Blank R, Stadek C, Chung WS, Harper JF (2010) Disruption of the vacuolar calcium-ATPase in Arabidopsis results in the activation of a salicylic acid-dependent programmed cell death pathway. Plant Physiol 154:1158–1171

Britton LM, Gonzales-Cope M, Zee BM, Garcia BA (2011) Breaking the histone code with quantitative mass spectrometry. Expert Rev Proteomics 8:631–643

Brodersen P, Petersen M, Bjorn Nielsen H, Zhu S, Newman MA, Shokat KM, Rietz S, Parker J, Mundy J (2006) *Arabidopsis* MAP kinase 4 regulates salicylic acid- and jasmonic acid/ethylene-dependent responses via EDS1 and PAD4. Plant J 47:532–546

Brooks DM, Bender CL, Kunkel BN (2005) The *Pseudomonas syringae* phytotoxin coronatine promotes virulence by overcoming salicylic acid-dependent defenses in *Arabidopsis thaliana*. Mol Plant Pathol 6:629–639

Camañes G, Pastor V, Cerezo M, Garcia-Andrade J, Vicedo B, Garcia-Agustin P, Flors V (2012) A deletion in *NRT2.1* attenuates *Pseudomonas syringae*-induced hormonal perturbation, resulting in primed plant defenses. Plant Physiol 158:1054–1066

Canet JV, Dobón A, Tornero P (2012) Non-recognition-of BTH4, an *Arabidopsis* mediator subunit homolog, is necessary for development and response to salicylic acid. Plant Cell 24:4220–4235

Cao H, Glazebrook J, Clarke J, Volko S, Dong X (1997) The Arabidopsis *npr1* gene that controls systemic acquired resistance encodes a novel protein containing ankyrin repeats. Cell 88:57–63

Cao X, Aufsatz W, Zilberman D, Mette MF, Jacobsen SE (2003) Role of the DRM and CMT3 methyltransferases in RNA-directed DNA methylation. Curr Biol 13:2212–2217

Carr JP, Lewsey MG, Palukaitis P (2010) Signaling in induced resistance. Adv Virus Res 76:57–121

Catinot J, Buchala A, Abou-Mansour E, Métraux JP (2008) Salicylic acid production in response to biotic and abiotic stress depends on isochorismate in *Nicotiana benthamiana*. FEBS Lett 582:473–478

Cevik V, Kidd BN, Zhang P, Hill C, Kiddle S, Denby KJ, Holub EB, Cahill DM, Manners JM, Schenk PM, Beynon J, Kazan K (2012) MEDIATOR25 acts as an integrative hub for the regulation of jasmonate-responsive gene expression in Arabidopsis. Plant Physiol 160:541–555

Chadha KC, Brown SA (1974) Biosynthesis of phenolic acids in tomato plants infected with *Agrobacterium tumefaciens*. Can J Bot 52:2041–2047

Chanda B, Xia Y, Mandal MK, Yu K, Sekine K-T, Gao Q-M, Selote D, Hu Y, Stromberg A, Navarre D, Kachroo A, Kachroo P (2011) Glycerol-3-phosphate is a critical mobile inducer of systemic immunity in plants. Nat Genet 43:421–427

Chaturvedi R, Krothapalli K, Makandar R, Nandi A, Sparks AA, Roth MR, Welti R, Shah J (2008) Plastid omega3-fatty acid desaturase-dependent accumulation of a systemic acquired resistance inducing activity in petiole exudates of *Arabidopsis thaliana* is independent of jasmonic acid. Plant J 54:106–117

Chaturvedi R, Venables B, Petros RA, Nalam V, Li M, Wang X, Takemoto LJ, Shah J (2012) An abietane diterpenoid is a potent activator of systemic acquired resistance. Plant J 71:161–172

Chellappan P, Vanitharani R, Ogbe F, Fauquet CM (2005) Effect of temperature on *Geminivirus-induced* RNA silencing in plants. Plant Physiol 138:1828–1841

Chen C, Chen Z (2002) Potentiation of developmentally regulated defense response by AtWRKY18, a pathogen-induced *Arabidopsis* transcription factor. Plant Physiol 129:706–716

Chen Z, Klessig DF (1991) Identification of a soluble salicylic acid-binding protein that may function in signal transduction in the plant disease resistance response. Proc Natl Acad Sci U S A 90:9533–9537

Chen ZX, Ricigliano JW, Klessig DF (1993) Purification and characterization of a soluble salicylic acid-binding protein that may function in signal from tobacco. Proc Natl Acad Sci U S A 90:9533–9537

Chen Z, Zheng Z, Huang J, Lai Z, Fan B (2009) Biosynthesis of salicylic acid in plants. Plant Signal Behav 4:493–496

Chen X, Hu Y, Zhou DX (2011) Epigenetic gene regulation by plant Jumonji group of histone demethylase. Biochim Biophys Acta 1809:421–426

Chen R, Jiang H, Li L, Zhai Q, Qi L, Zhou W, Liu X, Li H, Zheng W, Sun J, Li C (2012) The Arabidopsis mediator subunit MED25 differentially regulates jasmonate and abscisic acid signaling through interacting with MYC2 and ABI5 transcription factors. Plant Cell 24:2898–2916

Cheong YH, Moon BC, Kim JK, Kim CY, Kim MC, Kim IH, Park CY, Kim JC, Park BO, Koo SC, Yoon HW, Chung WS, Lim CO, Lee SY, Cho MJ (2003) BWMK1, a rice mitogen-activated protein kinase, locates in the nucleus and mediates pathogenesis-related gene expression by activation of a transcription factor. Plant Physiol 132:1961–1972

Chern M-S, Fitzgerald H, Yadav RC, Canlas PE, Dong X, Ronald PC (2001) Evidence for a disease-resistance pathway in rice similar to the *NPR1*-mediated signaling pathway in Arabidopsis. Plant J 27:101–113

Chern M, Fitzgerald HA, Canlas PE, Navarre DA, Ronald PC (2005) Overexpression of a rice NPR1 homolog leads to constitutive activation of defense response and hypersensitivity to light. Mol Plant-Microbe Interact 18:511–520

Chern M, Canlas PE, Ronald PC (2008) Strong suppression of systemic acquired resistance in *Arabidopsis* by NRR is dependent on its ability to interact with NPR1 and its putative repressive domain. Mol Plant 1:552–559

Chini A, Fonseca S, Fernández G, Adie B, Chico JM, Lorenzo O, Garcia-Casado G, López-Vidriero I, Lozano FM, Ponce MR, Micol JL, Solano R (2007) The JAZ family of repressors is the missing link in jasmonate signaling. Nature 448:666–671

Chini A, Fonseca S, Chico JM, Fernández-Calvo P, Solano R (2009) The ZIM domain mediates homo- and heteromeric interactions between Arabidopsis JAZ proteins. Plant J 59:77–87

Chivasa S, Murphy AM, Naylor M, Carr JP (1997) Salicylic acid interferes with tobacco mosaic virus replication as a novel salicylhydroxamic acid-sensitive mechanism. Plant Cell 9:547–557

Choi J, Huh SU, Kojima M, Sakakibara H, Paek K-H, Hwang I (2010) The cytokinin-activated transcription factor ARR2 promotes plant immunity via TGA3/NPR1-dependent salicylic acid signaling in *Arabidopsis*. Dev Cell 19:284–295

Choi J, Choi D, Lee R, Ryu CM, Hwang I (2011) Cytokinins and plant immunity: old foes or new friends. Trends Plant Sci 16:388–394

Chong J, Pierrel M-A, Atanassova R, Werck-Reichhart D, Fritig B, Saindrenan PS (2001) Free and conjugated benzoic acid in tobacco plants and cell cultures. Induced accumulation upon elicitation of defense responses and role as salicylic acid precursors. Plant Physiol 125:318–328

Chun HJ, Park HC, Koo SC, Lee JH, Park CY, Choi MS, Kang CH, Baek D, Cheong YH, Yun DJ, Chung WS, Cho MJ, Kim MC (2012) Constitutive expression of mammalian nitric oxide synthase in tobacco plants triggers disease resistance to pathogens. Mol Cells 34:463–471

Clark D, Durner J, Navarre DA, Klessig DF (2000) Nitric oxide inhibition of tobacco catalase and ascorbate peroxidase. Mol Plant-Microbe Interact 13:1380–1384

Clarke JD, Volko SM, Ledford H, Ausubel FM, Dong X (2000) Roles of salicylic acid, jasmonic acid, and ethylene in *cpr*-induced resistance in Arabidopsis. Plant Cell 12:2175–2190

Conaway RC, Conaway JW (2011a) Function and regulation of the Mediator complex. Curr Opin Genet Dev 21:225–230

Conaway RC, Conaway JW (2011b) Origin and activity of the Mediator complex. Semin Cell Dev Biol 22:729–734

Conrath U (2009) Priming of induced plant defense responses. Adv Bot Res 51:361–395

Conrath U (2011) Molecular aspects of defence priming. Trends Plant Sci 16:524–531

Conrath U, Chen Z, Ricigliano JR, Klessig DF (1995) Two inducers of plant defense responses, 2,6-dichloroisonicotinic acid and salicylic acid, inhibit catalase activity in tobacco. Proc Natl Acad Sci U S A 92:7143–7147

Conrath U, Klessig DF, Bachmair A (1998) Tobacco plants perturbed in the ubiquitin-dependent protein degradation system accumulate callose, salicylic acid, and pathogenesis-related protein 1. Plant Cell Rep 17:876–880

Conrath U, Thulke O, Katz V, Schwindling S, Kohler A (2001) Priming as a mechanism in induced systemic resistance of plants. Eur J Plant Pathol 107:113–119

Conrath U, Pieterse CMJ, Mauch-Mani B (2002) Priming in plant-pathogen interactions. Trends Plant Sci 7:210–216

Conrath U, Beckers GJM, Flors V, Garcia-Agustin P, Jakab G, Mauch F, Newman M-A, Pieterse CMJ, Poinssot B, Pozo MJ, Pugin A, Schaffrath U, Ton J, Wendehenne D, Zimmerli L, Mauch-Mani B (2006) Priming: getting ready for battle. Mol Plant-Microbe Interact 19:1062–1071

Cools HJ, Ishii H (2002) Pre-treatment of cucumber plants with acibenzolar-S-methyl systemically primes a phenylalanine ammonia lyase gene (*PAL1*) for enhanced expression upon attack with a pathogenic fungus. Physiol Mol Plant Pathol 61:273–280

Daudi A, Cheng Z, O'Brein JA, Mammarella N, Khan S, Ausubel FM, Bolwell GP (2012) The apoplastic oxidative burst peroxidase in *Arabidopsis* is a major component of pattern-triggered immunity. Plant Cell 24:275–287

de Jonge R, Thomma BPHJ (2009) Fungal LysM effectors: extinguishers of host immunity? Trends Microbiol 17:151–157

de Jonge R, van Esse HP, Kombrink A, Shinya T, Desaki Y, Bours R, van der Krol S, Shibuya N, Joosten MHAJ, Thomma BPHJ (2010) Conserved fungal LysM effector Ecp6 prevents chitin-triggered immunity in plants. Science 329:953–955

de Torres M, Mansfield JW, Grabov N, Brown IR, Ammouneh H, Tsiamis G, Forsyth A, Robatzek S, Grant M, Boch J (2006) *Pseudomonas syringae* effector AvrPtroB suppresses basal defence in Arabidopsis. Plant J 47:368–382

de Torres-Zabala M, Bennett MH, Truman WH, Grant MR (2009) Antagonism between salicylic and abscisic acid reflects early host-pathogen conflict and moulds plant defense responses. Plant J 59:375–386

De Vleeschauwer D, Van Buyten E, Satoh K, Balidion J, Mauleon R, Choi I-R, Vera-Cruz C, Kikuchi S, Höfte M (2012) Brassinosteroids antagonize gibberellin- and salicylate-mediated root immunity in rice. Plant Physiol 158:1833–1846

Dean JV, Mohammed LA, Fitzpatrick T (2005) The formation, vacuolar localization, and tonoplast transport salicylic acid glucose conjugates in tobacco cell suspension cultures. Planta 221:287–296
DebRoy S, Thilmony R, Kwack Y-B, Nomura K, He S-Y (2004) A family of conserved bacterial effectors inhibits salicylic acid-mediated basal immunity and promotes disease necrosis in plants. Proc Natl Acad Sci U S A 101:9927–9932
Deepak SA, Ishii A, Park P (2006) Acibenzolar-S-methyl primes cell wall strengthening genes and reactive oxygen species forming/scavenging enzymes in cucumber after fungal pathogen attack. Physiol Mol Plant Pathol 69:52–61
DeFalco TA, Bender KW, Snedden WA (2010) Breaking the code: Ca^{2+} sensors in plant signalling. Biochem J 425:27–40
Dellagi A, Helibronn J, Avrova AO, Montesano M, Palva ET, Stewart HE, Toth IK, Cooke DE, Lyon GD, Birch PR (2000) A potato gene encoding a WRKY-like transcription factor is induced in interactions with *Erwinia carotovora* subsp. *atroseptica* and *Phytophthora infestans* and is coregulated with a class I endochitinase expression. Mol Plant-Microbe Interact 13:1092–1101
Dempsey DA, Vlot AC, Wildermuth MC, Klessig DF (2011) Salicylic acid biosynthesis and metabolism. Arabidopsis Book 9:e0156. doi:10.1199/tab.0156
Denancé N, Sánchez-Vallet A, Goffner D, Molina A (2013) Disease resistance or growth: the role of plant hormones in balancing immune responses and fitness costs. Front Plant Sci 4:1–12, Article 155
Després C, DeLong C, Glaze S, Liu E, Fobert PR (2000) The Arabidopsis NPR1/NIM1 protein enhances the DNA binding activity of a subgroup of the TGA family of bZIP transcription factors. Plant Cell 12:279–290
Després C, Chubak C, Rochon A, Clark R, Bethune T, Desveaux D, Fobert PR (2003) The Arabidopsis *NPR1* disease resistance protein is a novel cofactor that confers redox regulation of DNA binding activity to the basic domain/leucine zipper transcription factor TGA1. Plant Cell 15:2181–2191
Desveaux D, Subramaniam R, Després C, Mess J-N, Lévesque C, Fobert PR, Dangl JL, Brisson N (2004) A 'Whirly' transcription factor is required for salicylic acid-dependent disease resistance in *Arabidopsis*. Dev Cell 6:229–240
Devadas SK, Enyedi A, Raina R (2002) The *Arabidopsis hrl1* mutation reveals novel overlapping roles for salicylic acid, jasmonic acid and ethylene signalling in cell death and defense against pathogens. Plant J 30:467–480
Diaz-Pendon JA, Li F, Li WX, Ding SW (2007) Suppression of antiviral silencing by cucumber mosaic virus 2b protein in *Arabidopsis* is associated with drastically reduced accumulation of three classes of viral small interfering RNAs. Plant Cell 19:2053–2063
Ding S-W (2010) RNA-based antiviral immunity. Nat Rev Immunol 10:632–644
Ding S-W, Voinnet O (2007) Antiviral immunity directed by small RNAs. Cell 130:413–426
Doares SH, Narvaez-Vasquez J, Conconi A, Ryan CA (1995) Salicylic acid inhibits synthesis of proteinase inhibitors in tomato leaves induced by systemin and jasmonic acid. Plant Physiol 108:1741–1746
Dodd AN, Kudla J, Sanders D (2010) The language of calcium signaling. Annu Rev Plant Biol 61:593–620
Dong J, Chen C, Chen Z (2003) Expression profiles of the Arabidopsis WRKY gene superfamily during plant defense response. Plant Mol Biol 51:21–37
Dreher K, Callis J (2007) Ubiquitin, hormones and biotic stress in plants. Ann Bot 99:787–822
Du H, Klessig DF (1997) Identification of a soluble, high-affinity salicylic acid-binding protein in tobacco. Plant Physiol 113:1319–1327
Du H, Zhang L, Liu L, Tang X-F, Yang W-J, Wu Y-M, Huang Y-B, Tang Y-X (2009) Biochemical and molecular characterization of plant MYB transcription factor family. Biochemistry (Moscow) 74:1–11
Du Q, Zhu W, Zhao Z, Qian X, Xu Y (2012) Novel benzo-1,2,3-thiadiazole-7-carboxylate derivatives as plant activators and the development of their agricultural applications. J Agric Food Chem 60:346–353

Durner J, Klessig DF (1995) Inhibition of ascorbate peroxidase by salicylic acid and 2,6-dichloroisonicotinic acid, two inducers of plant defense responses. Proc Natl Acad Sci U S A 92:11312–11316
Durner J, Klessig DF (1996) Salicylic acid is a modulator of tobacco and mammalian catalases. J Biol Chem 271:28492–28501
Durner J, Klessig DF (1999) Nitric oxide as a signal in plants. Curr Opin Plant Biol 2:369–374
Durner J, Shah J, Klessig DF (1997) Salicylic acid and disease resistance in plants. Trends Plant Sci 2:266–274
Durner J, Wendehenne D, Klessig DF (1998) Defense gene induction in tobacco by nitric oxide, cyclic GMP, and cyclic ADP ribose. Proc Natl Acad Sci U S A 95:10328–10333
Durrant WE, Dong X (2004) Systemic acquired resistance. Annu Rev Phytopathol 42:185–209
Eberharter A, Becker PB (2002) Histone acetylation: a switch between repressive and permissive chromatin. EMBO Rep 3:224–229
Edgar CI, McGrath KC, Dombrecht B, Manners JM, Maclean DC, Schenk PM, Kazan K (2006) Salicylic acid mediates resistance to the vascular wilt pathogen *Fusarium oxysporum* in the model host *Arabidopsis thaliana*. Australasian Plant Pathol 35:581–591
El Rahman TA, El Oirdi M, Gonzalez-Lamothe R, Bouarab K (2012) Necrotrophic pathogens use salicylic acid signaling pathway to promote disease development in tomato. Mol Plant-Microbe Interact 25:1584–1593
Enyedi AJ, Yalpani N, Silverman P, Raskin I (1992) Localization, conjugation, and function of salicylic acid in tobacco during the hypersensitive reaction to tobacco mosaic virus. Proc Natl Acad Sci U S A 89:2480–2484
Espunya C, De Michele R, Gomez-Cadenas AA, Martinez MC (2012) S-Nitrosoglutathione is a component of wound- and salicylic acid-induced systemic responses in *Arabidopsis thaliana*. J Exp Bot 63:3219–3227
Eulgem T, Rushton PJ, Schmelzer E, Hahlbrock K, Somssich IE (1999) Early nuclear events in plant defence signaling: rapid gene activation by WRKY transcription factors. EMBO J 18:4689–4699
Eulgem T, Rushton PJ, Robatzek S, Somssich IE (2000) The WRKY superfamily of plant transcription factors. Trends Plant Sci 5:199–206
Eulgem T, Weighman VJ, Chang H-S, McDowell JM, Holub EB, Glazebrook J, Zhu T, Dangl JL (2004) Gene expression signatures from three genetically separable resistance gene signaling pathways for downy mildew resistance. Plant Physiol 135:1129–1144
Fabro G, Di Rienzo JA, Voigt CA, Savchenko T, Dehesh K, Somerville S, Alvarez ME (2008) Genome-wide expression profiling Arabidopsis at the stage of *Golovinomyces cichoracearum* haustorium formation. Plant Physiol 146:1421–1439
Fan W, Dong X (2002) In vivo interaction between NPR1 and transcription factor TGA2 leads to salicylic acid-mediated gene activation in *Arabidopsis*. Plant Cell 14:1377–1389
Farmer EE, Almeras E, Krishnamurthy V (2003) Jasmonates and related oxylipins in plant responses to pathogenesis and herbivory. Curr Opin Plant Biol 6:372–378
Ferrari S, Plotnokova JM, De Lorenzo G, Ausubel FM (2003) *Arabidopsis* local resistance to *Botrytis cinerea* involves salicylic acid and camalexin and requires *EDS4* and *PAD2*, but not *SID2*, *EDS5* or *PAD4*. Plant J 35:193–205
Feys BJ, Moisan LJ, Newman MA, Parker JE (2001) Direct interaction between the Arabidopsis disease resistance signaling proteins, EDS1 and PAD4. EMBO J 20:5400–5411
Fitzgerald HA, Chern M-S, Navarre R, Ronald PC (2004) Overexpression of (*At*) *NPR1* in rice leads to a BTH- and environment-induced lesion-mimic/cell death phenotype. Mol Plant Microbe Interact 17:140–151
Fitzgerald HA, Canlas PE, Chern M-S, Ronald PC (2005) Alteration of TGA factor activity in rice results in enhanced tolerance to *Xanthomonas oryzae* pv. *oryzae*. Plant J 43:335–347
Flors V, Ton J, van Doorn R, Garcia-Agustin P, Mauch-Mani B (2008) Interplay between JA, SA, and ABA signalling during basal and induced resistance against *Pseudomonas syringae* and *Alternaria brassicicola*. Plant J 54:81–92

Foissner L, Wendehenne D, Langebartels C, Durner J (2000) In vivo imaging of an elicitor-induced nitric oxide burst in tobacco. Plant J 23:817–824

Forouhar F, Yang Y, Kumar D, Chen Y, Fridman E, Park SW, Chiang Y, Acton TB, Montelione GT, Pichersky E, Klessig DF, Tong L (2005) Structural and biochemical studies identify tobacco SABP2 as a methyl salicylate esterase and implicate it in plant innate immunity. Proc Natl Acad Sci U S A 102:1773–1778

Friedrich L, Lawton K, Reuss W, Masner P, Specker N, Gut Rella M, Meier B, Dincher S, Staub T, Uknes S, Metraux JP, Kessmann H, Ryals J (1996) A benzothiadiazole derivative induces systemic acquired resistance in tobacco. Plant J 10:61–70

Frye CA, Tang D, Innes RW (2001) Negative regulation of defense responses in plants by a conserved MAPKK kinase. Proc Natl Acad Sci U S A 98:373–378

Fu ZQ, Guo M, Jeong BR, Tian F, Elthon TE, Cerny RL, Staiger D, Alfano JR (2007) A type III effector ADP-ribosylates RNA-binding proteins and quells plant immunity. Nature 447:284–288

Fu DQ, Ghabrial S, Kachroo A (2009) *GmRAR1* and *Gm SGT1* are required for basal, *R* gene-mediated and systemic acquired resistance in soybean. Mol Plant-Microbe Interact 22:86–95

Fu L-J, Shi K, Gu M, Zhou Y-H, Dong D-K, Liang W-S, Song F-M, Yu J-Q (2010) Systemic induction and role of mitochondrial alternative oxidase and nitric oxide in a compatible tomato-Tobacco *mosaic virus* interaction. Mol Plant-Microbe Interact 23:39–48

Fu ZQ, Yang S, Saleh A, Wang W, Ruble J, Oka N, Mohan R, Spoel SH, Tada Y, Zheng N, Dong X (2012) NPR3 and NPR4 are receptors for the immune signal salicylic acid in plants. Nature 486:228–232

Fujiwara M, Umemura K, Kawasaki T, Shimamoto K (2006) Proteomics of Rac GTPase signaling reveals its predominant role in elicitor-induced defense response of cultured rice cells. Plant Physiol 140:734–745

Fung RWM, Gonzalo M, Fekete C, Kovacs LG, He Y, Marsh E, McIntyre LM, Schchtman DP, Qiu W (2008) Powdery mildew induces defense-oriented reprogramming of the transcriptome in a susceptible but not in a resistant grapevine. Plant Physiol 146:236–249

Gaille C, Kast P, Hass D (2002) Salicylate biosynthesis in *Pseudomonas aeruginosa*. Purification and characterization of PchB, a novel bifunctional enzyme displaying isochorismate pyruvate-lyase and chorismate mutase activities. J Biol Chem 277:21768–21775

Gaille C, Reimmann C, Haas D (2003) Isochorismate synthase (PchA), the first and rate-limiting enzyme in salicylate biosynthesis of *Pseudomonas aeruginosa*. J Biol Chem 278:16893–16898

Galon Y, Aloni R, Nachmias D, Snir O, Feldmesser E, Scrase-Field S, Boyce JM, Bouché N, Knight MR, Fromm H (2010) Calmodulin-binding transcription activator 1 mediates auxin signaling and responds to stresses in Arabidopsis. Planta 232:165–178

Gao M, Liu J, Zhang Z, Cheng F, Chen S, Zhang Y (2008) MEKK1, MKK1/MKK2 and MPK4 function together in a mitogen-activated protein kinase cascade to regulate innate immunity in plants. Cell Res 18:1190–1198

Gao Q, Liu Y, Wang M, Zhang J, Gai Y, Zhu C, Guo N (2009) Molecular cloning and characterization of an inducible RNA-dependent RNA polymerase gene, *GhRdRP*, from cotton (*Gossypium hirsutum* L.). Mol Biol Rep 36:47–56

Garcia-Brugger A, Lamotte O, Vandelle E, Bourque S, Lecourieux D, Poinssot B, Wendehenne D, Pugin A (2006) Early signaling events induced by elicitors of plant defenses. Mol Plant-Microbe Interact 19:711–724

Garcia-Ruiz H, Takeda A, Chapman EJ, Sullivan CM, Fahlgren N, Brempelis KJ, Carrington JC (2010) *Arabidopsis* RNA-dependent RNA polymerases and dicer-like proteins in antiviral defense and small interfering RNA biogenesis during *Turnip mosaic virus* infection. Plant Cell 22:481–496

Garcion C, Métraux J-P (2006) Salicylic acid. In: Hedden P, Thomas SG (eds) Plant hormone signaling, vol 24, Annual reviews. Blackwell Press, Oxford, pp 229–257

Garcion C, Lohmann A, Lamodiere E, Catinot J, Buchala A, Doermann P, Métraux J-P (2008) Characterization and biological function of the *ISOCHORISMATE SYNTHASE2* gene of the Arabidopsis. Plant Physiol 147:1279–1287

Gelli A, Higgins VJ, Blumwald E (1997) Activation of plant plasma membrane Ca^{2+}-permeable channels by race-specific fungal elicitors. Plant Physiol 113:269–279

Genger RK, Jurkowski GI, McDowell JM, Lu H, Jung HW, Greenberg JT, Bent AF (2008) Signaling pathways that regulate the enhanced disease resistance of *Arabidopsis* "Defense, No death" mutants. Mol Plant-Microbe Interact 21:1285–1296
Gilliland A, Singh DP, Hayward JM, Moore CA, Murphy AM, York CJ, Slator J, Carr JP (2003) Genetic modification of alternative respiration has differential effects on antimycin A-induced versus salicylic acid-induced resistance to *Tobacco mosaic virus*. Plant Physiol 132:1518–1528
Gimenez-Ibanez S, Solano R (2013) Nuclear jasmonate and salicylate signaling and crosstalk in defense against pathogens. Front Plant Sci 4:Article 72/1
Glazebrook J, Chen W, Estes B, Chang HS, Nawrath C, Metraux JP, Zhu T, Katagiri F (2003) Topology of the network integrating salicylate and jasmonate signal transduction derived from global expression phenotyping. Plant J 34:217–228
Göhre V, Spallek T, Häweker H, Mersmann S, Mentzel T, Boller T, de Torres M, Mansfield JW, Robatzek S (2008) Plant pattern-recognition receptor FLS2 is directed for degradation by the bacterial ubiquitin ligase AvrPtoB. Curr Biol 18:1824–1832
Greenberg MVC, Ausin I, Chan SWL, Cokus SJ, Cuperus JT, Feng S, Law JA, Chu C, Pellegrini M, Carrington JC, Jacobsen SE (2011) Identification of genes required for *de novo* DNA methylation in *Arabidopsis*. Epigenetics 6:344–354
Grennan AK (2008) Ethylene response factors in jasmonate signaling and defense response. Plant Physiol 146:1457–1458
Gudesblat GE, Torres PS, Vojnov AA (2009) *Xanthomonas campestris* overcomes Arabidopsis stomatal innate immunity through a DSF cell-to-cell signal-regulated virulence factor. Plant Physiol 149:1017–1027
Gupta V, Willits MG, Glazebrook J (2000) *Arabidopsis thaliana* EDS4 contributes to salicylic acid (SA)-dependent expression of defense responses: evidence for inhibition of jasmonic acid signaling by SA. Mol Plant-Microbe Interact 13:503–511
Hamada H, Kurusu T, Okuma E, Nokajima H, Kiyoduka M, Koyano T, Sugiyama Y, Okada K, Koga J, Saji H, Miyao A, Hirochika H, Yamane H, Murata Y, Kuchitsu K (2012) Regulation of a proteinaceous elicitor-induced Ca^{2+} influx and production of phytoalexins by a putative voltage-gated cation channel, OsTPC1, in cultured rice cells. J Biol Chem 287:9931–9939
Hamiduzzaman MM, Jakeb G, Barnavon L, Neuhaus JM, Mauch-Mani B (2005) β-Aminobutyric acid-induced resistance against downy mildew in grapevine acts through the potentiation of callose formation and jasmonic acid signaling. Mol Plant-Microbe Interact 18:819–829
Hammerschmidt R (2009) Systemic acquired resistance. Adv Bot Res 51:173–222
Harms K, Ramirez I, Peńa-Cortés H (1998) Inhibition of wound-induced accumulation of allene oxide synthase transcripts in flax leaves by aspirin and salicylic acid. Plant Physiol 118:1057–1065
Hashimoto K, Eckert C, Anschūtz U, Scholz M, Held K, Waadt R, Reyer A, Hippler M, Becker D, Kudla J (2012) Phosphorylation of calcineurin B-like (CBL) calcium sensor proteins by their CBL-interacting protein kinases (CIPKs) is required for full activity of CBL-CIPK complexes toward their target proteins. J Biol Chem 287:7956–7968
Hauck P, Thilmony R, He SY (2003) A *Pseudomonas syringae* type III effector suppresses cell wall-based extracellular defense in susceptible Arabidopsis plants. Proc Natl Acad Sci U S A 100:8577–8582
Henderson IR, Deleris A, Wong W, Zhong X, Chin HG, Horwitz GA, Kelly KA, Pradhan S, Jacobsen SE (2010) The *de novo* cytosine methyltransferase DRM2 requires intact UBA domains and a catalytically mutated paralog DRM3 during RNA-directed DNA methylation in *Arabidopsis thaliana*. PLoS Genet 6:e1001182
Hennig J, Malamy J, Grynkiewicz G, Indulski J, Klessig DF (1993) Interconversion of salicylic acid signal and its glucoside in tobacco. Plant J 4:593–600
Hermann M, Maier F, Masroor A, Hirth S, Pfitzner AJ, Pfitzner UM (2013) The Arabidopsis NIMIN proteins affect NPR1 differentially. Front Plant Sci 4:88
Hettenhausen C, Baldwin IT, Wu J (2012) Silencing *MPK4* in *Nicotiana attenuata* enhances photosynthesis and seed production but compromises abscisic acid-induced stomatal closure and

guard cell-mediated resistance to *Pseudomonas syringae* pv. *tomato* DC3000. Plant Physiol 158:759–776

Hirt H (2000) Connecting oxidative stress, auxin, and cell cycle regulation through a plant mitogen-activated protein kinase pathway. Proc Natl Acad Sci U S A 97:2405–2407

Hondo D, Hase S, Kanayama Y, Yoshikawa N, Takenaka S, Takahashi H (2007) The *LeATL6-associated* ubiquitin/proteasome system may contribute to fungal elicitor-activated defense response via the jasmonic acid-dependent signaling pathway in tomato. Mol Plant-Microbe Interact 20:72–81

Huang J, Wang H, Xie X, Zhang D, Liu Y, Guo G (2010) Roles of DNA methyltransferases in *Arabidopsis* development. Afr J Biotechnol 9:8506–8514

Hukkanen AT, Kokko HI, Buchala AJ, McDougall GJ, Stewart D, Kárenlampi SO, Karjalainen RO (2007) Benzothiadiazole induces the accumulation of phenolics and improves resistance to powdery mildew in strawberries. J Agric Food Chem 55:1862–1870

Ichimura K, Shinozaki K, Tena G, Sheen J, Henry Y (2002) Mitogen-activated protein kinase cascades in plants: a new nomenclature. Trends Plant Sci 7:301–308

Iwai T, Seo S, Mitsuhara I, Ohashi Y (2007) Probenazole-induced accumulation of salicylic acid confers resistance to *Magnaporthe grisea* in adult rice plants. Plant Cell Physiol 48:915–924

Jakob K, Kniskern JM, Bergelson J (2007) The role of pectate lyase and the jasmonic acid defense response in *Pseudomonas viridiflava* virulence. Mol Plant-Microbe Interact 20:146–158

Jakoby M, Weisshaar B, Droge-Laser W, Vicente-Carbajosa J, Tiedemann J, Kroj T, Parcy F (2002) bZIP transcription factors in *Arabidopsis*. Trends Plant Sci 7:106–111

Jaskiewicz M, Conrath U, Peterhänsel C (2011) Chromatin modification acts as a memory for systemic acquired resistance in the plant stress response. EMBO Rep 12:50–55

Jaubert M, Bhattacharjee S, Mello AF, Perry KL, Moffett P (2011) ARGONAUTE2 mediates RNA silencing anti-viral defenses against *Potato virus X* in Arabidopsis. Plant Physiol 156:1556–1564

Jelenska J, Yao N, Vinatzer BA, Wright CM, Brodsky JL, Greenberg JT (2007) A J domain virulence effector of *Pseudomonas syringae* remodels host chloroplasts and suppresses defenses. Curr Biol 17:499–508

Ji L-H, Ding S-W (2001) The suppressor of transgene RNA silencing encoded by *Cucumber mosaic virus* interferes with salicylic acid-mediated virus resistance. Mol Plant-Microbe Interact 14:715–724

Jing B, Xu S, Xu M, Li Y, Li S, Ding J, Zhang Y (2011) Brush and spray: a high-throughput systemic acquired resistance assay suitable for large-scale genetic screening. Plant Physiol 157:973–980

Jirage D, Tootle TL, Reuber TL, Frost LN, Feys BJ, Parker JE (1999) *Arabidopsis thaliana* PAD4 encodes a lipase-like gene that is important for salicylic acid signaling. Proc Natl Acad Sci U S A 96:13583–13588

Johnson C, Boden E, Arias J (2003) Salicylic acid and NPR1 induce the recruitment of *trans-activating* TGA factors to a defense gene promoter in Arabidopsis. Plant Cell 15:1846–1858

Johnson C, Mhatre A, Arias J (2008) NPR1 preferentially binds to the DNA-inactive form of *Arabidopsis* TGA2. Biochim Biophys Acta 1779:583–589

Journot-Catalino N, Somssich IE, Roby D, Kroj T (2006) The transcription factors WRKY11 and WRKY17 act as negative regulators of basal resistance in *Arabidopsis thaliana*. Plant Cell 18:3289–3302

Jovel J, Walker M, Sanfacon H (2011) Salicylic acid-dependent restriction of *Tomato ringspot virus* spread in tobacco is accompanied by a hypersensitive response, local RNA silencing, and moderate systemic resistance. Mol Plant-Microbe Interact 24:706–718

Jung HW, Tschaplinski TJ, Wang L, Glazebrook J, Greenberg JT (2009) Priming in systemic plant immunity. Science 324:89–91

Kachroo P, Yoshioka K, Shah J, Dooner HK, Klessig DF (2000) Resistance to turnip crinkle virus in *Arabidopsis* is regulated by two host genes and is salicylic acid dependent but NPR1, ethylene and jasmonate independent. Plant Cell 12:677–690

Kachroo P, Shanklin J, Shah J, Whittle EJ, Klessig DF (2001) A fatty acid desaturase modulates the activation of defense signaling pathways in plants. Proc Natl Acad Sci U S A 98:9448–9453
Kachroo P, Kachroo A, Lapchyk L, Hildebrand D, Klessig DF (2003a) Restoration of defective cross talk in *ssi2* mutants: role of salicylic acid, jasmonic acid, and fatty acids in *SSI2*-mediated signaling. Mol Plant-Microbe Interact 16:1022–1029
Kachroo A, Lapchyk L, Fukushige H, Hildebrand D, Klessig D, Kachroo P (2003b) Plastidial fatty acid signaling modulates salicylic acid- and jasmonic acid-mediated defense pathways in the Arabidopsis *ssi2* mutant. Plant Cell 15:2952–2965
Kachroo A, Venugopal SC, Lapchyk L, Falcone D, Hildebrand D, Kachroo P (2004) Oleic acid levels regulated by glycerolipid metabolism modulate defense gene expression in Arabidopsis. Proc Natl Acad Sci U S A 101:5152–5157
Kalde M, Barth M, Somssich IE, Lippok B (2003) Members of the *Arabidopsis* WRKY group III transcription factors are part of different plant defense signaling pathways. Mol Plant-Microbe Interact 16:295–305
Kallenbach M, Alagna F, Baldwin IT, Bonaventure G (2010) *Nicotiana attenuata* SIPK, WIPK, NPR1, and fatty acid-amino acid conjugates participate in the induction of jasmonic acid biosynthesis by affecting early enzymatic steps in the pathway. Plant Physiol 152:98–106
Kamakka RT, Biggins S (2005) Histone variants: deviants? Genes Dev 19:295–310
Kang CH, Jung WY, Kang YH, Kim JY, Kim DG, Jeong JC, Baek DW, Jin JB, Lee JY, Kim MO, Chung WS, Mengiste T, Koiwa H, Kwak SS, Bahk JD, Lee SY, Nam JS, Yun DJ, Cho MJ (2006) AtBAG6, a novel calmodulin-binding protein, induces programmed cell death in yeast and plants. Cell Death Differ 13:84–95
Katiyar-Agarwal S, Jin H (2010) Role of small RNAs in host-microbe interactions. Annu Rev Phytopathol 48:225–246
Katsir L, Schilmiller A, Staswick P, He SY, Howe GA (2008) COI1 is a critical component of a receptor for jasmonate and the bacterial virulence factor coronatine. Proc Natl Acad Sci U S A 105:7100–7105
Kesarwani M, Joo J, Dong X (2007) Genetic interactions of TGA transcription factors in the regulation of pathogenesis-related genes and disease resistance in *Arabidopsis*. Plant Physiol 144:336–346
Khang CH, Berruyer R, Giraldo MC, Kankanala P, Park SY, Czymmek K, Kang S, Valent B (2010) Translocation of *Magnaporthe oryzae* effectors into rice cells and their subsequent cell-to-cell movement. Plant Cell 22:1388–1403
Kidd BN, Cahill DM, Manners JM, Schenk PM, Kazan K (2011) Diverse roles of the Mediator complex in plants. Semin Cell Dev Biol 22:741–748
Kim HS, Delaney TP (2002) Overexpression of TGA5, which encodes a bZIP transcription factor that interacts with NIM1/NPR1, confers SAR-independent resistance in *Arabidopsis thaliana* to *Peronospora parasitica*. Plant J 32:151–163
Kim KC, Fan B, Chen Z (2006) Pathogen-induced Arabidopsis WRKY7 is a transcriptional repressor and enhances susceptibility to *Pseudomonas syringae*. Plant Physiol 142:1180–1192
Kim S-Y, Kim Y-C, Seong ES, Lee Y-H, Park M, Choi D (2007) The chili pepper CaATL1: an AT-hook motif-containing transcription factor implicated in defence responses against pathogens. Mol Plant Pathol 8:761–771
Kim K-C, Lai Z, Fan B, Chen Z (2008) Arabidopsis WRKY38 and WRKY62 transcription factors interact with histone deacetylase 19 in basal defense. Plant Cell 20:2357–2371
Kinkema M, Fan W, Dong X (2000) Nuclear localization of NPR1 is required for activation of PR gene expression. Plant Cell 12:2339–2350
Klessig DF, Durner J, Noad R, Navarre DA, Wendehenne D, Kumar D, Zhou JM, Shah J, Zhang S, Kachroo P, Trifa Y, Pontier D, Lam E, Silva H (2000) Nitric oxide and salicylic acid signalling in plant defense. Proc Natl Acad Sci U S A 97:8849–8855
Kloek AP, Verbsky ML, Sharma SB, Schoelz JE, Vogel J, Klessig DF, Kunkel BN (2001) Resistance to *Pseudomonas syringae* conferred by an *Arabidopsis thaliana* coronatine-insensitive (*coi*) mutation occurs through two distinct mechanisms. Plant J 26:509–522

Knoth C, Ringler J, Dangl JL, Eulgem T (2007) *Arabidopsis* WRKY70 is required for full *RPP4-mediated* disease resistance and basal defense against *Hyaloperonospora parasitica*. Mol Plant-Microbe Interact 20:120–128

Kobayashi M, Seo S, Hirai K, Yamamoto-Katou A, Katou S, Seto H, Meshi T, Mitsuhara I, Ohashi Y (2010) Silencing of WIPK and SIPK mitogen-activated protein kinases reduces *Tobacco mosaic virus* accumulation but permits systemic viral movement in tobacco possessing the *N* resistance gene. Mol Plant-Microbe Interact 23:1032–1041

Kohler A, Schwindling S, Conrath U (2002) Benzothiadiazole-induced priming for potentiated responses to pathogen infection, wounding, and infiltration of water into leaves requires the *NPR1/NIM1* gene in Arabidopsis. Plant Physiol 128:1046–1056

Koo YJ, Kim MA, Kim EH, Song JT, Jung CK, Moon JK, Kim JH, Seo HS, Song SL, Kim JK, Lee JS, Cheong JJ, Choi YD (2007) Overexpression of salicylic acid carboxyl methyltransferase reduces salicylic acid-mediated pathogen resistance in *Arabidopsis thaliana*. Plant Mol Biol 64:1–15

Koornneef A, Pieterse CMJ (2008) Cross talk in defense signaling. Plant Physiol 146:839–844

Korolev N, David DR, Elad Y (2008) The role of phytohormones in basal resistance in *Trichoderma-induced* systemic resistance to *Botrytis cinerea* in *Arabidopsis thaliana*. Biocontrol 53:667–683

Krečič-Stres C, Vučak C, Ravnikar M, Kovač M (2005) Systemic *Potato virus* Y^{NTN} infection and levels of salicylic and gentisic acids in different potato genotypes. Plant Pathol 54:441–447

Krinke O, Ruelland E, Valentová O, Vergnolle C, Renou J-P, Taconnat L, Flemr M, Burketova L, Zachowski A (2007) Phosphatidylinositol 4-kinase activation is an early response to salicylic acid in *Arabidopsis* suspension cells. Plant Physiol 144:1347–1359

Kudla J, Batistić O, Hashimoto K (2010) Calcium signals: the lead currency of plant information processing. Plant Cell 22:541–563

Kumar D, Klessig DF (2000) Differential induction of tobacco MAP kinases by the defense signals nitric oxide, salicylic acid, ethylene, and jasmonic acid. Mol Plant-Microbe Interact 13:347–351

Kumar D, Klessig DF (2003) High-affinity salicylic acid-binding protein 2 is required for plant innate immunity and has salicylic acid-stimulated lipase activity. Proc Natl Acad Sci U S A 100:16101–16106

Kumar D, Klessig DF (2008) The search for the salicylic acid receptor led to discovery of the SAR signal receptor. Plant Signal Behav 3:691–692

Kumar D, Gustafsson C, Klessig DF (2006) Validation of RNAi silencing specificity using synthetic genes: salicylic acid-binding protein 2 is required for innate immunity in plants. Plant J 45:863–868

Kwaaitaal M, Huisman R, Maintz J, Reinstädler A, Panstruga R (2011) Ionotropic glutamate receptor (iGluR)-like channels mediate MAMP-induced calcium influx in *Arabidopsis thaliana*. Biochem J 440:355–365

Kwon C (2010) Plant defense responses coming to shape. Plant Pathol J 26:115–120

Kwon S, Hamada K, Matsuyama A, Yasuda M, Nakashita H, Yamakawa T (2009) Biotic and abiotic stresses induce *AbSAMT1*, encoding *S*-adenosyl-L-methionine: salicylic acid carboxyl methyltransferase, in *Atropa belladonna*. Plant Biotechnol 26:207–215

Lamotte O, Gould K, Lecourieux D, Sequeira-Legrand A, Lebrun-Garcia A, Durner J, Pugin A, Wendehenne D (2004) Analysis of nitric oxide signaling functions in tobacco cells challenged by the elicitor cryptogein. Plant Physiol 135:516–529

Laquitaine L, Gomès E, Francois J, Marchive C, Pascal S, Hamdi S, Atanassova R, Delrot S, Coutos-Thévenot P (2006) Molecular basis of ergosterol-induced protection of grape against *Botrytis cinerea*: induction of type 1 LTP promoter activity, WRKY, and stilbene synthase gene expression. Mol Plant-Microbe Interact 19:1103–1112

Laudert D, Weiler EW (1998) Allene oxide synthase: a major control point in *Arabidopsis thaliana* octadecanoid signaling. Plant J 15:675–684

Latunde-Dada AO, Lucas JA (2001) The plant defense activator acibenzalor-S-methyl primes cowpea [*Vigna unguiculata* (L.) Walp.] seedlings for rapid induction of resistance. Physiol Mol Plant Pathol 58:199–208

Laurie-Berry N, Joardar V, Street JH, Kunkel BN (2006) The *Arabidopsis thaliana JASMONATE INSENSITIVE 1* gene is required for suppression of salicylic acid-dependent defenses during infection by *Pseudomonas syringae*. Mol Plant-Microbe Interact 19:789–800

Lawton K, Weymann K, Friedrich L, Vernooij B, Uknes S, Ryals J (1995) Systemic acquired resistance in *Arabidopsis* requires salicylic acid but not ethylene. Mol Plant Microbe Interact 8:863–870

Lecourieux D, Ranjeva R, Pugin A (2006) Calcium in plant defence-signalling pathways. New Phytol 171:249–269

Lee H-L, Raskin I (1999) Purification, cloning, and expression of a pathogen inducible UDP-glucose: salicylic acid glucosyltransferase from tobacco. J Biol Chem 274:36637–36642

Lee H-I, León J, Raskin I (1995) Biosynthesis and mechanism of salicylic acid. Proc Natl Acad Sci U S A 92:4076–4079

Lee DE, Lee IJ, Han O, Baik MG, Han SS, Back K (2004) Pathogen resistance of transgenic rice expressing mitogen-activated protein kinase 1, MK1, from Capsicum annuum. Mol Cells 17:81–85

Lee J, Nam J, Park HC, Na G, Miura K, Jin JB, Yoo CY, Baek D, Kim DH, Jeong JC, Kim D, Lee SY, Salt DE, Mengiste T, Gong Q, Ma S, Bohnert HJ, Kwak S-S, Bressan RA, Hasegawa PM, Yun D-J (2007) Salicylic acid-mediated innate immunity in *Arabidopsis* is regulated by SIZ1 SUMO E3 ligase. Plant J 49:79–90

Lee DH, Choi HW, Hwang BK (2011a) The pepper E3 ubiquitin ligase RING1 gene, *CaRING1*, is required for cell death and the salicylic acid-dependent defense response. Plant Physiol. doi:10.1104/pp. 111.177568 (On Line first)

Lee W-S, Fu S-F, Verchot-Lubicz J, Carr JP (2011b) Genetic modification of alternative respiration in *Nicotiana benthamiana* affects basal and salicylic acid-induced resistance to potato virus X. BMC Plant Biol 11:41

Lehtonen NT, Akita M, Frank W, Reski R, Valkonen JPT (2012) Involvement of a class III peroxidase and the mitochondrial protein TSPO in oxidative burst upon treatment of moss plants with a fungal elicitor. Mol Plant-Microbe Interact 25:363–371

Leibman D, Wolf D, Saharan V, Zelcer A, Arazi T, Yoel S, Gaba V, Gal-On A (2011) A high level of transgenic viral small RNA is associated with broad potyvirus resistance in cucurbits. Mol Plant-Microbe Interact 24:1220–1238

León J, Yalpani N, Raskin I, Lawton MA (1993) Induction of benzoic acid 2-hydroxylase in virus-inoculated tobacco. Plant Physiol 103:323–328

León J, Lawton MA, Raskin I (1995) Hydrogen peroxide stimulates salicylic acid biosynthesis in tobacco. Plant Physiol 108:1673–1678

Leon-Reyes A, Du Y, Koornneef A, Proietti S, Körbes AP, Memelink J, Pieterse CMJ, Ritsema T (2010a) Ethylene signaling renders the jasmonate response of *Arabidopsis* insensitive to future suppression by salicylic acid. Mol Plant-Microbe Interact 23:187–197

Leon-Reyes A, Van der Does D, De Lange ES, Delker C, Wasternack C, Van Wees SCM, Ritsema T, Pieterse CMJ (2010b) Salicylate-mediated suppression of jasmonate-responsive gene expression in *Arabidopsis* is targeted downstream of the jasmonate biosynthesis pathway. Planta 232:1423–1432

Levine A, Tenhaken R, Dixon R, Lamb C (1994) H_2O_2 from the oxidative burst orchestrates the plant hypersensitive disease resistance response. Cell 79:583–593

Lewsey MG, Murphy AM, MacLean D, Dalchau N, Westwood JH, Macaulay K, Bennett M, Moulin M, Hanke DE, Powell G, Smith AG, Carr JP (2010) Disruption of two signaling pathways by a viral RNA silencing suppressor. Mol Plant-Microbe Interact 23:835–845

Li J, Brader G, Palva ET (2004) The WRKY70 transcription factor: a node of convergence for jasmonate-mediated and salicylate-mediated signals in plant defense. Plant Cell 16:319–331

Li J, Brader G, Kariola T, Tapio Palva E (2006) WRKY70 modulates the selection of signaling pathways in plant defense. Plant J 46:477–491

Li C, Bonnema G, Che D, Dong L, Lindhout P, Visser R, Bai Y (2007) Biochemical and molecular mechanisms involved in monogenic resistance responses to tomato powdery mildew. Mol Plant-Microbe Interact 20:1161–1172

Li G, Meng X, Wang R, Mao G, Han L, Liu Y, Zhang S (2012) Dual-level regulation of ACC synthase activity by MPK3/MPK6 cascade and its downstream WRKY transcription factor during ethylene induction in Arabidopsis. PLoS Genet 8(6):e1002767. doi:10.1371/journal.pgen.1002767

Lim E-K, Doucet CJ, Li Y, Elias L, Worrall D, Spencer SP, Ross J, Bowles DJ (2002) Arabidopsis glycosyltransferases toward salicylic acid, 4-hydroxybenzoic acid and other benzoates. J Biol Chem 277:586–592

Lindermayr C, Sell S, Müller B, Leister D, Dumer J (2010) Redox regulation of NPR1-TGA1 system of Arabidopsis thaliana by nitric oxide. Plant Cell 22:2894–2907

Lippok B, Birkenbihl RP, Rivory G, Brummer J, Schmelzer E, Logemann E, Somssich IE (2007) Expression of *AtWRKY33* encoding a pathogen- or PAMP-responsive WRKY transcription factor is regulated by a composite DNA motif containing W box elements. Mol Plant-Microbe Interact 20:420–429

Liu J-J, Ekramoddoullah AKM, Yu X (2003) Differential expression of multiple PR10 proteins in western white pine following wounding, fungal infection and cold-hardening. Physiol Plant 119:544–553

Liu X, Bai X, Wang X, Chu C (2007) OsWRKY71, a rice transcription factor, is involved in rice defense response. J Plant Physiol 164:969–979

Liu P-P, Yang Y, Pichersky E, Klessig DF (2010) Altering expression of benzoic acid/salicylic acid carboxyl methyltransferase 1 compromises systemic acquired resistance and PAMP-triggered immunity in Arabidopsis. Mol Plant-Microbe Interact 23:82–90

Liu P-P, von Dahl CC, Klessig DF (2011a) The extent to which methyl salicylate is required for signaling systemic acquired resistance is dependent on exposure to light after infection. Plant Physiol 157:2216–2226

Liu P-P, von Dahl CC, Park S-W, Klessig DF (2011b) Interconnection between methyl salicylate and lipid-based long-distance signaling during the development of systemic acquired resistance in Arabidopsis and tobacco. Plant Physiol 155:1762–1768

Lorenc-Kukula K, Chaturvedi R, Roth M, Welti R, Shah J (2012) Biochemical and molecular-genetic characterization of SFD1's involvement in lipid metabolism and defense signaling. Front Plant Sci 3:26

Lorkovic ZJ (2009) Role of plant RNA-binding proteins in development, stress response and genome organization. Trends Plant Sci 14:229–236

Love AJ, Yun B-W, Laval V, Loake GJ, Milner JL (2005) *Cauliflower mosaic virus*, a compatible pathogen of Arabidopsis, engages three distinct defense-signaling pathways and activates rapid systemic generation of reactive oxygen species. Plant Physiol 139:935–948

Luan S, Kudla J, Rodriguez-Concepcion M, Yalovsky S, Gruissem W (2002) Calmodulins and calcineurin B-like proteins: calcium sensors for specific signal response coupling in plants. Plant Cell 14:S389–S400

Luna E, Bruce TJA, Roberts MR, Flors V, Ton J (2012) Next generation systemic acquired resistance. Plant Physiol 158:844–853

Ma W, Berkowitz GA (2007) The grateful dead: calcium and cell death in plant innate immunity. Cell Microbiol 9:2571–2585

Maeo K, Hayashi S, Kojima-Suzuki H, Morikami A, Nakamura D (2001) Role of conserved residues of the WRKY domain in the DNA-binding of tobacco WRKY family proteins. BioSci Biotechnol Biochem 65:2428–2436

Maier F, Zwicker S, Hückelhoven A, Meissner M, Funk J, Pfitzner AJ, Pfitzner UM (2011) NONEXPRESSOR OF PATHOGENESIS-RELATED PROTEINS1 (NPR1) and some NPR1-related proteins are sensitive to salicylic acid. Mol Plant Pathol 12:73–91

Malamy J, Hennig J, Klessig DF (1992) Temperature-dependent induction of salicylic acid and its conjugates during the resistance response to tobacco mosaic virus infection. Plant Cell 4:359–366

Maldonado AM, Doerner P, Dixon RA, Lamb CJ, Cameron RK (2002) A putative lipid transfer protein involved in systemic acquired resistance in potato. Mol Plant-Microbe Interact 23:82–90

Maleck K, Levine A, Eulgem T, Morgan A, Schmid J, Lawton KA, Dangl JL, Dietrich RA (2000) The transcriptome of *Arabidopsis thaliana* during systemic acquired resistance. Nat Genet 26:403–410

Malik SI, Hussain A, Yun BW, Spoel SH, Loake GJ (2011) GSNOR-mediated de-nitrosylation in the plant defence response. Plant Sci 181:540–544

Mandal B, Mandal S, Csinos AS, Martinez N, Culbreath AK, Pappu HR (2008) Biological and molecular analyses of the acibenzolar-S-methyl-induced systemic acquired resistance in flue-cured tobacco against *Tomato spotted wilt virus*. Phytopathology 98:196–204

Manosalva PM, Park S-W, Forouhar F, Tong L, Fry WE, Klessig DF (2010) *Methyl Esterase 1* (*StMES1*) is required for systemic acquired resistance in potato. Mol Plant-Microbe Interact 23:1151–1163

Mao P, Duan M, Wei C, Li Y (2007) WRKY62 transcription factor acts downstream of cytosolic NPR1 and negatively regulates jasmonate responsive gene expression. Plant Cell Physiol 48:833–842

March-Diaz R, Garcia-Domînguez M, Lozano-Juste J, León J, Florencio FJ, Reyes JC (2008) Histone H2A.Z and homologues of components of the SWR1 complex are required to control immunity in *Arabidopsis*. Plant J 53:475–487

Marchive C, Mzid R, Deluc L, Barrieu F, Pirrello J, Gauthier A, Corio-Costet M-F, Regard F, Cailleteau B, Hamdi S, Lauvergeat V (2007) Isolation and characterization of a *Vitis vinifera* transcription factor, VvWRKY1, and its effect on responses to fungal pathogens in transgenic tobacco plants. J Exp Bot 58:1999–2010

Mateo A, Muhlenbock P, Rustérucci C, Chang CC, Miszalski Z, Karpinska B, Parker JE, Mullineaux PM, Karpinski S (2004) *LESION SIMULATING DISEASE 1* is required for acclimation to conditions that promote excess excitation energy. Plant Physiol 136:2818–2830

Mauch-Mani B, Slusarenko AJ (1996) Production of salicylic acid precursors is a major function of phenylalanine ammonia-lyase in the resistance of Arabidopsis to *Peronospora parasitica*. Plant Cell 8:203–212

Maxwell DP, Wang Y, McIntosh L (1999) The alternative oxidase lowers mitochondrial reactive oxygen production in plant cells. Proc Natl Acad Sci U S A 96:8271–8276

Mayers CN, Lee KC, Moore CA, Wong SM, Carr JP (2005) Salicylic acid-induced resistance to *Cucumber mosaic virus* in squash and *Arabidopsis thaliana*: contrasting mechanisms of induction and antiviral action. Mol Plant-Microbe Interact 18:428–434

McAinsh MR, Pittman JK (2009) Shaping the calcium signature. New Phytol 181:275–294

McGrath KC, Dombrecht B, Manners JM, Schenk PM, Edgar CI, Maclean DJ, Scheible W-R, Udvardi MK, Kazan K (2005) Repressor- and activator-type ethylene response factors functioning in jasmonate signaling and disease resistance identified via a genome-wide screen of Arabidopsis transcription factor gene expression. Plant Physiol 139:949–959

Meldau S, Ullman-Zeunert L, Govind G, Bartram S, Baldwin IT (2012) MAPK-dependent JA and SA signalling in *Nicotiana attenuata* affects plant growth and fitness during competition with conspecifics. BMC Plant Biol 12(213):u7

Melotto M, Underwood W, Koczan J, Nomura K, He SY (2006) Plant stomata function in innate immunity against bacterial invasion. Cell 126:969–980

Mengiste T, Chen X, Salmeron J, Dietrich R (2003) The *Botrytis Susceptible 1* gene encodes an R2R3MYB transcription factor protein that is required for biotic and abiotic stress responses in Arabidopsis. Plant Cell 15:2551–2565

Menke FLH, Kang H-G, Chen Z, Mee Park J, Kumar D, Klessig DF (2005) Tobacco transcription factor WRKY1 is phosphorylated by the MAP kinase SIPK and mediates HR-like cell death in tobacco. Mol Plant-Microbe Interact 10:1027–1034

Mészáros T, Helfer A, Hatzimasoura E, Magyar Z, Serazetdinova L, Rios G, Bardóczy V, Teige M, Koncz C, Peck S, Bögre L (2006) The Arabidopsis MAP kinase kinase MKK1 participates in defence responses to the bacterial elicitor flagellin. Plant J 48:485–495

Metraux J-P (2002) Recent breakthroughs in the study of salicylic acid biosynthesis. Trends Plant Sci 7:332–334

Metz M, Dahlbeck D, Morales CQ, Al Sady B, Clark ET, Staskawicz BJ (2005) The conserved *Xanthomonas campestris* pv. *vesicatoria* effector protein XopX is a virulence factor and suppresses host defense in *Nicotiana benthamiana*. Plant J 41:801–814

Meur G, Budatha M, Gupta AD, Prakash S, Kirti PB (2006) Differential induction of NPR1 during defense responses in *Brassica juncea*. Physiol Mol Plant Pathol 68:128–137

Miao Y, Zentgraf U (2007) The antagonist function of *Arabidopsis* WRKY53 and ESR/ESP in leaf senescence is modulated by the jasmonic and salicylic acid equilibrium. Plant Cell 19:819–830

Midoh N, Iwata M (1996) Cloning and characterization of a probenazole-inducible gene for an intracellular pathogenesis-related protein in rice. Plant Cell Physiol 37:9–18

Mikami K, Sakamoto A, Iwabuchi M (1994) The HBP-1 family of wheat basic/leucine zipper proteins interacts with overlapping cis-acting hexamer motifs of plant histone genes. J Biol Chem 269:9974–9985

Mishina TE, Zeier J (2006) The Arabidopsis flavin-dependent monooxygenase FMO1 is an essential component of biologically induced systemic acquired resistance. Plant Physiol 141:1666–1675

Mitsuhara I, Iwai T, Seo S, Yanagawa Y, Kawahigasi H, Hirose S, Ohkawa Y, Ohashi Y (2008) Characteristic expression of twelve *PR1* family genes in response to pathogen infection, wounding, and defense-related signal compounds. Mol Genet Genomics 279:415–427

Moeder W, Urguhart W, Ung H, Yoshioka K (2011) The role of cyclic nucleotide-gated ion channels in plant immunity. Mol Plant 4:442–452

Mohr PG, Cahill DM (2007) Suppression by ABA of salicylic acid and lignin accumulation and the expression of multiple genes, in *Arabidopsis* infected with *Pseudomonas syringae* pv. *tomato*. Funct Integr Genomics 7:181–191

Momol MT, Olson SM, Funderburk JE, Stavisky J, Marois JJ (2004) Integrated management of tomato spotted wilt on field-grown tomatoes. Plant Dis 88:882–890

Moreau M, Tian M, Klessig DF (2012) Salicylic acid binds NPR3 and NPR4 to regulate NPR1-dependent defense responses. Cell Res. doi:10.1038/cr 2012.100

Mosher RA, Durrant WE, Wang D, Song J, Dong X (2006) A comprehensive structure-function analysis of *Arabidopsis* SNI1defines essential regions and transcriptional repressor activity. Plant Cell 18:1750–1765

Mou Z, Fan W, Dong X (2003) Inducers of plant systemic acquired resistance regulate NPR1 function through redox changes. Cell 113:935–944

Mucha E, Fricke I, Schaefer A, Wittinghofer A, Berken A (2011) Rho proteins of plants-functional cycle and regulation of cytoskeletal dynamics. Eur J Cell Biol 90:934–943

Mukherjee M, Larrimore KE, Ahmed NJ, Bedick TS, Barghouthi NT, Traw MB, Barth C (2010) Ascorbic acid deficiency in *Arabidopsis* induces constitutive priming that is dependent on hydrogen peroxide, salicylic acid, and the *NPR1* gene. Mol Plant-Microbe Interact 23:340–351

Mur LAJ, Bi YM, Darby RM, Firek S, Draper J (1997) Compromising early salicylic acid accumulation delays the hypersensitive response and increases viral dispersal during lesion development in TMV-infected tobacco. Plant J 12:1113–1126

Mur LAJ, Kenton P, Atzorn R, Miersch O, Wasternack C (2006) The outcomes of concentration-specific interactions between salicylate and jasmonate signaling include synergy, antagonism, and oxidative stress leading to cell death. Plant Physiol 140:249–262

Murphy AM, Chivasa S, Singh DP, Carr JP (1999) Salicylic acid-induced resistance to viruses and other pathogens: a parting of ways? Trends Plant Sci 4:155–160

Murphy AM, Gilliland A, York CJ, Hyman B, Carr JP (2004) High-level expression of alternative oxidase protein sequences enhances the spread of viral vectors in resistant and susceptible plants. J Gen Virol 85:3777–3786

Nakashita H, Yasuda M, Nishioka M, Hasegawa S, Arai Y, Uramoto M, Yoshida S, Yamaguchi I (2002) Chloroisonicotinamide derivative induces a broad range of disease resistance in rice and tobacco. Plant Cell Physiol 43:823–831

Nakashita H, Yasuda M, Okage R, Nishioka M, Arie T, Yoshida S (2003) A pyrazole derivative induces systemic acquired resistance with a new type of action. Plant Cell Physiol 44:S179–S179

Nandi A, Krothapalli K, Buseman C, Li M, Welti R, Enyedi A, Shah J (2003) The *Arabidopsis thaliana sid* mutants affect plastidic lipid composition and suppress dwarfing, cell death and the enhanced disease resistance phenotypes resulting from the deficiency of a fatty acid desaturase. Plant Cell 15:2383–2398

Nandi A, Welti R, Shah J (2004) The *Arabidopsis thaliana* dihydroxyacetone phosphate reductase gene SUPPRESSOR OF FATTY ACID DESATURASE DEFICIENCY1 is required for glycerolipid metabolism and for the activation of systemic acquired resistance. Plant Cell 16:465–477

Nandi A, Moeder W, Kachroo P, Klessig DF, Shah J (2005) Arabidopsis *ssi-2* conferred susceptibility to *Botrytis cinerea* is dependent on *EDS5* and *PAD4*. Mol Plant-Microbe Interact 18:363–370

Nandi D, Tahiliani P, Kumar A, Chandu D (2006) The Ubiquitin-Proteasome system. J Biosci 31:137–155

Naumann U, Daxinger L, Kanno T, Eun C, Long Q, Lorkovic ZJ, Matzke M, Matzke AJM (2011) Genetic evidence that DNA methyltransferase DRM2 has a direct catalytic role in RNA-directed DNA methylation in *Arabidopsis thaliana*. Genetics 187:977–979

Návarová H, Bernsdorff F, Döring A-C, Zeier J (2012) Pipecolic acid, an endogenous mediator of defense amplification and priming, is a critical regulator of inducible plant immunity. Plant Cell 24:5123–5141

Navarre DA, Mayo D (2004) Differential characteristics of salicylic acid-mediated signaling in potato. Physiol Mol Plant Pathol 64:179–188

Nawrath C, Heck S, Parinthawong N, Metraux JP (2002) EDS5, an essential component of salicylic acid-dependent signaling for disease resistance in *Arabidopsis*, is a member of MATE transporter family. Plant Cell 14:275–286

Naylor M, Murphy AM, Berry JO, Carr JP (1998) Salicylic acid can induce resistance in plant virus movement. Mol Plant Microbe Interact 11:860–868

Ndamukong I, Abdallat AA, Thurow C, Fode B, Zander M, Weigel R, Gatz C (2007) SA-inducible *Arabidopsis* glutaredoxin interacts with TGA factors and suppresses JA-responsive *PDF1.2* transcription. Plant J 50:128–139

Neill SJ, Desikan R, Clarke A, Hurst RD, Hancock JT (2002) Hydrogen peroxide and nitric oxide as signaling molecules in plants. J Exp Bot 53:1237–1247

Neill S, Bright J, Desikan R, Hancock J, Harrison J, Wilson I (2008) Nitric oxide evolution and perception. J Exp Bot 59:25–35

Nie X (2006) Salicylic acid suppresses *Potato virus Y* isolate N:O-induced symptoms in tobacco plants. Phytopathology 96:255–263

Niki T, Mitsuhara I, Seo S, Ohtsubo N, Ohasshi Y (1998) Antagonistic effect of salicylic acid and jasmonic acid on the expression of pathogenesis-related (PR) protein genes in wounded mature tobacco leaves. Plant Cell Physiol 39:500–507

Nishioka M, Nakashita H, Yasuda M, Yoshida S, Yamaguchi I (2005) Induction of resistance against rice bacterial blight by 3-chloro-1-methyl-1H-pyrazole-5-carboxylic acid. J Pest Sci 30:47–49

Nobuta K, Okrent RA, Stoutemyer M, Rodibaugh N, Kempema L, Wildermuth MC, Innes RW (2007) The GH3 acyl adenylase family member PBS3 regulates salicylic acid-dependent defense responses in Arabidopsis. Plant Physiol 144:1144–1156

Nürnberger T, Küfner I (2011) The role of the plant plasma membrane in microbial sensing and innate immunity: the plant plasma membrane. Plant Cell Monogr 19:471–483

O'Brein JA, Daudi A, Finch P, Butt VS, Whitelegge JP, Souda P, Ausubel FM, Bolwell GP (2012) A peroxidase-dependent apoplastic oxidative burst in cultured *Arabidopsis* cells functions in MAMP-elicited defense. Plant Physiol 158:2013–2027

Ogawa D, Nakajima N, Seo S, Mitsuhara I, Kamada H, Ohashi Y (2006) The phenylalanine pathway is the main route of salicylic acid biosynthesis in Tobacco mosaic virus-infected tobacco leaves. Plant Biotechnol 23:395–398

Oki K, Inaba N, Kitagawa K, Fujioka S, Kitano H, Fujisawa Y, Kato H, Iwasaki Y (2009) Function of the α subunit of rice heterotrimeric G protein in brassinosteroid signaling. Plant Cell Physiol 50:161–172

Olszak B, Malinovsky FG, Brodersen P, Grell M, Giese H, Petersen M, Mundy J (2006) A putative flavin-containing mono-oxygenase as a marker for certain defense and cell death pathways. Plant Sci 170:614–623

Oostendorp M, Kunz W, Dietrich B, Staub T (2001) Induced disease resistance in plants by chemicals. Eur J Plant Pathol 107:19–28

Ouyang J, Gill G (2009) SUMO engages multiple corepressors to regulate chromatin structure and transcription. Epigenetics 4:440–444

Pandey S, Wang R-S, Wilson L, Li S, Zhao Z, Gookin TE, Assmann SM, Albert R (2010) Boolean modeling of transcriptome data reveals novel modes of heterotrimeric G-protein action. Mol Syst Biol 6:372

Papaefthimiou I, Hamilton A, Denti M, Baulcombe D, Tsagris M, Tabler M (2001) Replicating *Potato spindle tuber viroid* RNA is accompanied by short RNA fragments that are characteristic of post-transcriptional gene silencing. Nucleic Acids Res 29:2395–2400

Park J, Choi HJ, Lee S, Lee T, Yang Z, Lee Y (2000) Rac-related GTP-binding protein in elicitor-induced reactive oxygen generation by suspension-cultured soybean cells. Plant Physiol 124:725–732

Park CY, Heo WD, Yoo JH, Lee JH, Kim MC, Chun HJ, Moon BC, Kim IH, Park HC, Choi MS, Ok HM, Cheong MS, Lee SM, Kim HS, Lee KH, Lim CO, Chung WS, Cho MJ (2004) Pathogenesis-related gene expression by specific calmodulin isoforms is dependent on NIM1, a key regulator of systemic acquired resistance. Mol Cells 18:207–213

Park SW, Kaimoyo E, Kumar D, Mosher S, Klessig DF (2007) Methyl salicylate is a critical mobile signal for plant systemic acquired resistance. Science 318:113–116

Park SW, Liu PP, Forouhar F, Vlot AC, Tong L, Tietjen K, Klessig DF (2009) Use of synthetic salicylic acid analog to investigate the roles of methyl salicylate and its esterases in plant disease resistance. J Biol Chem 284:7307–7317

Pastor V, Luna E, Mauch-Mani B, Ton J, Flors V (2013a) Primed plants do not forget. Environ Exp Bot 94:46–56

Pastor V, Luna E, Ton J, Cerezo M, Garcia-Agustin P, Flors V (2013b) Fine tuning of ROS homeostasis regulates primed immune responses in Arabidopsis. Plant-Microbe Interact 26:1334–1344

Pauwels L, Goossens A (2011) The JAZ proteins: a crucial interface in the jasmonate signaling cascade. Plant Cell 23:3089–3100

Pedley KF, Martin GB (2005) Role of mitogen-activated protein kinases in plant immunity. Curr Opin Plant Biol 8:541–547

Peňa-Cortés HT, Albrecht T, Prat S, Weiler EW, Willmitzer L (1993) Aspirin prevents wound-induced gene expression in tomato leaves by blocking jasmonic acid biosynthesis. Planta 191:123–128

Perchepied L, Balagué C, Riou C, Claudel-Renard C, Riviére N, Grezes-Besset B, Roby D (2010) Nitric oxide participates in the complex interplay of defense-related signaling pathways controlling disease resistance to *Sclerotinia sclerotiorum* in *Arabidopsis thaliana*. Mol Plant-Microbe Interact 23:846–860

Peters JM, Franke WW, Kleinschmidt JA (1994) Distinct 19S and 20S subcomplexes of the 26S proteasome and their distribution in the nucleus and the cytoplasm. J Biol Chem 269:7709–7718

Petersen M, Brodersen P, Naested H, Andreasson E, Lindhart U, Johansen B, Nielsen HB, Lacy M, Austin MJ, Parker JE, Sharma SB, Klessig DF, Martienssen R, Mattsson O, Jensen AB, Mundy J (2000) *Arabidopsis* MAP kinase 4 negatively regulates systemic acquired resistance. Cell 103:1111–1120

Petrov VD, van Breusegem F (2012) Hydrogen peroxide – a central hub for information flow in plant cells. AOB Plants 2012:pls014. doi:10.1093/aobpla/pls014

Pickart CM, Eddins MJ (2004) Ubiquitin: structures, functions, mechanisms. Biochim Biophys Acta 1695:55–72

Pieterse CMJ (2012) Prime time for transgenerational defense. Plant Physiol 158:545

Pieterse CMJ, Dicke M (2007) Plant interactions with microbes and insects: from molecular mechanisms to ecology. Trends Plant Sci 12:564–569
Pieterse CMJ, Leon-Reyes A, van der Does D, Verhage A, Koornneef A, van Pelt JA, van Wees SCM (2012) Networking by small-molecule hormones in plant immunity. Induced resistance against insects and diseases. IOBC-WPRS Bull 83:77–80
Pitzschke A, Djamei A, Bitton F, Hirt H (2009) A major role of the MEKK1-MKK1/2-MPK4 pathway in ROS signalling. Mol Plant 2:120–137
Potlakayala SD, Reed DW, Covello PS, Fobert PR (2007) Systemic acquired resistance in canola is linked with pathogenesis-related gene expression and requires salicylic acid. Phytopathology 97:794–802
Po-Wen C, Singh P, Zimmerli L (2013) Priming of the Arabidopsis pattern-triggered immunity response upon infection by necrotrophic *Pectobacterium carotovorum* bacteria. Mol Plant Pathol 14:58–70
Pré M, Atallah M, Champion A, De Vos M, Pieterse CMJ, Memelink J (2008) The AP2/ERF domain transcription factor ORA59 integrates jasmonic acid and ethylene signals in plant defense. Plant Physiol 147:1347–1357
Price MB, Jelesko J, Okumoto S (2012) Glutamine receptor homologs in plants: functions and evolutionary origins. Front Plant Sci 3:Article 235. doi:10.3389/fpls.2012.00235
Probst AV, Fransz PF, Paszkowski J, Mittelsten Scheid O (2003) Two means of transcriptional reactivation within heterochromatin. Plant J 33:743–749
Probst AV, Fagard M, Proux F, Mourrain P, Boutet S, Earley K, Lawrence RJ, Pikaard CS, Murfett J, Furner I, Vaucheret H, Scheid OM (2004) *Arabidopsis* histone deacetylase *HDA6* is required for maintenance of transcriptional gene silencing and determines nuclear organization of rDNA repeats. Plant Cell 16:1021–1034
Qi X, Bao FS, Xie Z (2009) Small RNA deep sequencing reveals role for *Arabidopsis thaliana* RNA-dependent RNA polymerases in viral siRNA biogenesis. PLoS One 4:e4971
Qi Y, Tsuda K, Joe A, Sato M, Nguyen LV, Glazebrook J, Alfano JR, Cohen JD, Katagiri F (2010) A putative RNA-binding protein positively regulates salicylic acid-mediated immunity in Arabidopsis. Mol Plant-Microbe Interact 23:1573–1583
Qiu D, Xiao J, Ding X, Xiong M, Cai M, Cao Y, Li X, Xu C, Wang S (2007) OsWRKY13 mediates rice disease resistance by regulating defense-related genes in salicylate- and jasmonate-dependent signaling. Mol Plant-Microbe Interact 20:492–499
Qiu JL, Fiil BK, Petersen K, Nielsen HB, Botanga CJ, Thorgrimsen S, Palma K, Suarez-Rodriguez MC, Sandbech-Clausen S, Lichota J, Brodersen P, Grasser KD, Mattson O, Glazebrook J, Mundy J, Petersen M (2008a) *Arabidopsis* MAP kinase 4 regulates gene expression through transcription factor release in the nucleus. EMBO J 27:2214–2221
Qiu JL, Zhou L, Yun B, Nielsen H, Fiil BK, Petersen K, MacKinlay J, Loake GJ, Mundy J, Morris PC (2008b) *Arabidopsis* mitogen-activated protein kinase kinases MKK1 and MKK2 have overlapping functions in defense signaling mediated by MEKK1, MPK4, and MKS1. Plant Physiol 148:212–222
Qu F, Morris TJ (2005) Suppressors of RNA silencing encoded by plant viruses and their role in viral infection. FEBS Lett 579:5958–5964
Quilis J, Peñas G, Messeguer J, Brugidou C, Segundo BS (2008) The *Arabidopsis AtNPR1* inversely modulates defense responses against fungal, bacterial, or viral pathogens while conferring hypersensitivity to abiotic stresses in transgenic rice. Mol Plant-Microbe Interact 21:1215–1231
Raffaele S, Rivas S, Roby D (2006) An essential role for salicylic acid in AtMYB30-mediated control of the hypersensitive cell death program in *Arabidopsis*. FEBS Lett 580:3498–3504
Rakshandehroo F, Takeshita M, Squires J, Palukaitis P (2009) The influence of RNA-dependent RNA polymerase 1 on potato virus Y infection and on other antiviral response genes. Mol Plant-Microbe Interact 22:1312–1318
Ranf S, Eschen-Lippold L, Pecher P, Lee J, Scheel D (2011) Interplay between calcium signaling and early signaling elements during defense responses to microbe- or damage-associated molecular patterns. Plant J 68:100–113

Reddy ASN, Ali GS, Celesnik H, Day IS (2011) Coping with stresses: roles of calcium- and calcium/calmodulin-regulated gene expression. Plant Cell 23:2010–2032

Ren D, Yang K-Y, Li G-J, Liu Y, Zhang S (2006) Activation of Ntf4, a tobacco mitogen-activated protein kinase, during plant defense and its involvement in hypersensitive response-like cell death. Plant Physiol 141:1482–1493

Ribnicky DM, Shulaev V, Raskin I (1998) Intermediates of salicylic acid biosynthesis in tobacco. Plant Physiol 118:565–572

Robert-Seilaniantz A, Grant MR, Jones JDG (2011) Hormone crosstalk in plant disease and defense: more than just jasmonate-salicylate antagonism. Annu Rev Phytopathol 49:2621–2627

Rocher F, Chollet J-F, Jousse C, Bonnemain J-L (2006) Salicylic acid, an ambimobile molecule exhibiting a high ability to accumulate in the phloem. Plant Physiol 141:1684–1693

Rochon A, Boyle P, Wignes T, Fobert PR, Després C (2006) The coactivator function of *Arabidopsis* NPR1 requires the core of its BTB/POZ domain and the oxidation of C-terminal cysteines. Plant Cell 18:3670–3685

Rushton PJ, Somssich IE, Ringler P, Shen QJ (2010) WRKY transcription factors. Trends Plant Sci 15:247–258

Rustérucci C, Aviv DH, Holt BF, Dangl JL, Parker JE (2001) The disease resistance signaling components *EDS1* and *PAD4* are essential regulators of the cell death pathway controlled by *LSD1* in Arabidopsis. Plant Cell 13:2211–2224

Ruthenburg AJ, Allis CD, Wysocka J (2007) Methylation of lysine 4 on histone H3: intricacy of writing and reading a single epigenetic mark. Mol Cell 25:15–30

Ryals JA, Weymann K, Lawton K, Friedrich L, Ellis D, Steiner H-Y, Johnson J, Delaney TP, Jesse T, Vos P, Uknes S (1997) The Arabidopsis NIM1 protein shows homology to the mammalian transcription factor inhibitor IkB. Plant Cell 9:425–439

Sagi M, Fluhr R (2006) Production of reactive oxygen species by plant NADPH oxidases. Plant Physiol 141:336–340

Sänchez G, Gerhardt N, Siciliano F, Vojnov A, Malcuit I, Marono MR (2010) Salicylic acid is involved in the *Nb*-mediated defense responses to *Potato virus X* in *Solanum tuberosum*. Mol Plant-Microbe Interact 23:394–405

Sanders D, Pelloux J, Brownlee C, Harper JF (2002) Calcium at the cross-roads of signaling. Plant Cell 14:S401–S417

Sano H, Seo S, Orudgev S, Youssefian K, Ishizuka K, Ohashi Y (1994) Expression of the gene for a small GTP binding protein in transgenic tobacco elevates endogenous cytokinin levels, abnormally induces salicylic acid in response to wounding, and increases resistance to tobacco mosaic virus infection. Proc Natl Acad Sci U S A 91:10556–10560

Sasaki Y, Asamizu E, Shibata D, Nakamura Y, Kaneko T, Awai K, Amagi M, Kuwata C, Tsugane T, Masuda T, Shimada H, Takamiya K, Ohta H, Tabata S (2001) Monitoring of methyl jasmonate-responsive genes in Arabidopsis by cDNA microarray: Self-activation of jasmonic acid biosynthesis and crosstalk with other phytohormone signaling pathways. DNA Res 8:153–161

Sawada H, Shim IS, Usui K (2006) Induction of benzoic acid 2-hydroxylase and salicylic acid biosynthesis-modulation by salt stress in rice seedlings. Plant Sci 171:263–270

Segonzac C, Zipfel C (2011) Activation of plant pattern-recognition receptors by bacteria. Curr Opin Microbiol 14:54–61

Seo PJ, Park C-M (2010) MYB96-mediated abscisic acid signals induce pathogen resistance response by promoting salicylic acid biosynthesis in *Arabidopsis*. New Phytol 186:471–483

Seo S, Okamoto M, Settee H, Ishizuka K, Sano H, Ohashi Y (1995) Tobacco MAP kinase: a possible mediator in wound signal transduction pathways. Science 270:1988–1992

Seo S, Katou S, Seto H, Gomi K, Ohashi Y (2007) The mitogen-activated protein kinases WIPK and SIPK regulate the levels of jasmonic and salicylic acids in wounded tobacco plants. Plant J 49:899–909

Seskar M, Shulaev V, Raskin I (1998) Endogenous methyl salicylate in pathogen-inoculated tobacco plants. Plant Physiol 116:387–392

Shah J (2009) Plants under attack: systemic signals in defence. Curr Opin Plant Biol 12:459–464
Shah J, Zeier J (2013) Long-distance communication and signal amplification in systemic acquired resistance. Front Plant Sci 4:30. doi:10.3389/fpls.2013.00030
Shapiro AD, Gutsche AT (2003) Capillary electrophoresis-based profiling and quantitation of total salicylic acid and related phenolics for analysis of early signaling in Arabidopsis disease resistance. Anal Biochem 320:223–233
Shen X, Yuan B, Liu H, Li X, Xu C, Wang S (2010) Opposite functions of a rice mitogen-activated protein kinase during the process of resistance against *Xanthomonas oryzae*. Plant J 64:86–99
Shi L, Fang Y (2011) Histone variants: making structurally divergent nucleosomes and linkers in chromatin. Front Biol 6:93–101
Shi J, An HL, Zhang L, Gao Z, Guo XQ (2010) GhMPK7, a novel multiple stress-responsive cotton group C MAPK gene, has a role in broad spectrum disease resistance and plant development. Plant Mol Biol 74:1–17
Shimono M, Sugano S, Nakayama A, Jiang C-J, Ono K, Toki S, Takatsuji H (2007) Rice WRKY45 plays a crucial role in benzothiadiazole-inducible blast resistance. Plant Cell 19:2064–2076
Shoresh M, Gal-On A, Leibman D, Chet I (2006) Characterization of a mitogen-activated protein kinase gene from cucumber required for *Trichoderma*-conferred plant resistance. Plant Physiol 142:1169–1179
Shulaev V, Silverman P, Raskin I (1997) Airborne signaling by methyl salicylate in plant pathogen resistance. Nature 385:718–727
Silverman P, Seskar M, Kanter D, Schweizer P, Métraux JP, Raskin I (1995) Salicylic acid in rice – biosynthesis, conjugation, and possible role. Plant Physiol 108:633–639
Sklodowska M, Gajewski E, Kuzniak E, Mikiciński A, Sobiczewski P (2010) BTH-mediated antioxidant system responses in apple leaf tissues. Scientia Hort 125:34–40
Slaughter A, Daniel X, Flors V, Luna E, Hohn B, Mauch-Mani B (2012) Descendants of primed *Arabidopsis* plants exhibit resistance to biotic stress. Plant Physiol 158:835–843
Slaymaker DH, Navarre DA, Clark D, del Pozo O, Martin GB, Klessig DF (2002) The tobacco salicylic acid-binding protein-3 (SABP3) is the chloroplast carbonic anhydrase, which exhibits antioxidant activity and plays a role in the hypersensitive defense response. Proc Natl Acad Sci U S A 99:11640–11645
Snedden WA, Fromm H (2001) Calmodulin as a versatile calcium signal transducer in plants. New Phytol 151:35–66
Song JT (2006) Induction of a salicylic acid glucosyltransferase, AtSGT1, is an early disease response in *Arabidopsis thaliana*. Mol Cells 22:233–238
Song F, Goodman RM (2001) Molecular biology of disease resistance in rice. Physiol Mol Plant Pathol 59:1–11
Song C, Yang B (2010) Mutagenesis of 18 type II effectors reveals virulence function of $XopZ_{Pxo99}$ in *Xanthomonas oryzae* pv *oryzae*. Mol Plant-Microbe Interact 23:893–902
Song JT, Lu H, Greenberg JT (2004a) Divergent roles in *Arabidopsis thaliana* development and defense of two homologous genes, aberrant growth and death2 and AGO2-LIKE DEFENSE RESPONSE PROTEIN1, encoding novel aminotransferases. Plant Cell 16:353–366
Song JT, Lu H, McDowell JM, Greenberg JT (2004b) A key role for *ALD1* in activation of local and systemic defenses in *Arabidopsis*. Plant J 40:200–212
Song JT, Koo YJ, Seo HS, Kim MC, Choi YD, Kim JH (2008) Overexpression of AtSGT1, an *Arabidopsis* salicylic acid glucosyltransferase, leads to increased susceptibility to *Pseudomonas syringae*. Phytochemistry 69:1128–1134
Spoel SH, Dong X (2008) Making sense of hormone crosstalk during plant immune response. Cell Host Microbe 3:348–351
Spoel SH, Dong X (2012) How do plants achieve immunity? Defence without specialized immune cells. Nat Rev Immunol 12:89–100
Spoel SH, Koornneef A, Claessens SMC, Korzelius JP, Van Pelt JA, Mueller MJ, Buchala AJ, Metraux J-P, Brown R, Kazan K, Van Loon LC, Dong X, Pieterse CMJ (2003) NPR1

modulates cross-talk between salicylate- and jasmonate-dependent defense pathways through a novel function in the cytosol. Plant Cell 15:760–770
Spoel SH, Johnson JS, Dong X (2007) Regulation of tradeoffs between plant defenses against pathogens with different lifestyles. Proc Natl Acad Sci U S A 104:18842–18847
Spoel SH, Mou Z, Tada Y, Spivey NW, Genschik P, Dong X (2009) Proteasome-mediated coactivator NPR1 plays dual roles in regulating plant immunity. Cell 137:860–872
Stael S, Wurzinger B, Mair A, Mehlmer N, Vothknecht UC, Teige M (2012) Plant organellar calcium signaling: an emerging field. J Exp Bot 63:1525–1542
Staswick PE, Tiryaki I (2004) The oxylipin signal jasmonic acid is activated by an enzyme that conjugates it to isoleucine in *Arabidopsis*. Plant Cell 16:2117–2127
Stein M, Dittgen J, Sanchez-Rodriquez C, Hou BH, Molina A, Schulze-Lefert P, Lipka V, Somerville S (2006) *Arabidopsis* PEN3/PDR8, an ATP binding cassette transporter, contributes to non-host resistance to inappropriate pathogens that enter by direct penetration. Plant Cell 18:731–746
Stracke R, Werber M, Weisshaar B (2001) The *R2R3-MYB* gene family in *Arabidopsis thaliana*. Curr Opin Plant Biol 4:447–456
Strawn MA, Marr SK, Inoue K, Inada N, Zubieta C, Wildermuth MC (2007) Arabidopsis isochorismate synthase functional in pathogen-induced salicylate biosynthesis exhibits properties consistent with a role in diverse stress responses. J Biol Chem 282:5919–5933
Suzuki N, Miller G, Morales J, Shulaev V, Torres MA, Mittler R (2011) Respiratory burst oxidases: the engines of ROS signaling. Curr Opin Plant Biol 14:691–699
Szczesny R, Buttner B, Escolar L, Schulze S, Seiferth A, Bonas U (2010) Suppression of the AvrBs1-specific hypersensitive response by the YopJ effector homolog AvrBsT from *Xanthomonas* depends on a SNF1-related kinase. New Phytol 187:1058–1074
Tada Y, Spoel SH, Pajerowska-Mukhtar K, Mou Z, Song J, Dong X (2008) Plant immunity requires conformational changes of NPR1 via S-nitrosylation and thioredoxin. Science 321:952–956
Takabatake R, Karita E, Seo S, Mitsuhara I, Kuchitsu K, Ohashi Y (2007) Pathogen-induced calmodulin isoforms in basal resistance against bacterial and fungal pathogens in tobacco. Plant Cell Physiol 48:414–423
Takahashi H, Miller J, Nozaki Y, Takeda M, Shah J, Hase S, Ikegami M, Ehara Y, Dinesh-Kumar SP, Sukamoto (2002) *RCY1*, an *Arabidopsis thaliana RPP8/HRT* family resistance gene, conferring resistance to cucumber mosaic virus requires salicylic acid, ethylene and a novel signal transduction mechanism. Plant J 32:655–657
Takahashi Y, Uehara Y, Berberich T, Ito A, Saitoh H, Miyazaki A, Terauchi R, Kusano T (2004) A subset of hypersensitive response marker genes, including HSR203J, is the downstream target of a spermine signal transduction pathway in tobacco. Plant J 40:586–595
Teige M, Scheikl E, Eulgem T, Doczi R, Ichimura K, Shinozaki K, Dangl JL, Hirt H (2004) The MKK2 pathway mediates cold and salt stress signaling in Arabidopsis. Mol Cell 15:141–152
Thaler JS, Bostock RM (2004) Interactions between abscisic acid-mediated responses and plant resistance to pathogens and insects. Ecology 85:48–58
Thibaud-Nissen F, Wu H, Richmond T, Redman JC, Johnson C, Green R, Arias J, Town CD (2006) Development of Arabidopsis whole-genome microarrays and their application to the discovery of binding sites for the TGA2 transcription factor in salicylic acid-treated plants. Plant J 47:152–162
Thines B, Katsir L, Melotto M, Niu Y, Mandaokar A, Liu G, Nomura K, He SY, Howe GA, Browse J (2007) JAZ repressor proteins are targets of the SCF^{COI1} complex during jasmonate signaling. Nature 448:661
Thomma B, Penninckx I, Broekaert WF, Cammue BPA (2001) The complexity of disease signaling in Arabidopsis. Curr Opin Immunol 13:63–68
Thomma BPHJ, Nürnberger T, Joosten MHAJ (2011) Of PAMPs and effectors: the blurred PTI-ETI dichotomy. Plant Cell 23:4–15
Tischner R, Koltermann M, Haesse H, Plath M (2010) Early responses of *Arabidopsis thaliana* to infection by *Verticillium longisporum*. Physiol Mol Plant Pathol 74:419–427
Torres MA, Jones JDG, Dangl JL (2006) Reactive oxygen species signaling in response to pathogens. Plant Physiol 141:373–378

Tripathi D, Jiang Y-L, Kumar D (2010) SABP2, a methyl salicylate esterase is required for the systemic acquired resistance induced by acibenzolar-S-methyl in plants. FEBS Lett 584:3458–3463

Truman G, Glazebrook J (2012) Coexpression analysis identifies putative targets for CBP60g and SARD1 regulation. BMC Plant Biol 12:216

Truman W, Bennett MH, Kubigsteltig I, Turnbull C, Grant M (2007) Arabidopsis systemic immunity uses conserved defense signaling pathways and is mediated by jasmonates. Proc Natl Acad Sci U S A 104:1075–1080

Tsuda K, Sato M, Glazebrook J, Cohen JD, Katagiri F (2008) Interplay between MAMP-triggered and SA-mediated defense responses. Plant J 53:763–775

Umemura K, Satou J, Iwata M, Uozumi N, Koga J, Kawano T, Koshiba T, Anzai H, Mitomi M (2009) Contribution of salicylic acid glucosyltransferase, OsSGT1, to chemically induced disease resistance in rice plants. Plant J 57:463–472

Uppalapati SR, Toyoda K, Yasuhiro I, Ichinose Y, Shiraishi T (2004) Differential regulation of MBP kinases by a glycoprotein elicitor and a polypeptide suppressor from *Mycosphaerella pinodes* in pea. Physiol Mol Plant Pathol 64:17–25

Uppalapati SR, Ishiga Y, Wangdi T, Kunkel BN, Anand A, Mysore KS, Bender CL (2007) The phytotoxin coronatine contributes to pathogen fitness and is required for suppression of salicylic acid accumulation in tomato inoculated with *Pseudomonas syringae* pv *tomato* DC3000. Mol Plant-Microbe Interact 20:955–965

Vaistij FE, Jones L (2009) Compromised virus-induced gene silencing in RDR6-deficient plants. Plant Physiol 149:1399–1407

Valent B, Khang CH (2010) Recent advances in rice blast effector research. Curr Opin Plant Biol 13:434–441

Vatsa P, Chiltz A, Bourgue S, Wendehenne D, Garcia-Brugger A, Pugin A (2011) Involvement of putative glutamate receptors in plant defence signaling and NO production. Biochimie 93:2095–2101

van den Burg HA, Takken FLW (2009) Does chromatin remodeling mark systemic acquired resistance? Trends Plant Sci 14:286–294

van den Burg HA, Takken FLW (2010) SUMO-, MAPK- and resistance protein-signaling converge at transcription complexes that regulate plant innate immunity. Plant Signal Behav 5:1507–1601

Van der Does D, Leon-Reyes A, Koornneef A, Van Verk MC, Rodenburg N, Pauwels L, Goossens A, Körbes AP, Memelink J, Ritsema T, Van Wees SCM, Pieterse CMJ (2013) Salicylic acid suppresses jasmonic acid signaling downstream of SCF^{COI1}-JAZ by targeting GCC promoter motifs via transcription factor ORA59. Plant Cell 25:744–751

Van der Ent S, Verhagen BWM, Van Doorn R, Bakker D, Verlaan MG, Pel MJC, Joosten RG, Proveniers MCG, Van Loon LC, Ton J, Pieterse CMJ (2008) *MYB72* is required in early signaling steps of rhizobacteria-induced systemic resistance in *Arabidopsis*. Plant Physiol 146:1293–1304

van Hulten M, Pelser M, van Loon LC, Pieterse CMJ, Ton J (2006) Costs and benefits of priming for defense in Arabidopsis. Proc Natl Acad Sci U S A 103:5602–5607

van Verk MC, Bol JF, Linthorst HJM (2011) WRKY transcription factors involved in activation of SA biosynthesis genes. BMC Plant Biol 11:89. doi:10.1186/1471-2229-11-89

van Verk MC, Pappaaioannou D, Neeleman L, Bol JF, Linthorst HJM (2008) A novel WRKY transcription factor is required for induction of *PR-1a* gene expression by salicylic acid and bacterial elicitors. Plant Physiol 140:1983–1995

Vernooij B, Friedrich L, Morse A, Resist R, Kolditz-Jawhar R, Ward E, Uknes S, Kessmann H, Ryals J (1994) Salicylic acid is not the translocated signal responsible for inducing systemic acquired resistance but is required in signal transduction. Plant Cell 6:959–965

Vicente J, Cascón T, Vicedo B, Garcia-Agustin P, Hamberg M, Castresana C (2012) Role of 9-lipoxygenase and α-dioxygenase oxylipin pathways as modulators of local and systemic defense. Mol Plant 5:914–928

Vidhyasekaran P (2007) Fungal pathogenesis in plants and crops: molecular biology and host defense mechanisms, 2nd edn. CRC Press/Taylor Francis Group, Boca Raton, p 510

Vidhyasekaran P (2014) PAMP signals in plant innate immunity: signal perception and transduction. Springer, Dordrecht, p 442

Vincill ED, Bieck AM, Spalding EP (2012) Ca^{2+} conduction by an amino acid-gated ion channel related to glutamate receptors. Plant Physiol 159:40–46

Vlot AC, Klessig DF, Park S-W (2008a) Systemic acquired resistance: the elusive signal(s). Curr Opin Plant Biol 11:436–442

Vlot AC, Liu P-P, Cameron RK, Park S-W, Yang Y, Kumar D, Zhou F, Padukkavidana T, Gustafsson C, Pichersky E, Klessig DF (2008b) Identification of likely orthologs of tobacco salicylic acid-binding protein 2 and their role in systemic acquired resistance in *Arabidopsis thaliana*. Plant J 56:445–456

Voinnet O (2008) Post-transcriptional RNA silencing in plant-microbe interactions: a touch of robustness and versatility. Curr Opin Plant Biol 11:464–470

Waller F, Müller A, Chung K-M, Yap Y-K, Nakamura K, Weiler E, Sano H (2006) Expression of a WIPK-activated transcription factor results in increase of endogenous salicylic acid and pathogen resistance in tobacco plants. Plant Cell Physiol 47:1169–1174

Wan D, Li R, Zou B, Zhang X, Cong J, Wang R, Xia Y, Li G (2012) Calmodulin-binding protein CBP60g is a positive regulator of both disease resistance and drought tolerance in *Arabidopsis*. Plant Cell Rep 31:1269–1281

Wang ZP, Yang PZ, Fan BF, Chen ZX (1998) An oligo selection procedure for identification of sequence-specific DNA-binding activities associated with the plant defense response. Plant J 16:515–522

Wang D, Amornsiripanitch N, Dong X (2006) A genomic approach to identify regulatory nodes in the transcriptional network of systemic acquired resistance in plants. PLoS Pathog 2:e123

Wang D, Pajerowska-Mukhtar K, Culler AH, Dong X (2007) Salicylic acid inhibits pathogen growth in plants through repression of the auxin signaling pathway. Curr Biol 17:1784–1790

Wang Y, Gao M, Li Q, Wang L, Jeon J-S, Qu N, Zhang Y, He Z (2008) *OsRAR1* and *OsSGT1* physically interact and function in rice basal disease resistance. Mol Plant-Microbe Interact 21:294–303

Wang L, Tsuda K, Sato M, Cohen JD, Katagiri F, Glazebrook J (2009) Arabidopsis CaM binding protein CBP60g contributes to MAMP-induced SA accumulation and is involved in disease resistance against *Pseudomonas syringae*. PLoS Pathog 5(2):e1000301. doi:10.1371/journal.ppat.1000301

Wang X-B, Wu Q, Ito T, Cillo F, Li W-X, Chen X, Yu J-L, Ding S-W (2010) RNAi-mediated viral immunity requires amplification of virus-derived siRNAs in *Arabidopsis thaliana*. Proc Natl Acad Sci U S A 107:484–489

Wang L, Tsuda K, Truman W, Sato M, le Nguyen V, Katagiri F, Glazebrook J (2011) CBP60g and SARD1 play partially redundant critical roles in salicylic acid signaling. Plant J 67:1029–1041

Wang W-H, Yi X-Q, Han A-D, Liu T-W, Chen J, Wu F-H, Dong X-J, He J-X, Pei Z-M, Pei Z-M, Zheng H-L (2012) Calcium-sensing receptor regulates stomatal closure through hydrogen peroxide and nitric oxide in response to extracellular calcium in *Arabidopsis*. J Exp Bot 63:177–190

Ward JM, Maser P, Schroeder JI (2009) Plant ion channels: gene families, physiology and functional genomic analyses. Annu Rev Physiol 71:59–82

Wasternack C (2007) Jasmonates: an update on biosynthesis, signal transduction and action in plant stress response, growth and development. Ann Bot (Lond) 100:681–697

Wathugala DL, Hemsley PA, Moffat CS, Cremelie P, Knight MR, Knight H (2012) The Mediator subunit SFR6/MED16 controls defence gene expression mediated by salicylic acid and jasmonate responsive pathways. New Phytol 195:217–230

Weigel RR, Pfitzner UM, Gatz C (2005) Interaction of NIMIN1 with NPR1 modulates PR gene expression in Arabidopsis. Plant Cell 17:1279–1291

Wendehenne D, Durner J, Chen Z, Klessig DF (1998) Benzothiadiazole, an inducer of plant defenses, inhibits catalase and ascorbate peroxidases. Phytochemistry 47:651–657

Wiermer M, Feys BJ, Parker JE (2005) Plant immunity: the EDS1 regulatory node. Curr Opin Plant Biol 8:383–389

Wildermuth MC, Dewdney J, Wu G, Ausubel FM (2001) Isochorismate synthesis is required to synthesize salicylic acid for plant defense. Nature 414:562–565

Willmann MR, Endres MW, Cook RT, Gregory BD (2011) The functions of RNA-dependent RNA polymerases in Arabidopsis. Arabidopsis Book 9:e0146. doi:10.1199/tab.0146

Wong CE, Carson RA, Carr JP (2002) Chemically induced virus resistance in *Arabidopsis thaliana* is independent of pathogenesis-related protein expression and the NPR1 gene. Mol Plant-Microbe Interact 15:75–81

Wu S, Lu D, Kabbage M, Wei H-L, Swingle B, Records AR, Dickman M, He P, Shan L (2011) Bacterial effector HopF2 suppresses *Arabidopsis* innate immunity at the plasma membrane. Mol Plant-Microbe Interact 24:585–593

Wu Y, Zhang D, Chu JY, Boyle P, Wang Y, Brindle ID, De Luca V, Despr½s C (2012) The Arabidopsis NPR1 protein is a receptor to the plant defense hormone salicylic acid. Cell Rep 1:639–647

Wypijewski K, Hornyik C, Shaw JA, Stephens J, Goraczniak R, Gunderson SI, Lacomme C (2009) Ectopic 5' splice sites inhibit gene expression by engaging RNA surveillance and silencing pathways in plants. Plant Physiol 151:955–965

Xia Y, Suzuki H, Borevitz J, Blount J, Guo Z, Patel K, Dixon RA, Lamb C (2004) An extracellular aspartic protease functions in *Arabidopsis* disease resistance signaling. EMBO J 23:980–988

Xiao S, Brown S, Patrick E, Brearley C, Turner J (2003) Enhanced transcription of the *Arabidopsis* disease resistance genes *RPW81* and *RPW82* via salicylic acid-dependent amplification circuit is required for hypersensitive cell death. Plant Cell 15:33–45

Xie Z, Fan B, Chen C, Chen Z (2001) An important role of an inducible RNA-dependent RNA polymerase in plant antiviral defense. Proc Natl Acad Sci U S A 98:6516–6521

Xie Z, Kasschau KD, Carrinton JC (2003) Negative feedback regulation of Dicer-Like1 in *Arabidopsis* by microRNA-guided mRNA degradation. Curr Biol 13:784–789

Xing D, Chen Z (2006) Effects of mutations and constitutive overexpression of *EDS1* and *PAD4* on plant resistance to different types of microbial pathogens. Plant Sci 171:251–262

Xing DH, Lai ZB, Zheng ZY, Vinod KM, Fan BF, Chen ZX (2008) Stress- and pathogen-induced *Arabidopsis* WRKY48 is a transcriptional activator that represses plant basal defense. Mol Plant 1:459–470

Xu X, Chen C, Fan B, Chen Z (2006) Physical and functional interactions between pathogen-induced *Arabidopsis* WRKY18, WRKY40, and WRKY60 transcription factors. Plant Cell 18:1310–1326

Xu J, Audenaert K, Hofte M, De Vleesschauwer D (2013) Abscisic acid promotes susceptibility to the rice blight pathogen *Xanthomonas oryzae* pv. *oryzae* by suppressing salicylic acid-mediated defenses. PLoS One 8(6):e67413

Yalowsky S, Baluska F, Jones A (2010) Integrated G proteins signaling in plants. Springer, Heidelberg

Yalpani N, Silverman P, Wilson TMA, Kleier DA, Raskin I (1991) Salicylic acid is a systemic signal and an inducer of pathogenesis-related proteins in virus-infected tobacco. Plant Cell 3:809–815

Yalpani N, León J, Lawton MA, Raskin I (1993) Pathway of salicylic acid biosynthesis in healthy and virus-inoculated tobacco. Plant Physiol 103:315–321

Yang Z (2002) Small GTPases: versatile signaling switches in plants. Plant Cell 14(Suppl):S375–S388

Yang D-L, Yang Y, He Z (2013) Roles of plant hormones and their interplay in rice immunity. Mol Plant 6:675–685

Yasuda M (2007) Regulation mechanisms of systemic acquired resistance induced by plant activators. J Pest Sci 32:281–282

Yasuda M, Nakashita H, Hasegawa S, Nishioka M, Arai Y, Uramoto M, Yamaguchi I, Yoshida S (2003) *N*-Cyanomethyl-2-chloroisonicotinamide induces systemic acquired resistance in *Arabidopsis* without salicylic acid accumulation. Biosci Biotechnol Biochem 67:322–328

Yoda H, Sano H (2003) Activation of hypersensitive response genes in the absence of pathogens in transgenic tobacco plants expressing a small GTPase. Planta 217:993–997

Yoda P, Ogawa M, Yamaguchi Y, Koizumi N, Kusano T, Sano H (2002) Identification of early-responsive genes associated with the hypersensitive response to tobacco mosaic virus and characterization of a WRKY-type transcription factor in tobacco plants. Mol Genet Genomics 267:154–161

Yoshioka K, Sugie K, Park HJ, Maeda H, Tsuda N, Kawakita K, Doke N (2001) Induction of plant gp91 phox homolog by fungal cell wall, arachidonic acid, and salicylic acid in potato. Mol Plant-Microbe Interact 14:725–736

Yu D, Chen C, Chen Z (2001) Evidence for an important role of WRKY DNA binding proteins in the regulation of *NPR1* gene expression. Plant Cell 13:1527–1539

Yu D, Fan B, MacFarlane SA, Chen Z (2003) Analysis of the involvement of an inducible *Arabidopsis* RNA-dependent RNA polymerase in antiviral defense. Mol Plant-Microbe Interact 16:206–216

Zago E, Morsa S, Dat JF, Alard P, Ferrarini A, Inzé D, Delledonne M, Van Breusegem F (2006) Nitric oxide- and hydrogen peroxide-responsive gene regulation during cell death induction in tobacco. Plant Physiol 141:404–411

Zahn LM (2009) Secondary messenger. Sci Signal 2:ec123

Zander M, La Camera S, Lamotte O, Metraux JP, Gatz C (2010) *Arabidopsis thaliana* class-II TGA transcription factors are essential activators of jasmonic acid/ethylene-induced defense responses. Plant J 61:200–210

Zander M, Chen S, Imkampe J, Thurow C, Gatz C (2012) Repression of the *Arabidopsis thaliana* jasmonic acid/ethylene-induced defense pathway by TGA-interacting glutaredoxins depends on their C-terminal ALWL motif. Mol Plant 5:831–840

Zhang X (2008) The epigenetic landscape of plants. Science 320:489–492

Zhang SQ, Klessig DF (1997) Salicylic acid activates a 48-kD MAP kinase in tobacco. Plant Cell 9:809–824

Zhang S, Liu Y (2001) Activation of salicylic acid-induced protein kinase, a mitogen-activated protein kinase, induces multiple defense responses in tobacco. Plant Cell 13:1877–1889

Zhang SQ, Du H, Klessig DF (1998) Activation of the tobacco SIP kinase by both a cell wall-derived carbohydrate elicitor and purified proteinaceous elicitins from *Phytophthora* spp. Plant Cell 10:435–449

Zhang Y, Fan M, Kinkema M, Li X, Dong X (1999) Interaction of NPR1 with basic leucine zipper protein transcription factors that bind sequences required for salicylic acid induction of the *PR1* gene. Proc Natl Acad Sci U S A 96:6523–6528

Zhang Y, Tessaro MJ, Lassner M, Li X (2003) Knockout analysis of *Arabidopsis* transcription factors *TGA2*, *TGA5*, and *TGA6* reveals their redundant and essential roles in systemic acquired resistance. Plant Cell 15:2647–2653

Zhang X, Dai Y, Xiong Y, DeFraia C, Li J, Dong X, Mou Z (2007a) Overexpression of Arabidopsis *MAP kinase kinase 7* leads to activation of plant basal and systemic acquired resistance. Plant J 52:1066–1079

Zhang Z, Li Q, Staswick PE, Wang M, Zhu Y, He Z (2007b) Dual regulation role of *GH3.5* in salicylic acid and auxin signaling Arabidopsis-*Pseudomonas syringae* interaction. Plant Physiol 145:450–464

Zhang Y, Xu S, Ding P, Wang D, Cheng YT, He J, Geo M, Xu F, Li Y, Zhu Z, Li X, Zhang Y (2010) Control of salicylic acid synthesis and systemic acquired resistance by two members of a plant-specific family of transcription factors. Proc Natl Acad Sci U S A 107:18220–18225

Zhang W, Jeon BW, Assmann SM (2011) Heterotrimeric G-protein regulation of ROS signalling and calcium currents in *Arabidopsis* guard cells. J Exp Bot 62:2371–2379

Zhang L, Li Y, Lu W, Meng F, Wu C, Guo X (2012a) Cotton *GhMKK5* affects disease resistance, induces HR-like cell death, and reduces the tolerance to salt and drought stress in transgenic *Nicotiana benthamiana*. J Exp Bot 63:3935–3952

Zhang X, Wang C, Zhang Y, Sun Y, Mou Z (2012b) The Arabidopsis mediator complex subunit16 positively regulates salicylate-mediated systemic acquired resistance and jasmonate/ethylene-induced defense pathways. Plant Cell 24:4294–4309

Zhang Z, Wu Y, Gao M, Zhang J, Kong Q, Liu Y, Ba H, Zhou J, Zhang Y (2012c) Disruption of PAMP-induced MAP kinase cascade by a *Pseudomonas syringae* effector activates plant immunity mediated by the NB-LRR protein SUMM2. Cell Host Microbe 11:253–263

Zheng Z, Qamar SA, Chen Z, Mengiste T (2006) Arabidopsis WRKY33 transcription factor is required for resistance to necrotrophic fungal pathogens. Plant J 48:596–605

Zheng Z, Mosher SL, Fan B, Klessig DF, Chen Z (2007) Functional analysis of Arabidopsis WRKY25 transcription factor in plant defense against *Pseudomonas syringae*. BMC Plant Biol 7:2

Zheng X, Spivey NW, Zeng W, Liu Po-Pu, Fu ZQ, Klessig DF, He SY, Dong X (2012) Coronatine promotes *Pseudomonas syringae* virulence in plants by activating a signaling cascade that inhibits salicylic acid accumulation. Cell Host Microbe 11:587–596

Zhou N, Tootle T, Tsui F, Klessig D, Glazebrook J (1998) *PAD4* functions upstream from salicylic acid to control defense responses to *Arabidopsis*. Plant Cell 10:1021–1030

Zhou F, Mosher S, Tian M, Sassi G, Parker J, Klessig DF (2008) The *Arabidopsis* gain-of-function mutant *ssi4* requires *RAR1* and *SGT1b* differentially for defense activation and morphological alterations. Mol Plant-Microbe Interact 21:40–49

Zhu YJ, Qiu X, Moore PH, Borth W, Hu J, Ferreira S, Albert HH (2003) Systemic acquired resistance induced by BTH in papaya. Physiol Mol Plant Pathol 63:237–248

Ziadi S, Barbedette S, Godard JF, Monot C, Le Corre D, Silue D (2008) Production of pathogenesis-related proteins in the cauliflower (*Brassica oleracea* var *botrytis*)-downy mildew (*Peronospora parasitica*) pathosystem treated with acibenzolar-S-methyl. Plant Pathol 50:579–586

Zoeller M, Stingl N, Krischke M, Fekete A, Waller F, Berger S, Mueller MJ (2012) Lipid profiling of the Arabidopsis hypersensitive response reveals specific lipid peroxidation and fragmentation processes: biogenesis of pimelic and azelaic acid. Plant Physiol 160:365–378

Zottini M, Costa A, De Michele R, Ruzzene M, Carimi F, Lo Schiavo F (2007) Salicylic acid activates nitric oxide synthesis in *Arabidopsis*. J Exp Bot 58:1397–1405

Zwicker S, Mast S, Stos V, Pfitzner AJP, Pfitzner UM (2007) Tobacco NIMIN2 proteins control *PR* gene induction through transient repression early in systemic acquired resistance. Mol Plant Pathol 8:385–400

Chapter 3
Jasmonate Signaling System in Plant Innate Immunity

Abstract The plant immune system uses several second messengers to encode information generated by the pathogen-associated molecular patterns (PAMPs) and deliver the information downstream of plant pattern recognition receptors (PRRs) to proteins which decode/interpret signals and initiate defense gene expression. Plant hormones play important role in intercellular and systemic signaling systems in plant immunity. Jasmonic acid (JA) system is a key component in the complex plant hormone signaling systems. The production of jasmonic acid (JA) in plants is a tightly regulated process. In healthy unstressed plants, JA content is very low and PAMP cues enhance the biosynthesis and accumulation of JA in the PAMP-treated plant tissues. PAMP elicitor signals activate G-proteins and the activated G-proteins further switch on calcium ion channels. Ca^{2+} influx and subsequent Ca^{2+} wave (calcium signature) may activate NADPH oxidase and H_2O_2 production. The Ca^{2+} influx-mediated NO plays important role in JA biosynthesis. Co-expression of MAPK genes with the genes involved in the JA biosynthesis pathway suggests that MAPK cascades may also be involved in JA biosynthesis. COI1 is as a key player in the jasmonate perception and signal transduction pathway. JA receptor is a three-molecule co-receptor complex, consisting of COI1, JAZ, and inositol pentakisphosphate, all of which are indispensable for high-affinity hormone binding. The bioactive jasmonate-isoleucine (JA-Ile) promotes physical interaction between JAZ1 and COI1. The JA receptor JAZ proteins have been identified as suppressors of jasmonate signaling. NINJA is an adaptor protein that interacts with the ZIM domain of most JAZs. NINJA contains an EAR (for ERF-associated amphiphilic repression) motif that recruits the corepressor TOPLESS (TPL). JAZs are a scaffold on which the NINJA–TPL corepressor complex is assembled. Repression of JA response genes involves binding of JAZ to -NINJA, which contains an EAR motif that recruits the corepressor TPL, which may silence gene expression. In the absence of JA signal, JAZ proteins actively repress the transcription factor MYC2, which binds to *cis*-acting elements of JA response genes. In response to cues that up-regulate JA-Ile synthesis, the hormone triggers the specific binding of JAZ proteins to COI1, leading to poly-ubiquitination. Subsequent degradation of JAZ by the 26S proteasome relieves repression of MYC2 and probably other transcription factors, permitting the expression of jasmonate-responsive genes. MEDIATOR25 subunit of the Mediator complex is a positive regulator of jasmonate-responsive gene expression in *Arabidopsis*. MED25 functions in association with transcriptional regulators of the JA pathway. Histone deacetylase may regulate JA-mediated signaling systems.

P. Vidhyasekaran, *Plant Hormone Signaling Systems in Plant Innate Immunity*, Signaling and Communication in Plants 2, DOI 10.1007/978-94-017-9285-1_3

JA and ET pathways appear to function cooperatively in modulating plant immune responses. Ethylene signaling may render JA signaling insensitive to subsequent suppression by SA. JA may inhibit SA signaling and SA may suppress the biosynthesis of JA. SA suppresses JA signaling system by targeting GCC-box motifs in JA-responsive promoters. JA signaling may be integrated into early ABA signaling and may affect ABA receptor complexes to regulate downstream common signal components. MicroRNA-directed RNA silencing system triggers JA signaling by specifically activating biosynthesis of JA. JA-Ile may be the mobile signaling cue involved in the induced systemic resistance (ISR). JA-Ile may be synthesized de novo and transported into systemic tissues to trigger ISR. JA-mediated priming plays an important role in the ISR.

3.1 Jasmonate Signaling System Is a Key Component in PAMP-Triggered Innate Immunity

Plant innate immune system is activated on the perception of the pathogen-associated molecular patterns (PAMPs) of invading pathogens (Nürnberger and Küfner 2011; Segonzac and Zipfel 2011). The plant immune system uses several second messengers to encode information generated by the PAMPs and deliver the information downstream of plant pattern recognition receptors (PRRs) to proteins which decode/interpret signals and initiate defense gene expression (Mersmann et al. 2010; Boudsocq et al. 2010; Hwang and Hwang 2011). Highly complex networks of signaling pathways are involved in the transmission of the signals to induce plant immune responses (Mur et al. 2006; Gfeller et al. 2010; Leon-Reyes et al. 2010a, b; Perchepied et al. 2010; Katagiri and Tsuda 2010; Alkan et al. 2012; Cheng et al. 2012). Plant hormones have been reported to play important role in intercellular and systemic signaling systems in plant immunity (Leon-Reyes et al. 2010a, b; An and Mou 2011; Robert-Seilaniantz et al. 2011; Pieterse et al. 2012a, b; Thaler et al. 2012; Van der Does et al. 2013; Yang et al. 2013).

Jasmonic acid (JA) system is a key component in the complex plant hormone signaling systems. Jasmonic acid and its derivatives are important signal molecules detected in a wide spectrum of plant species (Farmer et al. 2003; Kazan and Manners 2008; Wasternack and Hause 2013). Jasmonates are cellular signal molecules with key functions on the regulation of immune responses against necrotrophic pathogens (Pieterse et al. 2012a, b; Thaler et al. 2012; Wasternack and Hause 2013; Vidhyasekaran 2014). The production of jasmonic acid (JA) in plants is a tightly regulated process (Wasternack 2007; Balbi and Devoto 2008; Kazan and Manners 2008; Wasternack and Hause 2013). The concentrations of JA in healthy unperturbed plant tissues are very low. However, JA accumulates in plants infected by pathogens and/or treated with elicitors (Wang et al. 2000; Fabro et al. 2008). JA accumulates in *Arabidopsis* plants upon infection with *Botrytis cinerea* (Chehab et al. 2011). Several metabolites of JA have been reported to act as signal molecules in triggering plant immunity. Among them, (+)-7-iso-jasmonoyl-L-isoleucine

(JA-Ile) has been discovered as the major bioactive form of the hormone JA (Fonseca et al. 2009; Kombrink 2012; Wasternack and Hause 2013).

JA signaling systems modulate plant immune responses and confer resistance or susceptibility against different pathogens. Activation of JA signaling system induces resistance against *Botrytis cinerea* (Vicedo et al. 2009; Méndez-Bravo et al. 2011; Moffat et al. 2012) and *Alternaria brassicicola* (Thomma et al. 1998; Kidd et al. 2009). Suppression of JA signaling system results in the suppression of root-knot nematode infection in rice (Nahar et al. 2013). The maize *opr7opr8* double mutant shows dramatically reduced levels of JA, and these plants are highly susceptible to the root-rotting oomycetes *Pythium* spp. (Yan et al. 2012), suggesting that JA signaling is involved in conferring resistance against *Pythium* spp. JA signaling induces susceptibility against the bacterial pathogen *Pseudomonas syringae* in *Arabidopsis* (Laurie-Berry et al. 2006) and the wilt pathogen *Fusarium oxysporum* (Kidd et al. 2009). JA signaling but not JA biosynthesis has been shown to modulate plant immune responses in *Arabidopsis* in conferring resistance against the wilt pathogen *F. oxysporum* (Thatcher et al. 2009). Mutants of the impaired JA-perception gene *coi1*, but not JA biosynthesis mutants, were highly resistant to *F. oxysporum* in *Arabidopsis thaliana*. The *coi1*-mediated *F. oxysporum* resistance was independent of JA-dependent defense gene expression but correlated with compromised nondefensive aspects of the JA-dependent responses in this mutant (Thatcher et al. 2009). Collectively these studies suggest JA signaling system is an important component in the plant immune signaling system.

3.2 Biosynthesis of Jasmonates

Oxylipins are a large family of oxygenated fatty acids. Pathogen attack activates lipases that release unsaturated fatty acids, initiating the synthesis of oxylipins. One of the more widely known oxylipins is jasmonic acid. A critical step in oxylipin synthesis is the formation of fatty acid hydroperoxides, a reaction catalyzed by lipoxygenase (LOX) enzymes. *Arabidopsis* has six LOX genes, four of which encode 13-LOX enzymes. 13-LOX enzymes (numbered for the carbon at which oxygen is introduced) catalyze the first step in the synthesis of oxylipins such as jasmonic acid (Bertoni 2012).

Linolenic acid is mostly present in esterified glycerolipid form, and not in free fatty acid form. De novo JA biosynthesis begins with free α-linolenic acid, and release of linolenic acid from membranes is an important step in controlling JA synthesis (Wang et al. 2000). An increase in free linolenic acid is observed in cultured cells of several plant species after treatment with elicitors (Gundlach et al. 1992). Several phospholipase A1 (PLA1) enzymes have been reported to be involved in the release of linolenic acid (Ellinger et al. 2010). A phospholipase A (PLA) may mediate the release of linolenic acid from the membrane (Wasternack 2007). Two sn-1-specific acyl hydrolases, DEFECTIVE IN ANTHER DEHISCENCE1 (DAD1) and DONGLE (DGL), have been reported to be essential for jasmonate biosynthesis in *Arabidopsis thaliana* leaves (Ito et al. 2007; Hyun et al. 2008). The phospholipase A enzyme PLA1γ1 and still unidentified enzymes with sn-1 and sn-2 hydrolase activity were also involved in wound- and pathogen-induced jasmonate formation (Ellinger et al. 2010).

The released linolenic acid is oxygenated by lipoxygenases (LOXs) to hydroperoxide derivatives. Further reaction is catalyzed by allene oxide synthase (AOS) that converts 13-hydroperoxy-linolenic acid (13-HPOT) to an unstable allene oxide intermediate, which is in turn modified by an allene oxide cyclase (AOC) to form 12-oxo-phytodienoic acid (OPDA). OPDA is subsequently reduced by OPDA reductase 3 (OPR3) to 3-oxo-2-(2′-pentenyl)-cyclopentane-1-octanoic acid (OPC-8:0), which is followed by three rounds of β-oxidation to form 3-ketoacyl-CoA. Three key enzymes, such as acyl-CoA oxidase (ACX), multifunctional protein (MFP) possessing 2-trans enoyl-CoA hydratase and L-3-hydroxyacyl-CoA dehydrogenase activities, and 3-ketoacyl-CoA thiolase (KAT), are responsible for carrying out the fatty acid β-oxidation to form jasmonic acid (JA) (Fig. 3.1; Mei et al. 2006; Schilmiller et al. 2006; Delker et al. 2007; Vidhyasekaran 2007; Wasternack 2007; Balbi and Devoto 2008; Browse 2009; Wasternack and Kombrink 2010; Bonaventure et al. 2011; Sun et al. 2011; Kombrink 2012; Wasternack and Hause 2013).

3.3 Jasmonate Biosynthesis Intermediate OPDA in Defense Signaling

12-Oxo-phytodienoic acid (OPDA), an intermediate in JA biosynthesis pathway, triggers innate immune responses. OPDA exhibits individual signaling in plant defense reactions even in the absence of JA (Stintzi et al. 2001). OPDA has signaling properties different from JA (Böttcher and Pollmann 2009). It has been shown that OPDA triggers the expression of a subset of genes distinct from those induced by JA (Taki et al. 2005). While JA-induced genes function via COI1-dependent pathway, the OPDA-induced genes function via a COI1-independent pathway (Taki et al. 2005; Wasternack et al. 2013). OPDA does not fit into the binding pocket for JA-Ile (Sheard et al. 2010) and OPDA may function in a JA-COI1-independent pathway (Wasternack et al. 2013). OPDA specifically mediates resistance against *Botrytis cinerea* (Raacke et al. 2006).

3.4 JA Metabolites Involved in Defense Signaling

3.4.1 Methyl Jasmonate

JA can be metabolized to several derivatives and some of them are involved in defense signaling system. Among them, methyl jasmonate (MeJA) plays an important role in signaling system (Wasternack 2007). MeJA is a potent signal,

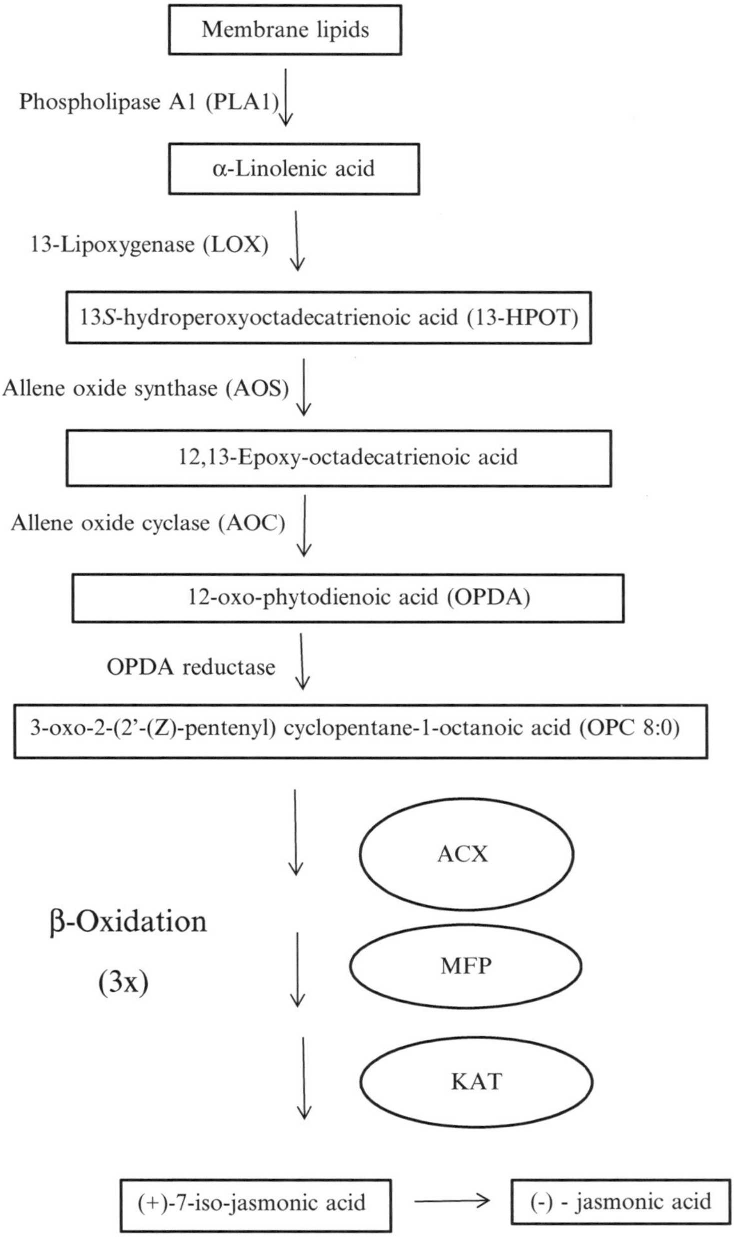

Fig. 3.1 Jasmonic acid biosynthesis pathway (Adapted from Balbi and Devoto 2008; Wasternack and Hause 2013)

functioning as a volatile in the atmosphere in interplant communication (Farmer and Ryan 1990). Jasmonic acid carboxyl methyltransferase (JMT) is involved in the conversion of JA to MeJA (Fig. 3.2; Seo et al. 2001). Enhanced pathogen resistance has been observed in *Arabidopsis* plants overexpressing a JMT, suggesting a defense

Fig. 3.2 Synthesis of methyl jasmonate from jasmonic acid (Adapted from Seo et al. 2001; Turner et al. 2002; Vidhyasekaran 2007; Wu et al. 2008)

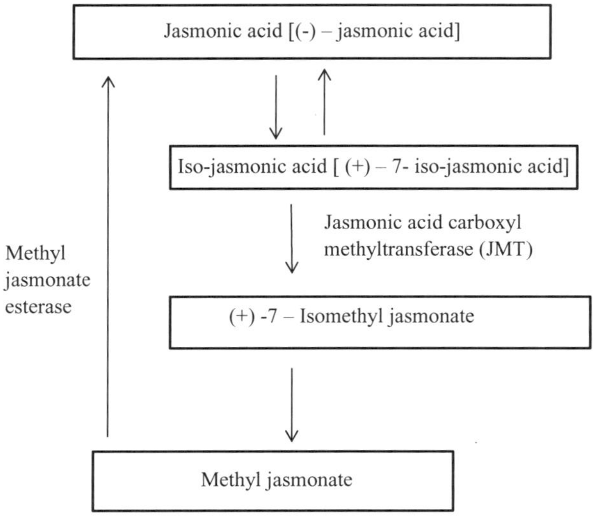

signaling role for MeJA (Seo et al. 2001). However, it has been shown that the biological activity of MeJA was only apparent when MeJA was converted to JA followed by its conjugation to JA-Ile (Stitz et al. 2011).

3.4.2 *Jasmonoyl-Isoleucine*

The JA amino acid conjugate (+)-7-*iso*-JA-Ile (jasmonoyl-isoleucine, JA-Ile) has been shown to be involved in defense signaling (Staswick et al. 1998; Staswick and Tiryaki 2004; Kang et al. 2006; Katsir et al. 2008; Gfeller et al. 2010; Kombrink 2012; Wasternack and Hause 2013). JAR1 (Jasmonate resistant1) is a JA–amino acid synthase and activates conjugation of JA to an amino acid (Staswick and Tiryaki 2004), and this enzyme may be involved in JA-Ile biosynthesis (Fig. 3.3; Hause et al. 2000; Méndez-Bravo et al. 2011; Westfall et al. 2012). In addition to Ile, the JAR family of related GH3 enzymes has the potential to conjugate other amino acids, such as Trp, Val, and Leu in tobacco. The JA-Trp, JA-Val, and JA-Leu may also participate in JA signaling pathway (Wang et al. 2008).

JA-Ile may be hydrolyzed to release JA and Ile (Kazan and Manners 2008). MeJA is hydrolyzed by a MeJA esterase to produce jasmonic acid (Wu et al. 2008). The ethylene precursor 1-aminocyclopropane-1-carboxylic acid (ACC) conjugated to JA is also found in plant tissue. An unidentified conjugating enzyme is suggested to be involved in this process (Staswick and Tiryaki 2004). Synthesis of JA-ACC provides a mechanism to coregulate the availability of JA and ACC for conversion to the active JA-Ile and ethylene, respectively (Staswick and Tiryaki 2004). Thus, the different forms of JAs are interconvertible.

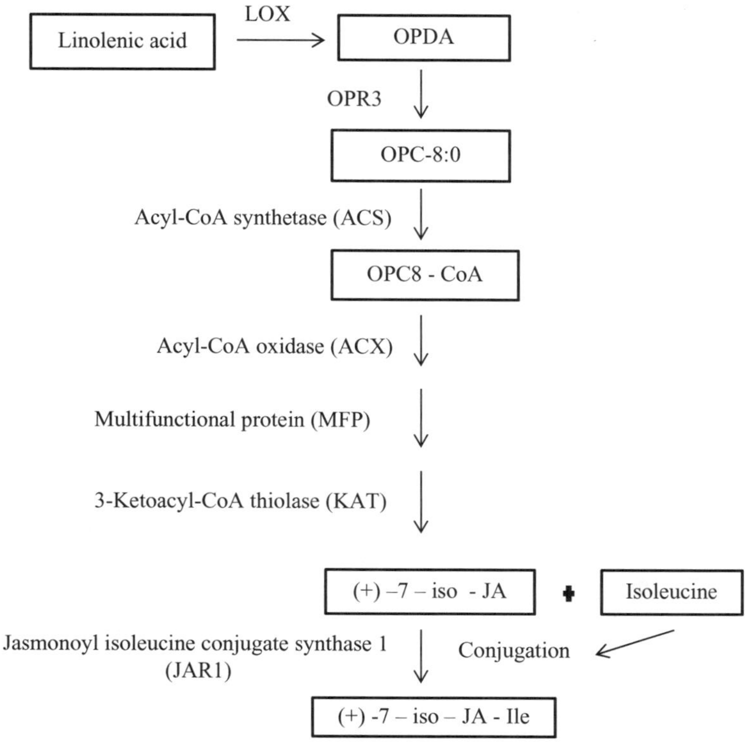

Fig. 3.3 Synthesis of JA-Ile (Adapted from Hause et al. 2000; Méndez-Bravo et al. 2011; Westfall et al. 2012; Wasternack et al. 2013)

3.5 Upstream of JA Biosynthesis

3.5.1 PAMP Triggers Enhanced Biosynthesis and Accumulation of JA

In healthy unstressed plants, JA content is very low and PAMP cues enhance the biosynthesis and accumulation of JA in the PAMP-treated plant tissues (Wang et al. 2000; Fabro et al. 2008; Chehab et al. 2011). The fungal PAMP chitosan treatment induced accumulation of jasmonic acid in tomato leaves (Doares et al. 1995). The PAMP also activates lipoxygenase, the key enzyme in JA-mediated signaling system (Bohland et al. 1997; Rakwal et al. 2002). Lipoxygenase activity significantly increased in chitosan-treated carrot plants (Jayaraj et al. 2009). The fungal PAMP chitin treatment induced lipoxygenase (LOX) (Lehtonen et al. 2012). The fungal PAMP β-1,3-glucan induces expression of *LOX* gene encoding lipoxygenase in grapevine (Aziz et al. 2003; Balbi and Devoto 2008) and tobacco cells (Klarzynski et al. 2000). The bacterial PAMP Flg22 induced the expression of *LOX3* and *LOX4* genes encoding lipoxygenases (LOX), *OPR3* gene encoding 12-oxophytodienoate reductase (OPR), and *ACX1* gene encoding acyl-CoA oxidase (ACX), and these enzymes are involved in biosynthesis of JA (Denoux et al. 2008). The PAMP Nep1 rapidly induces genes involved in JA biosynthesis (Bae et al. 2006). It triggered genes encoding lipoxygenases (*LOX*), 12-oxophytodienoate reductase, and allene oxide cyclase (*AOC2*), which are involved in JA biosynthesis (Bae et al. 2006). The host-associated molecular pattern (HAMP)/endogenous elicitor oligogalacturonates (OGs) triggered OPR3 and ACX1, the key enzymes involved in biosynthesis of JA (Denoux et al. 2008). These studies suggest that PAMPs initiate the JA biosynthesis pathway when the plants are invaded by pathogens.

3.5.2 G-Proteins in the Induction of JA Biosynthesis

Guanosine triphosphate (GTP)-binding proteins (G-proteins) are molecular switches in triggering various immune response signaling systems (Yalowsky et al. 2010; Zhang et al. 2011b, 2012a). Heterotrimeric G-proteins and small monomeric G-proteins (Ras/Ras-like small GTPases) are the two classes of signaling G-proteins reported in plants (Gu et al. 2004; Perfus-Barbeoch et al. 2004; Kiirika et al. 2012; Zhang et al. 2012a). The heterotrimeric G-proteins contain Gα-, Gβ-, and Gγ-subunits (Wang et al. 2008; Zhang et al. 2011b). Small G-proteins (small GTPases) are monomeric guanine nucleotide-binding proteins related to the α subunit of heterotrimeric G-proteins (Yang 2002; Gu et al. 2004; Nakashima et al. 2008; Kiirika et al. 2012). Both heterotrimeric G-proteins and small G-proteins play important roles in activating JA biosynthesis pathway (Gao et al. 2010a, b; Kiirika et al. 2012).

G-proteins may switch on jasmonate biosynthesis pathway. Inhibitors of G-proteins signaling pathway suppressed elicitor-induced increases in lipoxygenase activity, whereas activators of G-proteins signaling pathway increased lipoxygenase activity in Mexican cypress (*Cupressus lusitanica*) cell cultures (Zhao and Sakai 2003). Lipoxygenase is an important enzyme in the octadecanoid pathway leading to the biosynthesis of jasmonate (Schaller 2001; Zhao et al. 2004).

The early events in JA biosynthesis pathway involve activation of NADPH oxidase, which triggers production of ROS, which in turn activates lipoxygenase, the key enzyme in JA biosynthesis (Vandenabeele et al. 2003; Zago et al. 2006). The gene encoding lipoxygenase (*Lox1*) is induced by ROS (Vranová et al. 2002). The genes encoding lipoxygenase are activated in tobacco leaves accumulating ROS (Vandenabeele et al. 2003). These results suggest that ROS is involved in the induction of JA biosynthesis.

Both small monomeric G-proteins (Yang 2002; Schultheiss et al. 2003; Morel et al. 2004; Wong et al. 2007; Kiirika et al. 2012) and heterotrimeric G-proteins (Zhu et al. 2009; Zhao et al. 2010; Zhang et al. 2011b) are known to modulate the expression of NADPH oxidase. Small monomeric G-proteins are involved in the regulation of NADPH oxidase homologs of plants termed respiratory burst oxidase homolog (RBOH) (Agrawal et al. 2003; Kiirika et al. 2012). A small monomeric G-protein from rice, OsRac1, activates RBOH-mediated ROS signaling through direct binding of Rac1 to the catalytic subunits of the RBOH protein's N-terminal extension, which is specific for the plant RBOH proteins (Kawasaki et al. 2006; Wong et al. 2007; Nakashima et al. 2008). The small monomeric GTPase MtROP9 triggered the expression of *MtRBOH* gene involved in *Medicago truncatula* (Kiirika et al. 2012). Heterotrimeric G-proteins are also involved in the activation of NADPH oxidase. The *Arabidopsis* Gα subunit of heterotrimeric G-protein, GPA1, is involved in the activation of NADPH oxidase (Zhao et al. 2010). The *Arabidopsis gpa1* mutants have been shown to be disrupted in NADPH oxidase-mediated ROS production (Zhang et al. 2011b).

Numerous Rboh (for respiratory burst oxidase homolog) genes have been isolated in plants. All *rboh* genes identified to date possess a conserved N-terminal extension that contains two Ca^{2+}-binding EF-hand motifs. A substantial part of the N-terminal region of Rboh, including the two EF-hand motifs, is required for the Rac GTPases and the N-terminal extension interaction. Transient co-expression of *OsRac1* and *rbohB* enhanced NADPH oxidase-mediated ROS production in *Nicotiana benthamiana*, suggesting that direct Rac–Rboh interaction may activate NADPH oxidase activity in plants (Wong et al. 2007; Li et al. 2009; Zhao et al. 2010). The calcium ion sensor protein calmodulin activates GPA1, the Gα-subunit of heterotrimeric protein in *Arabidopsis*. The activated heterotrimeric G-protein in turn activates NADPH oxidases. GPA1 has been shown to function upstream of the NADPH oxidases AtrbohD and AtrbohF (Zhang et al. 2011b). Collectively these results suggest that both small monomeric G-proteins and heterotrimeric G-proteins are involved in the activation of NADPH oxidase (Fig. 3.4).

Another important enzyme involved in jasmonate biosynthesis is 12-oxo-10,15(Z)-phytodienoic acid (OPDA) reductase, which converts OPDA to jasmonate (Schaller 2001). Gβ mutants defective in Gβ activity showed lower induction of *OPR1* gene encoding *OPDA* in *Arabidopsis* (Trusov et al. 2006). The results suggest that the Gβ subunit of the heterotrimeric G-protein activates OPDA reductase and promotes biosynthesis of jasmonate (Fig. 3.4; Trusov et al. 2006).

3.5.3 G-Proteins-Activated Polyamine Synthesis in Triggering JA Biosynthesis

Polyamines are polycationic, ubiquitous aliphatic amines that occur in all plant cells (Tun et al. 2006; Ozawa et al. 2009, 2010; Szepesi et al. 2011; Nambeesan et al. 2012). G-proteins have been shown to activate polyamine biosynthesis. Arginine, ornithine, and S-adenosylmethionine are involved in the biosynthesis of the polyamine spermine. Arginine is converted to ornithine by arginase. The basic amino acids ornithine and arginine are decarboxylated by ornithine decarboxylase (ODC) and arginine decarboxylase (ADC), respectively (Walters 2003). Decarboxylation of ornithine by ODC or arginine by ADC leads to the synthesis of putrescine, which

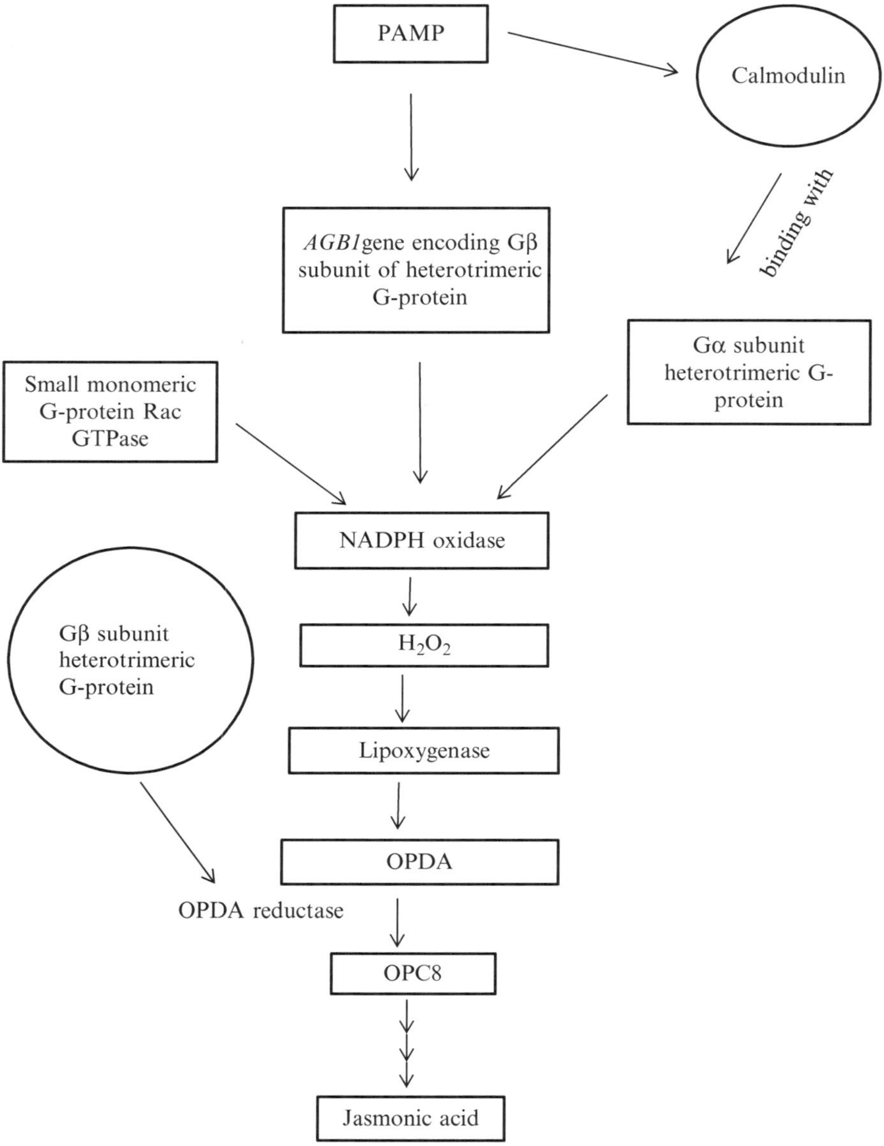

Fig. 3.4 Role of G-proteins in JA biosynthesis

is converted to spermidine by spermidine synthase. Spermidine, in turn, is then converted to spermine by spermine synthase (Nambeesan et al. 2012). In rice cells, the two key enzymes in biosynthesis of polyamines, arginase and spermidine synthase, are induced by the rice small monomeric G-protein OsRac1 (Fujiwara et al. 2006). S-adenosylmethionine decarboxylase is involved in decarboxylation of S-adenosylmethionine. OsRac1 also induces S-adenosylmethionine decarboxylase.

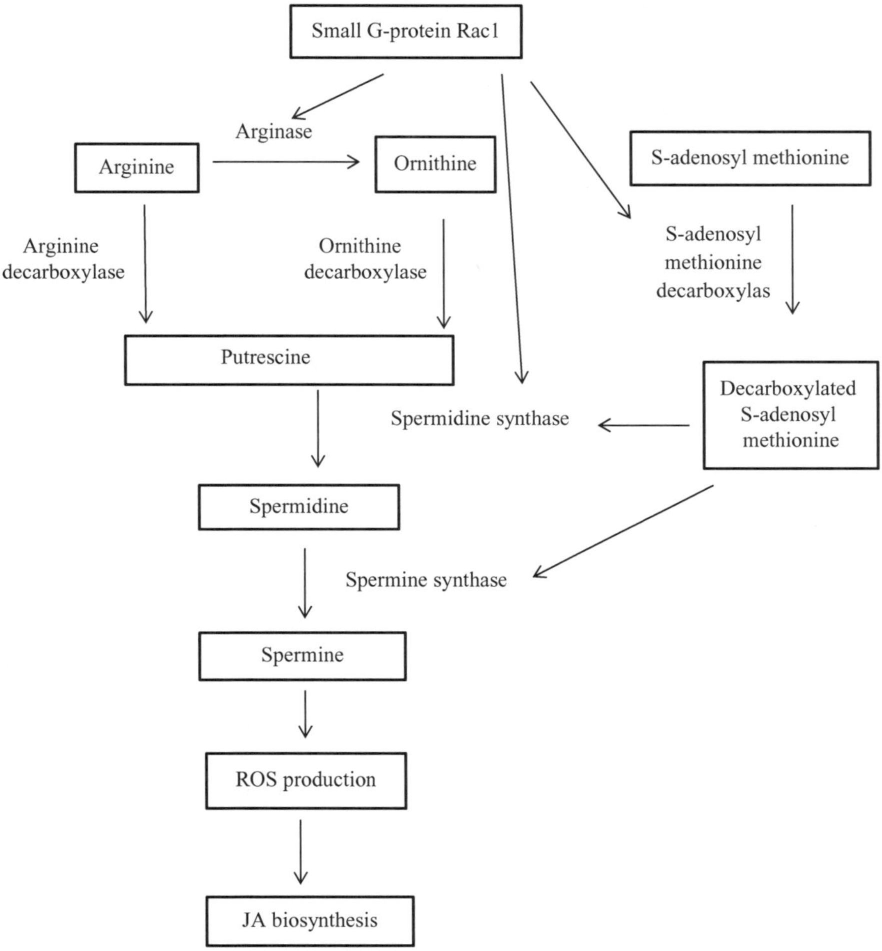

Fig. 3.5 Role of small monomeric G-protein Rac1 in polyamine-mediated JA biosynthesis (Adapted from Fujiwara et al. 2006; Kresge et al. 2007; Ozawa et al. 2009, 2010; Yoda et al. 2009; Nambeesan et al. 2012)

In these reactions, both sperimidine synthase and spermine synthase enzymes use aminopropyl residues derived from decarboxylated S-adenosylmethionine. The G-protein-triggered synthesis of polyamines is involved in the activation of JA biosynthesis (Fig. 3.5; Fujiwara et al. 2006; Kresge et al. 2007; Nambeesan et al. 2012).

Spermine treatment elicited biosynthesis of jasmonic acid in lima bean leaves (Fig. 3.6; Ozawa et al. 2009, 2010). Spermine treatment induced calcium influx and ROS production (Ozawa et al. 2010). Polyamines have been found to be a common source of hydrogen peroxide in host and nonhost hypersensitive response during pathogen infection (Yoda et al. 2009). ROS has been reported to activate lipoxygenase, which triggers biosynthesis of JA (Vranová et al. 2002; Vandenabeele et al. 2003; Zago et al. 2006).

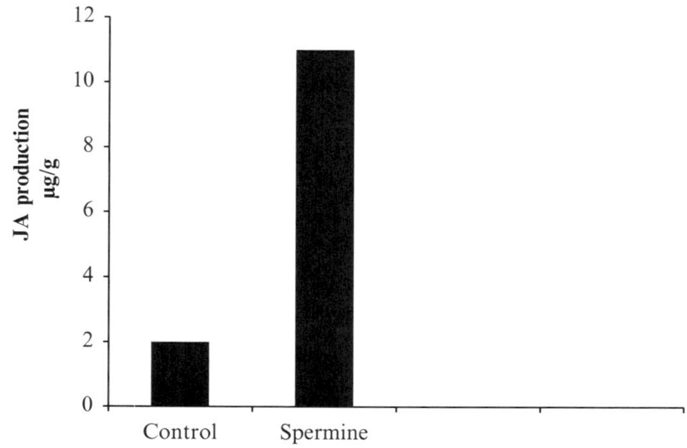

Fig. 3.6 Spermine triggers JA production (Adapted from Ozawa et al. 2009)

3.5.4 Calcium Signature Triggers JA Biosynthesis

One of the earliest upstream signaling events in JA biosynthesis pathway after PAMP perception by PRR is a rapid and transient rise in cytosolic Ca^{2+} levels through the function of plasma membrane-resident calcium ion channels (Kwaaitaal et al. 2011; Ranf et al. 2011). PAMP elicitor signals activate G-proteins and the activated G-proteins further switch on calcium ion channels (Zhao and Sakai 2003). Ca^{2+} influx and subsequent Ca^{2+} wave (calcium signature) may activate NADPH oxidase and H_2O_2 production (Zhao and Sakai 2003). Calcium ion influx-induced H_2O_2 production triggers increases in lipoxygenase activity (Zhao and Sakai 2003). Lipoxygenase is an important enzyme in the JA biosynthesis pathway (Schaller 2001; Vidhyasekaran 2007). Exogenous application of H_2O_2 has been found to induce the biosynthesis of endogenous jasmonate (Zhao and Sakai 2003). These studies suggest that Ca^{2+} influx may be involved in the activation of JA signaling system.

Glutamate receptors (GluRs) comprise a class of ion channel that might provide a calcium-permeable pathway across the plasma membrane (Vatsa et al. 2011; Price et al. 2012; Vincill et al. 2012; Li et al. 2013). The glutamate receptors are ligand-gated nonselective cation channels permeable to calcium (Kang et al. 2006; Tikhonov and Magazanik 2009; Vatsa et al. 2011). Glutamate receptor-like genes have been shown to form Ca^{2+} channels (Michard et al. 2011). The glutamate-gated Ca^{2+} channel is located in the plasma membrane of higher plants and plays a direct or indirect role in defense against pathogen infection by triggering JA biosynthesis (Kang et al. 2006). In *Arabidopsis thaliana* overexpressing a full-length cDNA clone (*RsGluR*) encoding a putative glutamate receptor from small radish, glutamate treatment triggered greater Ca^{2+} influx in the root cells of transgenic plants than in those of the wild type. JA biosynthetic genes were up-regulated in these transgenic plants overexpressing glutamate receptors (Kang et al. 2006). The glutamate receptors were found to be activated by the PAMP cryptogein, and they were involved in triggering Ca^{2+} influx (Vatsa et al. 2011). Kwaaitaal et al. (2011) showed that the initiation of innate immune responses upon the PAMPs flg22 and elf18 and chitin recognition involves apoplastic Ca^{2+} influx via glutamate receptor-like channels in *A. thaliana*. The downstream event in the glutamate receptor glutamate-gated Ca^{2+} channel-mediated signaling pathway included NO production (Vatsa et al. 2011). The Ca^{2+} influx-mediated NO plays important role in JA biosynthesis (del Rio et al. 2004; Xu et al. 2005; Grün et al. 2006; Zago et al. 2006; Palmieri et al. 2008). The downstream events in the glutamate receptor channel-mediated immune response signaling pathway also included mitogen-activated protein kinase (MAPK) cascade (Kwaaitaal et al. 2011). MAPK cascades may also be involved in JA biosynthesis (Seo et al. 1999; Lee et al. 2004; Kandoth et al. 2007). These results suggest that the calcium ion influx through glutamate receptor-gated channel is involved in JA biosynthesis pathway.

3.5.5 Role of ROS in JA Biosynthesis Pathway

The rapid and transient production of reactive oxygen species (ROS) is one of the most rapid defense responses observed in plants due to PAMP treatment (Faize et al. 2004; Asada 2006; Sagi and Fluhr 2006; Lehtonen et al. 2012). In ROS-elicited tobacco plants, a very rapid and sustained up-regulation of transcript tags that are involved in the production of JA signals was observed (Vandenabeele et al. 2003). The genes encoding lipase, lipoxygenase, and 12-oxophytodienoate reductase (12-ODPR), which are involved in JA biosynthesis, are activated in tobacco leaves accumulating ROS (Vandenabeele et al. 2003). Lipoxygenase is a key enzyme in JA biosynthesis (Rakwal et al. 2002; Grün et al. 2006; Zago et al. 2006; Gullner et al. 2010). The gene encoding lipoxygenase (*Lox1*) is induced by ROS (Vranová et al. 2002). These results suggest that ROS is involved in the induction of JA biosynthesis.

3.5.6 Role of NO in JA Biosynthesis Pathway

Nitric oxide is involved in triggering plant hormone signaling systems in plant innate immunity (Zago et al. 2006; Chun et al. 2012; Mur et al. 2012, 2013). NO is involved in the induction of biosynthesis of JA (Xu et al. 2005). JA accumulation was found to correlate with the rate of NO production in *Arabidopsis* (Mur et al. 2012). NO induces the key enzymes of the JA biosynthesis pathway (Fig. 3.7; del Rio et al. 2004; Palmieri et al. 2008). NO induces key enzymes of JA biosynthesis such as allene oxide synthase (AOS) and lipoxygenase (LOX2) in *A. thaliana* (Huang et al.

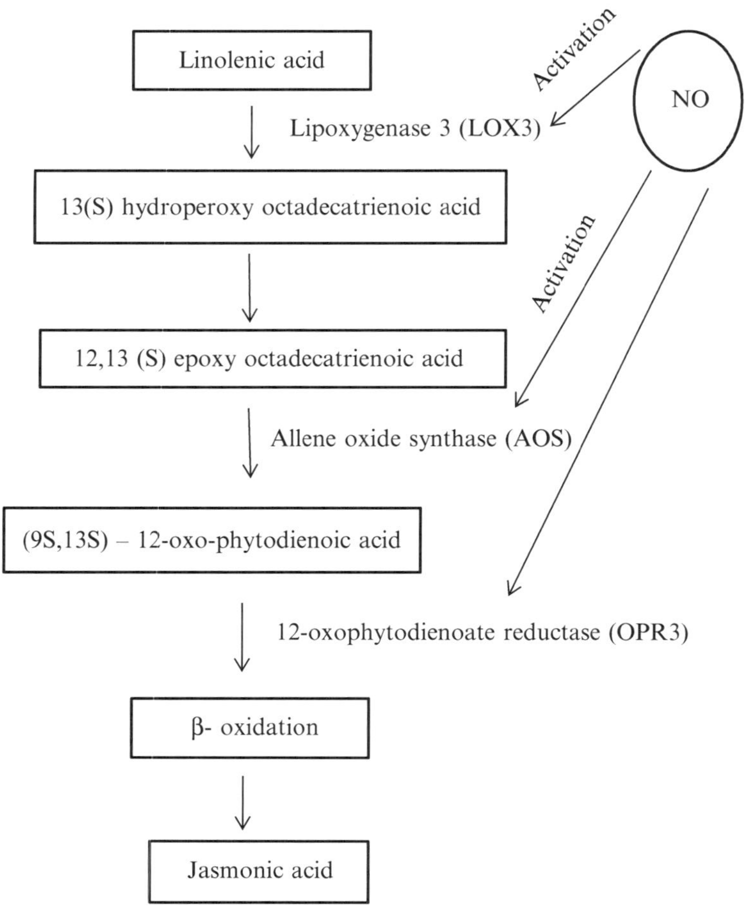

Fig. 3.7 Role of NO in activation of jasmonate biosynthesis (Adapted from Huang et al. 2004; Grün et al. 2006; Zago et al. 2006; Palmieri et al. 2008; Mur et al. 2013)

2004). Several genes of the JA biosynthetic pathway were up-regulated after NO treatment in *A. thaliana*. These included genes encoding three 12-oxophytodienoate reductases (OPR1, OPR2, and OPR3) and two lipoxygenases (LOX3 and a putative lipoxygenase protein) (Palmieri et al. 2008). NO induces allene oxide synthase gene (*AOS*) and 12-oxophytodienoate reductase gene *OPR3* (Grün et al. 2006); both are involved in JA biosynthesis (Vidhyasekaran 2007). Transcripts encoding lipoxygenase (LOX) and 12-oxophytodienoate reductase (12-OPR) involved in the biosynthesis of JA are up-regulated by NO (Grün et al. 2006; Zago et al. 2006). Four transcription factor-binding sites (TFBS elements) have been detected in promoter regions of the NO-regulated genes involved in JA biosynthesis (Palmieri et al. 2008).

3.5.7 *Mitogen-Activated Protein Kinases Functioning Upstream in JA Biosynthesis Pathway*

Mitogen-activated protein kinase (MAPK) cascades are major pathways downstream of PAMP/PRR complex that transduce extracellular stimuli into intracellular responses in plants (Hettenhausen et al. 2012; Zhang et al. 2012a). A typical MAPK signaling module consists of three interconnected protein kinases: an MAP kinase kinase kinase (MAPKKK or MEKK [for *M*APK/*e*xtracellular signal-regulated kinase *k*inase *k*inase]), an MAP kinase kinase (MAPKK or MKK), and an MAP kinase (MAPK or MPK) (Mészáros et al. 2006). Several distinct MAPK signaling three-kinase modules have been detected in plants (Cardinale et al. 2000; Pedley and Martin 2005; Mészáros et al. 2006). Some of the MAPK cascades may also be involved in JA biosynthesis (Seo et al. 1999; Lee et al. 2004). The MAPK modules MEKK1–MKK4–MPK3 and MEKK1–MKK1–MPK4 have been proposed to be responsible for JA biosynthesis in *Arabidopsis* (van Verk et al. 2011). A link between JA biosynthesis and the MAPK pathways has been reported in *Arabidopsis thaliana* (van Verk et al. 2011). MEK1 and MKK4 are co-expressed with *AOS* and *OPR3*, which are important genes in the JA biosynthesis pathway (Fig. 3.8; van Verk et al. 2011). The tomato MAPKs, LeMPK1 and LeMPK2, are activated in response to four different oligosaccharide elicitors (Holley et al. 2003), and these tomato MAPKs have been shown to function upstream of JA biosynthesis (Kandoth et al. 2007).

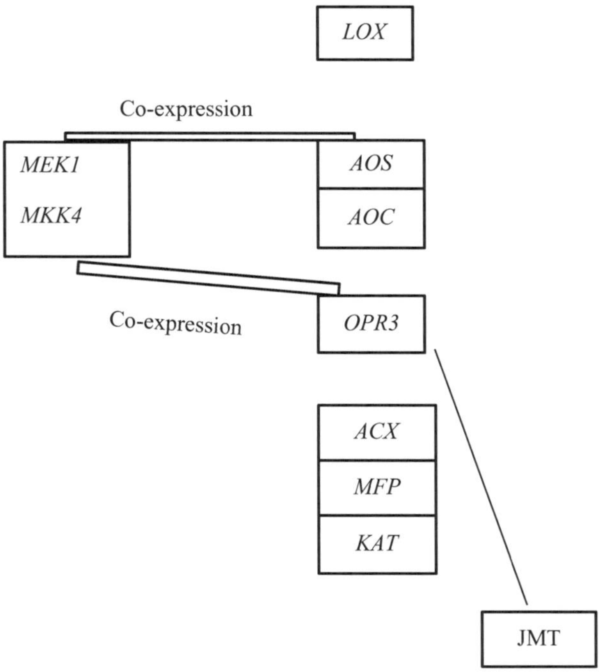

Fig. 3.8 Co-expression of *MAPK* genes with genes involved in the JA biosynthesis pathway (Adapted from van Verk et al. 2011)

Two mitogen-activated protein kinases, salicylic acid-induced protein kinase (SIPK) and wound-induced protein kinase (WIPK), are involved in plant immune responses (Ren et al. 2006; Kallenbach et al. 2010; Meldau et al. 2012). WIPK and SIPK play an important role in JA production (Seo et al. 2007). Silencing of WIPK/SIPK was correlated with a decrease in JA accumulation (Kobayashi et al. 2010), suggesting that WIPK/SIPK is involved in JA biosynthesis. The MKK2 overexpressor lines showed increased expression of lipoxygenase gene involved in JA biosynthesis (Teige et al. 2004). The MAPK WIPK-overproducing tobacco plants showed three- to fourfold higher levels of JA and methyl jasmonate than in the wild-type plants (Seo et al. 1999). Rice plants transformed with MK1, a homolog of WIPK from *Capsicum annuum*, showed a threefold higher level of JA than the wild type (Lee et al. 2004). Collectively these studies suggest a role for the MAPK in the production of JA.

MAPK cascades include both phosphorylation and dephosphorylation events resulting in transient increases in MAPK activity. MAP kinases are dephosphorylated and inactivated by MAPK phosphatases (MKPs), which are highly specific to MAPKs (Keyse 2000; Theodosiou and Ashworth 2002). MKPs are negative regulators of MAPKs (Katou et al. 2005; Lumbreras et al. 2010). MAPK phosphatases (MKPs) have been shown to negatively regulate JA biosynthesis in *A. thaliana* (Schweighofer et al. 2007).

3.5.8 Systemin Triggers JA Biosynthesis in Tomato

Systemin, an 18 amino acid polypeptide identified in tomato, has been reported to trigger the synthesis of JA. Systemin is processed from a larger prohormone protein, called prosystemin, by proteolytic cleavage (Ryan 2000). A cell surface receptor for systemin perception has been identified. Systemin induces a cascade of intracellular signaling events including depolarization of the plasma membrane, the opening of ion channels, an increase in Ca^{2+} influx, and activation of MAP kinase and phospholipase (Ryan 2000), leading to the release of linolenic acid from membrane lipids (Schaller and Ryan 1995; Bergey et al. 1996). Systemin triggers an increase in phospholipase A2 (PLA2) activity. PLA2 releases linolenic acid from membranes (Ryan 2000), and linoleic acid is the starting point in the lipoxygenase pathway leading to synthesis of JA (Li et al. 2005; Wasternack 2007; Balbi and Devoto 2008). Systemin activates JA biosynthetic enzymes such as allene oxide cyclase (AOC), and it leads to local rise in JA (Wasternack 2007). Systemin induces ROS (Orozco-Cárdenas et al. 2001) and ROS is involved in JA synthesis (Zhao and Sakai 2003). It also activates MAPKs and MAPK signaling system is involved in the biosynthesis of jasmonic acid in tomato (Kandoth et al. 2007). Co-silencing of the MAPKs *MPK1* and *MPK2* reduced MPK1/2 kinase activity, JA biosynthesis, and expression of JA-dependent defense genes. Application of methyl jasmonate restored the full defense response. These results show that MPK1 and MPK3 are essential components of the systemin signaling pathway, and the MAPKs function upstream of JA biosynthesis (Kandoth et al. 2007). Systemin is released into the vascular system of tomato plants and activates JA biosynthesis in the surrounding vascular tissues in which JA biosynthetic enzymes are located (Ryan 2000). Systemin was found only in tomato (Wasternack 2007).

It has been suggested that systemin may be the long-distance mobile signal for defense gene expression (McGurl et al. 1994; Ryan 2000). However, recent studies have revealed that systemin may not be the systemic signal. The *suppressor of prosystemin-mediated responses2* (*spr2*) mutation, which was isolated as suppressor of systemin-mediated signaling, impairs JA biosynthesis and the production of a long-distance signal (Sun et al. 2011). *Spr2* encodes fatty acid desaturase involved in JA biosynthesis (Li et al. 2003). The results suggest that systemin-induced activation of defense responses, as well as the production of the long-distance signal, is dependent on trienoic fatty acids that give rise to JA. The systemin action was found to be dependent on *Spr2*. The results suggest that JA, rather than systemin, may be the systemic signal, and the function of systemin may be in increasing JA biosynthesis to a level that is required for the systemic response (Sun et al. 2011).

3.6 Jasmonate Receptor Complex in JA Signal Perception

3.6.1 COI1, an F-Box Protein, Is a Jasmonate Receptor

Jasmonates are cellular signal molecules with key functions on the regulation of immune responses against necrotrophic pathogens (Pieterse et al. 2012a, b; Thaler et al. 2012; Wasternack and Hause 2013; Vidhyasekaran 2014). The discovery of coronatine-insensitive mutants in *Arabidopsis* led to the identification of COI1 (CORONATINE INSENSITVE1) as a key player in the jasmonate perception and signal transduction pathway. The *Arabidopsis* null mutant for *COI1* (*coronatine insensitive 1*) is fully insensitive to jasmonates (Feys et al. 1994; Xie et al. 1998). COI1 has been assigned a role in the direct perception of the jasmonate signal (Yan et al. 2009). COI1 has been identified as a receptor for jasmonates (Gfeller et al. 2010).

The gene *CoI1* is required for all the JA-mediated responses (Devoto et al. 2005; Balbi and Devoto 2008). The *Arabidopsis coi1* mutant displays defects in many, if not all, JA-dependent functions (Kazan and Manners 2008). Devoto et al. (2005) showed that COI1 was required for the expression of approximately 84 % of 212 genes induced by JA and COI1 was also required for the repression of 53 % of 104 genes whose expression was suppressed by JA. These results suggest that COI1 is a key regulator of genes involved in the methyl jasmonate-induced defense responses.

The COI1 locus encodes an F-box protein that is part of an E3 ubiquitin ligase complex involved in ubiquitin-mediated protein degradation (Balbi and Devoto 2008; Wager and Browse 2012). COI1 protein associates with CUL1 (*CUL*LIN1), RBX1 (*R*ING-*BOX* PROTEIN1), and the Skp1-like proteins ASK1 (*A*RABIDOPSIS

S-PHASE *K*INASE-ASSOCIATED PROTEIN*1*) and ASK2 to form the SCF^{COI1} (*S*KP1, *C*DC53p/CUL1, *F*-box^{COI1}) ubiquitin ligase complex in *A. thaliana* (Devoto et al. 2002; Xu et al. 2002; Pauwels and Goossens 2011; Zhang et al. 2012c). *COI1* gene has also been identified in soybean (Wang et al. 2005).

3.6.2 COI1–JAZ Receptor Complex

COI1 may not be the sole receptor in the JA-perception process, and it may be only one of the components in the jasmonate receptor complex (Katsir et al. 2008; Sheard et al. 2010). Another important component in the receptor complex belongs to JAZ (for *JA*SMONATE ZIM [Zinc-finger protein expressed in *I*nflorescence *M*eristem]-domain) family of transcriptional repressors. COI1–JAZ complex, rather than COI1 alone, functions as the high-affinity jasmonate receptor in a co-receptor form. JA-Ile binding to COI1 alone elicited <2 % binding signal relative to that of JA-Ile binding to COI1–JAZ (Sheard et al. 2010).

3.6.3 InsP5 Potentiates JA Perception by COI1–JAZ1 Complex

The structural and pharmacological studies revealed that the jasmonate receptor is a co-receptor complex, consisting of the F-box protein COI1, JAZ, and a newly discovered third component, inositol pentakisphosphate ($InsP_5$) (Sheard et al. 2010). InsP5 has been shown to potentiate JA perception by COI1–JAZ. $InsP_5$ binding has been shown to be crucial for the JA co-receptor to perceive the hormone with high sensitivity (Sheard et al. 2010).

It has been demonstrated that the JA receptor is a three-molecule co-receptor complex, consisting of COI1, JAZ, and inositol pentakisphosphate, all of which are indispensable for high-affinity hormone binding. JAZ boundaries show a unique bi-partite sequence that binds COI1 and directly participates in hormone recognition. The N-terminal clamp region of the JAZ1 that is critical for hormone binding is diverse among JAZ proteins (Sheard et al. 2010).

3.6.4 JA-Ile Promotes Physical Interaction Between JAZ1 and COI1

JA-Ile promotes physical interaction between JAZ1 and COI1. The carboxyl terminal Jas motif, but not the N-terminal domain or central ZIM domain of JAZ proteins, is critical for JA-Ile-dependent interactions with COI1. Two positively charged amino acid residues ($R^{205}R^{206}$ in JAZ1) in the Jas domain were found to be essential for COI1–JAZ interaction. These two positively charged residues are critical for JA signaling. Transgenic expression of JAZ1 carrying alanine substitutions of $R^{205}R^{206}$ conferred JA-insensitive phenotypes (Melotto et al. 2008).

3.7 JA Signaling Pathway

3.7.1 JAZ Proteins Suppress JA Signaling

The JA receptor JAZ proteins have been identified as suppressors of jasmonate signaling (Chini et al. 2007; Thines et al. 2007). JAZ proteins suppress the expression of the JA-responsive genes (Chini et al. 2007; Howe 2010; Wasternack and Hause 2013). JAZ family of transcriptional repressors has been identified as an important component in a receptor complex involved in the JA perception (Sheard et al. 2010). Twelve JAZ proteins have been identified in *Arabidopsis thaliana* (Chung and Howe 2009; Pauwels and Goossens 2011; Wager and Browse 2012).

The JAZ proteins contain a weakly conserved N-terminal domain, a highly conserved C-terminal Jas domain that mediates the interaction with the transcription factors, and the conserved ZIM (TIFY) domain responsible for JAZ dimerization and interaction with NINJA (for NOVEL INTERACTOR OF JAZ) (Chung and Howe 2009; Wasternack and Hause 2013). The Jas domain is required for the repressive activity of JAZ proteins (Chung and Howe 2009; Shyu et al. 2012). The conserved TIFY motif (TIFF/YXG) within the ZIM domain mediates homo- and heteromeric interactions between most *Arabidopsis thaliana* JAZs (Chini et al. 2009b; Chung and Howe 2009). There is a common occurrence of the ZIM and the Jas domain in all 12 JAZ proteins in *Arabidopsis* (Pauwels and Goossens 2011; Wasternack and Hause 2013).

NINJA is a member of the ABI-FIVE BINDING PROTEIN FAMILY, and it was identified as an adaptor protein that interacts with the ZIM domain of most JAZs (Pauwels et al. 2010). NINJA contains an EAR (for ERF-associated amphiphilic repression) motif that recruits the corepressor TOPLESS (TPL) (Pauwels et al. 2010). JAZs are a scaffold on which the NINJA–TPL corepressor complex is assembled (Howe 2010). *Arabidopsis* JAZ proteins recruit the Groucho/Tup1-type corepressor TPL and TPL-related proteins (TPRs) through the adaptor protein NINJA. NINJA acts as a transcriptional repressor whose activity is mediated by a functional TPL-binding EAR repression motif (Pauwels et al. 2010). JAZ proteins interact with several transcription factors including MYC2, MYC3, MYC4, MYB21, MYB24, EIN3, EIL, GL3, EGL3, TT8, GL1, and PAP (Pauwels and Goossens 2011; Wasternack and Hause 2013). The JAZ proteins actively repress the activity of transcription factor MYC2, and probably other transcription factors, which bind to *cis*-acting elements of JA response genes. Thus, the JAZ proteins interact with MYC2 (and other transcription factors) involved in the transcriptional regulation of JA-responsive gene expression (Lorenzo et al. 2004; Chini et al. 2007) and suppress their action in inducing expression of JA-responsive defense genes (Fig. 3.9; Lorenzo et al. 2004; Chini et al. 2007; Howe 2010; Wasternack and Hause 2013).

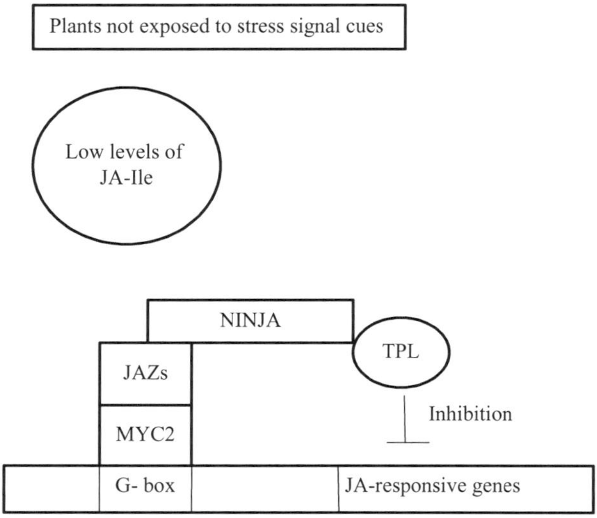

Fig. 3.9 Suppression of JA-responsive gene expression by JAZ proteins (Adapted from Howe 2010; Wasternack and Hause 2013)

When JA hormone levels are low in plant cells, JA-mediated responses are kept in repressed state by JAZ proteins (Wager and Browse 2012). In cells containing low JA JA-Ile levels, JAZ proteins bind to and inhibit transcription factors that promote the expression of JA response genes (Howe 2010; Woldemariam et al. 2012).

3.7.2 Role of COI1 Protein in the Degradation of JAZ Proteins by E3 Ubiquitin Ligase

The *Arabidopsis COI1* gene is required for the JA-mediated defense response against pathogens (Xie et al. 1998). COI1 protein contains an F-box motif and associates physically with AtCUL1, AtRbx1, and the Skp1-like proteins ASK1 and ASK2 to assemble SCF ^{COI1} ubiquitin ligase complexes (Skp1–cdc53–F-box protein) (Xu et al. 2002). SCF^{COI1} targets key regulators of JA signaling pathway for ubiquitination and subsequent degradation by the 26S proteasome (Kawamura et al. 2009).

Ubiquitin–proteasome-mediated proteolysis has been shown to be involved in jasmonate signaling system (Xu et al. 2002; Feng et al. 2003). *COI1* (*coronatine insensitive 1*) gene is required for jasmonate-regulated defense signaling in *Arabidopsis* (Xie et al. 1998). The *Arabidopsis* null mutant for *COI1-1* is fully insensitive to jasmonates, and the COI1 protein is required for all JA-dependent responses (Feys et al. 1994; Xie et al. 1998).

It has been shown that COI1 is required for removal of repressors of the JA signaling pathway (Xie et al. 1998; Balbi and Devoto 2008). The COI1 protein has been identified as an F-box protein, suggesting the involvement ubiquitin–proteasome-mediated protein degradation in JA signaling (Xie et al. 1998). *COI1* forms a functional E3 ubiquitin ligase, SCF^{COI1}, in plants (Devoto et al. 2002; Xu et al. 2002; Li et al. 2003). COI1 is present in the functional SKP–CULLIN–F-box-type E3 ubiquitin ligase complex (Devoto et al. 2002). COI1 associates physically with CUL1, RBX1, and SKP1-like proteins to assemble SCF^{COI1} complexes (Xu et al. 2002). A single amino acid substitution in the F-box motif of COI1 abolishes the formation of SCF^{COI1} complexes and affects JA-inducible gene expression. These observations suggest that SCF^{COI1} complex is important for JA signaling (Xu et al. 2002).

The COP9 signalosome (CSN) has been shown to be necessary for the function of SCF^{COI1} ubiquitin ligase. COP9 signalosome associates physically with SCF^{COI1} (Feng et al. 2003). Most of the COI1-dependent JA-responsive genes also required CSN function, and CSN abundance was shown to be important for JA responses (Feng et al. 2003). These observations suggest that both CSN and SCF^{COI1} work together to control genome expression and promote JA responses.

JAZ family of transcription repressors has been identified as SCF^{COI1} substrate targets, which associate with COI1 in a hormone-dependent manner (Chini et al. 2007; Thines et al. 2007; Yan et al. 2007). JAZ family of proteins has been found to be direct targets of the SCF^{COI1} E3 ubiquitin ligase, and jasmonate treatment induces their proteasome degradation (Chini et al. 2007). Degradation of JAZ proteins is dependent on activities of the SCF^{COI1} ubiquitin ligase and the 26S proteasome (Chini et al. 2007; Thines et al. 2007). The JA-dependent degradation of JAZs could be inhibited by a specific inhibitor of the 26S proteasome activity or in the *coi1* mutant background (Thines et al. 2007). The results suggest that COI1 is involved in the degradation of the repressors of the JA signaling pathway through SCF^{COI1}-dependent ubiquitination.

3.7.3 Role of JA-Ile in the JAZ Degradation by 26S Proteasome

JA-Ile, but not jasmonic acid itself or MeJA, promotes the interaction between SCF^{COI1}–JAZ complexes (Thines et al. 2007). This finding suggests that jasmonic acid may not be the signal directly responsible for the activation of the JA signaling pathway, but possibly it undergoes further modifications to be converted to a biologically active signal. *JAR1* (*JASMONATE RESISTANT1*) encodes a JA–amino synthetase, which activates conjugation of JA to Ile (Staswick and Tiryaki 2004). JA-Ile produced by JAR1 promotes the interaction between JAZ and SCF^{COI1} and takes part in the removal of repressors of JA signaling pathway (Kazan and Manners 2008; Melotto et al. 2008).

JA-Ile, but not jasmonic acid or other precursors of JA-Ile, stimulates physical interaction between COI1 and JAZ, which allows JAZs to be ubiquitinated by SCF^{COI1} and subsequently degraded by the 26S proteasome. The (*3R,7S*) stereoisomer of JA-Ile [also known as (+)-7-*iso*-JA-Ile] is much more active than the (*3R,7R*) isomer in promoting COI1–JAZ interactions and thus is considered the natural ligand for the receptor (Fonseca et al. 2009; Chung et al. 2010; Suza et al. 2010; Howe 2010).

JAZ1 degradation signal (degron) includes a short conserved LPIAR motif that seals JA-Ile in its binding pocket at the COI1–JAZ interface (Shyu et al. 2012). *Arabidopsis thaliana* JAZ8 lacks this motif and thus is unable to associate strongly with COI1 in the presence of JA-Ile. As a consequence, JAZ8 is stabilized against jasmonate (JA)-mediated degradation and, when ectopically expressed in *Arabidopsis*, represses JA-regulated defense responses. JAZ8-mediated repression depends on an LxLxL-type EAR (ERF-associated amphiphilic repression) motif at the JAZ8 N-terminus that binds the corepressor TOPLESS and represses transcriptional activation. JAZ8-mediated repression does not require the ZIM domain, which, in other JAZ proteins, recruits TOPLESS through the EAR motif-containing NINJA (Shyu et al. 2012). An alternatively spliced form (JAZ10.4) of JAZ10 that lacks the Jas domain, and as a consequence, is highly resistant to JA-induced degradation (Chung and Howe 2009).

Low JA-Ile levels permit the accumulation of JAZ proteins (with their Jas, ZIM/TIFY, and Domain1 regions) that bind to the bHLH-type transcription factor MYC2. Repression of JA response genes involves binding of JAZ to NINJA, which contains an EAR motif that recruits the corepressor TPL. TPL may silence gene expression through an unknown mechanism (Howe 2010). Low JA-Ile levels attenuate JA-mediated plant defenses (Woldemariam et al. 2012).

In response to stress cues that activate JA-Ile synthesis, high levels of JA-Ile promote SCF^{COI1}-mediated ubiquitination and subsequent degradation of JAZs by the 26S proteasome. JAZ degradation relieves TPL-mediated repression of gene expression (Fig. 3.10; Howe 2010; Wasternack and Hause 2013). In the absence of JA-Ile, the C-terminus of JAZ proteins binds to MYC2, repressing its transcription activity. When the concentration of JA-Ile reaches a threshold during wound or pest stress (Chung et al. 2008), COI1 begins to bind to the Jas domain of JAZ proteins, presumably in a ligand-dependent manner. The JA-Ile-dependent interaction between COI1 and the C-terminus of JAZ proteins would result in SCF^{COI1}-dependent ubiquitination and degradation of JAZ proteins through the 26S proteasome, thus derepressing JA signaling (Melotto et al. 2008). Through direct interaction with the transcription factor MYC2 and related transcription factors, JAZs target corepressor complexes to JA-responsive genes. Proteolytic removal of the JAZ scaffold by the JA-Ile/SCF^{COI1} pathway effectively uncouples the corepressor from its target genes (Howe 2010).

3.7.4 MYC2, MYC3, and MYC4 Transcription Factors Regulate JA-Responsive Gene Expression

MYC2 is a bHLH (basic helix-loop-helix) transcription factor (TF) belonging to the 133-member bHLH family characterized by its bHLH domain responsible for DNA binding and for the formation of homo- and/or heterodimers with other bHLH proteins (Wager and Browse 2012). MYC2 interacts with JAZ proteins and mediates JA

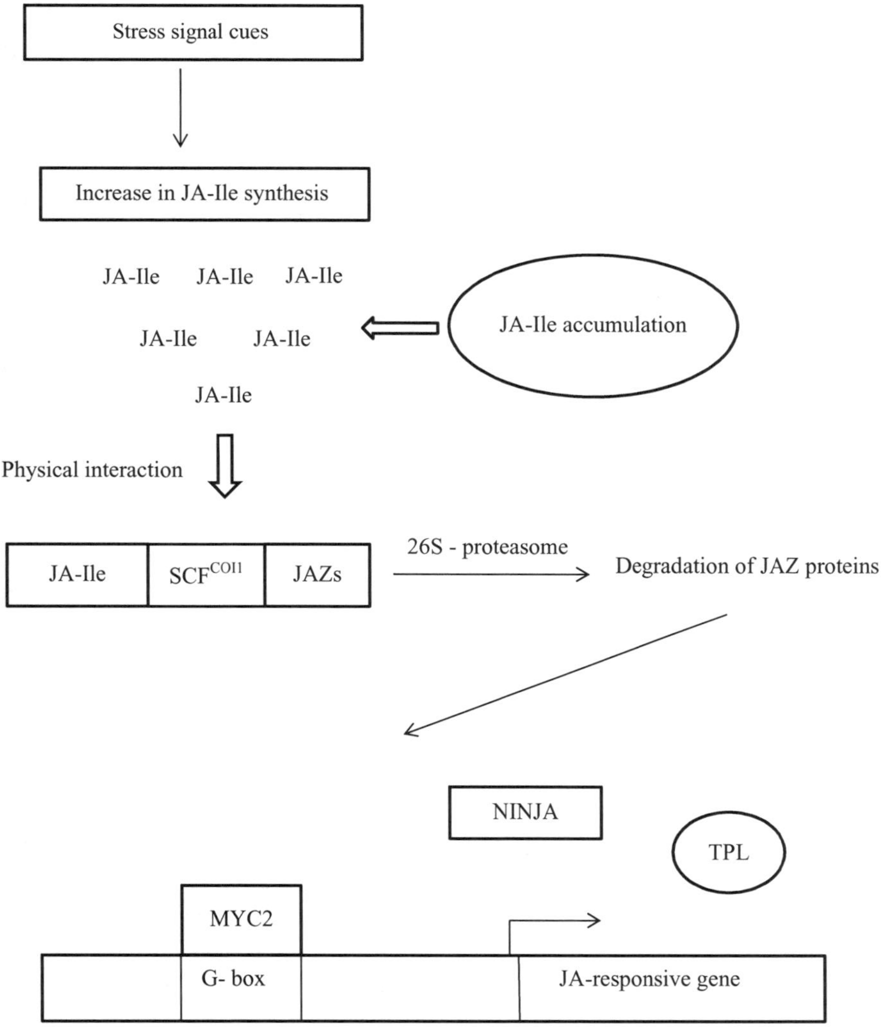

Fig. 3.10 Role of JA-Ile accumulation in SCF^{COI1} – mediated ubiquitination and JAZ degradation (Adapted from Howe 2010; Wasternack and Hause 2013)

responses. MYC2 is involved in the transcriptional regulation of JA-responsive gene expression (Lorenzo et al. 2004; Chini et al. 2007). It is suggested that, in the absence of a JA signal, JAZ proteins repress MYC2. Upon sensing of the JA signals, JAZ repressors are recruited to the SCF E3 complex for ubiquitination and subsequent degradation by the proteasome. The removal of these repressors then paves the way for MYC2 to regulate JA-dependent gene expression (Lorenzo et al. 2004; Chini et al. 2007, 2009a; Gfeller et al. 2010; Wager and Browse 2012). JAZs function as inhibitors of the transcription factor that, when free of JAZ inhibition, act to activate transcription from JA response genes (Wager and Browse 2012). The nuclear targeting of JAZ1 and JAZ9 has been shown to be dependent on physical interaction with MYC2 via a highly conserved region of the Jas domain (Withers et al. 2012).

MYC2 is involved in the expression of many JA-responsive genes (Dombrecht et al. 2007). MYC2-regulated genes include the JA-responsive pathogen defense genes *PDF1.2*, *CHIB/PR3*, and *HEL/PR4,* and they are negatively regulated by MYC2 (Anderson et al. 2004; Lorenzo et al. 2004; Dombrecht et al. 2007). MYC2 positively regulates the JA-responsive genes *VSP* and *LOX* (Lorenzo et al. 2004). *myc2* mutant plants show increased resistance to fungal pathogens such as *Plectosphaerella cucumerina*, *Botrytis cinerea*, and *Fusarium oxysporum* (Anderson et al. 2004; Lorenzo et al. 2004) and the bacterial pathogen *Pseudomonas syringae* (Nickstadt et al. 2004; Laurie-Berry et al. 2006).

Cheng et al. (2011) identified another bHLH transcription factor MYC3 which directly interacted with JAZs by virtue of its N-terminal region to regulate JA responses. The JAZ-interacting pattern and the JA-induced expression pattern of MYC3 were distinguishable from those of MYC2. MYC3 and MYC2 may have redundant but also distinguishable functions in the regulation of JA responses (Cheng et al. 2011). Fernández-Calvo et al. (2011) identified the TFs MYC3 and MYC4, phylogenetically closely related to MYC2. MYC3 and MYC4 interact in vitro and in vivo with JAZ repressors and also form homo- and heterodimers with MYC2 and among themselves. MYC3 and MYC4 are nuclear proteins that bind DNA with sequence specificity similar to that of MYC2. Loss-of-function mutations in any of these two TFs impair full responsiveness to JA and enhance the JA insensitivity of *myc2* mutants. The triple mutant *myc2myc3myc4* is as impaired as *coi1-1* in the activation of several, but not all, JA-mediated responses such as the defense against bacterial pathogens. MYC3 and MYC4 act additively with MYC2 to regulate specifically different JA-dependent transcriptional response sets (Fernández-Calvo et al. 2011). MYC3 and MYC4 share high sequence similarity with MYC2. MYC3 and MYC4 interact not only with JAZ1 but also with other JAZ proteins (JAZ3 and JAZ9). Similar to MYC2, MYC4 negatively regulates expression of *PDF1.2* (Niu et al. 2011).

MYC2 is a master regulator of the JA signaling pathway. MYC2 is required for induced systemic resistance (ISR) triggered by beneficial soil microbes (Kazan and Manners 2013). MYC2 function is targeted by pathogens during effector-mediated suppression of innate immunity in roots. MYC2 regulates crosstalk between the signaling pathways of JA and those of other phytohormones such as ABA, SA, GA, and auxin (Kazan and Manners 2013). MYC2 orthologs act as "master switches" that regulate JA-mediated biosynthesis of secondary metabolites (Kazan and Manners 2013).

In the absence of JA signal, JAZ proteins actively repress the transcription factor MYC2, which binds to *cis*-acting elements of JA response genes. In response to cues that up-regulate JA-Ile synthesis, the hormone triggers the specific binding of JAZ proteins to COI1, leading to poly-ubiquitination. Subsequent degradation of JAZ by the 26S proteasome relieves repression of MYC2 and probably other transcription factors, permitting the expression of jasmonate-responsive genes (Lorenzo et al. 2004; Chini et al. 2007, 2009a; Gfeller et al. 2010; Wager and Browse 2012). JAZ degradation relieves TPL-mediated repression of gene expression. This may alleviate passive expression by allowing MYC2 to engage RNA polymerase II (RNAP) and/or the Mediator complex that links RNAP to MYC2 (Howe 2010; Wasternack and Hause 2013).

3.8 Mediator Complex Regulates Transcription of JA-Responsive Genes by Interacting with Transcription Factors

Immune responses are activated upon the perception of PAMP signals. Activation of immune responses is accomplished by transcriptome reprogramming. During this process, transcription factors perceive the signals relayed by signaling molecules and translate into a functional response through recruiting or releasing RNA polymerase II (RNAPII) (Moore et al. 2011; An and Mou 2013). Mediator is a multiprotein cofactor of RNAPII, and it acts as a bridge between RNAPII- and DNA-binding transcription factors (Kidd et al. 2011; Cevik et al. 2012). The multiprotein Mediator complex acts as an adaptor/coregulator between sequence-specific transcription factor and RNAPII (An and Mou 2013). Mediator complex consists of about 20–30 subunits and the individual subunits may possess diverse roles (Kidd et al. 2011). The Mediator complex fine-tunes gene-specific transcriptional reprogramming by interacting with specific activators/repressors together with general transcription factors at the promoter site (Conaway and Conaway 2011a, b). The Mediator subunits are involved in mRNA, miRNA, and rRNA processing, as well as controlling DNA and protein stability (Kidd et al. 2011).

MEDIATOR25 subunit of the Mediator complex encoded by the PHYTOCHROME AND FLOWERING TIME1 gene (*PFT1*) is a positive regulator of jasmonate-responsive gene expression in *Arabidopsis* (Iñigo et al. 2012). MED25 functions in association with transcriptional regulators of the JA pathway (Cevik et al. 2012). MED25 interacts with several transcription factors (TFs) known to function in the control of JA-associated gene expression (Fig. 3.11; Cevik et al. 2012). These include three bHLH transcription factors MYC2, MYC3, and MYC4 and four members of group IX of the APETALA2 (AP2)/ETHYLENE RESPONSE FACTOR (AP2/ERF) transcription factors, ERF1, ERF59 (ORA59), TDR1/ERF98, and ERF15 (Cevik et al. 2012).

MED25 physically interacts with several transcriptional regulators of the JA signaling pathway, including MYC2, ORA59, and ERF1 (Cevik et al. 2012). MED25 physically associates with MYC2 in promoter regions of MYC2 target genes and exerts a positive effect in MYC2-regulated gene transcription (Chen et al. 2012). Transcription activation abilities of the MYC2, ERF1, and ORA59 were significantly reduced in the *med25* mutant leaves (Cevik et al. 2012). The results suggest that a functional MED25 is required for transcriptional activation by MYC2, ERF, and ORA59 transcription factors.

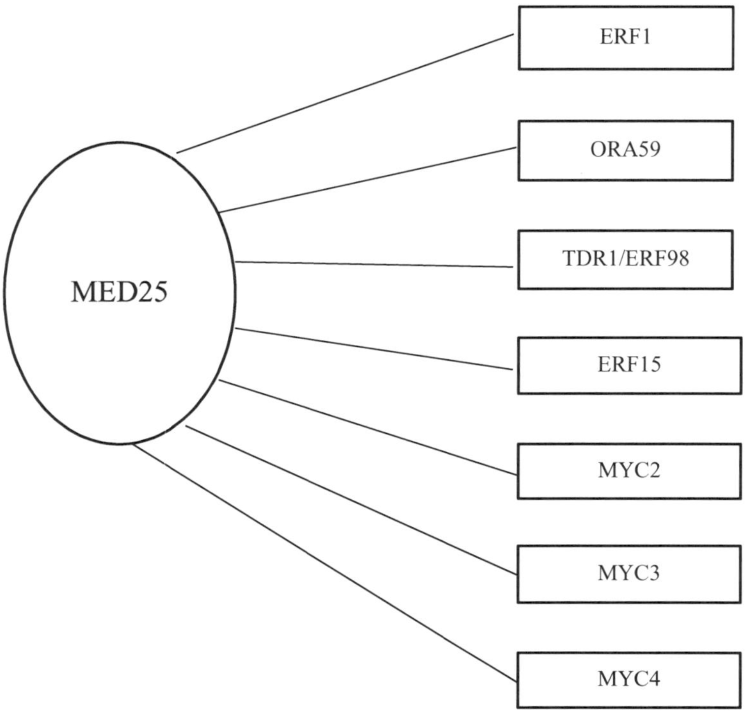

Fig. 3.11 MED25-interacting JA- and plant defense-associated transcription factors (Adapted from Cevik et al. 2012)

PLANT DEFENSIN1.2 (PDF1.2) is a pathogenesis-related protein (PR-12) involved in plant defense responses, and *PDF1.2* is induced by JA signaling (Vidhyasekaran 2007; Zander et al. 2010; Kravchuk et al. 2011; Po-Wen et al. 2013). Both ERF1 and ORA59 have been shown to activate expression of the *PDF1.2* gene, with the expression of *ORA59* being essential for *PDF1.2* activation in *Arabidopsis* (Lorenzo et al. 2003; Pré et al. 2008; Zander et al. 2010; Zarei et al. 2011; Cevik et al. 2012). Both ERF1 and/or ORA59 require a functional MED25 protein to activate *PDF1.2* expression (Cevik et al. 2012). The expression of either ERF1 or ORA59 significantly induced the expression of *PDF1.2* in wild-type plants. ORA59 activated *PDF1.2* expression to higher levels than ERF1 (Cevik et al. 2012). The *med25* mutant showed reduced expression of the *PDF1.2* gene (Kidd et al. 2009). When either *ORA59* or *ERF1* was expressed in the *med25* mutant plants, only partial activation of the *PDF1.2* gene expression was achieved (Cevik et al. 2012). Collectively, these studies suggest that MED25-dependent activation of both ERF1 and ORA59 is required for the activation of *PDF1.2* gene expression (Fig. 3.12).

In contrast to ERF1 and ORA59, MYC2 acts as a negative regulator of *PDF1.2* gene expression (Zander et al. 2010; Wager and Browse 2012; Wasternack and

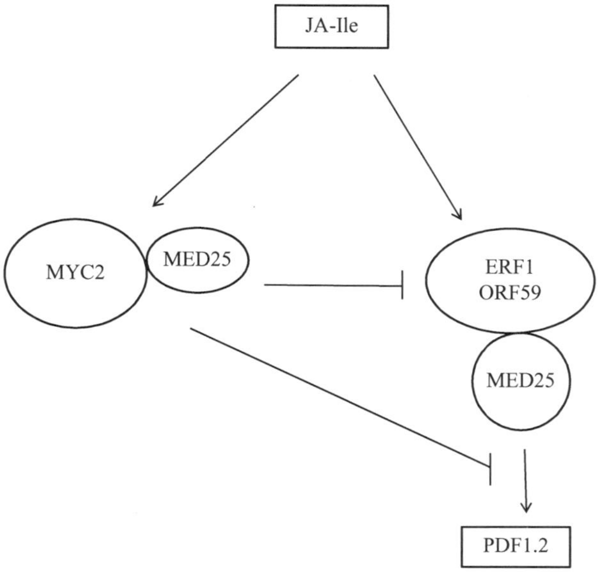

Fig. 3.12 Role of MED25 in regulation of *PDF1.2* expression (Adapted from Cevik et al. 2012)

Hause 2013). MYC2 inhibits the expression of *PDF1.2,* while ERF1 activates the expression of *PDF1.2* (Lorenzo et al. 2004). The results suggest that MYC2 differentially regulates the JA response through competitive interaction with ERF1. ERF1-, ORA59-, and MYC2-regulated genes were attenuated in *med25* mutant plants (Cevik et al. 2012). MED 25 has been shown to be required for the activity of MYC2 to repress the expression of *PDF1.2* gene. The expression of *PDF1.2* was reduced in the *med25* mutant (Kidd et al. 2009). In the *med25 myc2* double mutant, *PDF1.2* gene was attenuated and showed an expression level similar to that of the *med25* single mutant (Cevik et al. 2012). These results suggest that MED25 is essential for the activation of *PDF1.2* gene expression and that MYC2-mediated repression of the *PDF1.2* gene acts on a pathway that is dependent on MED25 (Fig. 3.12). After JA treatment, *ORA59* and *ERF1* expression was induced higher than the wild type in the *myc2* single mutant (Zander et al. 2010; Cevik et al. 2012). The results suggest that MYC2 negatively regulates both *ORA59* and *ERF1*. Both *ORA59* and *ERF1* showed attenuated expression in the *med25 myc2* double mutant, relative to the *myc2* single mutant (Cevik et al. 2012). These results suggest that MYC2 suppresses JA-associated genes through suppression of *ERF1* and *ORA59* and that this repression by MYC2 is dependent on a functional MED25 (Fig. 3.12).

MED25 is required for jasmonate-dependent defense gene expression and induces resistance against necrotrophic fungal pathogens *Alternaria brassicicola* and *Botrytis cinerea* in *Arabidopsis thaliana* (Kidd et al. 2009). Conversely, MED25 confers susceptibility to *Fusarium oxysporum*, a root-infecting hemibiotrophic fungal pathogen known to hijack jasmonate responses for disease development (Kidd et al. 2009; Thatcher et al. 2009).

Another Mediator complex subunit, MED16/SFR6 (SENSITIVE TO FREEZING6), has been shown to participate in JA signaling (Wathugala et al. 2012; Zhang et al. 2012b). *med16/sfr6* mutants were severely defective in the expression of plant defensin (*PDF*) genes in response to JA, suggesting that MED16 plays an important role in the induction of JA-responsive defense genes (Wathugala et al. 2012).

MED8 is another Mediator complex subunit involved in JA-dependent immune responses (Kidd et al. 2009). Similar to *med25/pft1*, a mutation in the *MED8* gene confers susceptibility to *A. brassicicola* but resistance against *F. oxysporum* in *Arabidopsis*. It suggests that MED8 has a role in JA-dependent response. The expression of *PDF1.2* was slightly lower in both methyl jasmonate (MeJA)-treated and MeJA-untreated *med8* plants than in similarly treated wild-type plants. However, in the *med25 med8* mutant, *PDF1.2* expression was greatly reduced, with the expression level in the double mutant being about 300- and 370-fold less in the MeJA-treated and untreated plants, respectively, than similarly treated *med25* mutant plants (Kidd et al. 2009). The results suggest MED8 and MED25 may act independently and their activity is additively enhanced in JA signaling pathway.

3.9 MAP Kinases May Regulate the Downstream Events in JA Signaling Pathway

MAP kinase cascades play an important role in JA signaling system, and they may positively or negatively regulate the JA signaling system. MAP kinases may regulate transcription factors by phosphorylation. The *Arabidopsis* MAP kinase MPK4 regulates downstream events in JA signaling system (Petersen et al. 2000; Liu and Zhang 2004; Gao et al. 2008; Qiu et al. 2008; Pitzschike et al. 2009). The induction of JA-responsive genes was blocked in *mpk4* mutant plants, suggesting that MPK4 positively regulates JA signaling. Inactivation of MPK4 in *mpk4* mutant plants resulted in suppression of JA-responsive genes (Petersen et al. 2000). Overexpression of an inactive form of MPK4 failed to complement *mpk4* phenotypes, indicating that kinase activity is required for MPK4 function (Petersen et al. 2000). The tobacco ortholog of the *Arabidopsis* MPK4, NtMPK4, is activated by SIPKK, a SIPK-interacting MAPK kinase. In NtMPK4-silenced tobacco plants, the induction of JA-responsive genes was inhibited, suggesting that NtMPK4 activates JA-responsive genes (Gomi et al. 2005). MAP KINASE4 SUBSTRATE 1 (MKS1) is the target of MPK4. MKS1 interacts with the WRKY transcription factors WRKY25 and WRKY33 and both of them can be phosphorylated by MPK4 (Andreasson et al. 2005). *wrky33* mutant plants showed reduced expression of JA-responsive genes, suggesting that WRKY33 after phosphorylation by MPK4 activates JA signaling system (Zheng et al. 2006).

The *Arabidopsis* MKK3–MPK6 cascade negatively regulates *MYC2* function (Takahashi et al. 2007). *MYC2* was shown to function as a downstream factor of the

MKK3–MPK6 cascade in JA signaling (Takahashi et al. 2007). *MYC2* encodes a nuclear localized basic helix-loop-helix leucine zipper-type transcription factor (Boter et al. 2004; Lorenzo et al. 2004). It plays a predominant role in JA signaling pathway, activating transcription of the JA-responsive *PDF1.2* encoding the PR-12 defensin protein (Lorenzo and Solano 2005; Takahashi et al. 2007). MYC2-regulated genes also include *CHIB/PR3* and *HEL/PR4*, and they are negatively regulated by MYC2 (Anderson et al. 2004; Lorenzo et al. 2004). MYC2 negatively regulates Trp and Trp-derived secondary metabolism such as indole glucosinolate biosynthesis (Dombrecht et al. 2007). It is suggested that MYC2 probably acts through the transcriptional orchestration of other transcription factors, which in turn regulate downstream JA response genes involved in diverse JA-dependent plant processes (Dombrecht et al. 2007).

A MAPKK encoding gene *GhMKK5* from cotton has been isolated and characterized. The expression of *GhMKK5* is significantly induced by methyl jasmonate (Zhang et al. 2012b). The expression of JA signaling system-inducible *PR-4* genes was greatly elevated in *GhMKK5*-overexpressing plants (Zhang et al. 2012b). The rice MAPK OsMPK6, an ortholog of AtMPK4, induces expression of JA signaling genes (Shen et al. 2010).

The MPK4 from soybean (*Glycine max*) GmMPK4 negatively regulates JA pathway. JA signaling is activated in *GmMPK4*-silenced plants. Significantly, a hallmark JA-responsive gene encoding defensin had 315-fold change in *GmMPK4*-silenced plants (Liu et al. 2011a). The MAPK WIPK-overproducing tobacco plants showed constitutive accumulation of PI-II (proteinase inhibitor II) transcript accumulation and *PI-II* is a JA-inducible PR gene (Seo et al. 1999). These observations suggest a role for the MAPK in the activation of JA-mediated signaling system.

EDR1 (ENHANCED DISEASE RESISTANCE 1) is an MAPKK kinase (MAPKKK), which functions at the top of an MAP kinase cascade. The *edr1* (*enhanced disease resistance 1*) gene encodes a putative MAPKKK. The PR-12 protein (PDF1.2) is an important pathogenesis-related protein involved in plant innate immune responses, and its expression is triggered by the JA signaling system (Jung et al. 2007; Oñate-Sánchez et al. 2007; Pré et al. 2008). The *PDF* genes are down-regulated in *edr1* mutants. *MYC2* encodes a basic helix-loop-helix leucine zipper transcription factor and negatively regulates JA-responsive defense genes (Lorenzo et al. 2004). *MYC2* is involved in the repression of *PDF1.2* expression, and *PDF1.2* was highly induced in *edr1myc2* double mutant (Hiruma and Takano 2011). It has been shown that EDR1 is critical for the expression of *PDF1.2* gene and the *MYC2*-encoded transcription factor represses *PDF1.2* expression. Inactivation of *MYC2* fully restored *PDF1.2* expression in *edr1* mutants (Hiruma and Takano 2011). It suggests that EDR1 cancels MYC2 function to regulate JA-inducible defense responses (Fig. 3.13). OsEDR1-suppressing knockout (KO) rice plants showed enhanced resistance against the bacterial blight pathogen *Xanthomonas oryzae* pv. *oryzae*. This resistance was associated with induced expression of JA-related defense genes (Shen et al. 2011). The results show that OsEDR1 negatively regulates JA signaling.

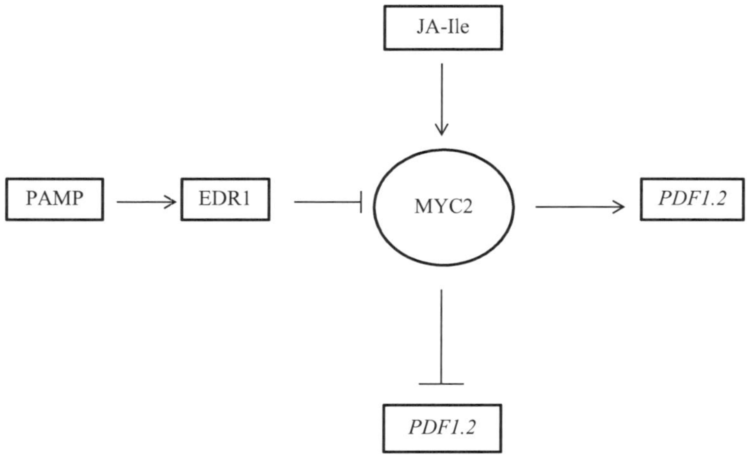

Fig. 3.13 EDR1 (MAPKKK) cancels MYC2 function to regulate JA-inducible *PDF1.2* gene expression (Adapted from Hiruma and Takano 2011)

3.10 Histone Acetylation May Regulate JA-Mediated Signaling Systems

Acetylation of the core histones in nucleosomes plays important role in gene regulation (Wu et al. 2008). Histone acetylation is modulated through the action of histone acetyltransferases and deacetylases (Zhou et al. 2005). Acetylation of the histones is often associated with increased gene activity, whereas deacetylation of histones is correlated with transcriptional repression. It suggests that the enzymatic modulation of histone acetylation is an integral component of transcription regulation (Berger 2002). Several histone deacetylatases (HDACs) have been identified in plants. The expression of a gene encoding a HDAC in *Arabidopsis*, *HDA19*, is induced by several signals (Zhou et al. 2005). Expression of jasmonic acid-regulated *PR2* and *PR3* genes was up-regulated in *35S:HDA19* transgenic plants but down-regulated in plants in which the expression of *HDA19* was silenced by RNA interference (RNAi) (Zhou et al. 2005). In another study, Wu et al. (2008) showed that the expression of the jasmonate-responsive genes, *PDF1.2*, *VSP2*, *JIN1*, and *ERF1*, was down-regulated in *HDA6* (another *HDAC* gene) mutant, *axe1-5*, and *HDA6*-RNAi *Arabidopsis* plants. Overexpression of *HDA19* in *35S:HDA19* plants decreased histone acetylation levels, whereas down-regulation of *HDA19* in *HDA19*-RNAi plants increased histone acetylation levels (Zhou et al. 2005). Similarly *Arabidopsis HDA6* mutant, *axe1-5*, and *HDA6*-RNAi plants displayed higher levels of deacetylated histone (Wu et al. 2008). These studies suggest that histone deacetylase may regulate JA-mediated signaling systems.

Arabidopsis HDA6 was shown to interact with COI1, an F-box protein which interacts with SKP1 and cullin proteins to form E3 ubiquitin ligases known as the SCF complexes that selectively recruit regulatory proteins targeted for ubiquitination (Devoto et al. 2002; Vierstra 2003). It suggests that COI1 may form a functional E3-type ubiquitin ligase in plants to regulate expression of jasmonate-responsive genes by targeted ubiquitination of a histone deacetylase (Devoto and Turner 2003; Zhou et al. 2005). Ubiquitin–proteasome-mediated proteolytic pathway may activate JA-inducible transcription factors. The COI1 activated four JA-inducible transcription factors, which include WRKY18, At1g74930, At3g53600, and AtMYC2 in *Arabidopsis*. These transcription factors positively regulate the jasmonate-mediated signaling system (Wang et al. 2008). These studies suggest that histone deacetylation may be involved in COI1-mediated activation of JA-inducible transcription factors.

3.11 JA-Induced Pep1 Peptide Amplifies JA Downstream Signaling to Induce JA-Responsive Genes

AtPep1 is a 23-aa peptide detected in *Arabidopsis*, and it activates downstream events leading to the transcription of the defensin *PDF1.2* gene, the marker gene activated in the JA signaling system (Huffaker et al. 2006). AtPep1 is cleaved from the C termini of precursor protein PROPEP1 that is induced by MeJA. MeJA induced a strong expression of the gene *PROPEP1* in *Arabidopsis* (Huffaker et al. 2006). AtPep1 induces the production of H_2O_2. The induction of both H_2O_2 and PDF1.2 by AtPep1 was blocked by first supplying the leaves with diphenyleneiodominium chloride (DPI), an inhibitor of NADPH oxidase (Huffaker et al. 2006). The NADPH oxidase-dependent H_2O_2 may induce the activation of the defensin *PDF1.2* gene. AtPep1 exhibits characteristics of an endogenous elicitor of defense response, and its major function is to amplify signaling that is initiated by JA (Huffaker et al. 2006). Orthologs of PROPEP1 have been identified in rice, wheat, barley, maize, potato, soybean, canola (*Brassica napus*), alfalfa (*Medicago sativa*), and poplar (*Populus balsamifera*) (Huffaker et al. 2006), suggesting that Pep1 peptide may participate in JA signaling system in a wide range of plants.

Both PEPR1 and PEPR2 were transcriptionally induced by the host-associated molecular pattern (HAMP) Pep peptides (Yamaguchi et al. 2010). Methyl jasmonate induced transcription of both *PEPR1* and *PEPR2* expression within 30 min, whereas methyl salicylate and 1-aminocyclopropan-1-carboxylic acid, an ethylene precursor, did not induce either PEPR1 or PEPR2 (Yamaguchi et al. 2010). The results suggest that JA signaling system may be involved in the activation of PEPRs. The receptors may be involved in the amplification of the innate immune response in *Arabidopsis* induced by Pep peptides (Yamaguchi et al. 2006).

3.12 Transcription Factors Acting Downstream of JA in Defense Signaling System

Several ERF, bHLH, WRKY, MYB, NAC, and bZIP transcription factors have been shown to act downstream of JA signaling system triggering expression of JA-responsive defense genes (Fig. 3.14; Memelink et al. 2001; Lorenzo et al. 2004; Lorenzo and Solano 2005; Zheng et al. 2006; Nurmberg et al. 2007; Wasternack 2007; Zander et al. 2010; Cheng et al. 2011; Le Hénanff et al. 2013). Transcription factors belonging to AP2/EREBP transcription factor family which are involved in inducing JA-responsive genes include ERF1, ERF2, ERF4, ERF5, ERF6, ERF14, ERF15, ERF59 (ORA59), ORA37, ORA47, and ERF98 (Gfeller et al. 2010; Zander et al.

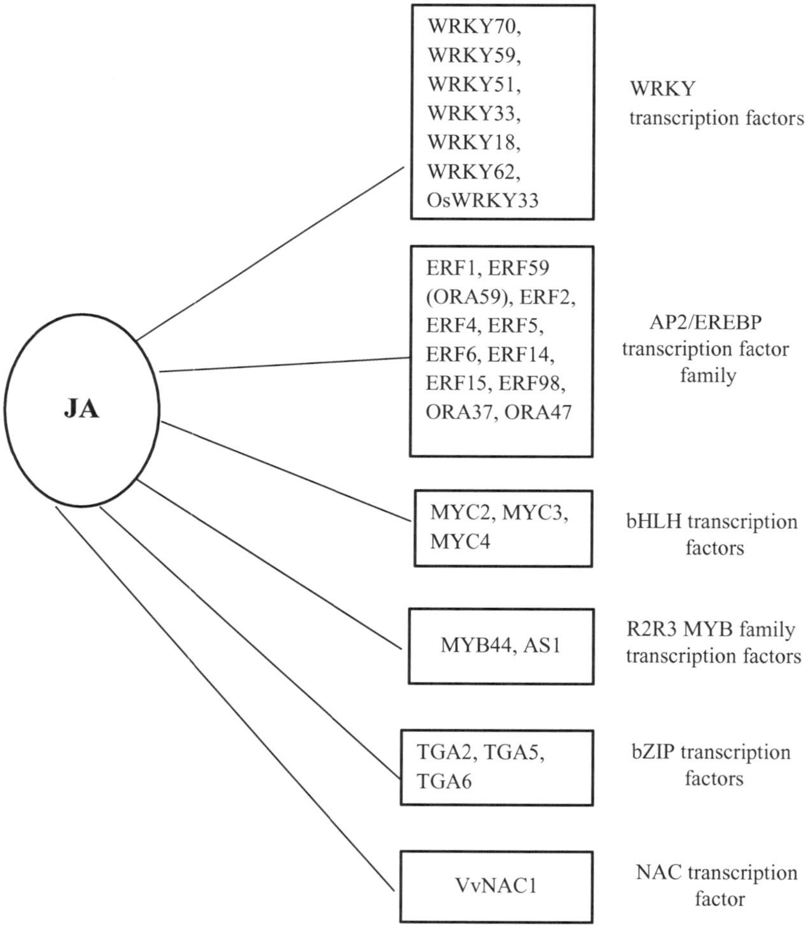

Fig. 3.14 Transcription factors acting downstream of JA in plant immune system

2010; Zarei et al. 2011; Cevik et al. 2012; Wager and Browse 2012; Wasternack and Hause 2013).

Ethylene response factor 1 (ERF1) is an important transcriptional factor regulating expression of defense genes in *Arabidopsis thaliana*. ERF1 is a downstream component of JA defense responses. Expression of *ERF1* was blocked in *coi1-1* mutant impaired in JA signal pathway (Berrocal-Lobo and Molina 2004). Other transcription factors, which act downstream of JA in defense signaling systems, include WRKY33, WRKY70, ORA59, and ORA37 (Memelink et al. 2001; Lorenzo et al. 2004; Lorenzo and Solano 2005; Zheng et al. 2006; Mao et al. 2007; Wasternack 2007). WRKY33 is a positive regulator of JA-mediated responses and confers resistance to necrotrophic fungal pathogens *Botrytis cinerea* and *Alternaria brassicicola* in *Arabidopsis*. *Arabidopsis* PAD4 is a regulatory gene inducing SA accumulation. The PAD4 expression and the SA-mediated PR genes expression are retarded by WRKY33, suggesting that WRKY33 is involved in the antagonism between SA- and JA-mediated signaling systems (Zheng et al. 2006). *Arabidopsis* transcription factor WRKY33 has been shown to mediate crosstalk between defense responses with different mechanisms of pathogenesis (Zheng et al. 2006). ORA47 is a COI1-dependent positive regulator of JA biosynthesis, whereas ORA59, ORA37, and WRKY70 act positively or negatively on different groups of defense genes (Wasternack 2007). The WRKY70 and WRKY53 transcription factors are involved in the crosstalk between JA and SA (Wang et al. 2006). Another transcription factor WRKY62 is involved in plant defense response (Kalde et al. 2003). Expression and function of the transcription factor WRKY62 was found to be induced by MeJA (Mao et al. 2007). The transcription factor from rice OsWRKY13 negatively regulates JA signaling system. It suppresses JA synthesis-related genes and JA-responsive genes (Qiu et al. 2007). A MYB transcription factor encoding gene *AS1* (for *ASYMMETRIC LEAVES1*) operates as negative regulator of JA-inducible defense genes by binding to the promoters of these genes (Nurmberg et al. 2007). The bHLH transcription factor MYC3 interacts with the jasmonate ZIM-domain proteins to mediate jasmonate response in *Arabidopsis* (Cheng et al. 2011).

The *Arabidopsis* class II TGA transcription factors TGA2, TGA5, and TGA6 are essential for the activation of JA- and ET-dependent defense mechanisms (Zander et al. 2010). Grapevine (*Vitis vinifera*) NAC1 transcription factor (*Vv*NAC1) expression is stimulated by *Botrytis cinerea* infection or by microbe-associated molecular pattern (MAMP) treatment. Methyl jasmonate treatment also activated the expression of *VvNAC1* and induced the expression of JA-responsive *PDF1.2* gene in grapevine plants (Le Hénanff et al. 2013).

Coronatine (COR) mimics the plant hormone jasmonic acid-isoleucine (JA-Ile). The expression of three homologous NAC (petunia *N*AM and *Arabidopsis* *A*TAF1, ATAF2, and *C*UC2) transcription factors is induced by COR (JA-Ile) directly through MYC2 in JA-Ile-COI1-JAZ-MYC2 module (Zheng et al. 2012). MYC2, MYC3, and MYC4 are JAZ-interacting transcription factors that regulate JA responses (Niu et al. 2011).

3.13 JA Signaling System-Activated Defense Genes

Several defense genes are induced by JA (Fig. 3.15). MeJA treatment induced eight defense genes consisting of *PR1.1*, *PR2* (β-1,3-glucanase), *PR3* (chitinase), *PR4* (wheatwin), *PR5* (thaumatin-like protein), *TaPERO* (peroxidase), *PR10*, and *TaGLP2a* (germin-like) in wheat (Desmond et al. 2006). JA induces β-1,3-glucanase

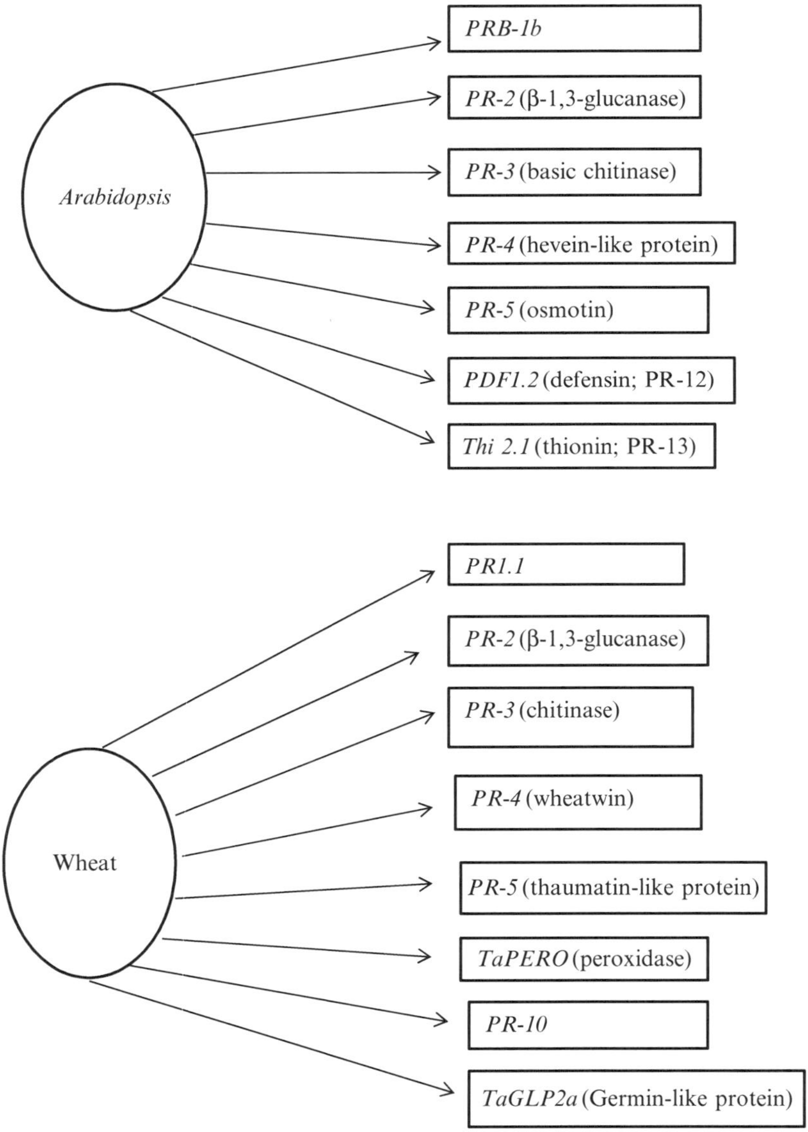

Fig. 3.15 JA-responsive genes in *Arabidopsis* and wheat

and thaumatin-like proteins in wheat (Jayaraj et al. 2004). JA/ET signaling systems activate the transcription of several defense-related *PR* genes including *PR-1*, *PRB-1b*, *PR-2* (β-1,3-glucanase), *ChiB* (PR-3; basic chitinase), *PR-4* (hevein-like protein), *PR-5* (osmotin), *PDF1.2* (PR-12; defensin), *Thi2.1* (PR-13; thionin), and germin-like proteins in *Arabidopsis* (Zhou et al. 2005; Cao et al. 2006; Johansson et al. 2006; Jung et al. 2007; Oñate-Sánchez et al. 2007; Pré et al. 2008). JA induced the defensin gene *PDF1.2* expression in *Arabidopsis* (Lee et al. 2007). Jasmonate is involved in the expression of *Thi2.1* gene encoding thionin and *CHI-B* gene encoding basic chitinase in *Arabidopsis* (Ellis and Turner 2001).

PR4 is induced by JA in rice (Agrawal et al. 2003) and maize (Bravo et al. 2003). Repka et al. (2004) reported that MeJA systemically induced accumulation of PR1, β-1,3-glucanase, chitinase, peroxidase, and extensin in the upper untreated grapevine leaves 14 days after treatment of lower leaves with MeJA. MeJA treatment induced *pin1* gene encoding a proteinase inhibitor in *Brachypodium distachyon* (Mur et al. 2004). Expression of the basic *PR-6* and *TPI-1* genes that encode proteinase inhibitor II and I, respectively, was found to be dependent on JA-mediated signaling pathway in tomato (Hondo et al. 2007). MeJA induced the thaumatin-like protein of the PR-5 family in *Pinus monticola*, while SA did not induce it (Piggott et al. 2004).

MeJA induces the expression of sesquiterpene cyclase genes involved in the biosynthesis of sesquiterpene phytoalexins in tobacco (Mandujano-Chávez et al. 2000). The biosynthesis of a phytoalexin β-thujaplicin, in *Cupressus lusitanica* cell cultures, is stimulated by MeJA (Zhao and Sakai 2003). JA signaling is involved in triggering the phytoalexin camalexin biosynthesis in *Arabidopsis* (Berr et al. 2010). MeJA induces stilbene accumulation in grapevine leaves and berries (Larronde et al. 2003). MeJA induced callose deposition in leaves and cell cultures of grapevine (Repka et al. 2004).

3.14 JA Signaling System Triggers Immune Responses Against Necrotrophic Pathogens

JA signaling system appears to confer resistance against diseases caused by necrotrophic pathogens. Application of N-isobutyl decanamide, the most highly active alkamide, significantly reduced necrosis caused by *Botrytis cinerea* in *Arabidopsis thaliana*. The alkamide enhanced the expression of genes encoding enzymes for jasmonic acid (JA) biosynthesis (Méndez-Bravo et al. 2011). The alkamide was ineffective in conferring resistance against the pathogen in the *Arabidopsis jar1* and *coi1* mutants which are altered in JA signaling (Méndez-Bravo et al. 2011). The *tga256* (the TGA transcription factors TGA2, TGA5 and TGA6) triple mutant is impaired in JA-induced *PDF1.2* and *b-CHI* gene expression, and this mutant showed higher susceptibility against *B. cinerea* (Zander et al. 2010). These results suggest that intact JA signaling system is necessary to induce resistance against the necrotrophic pathogen *B. cinerea.*

JA signaling system confers resistance to the necrotrophic fungal pathogens *Alternaria brassicicola* (Zheng et al. 2006; Flors et al. 2008) and *Botrytis cinerea*

(Korolev et al. 2008; Pré et al. 2008). JA signaling systems induced defense gene expression and conferred resistance against the necrotrophic wilt pathogens *Fusarium oxysporum* (Anderson et al. 2004) and *Verticillium longisporum* in *Arabidopsis* (Johansson et al. 2006). JA-mediated defense signaling system plays an important role in conferring resistance to *Sclerotinia sclerotiorum* in oilseed rape (Wang et al. 2009).

JASMONATE ZIM-DOMAIN (JAZ) proteins are negative regulators of JA signaling (Demianski et al. 2012). Jasmonyl-L-isoleucine (JA-Ile)-mediated processes contribute to *Pseudomonas syringae* disease susceptibility in Arabidopsis. The transcription factor JASMONATE INSENSITIVE1 (JIN1/AtMYC2) has been implicated in the regulation of JAZ gene expression. Eight of the 12 JAZ genes were induced during the infection of the bacterial pathogen of *A. thaliana*. A T-DNA insertion mutant and an RNA interference line disrupted for the expression of JAZ10 exhibited enhanced susceptibility and enhanced JA sensitivity. The results suggest that JAZ10 is a negative regulator of both JA signaling and disease development (Demianski et al. 2012).

A JA-dependent pathway has been shown to be effective against the necrotrophic bacterial pathogen *Erwinia carotovora* (Thomma et al. 2001). JA-mediated defense appears to be involved in conferring resistance against some oomycete pathogens. JA-mediated signaling system confers resistance against various *Pythium* spp. (Thomma et al. 2001) and against *Phytophthora capsici* in pepper (Ueeda et al. 2006). JA is involved in the induction of resistance against the damping-off oomycete pathogen *Pythium irregulare* in Arabidopsis. The insensitive *coi1* mutants in which JA signaling is suppressed show extreme susceptibility to *P. irregulare* (Adie et al. 2007).

The Mediator complex subunit 25 induces expression of JA-dependent defense genes and triggers resistance to the necrotrophic fungal pathogens *Alternaria brassicicola* and *Botrytis cinerea* in Arabidopsis (Kidd et al. 2009; An and Mou 2013). Suppression of a lipoxygenase gene (*GmLOX3*)-mediated JA signaling pathway in a *lox3* maize mutant induced enhanced susceptibility to the maize seed-infecting *Aspergillus* spp. (Gao et al. 2009). Collectively these studies suggest that JA signaling is involved in triggering resistance against necrotrophic pathogens.

3.15 JA and Ethylene Signaling Pathways May Operate Concomitantly in Plant Innate Immune System

3.15.1 Cooperative Function of JA and ET Signaling Pathways in Plant Innate Immunity

JA and ET pathways appear to function cooperatively in modulating plant immune responses (Anderson et al. 2004; Zander et al. 2010; Leon-Reyes et al. 2010b; Robert-Seilaniantz et al. 2011; Sharon et al. 2011; Martinez-Medina et al. 2013). Both JA and ET signaling are required for the expression of the defense-related gene PDF1.2 (Kunkel and Brooks 2002; Zhu et al. 2011; Po-Wen et al. 2013). JA and ET signaling pathways mediate various defense responses, either independently or collaboratively (Dong 1998). JA and ET pathways interact with each other coregulating expression of defense genes (Wang et al. 2002). JA and ET cooperatively regulate defense responses. JA stimulates proteinase inhibitor gene expression while Et alone does not, but JA and ET induce proteinase inhibitor genes synergistically in tobacco (O'Donnell et al. 1996). MeJA-induced β-thujaplicin accumulation in *Cupressus lusitanica* cell cultures can be partly blocked by inhibitors of ethylene biosynthesis and signaling, while blocking JA biosynthesis inhibits almost all ET-induced β-thujaplicin accumulation (Zhao et al. 2004). These results suggest that the ET and JA pathways interact in mediating β-thujaplicin production, with the jasmonate pathway working as a main control and the ET pathway as a fine modulator for β-thujaplicin accumulation.

ET positively regulates the induction of allene oxide synthase (AOS), which catalyzes the first step in the biosynthesis of JA (O'Donnell et al. 1996), while MeJA induced a rapid increase in 1-aminocyclopropane-1-carboxylic acid oxidase (ACO), resulting in increased synthesis of ET (Hudgins and Franceschi 2004). These results suggest that ET induces JA biosynthesis, while JA triggers ET biosynthesis and both JA and ET signaling pathways are interlinked.

3.15.2 ERF Transcription Factors May Concurrently Modulate JA and ET Signaling Pathways in Plant Immune System

Ethylene response factors (ERF) belonging to the APETELA2 (AP2)/ETHYLENE RESPONSIVE ELEMENT BINDING PROTEIN (EREBP) transcription factor family are the important group of transcription factors functioning downstream in ethylene signaling system (Berrocal-Lobo and Molina 2004; Gutterson and Reuber 2004; Grennan 2008). ERF genes have been shown to be responsive to both JA and ethylene (Lorenzo et al. 2003; Gutterson and Reuber 2004; McGrath et al. 2005). The transcription factor ERF, which regulates ET-mediated gene expression, also regulates JA-responsive gene expression (Brown et al. 2003; McGrath et al. 2005; Pieterse et al. 2009).

ERF1 is an important ERF transcription factor involved in JA and ET signaling pathways. JA signaling system confers resistance to necrotrophic fungal pathogens *Botrytis cinerea* (Korolev et al. 2008; Pré et al. 2008; Zander et al. 2010) and susceptibility to the hemibiotrophic bacterial pathogen *Pseudomonas syringae* (Demianski et al. 2012). Overexpression of *ERF1* enhances resistance against *B. cinerea* and other necrotrophic pathogens and increases susceptibility to the hemibiotroph *Pseudomonas syringae* (Berrocal-Lobo et al. 2002; Berrocal-Lobo and Molina 2004). The results suggest that ERF1 concurrently modulates the JA and ET signaling systems in the plant immune system. JA and ET pathways may converge in the transcriptional activation of *ERF1*, which encodes a transcription factor involved in the expression of defense genes. The expression of *ERF1* can be activated rapidly by ET or JA and can be activated synergistically by both hormones (Lorenzo et al. 2003). ERF1 is synergistically regulated by the ET and JA pathways (Gimenez-Ibanez and Solano 2013). In addition, both signaling pathways are required simultaneously to activate *ERF1*, because mutations that block any of them prevent *ERF1* induction by any of these hormones either alone or in combination (Lorenzo et al. 2003). Overexpression of *ERF1* expression could rescue the defense response defects of *coi1* and *ein2* mutants. ERF1 regulates in vivo the expression of a large number of genes responsive to both ET and JA (Lorenzo et al. 2003). These results suggest that ERF1 acts downstream of the intersection between ET and JA pathways. The transcription factor ERF1 acts as an integrator of JA and ET signaling pathways in *Arabidopsis* (Lorenzo et al. 2003). Constitutive overexpression of *ERF1* activates the expression of several JA-responsive defense-related genes, including *PDF1.2*, and basic chitinase (*ChiB*) (Lorenzo et al. 2003) and was shown to confer resistance to *B. cinerea* (Berrocal-Lobo et al. 2002).

The other ERF transcription factors involved in JA/ET signaling systems include ORA59, ERF5, ERF6, ERF2, and ERF4. ORA59 has been shown to be an essential integrator of the JA and ET signal transduction pathways. JA- and ET-responsive expression of several defense genes, including the defensin gene *PDF1.2*, depended on ORA59 (Pré et al. 2008). Constitutive overexpression of *AtERF2* or *AtERF4*, encoding ERF transcription factors, was reported to cause high levels of JA-responsive *PDF1.2* and *ChiB* gene expression in transgenic *Arabidopsis* plants (McGrath et al. 2005; Oñate-Sánchez et al. 2007). ERF5 and ERF6 play important roles as positive regulators of JA-/ET-mediated defense against *B. cinerea* in Arabidopsis (Moffat et al. 2012). Expression of JA-/ET-responsive genes and genes containing the GCC-box promoter motif was significantly up-regulated in both ERF5 and ERF6 transgenic plants, suggesting that ERF5 and ERF6 may act as positive regulators of JA-mediated defense (Moffat et al. 2012). Collectively these studies suggest that ERF transcription factors play important role in the concurrent expression of JA and ET signaling systems.

3.15.3 Role of Ethylene Transcription Factors EIN3 and EIL1 in JA/ET Signaling Synergy

Ethylene (ET), which is perceived by receptors, represses a Raf-like kinase CONSTITUTIVE ETHYLENE RESPONSE1 (CTR1). Downstream of CTR1 is ETHYLENE INSENSITIVE2 (EIN2), which is an essential positive regulator in ET signaling. ETHYLENE INSENSITIVE3 (EIN3) and its closest homolog EIN3-LIKE1 (EIL1) are the primary transcription factors downstream of EIN2, which are necessary and sufficient for the induction of the vast majority of ethylene response genes (An et al. 2010). In the absence of ET, EIN3 and EIL1 are quickly degraded through the action of two EIN3-binding F-box proteins 1 and 2 (EBF1 and EBF2).

ET enhances EIN3/EIL1 accumulation through the repression of EBF1/EBF2 (An et al. 2010).

EIN3 and EIL1 have been shown to be positive regulators of JA responses (Zhu et al. 2011). The JA induction of pathogenesis-related genes (*ERF1*, *ORA59*, and *PDF1.2*) was abolished in *ein3 eil1* mutant upon JA treatment (Zhu et al. 2011). JAZ proteins play important role in JA signaling (Chung and Howe 2009; Howe 2010; Sheard et al. 2010; Pauwels and Goossens 2011; Wager and Browse 2012; Wasternack and Hause 2013). JAZ-associated transcription factors are involved in integrating JA and ET signaling. Under normal conditions, EIN3 and EIL1 are repressed by JAZs, which recruit HDA6 as a corepressor to decrease EIN3/EIL1 protein stability (Guo et al. 2003; Zhou et al. 2005; An et al. 2010; Zhu et al. 2011). ET treatment enhances EIN3/EIL1 protein stability, while JA treatment leads to JAZ degradation and transcriptional derepression. When JA signaling is blocked in *coi1* mutant, JAZs highly accumulate and strongly repress EIN3/EIL1 function (Zhu et al. 2011). In JA signaling pathway, JAZ proteins are linked to the bHLH transcription factors MYC2, MYC3, and MYC4, while in ET signaling pathway, EIN3/EIL1 are JAZ-targeted transcription factors (Zhu et al. 2011). EIN3 and EIL1 physically interact with a number of JAZ proteins including JAZ1, JAZ3, and JAZ29, and JAZ proteins use the same module containing Jas domain to interact with EIN3/EIL1 and MYC2 (Chini et al. 2007; Zhu et al. 2011). These results suggest that EIN3/EIL1 may be the key integration node in JA and ET signaling synergy and the two ET transcription factors EIN3 and EIL1 integrate ET and JA signaling in the regulation of defense gene expression.

Crosstalk between ET and JA signaling occurs in many plant immune responses (Bari and Jones 2009; Grant and Jones 2009; Robert-Seilaniantz et al. 2011). Part of the crosstalk may be mediated through the connection of JAZ proteins with the transcription factors EIN3 and EIL1. The transcription regulators EIN3 and EIL1 are important positive regulators of ET signaling system (Zhu et al. 2011). JAZ1, JAZ3, and JAZ9 are known to bind EIN3 and EIL1 (Pauwels and Goossens 2011; Zhu et al. 2011). The JAZ proteins can repress the function of EIN3/EIL1, probably by suppressing the DNA binding of EIN3 (Zhu et al. 2011). ET is needed for EIN3/EIL1 stabilization and JA for EIN3/EIL1 release from the JAZ protein repression, and this may be responsible for the synergism between ET and JA signaling systems (Zhu et al. 2011).

3.15.4 Ethylene Signaling System May Protect JA Signaling System Against Its SA-Mediated Suppression

The action of JA and ET signaling is synergistic and mutually dependent in regulating defense responses against plant pathogens (Penninckx et al. 1996; Lorenzo et al. 2003; Gimenez-Ibanez and Solano 2013). Synergism between JA and ET signaling systems may fine-tune other signaling pathways to trigger defense responses against pathogens. Leon-Reyes et al. (2010a) showed that ethylene signaling may render JA signaling insensitive to subsequent suppression by SA. Exogenous application of SA is known to suppress expression of JA-responsive genes, such as *PDF1.2* and *VSP2* (Koornneef et al. 2008; Leon-Reyes et al. 2009). An *Arabidopsis* mutant *cev1*, which displays constitutive expression of JA and ET responses, appeared to be insensitive to SA-mediated suppression of the JA-responsive marker genes *PDF1.2* and *VSP2*. Similarly, induction of the JA and ET response upon inoculation with the necrotrophic pathogens *Botrytis cinerea* and *Alternaria brassicicola* resulted in insensitivity of the leaf tissue to the suppression of the JA response by SA (Leon-Reyes et al. 2010a). Application of high concentrations of ACC, the precursor of ET, neutralized the antagonistic effect of SA on JA-responsive gene expression. In the *PDF1.2*-overexpressing mutant *cev1*, *ORA59* was constitutively expressed. Overexpression of ORA59 in stable *ORA59* transformants resulted in constitutive *PDF1.2* expression, which could not be suppressed by SA treatment (Leon-Reyes et al. 2010a). These results suggest that ORA59 plays an important role in counteracting SA-mediated suppression of JA-/ET-responsive defense genes.

3.16 JA Signaling May Suppress SA Signaling System

SA signaling system is known to induce resistance against the bacterial pathogen *P. syringae* pv. *tomato* in *Arabidopsis* (Brooks et al. 2005; Jelenska et al. 2007). When *Arabidopsis* plants were inoculated with *P. syringae* pv. *tomato*, there was substantial increase in *WRKY33* transcript level (Dong et al. 2003; Zheng et al. 2006). WRKY33 induces JA-mediated signaling system and the induced JA signaling system reduces the SA-mediated signaling system by its antagonistic function (Zheng et al. 2006; van Verk et al. 2008). It suggests that the pathogen suppresses the SA signaling pathway, which is needed to induce resistance against the pathogen.

In the infection process, pathogen may suppress the particular signaling system which may be able to induce the expression of specific defense genes conferring resistance against the particular pathogen. Activation of SA-mediated pathway appears to confer resistance to *Pseudomonas syringae* pv. *tomato* DC3000. The pathogen produces the toxin coronatine, which induces JA signaling pathway while suppressing the SA-dependent pathway (Katsir et al. 2008). It suggests that the pathogen is able to suppress the signaling pathway which may suppress its pathogenesis and diverts the signals for the activation of an inefficient pathway resulting in defense response that cannot inhibit its pathogenesis (Spoel et al. 2003, 2007; Uppalapati et al. 2007). The PR-1 protein is required for the inhibition of *P. syringae* pv. *tomato* (Laurie-Berry et al. 2006), and induction of this pathogenesis-related protein is dependent on SA accumulation (van Verk et al. 2008). The toxin produced by the pathogen triggers JA signaling, which suppresses SA-mediated defenses. Thus, it appears that the bacterial pathogen utilizes the toxin to manipulate JA signaling both to suppress SA-mediated responses and to promote symptom development (Brooks et al. 2005; Laurie-Berry et al. 2006; Koornneef and Pieterse 2008).

Crosstalk between SA and JA/ET signaling systems has been reported in different plant–pathogen interactions. The tobacco basic PR protein gene *NtPRB1b* responds positively to JA and ET signaling systems but negatively to SA (Li et al. 2012). The MAP kinase MPK4 plays a role in the antagonism between SA and JA signaling systems in *Arabidopsis thaliana*. The *mpk4* knockout plants exhibit constitutive activation of SA-dependent signaling system but fail to induce JA-dependent signaling system (Petersen et al. 2000; Brodersen et al. 2006).

SA signaling system is known to induce resistance against the bacterial pathogen *P. syringae* pv. *tomato* in *Arabidopsis* (Brooks et al. 2005; Jelenska et al. 2007). When *Arabidopsis* plants were inoculated with *P. syringae* pv. *tomato*, there was substantial increase in *WRKY33* transcript level (Dong et al. 2003; Zheng et al. 2006). WRKY33 induces JA-mediated signaling system and the induced JA signaling system reduces the SA-mediated signaling system by its antagonistic function (Zheng et al. 2006; van Verk et al. 2008). It suggests that the pathogen suppresses the SA signaling pathway, which is needed to induce resistance against the pathogen.

JA may inhibit SA signaling. In tobacco, MeJA induced the transcription of a glucosyltransferase, a homolog of UDP-glucose/salicylic acid glucosyltransferase in tobacco (Lee and Raskin 1999) and *Arabidopsis* (Sasaki et al. 2001). It is suggested that MeJA inhibits salicylic acid signaling through the induction of the glucosyltransferase, because the sugar conjugate of salicylic acid is inactive (Sasaki et al. 2001).

3.17 Suppression of JA Signaling by SA Signaling System

3.17.1 SA Suppresses Biosynthesis of JA

Plants infected by SA-inducing biotrophic pathogens often suppress JA-dependent defense responses (Spoel et al. 2007). The elicitation of the JA pathway by pathogens can repress the SA response (Uppalapati et al. 2007). SA may suppress the biosynthesis of JA (Mur et al. 2006). SA has been suggested to suppress the expression of the JA biosynthetic enzymes lipoxygense (Spoel et al. 2003) and allene oxide synthase (Laudert and Weiler 2002). A large increase in JA levels has been reported in pathogen-challenged *Arabidopsis* NahG transgenic lines in which SA is suppressed (Spoel et al. 2003). The *Arabidopsis* mutants, which accumulate high levels of SA, show repression of JA-mediated pathway (Kachroo et al. 2003). SA strongly antagonized the JA signaling pathway, resulting in the down-regulation of a large set of JA-responsive genes, including the marker genes *PDF1.2* and *VSP2* in *Arabidopsis thaliana* (Leon-Reyes et al. 2010b). In the *Arabidopsis aos/dde2* mutant, the expression of the JA biosynthesis enzyme allene oxide synthase (AOS) was completely blocked. Mutant *aos/dde2* plants did not express the JA-responsive marker genes *PDF1.2* or *VSP2* in response to infection with the necrotrophic fungus *Alternaria brassicicola*. Bypassing JA biosynthesis by exogenous application of

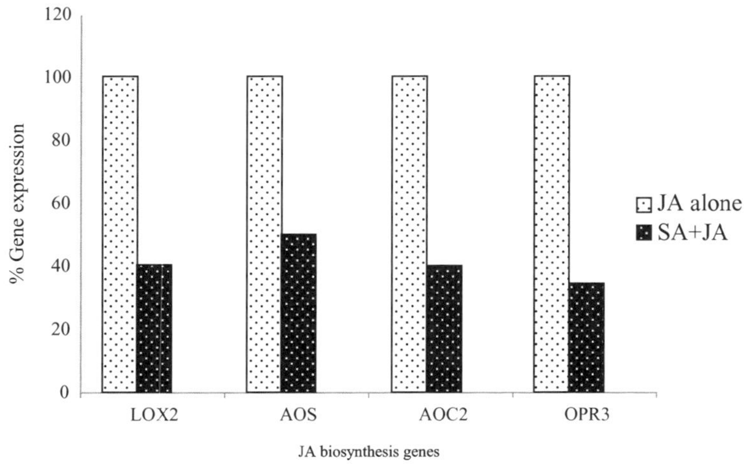

Northern blot analysis of transcript levels of the JA biosynthesis genes was carried out. Signal intensities of the Northern blots were quantified. The transcript levels in the single methyl jasmonate (JA) treatments were set to 100%.

Fig. 3.16 Suppression of JA biosynthesis genes by SA (Adapted from Leon Reyes et al. 2010b)

methyl jasmonate (MeJA) rescued this JA-responsive phenotype in *aos/dde2*. Application of SA suppressed MeJA-induced *PDF1.2* expression to the same level in the *aos/dde2* mutant as in wild-type plants (Leon-Reyes et al. 2010b). The results suggest that SA-mediated suppression of JA-responsive gene expression is targeted at a position downstream of the JA biosynthetic pathway.

Activation of genes encoding key enzymes in the JA biosynthesis pathway, such as *LOX29* (*LIPOXYGENASE2*), *AOS* (*ALLENE OXIDE SYNTHASE*), *AOC2* (*ALLENE OXIDE CYCLASE2*), and *OPR3* (12-OXO-PHYTODIENOATE REDUCTASE3), was repressed by SA (Fig. 3.16; Leon-Reyes et al. 2010b).

3.17.2 SA Suppresses JA Signaling System by Targeting GCC-Box Motifs in JA-Responsive Promoters

JA signaling pathway downstream of the COI1-JAZ receptor complex consists of two major branches: the MYC branch and the ERF branch. The MYC branch is controlled by MYC-type transcription factors that can bind to the G-box motif (CACGTG) and regulate expression of the marker gene VSP2 (Fernández-Calvo et al. 2011; Niu et al. 2011; Van der Does et al. 2013). The ERF branch is regulated by the members of the AP2/ERF family of transcription factors that bind to the GCC-box motif (AGCCGCC), such as ERF and ORA59, and regulate expression of the marker gene PDF1.2 encoding the pathogenesis-related defensin (PR-12

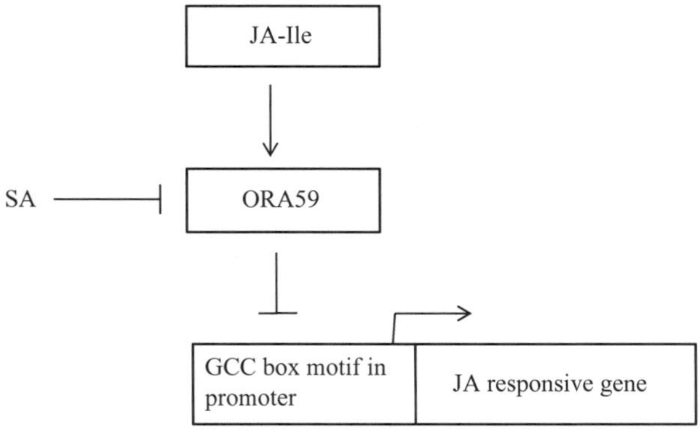

Fig. 3.17 SA suppresses JA signaling by targeting GCC-box motifs in JA-responsive promoters via a negative effect on the accumulation of ORA59 (Adapted from Van der Does et al. 2013)

protein) (Lorenzo et al. 2003; Pré et al. 2008; Pieterse et al. 2009; Gao et al. 2011; Zhu et al. 2011; Gimenez-Ibanez and Solano 2013; Wasternack and Hause 2013). The ERF branch is known to be associated with enhanced resistance to necrotrophic pathogens, whereas the MYC branch is associated with defense against insect pests (Verhage et al. 2011). SA has been reported to suppress the ERF branch pathway in the JA signaling system (Van der Does et al. 2013).

SA suppresses JA signaling downstream of the COI1–JAZ receptor complex by targeting GCC-box motifs in JA-responsive promoters via a negative effect on the accumulation of the AP2/ERF-type transcriptional activator ORA59. The GCC-box motif is overrepresented in JA-responsive promoters that are suppressed by SA, and this promoter motif is sufficient for SA-mediated suppression of JA-induced gene expression (Fig. 3.17; Van der Does et al. 2013). SA reduces the accumulation of the GCC-box binding TF ORA59, indicating that the antagonistic effect of SA on JA signaling is controlled at the level of transcriptional regulation, through the modulation of the TF levels (Van der Does et al. 2013).

3.17.3 *Role of WRKY62 Transcription Factor in the Suppression of JA Signaling by SA*

Salicylic acid has been shown to have a role in the suppression of JA signaling pathway. *NPR1* (*NONEXPRESSER OF PR GENES1*) is the transcription cofactor involved in the suppressive action of SA against JA signaling (Spoel et al. 2003, 2009). NPR1 forms a high-molecular-weight oligomeric complex in the cytosol. SA accumulation causes changes in cellular redox potential and it leads to the formation of NPR1 monomers that are translocated into the nucleus. However, the NPR1 homomeric complex remains in the cytoplasm (Mou et al. 2003). SA-induced suppression of the JA response is controlled by the function of NPR1 in the cytosol (Spoel et al. 2003; Koornneef and Pieterse 2008).

The NPR1 5'UTR contains several W-boxes, which are binding sites for WRKY transcription factors (Wasternack 2007). Some of the NPR1-activated transcription factors are involved in suppressing the activation of JA-responsive genes. The transcription factor WRKY62 is synergistically induced by MeJA and SA (Mao et al. 2007). MeJA inducibility of *WRKY62* is mediated by NPR1. Analysis of *wrky62* mutant and WRKY62-overexpressing plants showed that *WRKY62* suppresses JA-responsive gene expression (Mao et al. 2007). *WRKY62* may function downstream of cytosolic NPR1 and regulate SA-mediated inhibition of JA signaling (Mao et al. 2007).

3.17.4 Role of WRKY70 and MYB Transcription Factors in the Suppression of JA Signaling by SA

WRKY70 is a common transcription factor in SA- and JA-mediated signaling pathways. It acts downstream of SA and *NPR1* (Knoth et al. 2007) and inhibits JA-responsive genes (Li et al. 2004). *WRKY70* mRNA levels are enhanced in a *coi1* background, suggesting that JA represses its expression (Li et al. 2004). An R2R3 MYB transcription factor, AtMYB44, modulates antagonistic interaction between JA and SA signaling (Shim et al. 2013). AtMYB44 is induced by JA through COI1. AtMYB44 up-regulates WRKY70 expression, which is induced by SA. Induction of WRKY70 by SA is reduced in *atmyb44* and *npr1-1* mutants and is totally abolished in *atmyb44 npr1-1* double mutants, showing that WRKY70 is regulated through both NPR1 and AtMYB44. *AtMYB44* overexpression up-regulates SA-responsive *PR1* gene expression, while it suppresses JA-responsive *PDF1.2* expression (Shim et al. 2013). AtMYB44 overexpression down-regulated defense responses against the necrotrophic pathogen *Alternaria brassicicola*, but up-regulated *WRKY70* and SA-responsive *PR* genes, leading to enhanced resistance to the biotrophic pathogen *Pseudomonas syringae* pv. *tomato* DC3000 (Shim et al. 2013). The *wrky70* mutation suppressed AtMYB44-dependent up-regulation of SA-responsive *PR1* gene expression and down-regulation of JA-responsive *PDF1.2* expression. AtMYB44 binds to the *WRKY* promoter region, indicating that AtMYB44 acts as a transcriptional activator of WRKY70 by directly binding to a conserved sequence element in the *WRKY70* promoter. These results suggest that AtMYB44 modulates antagonistic interaction between SA and JA signaling systems through direct control of WRKY70 (Fig. 3.18; Shim et al. 2013).

3.17.5 WRKY50 and WRKY51 Transcription Factors May Modulate JA Signaling Suppression by SA

Several WRKY transcription factors have shown to be involved in the suppression of JA signaling by SA (Pieterse et al. 2012a, b). SA-mediated suppression of JA-induced *PDF1.2* gene expression was abolished in *wrky50* single and *wrky50 wrky51* double knockout mutants. Exogenous SA inhibited *PDF1.2* expression in

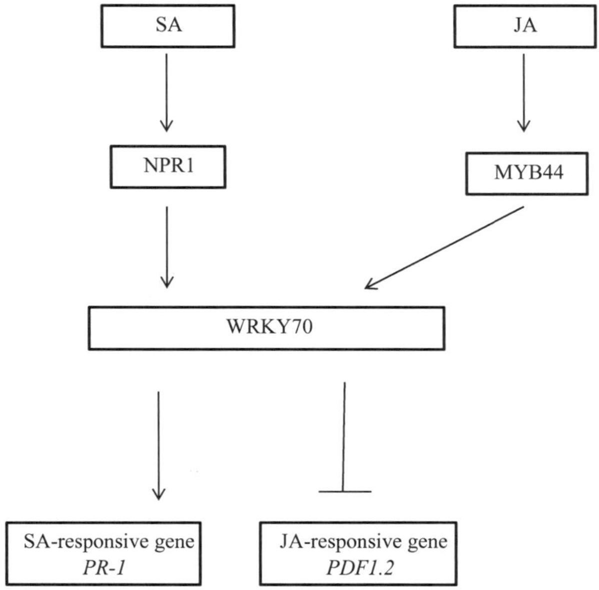

Fig. 3.18 Role of MYB44 and WRKY70 in regulation of suppression of JA signaling by SA signaling (Adapted from Shim et al. 2013)

the wild type but not in *wrky50* and *wrky51* mutant plants (Gao et al. 2011). These results suggest that the WRKY50 and WRKY51 proteins mediate SA-dependent repression of JA signaling.

3.17.6 Role of TGA Transcription Factors in the Suppression of JA Signaling by SA

Basic/leucine zipper-type transcription factors of the TGA family have been reported to regulate several SA-responsive genes (Gatz 2013). They interact with NPR1 to regulate SA-mediated suppression of JA. Expression of GRX480, a member of the glutaredoxin family, was found to be inducible by SA and the induction was dependent on NPR1. Glutaredoxins mediate redox regulation of TGA1 and NPR1 under inducing conditions. TGA transcription factors play important roles in SA-mediated gene expression. Arabidopsis plants lacking TGA factors TGA2, TGA5, and TGA6 were severely compromised in SA-responsive gene expression after treatment with the SA analog 2,6-dichloroisonicotinic acid (Zhang et al. 2003). Transgenic *Arabidopsis* plants ectopically expressing *GRX480* showed near wild-type expression of SA-inducible responses, while transcription of JA-responsive gene was suppressed in the transgenic plants. Suppression of JA-inducible gene depended on the presence of TGA transcription factors. The interaction between GRX480 and TGA transcription factors was essential for the GRX480-dependent suppression of JA signaling by SA. It is suggested that SA-activated NPR1 induces

GRX480, which in turn interacts with TGA transcription factors to suppress JA-responsive gene transcription (Ndamukong et al. 2007).

The bZIP transcription factors TGA2, TGA5, and TGA6 have been shown to be required for the SA-mediated suppression of JA signaling pathway in *A. thaliana* (Zander et al. 2010). JA-induced *PDF1.2* expression was strongly suppressed by SA in wild-type plants. No such negative effect of SA on PDF1.2 expression was observed in the *tga256 jin1* quadruple mutant. The results suggest that the antagonistic effects of TGA factors and JIN1/AtMYC2 on the JA pathway are necessary to evoke the SA-mediated suppression of JA-induced defense responses (Zander et al. 2010).

3.18 Interplay Between JA and Abscisic Acid Signaling Systems in Plant Immune Responses

A complex interplay between ABA and JA signaling pathways has been shown to regulate plant defense gene expression in *Arabidopsis* (Anderson et al. 2004). Mutations in the *ABA1* and *ABA2* genes, which encode enzymes involved in ABA biosynthesis, resulted in increased transcription of JA-responsive defense genes in *Arabidopsis* (Anderson et al. 2004). ABA is an essential signal for plant resistance to pathogens affecting the JA biosynthesis and activation of defenses in *Arabidopsis* (Adie et al. 2007). ABA treatment induced accumulation of JA, which resulted in reduced SA defense gene expression in Arabidopsis (Adie et al. 2007). MeJA stimulates ABA biosynthesis (Kim et al. 2009; Yoon et al. 2009). ABA suppressed JA-activated transcription of defense genes (Anderson et al. 2004). By contrast, ABA deficiency resulted in up-regulation of transcription of JA-responsive defense genes. Disruption of *AtMYC2*, which is a positive regulator of ABA signaling, resulted in elevated levels of transcription from JA-responsive defense genes in *Arabidopsis*. These results suggest the antagonistic interaction between ABA and JA signaling pathways (Anderson et al. 2004). *Pseudomonas syringae* pv. *tomato* infection-induced JA accumulation was reduced in ABA-deficient *aba3-1* mutant plants compared to the wild type (Fan et al. 2009). The results suggest that ABA promotes JA accumulation in Arabidopsis.

A tobacco gene from the PYR/PYL/RCAR family, *NtPYL4*, the expression of which is regulated by jasmonates, was found to encode a functional ABA receptor. NtPYL4 inhibited the type 2C protein phosphatases known to be key negative regulators of ABA signaling in an ABA-dependent manner (Lackman et al. 2011). Altered *PYL* expression affected the JA response (Lackman et al. 2011). These results suggest a connection between a component of the core ABA signaling pathway and the JA responses. JA signaling may be integrated into early ABA signaling and may affect ABA receptor complexes to regulate downstream common signal components in stomatal closure (Munemasa et al. 2011).

MeJA stimulates the expression of *9-Cis-epoxycarotenoid dioxygense3*, which encodes a crucial enzyme in ABA biosynthesis (Hossain et al. 2011). In the *aba2-2* mutant, MeJA did not induce JA-inducible *VSP1* (*Vegetative Storage Protein1*) gene expression. In wild-type leaves, an inhibitor of ABA biosynthesis, fluridon, inhibited *VSP1* expression (Hossain et al. 2011). These results suggest that endogenous ABA may be involved in JA biosynthesis and signal transduction (Hossain et al. 2011). ABA synthesis is required for JA production and the activation of plant defenses against *Pythium irregulare* (Adie et al. 2007).

Stomatal closure is an innate immune response involved in bacterial disease resistance (Hettenhausen et al. 2012). ABA induces stomatal closure immune response (Hubbard et al. 2012; Sugiyama et al. 2012; Uraji et al. 2012), and MeJA also induces stomatal closure in various plant species (Munemasa et al. 2010, 2011). MeJA as well as ABA stimulate stomatal closure in several plant species (Saito et al. 2008; Hossain et al. 2011; Munemasa et al. 2011). The induction of JA-responsive genes was blocked in *mpk4* mutant plants, suggesting that MPK4 positively regulates the JA pathway (Petersen et al. 2000; Gao et al. 2008; Qiu et al. 2008; Pitzschke et al. 2009). Silencing of *MPK4* in *Nicotiana attenuata* compromised abscisic acid-induced stomatal closure and guard cell-mediated resistance to *Pseudomonas syringae* pv. *tomato* DC3000 (Hettenhausen et al. 2012). The results suggest that MPK4-mediated JA signaling enhances ABA-mediated stomatal closure immune responses.

3.19 Crosstalk Between JA Signaling and Small RNA Signaling Systems

Small RNAs (sRNAs) are non-protein-coding RNAs of 20- to 30-nucleotide length (Havecker et al. 2010; Kulcheski et al. 2011; Wang et al. 2011). These small RNAs have been identified as regulatory RNAs that modulate gene expression at both the transcriptional and posttranscriptional levels (Zhao et al. 2012). They act as sequence-specific repressors of target gene expression, either at the transcriptional level through DNA and/or histone methylation or at the posttranscriptional level through transcriptional cleavage or translational inhibition (Ramachandran and Chen 2008). Small RNA molecules act as mobile signals that direct mRNA cleavage and DNA methylation in recipient cells (Ciomperlik et al. 2011; Molnár et al. 2011).

RNA silencing refers to a number of related cellular processes that employ the small RNAs to regulate the expression of genetic material in a sequence-specific manner (Qu et al. 2008; Jaubert et al. 2011; Zhao et al. 2012). RNA silencing protects plants against invading nucleic acids such as transgenes and viruses (Ellendorff et al. 2009). RNA silencing is a type of plant immune system conferring resistance against viruses and also against bacteria and fungi (Voinnet 2001; Katiyar-Agarwal and Jin 2010; Zhang et al. 2011a, c, d). Plants use RNA silencing as a surveillance mechanism to protect against fungal (Lu et al. 2007; Ellendorff et al. 2009), bacterial (Katiyar-Agarwal et al. 2006; Navarro et al. 2006, 2008; Agorio and Vera 2007;

Jin 2008; Li et al. 2010; Zhang et al. 2011a), and viral pathogens (Mlotshwa et al. 2002; Garcia-Ruiz et al. 2010).

RNA silencing is able to spread from cell to cell (Chitwood and Timmermans 2010) and spreads both to neighboring cells and systemically over long distances (Dunoyer et al. 2010). An RNA silencing signal has been shown to move through plasmodesmata and the phloem (Molnár et al. 2010). Both exogenous and endogenous siRNAs, as opposed to their precursor molecules, act as mobile silencing signals between plant cells (Dunoyer et al. 2010). The small RNA signaling system has been shown to play important role in host defense responses against viral, bacterial, and fungal pathogens.

Crosstalk between JA signaling and small RNA signaling systems has been reported. When a viral suppressor of small RNA-directed RNA silencing was overexpressed in transgenic tobacco plants, several genes involved in jasmonate signaling system were up-regulated. It up-regulated lipoxygenase genes (*LOX1* and *LOX2*) involved in JA biosynthesis (Endres et al. 2010). It also up-regulated several JA-responsive genes which include several transcription factors involved in JA signaling system (Endres et al. 2010). These results suggest that the viral suppressor, which suppresses RNA silencing, activates JA signaling system. The studies strongly suggest the existence of a crosstalk between these two different defense signaling pathways.

MicroRNAs are also involved in the modulation of JA signaling. One clade of the target genes of the miRNA miR319 have been identified as a clade of TCP (TEOSINTE BRANCHED/CYCLOIDEA/PCF) transcription factor genes (Schommer et al. 2008). TCPs positively regulate *LOX2* promoter activity. The miR319-targeted TCPs directly regulate expression of *LOX2* (Schommer et al. 2008). Lipoxygenase encoded by *LOX2* is the key enzyme in biosynthesis of JA (Vidhyasekaran 2007). The results suggest that the microRNA-directed RNA silencing system triggers JA signaling by specifically activating biosynthesis of JA through the action of the TCP transcription factor.

The targets of miRNAs have been found to include mostly transcription factors. The target sites of *Brassica napus* miRNA miR172 include MYB and WRKY transcription factors (Zhao et al. 2012). WRKY transcription factors are known to either positively or negatively regulate plant immune responses (Wang et al. 2006; Shen et al. 2007; Lai et al. 2008; Xing et al. 2008; Pandey and Somssich 2009). MYB transcription factors also play important role in plant innate immunity (Nurmberg et al. 2007; Van der Ent et al. 2008). A WRKY transcription factor, WIZZ (wound-induced leucine zipper zinc finger), is induced by JA, and expression of the mRNA of *WIZZ* was up-regulated by two times in the *HC-Pro* plants (Hara et al. 2000; Soitamo et al. 2011). The NAC transcription factor *Jasmonic acid 2*, which is involved in JA signaling, is down-regulated in *HC-Pro* transgenic plants (Soitamo et al. 2011).

The expression of the ethylene responsive factor 1 (*ERF1*) mRNA was up-regulated more than five times in the leaves of the *HC-Pro* expressing transgenic tobacco plants (Soitamo et al. 2011). ERF1 integrates signals from ethylene (ET) and JA signaling pathways (Lorenzo et al. 2003; Gutterson and Reuber 2004;

McGrath et al. 2005). Another ET-responsive gene, *ERF3* (Cao et al. 2006), was up-regulated by more than two times in *HC-Pro* transgenic plants (Soitamo et al. 2011). In addition, ERF4, an ET-responsive transcription factor and a negative regulator of JA-responsive defense related genes (McGrath et al. 2005), was down-regulated in these transgenic plants (Soitamo et al. 2011).

3.20 JA Signaling in Induced Systemic Immunity

3.20.1 JA Signaling Plays Major Role in Induced Systemic Resistance

Systemic immunity conferring resistance against a wide range of pathogens has been reported to be induced by several biotic and abiotic stresses in plants. Two distinctly different types of systemic induced resistance have been recognized. These include "systemic acquired resistance [SAR]" and "induced systemic resistance [ISR]." SAR is a systemic defense network that is usually induced in plants in response to attack by biotrophic and hemibiotrophic pathogens in plants. Localized attack by a necrotizing pathogen induces SAR to subsequent attack by a broad range of normally virulent pathogens (Liu et al. 2011b, c; Sharon et al. 2011). The necrotrophic bacterium *Pectobacterium carotovorum* has also been reported to induce SAR (Po-Wen et al. 2013). SAR is also induced by applying some abiotic substances (Flors et al. 2008; Mandal et al. 2008; Ziadi et al. 2008; Umemura et al. 2009; Jaskiewicz et al. 2011; Po-Wen et al. 2013). SA signaling has been shown to be required for the manifestation of SAR (Shah 2009; Du et al. 2012; Návarová et al. 2012; Shah and Zeier 2013). Without SA production or accumulation, SAR did not occur (Bari and Jones 2009).

ISR is triggered mostly by biocontrol agents (BCAs) and necrotrophic fungi. It is dependent mostly on jasmonic acid (Kravchuk et al. 2011; Niu et al. 2011; Weller et al. 2012; Zamioudis and Pieterse 2012; Bakker et al. 2013). Several chemicals and PAMPs/MAMPs have also been reported to trigger ISR (Kravchuk et al. 2011). *P. fluorescens* CHAOr strain induces defense responses in *Arabidopsis* (Iavicoli et al. 2003). CHAOr-mediated ISR effect was tested in various *Arabidopsis* mutants, such as *sid2-1* (nonproducing SA), *npr1-1* (nonexpressing NPR1 protein), *jar1-1* (insensitive to JA), *ein2-1* (insensitive to ET), *etr1-1* (insensitive to ET), and *eir1-1* (insensitive to ET in roots). Among them, only *jar1-1*, *eir1-1*, and *npr1* mutants were unable to undergo ISR. Sensitivity to JA and functional NPR1 and EIR1 proteins were found to be required for the full expression of CHAOr-mediated ISR (Iavicoli et al. 2003). These results suggest that *P. fluorescens* strain CHAOr activates defense response through JA/ET signaling pathway in *Arabidopsis*.

Pseudomonas fluorescens WCS374r strain elicits ISR against the rice blast fungal pathogen *Magnaporthe oryzae* through JA-regulated pathway (De Vleesschauwer et al. 2008). *P. fluorescens* strain 89B61 also elicited ISR against *Phytophthora infestans* in tomato through a SA-independent but JA-dependent signaling pathway (Yan et al. 2002). *Bacillus pumilus* SE34 induced resistance against *Phytophthora infestans* in tomato through a JA-dependent pathway (Yan et al. 2002).

In Arabidopsis JA has been reported as essential for *Trichoderma*-induced systemic resistance against *B. cinerea* and other pathogens (Korolev et al. 2008; Segarra et al. 2009; Brotman et al. 2012; Mathys et al. 2012). *Trichoderma virens* GV29-8 up-regulated allene oxide synthase (AOS), the key enzyme in lipoxygenase (LOX) pathway involved in JA biosynthesis. AOS gene was up-regulated by *Trichoderma virens* in maize leaves challenge inoculated with *Colletotrichum graminicola* (Djonovic et al. 2006). Another gene from the JA biosynthetic pathway is *OPR7* encoding 12-oxo-phytodienoic acid reductase, which was previously shown to be highly induced in maize by JA treatment but not by SA (Zhang et al. 2005), and *OPR7* was induced by *T. virens* GV29-8 (Djonovic et al. 2006). These results suggest that *T. virens* GV29-8 may activate the defense response through JA-dependent signaling pathway.

The elicitor/MAMP preparations (homogenate of mycelia) from *Pythium oligandrum* induced JA-dependent defense responses in tomato (Hase et al. 2008). The basic PR-6 protein is known to be induced through JA signaling pathway (Hondo et al. 2007). The expression of the JA-responsive gene for the basic PR-6 protein was induced in tomato cultivars treated with the elicitor from *P. oligandrum*. The elicitor induced resistance against the bacterial pathogen *Ralstonia solanacearum* and the induced resistance was compromised in the elicitor-treated *jai1-1* mutant plants defective in JA signaling (Hase et al. 2008). These results suggest that the JA-dependent signaling pathway is required for the elicitor-induced defense responses. When *A. thaliana* was treated with the elicitor, expression of the JA-responsive gene *PDF1.2* was up-regulated and the resistance to the bacterial pathogens *R. solanacearum* and *P. syringae* pv. *tomato* was enhanced. Such elicitor-induced defense responses were completely compromised in elicitor-treated *coi1-1* and *jar1-1* mutants with impaired JA signaling pathway (Kawamura et al. 2009).

Some chemical can also induce JA-dependent ISR. Root treatment with hexanoic acid induced resistance against foliar infection by *Botrytis cinerea* in tomato plants (Vicedo et al. 2009). Hexanoic acid-induced systemic resistance was blocked in the jasmonic acid-insensitive mutant *jai1* (a *coi1* homolog) (Vicedo et al. 2009). Soil drench treatments with hexanoic acid can effectively protect Arabidopsis plants against *B. cinerea* through a mechanism based on a stronger and faster accumulation of JA-dependent defenses. The impaired JA mutants, *jin1-2* and *jar1*, were unable to display hexanoic acid-triggered ISR against *B. cinerea* (Kravchuk et al. 2011). These studies demonstrated the importance of JA signaling in the hexanoic acid-induced systemic resistance. Repka et al. (2004) reported that MeJA systemically induced the accumulation of PR1, β-1,3-glucanase, chitinase, peroxidase, and extensin in the upper untreated grapevine leaves 14 days after treatment of lower leaves with MeJA. Collectively these studies show that various microbes, elicitors/MAMPs, and chemicals can induce JA signaling-dependent ISR.

3.20.2 *Mobile Signal Involved in Induced Systemic Resistance*

Jasmonoyl-isoleucine (JA-Ile) may be the mobile signal involved in the induction of disease resistance in distal leaves (Sato et al. 2011; Matsuura et al. 2012). Distal transport of JA and JA-Ile has been reported in response to leaf wounding in tobacco and tomato plants. Sato et al. (2011) could recover JA and JA-Ile in distal leaves after treating wounded leaves with JA or JA-Ile. JA-Ile had a higher mobility than JA. Application of exogenous JA-Ile to wounded leaves of *Nicotiana tabacum* led to a high accumulation of JA and JA-Ile in distal leaves compared with wounded control plants. Exudates from stem of tomato plants with damaged leaflets contained JA and JA-Ile at higher levels than in undamaged plant, and a significant difference in the levels of JA-Ile was observed within 30 min after wounding (Sato et al. 2011). Similar results were also obtained in *Arabidopsis thaliana* leaves (Koo et al. 2009). The wounding of *A. thaliana* leaves resulted in the rapid (within 5 min) accumulation of JA-Ile, in leaves distal to the wound site (Koo et al. 2009).

JA-Ile may be synthesized de novo and transported into systemic tissues. When [$^{13}C_6$]Ile and [$^{2}H_6$]JA were applied separately into the wounds on two different leaves, JA-[$^{13}C_6$]Ile and [$^{2}H_6$]JA-Ile were detected in the stem exudates but [$^{2}H_6$]JA-[$^{13}C_6$]Ile was not (Matsuura et al. 2012). The results indicate that JA is conjugated with Ile in the wounded leaf and the resulting JA-Ile was then transported into systemic tissues. Collectively, these studies suggest that JA-Ile may be the mobile signaling cue involved in ISR.

3.20.3 *Priming in Induced Systemic resistance*

JA signaling has been shown to be involved in ISR in plants (Bari and Jones 2009; Sharon et al. 2011; Martinez-Medina et al. 2013). Besides basal resistance responses that act at the site of pathogen infection, plants are also capable of developing systemic resistance that is effective against future pathogen attack (Bostock 2005). The induced resistance does not necessarily require a direct activation of defense responses but can also result from a faster and stronger expression of basal defense responses upon pathogen attack. This enhanced capacity to express infection-induced basal defenses is called "priming" (Conrath et al. 2006; Conrath 2009). The priming results in a faster and stronger induction of defense mechanisms after pathogen attack (Jung et al. 2009; Conrath 2011; Martinez-Medina et al. 2013; Po-Wen et al. 2013).

Several rhizobacterial strains trigger the plant immune response by activating JA signaling system. *Pseudomonas putida* strain BTP1 treatment primes tomato plants to activate two key enzymes of lipoxygenase (LOX) pathway, lipoxygenase (LOX) and lipid hydroperoxidase (LHP), after challenge inoculation with the pathogen *Botrytis cinerea* (Akram et al. 2008). The increase in the enzyme activities was observed only after challenge by the pathogen, and this priming effect appears to be almost invariably associated with the ISR phenomenon. The LOX pathway is involved in the JA signaling system (Delker et al. 2007; Vidhyasekaran 2007; Balbi and Devoto 2008), suggesting the role of JA signaling in the *P. putida* BTP1-induced systemic resistance.

Root colonization by selected *Trichoderma* isolates has been shown to activate in the plant a systemic response that is effective against a broad spectrum of plant pathogens (Martinez-Medina et al. 2013). In tomato, root colonization by *T. harzianum* rendered the leaves more resistant to *B. cinerea*. The systemic resistance induced in tomato against the necrophic pathogen was mainly based on priming JA-dependent responses (Martinez-Medina et al. 2013). The JA-deficient *defenseless1* (*def1*) mutants were unable to develop *Trichoderma*-induced resistance. The inability of *def1* plants to develop ISR correlated with a lack of priming of JA-responsive defense gene *Proteinase inhibitor II* expression (Martinez-Medina et al. 2013). Application of *Trichoderma hamatum* T382 to the roots of *A. thaliana* results in an increased resistance to subsequent leaf infections by the necrotrophic pathogen *B. cinerea* (Mathys et al. 2012). Treatment with *T. hamatum* T382 primes the plant (ISR-prime), resulting in an accelerated activation of defense responses against *B. cinerea* during ISR-boost in *A. thaliana* (Mathys et al. 2012).

Hexanoic acid-treated plants infected with *B. cinerea* showed priming in the expression of the JA-responsive genes *PDF1.2*, *PR-4*, and *VSP1* in *Arabidopsis*. JA signaling pathway showed dramatic changes after hexanoic acid treatment, mainly when the pathogen was present. The impaired JA mutants, *jin1-2* and *jar1*, were unable to display hexanoic acid-induced priming against *B. cinerea* (Kravchuk et al. 2011). JA and OPDA levels were primed at early stages by hexanoic acid. Hexanoic acid treatment could protect tomato plants against *B. cinerea* by priming OPDA and JA-Ile production (Kravchuk et al. 2011). Hexanoic treatment protected tomato plants against *B. cinerea* by priming defense responses by up-regulating proteinase inhibitor genes (Finiti et al. 2014). These studies showed that JA-mediated priming plays an important role in the induced systemic resistance.

References

Adie BA, Perez-Perez J, Godoy M, Sanchez-Serrano JJ, Schmelz EA, Solano R (2007) ABA is an essential signal for plant resistance to pathogens affecting JA biosynthesis and activation of defenses in *Arabidopsis*. Plant Cell 19:1665–1681

Agorio A, Vera P (2007) ARGONAUTE4 is required for resistance to *Pseudomonas syringae* in *Arabidopsis*. Plant Cell 19:3778–3790

Agrawal GK, Iwahashi H, Rakwal R (2003) Small GTPase 'Rop': molecular switch for plant defense responses. FEBS Lett 546:173–180

Akram A, Ongena M, Duby F, Dommes J, Thonart P (2008) Systemic resistance and lopoxygenase-related defence response induced in tomato by Pseudomonas putida strain BTP1. BMC Plant Biol 8:113

Alkan N, Fluhr R, Prusky D (2012) Ammonium secretion during *Colletotrichum coccodes* infection modulates salicylic and jasmonic acid pathways of ripe and unripe tomato fruit. Mol Plant-Microbe Interact 25:85–96

An C, Mou Z (2011) Salicylic acid and its function in plant immunity. J Integr Plant Biol 53:412–428

An C, Mou Z (2013) The function of the Mediator complex in plant immunity. Plant Signal Behav 8(3):e23182

An F, Zhao Q, Ji Y, Li W, Jiang Z, Yu X, Zhang C, Han Y, He W, Liu Y, Zhang S, Ecker JR, Guo H (2010) Ethylene-induced stabilization of ETHYLENE INSENSENSITIVE3 and EIN3-LIKE1 is mediated by proteasomal degradation of EIN3 binding F-box 1 and 2 that requires EIN2 in Arabidopsis. Plant Cell 22:2384–2401

Anderson JP, Badruzsaufari E, Schenk PM, Manners JM, Desmond OJ, Ehlert C, Maclean DJ, Ebert PR, Kazan K (2004) Antagonistic interaction between abscisic acid and jasmonate-ethylene signaling pathways modulates defense gene expression and disease resistance in *Arabidopsis*. Plant Cell 16:3460–3479

Andreasson E, Jenkins T, Brodersen P, Thorgrimsen S, Petersen NHT, Zhu S, Qiu J-L, Micheelsen P, Rocher A, Petersen M, Newman A, Nielsen HB, Hirt H, Somssich I, Mattsson O, Mundy J (2005) The MAP kinase substrate MKS1 is a regulator of plant defense responses. EMBO J 24:2579–2589

Asada K (2006) Production and scavenging of reactive oxygen species in chloroplasts and their functions. Plant Physiol 141:391–396

Aziz A, Poinssot B, Daire X, Adrian M, Bezier A, Lambert B, Joubert JM, Pugin A (2003) Laminarin elicits defense responses in grapevine and induces protection against *Botrytis cinerea* and *Plasmopara viticola*. Mol Plant-Microbe Interact 16:1118–1128

Bae H, Kim MS, Sicher RC, Bae H-J, Bailey BA (2006) Necrosis- and ethylene-inducing peptide from *Fusarium oxysporum* induces a complex cascade of transcripts associated with signal transduction and cell death in *Arabidopsis*. Plant Physiol 141:1056–1067

Bakker PAHM, Doornbos RF, Zamioudis C, Berendsen RL, Pieterse CMJ (2013) Induced systemic resistance and the rhizosphere microbiome. Plant Pathol J 29:136–143

Balbi V, Devoto A (2008) Jasmonate signalling network in *Arabidopsis thaliana*: Crucial regulatory nodes and new physiological scenarios. New Phytol 177:301–318

Bari R, Jones JDG (2009) Role of plant hormones in plant defence responses. Plant Mol Biol 69:473–488

Berger SL (2002) Histone modifications in transcriptional regulation. Curr Opin Genet Dev 2:142–148

Bergey DR, Howe GA, Ryan CA (1996) Polypeptide signalling for plant defensive genes exhibits analogies to defense signalling in animals. Proc Natl Acad Sci U S A 93:12053–12058

Berr A, McCallum EJ, Alioua A, Heintz D, Heitz T, Shen WH (2010) Arabidopsis histone methyltransferase SET DOMAIN GROUP8 mediates induction of the jasmonate/ethylene pathway genes in plant defense response to necrotrophic fungi. Plant Physiol 154:1403–1414

Berrocal-Lobo M, Molina A (2004) Ethylene response factor 1 mediates Arabidopsis resistance to the soilborne fungus *Fusarium oxysporum*. Mol Plant-Microbe Interact 17:763–770

Berrocal-Lobo M, Molina A, Solano R (2002) Constitutive expression of ETHYLENE-RESPONSE-FACTOR1 in *Arabidopsis* confers resistance to several necrotrophic fungi. Plant J 29:23–32

Bertoni G (2012) Oxylipins and plant palatability. Plant Cell 24:1305

Bohland C, Balkenhohl T, Loers G, Feussner I, Grambow HJ (1997) Differential induction of lipoxygenase isoforms in wheat upon treatment with rust fungus elicitor, chitin oligosaccharides, chitosan, and methyl jasmonate. Plant Physiol 114:679–685

Bonaventure G, Schuck S, Baldwin IT (2011) Revealing complexity and specificity in the activation of lipase-mediated oxylipin biosynthesis: a specific role of the *Nicotiana attenuata* GLA1 lipase in the activation of jasmonic acid biosynthesis in leaves and roots. Plant Cell Environ 34:1507–1530

Bostock RM (2005) Signal crosstalk and induced resistance: straddling the line between cost and benefit. Annu Rev Phytopathol 43:545–580

Boter M, Ruiz-Rivero O, Abdeen A, Prat S (2004) Conserved MYC transcription factors play a key role in jasmonate signalling both in tomato and *Arabidopsis*. Genes Dev 18:1577–1591

Böttcher C, Pollmann S (2009) Plant oxylipins: plant responses to 12-oxo-phytodienoic acid governed by its specific structural and functional properties. FEBS J 276:4693–4704

Boudsocq M, Willmann MR, McCormack M, Lee H, Shan L, He P, Bush J, Cheng SH, Sheen J (2010) Differential innate immune signalling via Ca^{2+} sensor protein kinases. Nature 464:418–422

Bravo JM, Campo S, Murillo J, Segundo BS (2003) Fungus and wound-induced accumulation of mRNA containing a class II chitinase of the pathogenesis-related protein 4 (PR-4) family of maize. Plant Mol Biol 52:745–759

Brodersen P, Petersen M, Bjorn Nielsen H, Zhu S, Newman MA, Shokat KM, Rietz S, Parker J, Mundy J (2006) *Arabidopsis* MAP kinase 4 regulates salicylic acid- and jasmonic acid/ethylene-dependent responses via EDS1 and PAD4. Plant J 47:532–546

Brooks DM, Bender CL, Kunkel BN (2005) The *Pseudomonas syringae* phytotoxin coronatine promotes virulence by overcoming salicylic acid-dependent defenses in *Arabidopsis thaliana*. Mol Plant Pathol 6:629–639

Brotman Y, Lisec J, Méret M, Chet I, Willmitzer L, Viterbo A (2012) Transcript and metabolite analysis of the *Trichoderma* – induced systemic resistance response to *Pseudomonas syringae* in *Arabidopsis thaliana*. Microbiology 158:139–146

Brown RL, Kazan K, McGrath KC, Maclean DJ, Manners JM (2003) A role for the GCC-box in jasmonate-mediated activation of the *PDF1.2* gene of *Arabidopsis*. Plant Physiol 132:1020–1032

Browse J (2009) Jasmonate passes muster: a receptor and targets for the defense hormone. Annu Rev Plant Biol 60:183–205

Cao Y, Wu Y, Zheng Z, Song F (2006) Overexpression of the rice EREBP-like gene *OsBIERF3* enhances disease resistance and salt tolerance in transgenic tobacco. Physiol Mol Plant Pathol 67:202–211

Cardinale F, Jonak C, Ligterink W, Niehaus K, Boller T, Hirt H (2000) Differential activation of four specific MAPK pathways by distinct elicitors. J Biol Chem 275:36734–36740

Cevik V, Kidd BN, Zhang P, Hill C, Kiddle S, Denby KJ, Holub EB, Cahill DM, Manners JM, Schenk PM, Beynon J, Kazan K (2012) MEDIATOR25 acts as an integrative hub for the regulation of jasmonate-responsive gene expression in Arabidopsis. Plant Physiol 160:541–555

Chehab EW, Kim S, Savchenko T, Kliebenstein D, Dehesh K, Braam J (2011) Intronic T-DNA insertion renders Arabidopsis *opr3* a conditioned jasmonic acid producing mutant. Plant Physiol 156:770–778

Chen R, Jiang H, Li L, Zhai Q, Qi L, Zhou W, Liu X, Li H, Zheng W, Sun J, Li C (2012) The Arabidopsis mediator subunit MED25 differentially regulates jasmonate and abscisic acid signaling through interacting with MYC2 and ABI5 transcription factors. Plant Cell 24:2898–2916

Cheng Z, Sun L, Qi T, Zhang B, Peng W, Liu Y, Xie D (2011) The bHLH transcription factor MYC3 interacts with the jasmonate ZIM-domain proteins to mediate jasmonate response in Arabidopsis. Mol Plant 4:279–288

Cheng B, Yu X, Ma Z, Dong S, Dou D, Wang Y, Zheng X (2012) *Phytophthora sojae* effector Avh331 suppresses the plant defence response by disturbing the MAPK signalling pathway. Physiol Mol Plant Pathol 77:1–9

Chini A, Fonseca S, Fernández G, Adie B, Chico JM, Lorenzo O, Garcia-Casado G, López-Vidriero I, Lozano FM, Ponce MR, Micol JL, Solano R (2007) The JAZ family of repressors is the missing link in jasmonate signaling. Nature 448:666–671

Chini A, Boter M, Solano R (2009a) Plant oxylipins: COI1/JAZs/MYC2 as the core jasmonic acid-signaling module. FEBS J 276:4682–4692

Chini A, Fonseca S, Chico JM, Fernández-Calvo P, Solano R (2009b) The ZIM domain mediates homo- and heteromeric interactions between Arabidopsis JAZ proteins. Plant J 59:77–87

Chitwood DH, Timmermans MCP (2010) Small RNAs are on the move. Nature 467:415–419

Chun HJ, Park HC, Koo SC, Lee JH, Park CY, Choi MS, Kang CH, Baek D, Cheong YH, Yun DJ, Chung WS, Cho MJ, Kim MC (2012) Constitutive expression of mammalian nitric oxide synthase in tobacco plants triggers disease resistance to pathogens. Mol Cells 34:463–471

Chung HS, Howe GA (2009) A critical role for the TIFY motif in repression of jasmonate signaling by a stabilized splice variant of the JASMONATE ZIM-domain protein JAZ10 in Arabidopsis. Plant Cell 21:131–145

Chung HS, Koo AJK, Gao X, Jayany S, Thines B, Jones AD, Howe GA (2008) Regulation and function of Arabidopsis JAZ genes in response to wounding and herbivory. Plant Physiol 146:952–964

Chung HS, Cooke TE, DePew CL, Patel LC, Ogawa N, Kobayashi Y, Howe GA (2010) Alternative splicing expands the repertoire of dominant JAZ repressors of jasmonate signaling. Plant J 63:613–622

Ciomperlik J, Omarov RT, Scholthof HB (2011) An antiviral RISC isolated from *Tobacco rattle virus*-infected plants. Virology 412:117–124

Conaway RC, Conaway JW (2011a) Function and regulation of the Mediator complex. Curr Opin Genet Dev 21:225–230

Conaway RC, Conaway JW (2011b) Origin and activity of the Mediator complex. Semin Cell Dev Biol 22:729–734

Conrath U (2009) Priming of induced plant defense responses. Adv Bot Res 51:361–395

Conrath U (2011) Molecular aspects of defence priming. Trends Plant Sci 16:524–531

Conrath U, Beckers GJM, Flors V, Garcia-Agustin P, Jakab G, Mauch F, Newman M-A, Pieterse CMJ, Poinssot B, Pozo MJ, Pugin A, Schaffrath U, Ton J, Wendehenne D, Zimmerli L, Mauch-Mani B (2006) Priming: getting ready for battle. Mol Plant-Microbe Interact 19:1062–1071

De Vleesschauwer D, Djavaheri M, Bakker PAHM, Höfte M (2008) *Pseudomonas fluorescens* WCS374r-induced systemic resistance in rice against *Magnaporthe oryzae* is based on pseudobactin-mediated priming for a salicylic acid-repressable multifaceted defense response. Plant Physiol 148:1996–2012

del Rio LA, Corpas FJ, Barroso JB (2004) Nitric oxide and nitric oxide synthase activity in plants. Phytochemistry 65:783–792

Delker C, Zolman BK, Miersch O, Wasternack C (2007) Jasmonate biosynthesis in *Arabidopsis thaliana* requires peroxisomal β-oxidation enzymes – Additional proof by properties of *pex6* and *aim1*. Phytochemistry 68:1542–1650

Demianski AJ, Chung KM, Kunkel BN (2012) Analysis of *Arabidopsis JAZ* gene expression during *Pseudomonas syringae* pathogenesis. Mol Plant Pathol 13:46–57

Denoux C, Galletti R, Mammarella N, Gopalan S, Werck D, De Lorenzo G, Ferrari S, Ausubel FM, Dewdney J (2008) Activation of defense response pathways by OGs and Flg22 elicitors in Arabidopsis seedlings. Mol Plant 1:423–445

Desmond OJ, Edgar CI, Manners JM, Maclean DJ, Schenk PM, Kazan K (2006) Methyl jasmonate induced gene expression in wheat delays symptom development by the crown rot pathogen *Fusarium pseudograminearum*. Physiol Mol Plant Pathol 67:171–179

Devoto A, Turner JC (2003) Regulation of jasmonate-mediated plant responses in Arabidopsis. Ann Bot 92:329–337

Devoto A, Nieto-Rostro M, Xie D, Ellis C, Harmston R, Patrick E, Davis J, Sherratt L, Coleman M, Turner JG (2002) COI1 links jasmonate signaling and fertility to the SCF ubiquitin-ligase complex in *Arabidopsis*. Plant J 32:457–466

Devoto A, Ellis C, Magusin A, Chang HS, Chilcott C, Zhu T, Turner JG (2005) Expression profiling reveals COI1 to be a key regulator of genes involved in wound- and methyl jasmonate-induced secondary metabolism, defence, and hormone interactions. Plant Mol Biol 58:497–513

Djonovic S, Pozo MJ, Dangott LJ, Howell CR, Kenerley CM (2006) Sm1, a proteinaceous elicitor secreted by the biocontrol fungus *Trichoderma virens* induces plant defense responses and systemic resistance. Mol Plant-Microbe Interact 19:838–853

Doares SH, Narvaez-Vasquez J, Conconi A, Ryan CA (1995) Salicylic acid inhibits synthesis of proteinase inhibitors in tomato leaves induced by systemin and jasmonic acid. Plant Physiol 108:1741–1746

Dombrecht B, Xue GP, Sprague SJ, Kirkegaard JA, Ross JJ, Reid JB, Fitt GP, Sewelam N, Schenk PM, Manners JM, Karzan K (2007) MYC2 differentially modulates diverse jasmonate-dependent functions in *Arabidopsis*. Plant Cell 19:2225–2245

Dong X (1998) SA, JA, ethylene, and disease resistance in plants. Curr Opin Plant Biol 1:316–323
Dong J, Chen C, Chen Z (2003) Expression profiles of the Arabidopsis WRKY gene superfamily during plant defense response. Plant Mol Biol 51:21–37
Du Q, Zhu W, Zhao Z, Qian X, Xu Y (2012) Novel benzo-1,2,3-thiadiazole-7-carboxylate derivatives as plant activators and the development of their agricultural applications. J Agric Food Chem 60:346–353
Dunoyer P, Schott G, Himber C, Meyer D, Takeda A, Carrington JC, Voinnet O (2010) Small RNA duplexes function as mobile silencing signals between plant cells. Science 328:912–916
Ellendorff U, Fradin EF, de Jonge R, Thomma BPHJ (2009) RNA silencing is required for *Arabidopsis* defence against *Verticillium* wilt disease. J Exp Bot 60:591–602
Ellinger D, Stingl N, Kubigsteltig II, Bals T, Juenger M, Pollmann S, Berger S, Schuenemann D, Mueller MJ (2010) DONGLE and DEFECTIVE IN ANTHER DEHISCENCE1 lipases are not essential for wound- and pathogen-induced jasmonate biosynthesis: redundant lipases contribute to jasmonate formation. Plant Physiol 153:114–127
Ellis C, Turner JG (2001) The Arabidopsis mutant *cev1* has constitutively active jasmonate and ethylene signal pathways and enhanced resistance to pathogens. Plant Cell 13:1025–1033
Endres MW, Gregory BD, Gao Z, Foreman AW, Mlotshwa S, Ge X, Pruss GJ, Ecker JR, Bowman LH, Vance V (2010) Two plant viral suppressors of silencing require the ethylene-inducible host transcription factor RAV2 to block RNA silencing. PLoS Pathog 6:e1000729
Fabro G, Di Rienzo JA, Voigt CA, Savchenko T, Dehesh K, Somerville S, Alvarez ME (2008) Genome-wide expression profiling Arabidopsis at the stage of *Golovinomyces cichoracearum* haustorium formation. Plant Physiol 146:1421–1439
Faize M, Faize L, Ishizaka M, Ishii H (2004) Expression of potential responses of Asian and European pears to infection with *Venturia nashicola*. Physiol Mol Plant Pathol 64:319–330
Fan J, Hill L, Crooks C, Doerner P, Lamb C (2009) Abscisic acid has a key role in modulating diverse plant-pathogen interactions. Plant Physiol 150:1750–1761
Farmer EE, Ryan CA (1990) Interplant communication: airborne methyl jasmonate induces synthesis of proteinase inhibitors in plant leaves. Proc Natl Acad Sci U S A 87:7713–7716
Farmer EE, Almeras E, Krishnamurthy V (2003) Jasmonates and related oxylipins in plant responses to pathogenesis and herbivory. Curr Opin Plant Biol 6:372–378
Feng S, Ma L, Wang X, Xie D, Dinesh-Kumar SP, Wei N, Deng XW (2003) The COP9 signalosome interacts physically with SCF^{COI1} and modulates jasmonate responses. Plant Cell 15:1083–1094
Fernández-Calvo P, Chini A, Fernández-Barbero G, Chico JM, Gimenez-Ibanez S, Geerinck J, Eeckhout D, Schweizer F, Godoy M, Franco-Zorrilla JM, Pauwels L, Witters E, Puga MI, Paz-Ares J, Goossens A, Reymond P, De Jaeger G, Solano R (2011) The Arabidopsis bHLH transcription factors MYC3 and MYC4 are targets of JAZ repressors and act additively with MYC2 in the activation of jasmonate responses. Plant Cell 23:701–715
Feys BJF, Benedetti CE, Penfold CN, Turner JG (1994) Arabidopsis mutants selected for resistance to the phytotoxin coronatine are male sterile, insensitive to methyl jasmonate, and resistant to a bacterial pathogen. Plant Cell 6:751–759
Finiti I, de la O Leyva M, Vicedo B, Gómez-Pastor R, López-Cruz J, Garcia-Agustin P, Real MD, González-Bosch C (2014) Hexanoic acid protects tomato plants against Botrytis cinerea by priming defence responses and reducing oxidative stress. Mol Plant Pathol 15:550–562
Flors V, Ton J, van Doorn R, Garcia-Agustin P, Mauch-Mani B (2008) Interplay between JA, SA, and ABA signalling during basal and induced resistance against *Pseudomonas syringae* and *Alternaria brassicicola*. Plant J 54:81–92
Fonseca S, Chini A, Hamberg M, Adie B, Porzel A, Kramell R, Miersch O, Wasternack C, Solano R (2009) (+)-7-*iso*-Jasmonoyl-L-isoleucine is the endogenous bioactive jasmonate. Nat Chem Biol 5:344–350
Fujiwara M, Umemura K, Kawasaki T, Shimamoto K (2006) Proteomics of Rac GTPase signaling reveals its predominant role in elicitor-induced defense response of cultured rice cells. Plant Physiol 140:734–745

Gao M, Liu J, Zhang Z, Cheng F, Chen S, Zhang Y (2008) MEKK1, MKK1/MKK2 and MPK4 function together in a mitogen-activated protein kinase cascade to regulate innate immunity in plants. Cell Res 18:1190–1198

Gao X, Brodhagen M, Isakeit T, Brown SH, Göbel C, Betran J, Feussner I, Keller NP, Kolomiets MV (2009) Inactivation of the lipoxygenase ZmLOX3 increases susceptibility of maize to *Aspergillus* spp. Mol Plant Microbe Interact 22:222–231

Gao Y, Li T, Liu Y, Ren C, Zhao Y, Wang M (2010a) Isolation and characterization of gene encoding G protein α subunit protein responsive to plant hormones and abiotic stresses in *Brassica napus*. Mol Biol Rep 37:3957–3965

Gao Y, Li T, Zhao Y, Ren C, Liu Y, Wang M (2010b) Cloning and characterization of a G protein β subunit gene responsive to plant hormones and abiotic stresses in *Brassica napus*. Plant Mol Biol Rep 28:450–459

Gao Q-M, Venugopal S, Navarre D, Kachroo A (2011) Low oleic acid-derived repression of jasmonic acid-inducible defense responses requires the WRKY50 and WRKY51 proteins. Plant Physiol 155:464–476

Garcia-Ruiz H, Takeda A, Chapman EJ, Sullivan CM, Fahlgren N, Brempelis KJ, Carrington JC (2010) *Arabidopsis* RNA-dependent RNA polymerases and dicer-like proteins in antiviral defense and small interfering RNA biogenesis during *Turnip mosaic virus* infection. Plant Cell 22:481–496

Gatz C (2013) From pioneers to team players: TGA transcription factors provide a molecular link between different stress pathways. Mol Plant-Microbe Interact 26:151–159

Gfeller A, Liechti R, Farmer EE (2010) Arabidopsis jasmonate signaling pathway. Sci Signal 3(109):cm4

Gimenez-Ibanez S, Solano R (2013) Nuclear jasmonate and salicylate signaling and crosstalk in defense against pathogens. Front Plant Sci 4: Article 72:1–11

Gomi K, Ogawa D, Katou S, Kamada H, Nakajima N, Saji H, Soyano T, Sasabe M, Machida Y, Mitsuhara I, Ohashi Y, Seo S (2005) A mitogen-activated protein kinase NtMPK4 activated by SIPKK is required for jasmonic acid signaling and involved in ozone tolerance via stomatal movement in tobacco. Plant Cell Physiol 46:1902–1914

Grant MR, Jones JDG (2009) Hormone (Dis)harmony moulds plant health and disease. Science 324:750–752

Grennan AK (2008) Ethylene response factors in jasmonate signaling and defense response. Plant Physiol 146:1457–1458

Grün S, Lindermayr C, Sell S, Durner J (2006) Nitric oxide and gene regulation in plants. J Exp Bot 57:507–516

Gu Y, Wang Z, Yang Z (2004) ROP/RAC GTPase: an old new master regulator for plant signaling. Curr Opin Plant Biol 7:527–536

Gullner G, Künstler A, Király L, Pogány M, Töbiás I (2010) Up-regulated expression of lipoxygenase and divinyl ether synthase genes in pepper leaves inoculated with Tobamoviruses. Physiol Mol Plant Pathol 74:387–393

Gundlach H, Muller MJ, Kutchan T, Zenk MH (1992) Jasmonic acid is a signal inducer in elicitor-induced plant cell cultures. Proc Natl Acad Sci U S A 89:2389–2393

Guo F-Q, Okamoto M, Crawford NM (2003) Identification of a plant nitric oxide synthase gene involved in hormonal signaling. Science 302:100–103

Gutterson N, Reuber TL (2004) Regulation of disease resistance pathways by AP2/ERF transcription factors. Curr Opin Plant Biol 7:465–471

Hara K, Yagi M, Kusano T, Sano H (2000) Rapid systemic accumulation of transcripts encoding a tobacco WRKY transcription factor upon wounding. Mol Gen Genet 263:30–37

Hase S, Takahashi S, Takenaka S, Nakaho K, Arie T, Seo S, Ohashi Y, Takahashi H (2008) Involvement of jasmonic acid signalling in bacterial wilt disease resistance induced by biocontrol agent *Pythium oligandrum* in tomato. Plant Pathol 57:870–876

Hause B, Stenzel I, Miersch O, Maucher H, Kramell R, Ziegler J, Wasternack C (2000) Tissue-specific oxylipin signature of tomato flowers: allene oxidase is highly expressed in distinct flower organs and vascular bundles. Plant J 24:113–126

Havecker ER, Wallbridge LM, Hardcastle TJ, Bush MS, Kelly KA, Dunn RM, Schwach F, Doonan JH, Baulcombe DC (2010) The *Arabidopsis* RNA-directed DNA methylation argonautes functionally diverge based on their expression and interaction with target loci. Plant Cell 22:321–334

Hettenhausen C, Baldwin IT, Wu J (2012) Silencing *MPK4* in *Nicotiana attenuata* enhances photosynthesis and seed production but compromises abscisic acid-induced stomatal closure and guard cell-mediated resistance to *Pseudomonas syringae* pv. *tomato* DC3000. Plant Physiol 158:759–776

Hiruma K, Takano Y (2011) Roles of EDR1 in non-host resistance of Arabidopsis. Plant Signal Behav 6:1831–1833

Holley SR, Yalamanchili RD, Moura DS, Ryan CA, Stratmann JW (2003) Convergence of signaling pathways induced by systemin, oligosaccharide elicitors, and ultraviolet-B radiation at the level of mitogen-activated protein kinases in *Lycopersicon peruvianum* suspension-cultured cells. Plant Physiol 132:1728–1738

Hondo D, Hase S, Kanayama Y, Yoshikawa N, Takenaka S, Takahashi H (2007) The *LeATL6-associated* ubiquitin/proteasome system may contribute to fungal elicitor-activated defense response via the jasmonic acid-dependent signaling pathway in tomato. Mol Plant-Microbe Interact 20:72–81

Hossain MA, Munemasa S, Uraji M, Nakamura Y, Mori IC, Murata Y (2011) Involvement of endogenous abscisic acid in methyl jasmonate-induced stomatal closure in *Arabidopsis*. Plant Physiol 156:430–438

Howe GA (2010) Ubiquitin ligase-coupled receptors extend their reach to jasmonate. Plant Physiol 154:471–474

Huang X, Stettmaier K, Michel C, Hutzler P, Mueller MJ, Durner J (2004) Nitric oxide is induced by wounding and influences jasmonic acid signaling in *Arabidopsis thaliana*. Planta 218:938–946

Hubbard KE, Siegel RS, Valerio G, Brandt B, Schroeder JI (2012) Abscisic acid and CO_2 signalling via calcium sensitivity in guard cells, new CDPK mutant phenotypes and a method for improved resolution of stomatal stimulus–response analyses. Ann Bot 109:5–17

Hudgins JW, Franceschi VR (2004) Methyl jasmonate-induced ethylene production is responsible for conifer phloem defense responses and reprogramming of stem cambial zone for traumatic resin duct formation. Plant Physiol 135:2134–2149

Huffaker A, Pearce G, Ryan CA (2006) An endogenous peptide signal in *Arabidopsis* activates components of the innate immune response. Proc Natl Acad Sci U S A 103:10098–10103

Hwang IS, Hwang BK (2011) The pepper mannose-binding lectin gene *CaMBL1* is required to regulate cell death and defense responses to microbial pathogens. Plant Physiol 155:447–463

Hyun Y, Choi S, Hwang H-J, Yu J, Nam SJ, Ko J, Park JY, Seo YS, Kim EY, Ryu SB, Kim WT, Lee YH, Kang H, Lee I (2008) Cooperation and functional diversification of two closely related galactolipase genes for jasmonate biosynthesis. Dev Cell 14:183–192

Iavicoli A, Boutet E, Buchala A, Metraux JP (2003) Induced systemic resistance in *Arabidopsis thaliana* in response to root inoculation with *Pseudomonas fluorescens* CHAO. Mol Plant-Microbe Interact 16:851–858

Iñigo S, Alvarez MJ, Strasser B, Califano A, Cerdán PD (2012) PFT1, the MED25 subunit of the plant Mediator complex, promotes flowering through CONSTANS dependent and independent mechanisms in Arabidopsis. Plant J 69:601–612

Ito T, Ng K-H, Lim T-S, Yu H, Meyerowitz EM (2007) The homeotic protein AGAMOUS controls late-stamen development by regulating a jasmonate biosynthetic gene in Arabidopsis. Plant Cell 19:3516–3529

Jaskiewicz M, Conrath U, Peterhänsel C (2011) Chromatin modification acts as a memory for systemic acquired resistance in the plant stress response. EMBO Rep 12:50–55

Jaubert M, Bhattacharjee S, Mello AF, Perry KL, Moffett P (2011) ARGONAUTE2 mediates RNA silencing anti-viral defenses against *Potato virus X* in Arabidopsis. Plant Physiol 156:1556–1564

Jayaraj J, Muthukrishnan S, Liang GH, Velazhahan R (2004) Jasmonic acid and salicylic acid induce accumulation of beta-1,3-glucanase and thaumatin-like proteins in wheat and enhance resistance against *Stagonospora nodorum*. Biol Plant 48:425–430

Jayaraj J, Rahman M, Wan A, Punja ZK (2009) Enhanced resistance to foliar fungal pathogens in carrot by application of elicitors. Ann Appl Biol 155:71–80

Jelenska J, Yao N, Vinatzer BA, Wright CM, Brodsky JL, Greenberg JT (2007) A J domain virulence effector of *Pseudomonas syringae* remodels host chloroplasts and suppresses defenses. Curr Biol 17:499–508

Jin H (2008) Endogenous small RNAs and antibacterial immunity in plants. FEBS Lett 582:2679–2684

Johansson A, Staal J, Dixelius C (2006) Early responses in the *Arabidopsis- Verticillium longisporum* pathosystem are dependent on *NDR1*, JA- and ET-associated signals via cytosolic NPR1 and *RFO1*. Mol Plant Microbe Interact 19:958–969

Jung C, Lyou SH, Yeu SY, Kim MA, Rhee S, Kim M, Lee JS, Choi YD, Cheong J-J (2007) Microarray-based screening of jasmonate-responsive genes in *Arabidopsis thaliana*. Plant Cell Rep 26:1053–1063

Jung HW, Tschaplinski TJ, Wang L, Glazebrook J, Greenberg JT (2009) Priming in systemic plant immunity. Science 324:89–91

Kachroo A, Lapchyk L, Fukushige H, Hildebrand D, Klessig D, Kachroo P (2003) Plastidial fatty acid signaling modulates salicylic acid- and jasmonic acid-mediated defense pathways in the Arabidopsis *ssi2* mutant. Plant Cell 15:2952–2965

Kalde M, Barth M, Somssich IE, Lippok B (2003) Members of the *Arabidopsis* WRKY group III transcription factors are part of different plant defense signaling pathways. Mol Plant-Microbe Interact 16:295–305

Kallenbach M, Alagna F, Baldwin IT, Bonaventure G (2010) *Nicotiana attenuata* SIPK, WIPK, NPR1, and fatty acid-amino acid conjugates participate in the induction of jasmonic acid biosynthesis by affecting early enzymatic steps in the pathway. Plant Physiol 152:98–106

Kandoth PK, Ranf S, Pancholi SS, Jayanty S, Walla MD, Miller W, Howe GA, Lincoln DE, Stratmann JW (2007) Tomato MAPKs LeMPK1, LeMPK2, and LeMPK3 function in the systemin-mediated defense response against herbivorous insects. Proc Natl Acad Sci U S A 104:12205–12210

Kang S, Kim HB, Lee H, Choi JY, Heu S, Oh CJ, Kwon SI, An CS (2006) Overexpression in Arabidopsis of a plasma membrane-targeting glutamate receptor from small radish increases glutamate-mediated Ca^{2+} influx and delays fungal infection. Mol Cells 21:418–427

Katagiri F, Tsuda K (2010) Understanding the plant immune system. Mol Plant-Microbe Interact 23:1531–1536

Katiyar-Agarwal S, Jin H (2010) Role of small RNAs in host-microbe interactions. Annu Rev Phytopathol 48:225–246

Katiyar-Agarwal S, Morgan R, Dahlbeck D, Borsani O, Villegas A, Zhu JK, Staskawicz BJ, Jin H (2006) A pathogen-inducible endogenous siRNA in plant immunity. Proc Natl Acad Sci U S A 103:18002–18007

Katou S, Karita E, Yamakawa H, Seo S, Mitsuhara I, Kuchitsu K, Ohashi Y (2005) Catalytic activation of the plant MAPK phosphatase NtMKP1 by its physiological substrate salicylic acid-induced protein kinase but not by calmodulins. J Biol Chem 280:39569–39581

Katsir L, Schilmiller A, Staswick P, He SY, Howe GA (2008) COI1 is a critical component of a receptor for jasmonate and the bacterial virulence factor coronatine. Proc Natl Acad Sci U S A 105:7100–7105

Kawamura Y, Hase S, Takenaka S, Kanayama Y, Yoshioka H, Kamoun S, Takahashi H (2009) INF1 elicitin activates jasmonic acid- and ethylene-mediated signaling pathways and induces resistance to bacterial wilt disease in tomato. J Phytopathol 157:287–297

Kawasaki T, Koita H, Nakatsubo T, Hasegawa K, Wakabayashi K, Takahashi H, Umemura K, Umezawa T, Shimamoto K (2006) Cinnamoyl-CoA reductase, a key enzyme in lignin biosynthesis, is an effector of small GTPase Rac in defense signaling in rice. Proc Natl Acad Sci U S A 103:230–235

Kazan K, Manners JM (2008) Jasmonate signaling: toward an integrated view. Plant Physiol 146:1459–1468

Kazan K, Manners JM (2013) MYC2: the master in action. Mol Plant 6(3):686–703. doi:10.1093/mp/sss128

Keyse SM (2000) Protein phosphatases and the regulation of mitogen-activated protein kinase signaling. Curr Opin Cell Biol 12:186–192

Kidd BN, Edgar CI, Kumar KK, Aitken EA, Schenk PM, Manners JM, Kazan K (2009) The Mediator complex subunit PFT1 is a key regulator of jasmonate-dependent defense in *Arabidopsis*. Plant Cell 21:2237–2252

Kidd BN, Cahill DM, Manners JM, Schenk PM, Kazan K (2011) Diverse roles of the Mediator complex in plants. Semin Cell Dev Biol 22:741–748

Kiirika LM, Bergmann HF, Schikowsky C, Wimmer D, Korte J, Schmitz U, Niehaus K, Colditz F (2012) Silencing of the Rac1 GTPase *MtROP9* in *Medicago truncatula* stimulates early mycorrhizal and oomycete root colonizations but negatively affects Rhizobial infection. Plant Physiol 159:501–516

Kim T-W, Guan S, Sun Y, Deng Z, Tang W, Shang JX, Sun Y, Burlingame AL, Wang ZY (2009) Brassinosteroid signal transduction from cell-surface receptor kinases to nuclear transcription factors. Nat Cell Biol 11:1254–1260

Klarzynski O, Plesse B, Joubert J-M, Yvin J-C, Kopp M, Kloareg B, Fritig B (2000) Linear β-1,3 glucans are elicitors of defense responses in tobacco. Plant Physiol 124:1027–1037

Knoth C, Ringler J, Dangl JL, Eulgem T (2007) *Arabidopsis* WRKY70 is required for full *RPP4-mediated* disease resistance and basal defense against *Hyaloperonospora parasitica*. Mol Plant-Microbe Interact 20:120–128

Kobayashi M, Seo S, Hirai K, Yamamoto-Katou A, Katou S, Seto H, Meshi T, Mitsuhara I, Ohashi Y (2010) Silencing of WIPK and SIPK mitogen-activated protein kinases reduces *Tobacco mosaic virus* accumulation but permits systemic viral movement in tobacco possessing the *N* resistance gene. Mol Plant-Microbe interact 23:1032–1041

Kombrink E (2012) Chemical and genetic exploration of jasmonate biosynthesis and signaling paths. Planta 236:1351–1366

Koo AJK, Gao XL, Jones AD, Howe GA (2009) A rapid wound signal activates the systemic synthesis of bioactive jasmonates in Arabidopsis. Plant J 59:974–986

Koornneef A, Pieterse CMJ (2008) Cross talk in defense signaling. Plant Physiol 146:839–844

Koornneef A, Leon-Reyes A, Ritsema T, Verhage A, Den Otter FC, Van Loon LC, Pieterse CMJ (2008) Kinetics of salicylate-mediated suppression of jasmonate signaling reveal a role for redox modulation. Plant Physiol 147:1358–1368

Korolev N, David DR, Elad Y (2008) The role of phytohormones in basal resistance in *Trichoderma-induced* systemic resistance to *Botrytis cinerea* in *Arabidopsis thaliana*. Biocontrol 53:667–683

Kravchuk Z, Vicedo B, Flors V, Camañes G, González-Bosch C, Garcia-Agustin P (2011) Priming for JA-dependent defenses using hexanoic acid is an effective mechanism to protect Arabidopsis against *B. cinerea*. J Plant Physiol 186:359–366

Kresge N, Simoni RD, Hill RL (2007) The biosynthesis of polyamines: the work of Herbert Tabor and Celia White Tabor. J Biol Chem 282:e26–e28

Kulcheski F, de Oliveira LFV, Molina LG, Almeräo MP, Rodrigues FA, Marcolino J, Barbosa JF, Stolf-Moreira R, Nepomuceno AL, Marcelino-Guimaräes FC, Abdelnoor RV, Nascimento LC, Carazzolle MF, Pereira GAG, Margis R (2011) Identification of novel soybean microRNAs involved in abiotic and biotic stresses. BMC Genomics 12:307

Kunkel BN, Brooks DM (2002) Cross talk between signaling pathways in pathogen defense. Curr Opin Plant Biol 5:325–331

Kwaaitaal M, Huisman R, Maintz J, Reinstädler A, Panstruga R (2011) Ionotropic glutamate receptor (iGluR)-like channels mediate MAMP-induced calcium influx in *Arabidopsis thaliana*. Biochem J 440:355–365

Lackman P, González-Guzmán M, Tilleman S, Carqueijeiro I, Cuéllar-Pérez A, Moses T, Seo M, Kanno Y, Häkkinen ST, Van Montagu MCE, Thevelein JM, Maaheimo H, Oksman-Caldentey K-M, Rodriguez PL, Rischer H, Goossens A (2011) Jasmonate signaling involves the abscisic acid receptor PYL4 to regulate metabolic reprogramming in *Arabidopsis* and tobacco. Proc Natl Acad Sci U S A 108:5891–5896

Lai Z, Vinod KM, Zheng Z, Fan B, Chen Z (2008) Roles of *Arabidopsis* WRKY3 and WRKY4 transcription factors in plant responses to pathogens. BMC Plant Biol 8:68

Larronde F, Gaudillère JP, Krisa S, Decendit A, Deffieux G, Mérillon JM (2003) Airborne methyl jasmonate induces stilbene accumulation in leaves and berries of grapevine plants. Am J Enol Vitic 54:63–66

Laudert D, Weiler EW (2002) Allene oxide synthase: a major control point in *Arabidopsis thaliana* octadecanoid signaling. Plant J 15:675–684

Laurie-Berry N, Joardar V, Street JH, Kunkel BN (2006) The *Arabidopsis thaliana JASMONATE INSENSITIVE 1* gene is required for suppression of salicylic acid-dependent defenses during infection by *Pseudomonas syringae*. Mol Plant Microbe Interact 19:789–800

Le Hénanff G, Profizi C, Courteaux B, Rabenoelina F, Gérard C, Clément C, Baillieul F, Cordelier S, Dhondt-Cordelier S (2013) Grapevine NAC1 transcription factor as a convergent node in developmental processes, abiotic stresses, and necrotrophic/biotrophic pathogen tolerance. J Exp Bot 64:4877. doi:10.1093/jxb/ert277

Lee H-L, Raskin I (1999) Purification, cloning, and expression of a pathogen inducible UDP-glucose: salicylic acid glucosyltransferase from tobacco. J Biol Chem 274:36637–36642

Lee DE, Lee IJ, Han O, Baik MG, Han SS, Back K (2004) Pathogen resistance of transgenic rice expressing mitogen-activated protein kinase 1, MK1, from *Capsicum annuum*. Mol Cells 17:81–85

Lee J, Nam J, Park HC, Na G, Miura K, Jin JB, Yoo CY, Baek D, Kim DH, Jeong JC, Kim D, Lee SY, Salt DE, Mengiste T, Gong Q, Ma S, Bohnert HJ, Kwak S-S, Bressan RA, Hasegawa PM, Yun D-J (2007) Salicylic acid-mediated innate immunity in *Arabidopsis* is regulated by SIZ1 SUMO E3 ligase. Plant J 49:79–90

Lehtonen NT, Akita M, Frank W, Reski R, Valkonen JPT (2012) Involvement of a class III peroxidase and the mitochondrial protein TSPO in oxidative burst upon treatment of moss plants with a fungal elicitor. Mol Plant Microbe Interact 25:363–371

Leon-Reyes A, Spoel SA, De Lange ES, Abe H, Kobayashi M, Tsuda S, Millenaar FF, Welschen RAM, Ritsema T, Pieterse CMJ (2009) Ethylene modulates the role of NONEXPRESSOR OF PATHOGENESIS-RELATED GENES1 in cross talk between salicylate and jasmonate signaling. Plant Physiol 149:1797–1809

Leon-Reyes A, Du Y, Koornneef A, Proietti S, Körbes AP, Memelink J, Pieterse CMJ, Ritsema T (2010a) Ethylene signaling renders the jasmonate response of *Arabidopsis* insensitive to future suppression by salicylic acid. Mol Plant-Microbe Interact 23:187–197

Leon-Reyes A, Van der Does D, De Lange ES, Delker C, Wasternack C, Van Wees SCM, Ritsema T, Pieterse CMJ (2010b) Salicylate-mediated suppression of jasmonate-responsive gene expression in *Arabidopsis* is targeted downstream of the jasmonate biosynthesis pathway. Planta 232:1423–1432

Li C, Liu G, Xu C, Lee GI, Bauer P, Ling HQ, Ganal MW, Howe GA (2003) The tomato *suppressor of prosystemin-mediated response2* gene encodes a fatty acid desaturase required for the biosynthesis of jasmonic acid and the production of a systemic wound signal for defense gene expression. Plant Cell 15:1646–1661

Li J, Brader G, Palva ET (2004) The WRKY70 transcription factor: a node of convergence for jasmonate-mediated and salicylate-mediated signals in plant defense. Plant Cell 16:319–331

Li C, Schilmiller AL, Liu G, Lee GI, Jayanty S, Sageman C, Vrebalov J, Giovannoni JJ, Yagi K, Kobayashi Y, Howe GA (2005) Role of β-oxidation in jasmonate biosynthesis and systemic wound signaling in tomato. Plant Cell 17:971–986

Li J-H, Liu Y-Q, Pin L, Lin H-F, Bai Y, Wang X-C, Chen Y-L (2009) A signaling pathway linking nitric oxide production to heterotrimeric G protein and hydrogen peroxide regulates extracellular calmodulin induction of stomatal closure in Arabidopsis. Plant Physiol 150:114–124

Li G, Meng X, Wang R, Mao G, Han L, Liu Y, Zhang S (2012) Dual-level regulation of ACC synthase activity by MPK3/MPK6 cascade and its downstream WRKY transcription factor during ethylene induction in Arabidopsis. PLoS Genet 8(6):e1002767. doi:10.1371/journal.pgen.1002767

Li F, Wang J, Ma C, Zhao Y, Wang Y, Hasi A, Qi Z (2013) Glutamate receptor-like channel3.3 is involved in mediating glutathione-triggered cytosolic calcium transients, transcriptional changes, and innate immunity responses in Arabidopsis. Plant Physiol 162:1497–1509

Li Y, Zhang Q, Zhang J, Wu L, Qi Y, Zhou J-M (2010) Identification of microRNAs involved in pathogen-associated molecular pattern-triggered plant innate immunity. Plant Physiol 152:2222–2231

Liu Y, Zhang S (2004) Phosphorylation of 1-aminocyclopropane-1-carboxylic acid synthase by MPK6, a stress-responsive mitogen-activated protein kinase, induces ethylene biosynthesis in *Arabidopsis*. Plant Cell 16:3386–3399

Liu J-Z, Horstman HD, Braun E, Graham MA, Zhang C, Navarre D, Qiu W-L, Lee Y, Nettleton D, Hill JH, Whitham SA (2011a) Soybean homologs of MPK4 negatively regulate defense responses and positively regulate growth and development. Plant Physiol 157:1363–1378

Liu P-P, von Dahl CC, Klessig DF (2011b) The extent to which methyl salicylate is required for signaling systemic acquired resistance is dependent on exposure to light after infection. Plant Physiol 157:2216–2226

Liu P-P, von Dahl CC, Park S-W, Klessig DF (2011c) Interconnection between methyl salicylate and lipid-based long-distance signaling during the development of systemic acquired resistance in *Arabidopsis* and tobacco. Plant Physiol 155:1762–1768

Lorenzo O, Solano R (2005) Molecular players regulating the jasmonate signaling network. Curr Opin Plant Biol 8:532–540

Lorenzo O, Piqueras R, Sánchez-Serrano JJ, Solano R (2003) ETHYLENE RESPONSE FACTOR 1 integrates signals from ethylene and jasmonate pathways in plant defense. Plant Cell 15:165–178

Lorenzo O, Chico JM, Sánchez-Serrano JJ, Solano R (2004) JASMONATE-INSENSITIVE1 encodes a MYC transcription factor essential to discriminate between different jasmonate-regulated defense responses in Arabidopsis. Plant Cell 16:1938–1950

Lu S, Sun Y-H, Amerson H, Chiang VL (2007) MicroRNAs in loblolly pine (*Pinus taeda* L.) and their association with fusiform rust gall development. Plant J 51:1077–1098

Lumbreras V, Vilela B, Irar S, Solé M, Capellades M, Valls M, Coca M, Pagés M (2010) MAPK phosphatase MKP2 mediates disease responses in *Arabidopsis* and functionally interacts with MPK3 and MPK6. Plant J 63:1017–1030

Mandal B, Mandal S, Csinos AS, Martinez N, Culbreath AK, Pappu HR (2008) Biological and molecular analyses of the acibenzolar-S-methyl-induced systemic acquired resistance in flue-cured tobacco against *Tomato spotted wilt virus*. Phytopathology 98:196–204

Mandujano-Chávez A, Schoenbeck MA, Ralston LF, Lozoya-Gloria E, Chappell J (2000) Differential induction of sesquiterpene metabolism in tobacco cell suspension cultures by methyl jasmonate and fungal elicitor. Arch Biochem Biophys 38:285–294

Mao P, Duan M, Wei C, Li Y (2007) WRKY62 transcription factor acts downstream of cytosolic NPR1 and negatively regulates jasmonate responsive gene expression. Plant Cell Physiol 48:833–842

Martinez-Medina A, Fernández I, Sánchez-Guzmán MJ, Jung SC, Pascual JA, Pozo MJ (2013) Deciphering the hormonal signalling network behind the systemic resistance induced by *Trichoderma harzianum* in tomato. Front Plant Sci 4:206

Mathys J, De Cremer K, Timmermans P, Van Kerckhove S, Lievens B, Vanhaecke M, Cammue BP, De Coninck B (2012) Genome-wide characterization of ISR induced in *Arabidopsis thaliana* by *Trichoderma hamatum* T382 against *Botrytis cinerea* infection. Front Plant Sci 3:108. doi:10.3389/fpls.2012.00108

Matsuura H, Takeishi S, Kiatoka N, Sato C, Sueda K, Masuta C, Nabeta K (2012) Transportation of de novo synthesized jasmonoyl isoleucine in tomato. Phytochemistry 83:25–33

McGrath KC, Dombrecht B, Manners JM, Schenk PM, Edgar CI, Maclean DJ, Scheible W-R, Udvardi MK, Kazan K (2005) Repressor- and activator-type ethylene response factors functioning in jasmonate signaling and disease resistance identified via a genome-wide screen of Arabidopsis transcription factor gene expression. Plant Physiol 139:949–959

McGurl B, Orozco-Cardenas M, Pearce G, Ryan CA (1994) Overexpression of the prosystemin gene in transgenic tomato plants generates a systemic signal that constitutively induces proteinase inhibitor synthesis. Proc Natl Acad Sci U S A 91:9799–9802

Mei C, Qi M, Sheng G, Yang Y (2006) Inducible overexpression of a rice allene oxide synthase gene increases the endogenous jasmonic acid level, PR gene expression, and host resistance to fungal infection. Mol Plant-Microbe Interact 19:1127–1137

Meldau S, Ullman-Zeunert L, Govind G, Bartram S, Baldwin IT (2012) MAPK-dependent JA and SA signalling in *Nicotiana attenuata* affects plant growth and fitness during competition with conspecifics. BMC Plant Biol 12:213

Melotto M, Mecey C, Niu Y, Chung HS, Katsir L, Yao J, Zeng W, Thines B, Staswick P, Browse J, Howe GA, He SY (2008) A critical role of two positively charged amino acids in the Jas motif of Arabidopsis JAZ proteins in mediating coronatine- and jasmonoyl isoleucine-dependent interactions with the COI1 F-box protein. Plant J 55:979–988

Memelink J, Verpoorte R, Kijne JW (2001) ORCAnization of jasmonate-responsive gene expression in alkaloid metabolism. Trends Plant Sci 6:212–219

Méndez-Bravo A, Calderón-Vázquez C, Ibarra-Laclette E, Raya-González J, Ramirez-Chávez E, Molina-Torres J, Guevara-Garcia AA, López-Bucio J, Herrera-Estrella L (2011) Alkamides activate jasmonic acid biosynthesis and signaling pathways and confer resistance to *Botrytis cinerea* in *Arabidopsis thaliana*. PLoS One 6(11):e27251

Mersmann S, Bourdais G, Rietz S, Robatzek S (2010) Ethylene signaling regulates accumulation of the FLS2 receptor and is required for the oxidative burst contributing to plant immunity. Plant Physiol 154:391–400

Mészáros T, Helfer A, Hatzimasoura E, Magyar Z, Serazetdinova L, Rios G, Bardóczy V, Teige M, Koncz C, Peck S, Bögre L (2006) The Arabidopsis MAP kinase kinase MKK1 participates in defence responses to the bacterial elicitor flagellin. Plant J 48:485–495

Michard E, Lima PT, Borges F, Silva AC, Portes MT, Carvalho JE, Gilliham M, Liu LH, Obermeyer G, Feijo JA (2011) Glutamate receptor-like genes form Ca^{2+} channels in pollen tubes and are regulated by pistil D-serine. Science 332:434–437

Mlotshwa S, Voinnet O, Florian Mette M, Matzke M, Vaucheret H, Ding SW, Pruss GT, Vance VB (2002) RNA silencing and the mobile silencing signal. Plant Cell 14:S.289–S.301

Moffat CS, Ingle RA, Wathugala DL, Saunders NJ, Knight H, Knight MR (2012) ERF5 and ERF6 play redundant roles as positive regulators of JA/Et-mediated defense against *Botrytis cinerea* in Arabidopsis. PLoS One 7(4):e35995. doi:10.137/journal.pone.0035995

Molnár A, Melnyk CW, Bassett A, Hardcastle TJ, Dunn R, Baulcombe DC (2010) Small silencing RNAs in plants are mobile and direct epigenetic modification in recipient cells. Science 328:872–875

Molnár A, Melnyk C, Baulcombe DC (2011) Silencing signals in plants: a long journey for small RNAs. Genome Biol 12:215

Moore JW, Loake CJ, Spoel SH (2011) Transcription dynamics in plant immunity. Plant Cell 23:2809–2820

Morel J, Fromentin J, Blein J-P, Simon-Plas F, Elmayan T (2004) RAC regulation of NtrbohD, the oxidase responsible for the oxidative burst in elicited tobacco cell. Plant J 37:282–293

Mou Z, Fan W, Dong X (2003) Inducers of plant systemic acquired resistance regulate NPR1 function through redox changes. Cell 113:935–944

Munemasa S, Hossain MA, Nakamura Y, Mori IC, Murata Y (2010) The Arabidopsis calcium-dependent protein kinase, CPK6, functions as a positive regulator of methyl jasmonate signaling in guard cells. Plant Physiol 155:553–561

Munemasa S, Mori IC, Murata Y (2011) Methyl jasmonate signaling and signal crosstalk between methyl jasmonate and abscisic acid in guard cells. Plant Signal Behav 6:939–941

Mur LAJ, Renlin X, Casson SA, Stoddart WM, Routledge APM, Draper J (2004) Characterization of a proteinase inhibitor from *Brachypodium distachyon* suggests the conservation of defense signalling pathways between dicotyledonous plants and grasses. Mol Plant Pathol 5:267–280

Mur LAJ, Kenton P, Atzorn R, Miersch O, Wasternack C (2006) The outcomes of concentration-specific interactions between salicylate and jasmonate signaling include synergy, antagonism, and oxidative stress leading to cell death. Plant Physiol 140:249–262

Mur LAJ, Sivakumaran A, Mandon J, Cristescu SM, Harren FJM, Hebelstrup KH (2012) Haemoglobin modulates salicylate and jasmonate/ethylene-mediated resistance mechanisms against pathogens. J Exp Bot 63:4375–4387

Mur LAJ, Prats E, Pierre S, Hall MA, Hebelstrup KH (2013) Integrating nitric oxide into salicylic acid and jasmonic acid/ethylene plant defense pathways. Front Plant Sci 4:215

Nahar K, Kyndt T, Hause B, Höfte M, Gheysen G (2013) Brassinosteroids suppress rice defense against root-knot nematodes through antagonism with jasmonate pathway. Mol Plant-Microbe Interact 26:106–115

Nakashima A, Chen L, Thao NP, Fujiwara M, Wong HL, Kuwano M, Umemura K, Shirasu K, Kawasaki T, Shimamoto K (2008) RACK1 functions in rice immunity by interacting with the Rac1 immune complex. Plant Cell 20:2265–2279

Nambeesan S, AbuQamar S, Laluk K, Mattoo AK, Mickelbart MV, Ferruzzi MG, Mengiste T, Handa AK (2012) Polyamines attenuate ethylene – mediated defense responses to abrogate resistance to *Botrytis cinerea* in tomato. Plant Physiol 158:1034–1045

Návarová H, Bernsdorff F, Döring A-C, Zeier J (2012) Pipecolic acid, an endogenous mediator of defense amplification and priming, is a critical regulator of inducible plant immunity. Plant Cell 24:5123–5141

Navarro L, Dunoyer P, Jay F, Arnold B, Dhamasiri N, Estelle M, Voinnet O, Jones JDG (2006) A plant miRNA contributes to antibacterial resistance by repressing auxin signaling. Science 312:436–439

Navarro L, Bari R, Achard P, Lison P, Nemri A, Harberd NP, Jones JDG (2008) DELLAs control plant immune responses by modulating the balance of jasmonic acid and salicylic acid signaling. Curr Biol 18:650–655

Ndamukong I, Abdallat AA, Thurow C, Fode B, Zander M, Weigel R, Gatz C (2007) SA-inducible *Arabidopsis* glutaredoxin interacts with TGA factors and suppresses JA-responsive *PDF1.2* transcription. Plant J 50:128–139

Nickstadt A, Thomma BPHJ, Feussner I, Kangasjarvi J, Zeier J, Loeffer C, Scheel D, Berger S (2004) The jasmonate-insensitive mutant *jin1* shows increased resistance to biotrophic as well as necrotrophic pathogens. Mol Plant Pathol 5:425–434

Niu Y, Figueroa P, Browse J (2011) Characterization of JAZ-interacting bHLH transcription factors that regulate jasmonate responses in Arabidopsis. J Exp Bot 62:2143–2154

Nurmberg PL, Knox KA, Yun B-W, Morris PC, Shafiei R, Hudson A, Loake GJ (2007) The developmental selector *AS1* is an evolutionarily conserved regulator of the plant immune response. Proc Natl Acad Sci U S A 104:18795–18800

Nürnberger T, Küfner I (2011) The role of the plant plasma membrane in microbial sensing and innate immunity: the plant plasma membrane. Plant Cell Monogr 19:471–483

O'Donnell PJ, Calvert R, Atzorn R, Wasternack C, Leyser HMO, Bowles DJ (1996) Ethylene as a signal mediating the wound response of tomato plants. Science 274:1914–1917

Oñate-Sánchez L, Anderson JP, Young J, Singh KB (2007) AtERF14, a member of the ERF family of transcription factors, plays a nonredundant role in plant defense. Plant Physiol 143:400–409

Orozco-Cárdenas ML, Narvaez-Vasquez J, Ryan CA (2001) Hydrogen peroxide acts as a second messenger for the induction of defense genes in tomato plants in response to wounding, systemin, and methyl jasmonate. Plant Cell 13:1047–1061

Ozawa R, Bertea CM, Foti M, Narayana R, Arimura G-I, Muroi A, Horiuchi JI, Nishioka T, Maffei ME, Takabayashi J (2009) Exogenous polyamines elicit herbivore-induced volatiles in lima bean leaves: involvement of calcium, H_2O_2 and jasmonic acid. Plant Cell Physiol 50:2183–2199

Ozawa R, Bertea CM, Foti M, Narayana R, Arimura G-I, Muroi A, Maffei ME, Takabayashi J (2010) Polyamines and jasmonic acid induce plasma membrane potential variations in Lima bean. Plant Signal Behav 5:308–310

Palmieri MC, Sell S, Huang X, Scherf M, Werner T, Durner J, Lindermayr C (2008) Nitric oxide-responsive genes and promoters in *Arabidopsis thaliana*: a bioinformatics approach. J Exp Bot 59:177–180

Pandey SP, Somssich IE (2009) The role of WRKY transcription factors in plant immunity. Plant Physiol 150:1648–1655

Pauwels L, Goossens A (2011) The JAZ proteins: a crucial interface in the jasmonate signaling cascade. Plant Cell 23:3080–3100

Pauwels L, Barbero GE, Geerinck J, Tilleman S, Grunewald W, Perez AC, Chico JM, Bossche RV, Sewell J, Gill E et al (2010) NINJA connects the co-repressor TOPLESS to jasmonate signaling. Nature 464:788–791

Pedley KF, Martin GB (2005) Role of mitogen-activated protein kinases in plant immunity. Curr Opin Plant Biol 8:541–547

Penninckx IAMA, Eggermont K, Terras FR, Thomma BP, De Samblanx GW, Buchala A, Metraux JP, Manners JM, Broekaert WF (1996) Pathogen-induced systemic activation of a plant defensin gene in Arabidopsis follows a salicylic acid-independent pathway. Plant Cell 8:2309–2323

Perchepied L, Balagué C, Riou C, Claudel-Renard C, Riviére N, Grezes-Besset B, Roby D (2010) Nitric oxide participates in the complex interplay of defense-related signaling pathways controlling disease resistance to *Sclerotinia sclerotiorum* in *Arabidopsis thaliana*. Mol Plant-Microbe Interact 23:846–860

Perfus-Barbeoch I, Jones AM, Assmann SM (2004) Plant heterotrimeric G protein function: insights from *Arabidopsis* and rice mutants. Curr Opin Plant Biol 7:719–731

Petersen M, Brodersen P, Naested H, Andreasson E, Lindhart U, Johansen B, Nielsen HB, Lacy M, Austin MJ, Parker JE, Sharma SB, Klessig DF, Martienssen R, Mattsson O, Jensen AB, Mundy J (2000) *Arabidopsis* MAP kinase 4 negatively regulates systemic acquired resistance. Cell 103:1111–1120

Pieterse CMJ, Leon-Reyes A, Van der Ent S, Van Wees SCM (2009) Networking by small-molecule hormones in plant immunity. Nat Chem Biol 5:308–316

Pieterse CMJ, Leon-Reyes A, van der Does D, Verhage A, Koornneef A, van Pelt JA, van Wees SCM (2012a) Networking by small-molecule hormones in plant immunity. Induced resistance against insects and diseases. IOBC-WPRS Bull 83:77–80

Pieterse CMJ, van der Does D, Zamioudis C, Leon-Reyes A, van Wees SCM (2012b) Hormonal modulation of plant immunity. Annu Rev Cell Dev Biol 28:489–521

Piggott N, Ekramoddoullah AKM, Liu J-J, Yu X (2004) Gene cloning of a thaumatin-like (PR-5) protein of western white pine (*Pinus monticola* D. Don) and expression studies of members of the PR-5 group. Physiol Mol Plant Pathol 64:1–8

Pitzschke A, Djamei A, Bitton F, Hirt H (2009) A major role of the MEKK1-MKK1/2-MPK4 pathway in ROS signaling. Mol Plant 2:120–137

Po-Wen C, Singh P, Zimmerli L (2013) Priming of the Arabidopsis pattern-triggered immunity response upon infection by necrotrophic *Pectobacterium carotovorum* bacteria. Mol Plant Pathol 14:58–70

Pré M, Atallah M, Champion A, De Vos M, Pieterse CMJ, Memelink J (2008) The AP2/ERF domain transcription factor ORA59 integrates jasmonic acid and ethylene signals in plant defense. Plant Physiol 147:1347–1357

Price MB, Jelesko J, Okumoto S (2012) Glutamine receptor homologs in plants: functions and evolutionary origins. Front Plant Sci 3:Article 235. doi:10.3389/fpls.2012.00235

Qiu D, Xiao J, Ding X, Xiong M, Cai M, Cao Y, Li X, Xu C, Wang S (2007) OsWRKY13 mediates rice disease resistance by regulating defense-related genes in salicylate- and jasmonate-dependent signaling. Mol Plant-Microbe Interact 20:492–499

Qiu JL, Fiil BK, Petersen K, Nielsen HB, Botanga CJ, Thorgrimsen S, Palma K, Suarez-Rodriguez MC, Sandbech-Clausen S, Lichota J, Brodersen P, Grasser KD, Mattson O, Glazebrook J, Mundy J, Petersen M (2008) *Arabidopsis* MAP kinase 4 regulates gene expression through transcription factor release in the nucleus. EMBO J 27:2214–2221

Qu F, Ye X, Morris TJ (2008) *Arabidopsis* DRB4, AGO1, AGO7, and RDR6 participate in a DCL4-initiated antiviral silencing pathway negatively regulated by DCL1. Proc Natl Acad Sci U S A 105:14732–14737

Raacke IC, Mueller MJ, Berger S (2006) Defects in allene oxide synthase and 12-oxo-phytodienoic acid reductase alter the resistance to *Pseudomonas syringae* and *Botrytis cinerea*. J Phyopathol 154:740–744

Rakwal R, Tamogani S, Agrawal GK, Iwahashi H (2002) Octadecanoid signalling component 'burst' in rice (*Oryza sativa* L) seedlings leaves upon wounding by cut and treatment with fungal elicitor chitosan. Biochem Biophys Res Commun 295:1041–1045

Ramachandran V, Chen X (2008) Small RNA metabolism in *Arabidopsis*. Trends Plant Sci 13:368–374

Ranf S, Eschen-Lippold L, Pecher P, Lee J, Scheel D (2011) Interplay between calcium signaling and early signaling elements during defense responses to microbe-or damage-associated molecular patterns. Plant J 68:100–113

Ren D, Yang K-Y, Li G-J, Liu Y, Zhang S (2006) Activation of Ntf4, a tobacco mitogen-activated protein kinase, during plant defense and its involvement in hypersensitive response-like cell death. Plant Physiol 141:1482–1493

Repka V, Fischerova I, Silharova K (2004) Metyl jasmonate is a potent elicitor of multiple defense responses in grapevine leaves and cell-suspension cultures. Biol Plant 48:273–283

Robert-Seilaniantz A, Grant M, Jones JDG (2011) Hormone crosstalk in plant disease and defense: more than just JASMONATE-SALICYLATE antagonism. Annu Rev Phytopathol 49:317–343

Ryan CA (2000) The systemin signaling pathway: differential activation of plant defensive genes. Biochim Biophys Acta 1447:112–121

Sagi M, Fluhr R (2006) Production of reactive oxygen species by plant NADPH oxidases. Plant Physiol 141:336–340

Saito N, Munemasa S, Nakamura Y, Shimoishi Y, Mori IC, Murata Y (2008) Roles of RCN1, regulatory A subunit of protein phosphatase 2A, in methyl jasmonate signaling and signal crosstalk between methyl jasmonate and abscisic acid. Plant Cell Physiol 49:1396–1401

Sasaki Y, Asamizu E, Shibata D, Nakamura Y, Kaneko T, Awai K, Amagi M, Kuwata C, Tsugane T, Masuda T, Shimada H, Takamiya K, Ohta H, Tabata S (2001) Monitoring of methyl jasmonate-responsive genes in Arabidopsis by cDNA microarray: self-activation of jasmonic acid biosynthesis and crosstalk with other phytohormone signaling pathways. DNA Res 8:153–161

Sato C, Aikawa K, Sugiyama S, Nabeta K, Masuta C, Matsuura H (2011) Distal transport of exogenously applied jasmonoyl-isoleucine with wounding stress. Plant Cell Physiol 52:509–517

Schaller F (2001) Enzymes of the biosynthesis of octadecanoid-derived signaling molecules. J Exp Bot 52:11–23

Schaller A, Ryan CA (1995) Systemin-a polypeptide defense signal in plants. Bioessays 18:27–33

Schilmiller AL, Koo AJK, Howe GA (2006) Functional diversification of acyl-coenzyme A oxidases in jasmonic acid biosynthesis and action. Plant Physiol 143:812–824

Schommer C, Palatnik JF, Aggarwal P, Chételat A, Cubas P, Farmer EE, Nath U, Weigel D (2008) Control of jasmonate biosynthesis and senescence by miR319 targets. PLoS Biol 6(9):e230

Schultheiss H, Dechert C, Kogel K-H, Hückelhoven R (2003) Functional analysis of barley RAC/ROP G-protein family members in susceptibility to the powdery mildew fungus. Plant J 36:589–601

Schweighofer A, Kazanaviciute V, Scheikl E, Teige M, Doczi R, Hirt H, Schwanninger M, Kant M, Schuurink R, Mauch F, Buchala A, Cardinale F, Meskiene I (2007) The PP2C-type phosphatase AP2C1, which negatively regulates MPK4 and MPK6, modulates innate immunity, jasmonic acid, and ethylene levels in *Arabidopsis*. Plant Cell 19:2213–2224

Segarra G, Van der Ent S, Trillas I, Pieterse CMJ (2009) MYB72, a node of convergence in induced systemic resistance triggered by a fungal and bacterial beneficial microbe. Plant Biol 11:90–96

Segonzac C, Zipfel C (2011) Activation of plant pattern-recognition receptors by bacteria. Curr Opin Microbiol 14:54–61

Seo S, Sano H, Ohashi Y (1999) Jasmonate-based wound signal transduction requires activation of WIPK, a tobacco mitogen-activated protein kinase. Plant Cell 11:289–298

Seo HS, Song JT, Cheong J-J, Lee Y-W, Hwang I, Lee JS, Choi YD (2001) Jasmonic acid carboxyl methyltransferase: a key enzyme for jasmonate-regulated plant responses. Proc Natl Acad Sci U S A 98:4788–4793

Seo S, Katou S, Seto H, Gomi K, Ohashi Y (2007) The mitogen-activated protein kinases WIPK and SIPK regulate the levels of jasmonic and salicylic acids in wounded tobacco plants. Plant J 49:899–909

Shah J (2009) Plants under attack: systemic signals in defence. Curr Opin Plant Biol 12:459–464

Shah J, Zeier J (2013) Long-distance communication and signal amplification in systemic acquired resistance. Front Plant Sci 4:Article 30. doi:10.3389/fpls.2013.00030

Sharon M, Freeman S, Sneh B (2011) Assessment of resistance pathways induced in *Arabidopsis thaliana* by hypovirulent *Rhizoctonia* spp. isolates. Phytopathology 101:828–838

Sheard LB, Tan X, Mao H, Withers J, Ben-Nissan G, Hinds TR, Kobayashi Y, Hsu FF, Sharon M, Browse J, He SY, Rizo J, Howe GA, Zheng N (2010) Jasmonate perception by inositol phosphate-potentiated COI1-JAZ co-receptor. Nature 468:400–405

Shen QH, Saijo Y, Mauch S, Biskup C, Bieri S, Keller B, Seki H, Ulker B, Somssich IE, Schulze-Lefert P (2007) Nuclear activity of MLA immune receptors links isolate-specific and basal disease resistance responses. Science 315:1098–1103

Shen X, Yuan B, Liu H, Li X, Xu C, Wang S (2010) Opposite functions of a rice mitogen-activated protein kinase during the process of resistance against Xanthomonas oryzae. Plant J 64:86–99

Shen X, Liu H, Yuan B, Li X, Xu C, Wang S (2011) OsEDR1 negatively regulates rice bacterial resistance via activation of ethylene biosynthesis. Plant Cell Environ 34:179–191

Shim JS, Jung C, Lee S, Min K, Lee YW, Choi Y, Lee JS, Song JT, Kim JK, Choi YD (2013) AtMYB44 regulates WRKY70 expression and modulates antagonistic interaction between salicylic acid and jasmonic acid signaling. Plant J 73:483–495

Shyu C, Figueroa P, Depew CL, Cooke TF, Sheard LB, Moreno JF, Katsir L, Zheng N, Browse J, Howe GA (2012) JAZ8 lacks a canonical degron and has an EAR motif that mediates transcriptional repression of jasmonate responses in Arabidopsis. Plant Cell 24:536–550

Soitamo AJ, Jada B, Lehto K (2011) HC-Pro silencing suppressor significantly alters the gene expression profile in tobacco leaves and flowers. BMC Plant Biol 11:68

Spoel SH, Koornneef A, Claessens SMC, Korzelius JP, Van Pelt JA, Mueller MJ, Buchala AJ, Metraux J-P, Brown R, Kazan K, Van Loon LC, Dong X, Pieterse CMJ (2003) NPR1 modulates cross-talk between salicylate- and jasmonate-dependent defense pathways through a novel function in the cytosol. Plant Cell 15:760–770

Spoel SH, Johnson JS, Dong X (2007) Regulation of tradeoffs between plant defenses against pathogens with different lifestyles. Proc Natl Acad Sci U S A 104:18842–18847

Spoel SH, Mou Z, Tada Y, Spivey NW, Genschik P, Dong X (2009) Proteasome-mediated coactivator NPR1 plays dual roles in regulating plant immunity. Cell 137:860–872

Staswick PE, Tiryaki I (2004) The oxylipin signal jasmonic acid is activated by an enzyme that conjugates it to isoleucine in *Arabidopsis*. Plant Cell 16:2117–2127

Staswick PE, Yuen GY, Lehman CC (1998) Jasmonate signaling mutants of Arabidopsis are susceptible to the soil fungus *Pythium irregulare*. Plant J 15:747–754

Stintzi A, Weber H, Reymond P, Browse J, Farmer EE (2001) Plant defense in the absence of jasmonic acid: the role of cyclopentenones. Proc Natl Acad Sci U S A 98:12837–12842

Stitz M, Baldwin IT, Gaquerel E (2011) Diverting the flux of the JA pathway in *Nicotiana attenuata* compromises the plant's defense metabolism and fitness in nature and glasshouse. PLoS One 8(10):e25925

Sugiyama Y, Uraji M, Watanabe-Sugimoto M, Okuma E, Munemasa S, Shimoishi Y, Nakamura Y, Mori IC, Iwai S, Murata Y (2012) FIA functions as an early signal component of abscisic acid signal cascade in *Vicia faba* guard cells. J Exp Bot 63:1357–1365

Sun J-Q, Jiang H-L, Li C-Y (2011) Systemin/jasmonate-mediated systemic defense signaling in tomato. Mol Plant 4:607–615

Suza WP, Rowe ML, Hamberg M, Staswick PE (2010) A tomato enzyme synthesizes (+)-7-*iso*-jasmonoyl-L-isoleucine in wounded leaves. Planta 231:717–728

Szepesi A, Gémes K, Orosz G, Petö A, Takács Z, Vorák M, Tari I (2011) Interaction between salicylic acid and polyamines and their possible roles in tomato hardening processes. Acta Biol Szegediensis 55:165–166

Takahashi F, Yoshida R, Ichimura K, Mizoguchi T, Seo S, Yonezawa M, Maruyama K, Yamaguchi-Shinozaki K, Shinozaki K (2007) The mitogen-activated protein kinase cascade MKK3-MPK6 is an important part of the jasmonate signal transduction pathway in Arabidopsis. Plant Cell 19:805–818

Taki N, Sasaki-Sekimoto Y, Obayashi T, Kikuta A, Kobayashi K, Ainai T, Yagi K, Sakurai N, Suzuki H, Masuda T, Takayama K-I, Shibata D, Kobayashi Y, Ohta H (2005) Oxo-phytodienoic acid triggers expression of a distinct set of genes and plays a role in wound-induced gene expression in Arabidopsis. Plant Physiol 139:1268–1283

Teige M, Scheikl E, Eulgem T, Doczi R, Ichimura K, Shinozaki K, Dangl JL, Hirt H (2004) The MKK2 pathway mediates cold and salt stress signaling in Arabidopsis. Mol Cell 15:141–152

Thaler JS, Humphrey PT, Whiteman NK (2012) Evolution of jasmonate and salicylate signal crosstalk. Trends Plant Sci 17:260–270

Thatcher LF, Manners JM, Kazan K (2009) *Fusarium oxysporum* hijacks COI1-mediated jasmonate signaling to promote disease development in *Arabidopsis*. Plant J 58:927–939

Theodosiou A, Ashworth A (2002) MAP kinase phosphatases. Genome Biol 3:1–10

Thines B, Katsir L, Melotto M, Niu Y, Mandaokar A, Liu G, Nomura K, He SY, Howe GA, Browse J (2007) JAZ repressor proteins are targets of the SCF^{COI1} complex during jasmonate signaling. Nature 448:661

Thomma BPHJ, Eggermont K, Penninckx IAMA, Mauch-Mani B, Vogelsang R, Cammue BPA, Broekaert WF (1998) Separate jasmonate-dependent and salicylate-dependent defense-response pathways in Arabidopsis are essential for resistance to distinct microbial pathogens. Proc Natl Acad Sci U S A 95:15107–15111

Thomma B, Penninckx I, Broekaert WF, Cammue BPA (2001) The complexity of disease signaling in *Arabidopsis*. Curr Opin Immunol 13:63–68

Tikhonov DB, Magazanik LG (2009) Origin and molecular evolution of ionotropic glutamate receptors. Neurosci Behav Physiol 39:763–773

Trusov Y, Rookes JE, Chakravorty D, Armour D, Schenk PM, Botella JR (2006) Heterotrimeric G proteins facilitate *Arabidopsis* resistance to necrotrophic pathogens and are involved in jasmonate signaling. Plant Physiol 140:210–220

Tun NN, Santa-Catarina C, Begum T, Silveira V, Handro W, Floh EIS, Scherer GFE (2006) Polyamines induce rapid biosynthesis of nitric oxide (NO) in *Arabidopsis thaliana* seedlings. Plant Cell Physiol 47:346–354

Turner JG, Ellis C, Devoto A (2002) The jasmonate signal pathway. Plant Cell 14:S153–S164

Ueeda M, Kubota M, Nishi K (2006) Contribution of jasmonic acid to resistance against *Phytophthora capsici* in *Capsicum annuum* cv SCM334. Physiol Mol Plant Pathol 67:149–154

Umemura K, Satou J, Iwata M, Uozumi N, Koga J, Kawano T, Koshiba T, Anzai H, Mitomi M (2009) Contribution of salicylic acid glucosyltransferase, OsSGT1, to chemically induced disease resistance in rice plants. Plant J 57:463–472

Uppalapati SR, Ishiga Y, Wangdi T, Kunkel BN, Anand A, Mysore KS, Bender CL (2007) The phytotoxin coronatine contributes to pathogen fitness and is required for suppression of salicylic acid accumulation in tomato inoculated with *Pseudomonas syringae* pv *tomato* DC3000. Mol Plant Microbe Interact 20:955–965

Uraji M, Katagiri T, Okuma E, Ye W, Hossain MA, Masuda C, Miura A, Nakamura Y, Mori IC, Shinozaki K, Murata Y (2012) Cooperative function of PLDδ and PLDα1 in abscisic acid-induced stomatal closure in Arabidopsis. Plant Physiol 159:450–460

Van der Does D, Leon-Reyes A, Koornneef A, Van Verk MC, Rodenburg N, Pauwels L, Goossens A, Körbes AP, Memelink J, Ritsema T, Van Wees SCM, Pieterse CMJ (2013) Salicylic acid suppresses jasmonic acid signaling downstream of SCF^{COI1}-JAZ by targeting GCC promoter motifs via transcription factor ORA59. Plant Cell 25:744–761

Van der Ent S, Verhagen BWM, Van Doorn R, Bakker D, Verlaan MG, Pel MJC, Joosten RG, Proveniers MCG, Van Loon LC, Ton J, Pieterse CMJ (2008) *MYB72* is required in early signaling steps of rhizobacteria-induced systemic resistance in *Arabidopsis*. Plant Physiol 146:1293–1304

van Verk MC, Pappaaioannou D, Neeleman L, Bol JF, Linthorst HJM (2008) A novel WRKY transcription factor is required for induction of *PR-1a* gene expression by salicylic acid and bacterial elicitors. Plant Physiol 140:1983–1995
van Verk MC, Bol JF, Linthorst HJM (2011) WRKY transcription factors involved in activation of SA biosynthesis genes. BMC Plant Biol 11:89. doi:101186/1471-2229-11-89
Vandenabeele S, Van Der Kelen K, Dat J, Gadjev I, Boonefaes T, Morsa S, Rottiers P, Slooten L, Van Montagu M, Zabeau M, Inzé D, Van Breusegem F (2003) A comprehensive analysis of hydrogen peroxide-induced gene expression in tobacco. Proc Natl Acad Sci U S A 100:16113–16118
Vatsa P, Chiltz A, Bourgue S, Wendehenne D, Garcia-Brugger A, Pugin A (2011) Involvement of putative glutamate receptors in plant defence signaling and NO production. Biochimie 93:2095–2101
Verhage A, Vlaardingerbroek I, Raaymakers C, Van Dam NM, Dicke M, Van Wees SCM, Pieterse CMJ (2011) Rewiring of the jasmonate signaling pathway in Arabidopsis during insect herbivory. Front Plant Sci 2:47. doi:10.3389/fpls.2011.00047
Vicedo B, Flors V, Leyva Mdl O, Finiti I, Kravchuk Z, Real MD, Garcia-Agustin P, González-Bosch C (2009) Hexanoic acid-induced resistance against *Botrytis cinerea* in tomato plants. Mol Plant-Microbe Interact 22:1455–1465
Vidhyasekaran P (2007) Fungal pathogenesis in plants and crops: molecular biology and host defense mechanisms, 2nd edn. CRC Press/Taylor Francis Group, Boca Raton, p 510
Vidhyasekaran P (2014) PAMP signals in plant innate immunity: signal perception and transduction. Springer, Dordrecht, p 442
Vierstra RD (2003) The ubiquitin/26S proteasome pathway, the complex last chapter in the life of many plant proteins. Trends Plant Sci 8:135–142
Vincill ED, Bieck AM, Spalding EP (2012) Ca^{2+} conduction by an amino acid-gated ion channel related to glutamate receptors. Plant Physiol 159:40–46
Voinnet O (2001) RNA silencing as a plant immune system against viruses. Trends Genet 17:449–459
Vranová E, Inzé D, Van Brueusegem F (2002) Signal transduction during oxidative stress. J Exp Bot 53:1227–1236
Wager A, Browse J (2012) Social network: JAZ protein interactions expand our knowledge of jasmonate signaling. Fron Plant Sci 3:Article 41
Walters D (2003) Resistance to plant pathogens: possible roles for free polyamines and polyamine catabolism. New Phytol 159:109–115
Wang C, Zien C, Afitlhile M, Welti R, Hildebrand DF, Wang X (2000) Involvement of phospholipase D in wound-induced accumulation of jasmonic acid in Arabidopsis. Plant Cell 12:2237–2246
Wang KLC, Li H, Ecker IP (2002) Ethylene biosynthesis and signaling networks. Plant Cell 14:S131–S155
Wang Z, Dai L, Jiang Z, Peng W, Zhang L, Wang G, Xie D (2005) *GmCOI1*, a soybean F-box protein gene, shows ability to mediate jasmonate-regulated plant defense and fertility in Arabidopsis. Mol Plant Microbe Interact 18:1285–1295
Wang D, Amonsiripanitch N, Dong X (2006) A genomic approach to identify regulatory nodes in the transcriptional network of systemic acquired resistance in plants. PLoS Pathog 2:e123
Wang L, Allmann S, Wu J, Baldwin IT (2008) Comparisons of LIPOXYGENASE3- and JASMONATE-RESISTANT4/6-silenced plants reveal that jasmonic acid-amino acid conjugates play different roles in herbivore resistance of *Nicotiana attenuata*. Plant Physiol 146:904–915
Wang Z, Mao H, Dong C, Ji R, Cai L, Fu H, Liu S (2009) Overexpression of *Brassica napus* MPK4 enhances resistance to *Sclerotinia sclerotiorum* in oilseed rape. Mol Plant Microbe Interact 22:235–244
Wang J, Liu C-Y, Zhang L-W, Wang J-L, Hu G-H, Ding J-J, Chen Q-S (2011) MicroRNAs involved in the pathogenesis of Phytophthora root rot of soybean (*Glycine max*). Agric Sci China 10:1150–1167

Wasternack C (2007) Jasmonates: an update on biosynthesis, signal transduction and action in plant stress response, growth and development. Ann Bot 100:681–697

Wasternack C, Hause B (2013) Jasmonates: biosynthesis, perception, signal transduction and action in plant stress response, growth and development. An update to the 2007 review in *Annals of Botany*. Ann Bot 111:1021–1058

Wasternack C, Kombrink E (2010) Jasmonates: structural requirements for lipid derived signals active in plant stress responses and development. ACS Chem Biol 5:63–77

Wasternack C, Forner S, Strnad M, Hause B (2013) Jasmonates in flower and seed development. Biochimie 95:79–85

Wathugala DL, Hemsley PA, Moffat CS, Cremelie P, Knight MR, Knight H (2012) The Mediator subunit SFR6/MED16 controls defence gene expression mediated by salicylic acid and jasmonate responsive pathways. New Phytol 195:217–230

Weller DM, Mavrodi DV, Van Pelt JA, Pieterse CMJ, Van Loon LC, Bakker PAHM (2012) Induced systemic resistance (ISR) in *Arabidopsis thaliana* against *Pseudomonas syringae* pv. tomato by 2,4-diacetylphloroglucinol-producing *Pseudomonas fluorescens*. Phytopathology 102:403–412

Westfall CS, Zubieta C, Herrmann J, Kapp U, Nanao MH, Jez JM (2012) Structural basis for pre-receptor modulation of plant hormones by GH3 proteins. Science 336:1708–1711

Withers J, Yao J, Mecey C, Howe GA, Melotto M, He SY (2012) Transcription factor-dependent nuclear localization of a transcriptional repressor in jasmonate hormone signaling. Proc Natl Acad Sci U S A 109:20148–20153

Woldemariam MG, Onkokesung N, Baldwin IT, Galis I (2012) Jasmonoyl-1-isoleucine hydrolase 1 (JIH1) regulates jasmonoyl-1-isoleucine levels and attenuates plant defenses against herbivores. Plant J 72:756–767

Wong HL, Pinontoan R, Hayashi K, Tabata R, Yaeno T, Hasegawa K, Kojima C, Yoshioka H, Iba K, Kawasaki T, Shimamoto K (2007) Regulation of rice NADPH oxidase by binding of Rac GTPase to its N-terminal extension. Plant Cell 19:4022–4034

Wu K, Zhang L, Zhou C, Yu C-W, Chalkam V (2008) HDA6 is required for jasmonate response, senescence and flowering in *Arabidopsis*. J Exp Bot 59:225–234

Xie DX, Feys BF, James S, Nieto-Rostro M, Turner JG (1998) *COI1*: An *Arabidopsis* gene required for jasmonate-regulated defense and fertility. Science 280:1091–1094

Xing DH, Lai ZB, Zheng ZY, Vinod KM, Fan BF, Chen ZX (2008) Stress- and pathogen-induced *Arabidopsis* WRKY48 is a transcriptional activator that represses plant basal defense. Mol Plant 1:459–470

Xu L, Liu F, Lechner E, Genschik P, Crosby WL, Ma H, Peng W, Huang D, Xie D (2002) The SCF(COI1) ubiquitin-ligase complexes are required for jasmonate response in *Arabidopsis*. Plant Cell 14:1919–1935

Xu M-J, Dong J-F, Zhu M-Y (2005) Nitric oxide mediates the fungal elicitor-induced hypericin production of *Hypericun perforatum* cell suspension cultures through a jasmonic-acid-dependent signal pathway. Plant Physiol 139:991–998

Yalowsky S, Baluska F, Jones A (2010) Integrated G proteins signaling in plants. Springer, Heidelberg

Yamaguchi Y, Huffaker A, Bryan AC, Tax FE, Ryan CA (2010) PEPR2 is a second receptor for the Pep1 and Pep2 peptides and contributes to defense responses in *Arabidopsis*. Plant Cell 22:508–522

Yamaguchi Y, Pearce G, Ryan CA (2006) The cell surface leucine-rich repeat receptor for *At*Pep1, an endogenous peptide elicitor in *Arabidopsis*, is functional in transgenic tobacco cells. Proc Natl Acad Sci U S A 103:10104–10109

Yan Z, Reddy MS, Ryu CM, McInroy JA, Wilson M, Kloepper JW (2002) Induced systemic protection against tomato late blight elicited by plant growth-promoting rhizobacteria. Phytopathology 92:1329–1333

Yan J, Tsuichihara N, Etoh T, Iwai S (2007) Reactive oxygen species and nitric oxide are involved in ABA inhibition of stomatal opening. Plant Cell Environ 30:1320–1325

Yan J, Zhang C, Gu M, Bai Z, Zhang W, Qi T, Cheng Z, Peng W, Luo H, Nam F, Wang Z, Xie D (2009) The Arabidopsis CORONATINE INSENSITIVE1 protein is a jasmonate receptor. Plant Cell 21:2220–2236

Yan Y, Christensen S, Isakeit T, Engelberth J, Meeley R, Hayward A, Emery RJ, Kolomiets MV (2012) Disruption of OPR7 and OPR8 reveals the versatile functions of jasmonic acid in maize development and defense. Plant Cell 24:1420–1436

Yang Z (2002) Small GTPases: versatile signaling switches in plants. Plant Cell 14(Suppl):S375–S388

Yang D-L, Yang Y, He Z (2013) Roles of plant hormones and their interplay in rice immunity. Mol Plant 6:675–685

Yoda H, Fujimura K, Takahashi H, Munemura I, Uchimiya H, Sano H (2009) Polyamines as a common source of hydrogen peroxide in host- and nonhost hypersensitive response during pathogen infection. Plant Mol Biol 70:103–112

Yoon JY, Hamayun M, Lee SK, Lee I-J (2009) Methyl jasmonate alleviated salinity stress in soybean. J Crop Sci Biotech 12:63–68

Zago E, Morsa S, Dat JF, Alard P, Ferrarini A, Inzé D, Delledonne M, Van Breusegem F (2006) Nitric oxide- and hydrogen peroxide-responsive gene regulation during cell death induction in tobacco. Plant Physiol 141:404–411

Zamioudis C, Pieterse CMJ (2012) Modulation of host immunity by beneficial microbes. Mol Plant Microbe Interact 25:139–150

Zander M, La Camera S, Lamotte O, Métraux JP, Gatz C (2010) *Arabidopsis thaliana* class-II TGA transcription factors are essential activators of jasmonic acid/ethylene-induced defense responses. Plant J 61:200–210

Zarei A, Körbes AP, Younessi P, Montiel G, Champion A, Memelink J (2011) Two GCC boxes and AP2/ERF-domain transcription factor ORA59 in jasmonate/ethylene-mediated activation of the PDF1.2 promoter in Arabidopsis. Plant Mol Biol 75:321–331

Zhang Y, Tessaro MJ, Lassner M, Li X (2003) Knockout analysis of *Arabidopsis* transcription factors *TGA2*, *TGA5*, and *TGA6* reveals their redundant and essential roles in systemic acquired resistance. Plant Cell 15:2647–2653

Zhang P, Vanderschuren H, Futterer J, Gruissem W (2005) Resistance to cassava mosaic disease in transgenic cassava expressing antisense RNAs targeting virus replication genes. Plant Biotechnol J 3:385–397

Zhang W, Gao S, Zhou X, Chellappan P, Chen Z, Zhou X, Zhang X, Fromuth N, Coutino G, Coffey M, Jin N (2011a) Bacteria-responsive microRNAs regulate plant innate immunity by modulating plant hormone networks. Plant Mol Biol l75:93–105

Zhang W, Jeon BW, Assmann SM (2011b) Heterotrimeric G-protein regulation of ROS signalling and calcium currents in Arabidopsis guard cells. J Exp Bot 62:2371–2379

Zhang X, Li H, Zhang J, Zhang C, Gong P, Ziaf K, Xiao F, Ye Z (2011c) Expression of artificial microRNAs in tomato confers efficient and stable virus resistance in a cell-autonomous manner. Transgenic Res 20:567–581

Zhang X, Zhao H, Gao S, Wang W-C, Katiyar-Agarwal S, Huang H-D, Raikhel N, Jin H (2011d) *Arabidopsis* Argonaute 2 regulates innate immunity via miRNA393*-mediated silencing of a Golgi-localized SNARE gene, MEMB12. Mol Cell 42:356–366

Zhang H, Gao Z, Zheng X, Zhang Z (2012a) The role of G-proteins in plant immunity. Plant Signal Behav 7:1284–1288

Zhang X, Wang C, Zhang Y, Sun Y, Mou Z (2012b) The Arabidopsis mediator complex subunit16 positively regulates salicylate-mediated systemic acquired resistance and jasmonate/ethylene-induced defense pathways. Plant Cell 24:4294–4309

Zhang X, Wu Q, Ren J, Qian W, He S, Huang K, Yu XC, Gao Y, Huang P, An C (2012c) Two novel RING-type ubiquitin ligases, RGLG3 and RGLG4, are essential for jasmonate-mediated responses in Arabidopsis. Plant Physiol 160:808–822

Zhao J, Sakai K (2003) Multiple signalling pathways mediate fungal elicitor-induced β-thujalplicin biosynthesis in *Cupressus lusitanica* cell cultures. J Exp Bot 54:647–656

Zhao J, Zheng S-H, Fujita K, Sakai K (2004) Jasmonate and ethylene signalling and their interaction are integral parts of the elicitor signalling pathway leading to β-thujaplicin biosynthesis in *Cupressus lusitanica* cell cultures. J Exp Bot 55:1003–1012

Zhao Z, Stanley BA, Zhang W, Assmann SM (2010) ABA-regulated G protein signaling in Arabidopsis guard cells: a proteomic perspective. J Proteome Res 9:1637–1647

Zhao Y-T, Wang M, Fu S-X, Yang W-C, Qi C-K, Wang X-J (2012) Small RNA profiling in two *Brassica napus* cultivars identifies microRNAs with oil production- and development-correlated expression and new small RNA classes. Plant Physiol 158:813–823

Zheng Z, Qamar SA, Chen Z, Mengiste T (2006) Arabidopsis WRKY33 transcription factor is required for resistance to necrotrophic fungal pathogens. Plant J 48:596–605

Zheng X, Spivey NW, Zeng W, Liu Po-Pu FZQ, Klessig DF, He SY, Dong X (2012) Coronatine promotes *Pseudomonas syringae* virulence in plants by activating a signaling cascade that inhibits salicylic acid accumulation. Cell Host Microbe 11:587–596

Zhou C, Zhang L, Duan J, Miki B, Wu K (2005) Histone deacetylase19 is involved in jasmonic acid and ethylene signaling of pathogen response in *Arabidopsis*. Plant Cell 17:1196–1204

Zhu H, Li G-J, Ding L, Cui X, Berg H, Assmann SM, Xia Y (2009) *Arabidopsis* extra large G-protein 2 (XLG2) interacts with the Gβ subunit of heterotrimeric G protein and functions in disease resistance. Mol Plant 2:513–525

Zhu Z, An F, Feng Y, Li P, Xue LAM, Jiang Z, Kim JM, To TK, Li W, Zhang X, Yu Q, Dong Z, Chen WQ, Seki M, Zhou JM, Guo H (2011) Derepression of ethylene-stabilized transcription factors (EIN3/EIL1) mediates jasmonate and ethylene signaling synergy in *Arabidopsis*. Proc Natl Acad Sci U S A 108:12539–12544

Ziadi S, Barbedette S, Godard JF, Monot C, Le Corre D, Silue D (2008) Production of pathogenesis-related proteins in the cauliflower (*Brassica oleracea* var *botrytis*)-downy mildew (*Peronospora parasitica*) pathosystem treated with acibenzolar-S-methyl. Plant Pathol 50:579–586

Chapter 4
Ethylene Signaling System in Plant Innate Immunity

Abstract Ethylene signaling is involved in the modulation of plant innate immunity. On perception of pathogen-associated molecular pattern (PAMP) signals, ethylene biosynthesis system is activated. It results in enhanced expression of ethylene biosynthesis genes. G-proteins act as molecular switches switching on the ethylene biosynthesis pathway. PAMP-induced signals induce calcium signaling, which is modulated by specific calcium signatures. Calcium sensors involved in ethylene signaling include Ca^{2+}-dependent protein kinases, which are involved in decoding calcium signals. The calcium waves activate ACC synthase, the key enzyme involved in the biosynthesis of ethylene. PAMP triggers ROS production within seconds or minutes, and an increase in H_2O_2 triggers transcription of genes involved in the biosynthesis of ethylene. Nitric oxide is involved in the generation of the biphasic pattern of ethylene production. MAP kinase cascades have been shown to play an important role in the induction of ethylene biosynthesis by PAMP signals. Ubiquitin- and proteasome-mediated degradation of enzymes involved in ethylene biosynthesis plays an important role in ethylene biosynthesis pathway. Ethylene signal transduction is initiated by the binding of ethylene by a family of five membrane-bound receptors. In all the receptors, the transmembrane domain at the N-terminus comprises the ethylene-binding domain, and the receptor N-terminus has three to four transmembrane domains that bind ethylene. The receptors interact with each other, and cooperativity for ethylene signaling exists among the receptors. Each member of the ethylene receptor family may have unique roles in ethylene signaling and may act cooperatively, rather than independently. CTR1 functions as a key mediator of ethylene signal transduction, acting just downstream of the receptors. EIN2 acts downstream of CTR1, and CTR1 directly phosphorylates the cytosolic C-terminal domain of EIN2. Upon ethylene perception by the receptors, CTR1 is inhibited and the inhibition of CTR1 upon ethylene perception is a signal for cleavage and nuclear localization of the EIN2 C-terminus, allowing the ethylene signal to reach the downstream transcription factors in the nucleus. In the nucleus, EIN2 activates the transcription factors EIN3 and EIL1. In the absence of ethylene, EIN3 is quickly degraded through a ubiquitin–proteasome pathway mediated by two F-box proteins, EBF1 and EBF2. Ethylene signaling suppresses the ubiquitin–proteasome-mediated proteolysis of EIN3. EIN3 becomes stabilized after perception of ethylene and induces ethylene-responsive gene expression. EIN3

P. Vidhyasekaran, *Plant Hormone Signaling Systems in Plant Innate Immunity*, Signaling and Communication in Plants 2, DOI 10.1007/978-94-017-9285-1_4

binds to the promoter sequence of the ethylene-inducible transcription factor ERF1. ERFs bind specifically to the GCC-box that is found in several promoters of the pathogenesis-related (PR) genes. Ethylene triggers increased expression of plant pattern recognition receptors (PRRs) and activates the plant immune responses. Ethylene and jasmonate signaling systems act concurrently in triggering plant immune responses.

4.1 Ethylene Signaling Is an Important Component in Plant Innate Immunity

Ethylene (ET) signaling system is an important component in plant innate immune system (Garaats et al. 2002, 2003; Cao et al. 2006; Iwai et al. 2006; Xu et al. 2007; Lin et al. 2008; Berr et al. 2010; Laluk et al. 2011; Nie et al. 2011; Shakeel et al. 2013). ET signaling system has been shown to induce resistance against the necrotrophic fungal pathogen *Botrytis cinerea* in *Arabidopsis* (Geraats et al. 2003; Moffat et al. 2012). Ethylene-insensitive *ein2* mutants showed enhanced susceptibility to *B. cinerea* in *Arabidopsis* (Nandi et al. 2003), suggesting that ethylene may play an important role in conferring resistance against *B. cinerea*. ET signaling pathway contributed for resistance against the blast pathogen *Magnaporthe oryzae* in rice (Iwai et al. 2006). The spermidine-accumulating transgenic tomato lines overexpressing yeast spermidine synthase (*ySpdSyn*) were more susceptible to *B. cinerea* than the wild-type plants. Exogenous application of ET precursors S-adenosyl-Met and 1- aminocyclopropane-1-carboxylic acid (ACC) reversed the response of the transgenic plants to *B. cinerea*. The increased susceptibility of the *ySpdSyn* transgenic tomato plants to *B. cinerea* was associated with down-regulation of gene transcripts involved in ethylene biosynthesis and signaling (Nambeesan et al. 2012). These results suggest that ET signaling is an important component in the plant immune system triggering defense responses against *B. cinerea*.

ET induced resistance against the necrotrophic pathogen *Cercospora nicotianae* and the bacterium *Erwinia carotovora*. In contrast, the ET system did not induce resistance against the biotrophic fungus *Oidium neolycopersici* and induced susceptibility against the oomycete *Peronospora tabacina* and *Tobacco mosaic virus* in tobacco (Geraats et al. 2003). Ethylene induced partial resistance in *Medicago truncatula* against the charcoal rot pathogen *Macrophomina phaseolina* (Gaige et al. 2010). Antisense suppression of *OsEIN2* confers ET insensitivity and increases susceptibility against the rice blast pathogen *M. oryzae* (De Vleesschauwer et al. 2008). ET biosynthesis has been shown to be an integral component of *R* gene-mediated blast resistance (Iwai et al. 2006). Ethephon treatment lowers susceptibility in compatible rice blast interactions (Singh et al. 2004). These results suggest that ET is involved in triggering immune responses against *M. oryzae* in rice.

ET has also been reported to be involved in systemic immunity. ET induces systemic resistance against *Turnip mosaic virus* (TuMV) infection in leafy mustard plants (Zhu et al. 2011a). Expression of ethylene biosynthesis-related genes (*ACS* and *ACO*) was enhanced in the upper uninoculated leaves at 6 days of postinoculation with *Turnip mosaic virus* (*TuMV*) in leafy mustard. Pretreatment with the ethylene precursor ACC resulted in reduced levels of *TuMV-CP*. Application of AVG (an ethylene biosynthesis inhibitor) substantially increased *TuMV* susceptibility in the upper leaves (Zhu et al. 2011a). These results suggest that ET is involved in triggering systemic immunity against the virus disease.

ET may also negatively regulate plant immune responses and confer susceptibility against some pathogens. In *Arabidopsis*, ETR1 is a membrane-bound receptor, and ethylene is perceived by the ETR receptor, which transmits the signal to downstream effectors (Wang et al. 2006). The ET signal receptor ETR1 negatively regulates plant immune responses against *Fusarium oxysporum* f. sp. *raphani* in *Arabidopsis*. The *etr1-1* mutant plants showed resistance against the wilt pathogen. These ET signaling-deficient plants showed reduced growth of the pathogen, suggesting the activation of defense responses against *F. oxysporum* (Pantelides et al. 2013). The results also suggest that *F. oxysporum* f. sp. *raphani* hijacks ETR1-mediated ET signaling to promote disease development in plants.

Exogenous ethephon (ET-releasing chemical) application enhances susceptibility, whereas genetic disruption of ET signaling renders plants less vulnerable to the brown spot pathogen *Cochliobolus miyabeanus* in rice. Feeding detached rice leaves with silver thiosulfate, an inhibitor of ET action induced resistance (De Vleesschauwer et al. 2010). These results suggest that ET signaling negatively regulates host defense responses against the fungal pathogen *C. miyabeanus* in rice.

OsEIN2 is a positive regulator component in ethylene signaling in rice. The antisense *OsEIN2* transgenics exhibit ET insensitivity and show a decreased expression of ET-responsive defense genes (Jun et al. 2004). Non-induced *OsEIN2* antisense seedlings showed significantly more resistance against *C. miyabeanus* than wild-type plants, indicating that ET action negatively interferes with basal resistance to *C. miyabeanus* (De Vleesschauwer et al. 2010). In fact, ET may act as a two-faced player in the rice immune response network, by inducing resistance against *M. oryzae* but promoting susceptibility to *C. miyabeanus* (De Vleesschauwer et al. 2008, 2010). Collectively, these studies suggest that ET is an important component in plant immune system.

4.2 Ethylene Biosynthesis in Plants

4.2.1 Enzymes Involved in Ethylene Biosynthesis

L-Methionine (L-Met) is the precursor of ethylene. It is converted to *S*-adenosylmethionine (*S*-AdoMet), and the formation of *S*-AdoMet from methionine is catalyzed by *S*-AdoMet synthetase (SAM synthetase) at the expense of one molecule of ATP per molecule of *S*-AdoMet synthesized (Wang et al. 2002). Nitric oxide (NO) induces SAM synthetase, which catalyzes the conversion of ATP and methionine into *S*-adenosyl-Met (Zago et al. 2006).

S-AdoMet is converted to ACC by ACC synthase (ACS). Several *ACS* genes have been identified, and transcriptional activation of these genes contributes to the increase in ACS isozymes in plants (Vogel et al. 1998; Chae et al. 2003; Peng et al. 2005). Phosphorylation of the ACC synthases appears to be essential for the activation of these enzymes (Spanu et al. 1991; Wang et al. 2002). Treatment with the protein kinase inhibitors staurosporine and K-252 inactivates the ACS activity, whereas protein phosphatase inhibitor calyculin A stimulated ACS activity (Spanu et al. 1991). In the absence of phosphorylation by mitogen-activated protein kinase, the newly synthesized ACC synthases are rapidly degraded through ubiquitin–proteasome pathway, resulting in no net increase in the ACS proteins (Liu and Zhang 2004). ACS isozymes have been shown to be substrates for E3 ligases (Dreher and Callis 2007). Phosphorylation results in ACS stability. Accumulation of ACS isozymes leads to increased synthesis of ACC (Liu and Zhang 2004). Finally ACC is oxidized by ACC oxidase (ACO) to form ethylene (Fig. 4.1; Wang et al. 2002; Vidhyasekaran 2007, 2014; Brisson et al. 2012).

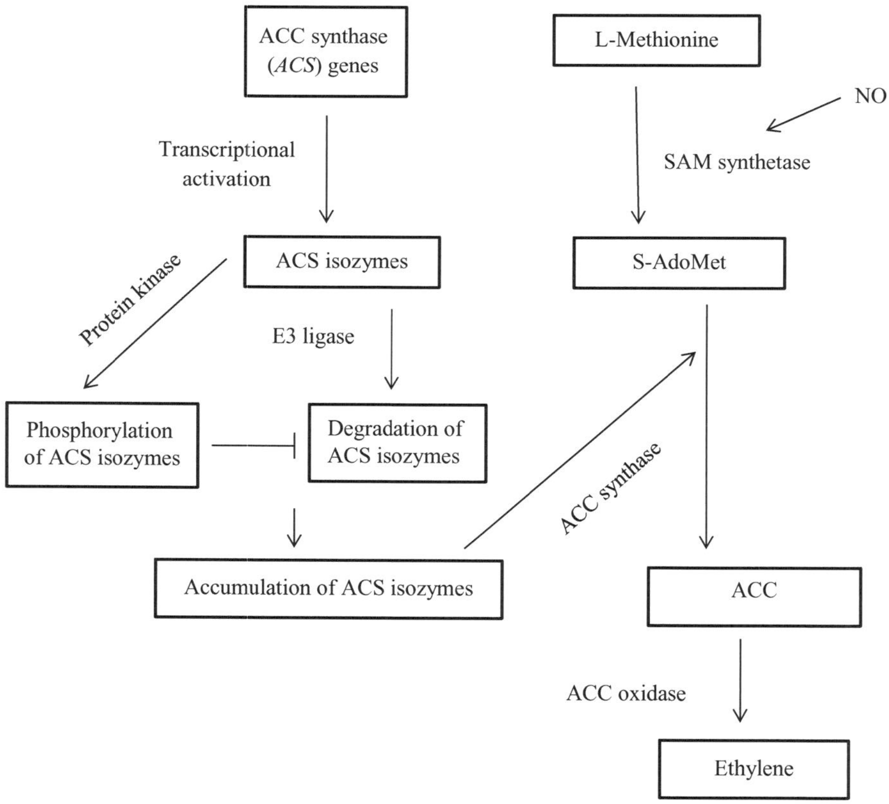

Fig. 4.1 Ethylene biosynthesis pathway (Adapted from Wang et al. 2002; Brisson et al. 2012; Vidhyasekaran 2014)

4.2.2 Pathogen Infection Triggers Enhanced Expression of Ethylene Biosynthesis Genes

Pathogen attack triggers ethylene biosynthesis in many host–pathogen interactions (Broekaert et al. 2006). Ethylene biosynthesis is often significantly increased during infection by fungal, bacterial, and viral pathogens (Bent et al. 1992; Chen et al. 2003; Mur et al. 2008; Zhu et al. 2011a; Nambeesan et al. 2012). Two periods of increased ethylene production were observed after inoculation of *Nicotiana tabacum* by *Colletotrichum destructivum*. Two different *ACS* genes showed increased expression after the first ethylene peak. Expression of an *ACO* gene increased during the first ethylene peak and then declined at the beginning of the second ET increase. A second *ACO* gene showed relatively little change in expression during infection. A third *ACO* gene showed a progressive decline in expression (Chen et al. 2003). ET production in tobacco following *Pseudomonas syringae* pathovars inoculation was also biphasic (Mur et al. 2008). A first transient rise in ET production was observed 1–4 h following inoculation with HR-eliciting, disease-forming, and nonpathogenic strains. A second rise was only seen with HR-eliciting strains (Mur et al. 2008). NO was shown to be required to generate the biphasic pattern of ET production (Mur et al. 2008). ET evolution occurs concomitantly with the progression of disease symptom development in tomato (Lund et al. 1998). The expression of the ET biosynthesis gene *ACS* increased 15-fold in response to *B. cinerea* infection in tomato (Nambeesan et al. 2012). Expression of *ACS* and *ACO* genes involved in ET biosynthesis was enhanced in mustard plants inoculated with *Turnip mosaic virus* (Zhu et al. 2011a).

4.2.3 PAMPs/HAMPs Induce Expression of ET Biosynthesis Genes and Trigger ET Biosynthesis

Pathogen-associated molecular patterns (PAMPs) and host-associated molecular patterns (HAMPs, endogenous elicitors) trigger ET biosynthesis and also enhance expression of ET biosynthesis genes (Qutob et al. 2006; Denoux et al. 2008; Vidhyasekaran 2014). A transient rise in ET production was observed following the bacterial PAMP flg22 treatment in tobacco (Mur et al. 2008). Flg22 induces ET production in *Arabidopsis* (Liu and Zhang 2004). Flg22 treatment induced the genes (*ACS2*, *ACS7*, and *ACS8*) encoding ACC synthase involved in ET biosynthesis (Denoux et al. 2008). The PAMP Nep1-like protein (NLP) isolated from the oomycete *Phytophthora parasitica* induced *ACS2* and *ACS7* genes involved in ET biosynthesis (Qutob et al. 2006). The oomycete PAMP cryptogein induced production of ET in tobacco cells (Milat et al. 1991). The *Pseudomonas* phytotoxin coronatine induces ethylene biosynthesis in *Arabidopsis* (Bent et al. 1992). *Zm*Pep1-treated maize leaves emitted a 5-fold increase in ethylene (ET). Expression of the gene encoding 1-aminocyclopropane-1-carboxylic acid oxidase (ACC oxidase) also responded to ZmPep1 treatment (Huffaker et al. 2011). The oligogalacturonide

(OG) elicitor induced the ET biosynthesis enzymes ACC synthase ACS7 and ACS8 (Denoux et al. 2008). These studies suggest that various PAMPs/elicitors trigger the activation of ET biosynthesis enzymes resulting in increased biosynthesis and accumulation of ethylene in plants.

4.2.4 G-Proteins May Trigger Ethylene Biosynthesis Pathway

G-proteins are known to act as molecular switches in signal transduction system (Zhang et al. 2011, 2012). These proteins are involved in ET biosynthesis in the plant immune system. The rice small G-protein OsRac1 induced methionine synthase and S-adenosylmethionine synthetase (also known as *S*-adenosyl-L-methionine synthase, methionine adenosyltransferase; EC 2.5.1.6) in rice cells (Fig. 4.2; Fujiwara et al. 2006). Methionine synthase is involved in the conversion of homocysteine to methionine, and S-adenosylmethionine synthase catalyzes the

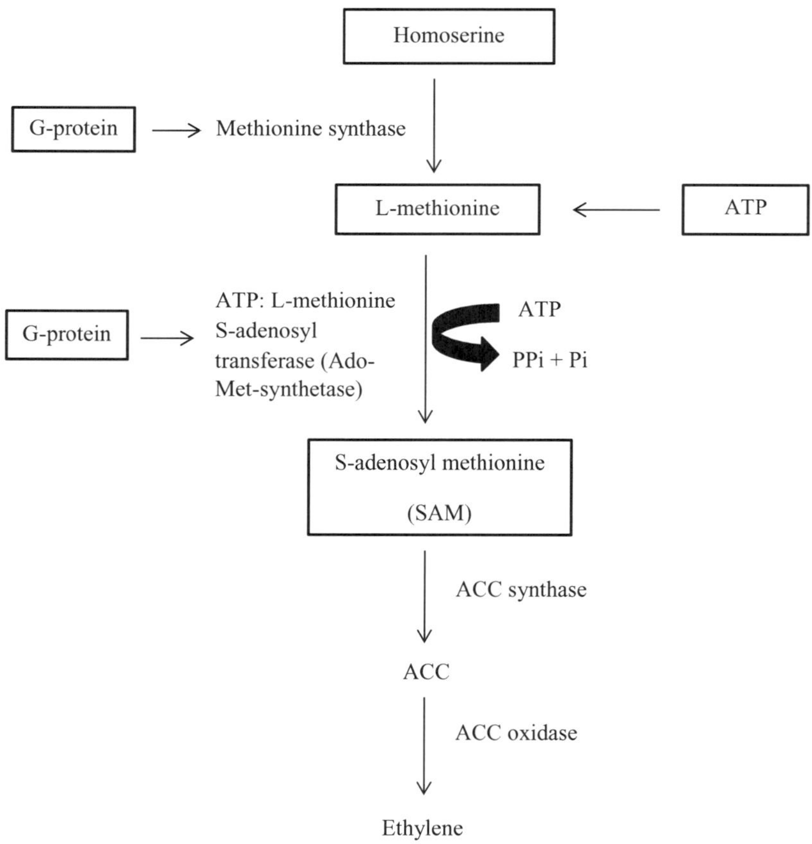

Fig. 4.2 G-protein-mediated ethylene biosynthesis pathway (Adapted from Fujiwara et al. 2006)

formation of S-adenosylmethionine by joining methionine and ATP (Zago et al. 2006). S-Adenosylmethionine is a precursor of ethylene, and it is converted into 1-aminocyclopropane-1-carboxylic acid (ACC) by the action of ACC synthase (Liu and Zhang 2004; Sebastiá et al. 2004; Peng et al. 2005). ACC is oxidized by ACC oxidase leading to the biosynthesis of ET (Gallardo et al. 1999; Wang et al. 2002; Vidhyasekaran 2014).

4.2.5 *Role of Ca^{2+} Influx-Mediated Ca^{2+} Signature in Ethylene Biosynthesis*

Ca^{2+} influx of extracellular Ca^{2+} through the plasma membrane is an early event in activating plant immune responses (Garcia-Brugger et al. 2006; Vadassery and Oelmūller 2009; Kwaaitaal et al. 2011; Vincill et al. 2012). Massive influx of Ca^{2+} is observed within 15–30 min after PAMP treatment in tobacco-cultured cells (Lecourieux-Ouaked et al. 2000). The initiation of innate immune responses upon the PAMPs recognition by their PRRs has been shown to be due to apoplastic Ca^{2+} influx via ion channels (Kwaaitaal et al. 2011). The Ca^{2+} influx has been shown to be involved in the activation of ACC oxidase, which is a key enzyme in the biosynthesis of ethylene (Fig. 4.3; Gallardo et al. 1999). In chickpea seeds, EGTA (ethylene glycol-bis (γβ-aminoethyl ether)-N,N,N′,N′-tetraacetic acid) (a selective extracellular calcium ion chelator) treatment strongly inhibited ACC oxidase activities as well as ethylene production (Gallardo et al. 1999). It indicates that the influx of Ca^{2+} is important for the ACC oxidase activity. The EGTA inhibition was restored by exogenous calcium ion treatment. Treatment of embryonic axes with either verapamil or $LaCl_3$ (both Ca^{2+} channel blockers) or TMB8 (an intracellular Ca^{2+} antagonist) provoked an inhibition of both ACC oxidase activity and ethylene production

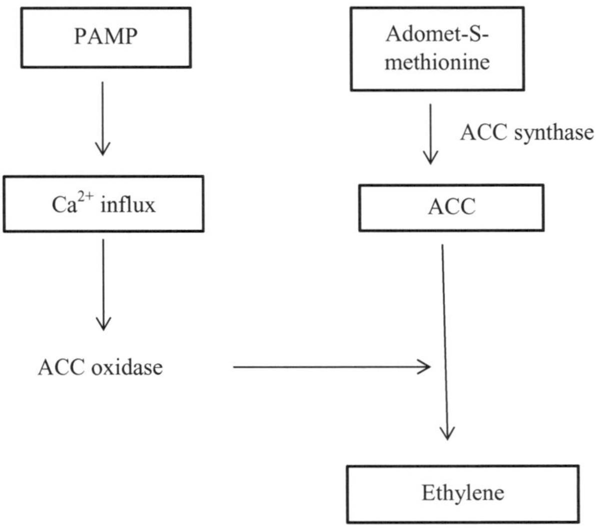

Fig. 4.3 Role of Ca^{2+} influx-triggered ACC oxidase in biosynthesis of ethylene (Adapted from Gallardo et al. 1999)

(Gallardo et al. 1999). Collectively these studies suggest that calcium ion fluxes and intracellular calcium levels may act as signals in inducing ethylene biosynthesis.

4.2.6 *Role of Calcium-Dependent Protein Kinase (CDPK) in Induction of Ethylene Biosynthesis*

Ca^{2+} influx through Ca^{2+} channels results in Ca^{2+} waves (calcium signature) in cytosol (McAinsh and Pittman 2009). Calcium signaling is modulated by specific calcium signatures (Lecourieux et al. 2006). The calcium signatures are recognized by different calcium sensors to transduce calcium-mediated signals into downstream events (Reddy et al. 2011a, b; Hashimoto et al. 2012; Wang et al. 2012). The Ca^{2+} sensors identified in plants include Ca^{2+}-dependent protein kinases. They are involved in decoding calcium signals (Luan 2009; Kudla et al. 2010; Hashimoto et al. 2012). The calcium wave-activated calcium-dependent protein kinase (CDPK) activates 1-aminocyclopropane-1-carboxylic acid (ACC) synthase, the key enzyme involved in the biosynthesis of ACC, a precursor for the biosynthesis of ethylene (Sebastiá et al. 2004) (Fig. 4.4).

4.2.7 *Reactive Oxygen Species May Trigger Transcription of Ethylene Biosynthesis Genes*

PAMPs trigger ROS production very rapidly, and the oxidative burst has been observed within seconds to few minutes after the PAMP treatment (Arnott and Murphy 1991; Bolwell et al. 1995; Sang et al. 2012). ROS may function in ethylene

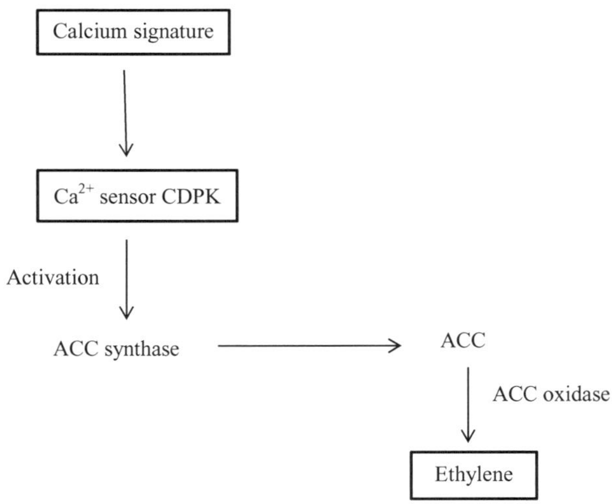

Fig. 4.4 Perception of Ca^{2+} signals by the Ca^{2+} sensor CDPK activates ACC synthase in ethylene biosynthesis pathway (Adapted from Sebastiá et al. 2004)

signaling system. An increase in H_2O_2 was able to trigger transcriptional changes of genes involved in the biosynthesis of ethylene. *S*-Adenosyl-L-methionine synthetase is the first enzyme in the ethylene biosynthesis pathway, and 1-aminocyclopropane-1-carboxylate (ACC) oxidase is the final stage enzyme involved in the conversion of ACC to ethylene. Transcript tags coding for both the enzymes were up-regulated within 2 h of increase in H_2O_2, followed by an increase in ethylene-responsive proteins, such as EREBP/AP2 domain proteins (Vandenabeele et al. 2003). The results suggest that H_2O_2 may be involved in the induction of ethylene biosynthesis genes (Fig. 4.5).

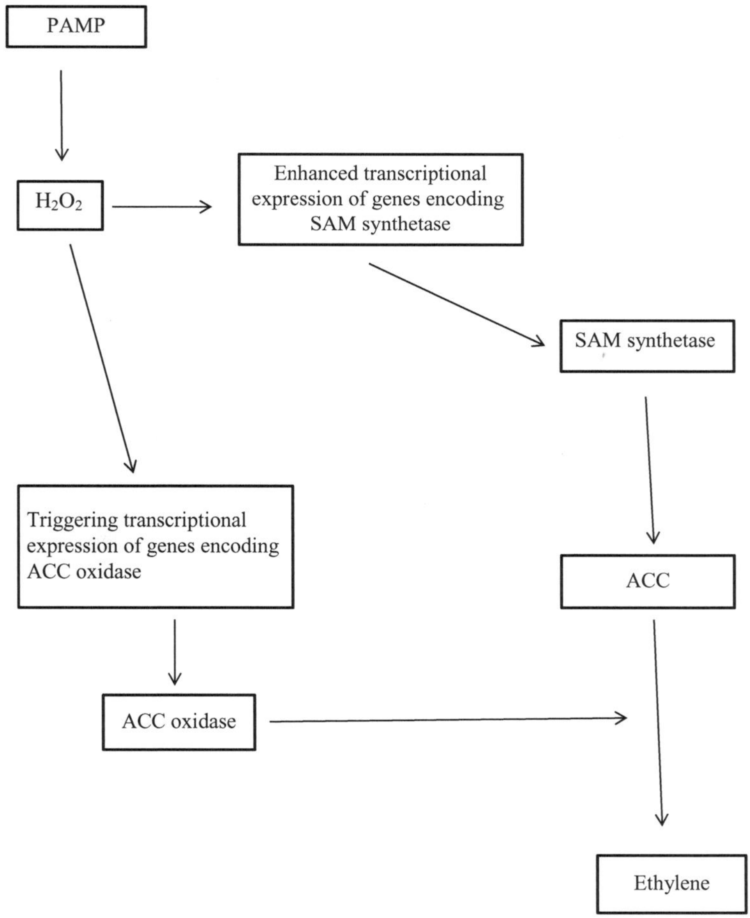

Fig. 4.5 Function of H_2O_2 in triggering ethylene biosynthesis (Adapted from Vandenabeele et al. 2003)

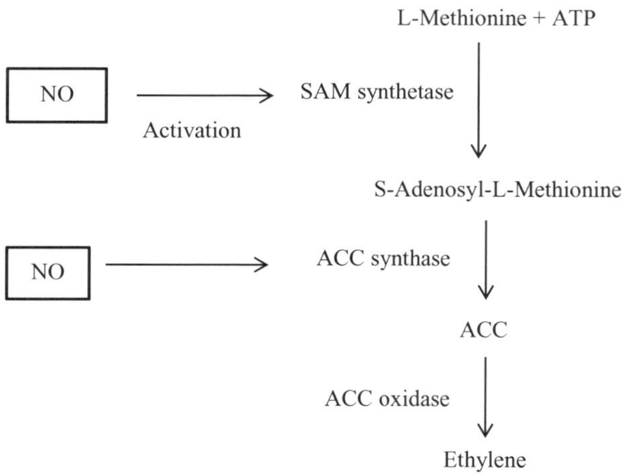

Fig. 4.6 Role of NO in induction of ethylene biosynthesis (Adapted from Lamotte et al. 2004; Zago et al. 2006)

4.2.8 *Nitric Oxide May Trigger Activation of Ethylene Biosynthesis Enzymes*

NO has been shown to be required to generate the biphasic pattern of ethylene production during the HR induced by *Pseudomonas syringae* pathovars in tobacco. The first transient rise in ethylene occurred at 1–4 h following inoculation with HR-eliciting, disease-forming, and nonpathogenic strains and also with the PAMP flg22 treatment. A second rise was observed approximately at 6 h after inoculation with HR-eliciting strains (Mur et al. 2008). NO induces *S*-adenosyl-L-methionine synthetase, which catalyzes the conversion of ATP and L-methionine into the ethylene precursor *S*-adenosyl-L-methionine (Fig. 4.6; Zago et al. 2006). It also induces ACC synthase involved in ethylene biosynthesis (Lamotte et al. 2004; Fig. 4.6). These results suggest that NO may function in the ethylene biosynthesis pathway (Lindermayr et al. 2005, 2006).

4.2.9 *MAP Kinase Cascades May Induce Biosynthesis of Ethylene*

Some of the MAP kinase cascades have been shown to be involved in inducing biosynthesis of ethylene in *Arabidopsis*, tobacco, and rice. In *Arabidopsis*, several MAP kinase cascades have been reported. A typical MAP kinase cascade is a three-kinase module consisting of three interconnected protein kinases: a MAP kinase kinase kinase (MAPKKK or MEKK [for *M*APK/*E*xtracellular signal-regulated kinase *K*inase *K*inase]), a MAP kinase kinase (MAPKK or MKK), and a MAP kinase (MAPK or MPK) (Mészáros et al. 2006). MAP kinase cascade involves sequence of phosphorylation events (Hirt 2000). MAPKs function at the bottom of the three-kinase cascade and are activated by MAPKKs through phosphorylation. The activity of MAPKKs is, in turn, regulated by MAPKKKs via phosphorylation. MAPKKKs receive signals from upstream PAMP/PRR signaling complex (Ichimura et al. 2002; Li et al. 2012; Singh and Jwa 2013). More than 80 putative MEKKS, 10 MKKs, and at least 20 MPKs have been reported in *Arabidopsis* (Ichimura et al. 2002; Jonak et al. 2002; Nakagami et al. 2005). Among the 20 MPKs, MPK3 and MPK6 have been reported to play an important role in triggering ethylene biosynthesis in *Arabidopsis* (Liu and Zhang 2004; Han et al. 2010; Li et al. 2012). In *Arabidopsis* six different MAP kinase cascades have been reported to activate MPK3/6. These include MEKK1–MKK4–MPK3, MEKK1–MKK5–MPK3, MEKK1–MKK4–MPK6, MEKK1–MKK5–MPK6, MEKK1–MKK1–MPK6, and MEKK1–MKK2–MPK6 pathways (Asai et al. 2002; Teige et al. 2004; Mészáros et al. 2006; Takahashi et al. 2007). The *Arabidopsis* CTR1 (Constitutive Triple Response 1) is a Raf-like kinase considered as a MAPKKK, and it controls MPK3/6 activation via MKK9 (MAPK kinase9). MKK4/5 also acts upstream of MPK3/6 in the MAPK signaling cascade. MPK3/6, together with its upstream MKK9 or MKK4/5, functions in the ethylene biosynthesis pathway (Liu and Zhang 2004; Menke et al. 2004; Joo et al. 2008; Xu et al. 2008).

Arabidopsis MPK3 and MPK6, which are involved in ethylene biosynthesis, are activated by pathogens, PAMP elicitors, and HAMPs (endogenous elicitors) (Asai et al. 2002; Bethke et al. 2012). The bacterial PAMP flg22 and the oligogalacturonides elicitor of host plant origin activated MPK3 and MPK6 in *Arabidopsis* (Galletti et al. 2011). The PAMP flg22 induces MAP kinases, which phosphorylate the ethylene biosynthetic enzymes ACC synthases 2 and 6, leading to their stabilization (Liu and Zhang 2004; Yoo et al. 2008). It is suggested that in the absence of flagellin, endogenous ethylene may ensure a constitutive level of the PRR FLS2 expression. On flg22 binding, FLS2 may activate MPK6 that, in turn, may phosphorylate ACS2/6, resulting in increased ethylene production (Boutrot et al. 2010).

ACS is a PAMP elicitor-inducible enzyme involved in the induction of biosynthesis of ethylene (Li et al. 2012). The *Arabidopsis* ACS isoforms ACS2 and ACS6 have been shown to be substrates of MPK3 and MPK6 (Liu and Zhang 2004; Han et al. 2010). The ACS isoforms were shown to be phosphorylated and stabilized by MPK3 and MPK6 functioning in the MAPK cascade consisting of MEKK1–MKK4/MKK5–MPK3/MPK6 (Han et al. 2010; Li et al. 2012). Phosphorylation of the two ACC synthases, ACS2 and ACS6, by MPK3 and MPK6 prevents rapid degradation of ACS2/ACS6 by the 26S proteasome pathway, resulting in an increase in cellular ACS activity (Han et al. 2010). The phosphorylation resulted in ACS stability and accumulation of ACS isozymes, which led to increased synthesis of ACC (Liu and Zhang 2004). Another enzyme involved in the biosynthesis of ethylene, ACC oxidase, oxidatively cleaves the accumulated ACC resulting in generation of ethylene (Wang et al. 2002). Collectively, these studies suggest that the PAMP signals activate the three-kinase signaling module and induce phosphorylation of the ET biosynthesis enzymes, resulting in the induction of ET biosynthesis (Fig. 4.7; Wang et al. 2002; Liu and Zhang 2004; Han et al. 2010; Boutrot et al. 2010; Li et al. 2012).

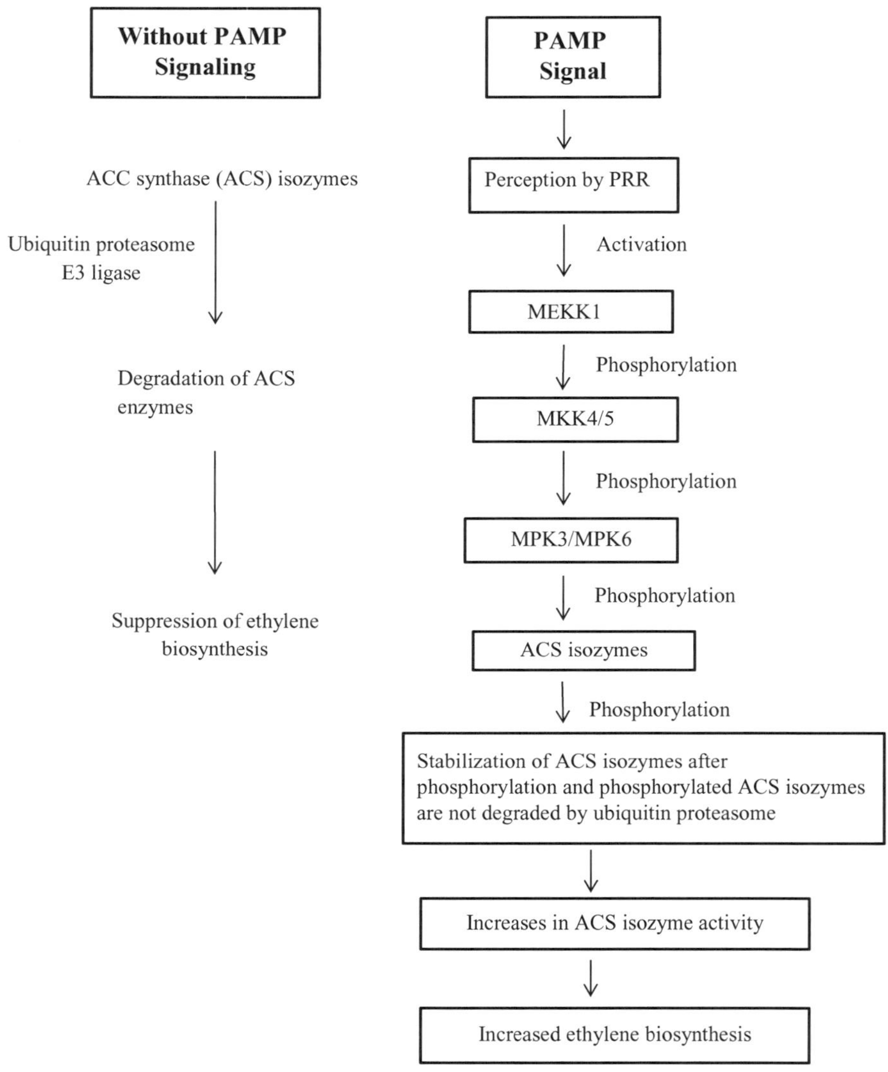

Fig. 4.7 Role of PAMP-induced MAP kinase cascade in inducing ethylene biosynthesis (Adapted from Wang et al. 2002; Liu and Zhang 2004; Han et al. 2010; Boutrot et al. 2010; Li et al. 2012)

MPK3 and MPK6 may also regulate the expression of *ACS* genes through another MPK3/MPK6 substrate, the WRKY transcription factor WRKY33 (Fig. 4.8; Li et al. 2012). WRKY33 binds to the W-boxes in the *ACS2/ACS6* promoters in vivo and is directly involved in MPK3/MPK6-induced *ACS2/ACS6* gene expression. These studies suggest that MPK3 and MPK6 not only function in the phosphorylation-induced stabilization of ACS2/ACS5 proteins but also signal the *ACS2* and *ACS6* gene activation in *Arabidopsis* (Li et al. 2012). Regulations of ACS isoforms at both transcriptional and posttranslational levels by MPK3 and MPK6 seem to contribute to the high level of ethylene production in plants (Li et al. 2012).

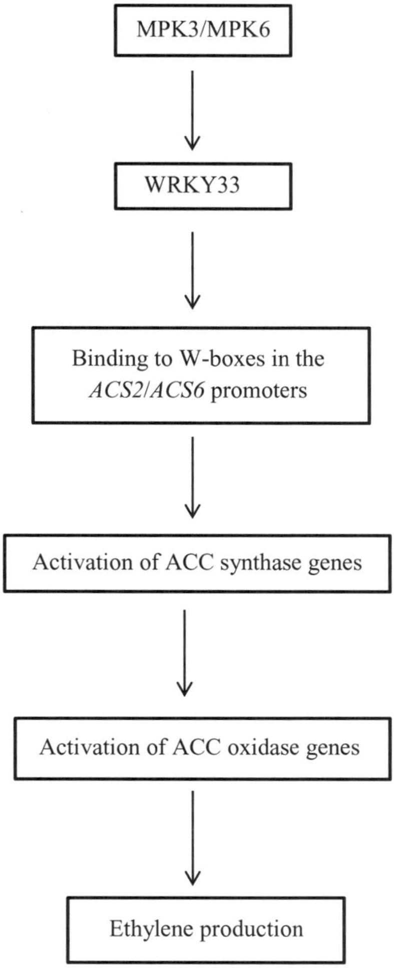

Fig. 4.8 Role of MPK3/MPK6 in triggering enhanced expression of ACC synthase genes in *Arabidopsis* (Adapted from Li et al. 2012)

In tobacco, activation of the MAP kinase cascade, which includes the NtMEK2 MAPKK and the SIPK and WIPK MAP kinases, has been reported to cause an increase in ethylene levels (Kim et al. 2003a; Ren et al. 2006). The activation of SIPK by the tobacco MAPK kinase, NtMEK2, resulted in a dramatic increase in ethylene production. The increase in ethylene production after the activation of SIPK coincided with a dramatic increase in ACC synthase (ACS) activity, which was followed by the activation of a subgroup of ACS and ACC oxidase (ACO) genes. *Tobacco mosaic virus* (TMV), which is known to activate the SIPK cascade and induce ethylene biosynthesis, also induced the same ACSs and ACOs (Kim et al. 2003a). These results suggest that ethylene is involved in plant defense responses mediated by the NtMEK2–SIPK/WIPK pathway.

In rice, *OsEDR1* encodes a MAP kinase kinase kinase (MAPKKK) (Kim et al. 2003b), which negatively regulates ethylene biosynthesis. *OsEDR1*-suppressing knockout (KO) rice plants showed suppression of 1-aminocyclopropane-carboxylic acid (ACC), the precursor of ethylene. Knockout of *OsEDR1* suppressed the ACC synthase (ACS) gene family, which encodes the enzymes of ethylene biosynthesis by catalyzing the formation of ACC (Shen et al. 2011).

4.2.10 Role of Ubiquitin–Proteasome in Ethylene Biosynthesis

Ubiquitin- and proteasome-mediated degradation of enzymes involved in ethylene biosynthesis plays an important role in ethylene biosynthesis pathway (Larsen and Cancel 2004; Dreher and Callis 2007). Proteasomes are large protein complexes located in the nucleus and the cytoplasm (Peters et al. 1994; Yao and Ndoja 2012; Zhang et al. 2012). The main function of the proteasome is to degrade proteins by proteolysis. Proteins are tagged for degradation by a small protein called ubiquitin (Pickart and Eddins 2004). E3 ligases play a key role in the ubiquitin–proteasome system (Lee et al. 2011). The E3 ubiquitin ligase recognizes the specific protein to be ubiquitinated and catalyzes the transfer of ubiquitin to this target protein for degradation (Smalle and Vierstra 2004; Dreher and Callis 2007).

ACC synthase (ACS) isozymes have been shown to be substrates for E3 ligases (Dreher and Callis 2007). ACC synthase is involved in the conversion of S-AdoMet to 1-aminocyclopropane-1-carboxylic acid (ACC), and ACC is oxidized by ACC oxidase to form ethylene. *Arabidopsis* ethylene overproduction mutants (*eto2* and *eto3*) are shown to bear mutations in the ACS genes, *ACS5* and *ACS9*, respectively (Vogel et al. 1998; Chae et al. 2003), showing that ACC synthase is the key enzyme involved in ET biosynthesis. Bostick et al. (2004) showed that silencing of two ubiquitin-related proteins, RUB1 and RUB2, which modify cullin RING ligases and regulate their activity, caused overproduction of ethylene in *Arabidopsis*. Similarly, mutation of the RUB-conjugating enzyme RCE1 also increased ethylene production (Larsen and Cancel 2004). These results suggest that ubiquitin protein-cullin RING ligases negatively regulate biosynthesis of ethylene.

4.3 Ethylene Signal Transduction Downstream of Ethylene Biosynthesis

4.3.1 Ethylene Signal Perception by Membrane-Bound Receptor Complex

Ethylene signal transduction is initiated by the binding of ethylene to its receptors (Zhao and Guo 2011; Binder et al. 2012; Shakeel et al. 2013). In *Arabidopsis*, ethylene is perceived by a family of five membrane-bound receptors (ETR1 [for Ethylene response1], ERS1 [for Ethylene response sensor1], ETR2, EIN4 [for Ethylene insensitive4], and ERS2) (Fig. 4.9). The receptors are localized to endoplasmic reticulum (ER) (Grefen et al. 2008). All the five members of the receptor family

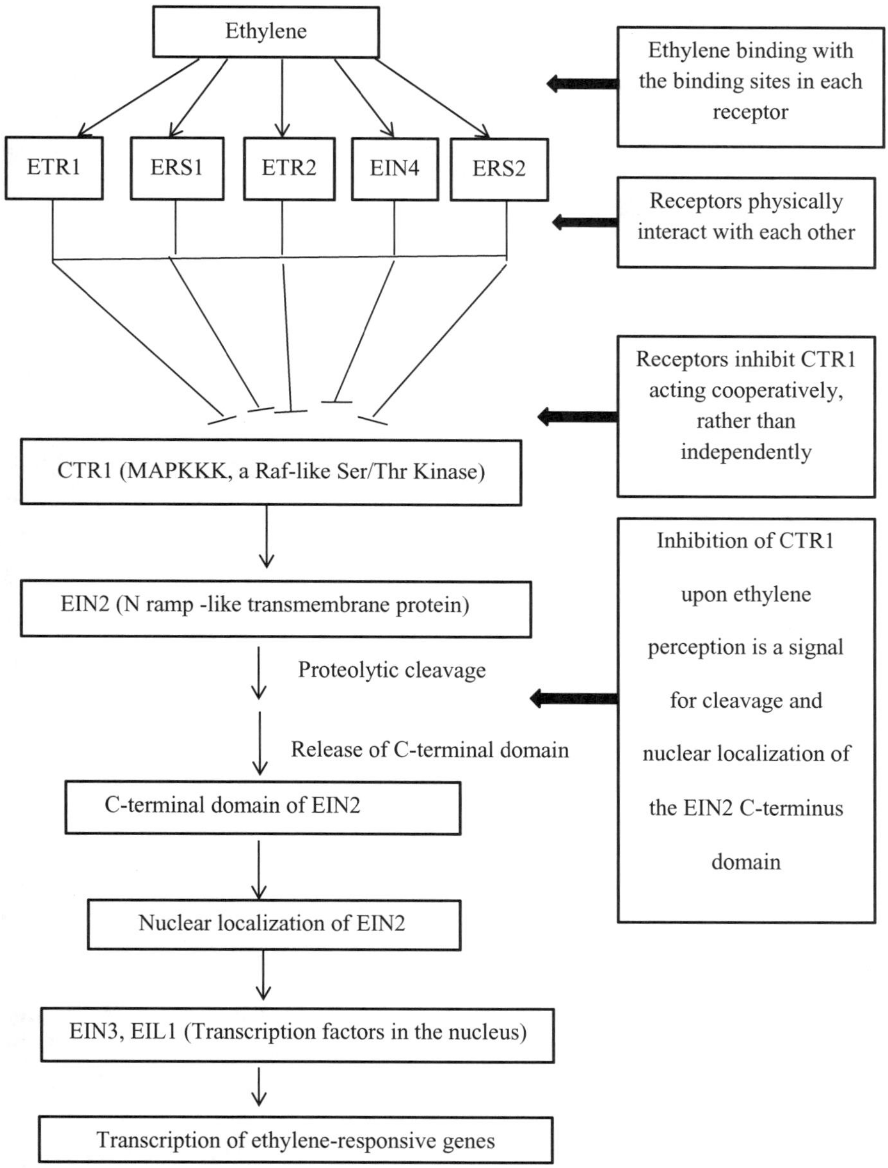

Fig. 4.9 Ethylene signaling pathway downstream of ethylene biosynthesis (Adapted from Ju et al. 2012; Qiao et al. 2012; Shakeel et al. 2013)

have a similar molecular structure. The ethylene receptors have similar overall structures with transmembrane domains near their N-termini and putative signaling motifs in their C-terminal halves. In all the receptors, the predicted transmembrane domain at the N-terminus comprises the ethylene-binding domain (EBD) (Grefen et al. 2008). The receptor N-terminus has three or four transmembrane domains that bind ethylene. The transmembrane region also encompasses the ethylene-binding site (O'Malley et al. 2005; Xie et al. 2006; Gao et al. 2008; Shakeel et al. 2013). The ethylene-binding site contains a copper cofactor, which is important for ethylene binding. The copper cofactor is made available to the receptors through the action of the copper transporter RAN1 (for Responsive-to-Antagonist1). RAN1 acts by delivering copper to create functional ethylene receptors (Hirayama et al. 1999; Rodriguez et al. 1999; Binder et al. 2010). This region is followed by a GAF (for cGMP-specific phosphodiesterases, adenylyl cyclases, and Fh1A) domain, which may be involved in protein–protein interaction (Gao et al. 2008; Grefen et al. 2008). The GAF domain mediates noncovalent receptor heterodimerization and may have a role in receptor cooperation (Gamble et al. 2002; Gao et al. 2008; Qiu et al. 2012). The C-terminal half of the receptors consists of a histidine (His) kinase domain and, in some cases (ETR1, ETR2, and EIN4), a receiver domain (Schaller et al. 2011).

The five receptors are classified into two subfamilies based on phylogenetic analysis and some shared structural features (Bleecker 1999; Wang et al. 2006). The subfamily 1 receptors ETR1 and ERS1 have a highly conserved histidine kinase domain containing all the required motifs essential for kinase functionality (Moussatche and Klee 2004). In the subfamily 2 members (ETR2, EIN4, and ERS2), the histidine kinase domain is less conserved, and it lacks most signature motifs essential for histidine kinase activity. They lack residues considered essential for histidine kinase activity and may act as serine/threonine kinases (Moussatche and Klee 2004; Qu et al. 2007; Wang et al. 2006). The subfamily 2 receptors also contain an extra transmembrane domain at the N-terminus, which may function as a signal sequence (Wang et al. 2006; Shakeel et al. 2013).

The subfamily 1 receptors play a greater role in signaling than the subfamily 2 receptors (Qu et al. 2007). Signaling by the subfamily 2 receptors may be partially dependent upon the subfamily 1 receptors (Cancel and Larsen 2002; Wang et al. 2003). The subfamily 2 members ETR1 and ERS1 exhibit a stronger interaction with CTR1 (for *C*onstitutive *T*riple *R*esponse1) than does subfamily 2 member ETR2 (Clark et al. 1998; Cancel and Larsen 2002). There may be a greater role for the receptors ETR1 and ERS1 (subfamily 1) in the regulation of ethylene signaling as compared to the other three members of the ethylene receptor family (subfamily 2). The predominant contribution of subfamily 1 receptors to ethylene signaling may arise from their greater effectiveness at maintaining CTR1 in an active state compared with the subfamily 2 receptors (Qu et al. 2007).

The five ethylene receptors interact with each other through noncovalent interactions, and the interactions among receptors may contribute to ethylene signal output (Gao et al. 2008). The five members of the ethylene receptor family can physically interact and form complexes, which implies that cooperativity for signaling may exist among the receptors (Fig. 4.9; Liu and Wen 2012). The receptors interact with each other, and this interaction may allow for increased sensitivity to ethylene through propagation of ethylene-induced conformational changes within a receptor cluster as well as facilitating transphosphorylation between His kinase domains and receiver domains of different receptors (Shakeel et al. 2013). The ethylene receptor gene mutations *etr1-1*$^{(C6_5Y)}$, *ers1-1*$^{(IC6_2P)}$, and *ers1*$^{(C6_5Y)}$ are dominant, and each confers ethylene insensitivity. It has been shown that *etr1-1*$^{(C6_5Y)}$, but not *ers1-1*$^{(IC6_2P)}$, substantially repressed various ethylene responses independent of other receptor genes (Liu and Wen 2012). The results suggest that *ETR1* and *ERS1* may differentially repress the ethylene response in combination with other ethylene receptor genes (Liu and Wen 2012). The lack of both *ETR1* and *EIN4* almost abolished the repression of ethylene responses by *ers1* ^{C6_5Y}, suggesting that ETR1 and EIN4 have synergistic effects on *ers1* ^{C6_5Y} functions (Liu and Wen 2012). Collectively, these studies suggest that cooperativity for signaling may exist among the ethylene receptors.

It has also been reported that there may be differences among the individual members of the subfamilies. In subfamily 1 of *Arabidopsis*, the receptor ETR1 contains a receiver domain, but ERS1 does not. EIN4 has retained the conserved His for phosphorylation, whereas the other subfamily 2 members have not (Shakeel et al. 2013). Although the receptor genes are genetically redundant, their functions in ethylene signaling appear to be irreplaceable by each other (Wang et al. 2003; O'Malley et al. 2005; Liu et al. 2010; Liu and Wen 2012). Each member of the ethylene receptor family may have unique roles in ethylene signaling and may act cooperatively, rather than independently (Liu and Wen 2012).

A positive regulator of ETR1 ethylene receptor has been identified in *Arabidopsis* by Resnick et al. (2008). *RTE1* (*REVERSION-TO-ETHYLENE SENSITIVITY1*), a novel gene conserved in plants and animals, was identified as a positive regulator of ETR1. The function of *RTE1* was found to be independent and distinct from that of *RAN1*, which encodes a copper transporter required for ethylene receptor function. It is suggested that *RTE1* promotes ETR1 signaling through a conformational effect on the ethylene-binding domain (Resnick et al. 2008).

4.3.2 Ethylene Receptors Physically Interact with CTR1 and Transmit the Ethylene Signal

The *Arabidopsis* CTR1 (Constitutive Triple Response 1) is a Raf-like Ser/Thr kinase considered as a MAPKKK, and it controls MPK3/MPK6 activation via MKK9. CTR1 functions as a key mediator of ethylene signal transduction, acting just downstream of the receptors. It negatively regulates ET signaling (Guo et al. 2004; Yoo et al. 2008). The five-member family of ethylene receptors has been shown to act as negative regulator in the ethylene signaling pathway (Hua and Meyerowitz 1998; Wang et al. 2003; Qu and Schaller 2004; Gao and Schaller 2009). All the five ethylene receptors physically interact with CTR1 (Fig. 4.9; Gao et al. 2003). CTR1 interacts more strongly with the subfamily 1 receptors ETR1 and ERS1 than it does with the subfamily 2 receptor ETR2 (Clark et al. 1998; Cancel and Larsen 2002). The major function of the receptors in ethylene signaling appears to be in the maintenance of CTR1 in an active state (Qu et al. 2007).

Downstream of CTR1 is ETHYLENE INSENSITIVE2 (EIN2), which is an essential positive regulator in ethylene signaling. The direct phosphorylation target of CTR1 is EIN2. EIN2 is an endoplasmic reticulum (ER)-bound protein with similarity to Nramp metal-ion transporters, which is maintained in an inactive state when phosphorylated by CTR1 (Ju et al. 2012; Qiao et al. 2012; Shakeel et al. 2013). CTR1 may directly phosphorylate EIN2 to inhibit the activity of EIN2. The significant sites of EIN2 phosphorylation may be on Ser645 and Ser924 (Chen et al. 2011; Ju et al. 2012; Qiao et al. 2012). Ser924 plays a predominant role in EIN2 regulation (Ju et al. 2012).

Upon ethylene binding to the ethylene receptors, the ethylene receptors inactivate CTR1, potentially through propagation of conformational changes in the receptor–CTR1 protein complex. EIN2 becomes dephosphorylated and this results in proteolytic cleavage and release of C-terminal domain of EIN2 (Fig. 4.9; Ju et al. 2012; Qiao et al. 2012; Wen et al. 2012; Shakeel et al. 2013). The C-terminal domain of EIN2 translocates to the nucleus, resulting in the activation of EIN3 and the transcriptional response to ethylene. Upon ethylene binding, the receptors inactivate CTR1, thereby relieving the suppression on the downstream signaling elements (Shakeel et al. 2013). As a result, EIN2 is proteolytically processed such that its C-terminal domain is released to migrate to the nucleus (Fig. 4.9; Qiao et al. 2012; Wen et al. 2012).

CTR1 interacts with and directly phosphorylates the cytosolic C-terminal domain of EIN2. Phosphorylation of EIN2 by CTR1 prevents EIN2 from signaling in the absence of ethylene. Upon ethylene perception by the receptors, CTR1 is inhibited and the inhibition of CTR1 upon ethylene perception is a signal for cleavage and nuclear localization of the EIN2 C-terminus, allowing the ethylene signal to reach the downstream transcription factors in the nucleus (Ju et al. 2012). In the nucleus, EIN2 either directly or indirectly activates the transcription factors EIN3 and EIN3-like1 (EIL1) to initiate the transcriptional response to ethylene (Fig. 4.9). In the absence of ethylene, the active receptors recruit CTR1 to associate with the membrane and thus become activated (Gao et al. 2003). This process represses the downstream signaling pathway mediated by EIN2 and EIN3 (Shakeel et al. 2013).

4.3.3 EIN2 Acts as the Central Regulator of Ethylene Signaling

EIN2 is a novel plant-specific protein of unknown function. EIN2 has two defined domains: an N-terminus with similarity to the Nramp family of metal-ion transporters and a unique hydrophilic C-terminus. Overexpression of the C-terminal domain is able to constitutively activate ethylene responses, suggesting that this domain is responsible for the transduction of the signal to downstream components (Alonso et al. 1999; Christians et al. 2008). Although the biochemical function of EIN2 is not understood, its critical role in ethylene signaling is manifested by the complete insensitivity of *EIN2* loss-of-function mutants (Love et al. 2005; Stepanova et al. 2005; Tang et al. 2005; Guo and Stotz 2007). EIN2 appears to be a key component involved in ethylene signaling and acts as the central regulator of ethylene signaling. EIN2 is a positive regulator of the ethylene signaling pathway in wheat, rice, and *Arabidopsis* (Travella et al. 2006). EIN2 may operate downstream of CTR1 and upstream of EIN3 (Fig. 4.9; Benavente and Alonso 2006; Yoo et al. 2008; Bisson and Groth 2011).

EIN2 is a membrane-spanning protein (An et al. 2010). EIN2 is recognized by two F-box proteins, EIN2-TARGETING PROTEIN1 (ETP1) and ETP2, for proteasomal degradation (Qiao et al. 2009). A nucleus-enriched multisubunit protein complex, called "COP9 signalosome (CSN)," is known to regulate ubiquitin–proteasome-mediated protein degradation (Feng et al. 2003). CSN is commonly present in plants. Several molecular studies have shown the presence of six loci, *COP9*, *FUS/COP11*, *FUS5*, *FUS4/COP8*, *FUS11*, and *FUS12*, encoding subunits of the COP9 signalosome, which are CSN8, CSN1, CSN7, CSN4, CSN3, and CSN2, respectively, in plants (Feng et al. 2003). CSN is associated with multiple SCF-type E3 ubiquitin ligases (Wang et al. 2003). CSN shows deneddylation activity toward the neddylated cullin subunit of SCF complexes, which is important for SCF ubiquitin ligase activity (Yang et al. 2002). EIN2 has been reported to interact with several components of the COP9 signalosome (Christians et al. 2008), a deneddylation enzyme required to sustain optimal activity of SCF-type E3 ligase. Mutation in a putative CSN subunit gene *ETHYLENE ENHANCED RESPONSE5* (*EER5*) leads to hypersensitivity to ethylene, suggesting that EIN2 might modulate CSN function to optimize the activity of SCF ligases in ethylene signaling (Christians et al. 2008).

4.3.4 Regulation of the Interaction of EIN2 and Ethylene Receptors by Protein Phosphorylation

The ethylene receptors may also associate with EIN2 (Fig. 4.10; Bisson et al. 2009; Bisson and Groth 2010; Ju and Chang 2012). EIN2 interacts tightly with all members of the *Arabidopsis* ethylene receptor family proteins that mark the starting point of the signaling pathway (Bisson and Groth 2011). The ethylene receptors, CTR1, and EIN2 are all localized to the endoplasmic reticulum (ER), and they interact with each other in the ethylene signal transduction pathway (Grefen et al. 2008; Bisson et al. 2009). The kinase domain of the receptors has been found to be essential for the formation of the EIN2–receptor complex. Phosphorylation appears to be a key mechanism in controlling the interaction of EIN2 and the ethylene receptors (Bisson and Groth 2011).

Protein phosphorylation is a key mechanism to control the interaction of EIN2 and the receptors. In vivo and in vitro fluorescence studies reveal that the kinase domain of the receptors is essential for the interaction (Bisson and Groth 2010; Scharein and Grath 2011; Hall et al. 2012). Cyanide, an ethylene agonist, which is known to reduce autophosphorylation of the ethylene receptor ETR1 or a mutation in the kinase domain of ETR1 that prevents autophosphorylation (the mutant H353A), increases the affinity of the receptors for EIN2. Mimicking permanent

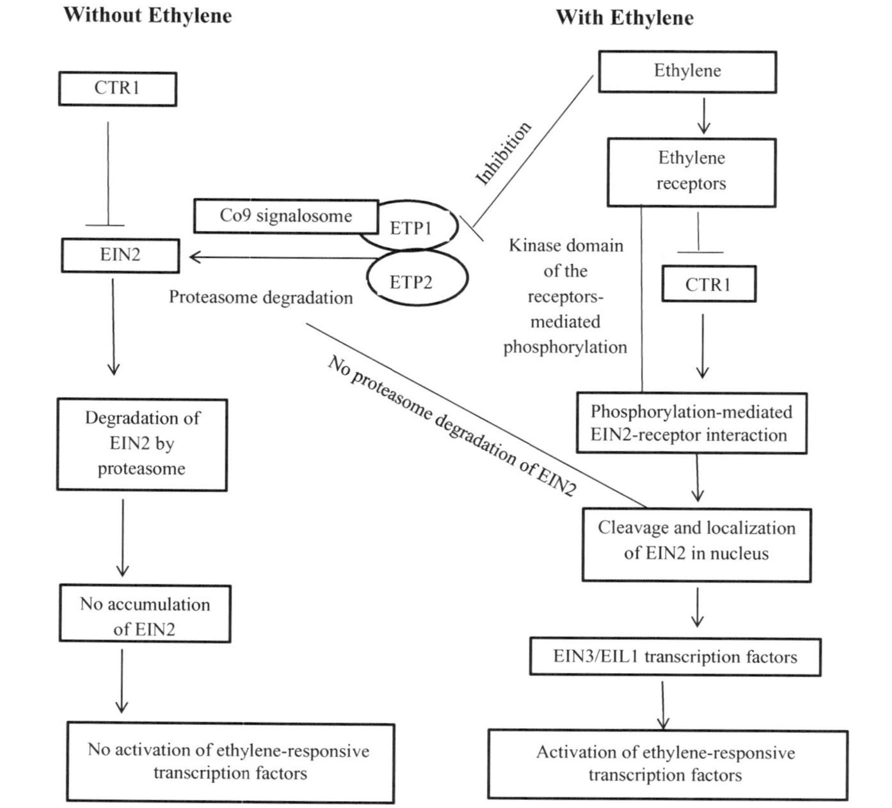

Fig. 4.10 Ethylene-triggered EIN2 C terminus cleavage to trigger transcription of ethylene-responsive transcription factors (Adapted from Gao et al. 2003; Qiao et al. 2009)

autophosphorylation of ETR1 as in the mutant H353E releases the EIN2–ETR1 interaction from the control by the plant hormone (Bisson and Groth 2010). These results suggest that autokinase activity of ETR1 and interaction with the soluble protein kinase CTR1 control the formation of signaling complexes of receptors and EIN2. Both the processes are regulated by ethylene. Without ethylene, binding receptor complexes are kept in their phosphorylated state due to their intrinsic autokinase activity (Voet-van-Vormizeele and Groth 2008), and they combine with CTR1, a negative regulator of ethylene signaling. Both processes distract the receptor complexes from interacting with the central regulator of ethylene signaling EIN2. Detached EIN2 is degraded by a proteasome pathway (Qiao et al. 2009). Almost no EIN2 accumulates in the absence of ethylene (Fig. 4.10). Upon ethylene binding, autokinase activity of the receptors is inhibited, converting them to their non-phosphorylated status (Voet-van-Vormizeele and Groth 2008), and CTR1 is inactivated by a conformational change or the release from the ER membrane (Gao et al. 2003). Both the mechanisms result in a tight interaction of the ethylene receptor complexes with EIN2, leading to accumulation of EIN2 at the ER membrane (Qiao et al. 2009).

4.3.5 EIN3/EIL Family of Proteins Functioning Downstream of EIN2 in Ethylene Signaling Pathway

A family of ETHYLENE INSENSITIVE3 (EIN3) and EIN3-like (EIL) proteins functions downstream of CTR1–EIN2 in the ethylene signaling pathway (Benavente and Alonso 2006). There are five EIN3 homologs (EIL1 to EIL5) in the *Arabidopsis* genome, and among them EIN3 and EIL1 play prominent roles in ethylene signaling (Binder et al. 2007; An et al. 2010). These plant-specific transcription factors are structurally and functionally conserved among several plant species (Chao et al. 1997; Lee and Kim 2003; Rieu et al. 2003). EIN3 and EIL1 are the primary transcription factors acting downstream of EIN2. In the nucleus, EIN2 either directly or indirectly activates the transcription factors EIN3 and EIL1 to initiate the transcriptional response to ethylene (Potuschak et al. 2003; van Loon et al. 2006; An et al. 2010).

EIN3 is expressed constitutively but is unable to accumulate as it is degraded continuously by the 26S proteasome-mediated proteolysis (Potuschak et al. 2003; Konishi and Yanagisawa 2008). The proteolysis of EIN3 is mediated by the E3 ubiquitin ligase $SCF^{EBF1/2}$ that contains either of the two closely related F-box proteins EBF1 (for *E*IN3-*B*inding *F* box 1) and EBF2 (Fig. 4.11; Guo and Ecker 2003; Yanagisawa et al. 2003; Gagne et al. 2004; Konishi and Yanagisawa 2008). The F-box proteins specifically recruit the target EIN proteins to an SCF ubiquitin ligase for degradation by the proteasome (Potuschak et al. 2003). EBF1 and EBF2

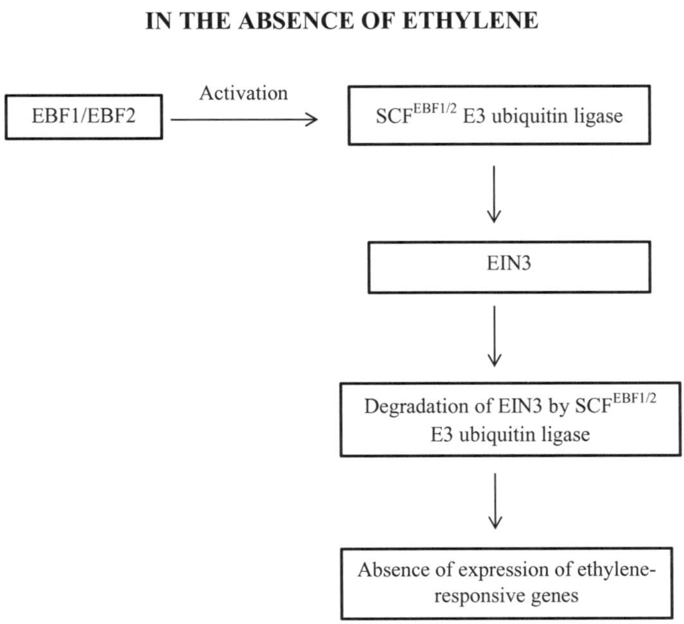

Fig. 4.11 Proteolysis of EIN3 mediated by the E3 ubiquitin ligase $SCF^{EBF1/2}$ (Adapted from Yanagisawa et al. 2003; Gagne et al. 2004; Konishi and Yanagisawa 2008)

have distinct roles in regulating EIN3 stability. EBF1 exerts its effect primarily in the absence of ethylene and during the initial phase of the response to ethylene treatment. By contrast, EBF2 plays a more prominent role during the later stages of the response to ethylene (Binder et al. 2007). The expression of EBF2 is transcriptionally induced by EIN3 that directly binds to the promoter of EBF2 (Konishi and Yanagisawa 2008).

The levels of EIN3 protein have been demonstrated to increase in response to the presence of ethylene (Olmedo et al. 2006). In the absence of ethylene, EIN3 is quickly degraded through a ubiquitin–proteasome pathway mediated by two F-box proteins, EBF1 and EBF2. EIN3 is short-lived that is degraded by the ubiquitin/26S proteasome pathway in the absence of ethylene. Upon ethylene treatment, EIN3 protein is quickly stabilized and accumulates in the nucleus (Guo and Ecker 2003, 2004; Potuschak et al. 2003; Yanagisawa et al. 2003; Gagne et al. 2004). EIN3 increases in response to increased levels of ethylene (Guo and Ecker 2003). EIN3 becomes stabilized after perception of ethylene and acts on its target promoters (Potuschak et al. 2003). Ethylene signaling suppresses the ubiquitin–proteasome-mediated proteolysis of EIN3, resulting in accumulation of EIN3. Thus, EIN3 becomes stabilized after perception of ethylene and induces ethylene-responsive gene expression (Fig. 4.12; Guo and Ecker 2003; Kepinski and Leyser 2003; Yanagisawa et al. 2003; Konishi and Yanagisawa 2008). These observations suggest that ethylene signaling action depends on EIN3 protein stabilization and proteolytic regulation of this protein may affect transcription of the defense-related genes (Fig. 4.12).

Konishi and Yanagisawa (2008) demonstrated feedback regulation of ethylene signaling that results from the direct up-regulation of the *EBF2* gene by EIN3. Although EBF1 and EBF2 show comparable activities as repressors of EIN3-dependent transcription, analysis of transgenic *Arabidopsis* plants has revealed that the *EBF2* promoter, but not the *EBF1* promoter, is activated by ethylene. Further studies revealed that EIN3 can bind and activate the *EBF2* promoter, indicating that EIN3 modulates *EBF2* gene expression. A sequence downstream of the EBF2 coding region has also been shown to be involved in modulation of both the EBF2 expression level and sensitivity to ethylene (Konishi and Yanagisawa 2008).

EBF1/EBF2-mediated proteolysis is repressed by ethylene (Gagne et al. 2004; An et al. 2010). Ethylene has been shown to enhance EIN3 and EIL1 stability by inducing EBF1/EBF2 proteasomal degradation (An et al. 2010). Ethylene stabilizes EIN3/EIL1 by promoting EBF1/EBF2 proteasomal degradation. It was also found that EIN2 is indispensable for mediating ethylene-induced EIN3/EIL1 accumulation and EBF1/2 degradation (An et al. 2010). Since EBF1 and EBF2 themselves are subject to proteasomal degradation, they may be the essential ethylene signal transducers. Collectively, these results suggest that EBF1 and EBF2 act as integrated ethylene signaling transducers downstream of CTR1 and EIN2 but upstream of EIN3/EIL1, and ethylene stimulates EIN3/EIL1 accumulation by promoting EBF1/EBF2 proteasomal degradation.

In *Arabidopsis thaliana*, EXORIBONUCLEASE4 (XRN4) has been identified as a component of the ethylene response pathway involved in the turnover of EIN3 (Potuschak et al. 2006). XRN4 has been shown to be allelic to the components

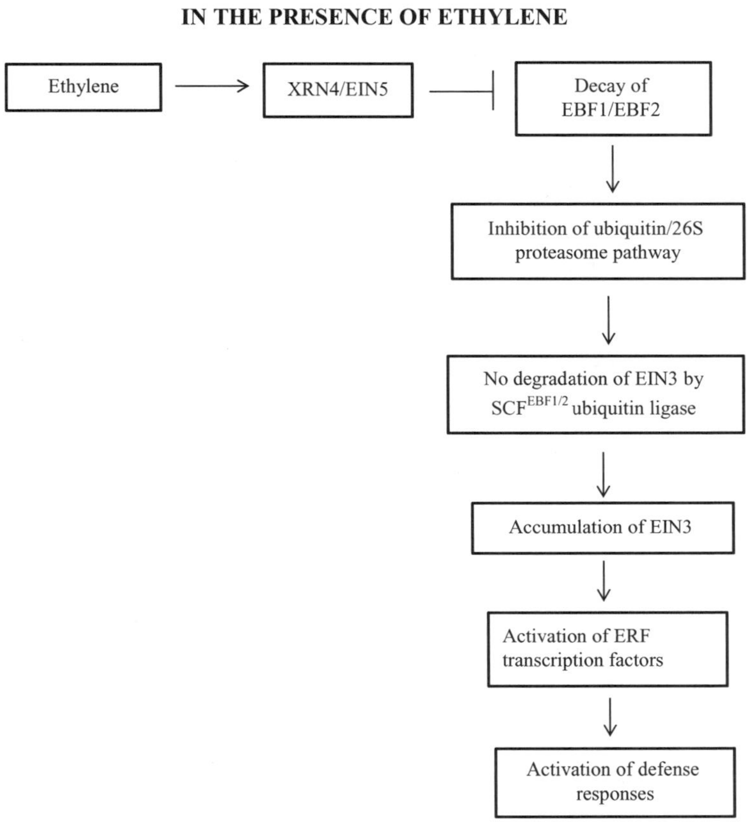

Fig. 4.12 Ethylene-induced suppression of ubiquitin-proteasome-mediated proteolysis of EIN3 promotes EIN3 accumulation to trigger defense responses (Adapted from Guo and Ecker 2003; Potuschak et al. 2003; Konishi and Yanagisawa 2008)

of the ethylene response pathway EIN5 (ETHYLENE INSENSITIVE5/ACC-INSENSITIVE1 (EIN5/AIN1) and EIN7 (Potuschak et al. 2006). EIN5 (ETHYLENE-INSENSITIVE5) has been identified as the 5′→3′ exoribonuclease XRN4. It is involved in the degradation of unstable mRNAs. EIN5/XRN4 acts downstream of CTR1 and upstream of EBF1/2 (Fig. 4.11; Potuschak et al. 2006). The *Arabidopsis xrn4* mutant seedlings are ethylene insensitive as a consequence of the up-regulation of EBF1 (EIN3 BINDING F-BOX PROTEIN 1) and EBF2 mRNA levels, which encode related F-box proteins involved in the turnover of EIN3 protein, an important transcriptional regulator of the ethylene response pathway. It was found that the ethylene insensitivity of *ein5* mutant plants was a consequence of the overaccumulation of EBF1 and EBF2 mRNAs resulting in the underaccumulatio n of EIN3 even in the presence of ethylene gas (Olmedo et al. 2006). These results suggest that the role of EIN5/XRN4 in ethylene perception is to antagonize the negative feedback regulation on EIN3 by promoting EBF1 and EBF2 mRNA decay, which consequently allows the accumulation of EIN3 protein to trigger the ethylene response.

4.3.6 ETR1–RTE1-Mediated CTR1-Independent Ethylene Signaling Pathway

A positive regulator of ETR1 ethylene receptor has been identified in *Arabidopsis* by Resnick et al. (2008). *RTE1* (*REVERSION-TO-ETHYLENE SENSITIVITY1*), a novel gene conserved in plants, animals, and some protists, was identified as a positive regulator of ETR1. The function of *RTE1* was found to be independent and distinct from that of *RAN1*, which encodes a copper transporter required for ethylene receptor function. It is suggested that *RTE1* promotes ETR1 signaling through a conformational effect on the ethylene-binding domain (Resnick et al. 2008).

RTE1 encodes a transmembrane protein that physically associates with ETR1 (Dong et al. 2008, 2010). The effects of RTE1 occur predominantly due to an isoform-specific interaction with ETR1 (Rivarola et al. 2009; Shakeel et al. 2013). Overexpression of *RTE1* results in reduced ethylene sensitivity, and this overexpression phenotype is largely dependent on the presence of ETR1 (Resnick et al. 2006; Zhou et al. 2007). RTE1 stabilizes or assists folding of ETR1 into the conformation it adopts in the absence of ethylene. It enhances the ability of ETR1 to repress ethylene signaling. The RTE1 transcript is up-regulated by ethylene, suggesting that RTE1 may facilitate the adaptation response of plants to ethylene (Resnick et al. 2006).

RTE1 and ETR1 may function together to mediate ethylene signaling through a CTR1-independent pathway (Qiu et al. 2012). Expression of the N-terminal half of ETR1 could partially suppress the constitutive ethylene response phenotype of a *ctr1* mutant (Qiu et al. 2012). The results suggest that there may be alternate pathway to the CTR1-dependent pathway. The results also suggest that the N-terminal half of ETR1 can mediate signal output through such an alternative CTR1-independent pathway. The role of RTE1 may be to maintain a particular conformation of ETR1 required for input into this alternative pathway (Shakeel et al. 2013).

RTE1 may directly promote ETR1 receptor signal output through the ETR1 N-terminus (Zhou et al. 2007; Qiu et al. 2012). ETR1 N-terminus does not have the CTR-1 interacting site. However, ETR1 N-terminus mediates the receptor signal to suppress constitutive ethylene responses. The ETR1 N-terminal signal may be mediated by a pathway independent of CTR1 (Gamble et al. 2002; Qu and Schaller 2004; Xie et al. 2006; Qiu et al. 2012). The presence of a CTR1-independent pathway has been suggested (Hua and Meyerowitz 1998; Cancel and Larsen 2002; Huang et al. 2003; Liu et al. 2010; Qiu et al. 2012). ETR1 may mediate the receptor signal output via its N-terminus. Overexpression of *RTE1* promotes ETR1 N-terminal signaling (Qiu et al. 2012). RTE1 was found to be essential for ETR1 N-terminal signaling independent of the CTR1 pathway. The ETR1 N-terminal signaling was not mediated by full-length ethylene receptors. The full-length ethylene receptors acted cooperatively with the ETR1 N-terminus to mediate the receptor signal independent of CTR1. ETR1 N-terminal signaling may involve RTE1, receptor cooperation, and negative regulation by the ETR1 C-terminus. The ETR1 C-terminus interacts with the N-terminus of CTR1 (Qiu et al. 2012). The ETR1 N-terminal signaling was not mediated via other full-length ethylene receptors, but the signal of full-length ethylene receptors could be mediated by the ETR1 N-terminus independent of CTR1. The ETR1 C-terminus may inhibit ETR1 N-terminal signaling, whereby deletion of the C-terminus facilitates N-terminal signaling (Qiu et al. 2012).

CTR1 plays a key role in mediating ethylene receptor signaling via its N-terminal interaction with the ethylene receptor C-terminal histidine kinase (HK) domain. Loss-of-function mutations of *CTR1* prevent ethylene receptor signaling, and corresponding *ctr1* mutants show a constitutive ethylene response phenotype (Xie et al. 2012). Expression of the truncated ethylene receptor ETR1 isoforms etr1$^{1\text{-}349}$ and dominant ethylene-insensitive etr1-1$^{1\text{-}349}$, lacking the C-terminal histidine kinase (HK) and receiver domains, both suppressed the *ctr1* mutant phenotype (Qiu et al. 2012). These results suggest that the ETR1 N-terminus is capable of receptor signaling independent of CTR1.

4.4 ERF Transcription Factors Functioning Downstream in Ethylene Signaling System

ERFs (Ethylene-responsive transcription factors) belonging to the APETELA2 (AP2)/ETHYLENE RESPONSIVE ELEMENT BINDING PROTEIN (EREBP) transcription factor family are the important group of transcription factors functioning downstream in ethylene signaling system (Gutterson and Reuber 2004; Grennan 2008). In *Arabidopsis*, 147 members of the AP2/EREBP gene family have been detected (Feng et al. 2005). The AP2/EREBP family is divided into the RAV, AP2, and EREBP subfamilies. The EREBP subfamily is further divided into DREB or A subgroup and the ERF or B subgroup. The ERF subgroup contains 65 ERF genes and contains all of the AP2/EREB genes that have been linked to disease resistance responses (Gutterson and Reuber 2004). ERF genes have been shown to be responsive to both JA and ethylene (Lorenzo et al. 2003; Gutterson and Reuber 2004; McGrath et al. 2005; Kim et al. 2011). The important ERF transcription factors involved in defense signaling system include AtERF1, AtERF2, AtERF4, and AtERF14 in *Arabidopsis* (McGrath et al. 2005; Grennan 2008); ERF2, ERF3, and ERF4 in tobacco (Koyama et al. 2003); ERF2 and ERF4 in canola (*Brassica napus*) (Yang et al. 2007); TaERF1 and TaERF3 in wheat (Xu et al. 2007; Zhang et al. 2007); CaPF1 and CaERFLP 1 in pepper (Lee et al. 2004; Yi et al. 2004); OsBIERF3 in rice (Cao et al. 2006); Pti4 and Pti5 in tomato (Zhou et al. 1997); and HvRAF in barley (Jung et al. 2007).

Functional studies of EIN3 have shown that EIN3 can bind to the promoter sequence of the ethylene-inducible transcription factor ERF1 (Solano et al. 1998). Up-regulation of *ERF1* gene expression in *Arabidopsis* by fungal infection is blocked in *ein2-5* mutant, impaired in the ethylene signal pathway. It indicates that ERF1 is a downstream component of ethylene signaling system (Berrocal-Lobo and Molina 2004). ERFs contain a single DNA-binding domain. ERFs have been shown to bind specifically to the GCC-box that is found in several promoters of the pathogenesis-related (PR) genes as well as ethylene- and jasmonate-inducible defense genes (Yamamoto et al. 1999; Gutterson and Reuber 2004). *Trichoderma viride* xylanase elicitor induced accumulation of mRNA for ERF2 in tobacco cells. The elicitor-inducible accumulation of the transcription factor was inhibited by staurosporine (an inhibitor of serine/threonine protein kinase) and calyculin A (an inhibitor of protein phosphatase). These results suggest that ERF2 might play a major role in the elicitor-induced GCC-box-mediated transcription of defense genes and that both protein kinase and protein phosphatase might be involved, as positive regulators in the signal transduction pathway that leads to expression of ERF2 and subsequent GCC-box-mediated transcription of defense genes (Yamamoto et al. 1999).

ERF proteins are grouped into three classes based on amino acid sequence identities within the ERF domain. Class I and class III ERFs act as activators, whereas class II ERFs act as repressors (Ohta et al. 2000; Koyama et al. 2003). Class II ERF repressors down-regulate the transactivation activity of class I and class II ERFs (Fujimoto et al. 2000). In tobacco, the *ERF3* gene coding for a class II repressor as well as genes for activators such as *ERF2* and *ERF4* were transcriptionally up-regulated in response to ethylene (Ohme-Takagi and Shinshi 1995; Kitajima et al. 2000). These *ERF* genes for both activators and a repressor were rapidly induced by a fungal elicitor treatment (Yamamoto et al. 1999).

Although most ERFs studied so far are positive regulators/activators, 14 *Arabidopsis* ERF proteins contain an ERF-associated amphiphilic repression (EAR) motif, which has been shown to function as a repression domain (Fujimoto et al. 2000; Ohta et al. 2001; McGrath et al. 2005; Nakano et al. 2006). In *Arabidopsis*, AtERF4, an EAR-containing ERF, negatively regulates the induction of *PDF1.2* and *CHIB* genes, while AtERF2 positively regulates the defense gene expression (Brown et al. 2003; McGrath et al. 2005). In tobacco, the *ERF3* gene acts as a repressor, while *ERF2* and *ERF4* act as positive regulators, and all three were transcriptionally up-regulated in response to ethylene (Ohme-Takagi and Shinshi 1995; Kitajima et al. 2000; Koyama et al. 2003). AtERF14 positively induces defense gene *ChiB* encoding the basic chitinase in *Arabidopsis* (Oñate-Sánchez et al. 2007).

Ubiquitin–proteasome system has been shown to be involved in the repression of class II ERFs (Koyama et al. 2003). A ubiquitin-conjugating enzyme (NtUBC2) was found to be involved in the repression activity of ERF3 (Koyama et al. 2003). The ubiquitin-conjugation activity of NtUBC2 may be involved in the regulation of repression activity of ERF3. The NtUBC2 interacted with ERF3 but not with ERF2 or ERF4. This suggests that the mechanism of regulation of the repression activity of ERF3 is distinct from that of the activation activity of ERF2 and ERF4. Since ERF repressor can suppress transactivation activity of ERF activators (Koyama et al. 2003), down-regulation of the repression activity of ERF3 by NtUBC2 may be operating for the induction of the GCC-box-mediated transcription of defense genes (Koyama et al. 2003). Thus, the interaction between ERF3 and NtUBC2 may be a critical step in activating transcription of various defense genes.

4.5 ROS and NO Signaling Systems Activate Transcription of Ethylene-Responsive Genes

On perception of pathogen-associated molecular patterns (PAMPs), plant pattern recognition receptors (PRRs) modulate signaling networks for defense responses that rely on rapid production of reactive nitrogen species (RNS) and reactive oxygen species (ROS) (Bellin et al. 2013). Both reactive oxygen species (H_2O_2) and reactive nitrogen species (NO) have been reported to trigger ethylene signaling system (Leshem et al. 1998; Desikan et al. 2001; Lindermayr et al. 2005, 2006). Ethylene receptor ETR1 has been shown to mediate H_2O_2 signaling in *Arabidopsis* (Desikan et al. 2005). ETR1 functions as an ROS sensor. Mutation of a Cys residue in the N-terminal region of ETR1 disrupts H_2O_2 signaling in plants (Desikan et al. 2005). The transcription factors induced by H_2O_2 include ethylene-responsive element-binding protein (EREBP) in *Arabidopsis* (Desikan et al. 2001). H_2O_2 up-regulates four ethylene-responsive element-binding proteins (EREBPs), the ethylene-responsive transcription factor (ERF1), and a CEO 1-like protein, which is a potential cofactor of EREBP transcription factors in tobacco (Fig. 4.12; Vandenabeele et al. 2003). NO signaling activates *EIN3*, which is involved in the activation of transcription of ethylene-responsive genes (Fig. 4.13; Chang and Stadler 2001).

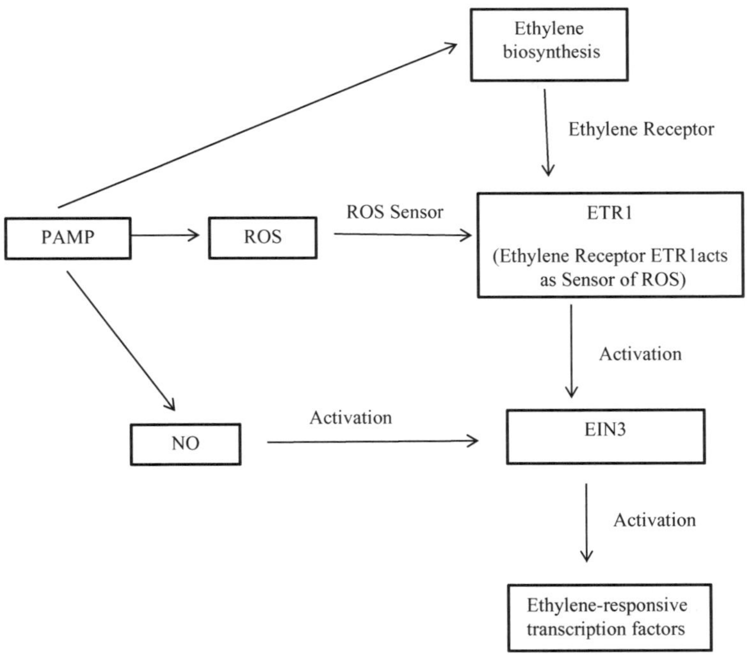

Fig. 4.13 ROS and NO signaling triggers ethylene signaling (Adapted from Chang and Stadler 2001; Vandenabeele et al. 2003; Desikan et al. 2005)

4.6 MAPK Cascade May Regulate Ethylene Signaling System

A MAPK cascade, MAPKKK (CTR1)–MKK9–MPK3/MPK6, has been shown to be an important downstream component in ethylene signaling system (Yoo et al. 2008). The *Arabidopsis* CTR1 (Constitutive Triple Response 1) is a Raf-like kinase considered as a MAPKKK, and it controls MPK3/6 activation via MKK9 (MAPK kinase9). It negatively regulates ET signaling (Guo et al. 2004; Yoo et al. 2008). CTR1 acts as an unconventional MAPKKK, blocking MKK9–MPK3/MPK6 activation and simultaneously enhancing EIN3 degradation by distinct MPK phosphorylation (Yoo et al. 2008).

Several pathogen-associated molecular patterns (PAMPs) are known to activate the phospholipase C (PLC)/diacylglycerol kinase (DAG kinase) pathway resulting in the biosynthesis of phosphatidic acid (PA) (Van der Luit et al. 2000; de Jong et al. 2004; Yamaguchi et al. 2005). PA is a second messenger involved in signal transduction system. It has been reported that PA binds CTR1 and inhibits its kinase activity. The kinase domain of CTR1 (CTR1-K) was found to bind PA directly. The PA-binding region was detected at the C-terminus of the kinase. Addition of PA resulted in almost complete inhibition of CTR1 kinase activity and disrupted the intramolecular interaction between CTR1-K and the CTR1 N-terminal regulatory domain. Further, PA blocked the interaction of CTR1 with the ethylene receptor ETR1 (Testerink et al. 2007). It is suggested that PA formed in response to PAMP/elicitor may induce ethylene responses through membrane recruitment and subsequent inhibition of kinase activity of the MAPKK kinase CTR1 (Testerink and Munnik 2005). The results suggest that CTR1-mediated MAP kinase cascade is involved in ethylene signaling.

The CTR1–MKK9–MPK3/MPK6 cascade positively modulates nuclear EIN3 stability. MPK3 or MPK6 activated by MKK9 in *Arabidopsis* protoplasts could directly phosphorylate EIN3 in vitro. These results suggest that direct phosphorylation by MKK9–MPK3/MPK6 in the nucleus may be a key step for EIN3 protein stabilization and ethylene signaling (Yoo et al. 2008). Binding of ethylene to the receptors (ETR1, ETR2, EIN4, ERS1, and ERS2) blocks the MKK9–MPK3/MPK6 cascade by failing to activate the MAPKKK protein (CTR1). This in turn allows EIN3 and its relatives to accumulate. The accumulated transcription factors induce the transcription of a variety of ethylene response genes such as the ERFs (Ouaked et al. 2003).

MKK9, one of the 10 *Arabidopsis* mitogen-activated protein kinase (MAPK) kinases, and its associated MAPKs, MPK3 and MPK6, are activated by ethylene, and the MAPK cascade acts positively in the ethylene signaling pathway downstream of CTR1 (Yoo et al. 2008). MKK9–MPK3/MPK6 module acts downstream of receptors/CTR1 and bypasses EIN2 to modulate EIN3 directly (Yoo et al. 2008). MKK9–MPK3/6 was reported to phosphorylate EIN3 and stabilize it; accordingly, the *mkk9* mutant displays a partial ethylene insensitivity phenotype (Yoo et al. 2008).

Search for putative MAPK phosphorylation sites and the conserved-docking sites for MAPK in EIN3 revealed that there were two MAPK phosphorylation sites

(174T175P and 592T593P), each coupled with a putative conserved-docking motif nearby, conserved in EIN3 and EIL1 (Alonso et al. 2003; Alonso and Stepanova 2004; Binder et al. 2007). The T174 mutation enhanced EIN3 degradation, whereas the T592 mutation enhanced EIN3 stability. The activated MPK6 phosphorylated $EIN3^{T592}$ but not $EIN3^{T174}$ in vivo. CTR1 degraded EIN3 but not $EIN3^{T592}$, supporting a direct link between T592 phosphorylation and degradation by the CTR1 pathway. It is suggested that the CTR1 pathway probably acted through MAPKs to phosphorylate T592 and promotes EIN3 degradation in the nucleus. The MKK9 phosphorylates T174 to promote EIN3 stability, whereas T592 is phosphorylated by a MAPK pathway mediated by CTR1 to promote EIN3 degradation (Yoo et al. 2008). Both the inhibition of CTR1 and the activation of MKK9 are induced by ethylene signaling. Without ethylene, CTR1 directly or indirectly inactivates MKK9–MPK3/MPK6 and probably activates downstream MAPKs to phosphorylate T592 to promote EIN3 degradation. Ethylene inactivates CTR1 for MKK9–MPK3/MPK6 activation and MKK9-induced T174 phosphorylation to stabilize EIN3 (Yoo et al. 2008). The role of MKK9–MPK3/MPK6 module in ethylene signaling is still being debated, and this pathway may act at least under certain conditions (Hahn and Harter 2009; Ju and Chang 2012; Ju et al. 2012).

The *Arabidopsis* MAP kinase MPK4 has been shown to be a requirement for induction of a subset of ET-regulated genes (Brodersen et al. 2006). MPK4 acts downstream or independently of ERF1 in ethylene signaling system (Brodersen et al. 2006). In *ctr1* mutant in which CTR1 activity is suppressed, the defense-related PDF1.2 protein accumulated to high levels. The high level of the PDF1.2 accumulation in the *ctr1* mutant was completely suppressed in the *ctr1/mpk4* double mutants, while *ctr1* and *ctr1/mpk4* both accumulated similar levels of *ERF1* mRNA (Brodersen et al. 2006). These results indicate that MPK4 functions downstream or independently of both CTR1 and ERF1 in triggering ethylene-responsive defense genes.

4.7 Ethylene Signaling Triggers Transcription of Plant Pattern Recognition Receptors (PRRs) in PAMP–PRR Signaling System

Pathogen-associated molecular patterns (PAMPs) are the signal molecules, which switch on the plant immune systems culminating in activation of defense genes (Thomma et al. 2011; Vidhyasekaran 2014). PAMPs are exclusively recognized as the molecules involved in triggering innate immunity. PAMPs are perceived as alarm/danger signals by cognate plant pattern recognition receptors (PRRs), and the PAMP–PRR complex activates the plant immune system (He et al. 2007; Wan et al. 2008; Iriti and Faoro 2009). Several receptors for the PAMPs have been recognized in plasma membrane of plant cells (Petutschnig et al. 2010; Schulze et al. 2010; Shinya et al. 2010; Segonzac and Zipfel 2011). Most of the PRRs are receptor-like kinases (RLKs) with a "receptor" and a "signaling domain" in one molecule

(Bi et al. 2010; Seifert and Blaukopf 2010; Lee et al. 2011). PAMPs bind with PRRs and induce conformational alteration in PRRs leading to their activation (Ali et al. 2007). PAMPs trigger increased transcription of PRR genes and accumulation of PRR proteins (Qutob et al. 2006; Lohmann et al. 2010). Sustained activation of plasma membrane-resident PRR signaling is important for mounting robust PAMP-triggered immunity (Saijo 2010). The plant immune system uses several second messengers to encode information generated by the PAMP/PRR signaling complex and deliver the information downstream of PRRs to proteins which decode/interpret signals and initiate defense gene expression (van Verk et al. 2008; Mersmann et al. 2010; Boudsocq et al. 2010; Hwang and Hwang 2011). The PRRs bind with PAMPs for their activation to trigger downstream immune signaling systems (Petutschnig et al. 2010; Boutrot et al. 2010). The binding of PAMP induces a conformational alteration in PRRs leading to their activation (Ali et al. 2007).

Boutrot et al. (2010) showed that ethylene is an integral part of PAMP-triggered immunity. Plants mutated in the key ethylene signaling protein EIN2 are impaired in all the PRR FLS2-mediated responses, correlated with reduced FLS2 transcription and protein accumulation. The EIN3 and EIN3-like transcription factors, which depend on EIN2 activity for their accumulation, directly control FLS2 expression. The results suggest a direct role for ethylene in the transcription of the PRR FLS2 (Boutrot et al. 2010).

Ethylene perception and signaling are crucial for *FLS2* gene transcription (Boutrot et al. 2010). FLS2 promoter revealed the presence of nine potential EIN3/EIL-binding sites (Boutrot et al. 2010), suggesting that EIN3 may bind to the promoter of the *FLS2* gene to influence its transcription. EIN3 binds to two positions in FLS2 promoter in *Arabidopsis* seedlings treated with ethylene. The results suggest that ethylene triggers the expression of the PRR gene through direct binding of the transcription factor EIN3 and potentially EIL1 to the PRR promoter (Boutrot et al. 2010). The ethylene-triggered increased expression of PRR genes might activate the plant immune responses.

4.8 Ethylene Triggers Ca^{2+} Influx in Downstream Ethylene Signaling System

Expression of several ethylene-inducible defense genes was found to be modulated by Ca^{2+}. Ca^{2+} influx may play an important role in ethylene signaling system (Raz and Fluhr 1992; Kwak and Lee 1997; Gallardo et al. 1999; Reddy et al. 2000). Blocking of Ca^{2+} fluxes inhibits ethylene-induced gene expression, and artificial elevation of $[Ca^{2+}]_{cyt}$ induces the expression of ethylene-regulated genes in the absence of ethylene (Kwak and Lee 1997; Gallardo et al. 1999). Expression of several ethylene-inducible defense genes was found to be modulated by Ca^{2+}. The induction of acidic and basic PR-1 protein families and chitinase accumulation were found to be inhibited by EGTA (ethylene glycol-bis (γβ-aminoethyl ether)-N,N,N′,N′-tetraacetic acid), the Ca^{2+} chelator.

Addition of calcium or application of Ca^{2+} ionophore induced the ethylene-inducible defense-related proteins (Raz and Fluhr 1992). These results suggest that downstream ethylene signaling system is triggered by Ca^{2+} signature.

Ethylene may also act upstream of Ca^{2+} influx from intracellular pools. Ethylene influenced Ca^{2+} influx into the segment tissues from pea epicotyls, and EGTA, a Ca^{2+} chelator, inhibited the ethylene response (Kwak and Lee 1997). These results suggest that Ca^{2+} influx acts at downstream of ethylene. Ca^{2+} from intracellular pools, but not Ca^{2+} from the apoplast, was the main player that interacted with ethylene signal transduction in pea (Petruzzelli et al. 2003). An ethylene-induced calmodulin-binding protein (EICBP, for *E*thylene-*I*nduced *CAM* *B*inding *P*rotein) has been isolated from ethylene-treated *Arabidopsis* seedlings (Reddy et al. 2000). The EICBP is a homolog of tobacco NtER1. The EICBP contains a CaM-binding domain. It contains, in addition to a CaM-binding domain, several features that are typical of transcription factors. These include a DNA-binding domain at the N-terminus, an acidic region at the C-terminus, and nuclear localization sequences. The expression of EICBP is induced by ethylene (Reddy et al. 2000). There seem to be several EICBP-related proteins in *Arabidopsis*, and EICBP-related proteins have also been detected in tomato and parsley (Reddy et al. 2000). These studies have revealed that these Ca^{2+}-regulated EICBPs may play a role in ethylene signaling system.

4.9 Ethylene and Jasmonate Signaling Interdependency in Triggering Plant Immune Responses

Ethylene (ET) and jasmonate (JA) signaling systems are known to act concurrently in triggering plant immune responses. JA/ET signaling systems concurrently activate transcription of several defense-related *PR* genes including *PR-1*, *PRB-1b*, *PR-2* (β-1,3-glucanase), *ChiB* (PR-3; basic chitinase), *PR-4* (hevein-like protein), *PR-5* (osmotin), *PDF1.2* (PR-12; defensin), *Thi2.1* (PR-13; thionin), and germin-like proteins in *Arabidopsis* (Zhou et al. 2005; Cao et al. 2006; Johansson et al. 2006; Jung et al. 2007; Oñate-Sánchez et al. 2007). Ethylene and JA signaling systems may induce the expression of the same defense genes; however, rate of the induced expression of the genes may vary. JA-induced expression of the defensin gene *PDF1.2* was significantly stronger than ET-induced expression of *PDF1.2* in *Arabidopsis* (Anderson et al. 2004).

JA and ET pathways appear to operate concomitantly in most cases (Anderson et al. 2004). JA and ET signaling pathways mediate various defense responses, either independently or collaboratively (Dong 1998). JA and ET pathways interact with each other coregulating expression of defense genes (Wang et al. 2002). JA and ET cooperatively regulate defense responses. JA stimulates proteinase inhibitor gene expression, while Et alone does not, but JA and ET induce proteinase inhibitor genes synergistically in tobacco (O'Donnell et al. 1996). MeJA-induced β-thujaplicin accumulation in *Cupressus lusitanica* cell cultures can be partly blocked by inhibitors of ethylene biosynthesis and signaling, while blocking JA biosynthesis inhibits almost all ET-induced β-thujaplicin accumulation (Zhao et al. 2004). These results suggest that the ET and JA pathways interact in mediating β-thujaplicin production, with the jasmonate pathway working as a main control and the ET pathway as a fine modulator for β-thujaplicin accumulation.

Ethylene signaling pathway activates JA biosynthesis. Ethylene positively regulates the induction of allene oxide synthase (AOS), a key enzyme involved in the biosynthesis of JA (O'Donnell et al. 1996). It also activates JA-inducible defense genes. The ethylene-responsive transcription factor ERF regulates JA-responsive gene expression (Brown et al. 2003; McGrath et al. 2005). ERF genes have been shown to be responsive to both JA and ethylene (Lorenzo et al. 2003; Gutterson and Reuber 2004; McGrath et al. 2005). The transcription factor ERF, which regulates ET-mediated gene expression, also regulates JA-responsive gene expression (Brown et al. 2003; McGrath et al. 2005; Pieterse et al. 2009). ERFs bind to the GCC-box and related elements in the promoters of JA-inducible *PR* genes, such as the defensin *PDF1.2*, basic chitinase (*ChiB*), and thionin (*Thi2.1*) (Fujimoto et al. 2000; Ohta et al. 2001; Berrocal-Lobo et al. 2002; Gu et al. 2002; Tournier et al. 2003; McGrath et al. 2005).

The expression of the important ERF transcription factor *ERF1* can be activated rapidly by ET or JA and can be activated synergistically by both ET and JA (Lorenzo et al. 2003). ERF1 is synergistically regulated by the ET and JA pathways (Gimenez-Ibanez and Solano 2013). In addition, both signaling pathways are required simultaneously to activate *ERF1*, because mutations that block any of them prevent *ERF1* induction by any of these hormones either alone or in combination (Lorenzo et al. 2003). Overexpression of *ERF1* expression could rescue the defense response defects of *coi1* and *ein2* mutants. ERF1 regulates in vivo the expression of a large number of genes responsive to both ET and JA (Lorenzo et al. 2003). Another ERF transcription factor, ORA59, has been shown to be an essential integrator of the JA and ET signal transduction pathways. JA- and ET-responsive expression of several defense genes, including the defensin gene *PDF1.2*, depended on ORA59 (Pré et al. 2008).

Constitutive overexpression of *AtERF2* or *AtERF4*, encoding ERF transcription factors, was reported to cause high levels of JA-responsive *PDF1.2* and *ChiB* gene expression in transgenic *Arabidopsis* plants (McGrath et al. 2005; Oñate-Sánchez et al. 2007). ERF5 and ERF6 play important roles as positive regulators of JA/ET-mediated defense against *B. cinerea* in *Arabidopsis* (Moffat et al. 2012). Expression of JA/ET-responsive genes and genes containing the GCC-box promoter motif was significantly up-regulated in both ERF5 and ERF6 transgenic plants, suggesting that ERF5 and ERF6 may act as positive regulators of JA-mediated defense (Moffat et al. 2012). Collectively these studies suggest that ERF transcription factors play important role in the concurrent expression of JA and ET signaling systems.

The molecular basis of ET/JA coaction and signaling interdependency has been described in Chap. 3 and will be briefly described here. The ethylene-stabilized transcription factors EIN3 and EIL1 and the jasmonate receptor JAZ (for JASMONATE

ZIM [Zinc-finger protein expressed in inflorescence Meristem]-domain) have been shown to take part in the concurrent activation of JA and ET signaling systems in plant immune responses (Zhu et al. 2011b). JAZ proteins physically interact with EIN3 and EIL1 (Chini et al. 2007; Zhu et al. 2011b). An EIN3 fragment containing amino residues 200 to 500 was sufficient for JAZ1 binding. The interaction between JAZ1 and an EIL1 fragment (amino acids 201–501) was also observed (Zhu et al. 2011b). JAZs strongly repress EIN3/EIL1 function (Zhu et al. 2011b). The JAZ proteins can repress the function of EIN3/EIL1, probably by suppressing the DNA binding of EIN3 (Zhu et al. 2011b). Without JA accumulation, EIN3 and EIL1 are repressed by JAZs, which recruit HDA6 as a corepressor to decrease EIN3/EIL1 protein stability (Zhou et al. 2005; An et al. 2010; Zhu et al. 2011b). JA treatment leads to JAZ degradation, while ET treatment enhances EIN3/EIL1 protein stability. ET is needed for EIN3/EIL1 stabilization and JA for EIN3/EIL1 release from the JAZ protein repression, and this may be responsible for the cooperative action of ET and JA signaling systems (Fig. 4.14; Zhu et al. 2011b). JAZ degradation activates the JA signaling system by activating MYC2 and other transcription factors, and it also activates ET signaling system by triggering the action of the ET-dependent transcription activators EIN3 and EIL1. Collectively, these results explain the molecular basis of interdependency of the two hormone signaling systems in plant immune responses.

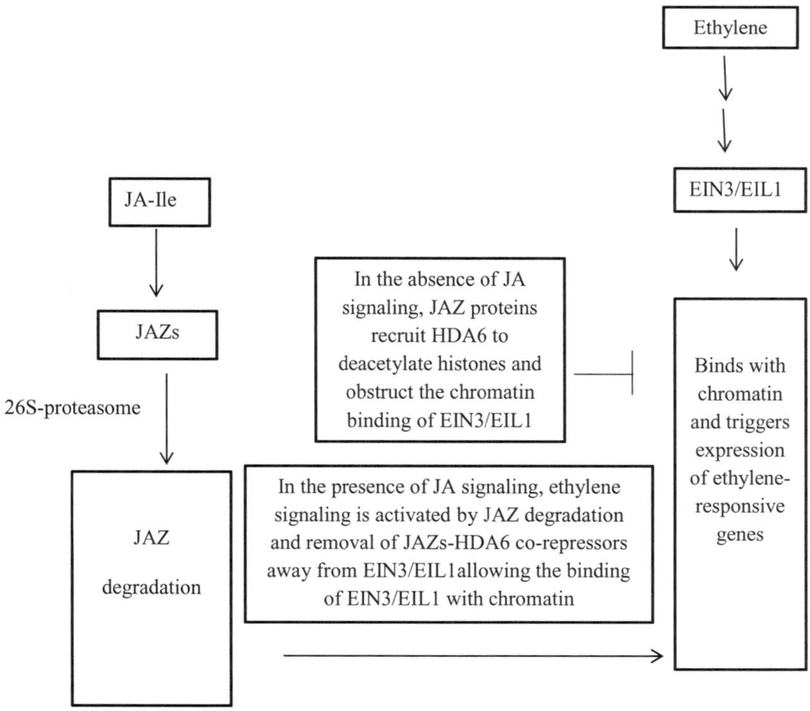

Fig. 4.14 Jasmonate activates ethylene-dependent defense responses (Adapted from Zhu et al. 2011b)

4.10 Ethylene Induces Transcription of Defense-Related Genes

The final stage of the ethylene signaling system includes induction of transcription of several defense genes (Fig. 4.15). The transcription factor ERF1 has been shown to induce transcription of several defense-related *PR* genes including *PR-1*, *PRB-1b*, *PR-2* (β-1,3-glucanase), *ChiB* (PR-3; basic chitinase), *PR-4* (hevein-like protein), *PR-5* (osmotin), *PDF1.2* (PR-12; defensin), *Thi2.1* (PR-13; thionin), and germin-like proteins in *Arabidopsis* (Sessa et al. 1995; Solano et al. 1998; Penninckx et al. 1998; Fujimoto et al. 2000; Frye et al. 2001; Lorenzo et al. 2003; Tang et al. 2005;

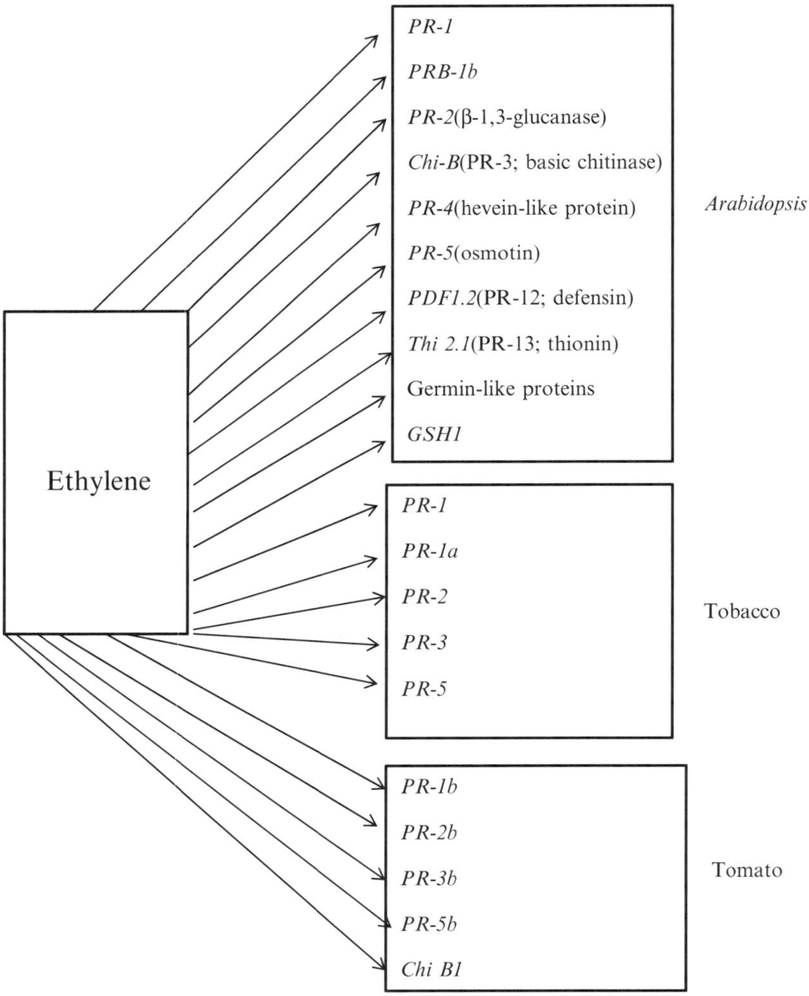

Fig. 4.15 Ethylene-inducible PR genes in *Arabidopsis*, tobacco, and tomato

Zhou et al. 2005; Johansson et al. 2006; Jung et al. 2007; Oñate-Sánchez et al. 2007). Overexpression of the *HvRAF* gene coding an ERF-type transcription factor from barley in *Arabidopsis* plants induced the activation of various defense genes such as *PDF1.2*, *PR1*, *PR5*, and *GSH1* (Jung et al. 2007). Ethylene induced expression of the *PR-1*, *PR-1a*, *PR-2*, *PR-3*, and *PR-5* genes in tobacco (Ohme-Takagi and Shinshi 1995; Shinshi et al. 1995; Yang et al. 1997; Park et al. 2001; Cao et al. 2006). An ethylene-responsive region has been detected in the promoter of the tobacco class 1 chitinase gene (Shinshi et al. 1995). Ethylene signaling system induces *PR1b1* and *ChiB* in tomato (Lin et al. 2008). PAMP treatment induced high expression of three basic ethylene-inducible defense-related genes (*PR-2b*, *PR-3b*, and *PR-5b*) in tomato (Hase et al. 2006). An ERF/AP2-type transcription factor (*CaPF1*) detected in pepper induced an increase in transcript levels of *PDF1.2* and *GST* (Yi et al. 2004).

4.11 Ethylene Signaling System Modulates Plant Immune Signaling System Triggering Resistance or Susceptibility Against Different Pathogens

Ethylene signaling system has been shown to induce resistance against some pathogens. It induced resistance against the necrotrophic pathogens *Botrytis cinerea* and *Cercospora nicotianae* and the bacterium *Erwinia carotovora*. In contrast, the ethylene system did not induce resistance against the biotrophic fungus *Oidium neolycopersici* and induced susceptibility against the oomycete *Peronospora tabacina* and *Tobacco mosaic virus* in tobacco (Geraats et al. 2003). The ethylene response factor 1 (ERF1) is a transcription factor involved in ethylene signaling system, and its overexpression in *Arabidopsis thaliana* enhances resistance to *Botrytis cinerea*, *Plectosphaerella cucumerina*, *Fusarium oxysporum* f. sp. *conglutinans*, and *F. oxysporum* f. sp. *lycopersici* (Berrocal-Lobo and Molina 2004). Overexpression of *ERF1* enhances resistance against *B. cinerea* and other necrotrophic pathogens, but increases susceptibility to the hemibiotroph *Pseudomonas syringae* (Berrocal-Lobo et al. 2002; Berrocal-Lobo and Molina 2004).

The rice *OsBIERF3* encodes a protein belonging to the ethylene-responsive element-binding protein (EREBP) transcription factor family (Cao et al. 2006). Overexpression of the *OsBIERF3* gene in transgenic tobacco plants resulted in elevation of defense genes and conferred resistance against infection by *Tomato mosaic virus* and bacterial wildfire pathogen *Pseudomonas syringae* pv. *tabaci* (Cao et al. 2006). Overexpression of NtERF5, a member of the tobacco ethylene response transcription factor family, enhances resistance of tobacco to *Tobacco mosaic virus* (Fischer and Dröge-Laser 2004). Tomato stress-responsive factor TSRF1 interacts with ethylene-responsive element GCC-box and regulates pathogen resistance to *Ralstonia solanacearum* (Zhang et al. 2004). The pepper ERF transcription factor *CaPF1* confers resistance to *P. syringae* pv. *tomato* in transgenic *Arabidopsis* and tobacco plants (Yi et al. 2004). Ethylene treatment induced resistance against the charcoal rot pathogen *Macrophomina phaseolina* in *Medicago truncatula* plants (Gaige et al. 2010). Ethylene confers resistance against *Botrytis cinerea* in tomato, and down-regulation of gene transcripts involved in ethylene biosynthesis and signaling was associated with increased susceptibility to the pathogen (Nambeesan et al. 2012). Overexpression of another pathogen-inducible ERF transcription factor, GbERF2, in tobacco enhances resistance to *Alternaria longipes* (Zuo et al. 2007). The barley ERF-type transcription factor HvRAF confers enhanced resistance to *Ralstonia solanacearum* (Jung et al. 2007). Overexpression of MtERF1-1 in *Medicago truncatula* roots increased resistance to the fungal pathogen *Rhizoctonia solani* and the oomycete pathogen *Phytophthora medicaginis* (Anderson et al. 2010).

ERF5 is an ethylene-responsive element-binding factor. It negatively regulates chitin signaling and plant defense against the fungal pathogen *Alternaria brassicicola* and positively regulates salicylic acid signaling and plant defense against the bacterial pathogen *Pseudomonas syringae* pv. *tomato* DC3000 (Son et al. 2012). It has been shown that 1-aminocyclopropane-1-carboxylic acid synthase (*ACS*)-silenced apple fruit that expresses a sense construct of *ACS* was more susceptible to *B. cinerea* than untransformed apple, suggesting that ethylene enhances fruit resistance to *B. cinerea* infection (Akagi et al. 2011).

OsEIN2 positively regulates ethylene signaling and triggers defense responses against the rice blast pathogen *Magnaporthe oryzae*. Antisense suppression of *OsEIN2* increases susceptibility to *M. oryzae* (De Vleesschauwer et al. 2008). Ethylene induces susceptibility against brown spot disease caused by *Cochliobolus miyabeanus* in rice plants (De Vleesschauwer et al. 2010). Symptoms caused by brown spot infection were more severe on rice seedlings pretreated with the ethylene-releasing chemical ethephon, implying a negative role in rice defense against *C. miyabeanus*. Disruption of the ethylene signaling pathway by antisense suppression of *OsEIN2*, a central transducer in the rice ethylene pathway, induced disease resistance. Infiltration of silver thiosulfate (STS), an inhibitor of ethylene action, induced disease resistance (De Vleesschauwer et al. 2010). These results suggest that ethylene may suppress host defense responses, resulting in susceptibility against *C. miyabeanus* in rice. In contrast, antisense suppression of *OsEIN2* induces heightened resistance against C. *miyabeanus* (De Vleesschauwer et al. 2010). These results suggest that ethylene may act as a two-faced player in the plant immune response network, triggering resistance or susceptibility against different pathogens (Fig. 4.16).

Ethylene signaling system induces susceptibility to *Clavibacter michiganensis* subsp. *michiganensis*, and increased synthesis of ethylene aids in disease development (Balaji et al. 2008). ET pathway induces susceptibility to *P. syringae* pv. *tomato* (Berrocal-Lobo et al. 2002).

Rice *OsERF922*, an AP2/ERF-type transcription factor, binds specifically to the GCC-box sequence and acts as a transcriptional activator in plant cells. Knockdown of *OsERF922* by means of RNAi enhanced resistance against the rice blast pathogen *Magnaporthe oryzae*. The elevated disease resistance was associated with increased expression of *PR1a*, *PR1b*, *PR10*, *PAL*, and other genes encoding phytoalexin biosynthetic enzymes (Liu et al. 2012). In contrast, *OsERF922*-overexpressing

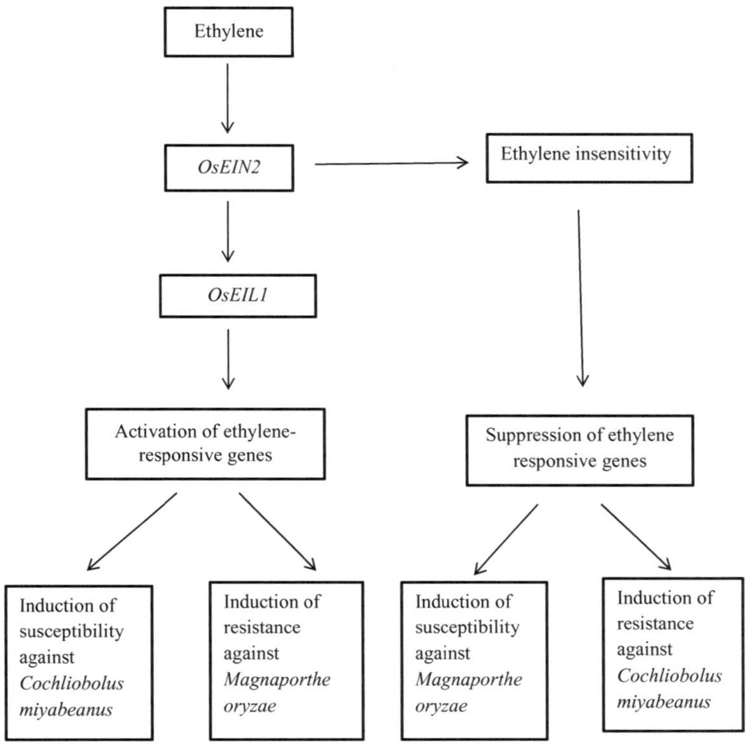

Fig. 4.16 Ethylene may induce resistance or susceptibility against different pathogens in rice (Adapted from De Vleesschauwer et al. 2008, 2010)

plants showed reduced expression of these defense-related genes and enhanced susceptibility to *M. oryzae* (Liu et al. 2012), suggesting that *OsERF922* is involved in conferring susceptibility against *M. oryzae*.

References

Akagi A, Dandekar AM, Stotz HU (2011) Resistance of *Malus domestica* fruit to *Botrytis cinerea* depends on endogenous ethylene biosynthesis. Phytopathology 101:1311–1321

Ali GS, Prasad KVSK, Day I, Reddy ASN (2007) Ligand-dependent reduction in the membrane mobility of FLAGELLIN SENSITIVE2, an Arabidopsis receptor-like kinase. Plant Cell Physiol 48:1601–1611

Alonso JM, Stepanova AN (2004) The ethylene signaling pathway. Science 306:1513–1515

Alonso JM, Hirayama T, Roman G, Nourizadeh S, Ecker JR (1999) EIN2, a bifunctional transducer of ethylene and stress responses in *Arabidopsis*. Science 284:2148–2152

Alonso JM, Stepanova AN, Solano R, Wisman E, Ferrari S, Ausubel FM, Ecker JR (2003) Five components of the ethylene-response pathway identified in a screen for weak ethylene-insensitive mutants in *Arabidopsis*. Proc Natl Acad Sci U S A 100:2992–2997

An F, Zhao Q, Ji Y, Li W, Jiang Z, Yu X, Zhang C, Han Y, He W, Liu Y, Zhang S, Ecker JR, Guo H (2010) Ethylene-induced stabilization of ETHYLENE INSENSENSITIVE3 and EIN3-LIKE1 is mediated by proteasomal degradation of EIN3 binding F-box 1 and 2 that requires EIN2 in Arabidopsis. Plant Cell 22:2384–2401

Anderson JP, Badruzsaufari E, Schenk PM, Manners JM, Desmond OJ, Ehlert C, Maclean DJ, Ebert PR, Kazan K (2004) Antagonistic interaction between abscisic acid and jasmonate-ethylene signaling pathways modulates defense gene expression and disease resistance in *Arabidopsis*. Plant Cell 16:3460–3479

Anderson JP, Lichtenzveig J, Gleason C, Oliver RP, Singh KB (2010) The B-3 ethylene response factor MtERF1-1 mediates resistance to a subset of root pathogens in *Medicago truncatula* without adversely affecting symbiosis with Rhizobia. Plant Physiol 154:861–873

Arnott T, Murphy TM (1991) A comparison of the effects of a fungal elicitor and ultraviolet radiation on ion transport and hydrogen peroxide synthesis by rose cells. Environ Exp Bot 31:209–216

Asai T, Tena G, Plotnikova J, Willmann MR, Chiu WL, Gomez-Gomez I, Boller T, Ausubel FM, Sheen J (2002) MAP kinase signaling cascade in *Arabidopsis* innate immunity. Nature 415:977–983

Balaji V, Mayrose M, Scherf O, Jacob-Hirsch J, Eichenlaub R, Iraki N, Manulis-Sasson S, Rechavi G, Barash I, Sessa G (2008) Tomato transcriptional changes in response to *Clavibacter michiganensis* reveal a role for ethylene in disease development. Plant Physiol 146:1797–1809

Bellin D, Asai S, Delledonne M, Yoshioka H (2013) Nitric oxide as a mediator for defense responses. Mol Plant Microbe Interact 26:271–277

Benavente LM, Alonso JM (2006) Molecular mechanisms of ethylene signaling in *Arabidopsis*. Mol BioSyst 2:165–173

Bent AF, Innes RW, Ecker JR, Staskawicz BJ (1992) Disease development in ethylene-insensitive *Arabidopsis thaliana* infected with virulent and avirulent *Pseudomonas* and *Xanthomonas* pathogens. Mol Plant Microbe Interact 5:372–378

Berr A, McCallum EJ, Alioua A, Heintz D, Heitz T, Shen WH (2010) Arabidopsis histone methyltransferase SET DOMAIN GROUP8 mediates induction of the jasmonate/ethylene pathway genes in plant defense response to necrotrophic fungi. Plant Physiol 154:1403–1414

Berrocal-Lobo M, Molina A (2004) Ethylene response factor 1 mediates *Arabidopsis* resistance to the soilborne fungus *Fusarium oxysporum*. Mol Plant Microbe Interact 17:763–770

Berrocal-Lobo M, Molina A, Solano R (2002) Constitutive expression of ETHYLENE-RESPONSE-FACTOR1 in *Arabidopsis* confers resistance to several necrotrophic fungi. Plant J 29:23–32

Bethke G, Pecher P, Eschen-Lippold L, Tsuda K, Katagiri F, Glazebrook J, Scheel D, Lee J (2012) Activation of the *Arabidopsis thaliana* mitogen-activated protein kinase MPK11 by the flagellin-derived elicitor peptide, flg22. Mol Plant Microbe Interact 25:471–480

Bi D, Cheng YT, Li X, Zhang Y (2010) Activation of plant immune responses by a gain-of-function mutation in an atypical receptor-like kinase. Plant Physiol 153:1771–1779

Binder BM, Rodrguez F, Bleecker AB (2010) The copper transporter RAN1 is essential for biogenesis of ethylene receptors in *Arabidopsis*. J Biol Chem 285:37263–37270

Binder BM, Walker JM, Gagne JM, Emborg TJ, Hemmann G, Bleecker AB, Vierstra RD (2007) The Arabidopsis EIN3 binding F-Box proteins EBF1 and EBF2 have distinct but overlapping roles in ethylene signaling. Plant Cell 19:509–523

Binder BM, Chang C, Schaller GE (2012) Perception of ethylene by plants – ethylene receptors. Annu Plant Rev 44:117–145

Bisson MMA, Groth G (2010) New insight in ethylene signaling: autokinase activity of ETR1 modulates the interaction of receptors and EIN2. Mol Plant 3:882–889

Bisson MMA, Groth G (2011) New paradigm in ethylene signaling: EIN2, the central regulator of the signaling pathway, interacts directly with the upstream receptors. Plant Signal Behav 6:164–166

Bisson MMA, Bleckmann A, Allekotte S, Groth G (2009) EIN2, the central regulator of ethylene signaling, is localized at the ER membrane where it interacts with the ethylene receptor ETR1. Biochem J 444:261–267

Bleecker AB (1999) Ethylene perception and signaling: an evolutionary perspective. Trends Plant Sci 4:269–274

Bolwell GP, Butt VS, Davies DR, Zimmerlin A (1995) The origin of the oxidative burst in plants. Free Radical Res 23:517–532

Bostick M, Lochhead SR, Honda A, Palmer S, Callis J (2004) Related to ubiquitin 1 and 2 are redundant and essential and regulate vegetative growth, auxin signaling, and ethylene production in *Arabidopsis*. Plant Cell 16:2418–2432

Boudsocq M, Willmann MR, McCormack M, Lee H, Shan L, He P, Bush J, Cheng SH, Sheen J (2010) Differential innate immune signalling via Ca^{2+} sensor protein kinases. Nature 464:418–422

Boutrot F, Segonzac C, Chang KN, Qiao H, Ecker JR, Zipfel C (2010) Direct transcriptional control of the *Arabidopsis* immune receptor FLS2 by the ethylene-dependent transcription factors EIN3 and EIL1. Proc Natl Acad Sci U S A 107:14502–14507

Brisson L, El-Bakkali-Taheri N, Giorgi M, Fadel A, Kaizer J, Rěglier M, Tron T, el Ajandouz H, Simaan AJ (2012) 1-Aminocyclopropane-1-carboxylic acid oxidase: insight into cofactor binding from experimental and theoretical studies. J Biol Inorg Chem 17:939–949

Brodersen P, Petersen M, Bjorn Nielsen H, Zhu S, Newman MA, Shokat KM, Rietz S, Parker J, Mundy J (2006) *Arabidopsis* MAP kinase 4 regulates salicylic acid- and jasmonic acid/ethylene-dependent responses via EDS1 and PAD4. Plant J 47:532–546

Broekaert WF, Delaure SI, De Bolle MFC, Commue BPA (2006) The role of ethylene in host-pathogen interactions. Annu Rev Phytopathol 44:393–416

Brown RL, Kazan K, McGrath KC, Maclean DJ, Manners JM (2003) A role for the GCC-box in jasmonate-mediated activation of the *PDF1.2* gene of *Arabidopsis*. Plant Physiol 132:1020–1032

Cancel JD, Larsen PB (2002) Loss-of function mutations in the ethylene receptor ETR1 cause enhanced sensitivity and exaggerated response to ethylene in Arabidopsis. Plant Physiol 129:1557–1567

Cao Y, Wu Y, Zheng Z, Song F (2006) Overexpression of the rice EREBP-like gene *OsBIERF3* enhances disease resistance and salt tolerance in transgenic tobacco. Physiol Mol Plant Pathol 202–211

Chae HS, Faure F, Kieber JJ (2003) The *eto1*, *eto2*, and *eto3* mutations and cytokinin treatment increase ethylene biosynthesis in Arabidopsis by increasing the stability of ACS protein. Plant Cell 15:545–559

Chang C, Stadler R (2001) Ethylene hormone receptor action in *Arabidopsis*. Bioessays 23:619–627

Chao Q, Rothenberg M, Solano R, Roman G, Terzaghi W, Ecker JR (1997) Activation of the ethylene gas response pathway in *Arabidopsis* by the nuclear protein Ethylene-Insensitive3 and related proteins. Cell 89:1133–1144

Chen N, Goodwin PH, Hsiang T (2003) The role of ethylene during the infection of *Nicotiana tabacum* by *Colletotrichum destructivum*. J Exp Bot 54:2449–2456

Chen R, Binder BM, Garrett WM, Tucker ML, Chang C, Cooper B (2011) Proteomic responses in *Arabidopsis thaliana* seedlings treated with ethylene. Mol Biosyst 7:2637–2650

Chini A, Fonseca S, Fernández G, Adie B, Chico JM, Lorenzo O, Garcia-Casado G, López-Vidriero I, Lozano FM, Ponce MR, Micol JL, Solano R (2007) The JAZ family of repressors is the missing link in jasmonate signaling. Nature 448:666–671

Christians MJ, Robles LM, Zeller SM, Larsen PB (2008) The *eer5* mutation, which affects a novel proteasome-related subunit, indicates a predominant role for the COP9 signalosome in resetting the ethylene-signaling pathway in Arabidopsis. Plant J 55:467–477

Clark KL, Larsen PB, Wang PB, Chang C (1998) Association of the Arabidopsis CTR1 raf-like kinase with the ETR1 and ERS ethylene receptors. Proc Natl Acad Sci U S A 95:5401–5406

de Jong CF, Laxalt AM, Bargmann BO, de Wit PJ, Joosten MH, Munnik T (2004) Phosphatidic acid accumulation is an early response in the *Cf-4/Avr4* interaction. Plant J 39:1–12

De Vleesschauwer D, Djavaheri M, Bakker PAHM, Höfte M (2008) *Pseudomonas fluorescens* WCS374r-induced systemic resistance in rice against *Magnaporthe oryzae* is based on pseudobactin-mediated priming for a salicylic acid-repressable multifaceted defense response. Plant Physiol 148:1996–2012

De Vleesschauwer D, Yang Y, Vera Cruz C, Höfte M (2010) Abscisic acid-induced resistance against the brown spot pathogen *Cochliobolus miyabeanus* in rice involves MAP kinase-mediated repression of ethylene signaling. Plant Physiol 152:2036–2052

Denoux C, Galletti R, Mammarella N, Gopalan S, Werck D, De Lorenzo G, Ferrari S, Ausubel FM, Dewdney J (2008) Activation of defense response pathways by OGs and Flg22 elicitors in Arabidopsis seedlings. Mol Plant 1:423–445

Desikan R, Mackerness S, Hancock JT, Neill SJ (2001) Regulation of the *Arabidopsis* transcriptome by oxidative stress. Plant Physiol 127:159–172

Desikan R, Hancock JT, Bright J, Harrison J, Weir I, Hooley R, Neill SJ (2005) A role for ETR1 in hydrogen peroxide signaling in stomatal guard cells. Plant Physiol 137:831–834

Dong X (1998) SA, JA, ethylene, and disease resistance in plants. Curr Opin Plant Biol 1:316–323

Dong CH, Rivarola M, Resnick JS, Maggin BD, Chang C (2008) Subcellular co-localization of Arabidopsis RTE1 and ETR1 supports a regulatory role for RTE1 in ETR1 ethylene signaling. Plant J 53:275–286

Dong CH, Jang M, Scharein B, Malach A, Rivarola M, Liesch J, Groth G, Hwang I, Chang C (2010) Molecular association of the *Arabidopsis* ETR1 ethylene receptor and a regulator of ethylene signaling, RTE1. J Biol Chem 285:40706–40713

Dreher K, Callis J (2007) Ubiquitin, hormones and biotic stress in plants. Ann Bot 99:787–822

Feng S, Ma L, Wang X, Xie D, Dinesh-Kumar SP, Wei N, Deng XW (2003) The COP9 signalosome interacts physically with SCF^{COI1} and modulates jasmonate responses. Plant Cell 15:1083–1094

Feng JX, Liu D, Pan Y, Gong W, Ma LG, Luo JC, Deng XW, Zhu YX (2005) An annotation update via cDNA sequence analysis and comprehensive profiling of developmental, hormonal or environmental responsiveness of the Arabidopsis AP2/EREBP transcription factor gene family. Plant Mol Biol 59:853–868

Fischer U, Dröge-Laser W (2004) Overexpression of *Nt*ERF5, a new member of the tobacco ethylene response transcription factor family enhances resistance to *Tobacco mosaic virus*. Mol Plant Microbe Interact 17:1162–1171

Frye CA, Tang D, Innes RW (2001) Negative regulation of defense responses in plants by a conserved MAPKK kinase. Proc Natl Acad Sci U S A 98:373–378

Fujimoto SY, Ohta M, Usui A, Shinshi H, Ohme-Takagi M (2000) Arabidopsis ethylene-responsive element binding factors act as transcriptional activators or repressors of GCC box-mediated gene expression. Plant Cell 12:393–404

Fujiwara M, Umemura K, Kawasaki T, Shimamoto K (2006) Proteomics of Rac GTPase signaling reveals its predominant role in elicitor-induced defense response of cultured rice cells. Plant Physiol 140:734–745

Gagne JF, Smalle J, Gingerich DJ, Walker JM, Yoo S-D, Yanagisawa S, Vierstra RD (2004) Arabidopsis EIN3-binding F-box 1 and 2 form ubiquitin-protein ligases that repress ethylene action and promote growth by directing EIN3 degradation. Proc Natl Acad Sci U S A 101:6803–6808

Gaige AR, Ayella A, Shuai B (2010) Methyl jasmonate and ethylene induce partial resistance in *Medicago truncatula* against the charcoal rot pathogen *Macrophomina phaseolina*. Physiol Mol Plant Pathol 74:412–416

Gallardo M, Gomez-Jimenez MC, Matilla A (1999) Involvement of calcium in ACC-oxidase activity from *Cicer arietinum* seed embryonic axes. Phytochemistry 50:373–376

Galletti R, Ferrari S, De Lorenzo G (2011) *Arabidopsis* MPK3 and MPK6 play different roles in basal and oligogalacturonide- or flagellin-induced resistance against *Botrytis cinerea*. Plant Physiol 157:804–814

Gamble RL, Qu X, Schaller GE (2002) Mutational analysis of the ethylene receptor ETR1: role of the histidine kinase domain in dominant ethylene insensitivity. Plant Physiol 128:1428–1438

Gao Z, Schaller GE (2009) The role of receptor interactions in regulating ethylene signal transduction. Plant Signal Behav 4:1152–1153

Gao Z, Chen YF, Randlett MD, Zhao XC, Findell JL, Kieber JJ, Schaller GE (2003) Localization of the Raf-like kinase CTR1 to the endoplasmic reticulum of Arabidopsis through participation in ethylene receptor signaling complexes. J Biol Chem 278:34725–34732

Gao Z, Wen C-K, Binder BM, Chen Y-F, Chang J, Chiang Y-H, Kerris RJ III, Chang C, Schaller GE (2008) Heteromeric interactions among ethylene receptors mediate signaling in *Arabidopsis*. J Biol Chem 283:23801–23810

Garcia-Brugger A, Lamotte O, Vandelle E, Bourque S, Lecourieux D, Poinssot B, Wendehenne D, Pugin A (2006) Early signaling events induced by elicitors of plant defenses. Mol Plant Microbe Interact 19:711–724

Geraats BPJ, Bakker PAHM, van Loon LC (2002) Ethylene insensitivity impairs resistance to soilborne pathogens in tobacco and *Arabidopsis thaliana*. Mol Plant Microbe Interact 15:1078–1085

Geraats BPJ, Bakker PAHM, Lawrence CB, Achuo EA, Höfte M, van Loon LC (2003) Ethylene-insensitive tobacco shows differentially altered susceptibility to different pathogens. Phytopathology 93:813–821

Gimenez-Ibanez S, Solano R (2013) Nuclear jasmonate and salicylate signaling and crosstalk in defense against pathogens. Front Plant Sci. 4: Article 72:1–11

Grefen C, Städele K, Ruzicka K, Obrdlik P, Harter K, Horák J (2008) Subcellular localization and in vivo interactions of the *Arabidopsis thaliana* ethylene receptor family members. Mol Plant 1:308–320

Grennan AK (2008) Ethylene response factors in jasmonate signaling and defense response. Plant Physiol 146:1457–1458

Gu YQ, Wildermuth MC, Chakravarthy S, Loh YT, Yang C, He X, Han Y, Martin GB (2002) Tomato transcription factors pti4, pti5, and pti6 activate defense responses when expressed in *Arabidopsis*. Plant Cell 14:817–831

Guo H, Ecker JR (2003) Plant responses to ethylene gas are mediated by $SCF^{(EBF1/EBF2)}$-dependent proteolysis of EIN3 transcription factor. Cell 115:667–677

Guo H, Ecker JR (2004) The ethylene signaling pathway; new insights. Curr Opin Plant Biol 7:40–49

Guo X, Stotz HU (2007) Defense against *Sclerotinia sclerotiorum* in *Arabidopsis* is dependent on jasmonic acid, salicylic acid, and ethylene signaling. Mol Plant Microbe Interact 20:1384–1395

Guo Z, Chen XJ, Wu XL, Ling JQ, Xu P (2004) Overexpression of the AP2/EREBP transcription factor OPBP1 enhances disease resistance and salt tolerance in tobacco. Plant Mol Biol 55:607–618

Gutterson N, Reuber TL (2004) Regulation of disease resistance pathways by AP2/ERF transcription factors. Curr Opin Plant Biol 7:465–471

Hahn A, Harter K (2009) Mitogen-activated protein kinase cascades and ethylene: signaling, biosynthesis, or both? Plant Physiol 149:1207–1210

Hall BP, Shakeel SN, Amir M, Haq NU, Qu X, Schaller GE (2012) Histidine kinase activity of the ethylene receptor ETR1 facilitates the ethylene response in *Arabidopsis*. Plant Physiol 159:682–695

Han L, Li GJ, Yang KY, Mao G, Wang R, Liu Y, Zhang S (2010) Mitogen-activated protein kinase 3 and 6 regulate *Botrytis cinerea*-induced ethylene production in *Arabidopsis*. Plant J 64:114–127

Hase S, Shimizu A, Nakaho K, Takenaka S, Takahashi H (2006) Induction of transient ethylene and reduction in severity of tomato bacterial wilt by *Pythium oligandrum*. Plant Pathol 55:537–543

Hashimoto K, Eckert C, Anschūtz U, Scholz M, Held K, Waadt R, Reyer A, Hippler M, Becker D, Kudla J (2012) Phosphorylation of calcineurin B-like (CBL) calcium sensor proteins by their CBL-interacting protein kinases (CIPKs) is required for full activity of CBL-CIPK complexes toward their target proteins. J Biol Chem 287:7956–7968

He P, Shan L, Sheen J (2007) Elicitation and suppression of microbe-associated molecular pattern-triggered immunity in plant-microbe interactions. Cell Microbiol 9:1385–1396

Hirayama T, Kieber JJ, Hirayama N, Kogan M, Guzman P, Nourizadeh S, Alonso JM, Dailey WP, Dancis A, Ecker JR (1999) RESPONSIVE-TO-ANTAGONIST1, a Menkes/Wilson disease-related copper transporter, is required for ethylene signaling in Arabidopsis. Cell 97:383–393

Hirt H (2000) Connecting oxidative stress, auxin, and cell cycle regulation through a plant mitogen-activated protein kinase pathway. Proc Natl Acad Sci U S A 97:2405–2407

Hua J, Meyerowitz EM (1998) Ethylene responses are negatively regulated by a receptor gene family in *Arabidopsis thaliana*. Cell 94:261–271

Huang Y, Li H, Hutchison CE, Laskey J, Keiber JJ (2003) Biochemical and functional analysis of CTR1, a protein kinase that negatively regulates ethylene signaling in *Arabidopsis*. Plant J 33:221–233

Huffaker A, Dafoe NJ, Schmelz EA (2011) ZmPep1, an ortholog of Arabidopsis elicitor peptide 1, regulates maize innate immunity and enhances disease resistance. Plant Physiol 155:1325–1338

Hwang IS, Hwang BK (2011) The pepper mannose-binding lectin gene *CaMBL1* is required to regulate cell death and defense responses to microbial pathogens. Plant Physiol 155:447–463

Ichimura K, Shinozaki K, Tena G, Sheen J, Henry Y (2002) Mitogen-activated protein kinase cascades in plants: a new nomenclature. Trends Plant Sci 7:301–308

Iriti M, Faoro F (2009) Chitosan as a MAMP, searching for a PRR. Plant Signal Behav 4:66–68

Iwai T, Miyasaka A, Seo S, Ohashi Y (2006) Contribution of ethylene biosynthesis for resistance to blast fungus infection in young rice plants. Plant Physiol 142:1202–1215

Johansson A, Staal J, Dixelius C (2006) Early responses in the *Arabidopsis-Verticillium longisporum* pathosystem are dependent on *NDR1*, JA- and ET-associated signals via cytosolic NPR1 and *RFO1*. Mol Plant Microbe Interact 19:958–969

Jonak C, Okrész L, Bogre L, Hirt H (2002) Complexity, cross-talk and integration of plant MAP kinase signaling. Curr Opin Plant Biol 5:415–424

Joo S, Liu Y, Leuth A, Zhang S (2008) MAPK phosphorylation-induced stabilization of ACS6 protein is mediated by the non-catalytic C-terminal domain, which also contains the *cis-determinant* for rapid degradation by the 26S proteasome pathway. Plant J 54:129–140

Ju C, Chang C (2012) Advances in ethylene signaling: protein complexes at the endoplasmic reticulum (ER) membrane. AoB PLANTS:pls031. doi:10.1093/aobpla/pls031

Ju C, Yoon GM, Shemansky JM, Lin DY, Ying ZI, Chang J, Gamett WM, Kessenbrock M, Groth G, Tucker ML, Cooper B, Kieber JJ, Chang C (2012) CTR1 phosphorylates the central regulator EIN2 to control ethylene hormone signaling from the ER membrane to the nucleus in *Arabidopsis*. Proc Natl Acad Sci U S A 109:19486–19491

Jun SH, Han MJ, Lee S, Seo YS, Kim WT, An GH (2004) OsEIN2 is a positive component in ethylene signaling in rice. Plant Cell Physiol 45:281–289

Jung J, Won SY, Suh SC, Kim H, Wing R, Jeong Y, Hwang I, Kim M (2007) The barley ERF-type transcription factor HvRAF confers enhanced pathogen resistance and salt tolerance in *Arabidopsis*. Planta 225:575–588

Kepinski S, Leyser O (2003) SCF-mediated proteolysis and negative regulation in ethylene signaling. Cell 115:647–648

Kim CY, Liu Y, Thorne ET, Yang H, Fukushige H, Gassmann W, Hildebrand D, Sharp RE, Zhang S (2003a) Activation of a stress-responsive mitogen-activated protein kinase cascade induces the biosynthesis of ethylene in plants. Plant Cell 15:2707–2718

Kim JA, Agrawal GK, Rakwal R, Han KS, Kim KN, Yun CH, Heu S, Park SY, Lee YH, Jwa NS (2003b) Molecular cloning and mRNA expression analysis of a novel rice (*Oryza sativa* L.) MAPK kinase kinase, OsEDR1, an ortholog of Arabidopsis AtEDR1, reveal its role in defense/stress signalling pathways and development. Biochem Biophys Res Commun 300:858–876

Kim H, Helmbrecht EE, Stalans MB, Schmitt C, Patel N, Wen C-K, Wang W, Binder BM (2011) Ethylene receptor ETHYLENE RECEPTOR1 domain requirements for ethylene response in Arabidopsis seedlings. Plant Physiol 156:417–429

Kitajima S, Koyama T, Ohme-Takagi M, Shinshi H, Sato F (2000) Characterization of gene expression of NsERFs, transcription factors of basic PR genes from *Nicotiana sylvestris*. Plant Cell Physiol 41:817–24

Konishi M, Yanagisawa S (2008) Ethylene signaling in Arabidopsis involves feedback regulation via the elaborate control of *EBF2* expression by EIN3. Plant J 55:821–831

Koyama T, Okada T, Kitajima S, Ohme-Takagi M, Shinshi H, Sato F (2003) Isolation of tobacco ubiquitin-conjugating enzyme cDNA in a yeast two-hybrid system with tobacco ERF3 as bait and its characterization of specific interaction. J Exp Bot 54:1175–1181

Kudla J, Batistić O, Hashimoto K (2010) Calcium signals: the lead currency of plant information processing. Plant Cell 22:541–563

Kwaaitaal M, Huisman R, Maintz J, Reinstädler A, Panstruga R (2011) Ionotropic glutamate receptor (iGluR)-like channels mediate MAMP-induced calcium influx in *Arabidopsis thaliana*. Biochem J 440:355–365

Kwak SH, Lee SH (1997) The requirements for Ca^{2+}, protein phosphorylation, and dephosphorylation for ethylene signal transduction in *Pisum sativum* L. Plant Cell Physiol 38:1142–1149

Laluk K, Luo H, Chai M, Dhawan R, Lai Z, Mengiste T (2011) Biochemical and genetic requirements for function of the immune response regulator BOTRYTIS-INDUCED KINASE1 in plant growth, ethylene signaling, and PAMP-triggered immunity in *Arabidopsis*. Plant Cell 23:2831–2849

Lamotte O, Gould K, Lecourieux D, Sequeira-Legrand A, Lebrun-Garcia A, Durner J, Pugin A, Wendehenne D (2004) Analysis of nitric oxide signaling functions in tobacco cells challenged by the elicitor cryptogein. Plant Physiol 135:516–529

Larsen PB, Cancel JD (2004) A recessive mutation in the RUB1-conjugating enzyme, RCE1, reveals a requirement for RUB modification for control of ethylene biosynthesis and proper induction of basic chitinase and PDF1.2 in *Arabidopsis*. Plant J 38:626–638

Lecourieux D, Ranjeva R, Pugin A (2006) Calcium in plant defence-signalling pathways. New Phytol 171:249–269

Lecourieux-Ouaked D, Pugin A, Lebrun-Garcia A (2000) Phosphoproteins involved in the signal transduction of cryptogein, an elicitor of defense reactions in tobacco. Mol Plant Microbe Interact 13:821–829

Lee JH, Kim WT (2003) Expression analysis of the tobacco EIN3 family members in relation to tissue-specific ethylene responses. J Exp Bot 54:2239–2244

Lee JH, Hong JP, Oh SK, Lee S, Choi D, Kim WT (2004) The ethylene-responsive factor like protein 1 (CaERFLP1) of hot pepper (*Capsicum annuum* L.) interacts in vitro with both GCC and DRE/CRT sequence with different binding affinities: possible biological roles of CaERFLP1 in response to pathogen infection and high salinity conditions in transgenic tobacco plants. Plant Mol Biol 42:335–344

Lee DH, Choi HW, Hwang BK (2011) The pepper E3 ubiquitin ligase RING1 gene, *CaRING1*, is required for cell death and the salicylic acid-dependent defense response. Plant Physiol 156:2011–2025

Leshem YY, Wills RBH, Veng-Va Ku V (1998) Evidence for the function of the free radical gas-nitric oxide (NO) – as an endogenous maturation and senescence regulating factor in higher plants. Plant Physiol Biochem 36:825–833

Li G, Meng X, Wang R, Mao G, Han L, Liu Y, Zhang S (2012) Dual-level regulation of ACC synthase activity by MPK3/MPK6 cascade and its downstream WRKY transcription factor during ethylene induction in Arabidopsis. PLoS Genet 8(6):e1002767. doi:10.1371/journal.pgen.1002767

Lin Z, Alexander L, Hackett R, Grierson D (2008) LeCTR2, a CTR1-like protein kinase from tomato, plays a role in ethylene signalling, development and defence. Plant J 54:1083–1093

Lindermayr C, Saalbach G, Durner J (2005) Proteomic identification of *S*-nitrosylated proteins in *Arabidopsis*. Plant Physiol 137:921–930

Lindermayr C, Saalbach G, Bahnweg G, Durner J (2006) Differential inhibition of *Arabidopsis* methionine adenosyltransferases by protein *S*-nitrosylation. J Biol Chem 281:4285–4291

Liu Q, Wen C-K (2012) Arabidopsis *ETR1* and *ERS1* differentially repress the ethylene response in combination with other ethylene receptor genes. Plant Physiol 158:1193–1207

Liu Y, Zhang S (2004) Phosphorylation of 1-aminocyclopropane-1-carboxylic acid synthase by MPK6, a stress-responsive mitogen-activated protein kinase, induces ethylene biosynthesis in *Arabidopsis*. Plant Cell 16:3386–3399

Liu Q, Xu C, Wen C-K (2010) Genetic and transformation studies reveal negative regulation of ERS1 ethylene receptor signaling in *Arabidopsis*. BMC Plant Biol 10:60

Liu D, Chen X, Liu J, Ye J, Guo Z (2012) The rice ERF transcription factor *OsERF922* negatively regulates resistance to *Magnaporthe oryzae* and salt tolerance. J Exp Bot 63:3899–3911

Lohmann GV, Shimoda Y, Nielsen W, Jørgensen FG, Grossmann C, Sandal N, Sørensen K, Thirup S, Madsen LH, Tabata S, Sato S, Stougaard J, Radutoiu S (2010) Evolution and regulation of the *Lotus japonicus* LysM receptor gene family. Mol Plant Microbe Interact 23:510–521

Lorenzo O, Piqueras R, Sánchez-Serrano JJ, Solano R (2003) ETHYLENE RESPONSE FACTOR 1 integrates signals from ethylene and jasmonate pathways in plant defense. Plant Cell 15:165–178

Love AJ, Yun B-W, Laval V, Loake GJ, Milner JL (2005) *Cauliflower mosaic virus*, a compatible pathogen of Arabidopsis, engages three distinct defense-signaling pathways and activates rapid systemic generation of reactive oxygen species. Plant Physiol 139:935–948

Luan S (2009) The CBL-CIPK network in plant calcium signaling. Trends Plant Sci 14:37–42

Lund S, Stall RE, Klee HJ (1998) Ethylene regulates the susceptible response to pathogen infection in tomato. Plant Cell 10:371–382

McAinsh MR, Pittman JK (2009) Shaping the calcium signature. New Phytol 181:275–294

McGrath KC, Dombrecht B, Manners JM, Schenk PM, Edgar CI, Maclean DJ, Scheible W-R, Udvardi MK, Kazan K (2005) Repressor- and activator-type ethylene response factors functioning in jasmonate signaling and disease resistance identified via a genome-wide screen of Arabidopsis transcription factor gene expression. Plant Physiol 139:949–959

Menke FLH, van Pelt JA, Pieterse CMJ, Klessig DF (2004) Silencing of the mitogen-activated protein kinase MPK6 compromises disease resistance in *Arabidopsis*. Plant Cell 16:897–907

Mersmann S, Bourdais G, Rietz S, Robatzek S (2010) Ethylene signaling regulates accumulation of the FLS2 receptor and is required for the oxidative burst contributing to plant immunity. Plant Physiol 154:391–400

Mészáros T, Helfer A, Hatzimasoura E, Magyar Z, Serazetdinova L, Rios G, Bardóczy V, Teige M, Koncz C, Peck S, Bögre L (2006) The Arabidopsis MAP kinase kinase MKK1 participates in defence responses to the bacterial elicitor flagellin. Plant J 48:485–495

Milat ML, Ricci P, Blein JP (1991) Capsidiol and ethylene production by tobacco cells in response to cryptogein, an elicitor from *Phytophthora cryptogea*. Phytochemistry 30:2171–2173

Moffat CS, Ingle RA, Wathugala DL, Saunders NJ, Knight H, Knight MR (2012) ERF5 and ERF6 play redundant roles as positive regulators of JA/Et-mediated defense against *Botrytis cinerea* in Arabidopsis. PLoS One 7(4):e35995. doi:10.1371/journal.pone.0035995

Moussatche P, Klee HJ (2004) Autophosphorylation activity of the Arabidopsis ethylene receptor multigene family. J Biol Chem 279:48734–48741

Mur LAJ, Laarhoven LJJ, Harren FJM, Hall MA, Smith AK (2008) Nitric oxide interacts with salicylate to regulate biphasic ethylene production during the hypersensitive response. Plant Physiol 148:1537–1546

Nakagami H, Pitzschke A, Hirt H (2005) Emerging MAP kinase pathways in plant stress signalling. Trends Plant Sci 10:339–346

Nakano T, Suzuki K, Fujimura T, Shinshi H (2006) Genome-wide analysis of the ERF gene family in *Arabidopsis* and rice. Plant Physiol 140:411–432

Nambeesan S, AbuQamar S, Laluk K, Mattoo AK, Mickelbart MV, Ferruzzi MG, Mengiste T, Handa AK (2012) Polyamines attenuate ethylene-mediated defense responses to abrogate resistance to *Botrytis cinerea* in tomato. Plant Physiol 158:1034–1045

Nandi A, Kachroo P, Fukushige H, Hildebrand DF, Klessig DF, Shah J (2003) Ethylene and jasmonic acid signaling pathways affect the NPR1-independent expression of defense genes without impacting resistance to *Pseudomonas syringae* and *Peronospora parasitica* in the *Arabidopsis ssi1* mutant. Mol Plant Microbe Interact 18:588–599

Nie H, Wu Y, Yao C, Tang D (2011) Suppression of *edr2*-mediated powdery mildew resistance, cell death and ethylene-induced senescence by mutations in *ALD1* in *Arabidopsis*. J Genet Genomics 38:137–148

O'Donnell PJ, Calvert R, Atzorn R, Wasternack C, Leyser HMO, Bowles DJ (1996) Ethylene as a signal mediating the wound response of tomato plants. Science 274:1914–1917

O'Malley RC, Rodriguez FI, Esch JJ, Binder BM, O'Donnel P, Klee HJ, Bleecker AB (2005) Ethylene-binding activity, gene expression levels, and receptor system output for ethylene receptor family members from *Arabidopsis* and tomato. Plant J 41:651–659

Ohme-Takagi M, Shinshi H (1995) Ethylene-inducible DNA binding proteins that interact with an ethylene-responsive element. Plant Cell 7:173–182

Ohta M, Ohme-Takagi M, Shinshi H (2000) Three ethylene-responsive transcription factors in tobacco with distinct transactivation functions. Plant J 22:29–38

Ohta M, Matsui K, Hiratsu K, Shinshi H, Ohme-Takagi M (2001) Repression domains of class II ERF transcriptional repressors share an essential motif for active repression. Plant Cell 13:1959–1968

Olmedo G, Guo H, Gregory BD, Nourizadeh SD, Aquilar-Henonin L, Li H, An F, Guzman P, Ecker JR (2006) ETHYLENE-INSENSITIVE5 encodes a $5' \rightarrow 3'$ exoribonuclease required for regulation of the EIN3-targeting F-box proteins EBF1/2. Proc Natl Acad Sci U S A 103: 13286–13293

Oñate-Sánchez L, Anderson JP, Young J, Singh KB (2007) AtERF14, a member of the ERF family of transcription factors, plays a nonredundant role in plant defense. Plant Physiol 143:400–409

Ouaked F, Rozhon W, Lecourieux D, Hirt H (2003) A MAPK pathway mediates ethylene signaling in plants. EMBO J 22:1282–1288

Pantelides S, Tjamos SE, Pappa S, Kargakis M, Paplomatas EJ (2013) The ethylene receptor ETR1 is required for *Fusarium oxysporum* pathogenicity. Plant Pathology 62:1302–1309

Park JM, Park CJ, Lee SB, Ham BK, Shin R, Paek KH (2001) Overexpression of the tobacco *Tsi1* gene encoding an EREBP/AP2-type transcription factor enhances resistance against pathogen attack and osmotic stress in tobacco. Plant Cell 13:1035–1046

Peng HP, Lin TY, Wang NN, Shih MC (2005) Differential expression of genes encoding 1-aminocyclopropane-1-carboxylate synthase in Arabidopsis during hypoxia. Plant Mol Biol 58:15–25

Penninckx IAMA, Thomma BPHJ, Buchala A, Metraux JP, Broekaert WF (1998) Concomitant activation of jasmonate and ethylene response pathways is required for induction of a plant defensin gene in Arabidopsis. Plant Cell 10:2103–2113

Peters JM, Franke WW, Kleinschmidt JA (1994) Distinct 19S and 20S subcomplexes of the 26S proteasome and their distribution in the nucleus and the cytoplasm. J Biol Chem 269:7709–7718

Petruzzelli L, Sturaro M, Mainieri D, Leubner-Metzger G (2003) Calcium requirement for ethylene-dependent responses involving 1-aminocyclopropane-1-carboxylic acid oxidase in radicle tissues of germinated pea seeds. Plant Cell Environ 26:661–671

Petutschnig EK, Jones AME, Serazetdinova L, Lipka U, Lipka V (2010) The Lysin motif receptor-like kinase (LysM-RLK) is a major chitin binding protein in *Arabidopsis thaliana* and subject to chitin-induced phosphorylation. J Biol Chem 285:28902–28911

Pickart CM, Eddins MJ (2004) Ubiquitin: structures, functions, mechanisms. Biochim Biophys Acta 1695:55–72

Pieterse CMJ, Leon-Reyes A, Van der Ent S, Van Wees SCM (2009) Networking by small-molecule hormones in plant immunity. Nat Chem Biol 5:308–316

Potuschak T, Lechner E, Parmentier Y, Yanagisawa S, Grava S, Koncz C, Genschik P (2003) EIN3-dependent regulation of plant ethylene hormone signaling by two *Arabidopsis* F box proteins EBF1 and EBF2. Cell 115:679–689

Potuschak T, Vansiri A, Binder BM, Lechner E, Vierstra RD, Genschik P (2006) The exoribonuclease XRN4 is a component of the ethylene response pathway in *Arabidopsis*. Plant Cell 18:3047–3057

Pré M, Atallah M, Champion A, De Vos M, Pieterse CMJ, Memelink J (2008) The AP2/ERF domain transcription factor ORA59 integrates jasmonic acid and ethylene signals in plant defense. Plant Physiol 147:1347–1357

Qiao H, Chang KN, Yazaki J, Ecker JR (2009) Interplay between ethylene, ETP1/ETR2 F-box proteins, and degradation of EIN2 triggers ethylene responses in Arabidopsis. Genes Dev 23:512–521

Qiao H, Shen ZS, Huang C, Schmitz RJ, Urich MA, Briggs SP, Ecker JR (2012) Processing and subcellular trafficking of ER-tethered EIN2 control response to ethylene gas. Science 338:390–393

Qiu L, Xie F, Yu J, Wen CK (2012) Arabidopsis RTE1 is essential to ethylene receptor ETR1 amino-terminal signaling independent of CTR1. Plant Physiol 159:1263–1276

Qu X, Schaller GE (2004) Requirement of the histidine kinase domain for signal transduction by the ethylene receptor ETR1. Plant Physiol 136:2961–2970

Qu X, Hall BP, Gao Z, Schaller GE (2007) A strong constitutive ethylene-response phenotype conferred on Arabidopsis plants containing null mutations in the ethylene receptors *ETR1* and *ERS1*. BMC Plant Biol 7:3

Qutob D, Kemmerling B, Brunner F, Küfner I, Engelhardt S, Gust AA, Luberacki B, Seitz HU, Stahl D, Rauhut T, Glawischnig E, Schween G, Lacombe B, Watanabe N, Lam E, Schlichting R, Scheel D, Nau K, Dodt G, Hubert D, Gijzen M, Nürnberger T (2006) Phytotoxicity and innate immune responses induced by NEP1-like proteins. Plant Cell 18:3721–3744

Raz V, Fluhr A (1992) Calcium requirement for ethylene-dependent responses. Plant Cell 4:1123–1130

Reddy ASN, Reddy VS, Golovkin M (2000) A calmodulin-binding protein from *Arabidopsis* is induced by ethylene and contains a DNA-binding motif. Biochem Biophys Res Commun 279:762–769

Reddy ASN, Ali GS, Celesnik H, Day IS (2011a) Coping with stresses: roles of calcium- and calcium/calmodulin-regulated gene expression. Plant Cell 23:2010–2032

Reddy ASN, Ben-Hur A, Day IS (2011b) Experimental and computational approaches for the study of calmodulin interactions. Phytochemistry 72:1007–1019

Ren D, Yang K-Y, Li G-J, Liu Y, Zhang S (2006) Activation of Ntf4, a tobacco mitogen-activated protein kinase, during plant defense and its involvement in hypersensitive response-like cell death. Plant Physiol 141:1482–1493

Resnick JS, Wen CK, Shockey JA, Chang C (2006) REVERSION-TO-ETHYLENE SENSITIVITY1, a conserved gene that regulates ethylene receptor function in *Arabidopsis*. Proc Natl Acad Sci U S A 103:7917–7922

Resnick JS, Rivarola M, Chang C (2008) Involvement of *RTE1* in conformational changes promoting ETR1 ethylene receptor signaling in Arabidopsis. Plant J 56:423–431

Rieu I, Mariani C, Weterings K (2003) Expression analysis of five tobacco EIN3 family members in relation to tissue-specific ethylene responses. J Exp Bot 54:2239–2244

Rivarola M, McClellan CA, Resnick JS, Chang C (2009) ETR1-specific mutations distinguish ETR1 from other *Arabidopsis* ethylene receptors as revealed by genetic interaction with RTE1. Plant Physiol 150:547–551

Rodriguez FI, Esch JJ, Hall AE, Binder BM, Schaller GE, Bleecker AB (1999) A copper cofactor for the ethylene receptor ETR1 from *Arabidopsis*. Science 283:996–998

Saijo Y (2010) ER quality control of immune receptors and regulators in plants. Cell Microbiol 12:716–724

Sang S, Li X, Gao R, You Z, Lü B, Liu P, Ma Q, Dong H (2012) Apoplastic and cytoplasmic location of harpin protein $Hpa1_{Xoo}$ plays different roles in H_2O_2 generation and pathogen resistance in *Arabidopsis*. Plant Mol Biol 79:375–391

Schaller GE, Shiu SH, Armitage JP (2011) Two-component systems and their co-option for eukaryotic signal transduction. Curr Biol 21:R320–R330

Scharein B, Grath G (2011) Phosphorylation alters the interaction of the *Arabidopsis* phosphotransfer protein AHP1 with its sensor kinase ETR1. PLoS One 6:e24173

Schulze B, Mentzel T, Jehle A, Mueller K, Beeler S, Boller T, Felix G, Chinchilla D (2010) Rapid heteromerization and phosphorylation of ligand-activated plant transmembrane receptors and their associated kinase BAK1. J Biol Chem 285:9444–9451

Sebastiá CH, Hardin SC, Clouse SD, Kieber JJ, Huber SC (2004) Identification of a new motif for CDPK phosphorylation in vitro that suggests ACC synthase may be a CDPK substrate. Arch Biochem Biophys 428:81–91

Segonzac C, Zipfel C (2011) Activation of plant pattern-recognition receptors by bacteria. Curr Opin Microbiol 1(4):54–61

Seifert GJ, Blaukopf C (2010) Irritable walls: the plant extracellular matrix and signaling. Plant Physiol 153:467–478

Sessa G, Meller Y, Fluhr R (1995) A GCC element and a motif participate in ethylene-induced expression of the *PRB-1b* gene. Plant Mol Biol 28:145–153

Shakeel SN, Wang X, Binder BM, Eric Schaller G (2013) Mechanisms of signal transduction by ethylene: overlapping and non-overlapping signalling roles in a receptor family. AoB PLANTS 5:plt010. doi:10.1093/aobpla/plt010

Shen X, Liu H, Yuan B, Li X, Xu C, Wang S (2011) OsEDR1 negatively regulates rice bacterial resistance via activation of ethylene biosynthesis. Plant Cell Environ 34:179–191

Shinshi H, Usami S, Ohme-Takagi M (1995) Identification of an ethylene responsive region in the promoter of a tobacco class I chitinase gene. Plant Mol Biol 27:923–932

Shinya T, Osada T, Desaki Y, Hatamoto M, Yamanaka Y, Hirano H, Takai R, Che F-S, Kaku H, Shibuya N (2010) Characterization of receptor proteins using affinity cross-linking with biotinylated ligands. Plant Cell Physiol 51:262–270

Singh R, Jwa NS (2013) The rice MAPKK-MAPK interactome: the biological significance of MAPK components in hormone signal transduction. Plant Cell Rep 32:923–931

Singh MP, Lee FN, Counce PA, Gibbons JH (2004) Mediation of partial resistance to rice blast through anaerobic induction of ethylene. Phytopathology 94:819–825

Smalle J, Vierstra RD (2004) The ubiquitin 26S proteasome proteolytic pathway. Annu Rev Plant Biol 55:555–590

Solano R, Stepanova A, Chao Q, Ecker JR (1998) Nuclear events in ethylene signaling: a transcriptional cascade mediated by ETHYLENE-INSENSITIVE3 and ETHYLENE-RESPONSE FACTOR1. Gen Dev 12:3703–3714

Son GH, Wan J, Kim HJ, Nguyen XC, Chung W-S, Hong JC, Stacey G (2012) Ethylene-responsive element-binding factor 5, ERF5, is involved in chitin-induced innate immunity response. Mol Plant Microbe Interact 25:48–60

Spanu P, Grosskopf DG, Felix G, Boller T (1991) The apparent turnover of 1-aminocyclopropane-1-carboxylate synthase in tomato cells is regulated by protein-phosphorylation and dephosphorylation. Plant Physiol 106:529–535

Stepanova AN, Hoyt JM, Hamilton AA, Alonso JM (2005) A link between ethylene and auxin uncovered by the characterization of two root-specific ethylene-insensitive mutants in *Arabidopsis*. Plant Cell 17:2230–2242

Takahashi Y, Nasir KHB, Ito A, Kanzaki H, Matsumura H, Saitoh H, Fujisawa S, Kamoun S, Terauchi R (2007) A high-throughput screen of cell-death-inducing factors in *Nicotiana benthamiana* identifies a novel MAPKK that mediates INF1-induced cell death signaling and non-host resistance to *Pseudomonas cichorii*. Plant J 49:1030–1040

Tang D, Christiansen KM, Innes RW (2005) Regulation of plant disease resistance, stress responses, cell death, and ethylene signaling in Arabidopsis by the EDR1 protein kinase. Plant Physiol 138:1018–1026

Teige M, Scheikl E, Eulgem T, Doczi R, Ichimura K, Shinozaki K, Dangl JL, Hirt H (2004) The MKK2 pathway mediates cold and salt stress signaling in Arabidopsis. Mol Cell 15:141–152

Testerink C, Larsen PB, van der Does D, van Himbergen JAJ, Munnik T (2007) Phosphatidic acid binds to and inhibits the activity of *Arabidopsis* CTR1. J Exp Bot 58:3905–3914

Testerink C, Munnik T (2005) Phosphatidic acid: a multifunctional stress signaling lipid in plants. Trends Plant Sci 10:368–375

Thomma BPHJ, Nürnberger T, Joosten MHAJ (2011) Of PAMPs and effectors: The blurred PTI-ETI dichotomy. Plant Cell 23:4–15
Tournier B, Sanchez-Ballesta MT, Jones B, Pesquet E, Regad F, Latché A, Pech JC, Bouzayen M (2003) New members of the tomato ERF family show specific expression pattern and diverse DNA-binding capacity to the GCC box element. FEBS Lett 550:149–154
Travella S, Klimm TE, Keller B (2006) RNA interference-based gene silencing as an efficient tool for functional genomics in hexaploid bread wheat. Plant Physiol 142:6–20
Vadassery J, Oelmūller R (2009) Calcium signaling in pathogenic and beneficial plant microbe interactions. What can we learn from the interaction between *Piriformospora indica* and *Arabidopsis thaliana*. Plant Signal Behav 411:1024–1027
van der Luit AH, Piatti T, van Doorn A, Musgrave A, Felix G, Boller T, Munnik T (2000) Elicitation of suspension-cultured tomato cells triggers the formation of phosphatidic acid and diacylglycerol pyrophosphate. Plant Physiol 123:1507–1515
van Loon LC, Rep M, Pieterse C (2006) Significance of inducible defense-related proteins in infected plants. Annu Rev Phytopathol 44:71–728
van Verk MC, Pappaaioannou D, Neeleman L, Bol JF, Linthorst HJM (2008) A novel WRKY transcription factor is required for induction of *PR-1a* gene expression by salicylic acid and bacterial elicitors. Plant Physiol 140:1983–1995
Vandenabeele S, Van Der Kelen K, Dat J, Gadjev I, Boonefaes T, Morsa S, Rottiers P, Slooten L, Van Montagu M, Zabeau M, Inzé D, Van Breusegem F (2003) A comprehensive analysis of hydrogen peroxide-induced gene expression in tobacco. Proc Natl Acad Sci U S A 100: 16113–16118
Vidhyasekaran P (2007) Fungal pathogenesis in plants and crops: molecular biology and host defense mechanisms II edition. CRC Press/Taylor Francis Group, Boca Raton, p 510
Vidhyasekaran P (2014) PAMP signals in plant innate immunity: signal perception and transduction. Springer, Dordrecht, p 442
Vincill ED, Bieck AM, Spalding EP (2012) Ca^{2+} conduction by an amino acid-gated ion channel related to glutamate receptors. Plant Physiol 159:40–46
Voet-van-Vormizeele J, Groth G (2008) Ethylene controls autophosphorylation of the histidine kinase domain in ethylene receptor ETR1. Mol Plant 1:380–387
Vogel JP, Woeste KE, Theologis A, Kicher JJ (1998) Recessive and dominant mutations in the ethylene biosynthetic gene *ACS5* of arabidopsis confer cytokinin insensitivity and ethylene overproduction, respectively. Proc Natl Acad Sci U S A 95:4766–4771
Wan J, Zhang X-C, Neece D, Ramonell KM, Clough S, Kim S-Y, Stacey MG, Stacey G (2008) A LysM receptor-like kinase plays a critical role in chitin signaling and fungal resistance in *Arabidopsis*. Plant Cell 20:471–481
Wang KLC, Li H, Ecker IP (2002) Ethylene biosynthesis and signaling networks. Plant Cell 14:S131–S155
Wang W, Hall AE, O'Malley R, Bleecker AB (2003) Canonical histidine kinase activity of the transmitter domain of the ETR1 ethylene receptor from *Arabidopsis* is not required for signal transmission. Proc Natl Acad Sci U S A 100:352–357
Wang W, Esch JJ, Shiu S-H, Agula H, Binder BM, Chang C, Patterson S, Bleecker AB (2006) Identification of important regions for ethylene binding and signaling in the transmembrane domain of the ER1 ethylene receptor of *Arabidopsis*. Plant Cell 18:3429–3442
Wang W-H, Yi X-Q, Han A-D, Liu T-W, Chen J, Wu F-H, Dong X-J, He J-X, Pei Z-M, Pei Z-M, Zheng H-L (2012) Calcium-sensing receptor regulates stomatal closure through hydrogen peroxide and nitric oxide in response to extracellular calcium in *Arabidopsis*. J Exp Bot 63:177–190
Wen X, Zhang C, Ji Y, Zhao Q, He W, An F, Jiang L, Guo H (2012) Activation of ethylene signaling is mediated by nuclear translocation of the cleaved EIN2 carboxyl terminus. Cell Res 22:1613–1616
Xie F, Liu Q, Wen CK (2006) Receptor signal output mediated by the ETR1 N terminus is primarily subfamily 1 receptor dependent. Plant Physiol 142:482–508
Xie F, Qiu L, Wen C-K (2012) Possible modulation of Arabidopsis ETR1 N-terminal signaling by CTR1. Plant Signal Behav 7:1243–1245

Xu Z-S, Xia L-Q, Chen M, Cheng X-G, Zhang R-Y, Li L-C, Zhao Y-X, Lu Y, Ni Z-Y, Liu L, Qiu Z-G, Ma Y-Z (2007) Isolation and molecular characterization of the *Triticum aestivum* L ethylene-responsive factor1 (*TaERF1*) that increases multiple stress tolerance. Plant Mol Biol 65:719–732

Xu J, Li Y, Wang Y, Liu H, Lei L, Yang H, Liu G, Ren D (2008) Activation of MAPK kinase 9 induces ethylene and camalexin biosynthesis and enhances sensitivity to salt stress in *Arabidopsis*. J Biol Chem 283:26996–27006

Yamaguchi T, Minami E, Ueki J, Shibuya N (2005) Elicitor-induced activation of phospholipases plays an important role for the induction of defense responses in suspension-cultured rice cells. Plant Cell Physiol 46:579–587

Yamamoto S, Suzuki K, Shinshi H (1999) Elicitor-responsive, ethylene-independent activation of GCC box-mediated transcription that is regulated by both protein phosphorylation and dephosphorylation in cultured tobacco cells. Plant J 20:571–579

Yanagisawa S, Yoo SD, Sheen J (2003) Differential regulation of EIN3 stability by glucose and ethylene signalling in plants. Nature 425:521–525

Yang Y, Shah J, Klessig DF (1997) Signal perception and transduction in plant defense responses. Genes & Dev 11:1621–1639

Yang X, Menon S, Lykke-Andersen K, Tsuge T, Di X, Wang X, Rodriguez-Suarez-Suarez RJ, Zhang H, Wei N (2002) The COP9 signalosome inhibits $p27^{klp1}$ degradation and impedes G1-S phase progression via deneddylation of SCF Cul1. Curr Biol 12:667–672

Yang B, Srivastava S, Deyholos MK, Kav NNV (2007) Transcriptional profiling of canola (*Brassica napus* L) responses to the fungal pathogen *Sclerotinia sclerotiorum*. Plant Sci 173:156–171

Yao T, Ndoja A (2012) Regulation of gene expression by the ubiquitin-proteasome system. Semin Cell Dev Biol 23:523–529

Yi SY, Kim JH, Joung YH, Lee S, Kim WT, Yu SH, Choi D (2004) The pepper transcription factor *CaPF1* confers pathogen and freezing tolerance in Arabidopsis. Plant Physiol 136:2862–2874

Yoo S-D, Cho Y-H, Tena G, Xiong Y, Sheen J (2008) Dual control of nuclear EIN3 by bifurcate MAPK cascades in C_2H_4 signalling. Nature 451:789–795

Zago E, Morsa S, Dat JF, Alard P, Ferrarini A, Inzé D, Delledonne M, Van Breusegem F (2006) Nitric oxide- and hydrogen peroxide-responsive gene regulation during cell death induction in tobacco. Plant Physiol 404–411

Zhang H, Zhang D, Chen J, Yang Y, Huang Z, Huang D, Wang X-C, Huang R (2004) Tomato stress-responsive factor TSRF1 interacts with ethylene responsive element GCC box and regulates pathogen resistance to *Ralstonia solanacearum*. Plant Mol Biol 55:825–834

Zhang Z, Yao W, Dong N, Liang H, Liu H, Huang R (2007) A novel ERF transcription activator in wheat and its induction kinetics after pathogen and hormone treatments. J Exp Bot 58:2993–3003

Zhang W, Jeon BW, Assmann SM (2011) Heterotrimeric G-protein regulation of ROS signalling and calcium currents in *Arabidopsis* guard cells. J Exp Bot 62:2371–2379

Zhang H, Gao Z, Zheng X, Zhang Z (2012) The role of G-proteins in plant immunity. Plant Signal Behav 7:1284–1288

Zhao Q, Guo HW (2011) Paradigms and paradox in the ethylene signaling pathway and interaction network. Mol Plant 4:626–634

Zhao J, Zheng S-H, Fujita K, Sakai K (2004) Jasmonate and ethylene signalling and their interaction are integral parts of the elicitor signalling pathway leading to β-thujaplicin biosynthesis in *Cupressus lusitanica* cell cultures. J Exp Bot 55:1003–1012

Zhou J, Tang T, Martin GB (1997) The Pto kinase conferring resistance to tomato bacterial speck disease interacts with proteins that bind a *cis*-element of pathogenesis-related genes. EMBO J 16:3207–3218

Zhou C, Zhang L, Duan J, Miki B, Wu K (2005) Histone deacetylase19 is involved in jasmonic acid and ethylene signaling of pathogen response in *Arabidopsis*. Plant Cell 17:1196–1204

Zhou X, Liu Q, Xie F, Wen C-K (2007) RTE1is a Golgi-associated and ETR1-dependent negative regulator of ethylene responses. Plant Physiol 145:75–86

Zhu L, Li Y, Li L, Yang J, Zhang M (2011a) Ethylene is involved in leafy mustard systemic resistance to *Turnip mosaic virus* infection through the mitochondrial alternative oxidase pathway. Physiol Mol Plant Pathol 76:166–172

Zhu Z, An F, Feng Y, Li P, Xue LAM, Jiang Z, Kim JM, To TK, Li W, Zhang X, Yu Q, Dong Z, Chen WQ, Seki M, Zhou JM, Guo H (2011b) Derepression of ethylene-stabilized transcription factors (EIN3/EIL1) mediates jasmonate and ethylene signaling synergy in *Arabidopsis*. Proc Natl Acad Sci U S A 108:12539–12544

Zuo K-J, Qin J, Zhao J-Y, Ling H, Zhang L-D, Cao Y-F, Tang K-X (2007) Over-expression of *GbERF2* transcription factor in tobacco enhances brown spots disease resistance by activating expression of downstream genes. Gene 391:80–90

Chapter 5
Abscisic Acid Signaling System in Plant Innate Immunity

Abstract Abscisic acid is a plant hormone involved in immune signal transduction system, and it positively or negatively regulates the plant immune system. On the perception of pathogen-associated molecular pattern (PAMP) signals, ABA biosynthesis pathway is activated and ABA accumulates in the infected tissues. The ABA signaling complex includes PYR/PYL/RCAR (an ABA receptor), the type 2C protein phosphatase (PP2C, a negative regulator), and SNF1-related protein kinase (SnRK2, a positive regulator). Binding of ABA to the receptor induces a conformational change in the receptor that allows it to sequester the PP2Cs. This excludes the PP2Cs from the negative regulation of the downstream SnRK2-activated kinases which phosphorylate and activate downstream transcription factors. Phosphorylation of the transcription factors initiates the transcription of ABA-responsive promoter elements. ABA regulates the expression of several genes involved in host defense responses. The *cis*-regulatory elements responsible for the ABA regulation of gene expression share a conserved motif, ACGTGGC, which is known as ABA-responsive element (ABRE). ABRE appears in the promoters of many defense genes. The bZIP-type transcription factors bind ABRE and transactivate downstream gene expression. Transcription factors perceive signals relayed by signaling molecules and translate into a functional response through recruiting or releasing RNA polymerase II (RNAPII). MEDIATOR25 (MED25), a cofactor of RNAPII, acts as a bridge between RNAPII and DNA-binding transcription factors and regulates ABA-triggered gene transcription. ABA has been shown to be a complex modulator of various plant hormone signaling systems in triggering or suppressing host defense responses. Synergistic and antagonistic interaction between ABA and JA, SA, and ET in the modulation of plant immune system has been widely reported. ABA and JA signaling pathways may be interconnected, and they may act cooperatively in the induction of defense genes. ABA may also suppress JA-activated transcription of defense genes in some plant–pathogen interactions. ABA may suppress SA biosynthesis and SA-mediated defense responses. ABA signaling system suppresses SA signaling system and vice versa. In some plant–pathogen interactions, ABA signaling may act synergistically with SA signaling in triggering plant immune responses. ABA activates ethylene biosynthesis, while ethylene activates ABA biosynthesis. Synergistic and antagonistic interaction between ABA and ethylene signaling systems has also been reported. ABA induces resistance against many necrotrophic

P. Vidhyasekaran, *Plant Hormone Signaling Systems in Plant Innate Immunity*, Signaling and Communication in Plants 2, DOI 10.1007/978-94-017-9285-1_5

pathogens, while it negatively regulates defense responses against biotrophic pathogens (with a few exceptions). ABA has been shown to promote susceptibility or resistance against bacterial pathogens, and ABA plays positive roles during early stages and negative roles at later stages of bacterial infection. ABA signaling induces resistance or susceptibility against virus diseases. Different ABA signaling pathway components may decide induction or suppression of defense responses against virus diseases. ABA may regulate host defense responses either positively or negatively through its direct or indirect action on the plant innate immune system. ABA may suppress plant immune responses and induce susceptibility. Pathogens may hijack ABA signaling pathway to cause disease. Pathogens may secrete toxins/effectors and suppress ABA-dependent defense responses.

5.1 Abscisic Acid as a Multifaceted Plant Hormone Signal Triggering or Suppressing Plant Defense Responses

The phytohormone abscisic acid (ABA) plays a multifaceted role in plant immunity (Cao et al. 2011; Alazem et al. 2014). ABA induces defense responses (Asselbergh et al. 2008b; Fan et al. 2009; Garcia-Andrade et al. 2011) or suppresses the immune responses depending upon the type of plant–pathogen interactions (Sánchez-Vallet et al. 2012; Yazawa et al. 2012). ABA may induce resistance against necrotrophic pathogens such as *Botrytis cinerea* (Garcia-Andrade et al. 2011), *Alternaria brassicicola* (Fan et al. 2009), *Plectosphaerella cucumerina* (Ton and Mauch-Mani 2004; Garcia-Andrade et al. 2011), and *Pythium irregulare* (Adie et al. 2007) in many host plants. It has been widely reported that ABA is a signal required for resistance against necrotrophic pathogens (Adie et al. 2007). However, ABA may not be a positive signal for plant defense against all necrotrophs. *Arabidopsis* ABA-deficient mutants (*aba2* and *aao3*) are more resistant than wild-type plants to the necrotrophic pathogen *B. cinerea* and to the vascular fungal pathogen *Fusarium oxysporum* (Audenaert et al. 2002; L'Haridon et al. 2011). In tomato, a negative regulatory role of ABA in resistance to necrotrophic pathogens, such as the fungus *B. cinerea* and the bacterial pathogen *Dickeia dadantii*, has been reported. These pathogens are less virulent on the tomato *sitiens* mutant, in which ABA biosynthesis is disrupted (Audenaert et al. 2002; Asselbergh et al. 2007, 2008a, b).

A positive role of ABA in resistance to *Plectosphaerella cucumerina* has been reported (Garcia-Andrade et al. 2011). The *aba2* mutant was shown to be hypersusceptible to the pathogen, and this function was associated with the regulation of callose deposition upon infection (Garcia-Andrade et al. 2011). However, it has also been reported that ABA plays a negative regulatory function in balancing *Arabidopsis* resistance response to *P. cucumerina*, a necrotrophic pathogen (Sánchez-Vallet et al. 2012). The *aba1-6* mutant defective in ABA biosynthesis and the quadruple *pyr/pyl* mutant impaired in four ABA receptors were more resistant to *P. cucumerina* than wild-type plants, whereas the triple *hab1-1abi1-2abi2-1* mutant, which has a constitutive activation of ABA signaling due to the impairment of three PP2C phosphatases (from clade A) that negatively regulate ABA signaling, displayed an enhanced susceptibility to *P. cucumerina* (Sánchez-Vallet et al. 2012).

ABA has been shown to negatively regulate defense responses and confer susceptibility against biotrophic pathogens. ABA accumulation compromised resistance to the biotrophic oomycete pathogen *Hyaloperonospora arabidopsidis* in *Arabidopsis thaliana* (Fan et al. 2009). In mutants deficient in ABA biosynthesis, the resistance to the biotrophic pathogens *Blumeria graminis* f. sp. *hordei* and *Hyaloperonospora parasitica* is enhanced in *Arabidopsis* (Mohr and Cahill 2003; Jensen et al. 2008). Transcriptional regulation by an NAC (NAM-ATAF1,2-CUC2) transcription factor attenuates ABA signaling for efficient basal defence toward *Blumeria graminis* f. sp. *hordei* in *Arabidopsis* (Jensen et al. 2008). The results suggest that ABA negatively regulates resistance against the powdery mildew pathogen.

ABA may induce resistance against a fungal pathogen and induce susceptibility against another fungal pathogen in the same host plant. ABA is the key factor for suppressing rice blast disease resistance and making rice plants susceptible to *Magnaporthe oryzae* (Koga et al. 2004). Exogenous application of abscisic acid promoted the blast pathogen *M. oryzae* infection (Jiang et al. 2010). ABA promoted *M. oryzae* infection of rice during early stages of infection (Yazawa et al. 2012). Decreased ABA levels or inhibition of ABA signaling by expressing *OsABABox1* or *OsABI(G-to-A)* genes before, but not after, *M. oryzae* inoculation decreased lesion numbers (Yazawa et al. 2012). Expression of ABA biosynthetic genes and accumulation of ABA were decreased in the *OsERF922* RNAi plants, in line with enhanced resistance against *M. oryzae*, whereas *OsERF922* overexpressing plants were more susceptible (Liu et al. 2012). In contrast, ABA has been shown as an activator of resistance against the rice brown spot pathogen *Cochliobolus miyabeanus* (De Vleesschauwer et al. 2010). The results suggest that ABA induces resistance against the brown spot fungal pathogen, while it enhances host susceptibility against another fungal pathogen *M. oryzae* in rice plants.

ABA has been shown to be involved in conferring susceptibility against bacterial diseases. ABA treatment stimulated susceptibility to *Pseudomonas syringae* pv. *tomato* (Mohr and Cahill 2003). ABA has been shown to promote susceptibility to the rice bacterial blight pathogen *Xanthomonas oryzae* pv. *oryzae* by suppressing SA-mediated defenses (Xu et al. 2013). Activation of NCED5, a key enzyme regulating de novo ABA synthesis, led to elevated endogenous ABA levels, and it enhanced susceptibility to infection by the hemibiotrophic bacterial pathogen *Pseudomonas syringae* in *Arabidopsis* (Fan et al. 2009). In tomato, a negative regulatory role of ABA in resistance to the bacterial pathogen *Dickeia dadantii* has been reported. The bacterial pathogen was less virulent on the tomato *sitiens* mutant, in which ABA biosynthesis is disrupted, than in wild-type plants (Audenaert et al. 2002; Asselbergh et al. 2007, 2008a, b). ABA deficiency leads to rapid activation of tomato defense responses upon infection with *Erwinia chrysanthemi*, suggesting that ABA negatively regulates resistance against *E. chrysanthemi* in tomato (Asselbergh et al. 2008a). ABA treatment induces *ABR1* (*ABA-responsive1*) in pepper plants and *ABR1* is a negative regulator of ABA signaling. Heterologous transgenic expression of pepper ABR1 in *Arabidopsis* conferred enhanced resistance to *Pseudomonas syringae* (Choi and Hwang 2012). The results suggest that ABA signaling negatively regulates disease resistance against *P. syringae*. Application of ABA decreased resistance to *Ralstonia solanacearum* in tobacco plants, when the bacterial pathogen was inoculated at a concentration of 10^7 and 10^8 cfu ml^{-1} (Zhou et al. 2008). In contrast, a resistance response was seen after ABA treatment when the bacteria were inoculated at a concentration of 10^{-4} cfu ml^{-1} (Zhou et al. 2008). These results suggest that ABA may induce resistance or susceptibility depending on different inoculation concentrations of the pathogen. ABA may also be involved in induced systemic resistance against bacterial pathogens. Mutants in a large cluster of genes (*hrp*) involved in the pathogenicity of *Ralstonia solanacearum* were used to control the bacterial wilt caused by *R. solanacearum* in *Arabidopsis*. High protection rates associated with a significant decrease in the multiplication of *R. solanacearum* were observed in plants pre-inoculated with the Δ*hrpB* mutant strain. The induced resistance triggered by the mutant strain was associated with the up-regulation of genes involved in ABA biosynthesis and signaling (Feng et al. 2012).

ABA signaling has been shown to be involved in both resistance and susceptibility against virus diseases (Alazem et al. 2014). ABA pathway has multifaceted effects on the accumulation of *Bamboo mosaic virus* (BAMV) and *Cucumber mosaic virus* (CMV) in two different hosts: *Arabidopsis thaliana* and *Nicotiana benthamiana* (Alazem et al. 2014). *ABA2* plays a critical role in accumulation of BAMV and CMV. Mutants downstream of *ABA2* (*aao3*, *abi-1*, *abi3-1*, and *abi4-1*) were susceptible to BAMV, indicating that the ABA pathway downstream of ABA2 is essential for BaMV resistance. The *aba2-1* mutant decreased the accumulation of BaMV (+)RNA, (−)RNA, and coat protein (Alazem et al. 2014). The results suggest the ABA2 is involved in inducing susceptibility, while downstream genes are involved in conferring resistance (Alazem et al. 2014). *ABA2* appears to divide the ABA pathway into upstream genes required for BaMV accumulation and downstream genes required for plant resistance. BaMV titers were decreased in the ABA biosynthesis mutant *aba2-1* and were affected to a lesser extent in a mutant of the upstream gene *NCED3* (Alazem et al. 2014). Exogenous application of ABA reduced the BaMV titer in wild-type plants. Viral titers were increased in signaling pathway mutants (*abi1-1*, *abi3-1*, and *abi4-1*) in a similar manner as in *aao3* mutant (Alazem et al. 2014). Although ABA was induced in BaMV-infected *abi1-1* and *abi4-1* lines, disruption of the signaling pathway prevented ABA-mediated resistance to BaMV. *abi1-1* is a gain-of-function mutant of the negative regulator ABI1, in which the PP2C–SnRK2 complex is insensitive to the ABA; thus, SnRK2 will not be released from the complex, nor will it phosphorylate or be able to activate ABA-responsive elements (Sheard and Zheng 2009). Therefore, no ABA defense mechanism will be activated in this mutant. Similarly, *abi3-1* and *abi4-1* are loss-of-function mutants of the positive regulators ABI3 and ABI4, in which the ABA signaling pathway is impaired and results, likewise, in no activation of ABA-related defense. *ABI3* and *ABI4* encode transcription factors that regulate ABA-responsive genes (Alazem et al. 2014). These results suggest that the different ABA signaling pathway components may decide induction or suppression of defense responses.

ABA may regulate host defense responses either positively or negatively through its direct or indirect action on the plant immune system. ABA accumulation directly potentiated callose deposition and enhanced disease resistance against *Botrytis cinerea* and *Plectosphaerella cucumerina* (Garcia-Andrade et al. 2011). ABA triggers callose deposition upon *Alternaria brassicicola* infection, a defense mechanism associated with restriction of colonization by the fungus (Flors et al. 2008). Basal tomato defenses to *Botrytis cinerea* include abscisic acid-dependent callose formation (Asselbergh and Höfte 2007). ABA was rapidly synthesized in plants following inoculation with the necrotrophic fungal pathogens, and it induced callose deposition (Sánchez-Vallet et al. 2012). ABA was found to mediate plant defense through increasing callose deposition and thereby limiting virus movement. ABA plays an important role in mediating callose deposition and restricting movement of *Tobacco necrosis virus* (TNV) in *Phaseolus vulgaris*; these latter effects were not evident in *P. vulgaris* plants treated with an ABA inhibitor (Iriti and Faoro 2008).

ABA may negatively regulate the host defense responses. A mutant of tomato with reduced ABA production (*sitiens*) exhibits increased resistance to the necrotrophic fungus *Botrytis cinerea*. This resistance was correlated with a rapid and strong hydrogen peroxide-driven cell wall fortification response in epidermis cells that was absent in tomato with normal ABA production (Curvers et al. 2010). Enhanced resistance of *Arabidopsis* ABA-deficient mutants (*aba2*) has been suggested to be due to increased cuticle permeability, which may result in stronger and faster accumulation of ROS following fungal infection (Curvers et al. 2010; L'Haridon et al. 2011). ROS signaling system has been shown to be involved in the activation of several defense genes. ROS signaling induced the PR protein PRB1-b and the antimicrobial protein chitinase class 4 in tobacco (Vranová et al. 2002). Thaumatin-like protein (PR-5 protein) is induced by H_2O_2 (Desikan et al. 2001). H_2O_2 induces PR-1 protein accumulation in tobacco (Neill et al. 2002). The ROS stress induced various enzymes involved in phytoalexins synthesis in tobacco (Vranová et al. 2002). ROS is known to induce the expression of phenylalanine ammonia lyase (Neill et al. 2002). Lignin biosynthesis is also activated by ROS (Vranová et al. 2002).

ABA may suppress the expression of several defense genes and thus it may negatively regulate ABA-induced defense responses. ABA modifies the expression of ethylene-responsive GCC-box-containing genes such as *PRB 1b*, *osmotin*, *CHN50*, and *GL4*. Many PR genes, such as *PR1*, *PR2*, *PR3*, and *PR5*, have GCC-box in their promoters (Zhou et al. 2008). ABA decreases the expression of GCC-box-containing genes (Zhou et al. 2008). ABA suppresses phenylalanine ammonia lyase (PAL) and resistance of soybeans to *Phytophthora megasperma* f. sp. *glycinea* (Ward et al. 1989). ABA treatment suppresses phytoalexin synthesis. It inhibits the activity and transcript accumulation of phenylalanine ammonia lyase (Henfling et al. 1980; Ward et al. 1989; McDonald and Cahill 1996; Mauch-Mani and Mauch 2005). Among the defensive proteins encoded by constitutively up-regulated genes in *aba1-6* mutant in which ABA biosynthesis pathway is impaired were antimicrobial proteins (PDF1.2, PR-1, and LTP3); enzymes involved in the biosynthesis and target delivery of Trp-derived secondary metabolites, such as indole glucosinolates and the phytoalexin camalexin (PEN3, PAD3, CYP81F2); and the ATRBOHD and PRX33/PRX34 proteins that are involved in ROS generation (Sánchez-Vallet et al. 2012). ROS accumulation has been linked to the enhanced resistance phenotype of some ABA mutants (Asselbergh et al. 2007; L'Haridon et al. 2011). Some transcriptional factors such as WRKY33 and ORA59, which regulate *Arabidopsis* immune responses, were up-regulated in unchallenged *aba1-6* and quadruple *pyr/pyl* mutants (Sánchez-Vallet et al. 2012). The *cis*-regulatory elements of the *aba1-6* up-regulated genes were enriched in W-box and GCC-box sequences that bind transcriptional factors such as WRKY33 and ORA59, respectively (Sánchez-Vallet et al. 2012).

ABA may also act indirectly in triggering or suppressing host plant defense responses by interacting with other signaling pathways involved in plant innate immunity. ABA has been shown to be complex modulator of various signaling systems involved in plant innate immunity (Asselbergh et al. 2008b; Feng et al. 2012; Sánchez-Vallet et al. 2012; Denancĕ et al. 2013). ABA strongly antagonizes many hormone signaling pathways, including SA (Audenaert et al. 2002; Soosaar et al. 2005; Yasuda et al. 2008), JA (Broekaert et al. 2006; Sánchez-Vallet et al. 2012), and ethylene (Ghassemian et al. 2000; Cheng et al. 2009), and the synergistic ET/JA pathway (Broekaert et al. 2006). Antagonism between ABA and ethylene in *Arabidopsis* acts in parallel with the reciprocal regulation of their metabolism and signaling pathways (Cheng et al. 2009).

ABA promotes infection of the rice blast pathogen *Magnaporthe oryzae* (Yazawa et al. 2012). The blast fungus *M. oryzae* was found to produce and secrete ABA and cause disease (Jiang et al. 2010). Exogenous application of ABA compromises rice resistance to the blast pathogen *M. oryzae* (Jiang et al. 2010). It has been shown that ABA interacts antagonistically with SA signaling pathway and confers susceptibility against the rice blast pathogen *Magnaporthe oryzae* (Jiang et al. 2010). ABA induces susceptibility to *B. cinerea* by suppressing SA-dependent signaling mechanisms in tomato (Audenaert et al. 2002). ABA determines basal susceptibility of tomato to *B. cinerea* and suppresses salicylic acid-dependent signaling system (Audenaert et al. 2002). Exogenous application of ABA to Arabidopsis resulted in an enhanced susceptibility to *Pseudomonas syringae* pv. *tomato* DC3000 that was correlated with reduced SA accumulation (Anderson et al. 2004; Mohr and Cahill 2007). The AvrPtoB effector of *P. syringae* pv. *tomato* DC3000 induces the accumulation of ABA in *Arabidopsis* to suppress the induction of basal defense by down-regulating SA biosynthesis and signaling pathways (de Torres et al. 2006). *Nicotiana benthamiana* infected with BAMV or CMV showed increased ABA contents, and the infection activated the SA and ABA pathways (Alazem et al. 2014). In *N. benthamiana*, infection with CMV or BaMV induced ABA accumulation and up-regulated the mRNA levels of several genes (*EDS1*, *SID2*, *NPR1*, *PR* genes) in the SA and (*ABA1, ABA2, AAO3, ABI3, ABI5, RAB18* genes) in the ABA pathways (Alazem et al. 2014). *NPR1* shows a negative effect over ABA biosynthesis genes *ABA1* and *ABA2*, as well as over some ABA-responsive genes such as *RAB18*, while *AAO3* suppresses the accumulation of SID2 in SA biosynthesis pathway (Alazem et al. 2014).

ABA may act cooperatively with JA in triggering resistance against the necrotrophic fungal pathogens *B. cinerea* and *Plectosphaerella cucumerina* in *Arabidopsis* (Asselbergh and Höfte 2007; Garcia-Andrade et al. 2011). ABA synthesis has been shown to be required for JA production and the activation of plant defenses against *Pythium irregulare* in *Arabidopsis* (Adie et al. 2007). ABA-induced resistance against the brown spot pathogen *Cochliobolus miyabeanus* in rice involves MAP kinase-mediated repression of ethylene-mediated repression of ethylene signaling (De Vleesschauwer et al. 2010).

The multifaceted role of ABA in plant defense has been suggested to be phase specific. It plays positive roles during the early stages of infection (by inducing stomatal closure and callose deposition) and negative roles at later stages (which include suppressing ROS accumulation and SA induction) (Munemasa et al. 2007; Asselbergh et al. 2008b; Ton et al. 2009). It has also been suggested that the dual role of ABA inducing resistance and susceptibility may be due to the differential action of upstream and downstream components in ABA signaling pathway (Alazem et al. 2014).

5.2 ABA Biosynthesis in Innate Immune Responses

5.2.1 Pathogen/PAMP Triggers Biosynthesis and Accumulation of ABA

Pathogen/pathogen-associated molecular patterns (PAMPs) induced increase in ABA accumulation inducing disease resistance (Whenham et al. 1986), or susceptibility (Mohr and Cahill 2003; Koga et al. 2004; Schmidt et al. 2008). ABA was found to accumulate during pathogenesis in plants (Alazem et al. 2014). ABA was rapidly synthesized in plants following inoculation with necrotrophic fungal pathogens (Sánchez-Vallet et al. 2012). ABA was found to accumulate in tobacco and tomato infected with *Tobacco mosaic virus* (Whenham et al. 1986; Fraser and Whenham 1989). The concentration of ABA increased in sugar beet (*Beta vulgaris*) leaves during *Cercospora beticola* infection. After an initial burst of ABA induced by inoculation, elevated ABA concentrations were detected during the fungal penetration and colonization phases (Schmidt et al. 2008).

5.2.2 ABA Biosynthesis Pathway

The early precursor of ABA is isopentenyl diphosphate and is synthesized from glyceraldehyde 3-phosphate and pyruvate via the methylerythritol phosphate pathway. This leads to the successive production of phytoene and lycopene as the intermediates. The lycopene is cyclized and hydroxylated to form zeaxanthin, the first oxygenated carotenoid. Zeaxanthin is then oxidized to violaxanthin by the action of *Arabidopsis* ABA1/tobacco ABA2 and rearranged to 9′-*cis*-neoxanthin (Marin et al. 1996; Milborrow 2001; Wasilewska et al. 2008).

ABA is derived from oxidative cleavage of the 11,12 double bond of 9-*cis* epoxy carotenoids (neoxanthin and/or violaxanthin). The 9-*cis*-epoxycarotenoid dioxygenase (NCED) enzyme cleaves 9-*cis* xanthophylls to xanthoxin (Tan et al. 2003). In *Arabidopsis*, both violaxanthin and neoxanthin are alternative in vivo substrates of NCED to produce xanthoxin (Wasilewska et al. 2008). Six genes encoding NCED have been identified. The T-DNA insertion adjacent to *9-cis-epoxycarotenoid dioxygenase5* (*NCED5*) caused a massive increase in transcript level and enhanced ABA levels (Fan et al. 2009). *NCED2*, *NCED3*, and *NCED5* were induced, and ABA accumulated strongly following compatible *Pseudomonas syringae* infection (Fan et al. 2009). *NCED3* encodes a member of the 9-*cis*-epoxycarotenoid dioxygenase family, responsible for converting 9-*cis*-violaxanthin and 9-*cis*-neoxanthin into xanthoxin (Schwartz et al. 2003). *ABA2* encodes a short-chain alcohol dehydrogenase, which catalyzes the conversion of xanthoxin into ABA aldehyde (Gonzalez-Guzman et al. 2002). AAO3 (Abscisic Acid Oxidase3) catalyzes the conversion of ABA aldehyde to ABA (Seo et al. 2000; Fig. 5.1). ABA2 converts xanthoxin to ABA aldehyde, which is used by AAO3 and cofactor ABA3 to produce ABA (Hauser et al. 2011).

Xanthoxin is the first cytoplasmic precursor for the biosynthesis of ABA. It is converted to abscisic aldehyde by the action of a dehydrogenase/reductase encoded by *ABA2*. Abscisic aldehyde is converted to ABA by the action of two enzymes, ABA3 and AAO3. *ABA3* encodes the molybdenum cofactor sulfurase that adds the sulfur atom to the molybdenum center. The *AAO3* encodes an aldehyde oxidase that requires molybdenum cofactor for its activity. Abscisic aldehyde may also be converted to abscisic alcohol, which in turn is converted to ABA through a shunt pathway (Wasilewska et al. 2008).

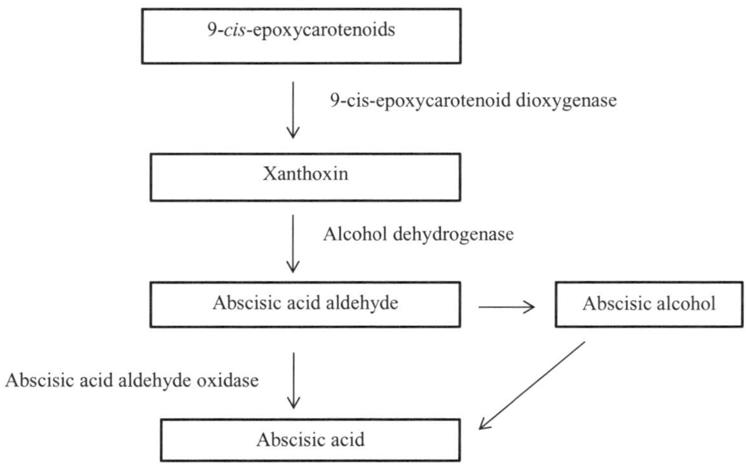

Fig. 5.1 Abscisic acid biosynthesis pathway (Adapted from Gonzalez-Guzman et al. 2002; Wasilewska et al. 2008; Ton et al. 2009; Hauser et al. 2011)

5.2.3 G-Proteins May Be Involved in ABA Biosynthesis

G-proteins are the regulatory GTPases acting as molecular switches in signal transduction system (Yalowsky et al. 2010; Zhang et al. 2011, 2012). G-proteins physically couple the recognition of extracellular signals with specific cell-surface receptors called G-protein-coupled receptors (GPCRs) (Hu et al. 2010; Tesmer 2010). Heterotrimeric G-proteins are composed of α-, β-, γ-subunits, which exist as associated heterotrimers in their inactive state. Once activated by the GPCR, the Gα protein, which possesses a GDP/GTP-nucleotide-binding site and GTP-hydrolase activity, changes its form to a structure that allows exchange of GDP for GTP (Pandey et al. 2010). The GPCR works as a guanine exchange factor (GEF) for Gα and facilitates the Gα subunit to exchange GDP for GTP and become active (Oki et al. 2009). GTP binding is accompanied by structural rearrangements that disengage the Gβγ interaction and result in heterotrimer dissociation. The free subunits then relay signals by interacting with downstream proteins called effectors. The GTP-bound Gα separates from the associated Gβγ dimer, and the freed Gα and Gβγ proteins can then interact separately with a variety of downstream effector molecules to transduce the signal (Pandey et al. 2010). Subsequent to signal propagation, the intrinsic GTPase activity of Gα eventually results in hydrolysis of bound GTP to GDP, which inactivates Gα and allows its reassociation with the Gβγ dimer to reform the inactive G-protein complex (Liu et al. 2010; Pandey et al. 2010; Trusov et al. 2012). The GTP form exhibits a weak intrinsic GTPase activity for GTP hydrolysis, requiring specific GTPases-accelerating proteins (GAPs) with unique domain composition for efficient deactivation (Yang 2002; Mucha et al. 2011).

In *Arabidopsis*, the G-protein-mediated signal transduction chain consists of many components, including α-, β-, γ-subunits of the G-protein heterotrimer, G-protein-coupled receptors (GPCRs), regulator of G-protein signaling (RGS), and many other downstream effectors (Chen et al. 2006). RGS functions as a Gα subunit-accelerating protein (GAP) in *Arabidopsis* (Chen et al. 2006). RGS proteins act as an effective desensitizer of G-protein-mediated signaling by accelerating the deactivation of G-protein α-subunits from the G-protein-GTP binding form to the G-protein-GDP binding form (McCudden et al. 2005). *Arabidopsis* genome contains only one member of the RGS proteins, AtRGS1 (Chen et al. 2003). AtRGS1 strongly interacts with the Gα subunit (AtGPA1) and has a GTPase activity, implying that AtRGS1 protein may be involved in G-protein-mediated signaling (Chen et al. 2003). AtRGS1 functions as a Gα subunit-accelerating protein (GAP) (Chen et al. 2006).

Transgenic *Arabidopsis* plants overexpressing AtRGS1, the regulator of G-protein signaling protein, were developed, and AtRGS1 overexpression significantly stimulated the expression of *NCED* encoding the 9-*cis*-epoxycarotenoid dioxygenase (NCED) enzyme, which cleaves 9-*cis* xanthophylls to xanthoxin. The NCED is the first committed step for ABA synthesis (Schwartz et al. 2003; Tan et al. 2003; Chen et al. 2006; Wasilewska et al. 2008; Fan et al. 2009). The transgenic plants overexpressing AtRGS1 also showed increased expression of *ABA2* gene (Chen et al. 2006), which encodes dehydrogenase/reductase involved in conversion

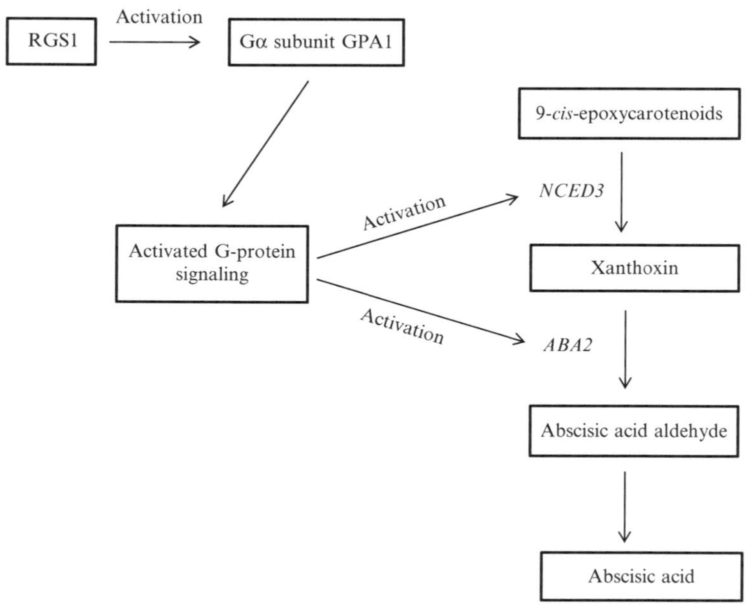

Fig. 5.2 Role of G-protein signaling in triggering ABA biosynthesis (Adapted from Chen et al. 2006)

of xanthoxin to abscisic acid aldehyde, which is used by AAO3 and cofactor ABA3 to produce ABA (Gonzalez-Guzman et al. 2002; Wasilewska et al. 2008; Hauser et al. 2011). These results indicate that RGS1 is involved in the biosynthesis of ABA (Fig. 5.2; Chen et al. 2006).

5.3 ABA Perception and Signal Transduction

5.3.1 ABA Signaling Pathway

Three major components are involved in the ABA signal perception and transduction pathway. These components include PYR/PYL/RCAR (an abscisic acid receptor), type 2C protein phosphatase (PP2C; a negative regulator), and SNF1-related protein kinase (SnRK2; a positive regulator) (Schweighofer et al. 2004; Umezawa et al. 2010, 2013). Cellular ABA levels are generally low in unstressed normal plants, and in these plants, PP2Cs bind to SnRK2s and dephosphorylate the SnRK2s to keep the SnRK2s in an inactive state (Guo et al. 2011). Pathogen infection or PAMP treatment triggers ABA accumulation (Sánchez-Vallet et al. 2012; Alazem et al. 2014; Vidhyasekaran 2014). In the presence of accumulated ABA, the PYR/PYL/RCAR receptor proteins disrupt the interaction between the SnRK2s and PP2Cs, thus preventing the PP2Cs-mediated dephosphorylation of the SnRK2s and resulting in the activation of the SnRK2 kinases (Fujii et al. 2009). ABA binds with the receptor and the binding of ABA to the receptor induces a conformational change in the receptor that allows it to sequester the PP2Cs. This excludes the PP2Cs from the negative regulation of the downstream ABA-activated SnRK protein kinases (Miyazono et al. 2009; Umezawa et al. 2009; Vlad et al. 2009; Raghavendra et al. 2010). The recognition of ABA by the PYR/PYL/RCAR receptor results in the dissociation of the SnRK2/PP2C complex and the subsequent release of SnRK2s. The latter phosphorylate and activate downstream transcription factors, which initiate transcription at ABA-responsive promoter elements (Sheard and Zheng 2009; Fig. 5.3). Several b-ZIP transcription factors, ion channels (SLAC1, KAT1), and a NADPH oxidase (AtrbohF) are the important downstream targets of ABA (Joshi-Saha et al. 2011).

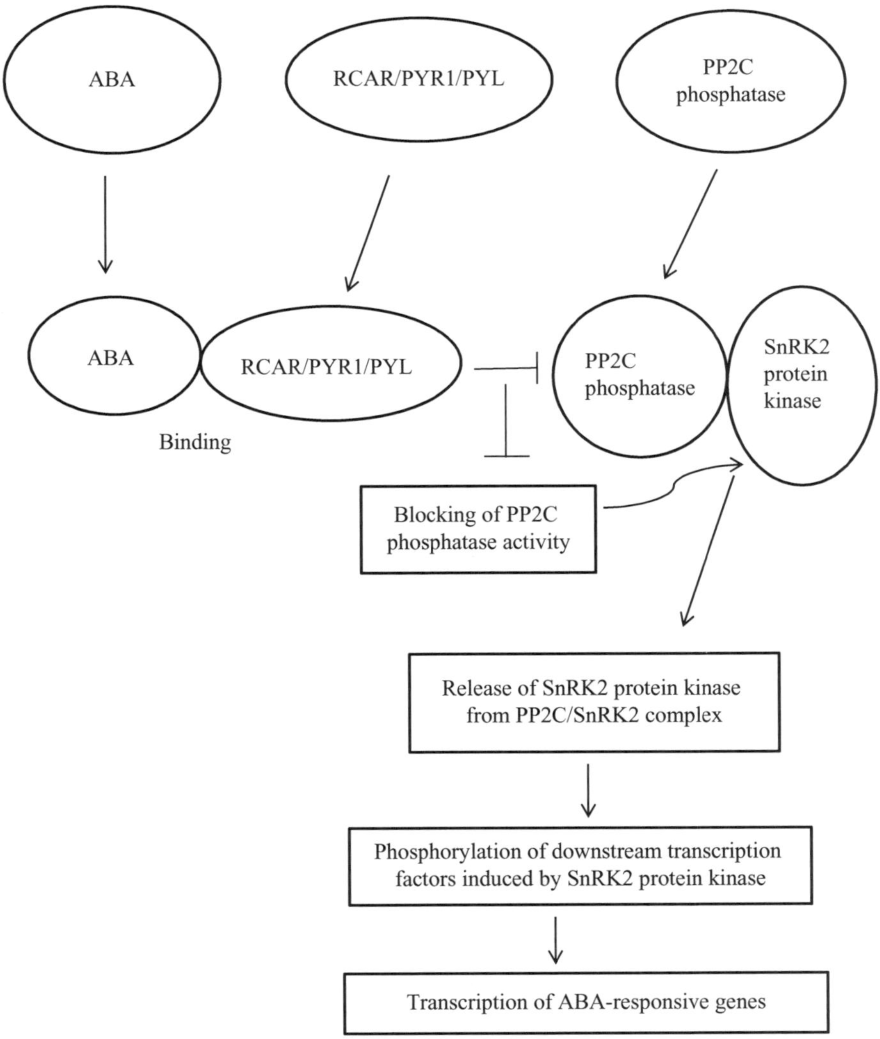

Fig. 5.3 ABA signaling pathway (Fujii et al. 2009; Ma et al. 2009; Park et al. 2009; Weiner et al. 2010; Joshi-Saha et al. 2011)

5.3.2 ABA Receptors

5.3.2.1 Early Search for ABA Receptors

Although the role of ABA is known since several decades ago, the mechanisms of ABA signaling have remained largely obscure (Hirayama and Shinozaki 2007; Cutler et al. 2010; Hubbard et al. 2010; Kline et al. 2010; Joshi-Saha et al. 2011). Several attempts have been made for over a decade to detect receptor sites for ABA signals. These studies have revealed the existence of perception sites located on the "inside" as well as the "outside" of cells. However, the corresponding proteins could not be identified (Wasilewska et al. 2008). Three distinct loci encoding ABA receptors have been identified, and these include FCA (for *f*lowering time *c*ontrol protein *A*) (Razem et al. 2006), ABAR/CHLH (for putative *ABA r*eceptor/magnesium-protoporphyrin IX *chel*atase *H* unit) (Shen et al. 2006), and GCR2 (for G-protein-coupled receptor 2) (Liu et al. 2007a). Another class of GPCRs (G-protein-coupled receptors), GTGs (GPCR-type G-proteins) has been proposed as ABA receptors (Pandey et al. 2009).

5.3.2.2 FCA May Not Be the ABA Receptor

FCA is an ABA receptor that controls ABA-dependent flowering (Razem et al. 2006). It is a nuclear RNA-binding protein (Wang and Zhang 2008; Wasilewska et al. 2008). Razem et al. (2006) reported that FCA functions as an ABA receptor. However, Risk et al. (2008) questioned the binding property of FCA to ABA. In fact, the original article claiming FCA as an ABA receptor was retracted (Razem et al. 2008).

5.3.2.3 Controversial Role of ABAR/CHLH as ABA Receptor

An ABA-specific binding protein was isolated from broad bean leaves, and it was named ABAR. Based on sequencing information of the purified protein, a cDNA fragment was isolated and found to encode the C-terminal half of about 770 amino acids of the putative H subunit (CHLH) of the magnesium-protoporphyrin IX chelatase (Zhang et al. 2002). The cDNA fragment encoding the broad bean ABAR/CHLH could specifically bind ABA. CHLH plays a key role in mediating plastid-to-nucleus retrograde signaling, in which plastids send signals to the nucleus to control the expression of nuclear genes (Wang and Zhang 2008). ABA binding to ABAR/CHLH specifically depended on physiologically active ABA and the ABA binding displayed typical ligand-receptor binding characteristics (Wang and Zhang 2008). The mechanism of ABAR perception of ABA and downstream signaling events are still unknown. Further, both the FCA and ABAR receptors have been shown to encode genes involved in floral timing and chlorophyll biosynthesis, which are not considered traditional ABA responses. Hence the role of these proteins as ABA receptor for triggering defense ABA signaling is doubtful (McCourt and Creelman 2008). Only conflicting experimental evidence has been presented to show that ABAR could function as an ABA receptor (Muller and Hansson 2009; Shang et al. 2010).

5.3.2.4 Doubtful Role of GPCR as ABA Receptor

A G-protein-coupled receptor (GPCR), GCR2, has been identified as a plasma membrane-localized receptor for ABA (Liu et al. 2007a). GCR2 is a membrane protein with seven transmembrane helices. The GCR2 genetically and physically interacts with the G-protein α-subunit GPA1 to mediate all known ABA responses in *Arabidopsis*. This receptor binds ABA with high affinity at physiological concentration. The GCR2 interacts with the Gαβγ complex. Binding of ABA to GCR2 results in the release of Gα and Gβγ dimer to activate downstream ABA signaling events (Liu et al. 2007a).

However, there is a debate on the role of the GCR2 as an ABA receptor. Johnston et al. (2007) reported that GCR2 is neither a transmembrane protein nor a G-protein-coupled receptor, but rather a plant homolog of bacterial lanthionine synthetase (LanC) proteins. Gao et al. (2007) reported that genetic characterization revealed no role for the reported ABA receptor, GCR2, in ABA control of seed germination and early seedling development in *Arabidopsis*. However, Liu et al. (2007b) studied only the role of ABA-related stomatal regulation, and not the role of ABA control of seed germination. They provided additional evidences to show that GCR2 is biochemically distinct from human LanC-like (LANCL) proteins, including the difference in membrane association properties between GCR2 and LANCL1. Further, LANCL1 does not interact with G-protein, whereas GCR2 interacted with Gα and Gβ subunits. They suggested that GCR2 is an ABA receptor and interacts with G-protein α- and β-subunits. They also argued that although other properties of GCR2 may not be consistent with classic GPCRs, this protein may be a new type of nonclassical GPCR (Liu et al. 2007b). However, Risk et al. (2009) showed that GCR2 does not bind ABA.

5.3.2.5 GTG1/GTG2

Another novel class of GPCRs, GTGs (GPCR-type G proteins), has been suggested as ABA receptors (Pandey et al. 2009). GTG proteins contain GTP binding/GTPase activity. GTGs interact with the heterotrimeric G-protein α-subunit, GPA1, and the GTP-bound form of GPA1 inhibits the GTPase activity of GTG proteins. The GDP-bound GTGs bind ABA stronger than GTP-bound form; thus, it might be the active form for perceiving and transducing ABA signal (Guo et al. 2011). However, the ABA-induced responses were not abolished in the *Arabidopsis gtg1/gtg2* double mutant. It suggests that besides GTG1/GTG2, there may be some other ABA receptors involved in ABA signaling (Christmann and Grill 2009; Pandey et al. 2009; Risk et al. 2009). It has also been reported that the binding activity of GTGs is also very low (Guo et al. 2011).

5.3.2.6 PYR/PYL/RCAR Proteins May Be the Potential ABA Receptors

Although some candidate receptors have been put forward during the last decade, none of them at the time has fully satisfied the criteria that characterize the key receptor (Joshi-Saha et al. 2011). In the year 2009, a breakthrough in the search for the ABA receptor has occurred (Ma et al. 2009; Park et al. 2009). A new class of START (for *ST*EROIDOGENIC *A*CUTE *R*EGULATORY-RELATED LIPID *T*RANSFER) proteins has been identified as ABA receptors in *Arabidopsis thaliana* (Ma et al. 2009; Melcher et al. 2009; Park et al. 2009; Yin et al. 2009). These receptors were identified by resistance to pyrabactin, a synthetic inhibitor of seed germination (Melcher et al. 2010a, b). These proteins are called "REGULATORY COMPONENT OF ABA RECEPTORS (RCARs)" (Ma et al. 2009) or PYRABACTIN RESISTANCE 1 (PYR1) and PYR1-LIKE PROTEINS (PYLs) (Park et al. 2009). A family of soluble proteins with at least 14 known members in *Arabidopsis thaliana* has been suggested to be the ABA receptors (Ma et al. 2009; Melcher et al. 2009; Park et al. 2009; Yin et al. 2009). These proteins belong to the pathogenesis-related protein class 10 (PR-10). These proteins of 150–200 amino acids share a conserved START proteins domain (Joshi-Saha et al. 2011). These proteins interact with ABA and regulate the activity of clade A-group protein phosphatase 2C (PP2C) (Park et al. 2009; Yin et al. 2009; Umezawa et al. 2010; Guo et al. 2011; Antoni et al. 2012). PYR/PYL/RCAR proteins can directly bind to ABA (Ma et al. 2009; Park et al. 2009; Santiago et al. 2009a, b; Szostkiewicz et al. 2010) and regulate the activity of PP2C (Ma et al. 2009; Park et al. 2009; Guo et al. 2010).

5.3.3 PYR/PYL/RCAR Negatively Regulates PP2C

Plant PP2Cs, such as ABI1 and ABI2, are known core components in the ABA signaling pathway, repressing ABA signaling at an early step (Yoshida et al. 2006). ABA binds PYR1, and the ABA-bound PYR1 then binds to PP2C and inhibits its activity, supporting the view that PYR/PYLs function as ABA receptors through inhibiting the phosphatase activity of PP2C (Fig. 5.3; Guo et al. 2011). The ABA-binding induces conformational changes of PYR/PYL/RCARs, creating a new surface for PP2Cs to interact with (Guo et al. 2011). The conformational change in PYR/PYL/RCARs allows the proteins to sequester the PP2Cs (Joshi-Saha et al. 2011). The interacting interface between ABA-bound PYR/PYL/RCARs and PP2Cs is located at the enzymatic active site of the PP2Cs, and therefore the interaction inhibits the phosphatase activity of PP2Cs by blocking the access of their substrates to the catalytic center (Guo et al. 2011).

5.3.4 ABA-Bound PYR/PYL/RCAR Can Shift ABA Signaling Status to "Active" State

PYR/PYL/RCARs have been shown to be ABA receptors and these proteins consist of homodimers with one subunit binding ABA. The binding of ABA may dissociate the dimer. The binding cavity of PYR/PYL/RCARs is flanked by the gate and latch loops, which upon binding of an ABA molecule fold over and enclose it inside (Melcher et al. 2009; Nishimura et al. 2009; Yin et al. 2009; Danquash et al. 2013). The crystal structures of apo and ABA-bound receptors as well as a ternary PYL2-ABA-PP2C complex revealed that the apo receptors contain an open ligand-binding pocket flanked by a gate that closes in response to ABA by way of conformational changes in two highly conserved β-loops that serve as a gate and a latch. The ABA-induced closure of the gate creates a surface that enables the receptor to dock into and competitively inhibit the PP2C active site. A conserved tryptophan in the PP2C inserts directly between the gate and latch, which functions to further lock the receptor in a closed conformation. These studies suggest there may be a conserved gate–latch–lock mechanism underlying ABA signaling (Melcher et al. 2009).

Yin et al. (2009) reported that apo-PYL2 contains a pocket surrounded by four highly conserved surface loops. In response to ABA binding, loop CL2 closes into the pocket, creating a surface that recognizes ABI1. In the ternary complex, the CL2 loop is located near the active site of ABI1, blocking the entry of substrate proteins. The results suggest that ABA binding induces a conformational change in PYLs that allows subsequent binding of PP2Cs and ABA-bound PYLs inhibit PP2Cs by blocking entry of substrate proteins. PYR/PYL/RCAR family members have been found to be major in vivo ABI1 protein phosphatase 2C-interacting proteins in *Arabidopsis* (Nishimura et al. 2010). Because PP2C is a major negative regulator of ABA signaling, it seems that ABA-bound PYR/PYL/RCAR can shift ABA signaling status to "active" state (Nishimura et al. 2010).

5.3.5 ABA-Induced PP2C Phosphatase Inhibition Leads to SnRK2 Protein Kinase-Activated Phosphorylation of ABA-Responsive Genes

ABA binding to the PYR/PYL/RCAR receptors increases their ability to bind and inhibit type 2C protein phosphatases such as ABI1, ABI2, and HAB1. In the absence of ABA, these PP2Cs bind and inactivate subfamily 2 members of SNF1-related protein kinases (SnRK2 kinases) by dephosphorylating serine and threonine residues in their activation loop (Yoshida et al. 2006; Umezawa et al. 2009). The results suggest that the activity of PP2Cs is regulated by ABA-bound PYR/PYL/RCARs.

Cellular ABA levels are generally low in unstressed normal plants, and in these plants, PP2Cs bind to SnRK2s and dephosphorylate the SnRK2s to keep the SnRK2s in an inactive state. In contrast, when plants are exposed to biotic or abiotic stress conditions, the ABA levels are elevated. The elevated cellular ABA binds PYR/

PYL/RCARs, which in turn may bind and deactivate phosphatase activity of PP2Cs. The SnRK2s may then be released from PP2Cs and self-activated via autophosphorylation. The activated SnRK2s may phosphorylate different downstream targets in the ABA signaling complex (Guo et al. 2011).

ABA-induced PP2C inhibition leads to SnRK2 activation by autophosphorylation and/or phosphorylation by upstream kinases. Activated SnRK2 in turn phosphorylate and activate downstream effectors such as the basic leucine zipper transcription factors called ABFs/AREBs to switch on defense responses (Yoshida et al. 2006; Furihata et al. 2006; Melcher et al. 2009, 2010a, b).

The SnRK2 in *Arabidopsis* is a family of 10, with SnRK2.2, SnRK2.3, and OST1/SnRK2.6/SRK2E (from the subclass III) bearing a particular Domain II motif near their C-terminus that constitutes the binding site for ABI1 (Fujita et al. 2009). ABA regulates several membrane proteins. SLOW ANION CHANNEL-ASSOCIATED (SLAC) is the slow (S)-type anion channel that triggers membrane depolarization required to initiate stomatal closure (Chen et al. 2010; Joshi-Saha et al. 2011). OST1 phosphorylation may alter the gating properties of SLAC1 (Chen et al. 2010). The major guard cell K^+ current rectifier KAT1 is down-regulated by OST1 phosphorylation (Sato et al. 2009). OST1 directly phosphorylates AtrhobF, one of the two major NADPH oxidases responsible for H_2O_2 production in guard cells (Sirichandra et al. 2009). SnRK2 kinases phosphorylate basic leucine zipper domain (b-ZIP) transcription factors to activate ABA-responsive genes (Joshi-Saha et al. 2011).

5.3.6 Phosphatases in ABA Signaling Network

Protein phosphatases are key components in ABA signaling (Moes et al. 2008). The members of the type 2C Ser/Thr phosphatase (PP2C) superfamily, ABI1, ABI2, PP2C-HAB and PP2CA, are key negative regulators of ABA-mediated responses (Merlot et al. 2001; Murata et al. 2001; Saez et al. 2004, 2006; Kuhn et al. 2006). The loss of ABI1 and ABI2 PP2C activity in intragenic revertants of *abi1-1* or *abi2-1* leads to an enhanced response to ABA (Gosti et al. 1999; Merlot et al. 2001). Antisense inhibition of AtPP2CA results in the enhancement of ABA-induced gene expression (Tahtiharju and Palva 2001). Overexpression of ABI1 or AtPP2CA blocks the expression of ABA-inducible genes in *Arabidopsis* protoplasts (Sheen 1998). These results suggest that the protein phosphatases act as negative regulators of ABA signaling system. In contrast to the PP2Cs ABI1 and ABI2, a PP2A, RCN1, has been identified as a positive regulator of ABA response involved in early events of ABA signaling (Kwak et al. 2002).

Two distinct PP2Cs encoded by *abi1-1* and *abi2-1* have been detected in *Arabidopsis* cells. In *abi1-1* mutant, ABA did not induce ROS production, but H_2O_2 activation of calcium influx (I_{Ca}) channels was not disturbed, suggesting that *abi1-1* impairs ABA signaling between ABA reception and ROS production.

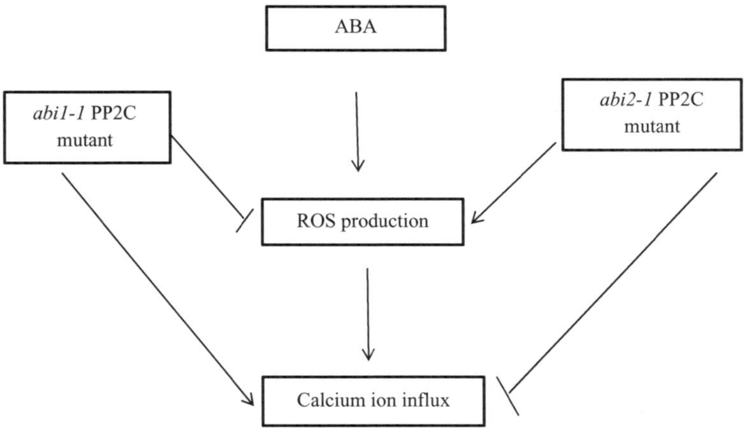

Fig. 5.4 Two different PP2Cs take part in the ABS signaling at two different stages (Adapted from Murata et al. 2001)

In contrast to *abi1-1*, *abi2-1* elicited ROS production, but impaired H_2O_2 activation of I_{Ca} (Murata et al. 2001). These results suggest that the two PP2Cs may take part in the ABS signaling network at two different stages (Fig. 5.4).

5.3.7 Role of SnRK2 Protein Kinase in ABA Signaling

Protein phosphorylation is the important mechanism for ABA signaling (He and Li 2008). Specific protein kinases are activated in response to ABA and they play a positive role in ABA signaling (Li et al. 2000; Lu et al. 2002; Fujii et al. 2007; Kline et al. 2010; Guo et al. 2011). ABA response element-binding factors (ABFs) have been shown to be activated by phosphorylation by protein kinases (Uno et al. 2000). The plant AMP-activated protein kinase (AMPK)/sucrose-non-fermentation 1 (SNF1)-related kinase subfamily 2 (SnRK2) is a positive regulator of ABA signaling (Nakashima et al. 2009). SnRK family of protein kinases, belonging to CaMK group of protein kinases, is involved in the phosphorylation of ABF proteins (Fujii et al. 2007; Vidhyasekaran 2014). The *Arabidopsis* genome contains 38 SnRKs, of which 10 are SnRK2s (Hrabak et al. 2003). Five of these SnRK2s were shown to be activated by ABA, and they phosphorylated ABI1, ABF2, and ABF4 (Furihata et al. 2006; Yoshida et al. 2006). SnRK2.2 and SnRK2.3 are the two protein kinases, and the effect of these protein kinases is mediated by phosphorylating ABFs and regulating ABA-responsive genes (Fujii et al. 2007).

Umezawa et al. (2013) identified 35 peptides that were differentially phosphorylated in *Arabidopsis* plants treated with ABA. SnRK2 promoted the ABA-induced activation of the MAP kinases AtMPK1 and AtMPK2. SnRK2 is involved in phosphorylation of Ser45 in the bZIP transcription factor AREB1 (ABA-responsive

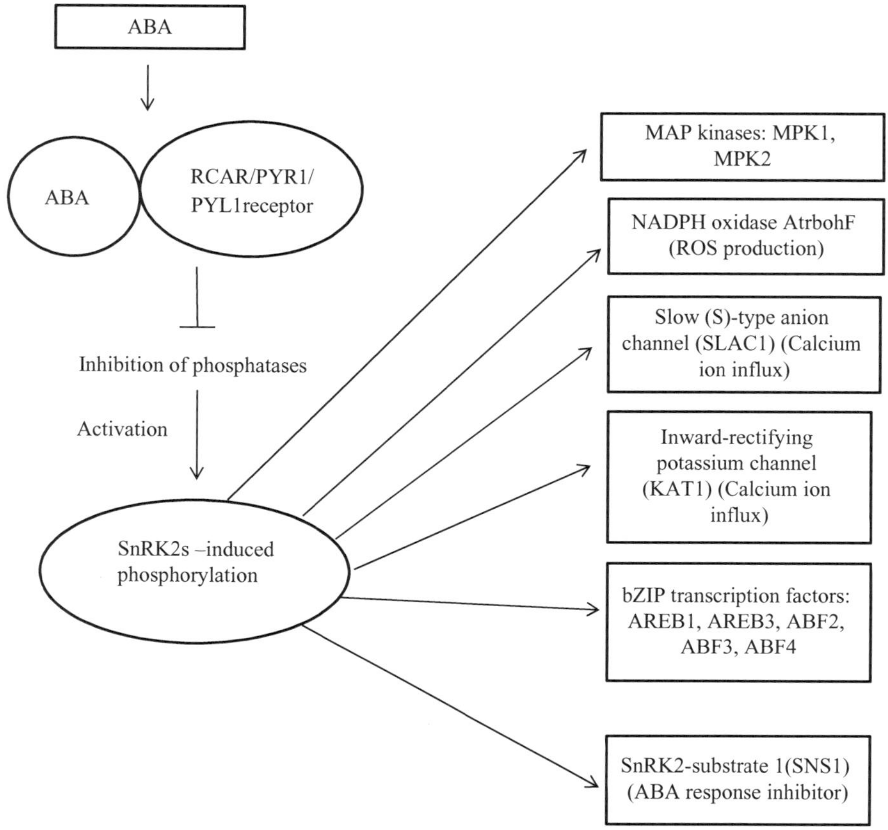

Fig. 5.5 ABA-induced SnRK2-mediated protein phosphorylation network in ABA signaling pathway (Adapted from Kline et al. 2010; Umezawa et al. 2013)

element-binding protein 1), which stimulated ABA-responsive gene expression. The bZIP transcription factors bind ABA-responsive elements (ABRE) in the promoter of ABA responsive genes, including *AREB1* (*ABRE binding protein 1*), *AREB2*, *AREB3*, and *ABI5* (*ABA insensitive 5*) (Fujii et al. 2009; Yoshida et al. 2010; Guo et al. 2011). It is suggested that SnRK2s can phosphorylate AREB1 and the phosphorylated AREB1 can then activate the expression of ABA responsive genes (Furihata et al. 2006; Fujii et al. 2009). ABA-activated SnRK2 phosphorylated in vivo SNS1 (SnRK2-substrate 1, SNS1), which was shown to inhibit ABA responses in *Arabidopsis* (Fig. 5.5; Kline et al. 2010; Umezawa et al. 2013).

SnRK2 phosphorylates not only basic domain leucine zipper (bZIP) transcription factors, but also phosphorylates various membrane proteins (Fig. 5.5; Umezawa et al. 2010; Guo et al. 2011). Several membrane proteins have been identified as SnRK2 substrates. SLAC1, a slow anion channel, is phosphorylated and activated by SRK2E/OST1/SnRK2 under the control of PP2C (Geiger et al. 2009; Lee et al. 2009). SLAC1 has a central role in guard cells involved in stomatal closure (Negi et al. 2008; Kim et al. 2010). KAT1, an inward-rectifying potassium channel, is another target of SnRK2 (Umezawa et al. 2010). The SnRK2s-mediated phosphorylation activates SLAC1 and deactivates KAT1 and consequently leads to the depolarization of plasma membrane and activation of the outward K^+ channel. The leaking of anions and K^+ may cause loss of turgor in guard cells, leading to stomatal closure (Geiger et al. 2009; Lee et al. 2009). SnRK2E/OST1 phosphorylates Thr306 of KAT1, and the modification of Thr306 greatly affects KAT1 activity (Sato et al. 2009). ABA-activated SnRK2 may inhibit KAT1 by phosphorylation to promote stomatal closure (Umezawa et al. 2010). ABA-induced stomatal closure has been suggested to be an important plant innate immune response against bacterial pathogens (Melotto et al. 2006; Takahashi et al. 2007; Ton et al. 2009; Uraji et al. 2012). SnRK2 can also phosphorylate an NADPH oxidase, AtrbohF, involved in ROS signaling (Sirichandra et al. 2009, 2010).

In rice, SAPK8, SAPK9, and SAPK10, which are homologous with SnRK2.2, SnRK2.3, and SnRK2.6 of *Arabidopsis*, were activated by ABA (Kobayashi et al. 2004). These SAPKs phosphorylated TRAB1, which is a rice ortholog of the *Arabidopsis* ABFs (Kobayashi et al. 2005). ABA induced the wheat SnRK, PKABA1, at the transcript level. The ABA-induced PKAB1 phosphorylated TaABF, the wheat bZIP transcription factor (Johnson et al. 2002).

Another SNF1-related protein kinase, AAPK (for ABA-activated protein kinase) was detected in broad bean (*Vicia faba*). AAPK is stimulated by ABA, and it is a positive regulator of ABA response (Li et al. 2000, 2002; Johnson et al. 2002). Takahashi et al. (2007) showed that ABA induced binding of a 14-3-3 protein to proteins with molecular masses of 61, 43, and 39 kDa. Autophosphorylation of AAPK, which mediates anion channel activation and ABA-induced phosphorylation of the 61 kDa protein, showed similar time courses. AAPK elicits the binding of the 14-3-3 protein to the 61-kDa protein in vitro when AAPK was activated by ABA. It has also been suggested that the 61 kDa protein may be a substrate for AAPK and the 61 kDa protein is located upstream of H_2O_2 and Ca^{2+} or on Ca^{2+}-independent signaling pathway (Takahashi et al. 2007).

An ABA-insensitive *Vicia faba* mutant, *fia* (faba bean impaired in ABA-induced stomatal closure), had been isolated. Unlike ABA, H_2O_2, and nitric oxide (NO), induced stomatal closure in the *fia* mutant. ABA did not induce production of either reactive oxygen species or NO in the mutant. ABA also did not suppress inward-rectifying K^+ currents or activate AAPK in mutant guard cells. These results suggest that FIA functions as an early signal component upstream of AAPK activation in ABA signaling in guard cells of *Vicia faba* (Sugiyama et al. 2012).

Arabidopsis OST1/SRK2E, an *Arabidopsis* ortholog of *Vicia faba* AAPK, which is an ABA-activated, Ca^{2+}-independent protein kinase, is involved in ABA signaling, and it acts upstream of ROS production (Mustilli et al. 2002; Yoshida et al. 2002). Phosphorylation has been shown to function in ABA-induced ROS production and I_{ca} channel activation (Murata et al. 2001; Köhler and Blatt 2002).

5.3.8 Phospholipase D in ABA Signaling Pathway

Phospholipase D (PLD) is a phosphatidylcholine (PC)-hydrolyzing enzyme that generates phosphatidic acid (PA), a lipid second messenger that modulates diverse intercellular signaling (Jang et al. 2012). Phospholipase D-mediated phosphatidic acid (PA) production has been shown to promote ABA-induced gene expression (Fig. 5.6; Zhang et al. 2004b). Phospholipase Dα-1-derived PA regulates a protein phosphatase 2C, ABI1, which is a negative regulator of ABA responses in

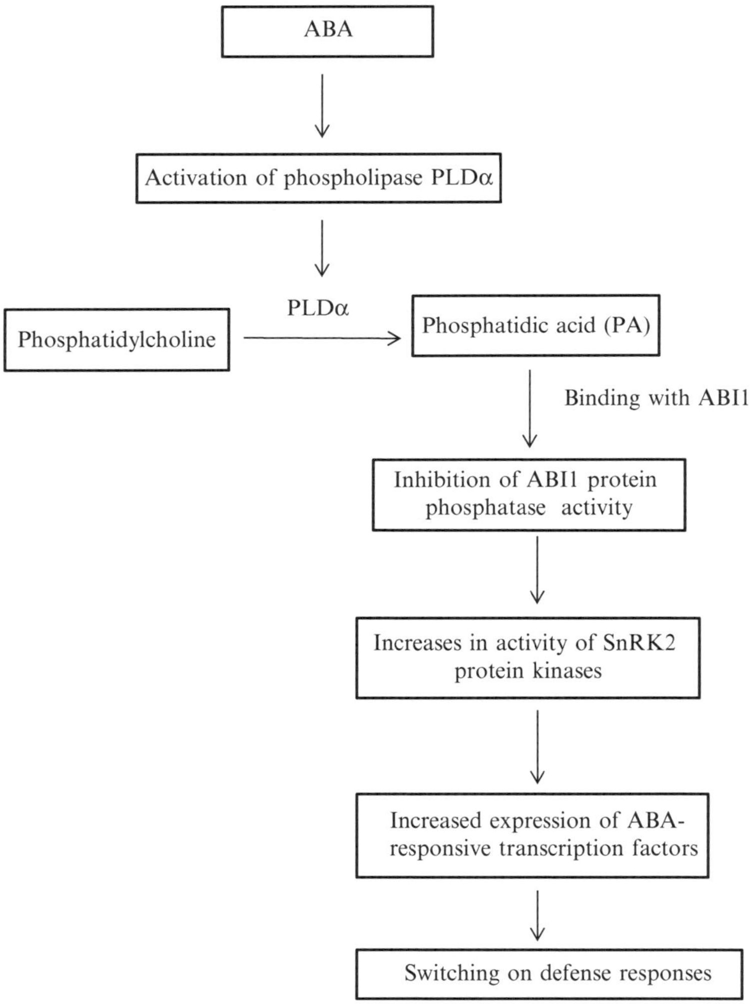

Fig. 5.6 Role of ABA-induced phosphatidic acid production in triggering ABA-responsive gene transcription (Adapted from Zhang et al. 2004b; Wang 2005)

Arabidopsis (Zhang et al. 2004b, 2005, 2009). ABA treatment activates PLD activity and promotes an increase in PA from phosphatidylcholine (Fig. 5.6; Zhang et al. 2004b). The PA binds to ABI1 (ABA insensitive1), a protein phosphatase 2C (PP2C) that is a negative regulator of ABA in *Arabidopsis,* and arginine 73 in ABI1 is essential for PA-ABI1 binding. PA binds with ABI1. The PA binding decreases PP2C activity and also appears to reduce its translocation to nuclei in response to ABA by tethering ABI1 to the plasma membrane (Zhang et al. 2004b; Wang 2005; Mishra et al. 2006). Binding of PA to ABI1 protein results in anchorage of ABI1 to the plasma membrane and a decrease in ABI1 PP2C activity. This membrane tethering reduces the movement of ABI1 from the cytosol into the nucleus (Zhang et al. 2004b). It has been shown that there is a functional link between the two signaling enzymes, phospholipase and phosphatase (Zhang et al. 2004b). These results suggest that PA produced by PLDα1 inhibits the function of the negative regulator ABI1, thus promoting ABA signaling. In the ABA signaling pathway, the inhibition of type 2C protein phosphatase PP2C (a negative regulator) results in activation of SNF1-related protein kinase (SnRK2; a positive regulator) (Umezawa et al. 2010). The activated SnRK2s are known to phosphorylate many different downstream targets in the ABA signaling complex (Guo et al. 2011). Inhibition of ABI1 phosphatase leads to SnRK2 activation by autophosphorylation and/or phosphorylation by upstream kinases. Activated SnRK2 in turn phosphorylates and activates the downstream transcription factors called ABFs/AREBs to switch on defense responses (Yoshida et al. 2006; Melcher et al. 2010b; Umezawa et al. 2013).

5.4 ABA Signaling Events Downstream of PYR/PYL/RCAR-PP2C–SNRK2 Signaling Complex

5.4.1 Role of G-Proteins in ABA Downstream Signaling

5.4.1.1 Switching on ROS and Ca^{2+} Signaling Systems

G-proteins act as molecular switches in signal transduction system (Yalowsky et al. 2010; Zhang et al. 2011, 2012). Plant cells contain many G-proteins, which are classified into two classes, such as heterotrimeric G-proteins and small monomeric G-proteins (Ras/Ras-like small GTPases) (Gu et al. 2004; Perfus-Barbeoch et al. 2004; Yong et al. 2010; Zhang et al. 2012). The G-proteins have been shown to be induced by ABA signaling, and they regulate downstream events in the ABA signaling system.

ABA treatment triggered an increase in the expression of *BnGβ1* encoding a putative heterotrimeric G-protein β-subunit (Gβ) in *Brassica napus*. The expression of *BnGβ1* was induced by low ABA concentrations, but high ABA concentrations inhibited the expression, suggesting that there might be an ABA-mediated feedback expression of *BnGβ1* expression (Gao et al. 2010b). In contrast, the expression of *BnGA1* encoding Gα G-protein subunit in *B. napus* was significantly induced by the high concentrations of ABA (Gao et al. 2010a).

The Gα subunit GPA1, the Gβ subunit (AGB1), and the G-protein-coupled receptor (GCR1) regulate the downstream events in the ABA signaling in *Arabidopsis* (Pandey et al. 2006). GCR1 may interact with GPA1 and the ligand binding to GCR1 may regulate heterotrimeric G-protein signaling via GPA1 and the βγ dimer (Eckardt 2004). GPA1 acts upstream of the small GTPase Rac in inducing disease resistance response in rice (Suharsono et al. 2002).

The *Arabidopsis* GPA1 has been demonstrated to be involved in the regulation of Ca^{2+}-permeable channels including inward K^+ channels and S-type anion channels activated by ABA (Wang et al. 2001). *Arabidopsis* mutants deficient in GPA1 were found to be impaired in ABA inhibition of K^+ influx channels and in pH-independent activation of anion efflux channels. The *gpa1* mutants were disrupted both in ABA-induced Ca^{2+}-channel activation and in production of reactive oxygen species (ROS) in response to ABA. These studies further suggested that absence of GPA1 interrupts ABA signaling between ABA reception and ROS production, with a consequent impairment in Ca^{2+}-channel activation (Fig. 5.7; Zhang et al. 2011). These results strongly suggest that the heterotrimeric G-protein is involved in ABA signaling system. The inward K^+ channels and slow anion channels are regulated by Gα-dependent pathways. These channels are activated by ABA, and this activation is lost in *gpa1* mutant lines (Zhang et al. 2011). The results suggest that ABA activation

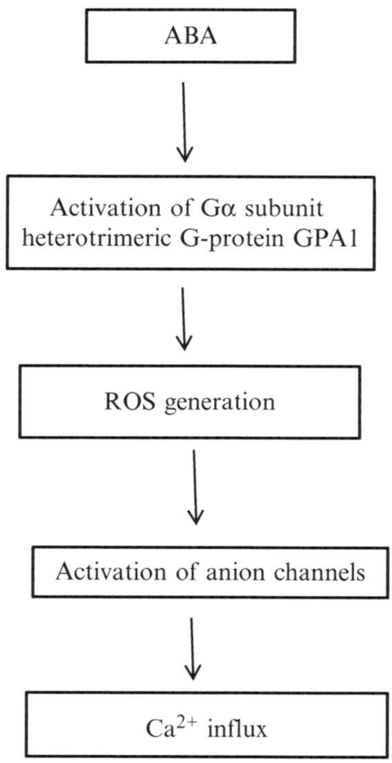

Fig. 5.7 Role of ABA-induced G-protein in activation of ROS and Ca^{2+} signaling systems (Adapted from Wang et al. 2001; Gao et al. 2010a; Zhang et al. 2011)

of these channels is dependent on G-proteins. The early immune signaling events induced by the PAMPs flg22 and elf18 were found to involve Ca^{2+}-associated opening of plasma membrane anion channels (Jeworutzki et al. 2010). The elicitor-induced Ca^{2+} influx is inhibited by different anion-channel blockers (Ward et al. 1995; Ebel and Mithofer 1998). It suggests that anion flux precedes and controls Ca^{2+} influx. Collectively these studies suggest that ABA-activated G-proteins modulate ROS and calcium ion signaling systems.

ABA has been shown to activate plasma membrane Ca^{2+}-permeable channels of *Arabidopsis* guard cells in a GPA1-dependent manner (Hamilton et al. 2000; Pei et al. 2000; Zhang et al. 2011). ABA stimulates ROS production in guard cells and the ROS H_2O_2 activates plasma membrane Ca^{2+}-permeable currents in guard cells (Pei et al. 2000; Zhang et al. 2011). Following ABA application, increasing ROS production was observed in wild-type guard cells, while no significant ROS production was elicited by ABA in *gpa1* mutant cells (Zhang et al. 2011). The results suggest that ROS function downstream of GPA1 in ABA-signaling cascade in activating Ca^{2+}-permeable currents. ABA activates R-type (rapid) and S-type anion channels (Roelfsema and Hedrich 2004). ABA activation of slow anion channels may be dependent on the G-protein GPA1 (Fan et al. 2008). The activated anion channels further intensify Ca^{2+} influx which will pave the way for triggering Ca^{2+}-mediated signaling system. K^+ channels also play important role in modulation of Ca^{2+} signaling system (Maathuis et al. 1997). These studies suggest that the G-proteins switch on both ROS and Ca^{2+} signaling systems downstream of ABA signaling system.

5.4.1.2 Modulation of ABA-Induced Stomatal Closure Immune Response by G-Proteins

Stomata are entry gates for pathogens, and the PAMP-mediated regulation of stomatal apertures is an important immune response particularly against bacterial pathogens (Melotto et al. 2006; Jammes et al. 2011; Uraji et al. 2012). Various PAMPs have been shown to trigger stomatal closure or prevent light-induced stomatal opening (Zeng et al. 2010). Arabidopsis recognizes the invasion through stomata by the bacterial pathogen *Pseudomonas syringae* pv. *tomato* and triggers ABA signaling leading to stomatal closure to prevent further bacterial infection (Melotto et al. 2006). ABA is known to be involved in stomatal pore closure which is a key component in plant immune responses against bacterial pathogens (Takahashi et al. 2007; Hettenhausen et al. 2012; Uraji et al. 2012). G-proteins may be involved in the modulation of ABA-induced stomatal closure immune response. The *Arabidopsis* small GTPase protein AtRac1 has been shown as a central component in the ABA-mediated stomatal closure (Lemichez et al. 2001). ABA treatment induced inactivation of AtRac GTPases, and this ABA-treatment-induced AtRac inactivation was not observed in the *Arabidopsis* ABA-insensitive mutant *abi1-1*, which is impaired in stomatal closure. Expression of a dominant-positive mutant of *AtRac1* blocked the ABA-mediated stomatal closure in wild-type plants, whereas expression of a dominant-negative *AtRac1* mutant recapitulated the ABA effects in the absence of the hormone. It was also observed that the dominant-negative form of *AtRac1* could restore stomatal closure in the *abi-1* mutant (Lemichez et al. 2001). These results suggest that a small monomeric G-protein plays a key role in the stomatal closure immune response induced by ABA.

5.4.2 Role of Ca^{2+} Signaling System in ABA Downstream Signaling

5.4.2.1 ABA Modulates Ca^{2+} Influx-Mediated Calcium Signature

One of the earliest detectable signaling events after ABA treatment is a rapid and transient rise in cytosolic Ca^{2+} levels through the function of plasma membrane-resident Ca^{2+} channels (Schroeder and Hagiwara 1990; Staxén et al. 1999; Hamilton et al. 2000; Pei et al. 2000; Klüsener et al. 2002; Kwak et al. 2003; Brault et al. 2004). In plant cells, the calcium ion is a ubiquitous intracellular second messenger involved in numerous signaling pathways (DeFalco et al. 2010; Dodd et al. 2010; Hamada et al. 2012; Stael et al. 2012). Ca^{2+} is a master regulator of gene expression in plants (Galon et al. 2010a, b). Calcium signaling is modulated by specific calcium signatures. Spatial and temporal changes in cytosolic calcium ($[Ca^{2+}]_{cyt}$) such as single calcium transients, oscillations, or repeated spikes/waves are referred to as "calcium signature" (Luan et al. 2002; Lecourieux et al. 2006). Ca^{2+} signatures are generated in the cytosol and in noncytosolic locations including the nucleus and chloroplast through the coordinated action of Ca^{2+} influx and efflux pathways (McAinsh and Pittman 2009). Specific calcium signatures are recognized by different calcium sensors to transduce calcium-mediated signals into downstream events (Reddy et al. 2011a, b; Wang et al. 2012; Hashimoto et al. 2012).

Several studies have revealed that Ca^{2+} signaling is activated by ABA. ABA produced transient repetitive increases in the cytosolic-free Ca^{2+} concentration in *Vicia faba* cells (Schroeder and Hagiwara 1990; Kwak et al. 2003). Free cytosolic Ca^{2+} concentration ($[Ca^{2+}]_{cyt}$) regulates plasma membrane and vacuolar ion channels (Hubbard et al. 2012). The Ca^{2+} transients were accompanied by concomitantly occurring increases in an inward-directed ion current. Depolarization of the membrane terminated both repetitive elevations in cytosolic Ca^{2+} and inward-directed ion currents, suggesting that ABA-mediated Ca^{2+} transients were produced by passive influx of Ca^{2+} from the extracellular space through Ca^{2+}-permeable channels (Schroeder and Hagiwara 1990). ABA may activate hyperpolarization-activated Ca^{2+}-permeable channels. In *Vicia faba* cells, intracellular ABA transiently enhanced the activity of hyperpolarization-activated Ca^{2+} channels (Hamilton et al. 2000). In *Arabidopsis*, ABA induced hyperpolarization-activated Ca^{2+}-permeable channels (Pei et al. 2000). ROS production precedes the activation of Ca^{2+} influx, and H_2O_2 activates plasma membrane Ca^{2+} channels (Coelho et al. 2002).

ABA triggered a rapid plasma membrane depolarization in *Arabidopsis* suspension cells. The ABA-induced depolarization was suggested to be due to two different processes: activation of anion channels and inhibition of H^+-ATPases. Both these processes were shown to be dependent on the $[Ca^{2+}]_{cyt}$ increase induced by ABA (Brault et al. 2004). Cytosolic Ca^{2+} elevation induced by ABA activates slow (S-type) anion channels (Schroeder and Hagiwara 1989; Brault et al. 2004). ABA activates the plasma membrane anion channels, and in several species, this response is associated with changes in the cytoplasmic Ca^{2+} concentration (Marten et al. 2006). However, in *Vicia faba*, anion cells are activated by ABA in a Ca^{2+}-independent manner. In *Nicotiana tabacum*, ABA could activate anion channels independent of a rise in Ca^{2+} concentration, but activated by elevated Ca^{2+} in an ATP-dependent manner (Marten et al. 2007). ABA down-regulates inward K^+ channels (Schroeder and Hagiwara 1989).

ABA induces oscillations in $[Ca^{2+}]_{cyt}$ by inducing both Ca^{2+} release from intracellular stores and Ca^{2+} influx from the extracellular space (Staxén et al. 1999; Hamilton et al. 2000; Pei et al. 2000; Klüsener et al. 2002). ABA-induced Ca^{2+} release from internal stores is mediated by inositol 1,4,5-trisphosphate in *V. faba* (Lee et al. 1996). Phosphoinositide-specific phospholipase C is involved in ABA-induced oscillations in cytosolic-free calcium (Staxén et al. 1999). ABA-induced Ca^{2+} release into cytosol is also mediated by cyclic adenosine 5′-diphosphoribose (cADPR) (Leckie et al. 1998).

5.4.2.2 CDPKs Are Involved in ABA Downstream Signaling

The calcium signature is perceived by different Ca^{2+}-binding proteins, which are also known as Ca^{2+} sensors (Kudla et al. 2010). The changes in $[Ca^{2+}]_{cyt}$ concentrations are monitored by the Ca^{2+} sensors, and the Ca^{2+} signals are subsequently decoded and propagated downstream to activate plant defense responses (Snedden and Fromm 2001). The Ca^{2+}-dependent protein kinases (CDPKs) are the important Ca^{2+} sensors (Luan et al. 2002; Geiger et al. 2010). CDPKs function as sensors of fluctuations in cytosolic Ca^{2+} and initiate downstream signaling events (Harmon et al. 2000; Hrabak et al. 2003; Kobayashi et al. 2007; Ito et al. 2010).

ABA is known to up-regulate the expression of CDPKs. The expression of various CDPKs of tobacco was up-regulated by ABA (Ludwig et al. 2004). *CaCDPK3* gene encoding the *Capsicum annuum* calcium-dependent protein kinase 3 was induced by ABA in pepper (Chung et al. 2004). Several CDPKs have been shown to be involved in ABA signaling (Sheen 1996; Romeis et al. 2001; Choi et al. 2005). Two *Arabidopsis* CDPKs, AtCPK10, and AtCPK30 activate an ABA-inducible promoter in maize leaf protoplasts (Sheen 1996). Another CDPK in *Arabidopsis*, AtCPK32, interacts with C2–C3 conserved region of ABF4 (Choi et al. 2005). AtCPK32 has autophosphorylation activity and phosphorylates ABF4 in vitro. The CDPK has been shown to be an ABA signaling component that positively modulates ABF4 function (Choi et al. 2005).

Two calcium-dependent protein kinases, CPK4 and CPK11, regulate ABA signal transduction in *Arabidopsis* (Zhu et al. 2007). The CPK4 and CPK11 kinases both phosphorylated two ABA-responsive transcription factors, ABF1 and ABF4, in vitro, suggesting that the two kinases may regulate ABA signaling through these transcription factors (Zhu et al. 2007).

ABA is involved in the activation of CDPKs to trigger the downstream events in the plant immune response. A grape berry 58-kDa ABA-stimulated calcium-dependent protein kinase, designated ACPK1 (for ABA-stimulated calcium-dependent protein kinase1), has been isolated from grape berry (*Vitis vinifera* × *V. labrusca*). Expression of the grape CDPK, ACPK1, in *Arabidopsis thaliana* activates ABA downstream signaling (Yu et al. 2007). Analysis of transcriptome changes in *Arabidopsis* revealed 230 calcium-responsive genes, of which 162 were up-regulated and 68 were down-regulated. Analysis of upstreams revealed, exclusively in the up-regulated genes, a highly significant occurrence of a consensus sequence comprising two ABA-specific *cis* elements: the ABA-responsive element (ABRE, CACGTG [T/C/G]) and its coupling element ([C/A] ACGCG[T/C/A]). A tetramer of the ABRE *cis* element was sufficient to confer transcriptional activation in response to cytosolic Ca^{2+} transients (Yu et al. 2007). Thus, for some specific Ca^{2+} transients and motif combinations, ABREs function as Ca^{2+}-responsive *cis* elements (Kaplan et al. 2006; Yu et al. 2007). The link between Ca^{2+}-responsive transcription factors with ABA-responsive *cis*-elements suggests a link between Ca^{2+} signaling and ABA signaling systems. Phenylalanine ammonia lyase (PAL), the key enzyme in phenylpropanoid pathway involved in biosynthesis of phytoalexins, has been shown to be a substrate of a specific constitutively active *Arabidopsis* CDPK expressed in corn protoplasts (Cheng et al. 2001).

Interaction between Ca^{2+} signaling and ABA signaling systems was further demonstrated by showing that CDPKs regulate ABA signal transduction in *Arabidopsis* (Mori et al. 2006; Zhu et al. 2007; Geiger et al. 2010). ABA stimulated two homologous CDPKs (CPK4 and CPK11) in *A. thaliana*. Loss-of-function mutations of *CPK4* and *CPK11* resulted in ABA insensitive phenotypes. The CPK4 and CPK11 kinases both phosphorylated two ABA-responsive transcription factors, ABF1 and ABF4, in vitro, suggesting that the two kinases may regulate ABA signaling through these transcription factors (Zhu et al. 2007). These results suggest that CDPK may regulate ABA signaling pathway.

A calcineurin B-like (CBL) protein kinase CIPK15 interacts with the calcium-modulated protein phosphatases ABI1 and ABI2 and a CBL Ca^{2+}-binding protein ScaBP5. CIPK15 and one of its homologs CIPK2 are involved in ABA signaling as negative regulators (Guo et al. 2002; Kim et al. 2003). The kinase substrate of CIPK15 has been identified as an AP2 transcription factor AtERT7 that negatively regulates ABA signaling (Song et al. 2005).

5.4.3 ABA Activates ROS Signaling System Downstream of ABA Signaling System

ABA has been shown to induce the production of ROS (Pei et al. 2000; An et al. 2008; Hung et al. 2008; Sakamoto et al. 2008; Wang and Song 2008; Zhang et al. 2011; Munemasa et al. 2013). H_2O_2 is synthesized in response to exogenous ABA (Guan et al. 2000; Pei et al. 2000; Gayatri et al. 2013). *AtrbohD* and *AtrbohF* genes encoding NADPH oxidase have been shown to function in ROS-dependent ABA signaling in *Arabidopsis* (Kwak et al. 2003; Wang et al. 2006a, b). Calcium-permeable channels are activated by ROS, and ABA signaling network is involved in the ROS-induced activation of calcium-permeable channels (Pei et al. 2000; Murata et al. 2001; Trouverie et al. 2008; Zhang et al. 2011). Activation of these channels requires the presence of NAD(P)H in the cytosol. Type 2C protein phosphatases (PP2Cs) also play a role in the ABA-induced ROS-dependent activation of calcium channels (Murata et al. 2001). Anion channel activation and proton pumping inhibition involved in the plasma membrane depolarization induced by ABA in *A. thaliana* suspension cells were shown to be ROS dependent (Trouverie et al. 2008). Abscisic acid (ABA) promotes ROS production, activates plasma membrane Ca^{2+}-permeable channels, and triggers cytosolic Ca^{2+} increases in *Arabidopsis thaliana*. Disruption of NADPH oxidase catalytic subunits genes, *AtrbohD* and *AtrbohF*, has been shown to impair ABA signaling, ABA promotion of ROS production, ABA-induced cytosolic Ca^{2+} increases, and ABA activation of plasma membrane Ca^{2+}-permeable channels in guard cells (Kwak et al. 2003).

5.4.4 Nitric Oxide (NO) Acts Downstream of H_2O_2 in ABA Signaling System

ABA treatment triggers a marked increase of NO in guard cells of *Arabidopsis*, *Vicia*, and pea (Gayatri et al. 2013). NO has been shown to act downstream of H_2O_2 in ABA signaling system (Lü et al. 2005; Gayatri et al. 2013). ABA-mediated NO generation has been shown to be dependent on ABA-induced H_2O_2 production in *Vicia faba* and *Arabidopsis* (Lü et al. 2005; Bright et al. 2006). Nitrate reductase, but not nitric oxide synthase (NOS), was involved in the ABA-mediated NO production and stomatal closure (Bright et al. 2006). The existence of true NOS in plant tissues and its role in NO production are still a matter of intense debate (Gayatri et al. 2013). However, Yan et al. (2007) showed that ABA-induced NO is synthesized by a NOS-like enzyme in *V. faba*. NOS-like enzyme and nitrate reductase have been shown to be involved in NO production in guard cells (Gayatri et al. 2013). ABA-induced NO is involved in ABA-induced stomatal closure; stomatal closure is part of plant immune response to restrict bacterial invasion (Melotto et al. 2006). It is suggested that both H_2O_2 and NO are synthesized in parallel, in response to ABA (Wang and Song 2008).

5.4.5 MAP Kinases Function Downstream of ABA Signaling System

It has been widely reported that ABA activates mitogen-activated protein kinase (MAPK)-mediated signaling system (Gomi et al. 2005; Wang and Song 2008). An *Arabidopsis* MAPK, AtMAPK3, and a rice MAPK, OsMAPK5, have been reported as ABA-activated MAPKs (Lu et al. 2002; Xiong and Yang 2003). It has been found that ABA and H_2O_2 can activate the same MAPK (Desikan et al. 2004). ROS may act upstream of the MAPK cascade in the ABA signaling system in maize leaves (Zhang et al. 2006). ABA-induced activation of MAPK was almost fully arrested by pretreatment with inhibitors of ROS production, suggesting that ABA-induced H_2O_2 production activates MAPK (Zhang et al. 2006).

Two MAP kinases, *MPK9* and *MPK12*, are highly expressed in guard cells and function as positive regulators of ROS-mediated ABA signaling in guard cells (Jammes et al. 2011). It has been shown that *MPK9* and *MPK12* act downstream of ROS and cytosolic Ca^{2+} and upstream of anion channels in the guard cell ABA signaling cascade (Jammes et al. 2011). The MAP kinases MPK9 and MPK12 have been shown to be preferentially expressed in stomata guard cells and positively regulate ROS-mediated ABA signaling (Jammes et al. 2009). The MAPKs act upstream of anion channels in guard cell ABA signaling. MPK9 and MPK12 function downstream of ROS to regulate ABA signaling (Jammes et al. 2009). ABA-induced stomatal closure plays an important role in bacterial disease resistance. The *mpk9/mpk12* double mutants have been shown to be highly susceptible to *P. syringae* pv. *tomato* DC3000 compared to wild-type *Arabidopsis thaliana* plants (Jammes et al. 2011). The results suggest that the regulation of stomatal apertures by MPK9 and MPK12 contributes to the first line of defense against pathogens.

5.4.6 ABA Regulates the Expression of Several Transcription Factors

ABA regulates the expression of several genes (Choi et al. 2005). The *cis*-regulatory elements responsible for the ABA regulation of gene expression share a conserved motif, ACGTGGC, which is known as ABA-responsive element (ABRE). ABRE characterized by ACGTGGC (Kaplan et al. 2006) appears in the promoters of many of the defense genes (Adie et al. 2007). The ABA-induced gene regulation is mediated by a subfamily of basic leucine zipper class transcription factors referred to as ABA response element (ABRE) binding factors (ABFs; also referred to as AREBs) (Choi et al. 2005; Furihata et al. 2006). A large number of basic leucine zipper-type (bZIP) class transcription factors have been isolated from *Arabidopsis thaliana* based on their in vitro interaction with the ABRE. The ABF subfamily members, ABF1, ABF2 (AREB1), ABF3, ABF4 (AREB2), and ABI5, and their homologs have nearly identical bZIP domains in their C-terminal regions and three highly conserved regions in their N-terminal halves. The N-terminal-conserved regions invariably possess one to two Ser or Thr residues, which are potential substrate sites for various protein kinases (Choi et al. 2005). Orthologs of AREB (ABF) subfamily transcription factors have been reported in rice (TRABI) (Hobo et al. 1999), barley (HvABI5) (Casaretto and Ho 2003), wheat (TaABF) (Johnson et al. 2002), and sunflower (DPBF1, DPBF2, DPBF3) (Kim et al. 1997; Kim and Thomas 1998).

bZIP-type transcription factors AREBs/ABFs bind ABRE and transactivate downstream gene expression in *Arabidopsis* (Furihata et al. 2006). Overexpression of AREB1 could not induce downstream gene expression, and it has been shown that activation of AREB1 requires ABA-dependent posttranscriptional modification. ABA activated 42-kDa kinase activity, which in turn, phosphorylated Ser/Thr residues of R-X-X-S/T sites in the conserved regions of AREB1. Substitution of the Ser/Thr residues to Asp resulted in high transactivation activity without exogenous ABA application. A phosphorylated, transcriptionally active form was achieved by the substitution of Ser/Thr in all conserved R-X-X-S/T sites to Asp. Transgenic plants overexpressing the phosphorylated active form of AREB1 expressed many ABA-inducible genes. These results suggest that the ABA-dependent multisite phosphorylation of AREB1 regulates its own activation in plants (Furihata et al. 2006).

AtMYC2 and AtMYB2 are transcriptional activators of ABA-inducible gene expression. Expression of several ABA-inducible genes was up-regulated in *Arabidopsis* plants overexpressing *AtMYC2* and/or *AtMYB2* cDNAs (Abe et al. 2003). AtMYC2 is a positive regulator of ABA signaling (Anderson et al. 2004). The *Arabidopsis* MYB-related protein BOTRYTIS SUSCEPTIBLE1 (BOS1) shows high sequence similarity to AtMYB2. The promoter of BOS1 contained ABA-responsive elements. BOS1 induces the expression of ABA-inducible genes (Mengiste et al. 2003).

Functional analyses of the rice WRKY gene superfamily reveal positive and negative regulators of abscisic acid signaling (Xie et al. 2005). There are at least four OsWRKY genes involved in ABA signaling, with two functioning as positive and two as negative regulators (Xie et al. 2005). A WRKY gene, *LtWRKY21*, from creosote bush (*Larrea tridentata*) encodes an activator of the ABA signaling pathway (Zou et al. 2004). The LtWRKY21 protein is localized in the nucleus and activates the promoter of an ABA-inducible gene, HVA22. The transactivating activity of LtWRKY21 relies on the C-terminal sequence containing the WRKY domain and a N-terminal motif that is essential for the repression activity of some regulators in ethylene signaling. It is suggested that LtWRKY21 may function downstream of ABI1 to control ABA-regulated expression of genes (Zou et al. 2004).

5.5 Systemic Movement of ABA and Intercellular ABA Signaling Pathway

5.5.1 AtABCG25 Is Involved in the Intercellular Transport of ABA in ABA Signaling Pathway

ABA is predominantly biosynthesized and metabolized in vascular tissues but acts in the stomatal responses of distant guard cells (Wilkinson and Davies 2010). Under stress conditions, ABA concentrations increase in the apoplast, leading to stomatal closure in the leaves (Schachtman and Goodger 2008; Wilkinson and Davies 2010).

Systemic changes in gene expression in response to ABA have also been reported (Christmann et al. 2007; Endo et al. 2008). These results suggest that ABA synthesized in vasculature is transported to distant tissues. Recently, molecular basis of ABA transport has been studied. One of the ATP-binding cassette (ABC) transporter genes, AtABCG25, was found to encode a protein responsible for ABA transport and response in *Arabidopsis* (Kuromori et al. 2010). The AtABCG25 was expressed mainly in vascular tissues, where ABA is predominantly biosynthesized. The protein was found to be localized at the plasma membrane in plant cells (Kuromori et al. 2010). The results suggest that AtABCG25 is an exporter of ABA through the plasma membrane and is involved in the intercellular ABA signaling pathway (Umezawa et al. 2010).

5.5.2 AtABCG40 Is Involved in Intercellular Transport of ABA in ABA Signaling Pathway

AtABCG40 is another ABC transporter in *Arabidopsis*, which functions as an ABC transporter in plant cells (Kang et al. 2010). AtABCG40 is expressed in the leaves and the expression is very high in guard cells (Kang et al. 2010). It was found to be localized in plasma membrane in guard cells. In loss of function *atabcg40* mutants, the stomata closed more slowly in response to ABA. In response to exogenous ABA, the up-regulation of ABA-inducible genes was strongly delayed in *atabcg40* plants (Kang et al. 2010). These results suggest that AtABCG40 is an importer of ABA through plasma membranes and integrates ABA-dependent signaling and transport processes. It is also proposed that ABA is exported from ABA-biosynthesizing cells to the apoplastic area and then imported from the apoplast into guard cells through the action of ABA transport proteins (Kang et al. 2010; Kuromori et al. 2010; Umezawa et al. 2010).

5.6 Interplay Between ABA and JA Signaling Systems

5.6.1 ABA Signaling and JA Signaling Pathways May Be Interconnected

Interplay of ABA with JA signaling system has been widely reported (Broekaert et al. 2006; Lackman et al. 2011; Sánchez-Vallet et al. 2012). Both the ABA and JA signaling pathways may be interconnected. A tobacco gene from the PYR/PYL/RCAR family, *NtPYL4*, the expression of which is regulated by jasmonates, was found to encode a functional ABA receptor. NtPYL4 inhibited the type 2C protein phosphatases known to be key negative regulators of ABA signaling in an ABA-dependent manner (Lackman et al. 2011). Altered *PYL* expression affected the JA response (Lackman et al. 2011). These results suggest a connection between a component of the core ABA signaling pathway and the JA responses. JA signaling may be integrated into early ABA signaling and may affect ABA receptor complexes to regulate downstream common signal components in stomatal closure (Munemasa et al. 2011). *Botrytis susceptible1* (*BOS1*) gene is a MYB transcription factor. The expression of (*BOS1*) is induced by *B. cinerea* via the jasmonic acid pathway, and its promoter contained ABA-responsive elements (Mengiste et al. 2003).

ABA has been shown to play an important role in JA biosynthesis. JA and its precursor 12-oxo-phytodienoic acid (OPDA) levels increased following infection with *Pythium irregulare* both in wild-type and in ABA biosynthesis mutant (*aba2-12*) *Arabidopsis* plants (Fig. 5.8; Adie et al. 2007). *ABA2* encodes a short-chain alcohol dehydrogenase, which catalyzes the conversion of xanthoxin to ABA-aldehyde, which is used by another enzyme AAO3 and cofactor ABA3 to produce ABA (see Fig. 5.1; Hauser et al. 2011). The increase in JA and OPDA levels following infection was much lower in *aba2-12* mutants than in wild-type plants (Fig. 5.8; Adie et al. 2007). The results indicate that ABA synthesis is required for JA production. *Pseudomonas syringae* pv. *tomato* infection-induced JA accumulation was reduced in ABA-deficient *aba3-1* mutant plants compared to the wild-type (Fan et al. 2009). The results suggest that ABA promotes JA accumulation in *Arabidopsis*.

It has also been shown that JA stimulates ABA biosynthesis (Kim et al. 2009; Yoon et al. 2009). MeJA stimulates the expression of *9-cis-epoxycarotenoid dioxygense3*, which encodes a crucial enzyme in ABA biosynthesis (Hossain et al. 2011). In the *aba2-2* mutant, MeJA did not induce JA-inducible *VSP1* (*Vegetative Storage Protein1*) gene expression. In wild-type leaves, an inhibitor of ABA biosynthesis,

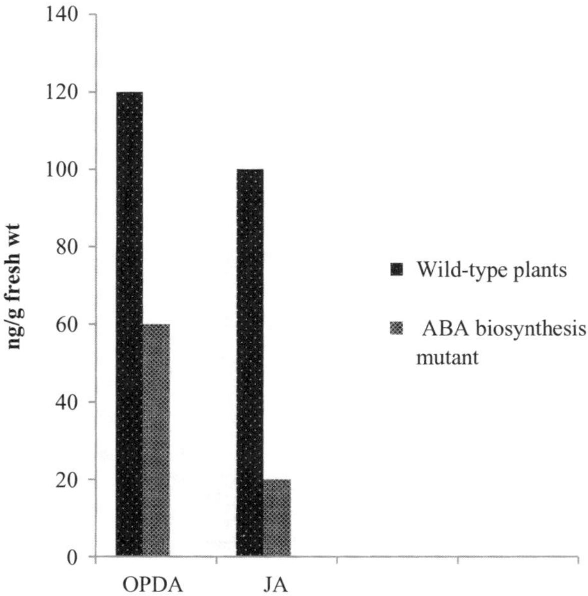

Fig. 5.8 Changes in JA and OPDA levels in wild-type and ABA biosynthesis-mutant (*aba2-12*) *Arabidopsis* plants 12 h after infection with *Pythium irregulare* (Adapted from Adie et al. 2007)

fluridon, inhibited *VSP1* expression (Hossain et al. 2011). These results suggest that endogenous ABA may be involved in JA biosynthesis and signal transduction (Hossain et al. 2011).

Many signaling components are commonly involved in both ABA and JA signaling systems. A regulatory A-subunit of protein phosphatase type 2A (PP2A), RCN1, is involved in both JA and ABA signaling upstream of ROS and NO production (Saito et al. 2008, 2009). Myrosinases, TGG1 and TGG2, function downstream of ROS production in guard cell JA and ABA signaling (Islam et al. 2009). CPK6 is required for the activation of I_{ca2} channels and S-type anion channels in both JA and ABA signaling (Munemasa et al. 2011).

5.6.2 ABA and JA May Act Cooperatively in the Induction of Defense Genes

ABA has also been shown to be involved in the cooperative activation of JA-regulated defense genes in *Arabidopsis*. *Pythium irregulare* infection regulated 38 ABA-dependent genes in wild-type plants. Two major classes of genes, ABA-regulated genes and JA-regulated genes, could be identified within this ABA-dependent group of genes in the wild-type plants infected with *P. irregulare*. About half of the ABA-regulated genes were also regulated by JA. In contrast, there was deficiency in the activation of JA-induced genes in the ABA-biosynthesis mutant *aba2-12* infected with the pathogen (Adie et al. 2007). ABA levels increased rapidly in the wild-type plants, and not in *aba2-12* mutants, when infected with *P. irregulare* (Adie et al. 2007).

5.6.3 ABA May Suppress JA-Activated Defense Responses

Antagonism between ABA and JA signaling systems has also been reported. Antagonistic interaction between ABA and JA signaling pathways modulates defense gene expression and disease resistance in *Arabidopsis*. ABA suppressed JA-activated transcription of defense genes in *Arabidopsis* (Anderson et al. 2004). ABA deficiency resulted in up-regulation of transcription of JA-responsive defense genes. ABA is a negative regulator of JA-responsive defense genes (Anderson et al. 2004). AtMYC2 is a positive regulator of ABA signaling. Disruption of *AtMYC2* resulted in elevated levels of transcription from JA-responsive defense genes in *Arabidopsis*. These results suggest the antagonistic interaction between ABA and JA signaling pathways (Anderson et al. 2004). Mutations in the *ABA1* and *ABA2* genes, which encode enzymes involved in ABA biosynthesis, resulted in increased transcription of JA-responsive defense genes in *Arabidopsis* (Anderson et al. 2004).

AtMYC2 and AtMYB2 are transcriptional activators of ABA-inducible gene expression (Abe et al. 2003). The expression of several ABA-inducible genes was up-regulated in the plants overexpressing *AtMYC2* or *AtMYB2*. The ABA-responsive gene *rd22* contains MYC- and MYB-recognition sites in its promoter region that bind to AtMYC2 and AtMYB2 (Abe et al. 2003). AtMYC2 is a negative regulator of JA-regulated defense genes, such as *PDF1.2* and chitinase (Lorenzo et al. 2004).

5.6.4 Role of Mediator Subunit MED25 in ABA and JA Signaling Interplay

Transcription factors perceive the signals relayed by signaling molecules in the plant immune system and translate into a functional response through recruiting or releasing RNA polymerase II (RNAPII) (Moore et al. 2011; An and Mou 2013). Mediator is a multiprotein cofactor of RNAPII, and it acts as a bridge between RNAPII and DNA binding transcription factors (Kidd et al. 2011; Cevik et al. 2012). The multisubunit Mediator complex acts as bridge between gene-specific transcription factors and the RNA polymerase machinery to regulate transcription (Chen et al. 2012; An and Mou 2013).

MEDIATOR25 (MED25) subunit of the Mediator complex encoded by the PHYTOCHROME AND FLOWERING TIME1 gene (*PFT1*) has been reported to regulate ABA and JA-triggered gene transcription (Chen et al. 2012). MED25 interacts with several transcription factors (MYC2, MYC3, MYC4, ERF1, ERF59 [ORA59], TDR1/ERF98, and ERF15) functioning in the JA signaling system (Cevik et al. 2012). MED25 physically associates with MYC2 in promoter regions of MYC2 target genes and exerts a positive effect on MYC2-regulated gene transcription in JA signaling system (Cevik et al. 2012; Chen et al. 2012). It has also been reported that MED 25 physically associates with ABI5 (ABA-Insensitive5) in promoter regions of ABI5 target genes and shows a negative effect on ABI5-regulated gene transcription in ABA signaling system (Chen et al. 2012). The conserved activator-interacting domain (ACID) of MED25 was found to be sufficient for its interaction with ABI5 but not sufficient for its interaction with MYC2. The putative transcriptional activation domain (TAD) of MYC2 was involved in its interaction with MED25; the TAD domain of ABI5 was not involved in its interaction with MED25 (Chen et al. 2012). MED25 acts genetically downstream of MYC2 in JA signaling, whereas MED25 acts genetically downstream of ABI5 in ABA signaling. In the absence of ABA, ABI5 accumulation at its promoter site is low, whereas MED25 accumulation at the same promoter region is high (Chen et al. 2012).

MED25 has been suggested to be an interaction node between ABA and JA signaling systems. It positively regulates JA signaling, whereas it negatively regulates ABA signaling. Both antagonism and synergism between ABA and JA signaling have been reported (Chen et al. 2012). Both JA and ABA induce the expression of MYC2. Induction of MYC2 by ABA depends on the JA receptor COI1 and ABA activates the expression of MYC2 by the JA signaling pathway (Lorenzo et al. 2004). *Arabidopsis* mutants defective in ABA biosynthesis or signaling show increased resistance to *Fusarium oxysporum* (Anderson et al. 2004). In contrast, ABA activates JA biosynthesis and induces resistance of *Arabidopsis* plants to *Pythium irregulare* (Adie et al. 2007).

5.7 Interplay Between ABA and SA Signaling Systems

5.7.1 ABA May Suppress SA Biosynthesis

ABA has been shown to suppress SA signaling system by suppressing SA biosynthesis. ABA suppresses SA accumulation in *Arabidopsis* infected with *Pseudomonas syringae* pv. *tomato* (Mohr and Cahill 2007). Phenylalanine ammonia lyase (PAL) is a key enzyme catalyzing the first committed step of the phenylpropanoid biosynthetic pathway, leading to the production of SA (Kim and Hwang 2014). ABA suppresses PAL activity and reduces SA accumulation in *Arabidopsis* (Mohr and Cahill 2007; Fig. 5.9). Liu et al. (2012) found that PAL expression was elevated and further increased by inoculation with *Magnaporthe oryzae* in *OsERF922* RNAi rice plants coincident with the low level of ABA accumulation in those plants. In contrast, the pathogen-inducible PAL expression was suppressed in the *OsERF922* overexpression lines. OsERF922-overexpressing rice lines show increased accumulation of ABA. These results suggest that increased ABA biosynthesis suppresses PAL activity and SA biosynthesis. The AvrPtoB effector of *P. syringae* pv. *tomato* DC3000 induces the accumulation of ABA

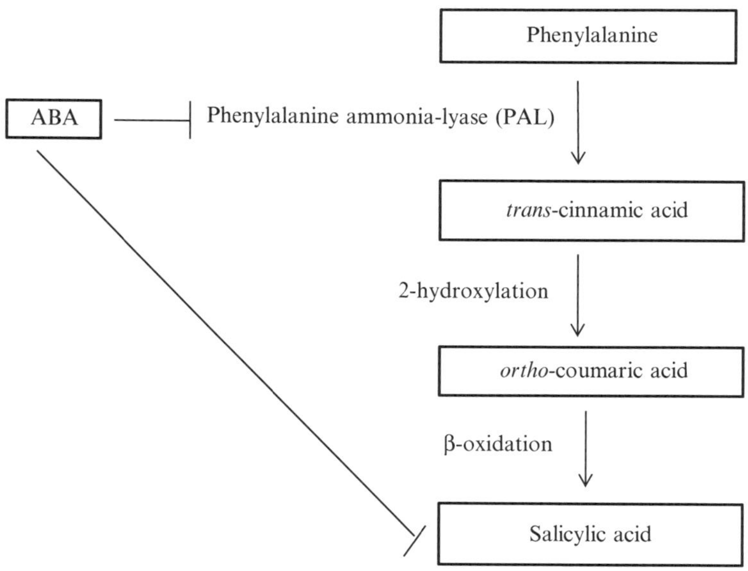

Fig. 5.9 ABA reduces SA accumulation by suppressing phenylalanine ammonia-lyase (PAL) activity (Adapted from Mohr and Cahill 2007)

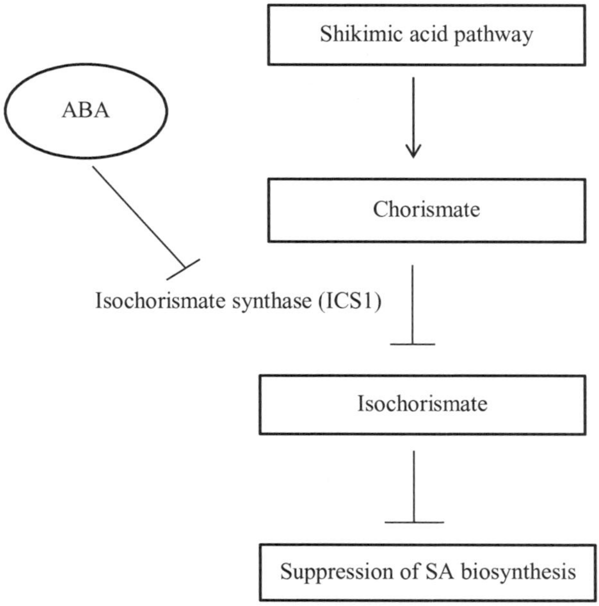

Fig. 5.10 Suppression of SA biosynthesis isochorismate pathway by ABA in *Arabidopsis thaliana* (Adapted from Yasuda et al. 2008)

in *Arabidopsis* to suppress the induction of basal defense by down-regulating SA biosynthesis and signaling pathways (de Torres et al. 2006).

Treatment of *Arabidopsis* with ABA blocked the activation of isochorismate synthase (ICS) (Yasuda et al. 2008). ICS is involved in the conversion of chorismate to isochorismate, which in turn leads to the SA biosynthesis in the isochorismate pathway (Wildermuth et al. 2001). The results suggest that ABA may suppress isochorismate pathway resulting in the suppression of SA accumulation (Fig. 5.10; Yasuda et al. 2008). ABA suppresses inducible innate immune responses by down-regulating SA biosynthesis and SA-mediated defenses (de Torres et al. 2006).

5.7.2 Suppression of SA Signaling System by ABA

ABA negatively regulates SA-dependent immune responses, which are essential for *Arabidopsis* resistance to the necrotrophic pathogen *Plectosphaerella cucumerina* (Sánchez-Vallet et al. 2012). Exogenous application of ABA to *Arabidopsis* resulted in an enhanced susceptibility to *Pseudomonas syringae* pv. *tomato* DC3000 that was correlated with reduced SA accumulation, inhibition of lignin biosynthesis, and suppression of several defense-related genes (Anderson et al. 2004; Mohr and Cahill 2007). Treatment of *Arabidopsis* with ABA blocked the activation of systemic acquired resistance both upstream and downstream of SA biosynthesis (Yasuda et al. 2008). Abscisic acid suppresses salicylic acid-dependent signaling mechanisms (Audenaert et al. 2002). The ABA-deficient mutant of tomato had greater

SA-mediated responses, suggesting the antagonistic interaction between ABA and SA signaling systems (Thaler and Bostock 2004). ABA reduced SA defense gene expression in *Arabidopsis* (Adie et al. 2007). ABA inhibits the accumulation of SA and the expression of SA-mediated defense genes (Mohr and Cahill 2007; Spoel and Dong 2008). ABA suppresses inducible innate immune responses by down-regulating SA biosynthesis and SA-mediated defenses in *Arabidopsis* (de Torres et al. 2006). ABA has been shown to promote susceptibility to the rice bacterial blight pathogen *Xanthomonas oryzae* pv. *oryzae* by suppressing SA-mediated defenses (Xu et al. 2013).

5.7.3 Reciprocal Antagonistic Interaction Between ABA and SA Signaling Systems

Interplay between ABA and SA signaling systems is highly complex. ABA signaling system suppresses SA signaling system and vice versa (Fig. 5.11; Alazem et al. 2014). It has been shown that SA signaling system rapidly disrupts ABA signaling transduction (Kim et al. 2011). *Nicotiana benthamiana* infected with *Bamboo mosaic virus* (BAMV) or *Cucumber mosaic virus* (CMV) showed increased ABA contents, and the infection activated the SA and ABA pathways (Alazem et al.

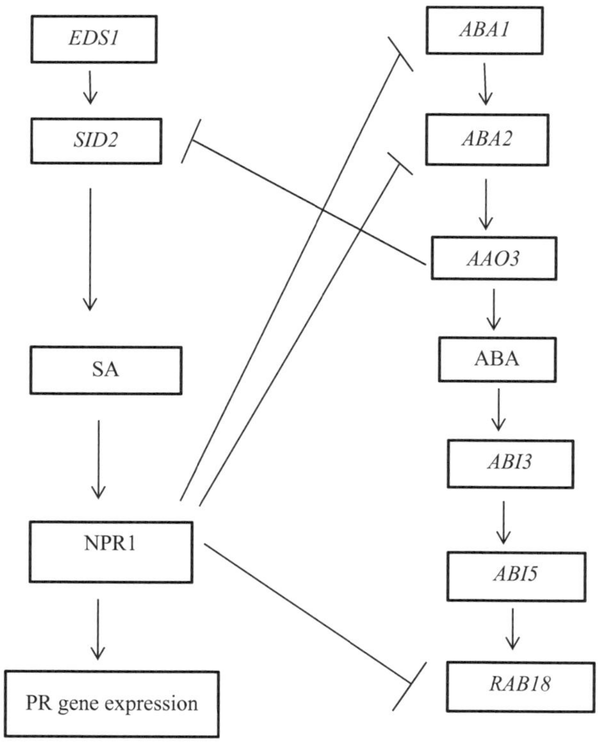

Fig. 5.11 Interplay between ABA and SA signaling pathways in *Nicotiana benthamiana* (Adapted from Alazem et al. 2014)

2014). Infection with CMV or BaMV induced ABA accumulation and up-regulated the mRNA levels of several genes (*EDS1*, *SID2*, *NPR1*, *PR* genes) in the SA and (*ABA1, ABA2, AAO3, ABI3, ABI5, RAB18* genes) in the ABA pathways (Alazem et al. 2014). *NPR1* (for *nonexpresser of PR gene1*; also known as *NIM1* [*nonimmunity1*] and *SAI1* [*salicylic acid inducible1*]) gene is a master regulator of the SA-mediated induction of defense genes. NPR1 is an important regulator of responses downstream of SA (Mou et al. 2003; Zhang et al. 2003; Chern et al. 2008; Wu et al. 2012). *NPR1* shows a negative effect over ABA biosynthesis genes *ABA1* and *ABA2*, as well as over some ABA-responsive genes such as *RAB18*. *AAO3* from ABA biosynthesis pathway suppresses the expression of *SID2* gene in the SA biosynthesis pathway in *N. benthamiana*. These results suggest that SA signaling negatively regulates ABA signaling, which in turn suppresses the SA signaling system (Fig. 5.11).

5.7.4 Synergistic Interaction Between ABA and SA Signaling Systems

In some plant-pathogen interactions, ABA signaling may act synergistically with SA signaling in triggering plant immune responses. ABA triggers SA biosynthesis and signaling in *Arabidopsis*. The MYB 96 transcription factor, a R2R3-type MYB member, has been shown to serve as a signaling link that integrates ABA and SA signals and regulates a synergistic interaction between the two hormones. ABA-mediated stress signals regulate SA biosynthesis by inducing the SA biosynthetic enzyme gene, *SID2* (Fig. 5.12; Seo and Park 2010). *MYB96*-overexpressing

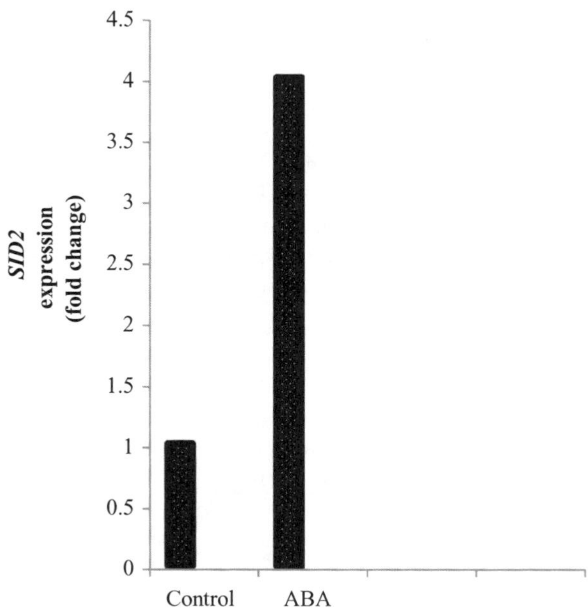

Fig. 5.12 Abscisic acid triggers the expression of the SA biosynthetic gene *SID2* (Adapted from Seo and Park 2010)

myb96-1d activation-tagging *Arabidopsis* line showed enhanced resistance to *P. syringae* infection. By contrast, the T-DNA insertional *myb96-1* mutant was susceptible to pathogen infection. The *SID2* gene was up-regulated, and the concentrations of endogenous SA were elevated in the activation-tagging *myb96-1d* line, indicating that the enhanced pathogen resistance of the mutant is derived from increased SA biosynthesis. ABA induced the *SID2* gene. However, the inductive effects of ABA were reduced in the *myb96-1* mutant, indicating that the *MYB96* gene is required for the *SID2* induction by ABA-mediated signals (Seo and Park 2010).

5.8 Interplay Between ABA and Ethylene Signaling Systems

5.8.1 ABA Activates Ethylene Biosynthesis and Ethylene Signaling Pathway

Ethylene response factor (ERF) proteins play important role in ET signaling system. ERF proteins are regulated by the important components in ET signaling system, ETR1, CTR1, EIN2, and EIN3 (Beaudoin et al. 2000; Alonso and Stepanova 2004; Anderson et al. 2004; Guo and Ecker 2004). Most of the ERF proteins bind to the GCC-box and modulate the expression of pathogenesis-related genes that function in plant immune responses (Solano et al. 1998; Fujimoto et al. 2000; Gu et al. 2002; Zhou et al. 2008; Liu et al. 2012). The ERF proteins Ts6-14i (Park et al. 2001) and TaERF1 (Zhang et al. 2007) regulate the expression of GCC-box-containing genes. An ERF protein, TSRF1, significantly enhanced resistance to *Ralstonia solanacearum*, *Pseudomonas syringae*, and *Botrytis cinerea* in both tobacco and tomato (Zhang et al. 2004a; Zhang et al. 2008). The transgenic tobacco plants expressing *TSRF1* showed increased production of both ABA and ethylene (Zhou et al. 2008). Methionine (L-Met) is the precursor of ethylene. It is converted to *S*-adenosylmethionine (*S*-AdoMet). by *S*-AdoMet synthetase (SAM synthetase). *S*-AdoMet is converted to 1-aminocyclopropane-1-carboxylic acid (ACC) by ACC synthase (ACS). Accumulation of ACS isozymes leads to increased synthesis of ACC (Liu and Zhang 2004). Finally ACC is oxidized by ACC oxidase (ACO) to form ethylene. The ethylene biosynthetic gene *ACO1* expression was enhanced in the transgenic plants expressing the ERF gene TSRF1. The ethylene production and the expression of *ACO1* were strengthened by ABA application (Zhou et al. 2008). These studies suggest that ABA stimulates the expression of *ACO1* (Fig. 5.13; Zhou et al. 2008) and production of ethylene (Zhou et al. 2008). ABA modifies the expression of ethylene-responsive GCC-box-containing genes such as *PRB 1b*, *osmotin*, *CHN50*, and *GL4*. Many PR genes, such as *PR1*, *PR2*, *PR3*, and *PR5* have GCC-box in their promoters (Zhou et al. 2008). ABA decreases the expression of GCC-box-containing genes (Zhou et al. 2008).

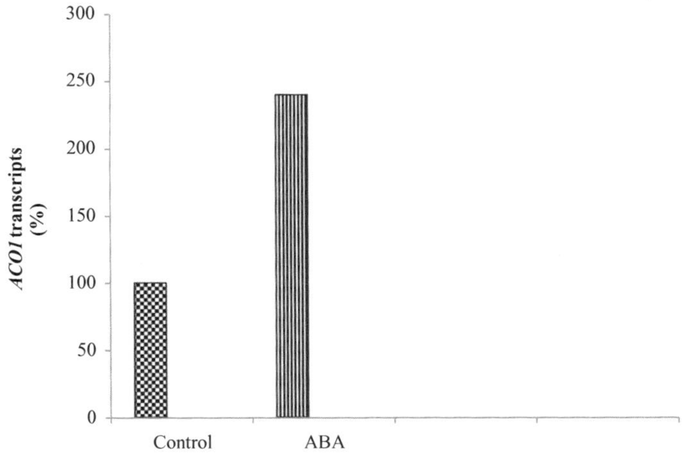

Fig. 5.13 ABA triggers expression of ethylene biosynthetic gene *ACO1* in tobacco plants (Adapted from Zhou et al. 2008)

5.8.2 Ethylene Signaling Triggers ABA Biosynthesis

In the ethylene signaling pathway, EIN2 either directly or indirectly activates the transcription factors EIN3 and EIL1 to initiate the transcriptional response to ethylene (Potuschak et al. 2003; van Loon et al. 2006; An et al. 2010). ERFs (ethylene-responsive transcription factors) are the important group of transcription factors functioning downstream in ethylene signaling system (Grennan 2008). EIN3 and EIL proteins transduce the signal from the upstream component EIN2 to downstream components including the ERF transcription factors. ERFs contain a single DNA-binding domain. ERFs have been shown to bind specifically to the GCC-box that is found in several promoters of the pathogenesis-related (PR) genes (Yamamoto et al. 1999; Gutterson and Reuber 2004).

The rice ERF, OsERF922, acts downstream of OsEIL1 signal transduction in rice (Bailey et al. 2009; Yoo et al. 2009; Liu et al. 2012). *OsERF922* overexpressing *Arabidopsis* lines showed increased expression of 9-*cis*-epoxycarotenoid dioxygenase (Os*NCED*3 and *OsNCED4*) genes (Liu et al. 2012). NCED catalyses the conversion of 9-*cis*-epoxycarotenoids to xanthoxin in the ABA biosynthesis pathway (Wasilewska et al. 2008; Hauser et al. 2011; Liu et al. 2012). The results suggest that ethylene signaling pathway triggers ABA biosynthesis pathway (Fig. 5.14). Transgenic rice plants overexpressing *OsERF922* showed increased accumulation of ABA and knockdown of OsERF922 by means of RNAi suppressed ABA resulted in the suppression of ABA accumulation (Fig. 5.15; Liu et al. 2012). The results

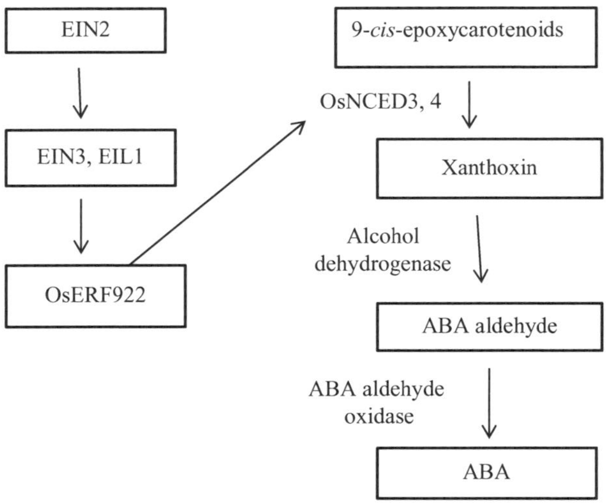

Fig. 5.14 ERF in ethylene signaling pathway triggers ABA biosynthesis pathway in rice (Adapted from Wasilewska et al. 2008; Hauser et al. 2011; Liu et al. 2012)

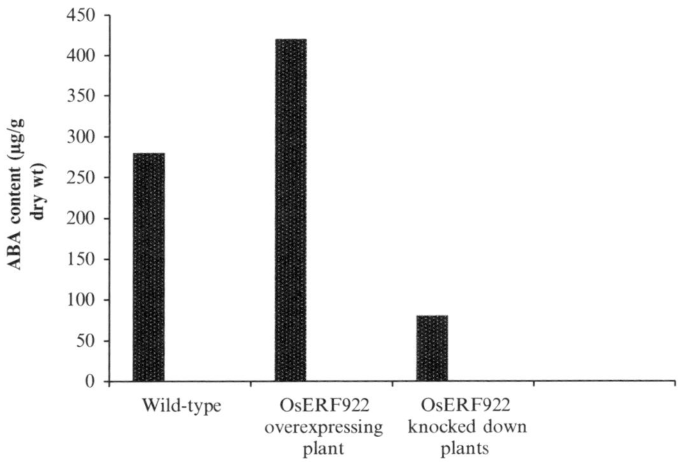

Fig. 5.15 Role of *OsERF922* in the ethylene signaling pathway in triggering ABA biosynthesis (Adapted from Liu et al. 2012)

suggest that the OsERF922 in the ethylene signaling pathway triggers accumulation of ABA. Conversely the rice ERF *OsERF922* is induced rapidly and strongly in response to ABA and in response to *Magnaporthe oryzae* infection in rice (Liu et al. 2012). Collectively these results suggest a reciprocal interaction between ABA and ethylene signaling pathways.

5.8.3 Synergistic and Antagonistic Interaction Between ABA and Ethylene Signaling Systems

Synergistic interaction between ABA and ethylene signaling systems has also been reported. The *ein2-1 aba1-6* double mutant was more susceptible to *P. cucumerina* than *aba1-6* plants (Sánchez-Vallet et al. 2012). Antagonism between ABA and ethylene in *Arabidopsis* acts in parallel with the reciprocal regulation of their signaling pathways (Cheng et al. 2009). An ACC oxidase involved in ethylene biosynthesis pathway was significantly up-regulated in an *Arabidopsis aba2* mutant. Conversely the *NCED3* gene was up-regulated in an *ein2* mutant. *ABI1* gene in the *etr1-1* (ethylene mutant) mutant was down-regulated (Cheng et al. 2009). These studies suggest that ABA and ethylene may antagonistically regulate the biosynthesis and signaling of each other.

5.9 ABA Signaling System May Trigger Defense Responses Against Pathogens

5.9.1 ABA Signaling Is Involved in Conferring Resistance Against a Wide Range of Pathogens

ABA signaling system has been shown to be involved in conferring resistance against fungal (Ton and Mauch-Mani 2004; Asselbergh and Höfte 2007; Kaliff et al. 2007), oomycete (Adie et al. 2007), bacterial (Hernández-Blanco et al. 2007; Zhou et al. 2008), and viral (Rezzonico et al. 1998; Mauch-Mani and Mauch 2005) pathogens. ABA-dependent signaling pathway has been shown to confer resistance against *Alternaria brassicicola* and *Plectosphaerella cucumerina* in *Arabidopsis* (Ton and Mauch-Mani 2004). ABA signaling is required for resistance to *Leptosphaeria maculans* in *Arabidopsis* (Kaliff et al. 2007) and *Botrytis cinerea* in tomato (Asselbergh and Höfte 2007). ABA induced resistance against the rice brown spot pathogen *Cochliobolus miyabeanus* (De Vleesschauwer et al. 2010). ABA-induced defense responses against the oomycete pathogen *Pythium irregulare* in *Arabidopsis* (Adie et al. 2007). ABA induces resistance against the bacterial pathogen *Ralstonia solanacearum* in tobacco and tomato plants (Zhou et al. 2008). ABA has a direct role in resistance to *R. solanacearum* in *Arabidopsis* (Hernández-Blanco et al. 2007). ABA induced resistance against crucifer-infecting *Tobacco mosaic virus* (TMV-cg). Exogenous applications of ABA reduced the systemic accumulation of TMV-cg (Chen et al. 2013). The enhanced and constitutive accumulation of ABA in the Arabidopsis *irxi-6* mutant, impaired in a secondary cell wall cellulose synthase (CESA8), and in the *ocp3* mutant, defective in a transcriptional factor of the homeodomain family, has been suggested to contribute to increased resistance to several pathogens (Hernández-Blanco et al. 2007; Garcia-Andrade et al. 2011). The results suggest a positive role for ABA in the regulation of *Arabidopsis* basal resistance to certain pathogens.

ABA may also have a negative role in conferring resistance against some pathogens. In tomato, a negative regulatory role of ABA in resistance to necrotrophic pathogens, such as the fungus *B. cinerea* and the bacterial pathogen *Dickeia dadantii*, has been reported. These pathogens are less virulent on the tomato *sitiens* mutant, in which ABA biosynthesis is disrupted, than in wild-type plants (Audenaert et al. 2002; Asselbergh et al. 2007, 2008a, b). ABA-deficient *aba2-1* mutant of *Arabidopsis* shows enhanced resistance to *Fusarium oxysporum* (Anderson et al. 2004) and *Hyaloperonospora parasitica* (Mohr and Cahill 2003). ABA deficiency in the *sitiens* mutant of tomato confers resistance to *Erwinia chrysanthemi* (Asselbergh et al. 2008a), *Botrytis cinerea* (Audenaert et al. 2002), *Oidium neolycopersici* (Achuo et al. 2006), and *Pseudomonas syringae* (Thaler and Bostock 2004). Decreased ABA levels or inhibition of ABA signaling decreased blast lesion numbers induced by *Magnaporthe oryzae* in rice plants (Yazawa et al. 2012). Abscisic acid deficiency leads to rapid activation of tomato defense responses upon infection with *Erwinia chrysanthemi* (Asselbergh et al. 2008a). Increased susceptibility to the necrotrophic fungus *Alternaria brassicicola*, the vascular oomycete *Pythium irregulare*, and the necrotrophic fungus *Plectosphaerella cucumerina* was observed in ABA-deficient mutants (Adie et al. 2007; Flors et al. 2008; Garcia-Andrade et al. 2011).

5.9.2 ABA Signaling System Triggers Callose Deposition and Confers Disease Resistance

ABA signaling system is involved in the induction of callose (β-1,3-glucan) deposition (Yazawa et al. 2012). The *Arabidopsis* mutant defective in the ABA-biosynthetic zeaxanthin epoxidase gene showed decreased callose deposition (Ton et al. 2005). β-aminobutyric acid (BABA)-induced resistance against necrotrophic pathogens in *Arabidopsis* is shown to be due to primed callose accumulation, which is controlled by an ABA-dependent defense pathway (Ton and Mauch-Mani 2004). In tomato, callose formation is at least in part ABA dependent (Asselbergh and Höfte 2007). ABA interacts with callose deposition allowing an early and efficient build up of papillae conferring resistance to pathogen attack (Flors et al. 2005). ABA-induced callose deposition plays a role in defense against *Pythium irregulare* in *Arabidopsis* (Adie et al. 2007).

Callose production is a secretory process and the fusion of the involved vesicles with the plasma membrane is mediated by SNARE (soluble *N*-ethyl-maleimide-sensitive fusion protein attachment receptor proteins) (Mauch-Mani and Mauch 2005). During cell-plate formation, callose synthase is transported in vesicles to the location where its function, the synthesis of callose, is required (Verma and Hong 2001). SNAREs have been implicated in ABA-dependent disease resistance responses (Collins et al. 2003). It is suggested that ABA is involved in callose deposition through the regulation of specific SNAREs (Mauch-Mani and Mauch 2005).

ABA increased virus resistance in tobacco. ABA inhibited the transcription of a basic β-1,3-glucanase that can degrade the β-1,3-glucan callose (Rezzonico et al. 1998). The inhibition of callose degradation by ABA would result in forming a physical barrier to viral spread through plasmodesmata (Mauch-Mani and Mauch 2005). The down-regulation of β-1,3-glucanase by ABA might have resulted in callose accumulation and conferred resistance to virus spread.

5.9.3 ABA Signaling Cascade May Trigger Stomatal Closure Immune Responses

Stomatal closure is an innate immune response involved in bacterial disease resistance (Hettenhausen et al. 2012). Stomata serve as passive ports of bacterial entry during infection. Bacterial entry into host tissue is a critical first step in causing infection by bacterial pathogens in plants (Melotto et al. 2006; Gudesblat et al. 2009). The stomata constitute one entry point for bacteria, which need to reach apoplastic spaces to multiply and cause disease (Nicaise et al. 2009; Hubbard et al. 2010). The stomata in the *Arabidopsis* leaf epidermis have been shown to act as innate immunity gates to actively prevent bacteria from entering the plant leaf (Melotto et al. 2006, 2008). The bacterial PAMP flg22 triggers closure of stomata which occurs within the first hour of contact with plant tissue (Melotto et al. 2006). ABA induces stomatal closure immune response (Saito et al. 2008; Hossain et al. 2011; Munemasa et al. 2011; Hubbard et al. 2012; Sugiyama et al. 2012; Uraji et al. 2012). ABA signaling system has been reported to be involved in the stomatal closure (Hubbard et al. 2010). ABA increase was the critical early event in stomatal closure induced by flg22 (Melotto et al. 2006).

Phospholipase D (PLD)-induced phosphatidic acid (PA) production has been shown to be involved in ABA signaling-mediated stomatal closure immune response (Uraji et al. 2012). Two *Arabidopsis* PLDs (PLDα1 and PLDδ) have been reported to be involved in ABA-induced stomatal closure signaling in guard cells. ABA induced stomatal closure was suppressed in the *pldα1 pldδ* double mutant but not in the *pld* single mutants. The *pldα1* and *pldδ* mutations reduced ABA-induced phosphatidic acid production. During ABA-induced stomatal closure, wild-type guard cells showed increased production of ROS and NO and also showed increased cytosolic alkalization. These changes were reduced in guard cells of the *pldα1pldδ* double mutant plants. Inward-rectifying K^+ channel currents of guard cells were inhibited by ABA in the wild-type but not in *pldα1pldδ* mutant plants (Uraji et al. 2012). ABA signaling system has been shown to involve ROS, NO, and PA, the product of PLD activity, in *Arabidopsis* (Distéfano et al. 2012). These studies suggest that ABA induces production of phosphatidic acid through the action of phospholipase D and the phosphatidic acid acting as a second messenger triggers ROS, NO, and Ca^{2+} signaling systems triggering stomatal closure immune response.

The MAP kinase MPK4 has been shown to be involved in abscisic acid-induced stomatal closure and guard cell-mediated resistance to *Pseudomonas syringae* DC3000 in *Nicotiana attenuata*. Silencing *MPK4* gene compromised ABA-induced stomatal closure immune response against the bacterial pathogen (Hettenhausen et al. 2012).

5.9.4 ABA Signaling May Modulate Other Hormone Signaling Systems and Trigger Defense Responses

ABA may modulate other plant hormone signaling systems and trigger defense responses. ABA has been shown to interact with JA signaling system and trigger defense responses. ABA is required for JA biosynthesis (Hossain et al. 2011) and induction of JA-regulated defense genes, which are required for *Arabidopsis* resistance to *Pythium irregulare* (Adie et al. 2007). *Pseudomonas syringae* pv. *tomato* infection-induced JA accumulation was reduced in ABA-deficient *aba3-1* mutant plants compared to the wild type (Fan et al. 2009). The results suggest that ABA promotes JA accumulation in *Arabidopsis*. JA signaling may be integrated into early ABA signaling and may affect ABA receptor complexes to regulate downstream common signal components in stomatal closure-related immune responses (Munemasa et al. 2011).

ABA-induced resistance against the brown spot pathogen *Cochliobolus miyabeanus* in rice involves repression of ethylene signaling (De Vleesschauwer et al. 2010). Exogenously administered ABA enhanced basal resistance of rice against *C. miyabeanus*. The ABA-induced resistance did not rely on boosted expression of SA- or JA-dependent resistance mechanisms, but it was dependent on repression of ethylene signaling system, which suppresses host defense responses. Exogenous application of ethylene enhanced susceptibility and ABA induces resistance against the pathogen by suppressing the action of ethylene (De Vleesschauwer et al. 2010).

ABA signaling may regulate plant defense responses by its interplay with SA, JA, and ET signaling systems. An extensive and complex crosstalk among different signaling pathways has been reported in ABA-mediated defense responses. Hierarchical clustering of genes differentially regulated in *aba1-6* mutant, which negatively regulates plant defense responses, showed that two of the *aba1-6* clusters contained up-regulated genes whose expression was positively modulated by either JA or SA treatment. Genetic disruption of these pathways by generating *coi1-1 aba1-6*, *aos aba1-6*, and *sid2-1 aba1-6* double mutants led to partial impairment of the *aba1-6* resistance phenotype. The *ein2-1 aba1-6* double mutant was more susceptible to *P. cucumerina* than *aba1-6* plants (Sánchez-Vallet et al. 2012). These results suggest that ABA, JA, ET, and SA signaling pathways cooperatively contribute to *Arabidopsis* resistance to *P. cucumerina* (Sánchez-Vallet et al. 2012).

5.10 ABA Signaling System May Confer Susceptibility Against Pathogens

5.10.1 *ABA Induces Susceptibility Against Fungal and Bacterial Pathogens*

ABA has been shown to induce susceptibility against many pathogens. ABA promotes infection of the rice blast pathogen *Magnaporthe oryzae* (Yazawa et al. 2012). The blast pathogen *M. oryzae* was found to produce and secrete ABA and cause disease (Jiang et al. 2010). Exogenous application of ABA induces susceptibility in rice against *M. oryzae* (Jiang et al. 2010). ABA is the key factor for suppressing rice blast disease resistance and making rice plants susceptible to *M. oryzae* (Koga et al. 2004). ABA signaling pathway induces susceptibility to *Fusarium oxysporum* in *Arabidopsis* (Anderson et al. 2004). ABA determines basal susceptibility to *Botrytis cinerea* in tomato (Audenaert et al. 2002). ABA promotes susceptibility to the rice bacterial blight pathogen *Xanthomonas oryzae* pv. *oryzae* (Xu et al. 2013). Exogenous application of ABA to *Arabidopsis* resulted in an enhanced susceptibility to *Pseudomonas syringae* pv. *tomato* DC3000 (Anderson et al. 2004; Mohr and Cahill 2007). ABA signaling pathway negatively regulates resistance to various pathogens. It has been shown that ABA deficiency leads to rapid activation of tomato defense responses upon infection with *Erwinia chrysanthemi*, suggesting that ABA negatively regulates resistance against *E. chrysanthemi* in tomato (Asselbergh et al. 2008a).

5.10.2 *ABA May Suppress Plant Immune Responses and Induce Susceptibility*

ABA may induce disease incidence by suppressing the host defense mechanisms. ABA treatment suppresses phytoalexin synthesis. The production of the phytoalexin glyceollin is reduced in an ABA-dependent manner in soybean upon infection with *Phytophthora sojae* (Mohr and Cahill 2003). Expression of *OsCPS2* and *OsCPS4*, the two cyclase genes leading to the synthesis of the diterpenoid phytoalexins, was suppressed significantly in *OsERF922* overexpressing plants. OsERF922 overexpressing plants show elevated ABA levels (Liu et al. 2012). In mutants deficient in ABA biosynthesis, the synthesis of capsidiol, a sesquiterpene phytoalexin in tobacco is induced higher than wild-type plants when infected with *Botrytis cinerea*, or elicited with elicitors, while the application of exogenous ABA reversed the elevated effect (Miyaloudama et al. 2009). ABA inhibits the activity and transcript accumulation of phenylalanine ammonia lyase involved in phytoalexin and phenolics biosynthesis (Henfling et al. 1980; Ward et al. 1989; McDonald and Cahill 1996; Mauch-Mani and Mauch 2005). ABA suppresses PAL and resistance of soybeans to

Phytophthora megasperma f. sp. *glycinea* (Ward et al. 1989). ABA accumulation suppressed the expression of defense genes and enhanced the susceptibility of *A. thaliana* to *P. syringae* pv. *tomato* (de Torres-Zabala et al. 2007).

A mutant of tomato with reduced ABA production (*sitiens*) exhibits increased resistance to the necrotrophic fungus *Botrytis cinerea*. This resistance is correlated with a rapid and strong hydrogen peroxide-driven cell wall fortification response in epidermis cells that is absent in tomato with normal ABA production. Basal expression of defense genes is higher in the mutant compared with the wild-type tomato (Curvers et al. 2010). The results suggest ABA suppresses peroxide-derived cell wall fortification. The cell wall fortification plays important role in suppressing disease development (Vidhyasekaran 2007). The cell wall fortification involves lignin deposition (Kang and Buchenauer 2000). ABA-deficient mutants of *Arabidopsis*, such as *aba1-1*, were resistant to *Hyaloperonospora parasitica*. ABA treatment induced susceptibility to avirulent isolates of *Pseudomonas syringae* pv. *tomato* in *Arabidopsis* (Mohr and Cahill 2004, 2007). The results suggest that ABA plays an important role in determining either susceptibility or resistance to pathogen attack. It was found that the production of the cell wall strengthening compound, lignin, was increased during resistant interactions of *aba1-1* but was suppressed in ABA-induced susceptible interactions. ABA down-regulated key genes of phenylpropanoid pathway, especially of those genes involved directly in lignin biosynthesis (Mohr and Cahill 2003, 2004, 2007). These results suggest that ABA suppresses lignin biosynthesis and the suppression of lignin biosynthesis induces susceptibility against pathogens.

5.10.3 ABA May Modulate JA, SA, and ET Signaling Pathways and Confer Susceptibility Against Pathogens

ABA may confer susceptibility by modulating JA or SA or ET, or JA/ET signaling pathways. ABA negatively regulates SA-/ET-/JA-dependent immune responses, which are essential for *Arabidopsis* resistance to the necrotrophic pathogen *Plectosphaerella cucumerina* (Sánchez-Vallet et al. 2012). The rice blast pathogen *M. oryzae* was found to produce and secrete ABA, which antagonizes the SA-signaling pathway to attenuate host defense responses (Jiang et al. 2010). ABA suppresses SA signaling system and it markedly suppressed the transcriptional up-regulation of WRKY45 and OsNPR1, the two key components of the SA signaling pathway in rice, induced by SA (Jiang et al. 2010). ABA promotes susceptibility to the rice bacterial blight pathogen *Xanthomonas oryzae* pv. *oryzae* by suppressing salicylic acid-mediated defense responses (Xu et al. 2013).

Treatment of *Arabidopsis* with ABA blocked the activation of SA-activated systemic acquired resistance both upstream and downstream of SA biosynthesis (Yasuda et al. 2008). SAR activation also disrupted the up-regulation of ABA biosynthesis and signaling in plants (Yasuda et al. 2008). The results suggest that ABA suppresses SA signaling system and in turn induces susceptibility. Exogenous application of ABA to *Arabidopsis* resulted in an enhanced susceptibility to *Pseudomonas syringae* pv. *tomato* DC3000 that was correlated with reduced SA accumulation and suppression of several defense-related genes, including some regulated by SA and JA/ET signaling pathways (Anderson et al. 2004; Mohr and Cahill 2007).

5.11 Pathogens May Suppress Host Defense Mechanisms by Activating ABA Signaling System to Cause Disease

The concentration of ABA increased in sugar beet (*Beta vulgaris*) leaves during *Cercospora beticola* infection. After an initial burst of ABA induced by inoculation, elevated ABA concentrations were detected during the fungal penetration and colonization phases. Expression of the basic leucine zipper transcription factor AREB1 was induced by the fungal infection. Activation of the ABA-dependent signaling cascade resulted in reduced expression of the phenylalanine ammonia lyase (*PAL*) gene. These results suggest that activation of ABA-dependent signaling cascade may be the primary cause of suppression of the important defense gene *PAL* (Schmidt et al. 2008). ABA suppresses the deposition of lignin, which reinforces the cell wall to prevent the bacterial invasion (Mohr and Cahill 2007).

5.12 Pathogens May Hijack ABA Signaling Pathway to Cause Disease

Pathogens may induce specific signaling systems, which may favor disease development. *Pseudomonas syringae* pv. *tomato* has been shown to hijack the *Arabidopsis* ABA signaling pathway to cause disease (de Torres-Zabala et al. 2007). The major target for effectors secreted by *Pseudomonas syringae* is the ABA signaling pathway. The bacterial effector delivered into the plant cell via a type III protein secretion system interferes with the host defense signaling. In a compatible interaction leading to disease, *NCED*, the gene encoding 9-*cis*-epoxycarotenoid dioxygenase, the key enzyme of ABA biosynthesis, and the *abscisic acid insensitive 1* (*ABI1*) clade of genes encoding protein phosphatases type 2C (PP2C) involved in the ABA signaling were strongly induced. These events led to the accumulation of ABA. ABA accumulation suppressed the expression of defense genes and enhanced the susceptibility of *A. thaliana* to *P. syringae* pv. *tomato* (de Torres-Zabala et al. 2007).

The rice bacterial blight pathogen *Xanthomonas oryzae* pv. *oryzae* has been shown to hijack the rice ABA machinery to cause disease (Xu et al. 2013). ABA negatively regulated resistance to the bacterial pathogen. Exogenously applied ABA rendered rice hypersusceptible to infection, whereas chemical and genetic disruption of ABA biosynthesis and signaling led to enhanced resistance to the pathogen. Thus, ABA was found to suppress the host immune responses. This immune-suppressive effect of ABA may be due to suppression of SA-mediated defenses that normally serve to limit pathogen growth. Successful *X. oryzae* pv. *oryzae* infection was found to be associated with extensive reprogramming of ABA biosynthesis and response genes, suggesting that ABA functions as a virulence factor for the pathogen (Xu et al. 2013). These results suggest that the pathogen hijacks ABA signaling system to cause the disease.

5.13 Pathogen Produces Toxins/Effectors and Suppresses ABA-Dependent Defenses

ABA increase was the critical early event in stomatal closure immune response induced by PAMPs and the toxin coronatine inhibited PAMP- or ABA-induced stomatal closure (Melotto et al. 2006). The bacterial pathogen *Pseudomonas syringae* pv. *tomato* reopens the closed stomata and counteracts the PAMP-induced/SA and ABA-mediated stomatal closure, by producing virulence factors, specifically the phytotoxin coronatine. The toxin overcomes or suppresses ABA-dependent defenses (Brooks et al. 2005) and probably suppresses the stomatal defense. Collectively, these results suggest that ABA signaling systems are involved in host defense mechanism against *P. syringae* pv. *tomato* in *Arabidopsis* and the virulent pathogen suppresses both SA and ABA signaling systems and cause disease. ABA accumulation suppressed the expression of defense genes and enhanced the susceptibility of *A. thaliana* to *P. syringae* pv. *tomato* (de Torres-Zabala et al. 2007). The AvrPtoB effector of *P. syringae* pv. *tomato* DC3000 induces the accumulation of ABA in *Arabidopsis* and suppresses the induction of basal defense by down-regulating SA biosynthesis and signaling pathways (de Torres-Zabala et al. 2007, 2009).

Oxalic acid is a virulence factor of *Sclerotinia sclerotiorum*, which causes foliar wilting of *Vicia faba* (Guimarães and Stotz 2004). Oxalate causes foliar dehydration by disturbing guard cell function. It triggers stomatal opening, which causes increase in transpiration and wilting. ABA signaling plays important role in stomatal function and confers resistance to the pathogen. It induces stomatal closure, thus preventing excess transpiration, dehydration, and foliar wilting. For successful pathogenesis, the pathogen produces oxalate, which inhibits the ABA signaling system-induced stomatal closure and keeps the stomata open. The induction of stomatal opening favors disease development (Guimarães and Stotz 2004). The results suggest that the virulent pathogen suppresses ABA-signaling system and causes disease.

References

Abe H, Urao T, Seki M, Shinozaki K, Yamaguchi-Shinozaki Y (2003) Arabidopsis AtMYC2 (bHLH) and AtMYB2 (MYB) function as transcriptional activators in abscisic acid signaling. Plant Cell 15:63–78

Achuo EA, Prinsen E, Höfte M (2006) Influence of drought, salt stress and abscisic acid in the resistance of tomato to *Botrytis cinerea* and *Oidium neolycopersici*. Plant Pathol 55:179–186

Adie BA, Perez-Perez J, Godoy M, Sanchez-Serrano JJ, Schmelz EA, Solano R (2007) ABA is an essential signal for plant resistance to pathogens affecting JA biosynthesis and activation of defenses in *Arabidopsis*. Plant Cell 19:1665–1681

Alazem M, Lin K-Y, Lin N-S (2014) The abscisic acid pathway has multifaceted effects on the accumulation of *Bamboo mosaic virus*. Mol Plant Microbe Interact 27:177–189

Alonso JM, Stepanova AN (2004) The ethylene signaling pathway. Science 306:1513–1515

An C, Mou Z (2013) The function of the Mediator complex in plant immunity. Plant Signal Behav 8:e23182

An Z, Jing W, Liu Y, Zhang W (2008) Hydrogen peroxide generated by copper amine oxidase is involved in abscisic acid-induced stomatal closure in *Vicia faba*. J Exp Bot 59:815–825

An F, Zhao Q, Ji Y, Li W, Jiang Z, Yu X, Zhang C, Han Y, He W, Liu Y, Zhang S, Ecker JR, Guo H (2010) Ethylene-induced stabilization of ETHYLENE INSENSENSITIVE3 and EIN3-LIKE1 is mediated by proteasomal degradation of EIN3 binding F-box 1 and 2 that requires EIN2 in *Arabidopsis*. Plant Cell 22:2384–2401

Anderson JP, Badruzsaufari E, Schenk PM, Manners JM, Desmond OJ, Ehlert C, Maclean DJ, Ebert PR, Kazan K (2004) Antagonistic interaction between abscisic acid and jasmonate-ethylene signaling pathways modulates defense gene expression and disease resistance in *Arabidopsis*. Plant Cell 16:3460–3479

Antoni R, Gonzalez-Guzman M, Rodriguez L, Rodrigues A, Pizzio GA, Rodriguez PL (2012) Selective inhibition of clade A phosphatases type 2C by PYR/PYL/RCAR abscisic acid receptors. Plant Physiol 158:970–980

Asselbergh B, Höfte M (2007) Basal tomato defences to *Botrytis cinerea* include abscisic acid-dependent callose formation. Physiol Mol Plant Pathol 71:33–40

Asselbergh B, Curvers K, Franca SC, Audenaert K, Vuylsteke M, Van Breusegem F, Höfte M (2007) Resistance to *Botrytis cinerea* in *sitiens*, an abscisic acid-deficient tomato mutant, involves timely production of hydrogen peroxide and cell wall modifications in the epidermis. Plant Physiol 144:1863–1877

Asselbergh B, Achuo AE, Höfte M, Van Gijsegem F (2008a) Abscisic acid deficiency leads to rapid activation of tomato defence responses upon infection with *Erwinia chrysanthemi*. Mol Plant Pathol 9:11–24

Asselbergh B, De Vleesschauwer D, Höfte M (2008b) Global switches and fine-tuning ABA modulates plant pathogen defense. Mol Plant Microbe Interact 21:709–719

Audenaert K, De Meyer GB, Höfte MM (2002) Abscisic acid determines basal susceptibility of tomato to *Botrytis cinerea* and suppresses salicylic acid-dependent signaling mechanisms. Plant Physiol 128:491–501

Bailey TA, Zhou X, Chen J, Yang Y (2009) Role of ethylene, abscisic acid and MAP kinase pathways in rice blast resistance. In: Wang GL, Valent B (eds) Advances in genetics, genomics and control of rice blast disease. Springer, Dordrecht, pp 185–190

Beaudoin N, Serizet C, Gosti F, Giraudat J (2000) Interactions between abscisic acid and ethylene signaling cascades. Plant Cell 12:1103–1115

Brault M, Amiar Z, Pennarun A-M, Monestiez M, Zhang Z, Cornel D, Dellis O, Knight H, Bouteau F, Rona J-P (2004) Plasma membrane depolarization induced by abscisic acid in *Arabidopsis* suspension cells involves reduction of proton pumping in addition to anion channel activation, which are both Ca^{2+} dependent. Plant Physiol 135:231–243

Bright J, Desikan R, Hancock JT, Weir IS, Neill SJ (2006) ABA-induced NO generation and stomatal closure in *Arabidopsis* are dependent on H_2O_2 synthesis. Plant J 45:113–122

Broekaert WF, Delaure SI, De Bolle MFC, Commue BPA (2006) The role of ethylene in host-pathogen interactions. Annu Rev Phytopathol 44:393–416

Brooks DM, Bender CL, Kunkel BN (2005) The *Pseudomonas syringae* phytotoxin coronatine promotes virulence by overcoming salicylic acid-dependent defenses in *Arabidopsis thaliana*. Mol Plant Pathol 6:629–639

Cao FY, Yoshioka K, Desveaux D (2011) The roles of ABA in plant-pathogen interactions. J Plant Res 124:489–499

Casaretto J, Ho TD (2003) The transcription factors HvABI5 and HvVP1 are required for the abscisic acid induction of gene expression in barley aleurone cells. Plant Cell 15:271–284

Cevik V, Kidd BN, Zhang P, Hill C, Kiddle S, Denby KJ, Holub EB, Cahill DM, Manners JM, Schenk PM, Beynon J, Kazan K (2012) MEDIATOR25 acts as an integrative hub for the regulation of jasmonate-responsive gene expression in Arabidopsis. Plant Physiol 160:541–555

Chen N, Goodwin PH, Hsiang T (2003) The role of ethylene during the infection of *Nicotiana tabacum* by *Colletotrichum destructivum*. J Exp Bot 54:2449–2456

Chen Y, Fangfang JI, Xie H, Liang J (2006) Overexpression of the regulator of G-protein signalling protein enhances ABA-mediated inhibition of root elongation and drought tolerance in *Arabidopsis*. J Exp Bot 57:2101–2110

Chen Y-H, Hu L, Punta M, Bruni R, Hillerich B, Kloss B, Rost B, Love J, Siegelbaum SA, Hendrickson WA (2010) Homologue structure of the SLAC1 anion channel for closing stomata in leaves. Nature 467:1074–1080

Chen R, Jiang H, Li L, Zhai Q, Qi L, Zhou W, Liu X, Li H, Zheng W, Sun J, Li C (2012) The Arabidopsis mediator subunit MED25 differentially regulates jasmonate and abscisic acid signaling through interacting with MYC2 and ABI5 transcription factors. Plant Cell 24:2898–2916

Chen L, Zhang L, Li D, Wang F, Yu D (2013) WRKY8 transcription factor functions in the TMV-cg defense response by mediating both abscisic acid and ethylene signaling in Arabidopsis. Proc Natl Acad Sci U S A 110(21):E1983–1971. doi:10.1073/pnas.1221347110

Cheng S-H, Sheen J, Gerrish C, Bolwell GP (2001) Molecular identification of phenylalanine ammonia-lyase as a substrate of a specific constitutively active *Arabidopsis* CDPK expressed in maize protoplasts. FEBS Lett 503:185–188

Cheng WH, Chiang MH, Hwang SG, Lin PC (2009) Antagonism between abscisic acid and ethylene in *Arabidopsis* acts in parallel with the reciprocal regulation of their metabolism and signaling pathways. Plant Mol Biol 71:61–80

Chern M, Canlas PE, Ronald PC (2008) Strong suppression of systemic acquired resistance in *Arabidopsis* by NRR is dependent on its ability to interact with NPR1 and its putative repressive domain. Mol Plant 1:552–559

Choi DS, Hwang BK (2012) Proteomics and functional analyses of pepper *Abscisic acid-responsive1* (*ABR1*), which is involved in cell death and defense signalling. Plant Cell 23:823–842

Choi H-L, Park H-J, Park JJ, Kim S, Im M-Y, Seo H-H, Kim Y-W, Hwang I, Kim SY (2005) Arabidopsis calcium-dependent protein kinase AtCPK32 interacts with ABF4, a transcriptional regulator of abscisic acid-responsive gene expression, and modulates its activity. Plant Physiol 139:1750–1761

Christmann A, Grill E (2009) Are GTGs ABA's biggest fans? Cell 136:21–23

Christmann A, Weiler EW, Steudle E, Grill E (2007) A hydraulic signal in root-to-shoot signalling of water shortage. Plant J 52:167–174

Chung F, Park JM, Oh SK, Joung YH, Lee S, Choi D (2004) Molecular and biochemical characterization of the *Capsicum annuum* calcium-dependent protein kinase 3 (*CaCDPK3*) gene induced by abiotic and biotic stresses. Planta 220:286–295

Coelho SM, Taylor AR, Ryan KP, Sousa-Pinto I, Brown MT, Brownlee C (2002) Spatiotemporal patterning of reactive oxygen production and Ca^{2+} wave propagation in *Fucus* rhizoid cells. Plant Cell 14:2369–2381

Collins NC, Thordal-Christensen H, Lipka V, Bau S, Kombrink E, Qiu JL, Huckelhoven R, Stein M, Frelaldenhoven A, Somerville SC, Schulze-Lefert P (2003) SNARE-protein-mediated disease resistance at the plant cell wall. Nature 425:973–977

Curvers K, Seifi H, Mouille G, de Rycke R, Asselbergh B, Van Hecke A, Vanderschaeghe D, Höfte H, Callewaert N, Van Brueusegem F, Höfte M (2010) Abscisic acid deficiency causes changes in cuticle permeability and pectin composition that influence tomato resistance to *Botrytis cinerea*. Plant Physiol 154:847–860

Cutler SR, Rodríguez PL, Finkelstein RR, Abrams SR (2010) Abscisic acid: emergence of a core signaling network. Annu Rev Plant Biol 61:651–679

Danquash A, de Zelicourt A, Colcombet J, Hirt H (2013) The role of ABA and MAPK signaling pathways in plant abiotic stress responses. Biotechnol Adv. doi:10.1016/j.biotechadv.2013.09.006

DeFalco TA, Bender KW, Snedden WA (2010) Breaking the code: Ca^{2+} sensors in plant signalling. Biochem J 425:27–40

Denancě N, Sánchez-Vallet A, Goffner D, Molina A (2013) Disease resistance or growth: the role of plant hormones in balancing immune responses and fitness costs. Front Plant Sci 4; Article 155:1–12. doi:10.3389/fpls 2013.00155

Desikan R, Mackerness S, Hancock JT, Neill SJ (2001) Regulation of the *Arabidopsis* transcriptome by oxidative stress. Plant Physiol 127:159–172

Desikan R, Cheung M-K, Bright J, Henson D, Hancock JT, Neill SJ (2004) ABA, hydrogen peroxide and nitric oxide signaling in stomatal guard cells. J Exp Bot 55:205–212

de Torres M, Mansfield JW, Grabov N, Brown IR, Ammouneh H, Tsiamis G, Forsyth A, Robatzek S, Grant M, Boch J (2006) *Pseudomonas syringae* effector AvrPtroB suppresses basal defence in Arabidopsis. Plant J 47:368–382

de Torres-Zabala M, Truman W, Bennett MH, Lafforgue G, Mansfield JW, Rodriguez Egea P, Bogre L, Grant M (2007) *Pseudomonas syringae* pv. *tomato* hijacks the *Arabidopsis* abscisic acid signalling pathway to cause disease. EMBO J 26:1434–1443

de Torres-Zabala M, Bennett MH, Truman WH, Grant MR (2009) Antagonism between salicylic and abscisic acid reflects early host-pathogen conflict and moulds plant defense responses. Plant J 59:375–386

De Vleesschauwer D, Yang Y, Vera Cruz C, Höfte M (2010) Abscisic acid –induced resistance against the brown spot pathogen *Cochliobolus miyabeanus* in rice involves MAP kinase-mediated repression of ethylene signaling. Plant Physiol 152:2036–2052

Distéfano AM, Scuffi D, Garcia-Mata C, Lamattina L, Laxalt AM (2012) Phospholipase Dδ is involved in nitric oxide-induced stomatal closure. Planta 236:1899–1907

Dodd AN, Kudla J, Sanders D (2010) The language of calcium signaling. Annu Rev Plant Biol 61:593–620

Ebel J, Mithofer A (1998) Early events in the elicitation of plant defence. Planta 206:335–348

Eckardt NA (2004) Abscisic acid signal transduction: function of G protein-coupled receptor 1 in *Arabidopsis*. Plant Cell 16:1353–1354

Endo A, Koshiba T, Kamiya Y, Nambara E (2008) Vascular system is a node of systemic stress responses. Plant Signal Behav 3:1138–1140

Fan J, Hill L, Crooks C, Doerner P, Lamb C (2009) Abscisic acid has a key role in modulating diverse plant-pathogen interactions. Plant Physiol 150:1750–1761

Fan LM, Zhang W, Chen J-G, Taylor JP, Jones AM, Assmann SM (2008) Abscisic acid regulation of guard cell K^+ and anion channels in Gβ and RGS-deficient *Arabidopsis* lines. Proc Natl Acad Sci U S A 105:8476–8481

Feng DX, Tasset C, Hanemian M, Barlet X, Hu J, Trèmousaygue D, Deslandes L, Marco Y (2012) Biological control of bacterial wilt in *Arabidopsis thaliana* involves abscisic acid signalling. New Phytol 194:1035–1045

Flors V, Ton J, Jakab G, Mauch-Mani B (2005) Abscisic acid and callose: team players in defence against pathogens? J Phytopathol 153:377–383

Flors V, Ton J, van Doorn R, Garcia-Agustin P, Mauch-Mani B (2008) Interplay between JA, SA, and ABA signalling during basal and induced resistance against *Pseudomonas syringae* and *Alternaria brassicicola*. Plant J 54:81–92

Fraser RSS, Whenham RJ (1989) Abscisic-acid metabolism in tomato plants infected with tobacco mosaic virus – relationships with growth, symptoms and the *Tm-1* gene for TMV resistance. Physiol Mol Plant Pathol 34:212–226

Fujii H, Verslues PE, Zhu J-K (2007) Identification of two protein kinases required for abscisic acid regulation of seed germination, root growth, and gene expression in *Arabidopsis*. Plant Cell 19:485–494

Fujii H, Chinnusamy V, Rodrigues A, Rubio S, Antoni R, Park SY, Cutler SR, Sheen J, Rodriguez PL, Zhu JK (2009) *In vitro* reconstitution of an abscisic acid signalling pathway. Nature 462:660–664

Fujimoto SY, Ohta M, Usui A, Shinshi H, Ohme-Takagi M (2000) Arabidopsis ethylene-responsive element binding factors act as transcriptional activators or repressors of GCC box-mediated gene expression. Plant Cell 12:393–404

Fujita Y, Nakashima K, Yoshida T, Katagiri T, Kidokro S, Kanamori N, Umezawa T, Fujita M, Maruyama K, Ishiyama K, Kobayashi M, Nakasone S, Yamada K, Ito I, Shinozaki K, Yamaguchi-Shinozaki K (2009) Three SnRK2 kinases are the main positive regulators of abscisic acid signalling in response to water stress in *Arabidopsis*. Plant Cell Physiol 50:2123–2132

Furihata T, Manuyama K, Fujita Y, Umezawa T, Yoshida R, Shinozaki K, Yamaguchi-Shinozaki K (2006) Abscisic acid-dependent multisite phosphorylation regulates the activity of a transcription activator AREB1. Proc Natl Acad Sci U S A 103:1988–1993

Galon Y, Aloni R, Nachmias D, Snir O, Feldmesser E, Scrase-Field S, Boyce JM, Bouché N, Knight MR, Fromm H (2010a) Calmodulin-binding transcription activator 1 mediates auxin signaling and responds to stresses in Arabidopsis. Planta 232:165–178

Galon Y, Finkler A, Fromm H (2010b) Calcium-regulated transcription in plants. Mol Plant 3:653–669

Gao Y, Zeng Q, Guo J, Cheng J, Ellis BE, Jin-Gui C (2007) Genetic characterization reveals no role for the reported ABA receptor, GCR2, in ABA control of seed germination and early seedling development in Arabidopsis. Plant J 52:1001–1013

Gao Y, Li T, Liu Y, Ren C, Zhao Y, Wang M (2010a) Isolation and characterization of gene encoding G protein α subunit protein responsive to plant hormones and abiotic stresses in *Brassica napus*. Mol Biol Rep 37:3957–3965

Gao Y, Li T, Zhao Y, Ren C, Liu Y, Wang M (2010b) Cloning and characterization of a G protein β subunit gene responsive to plant hormones and abiotic stresses in *Brassica napus*. Plant Mol Biol Rep 28:450–459

Garcia-Andrade J, Ramirez V, Flors V, Vera P (2011) Arabidopsis *ocp3* mutant reveals a mechanism linking ABA and JA to pathogen-induced callose deposition. Plant J 67:783–794

Gayatri D, Agurla S, Raghavendra AS (2013) Nitric oxide in guard cells as an important secondary messenger during stomatal closure. Front Plant Sci 4:425

Geiger D, Scherzer S, Mumm P, Stange A, Marten I, Bauer H, Ache P, Matschi S, Liese A, Al-Rasheid KA, Romeis T, Hedrich R (2009) Activity of guard cell anion channel SLAC1 is controlled by drought-stress signaling kinase-phosphatase pair. Proc Natl Acad Sci U S A 106:21425–21430

Geiger D, Scherzer S, Mumm P, Marten I, Ache P, Matschi S, Liese A, Wellmann C, Al-Rasheid KAS, Grill E, Romeis T, Hedrich R (2010) Guard cell anion channel SLAC1 is regulated by CDPK protein kinases with distinct Ca^{2+} affinities. Proc Natl Acad Sci U S A 107: 8023–8028

Ghassemian M, Nambara E, Cutler S, Kawaide H, Kamiya Y, McCourt P (2000) Regulation of abscisic acid signaling by the ethylene response pathway in Arabidopsis. Plant Cell 12:1117–1126

Gomi K, Ogawa D, Katou S, Kamada H, Nakajima N, Saji H, Soyano T, Sasabe M, Machida Y, Mitsuhara I, Ohashi Y, Seo S (2005) A mitogen-activated protein kinase NtMPK4 activated by SIPKK is required for jasmonic acid signaling and involved in ozone tolerance via stomatal movement in tobacco. Plant Cell Physiol 46:1902–1914

Gonzalez-Guzman M, Apostolova N, Belles JM, Barrero JM, Barrero JM, Piqueras P, Ponce MR, Micol JL, Serrano R, Rodriguez FL (2002) The short-chain alcohol dehydrogenase ABA2 catalyzes the conversion of xanthoxin to abscisic aldehyde. Plant Cell 14:1833–1846

Gosti F, Beaudoin N, Serizet C, Webb AAR, Vattanian N, Giraudat J (1999) ABI1 protein phosphatase 2C is a negative regulator of abscisic acid signaling. Plant Cell 11:1897–1909

Grennan AK (2008) Ethylene response factors in jasmonate signaling and defense response. Plant Physiol 146:1457–1458

Gu YQ, Wildermuth MC, Chakravarthy S, Loh YT, Yang C, He X, Han Y, Martin GB (2002) Tomato transcription factors pti4, pti5, and pti6 activate defense responses when expressed in *Arabidopsis*. Plant Cell 14:817–831

Gu Y, Wang Z, Yang Z (2004) ROP/RAC GTPase: an old new master regulator for plant signaling. Curr Opin Plant Biol 7:527–536

Guan LM, Zhao J, Scandalios JG (2000) *Cis*-elements and *trans*-factors that regulate expression of the maize *Cat1* antioxidant gene response to ABA and osmotic stress: H_2O_2 is the likely intermediary signaling molecule for the response. Plant J 22:87–95

Gudesblat GE, Torres PS, Vojnov AA (2009) *Xanthomonas campestris* overcomes Arabidopsis stomatal innate immunity through a DSF cell-to-cell signal-regulated virulence factor. Plant Physiol 149:1017–1027

Guimarães RL, Stotz HU (2004) Oxalate production by *Sclerotinia sclerotiorum* deregulates guard cells during infection. Plant Physiol 136:3703–3711

Guo H, Ecker JR (2004) The ethylene signaling pathway; new insights. Curr Opin Plant Biol 7:40–49

Guo Y, Xiong L, Song CP, Gong D, Halfter U, Zhu JK (2002) A calcium sensor and its interacting protein kinase are global regulators of abscisic acid signaling in *Arabidopsis*. Dev Cell 3:233–244

Guo L, Yu Y, Law JA, Zhang X (2010) SET DOMAIN GROUP2 is the major histone H3 lysine 4 trimethyltransferase in *Arabidopsis*. Proc Natl Acad Sci U S A 107:18557–18562

Guo J, Yang X, Weston DJ, Chen J-G (2011) Abscisic acid receptors: past, present and future. J Integr Plant Biol 53:469–479

Gutterson N, Reuber TL (2004) Regulation of disease resistance pathways by AP2/ERF transcription factors. Curr Opin Plant Biol 7:465–471

Hamada H, Kurusu T, Okuma E, Nokajima H, Kiyoduka M, Koyano T, Sugiyama Y, Okada K, Koga J, Saji H, Miyao A, Hirochika H, Yamane H, Murata Y, Kuchitsu K (2012) Regulation of a proteinaceous elicitor-induced Ca^{2+} influx and production of phytoalexins by a putative voltage-gated cation channel, OsTPC1, in cultured rice cells. J Biol Chem 287:9931–9939

Hamilton DWA, Hills A, Köhler B, Blatt MR (2000) Ca^{2+} channels at the plasma membrane of stomatal guard cells are activated by hyperpolarization and abscisic acid. Proc Natl Acad Sci U S A 97:4967–4972

Harmon AC, Gribskov M, Harper JF (2000) CDPKs – a kinase for every Ca^{2+} signal? Trends Plant Sci 5:154–159

Hashimoto K, Eckert C, Anschūtz U, Scholz M, Held K, Waadt R, Reyer A, Hippler M, Becker D, Kudla J (2012) Phosphorylation of calcineurin B-like (CBL) calcium sensor proteins by their CBL-interacting protein kinases (CIPKs) is required for full activity of CBL-CIPK complexes toward their target proteins. J Biol Chem 287:7956–7968

Hauser SY, Waadt R, Schroeder JJL (2011) Evolution of abscisic acid synthesis and signalling mechanism. Curr Biol 21:R346–R355

He H, Li J (2008) Proteomic analysis of phosphoproteins regulated by abscisic acid in rice leaves. Biochem Biophys Res Commun 371:883–888

Henfling JWDM, Bostock R, Kuc J (1980) Effect of abscisic acid on rishitin and lubimin accumulation and resistance to *Phytophthora infestans* and *Cladosporium cucumerinum* in potato tuber tissue slices. Phytopathology 70:1074–1078

Hernández-Blanco C, Feng DX, Hu J, Sanchez-Vallet A, Deslandes L, Llorente F, Berrocal-Lobo M, Keller H, Barlet X, Sánchez-Rodriguez C, Anderson LK, Somerville S, Marco Y, Molina A (2007) Impairment of cellulose synthase required for *Arabidopsis* secondary cell wall formation enhances disease resistance. Plant Cell 19:890–903

Hettenhausen C, Baldwin IT, Wu J (2012) Silencing *MPK4* in *Nicotiana attenuata* enhances photosynthesis and seed production but compromises abscisic acid-induced stomatal closure and guard cell-mediated resistance to *Pseudomonas syringae* pv. *tomato* DC3000. Plant Physiol 158:759–776

Hirayama T, Shinozaki K (2007) Perception and transduction of abscisic acid signals: keys to the function of the versatile plant hormone ABA. Trends Plant Sci 12:343–351

Hobo T, Kowyama Y, Hattori T (1999) A bZIP factor, TRAB1, interacts with VP1 and mediates abscisic acid-induced transcription factor. Proc Natl Acad Sci U S A 96:15348–15353

Hossain MA, Munemasa S, Uraji M, Nakamura Y, Mori IC, Murata Y (2011) Involvement of endogenous abscisic acid in methyl jasmonate-induced stomatal closure in *Arabidopsis*. Plant Physiol 156:430–438

Hrabak EM, Chan CW, Gribskov M, Harper JF, Choi JH, Halford N, Kudla J, Luan S, Nimmo HG, Sussman MR, Thomas M, Walker-Simmons K, Zhu J-K, Harmon AC (2003) The Arabidopsis CDPK-SnRK superfamily of protein kinases. Plant Physiol 132:666–680

Hu J, Wang Y, Zhang X, Lloyd JR, Li JH, Karpiak J, Costanzi S, Wess J (2010) Structural basis of G protein-coupled receptor – G protein interactions. Nat Chem Biol 6:541–548

Hubbard KE, Nishimura N, Hitomi K, Getzoff ED, Schroeder JI (2010) Early abscisic acid signal transduction mechanisms: newly discovered components and newly emerging questions. Genes Dev 24:1695–1708

Hubbard KE, Siegel RS, Valerio G, Brandt B, Schroeder JI (2012) Abscisic acid and CO_2 signaling via calcium sensitivity in guard cells, new CDPK mutant phenotypes and a method for improved resolution of stomatal stimulus–response analyses. Ann Bot 109:5–17

Hung KT, Cheng DG, Hsu YT, Kao CH (2008) Abscisic acid-induced hydrogen peroxide is required for anthocyanin accumulation in leaves of rice seedlings. J Plant Physiol 165:1280–1287

Iriti M, Faoro F (2008) Abscisic acid is involved in chitosan-induced resistance to *Tobacco necrosis virus* (TNV). Plant Physiol Biochem 46:1106–1111

Islam MM, Tani C, Watanabe-Sugimoto M, Uraji M, Jahan MS, Masuda C, Nakamura Y, Mori IC, Murata Y (2009) Myrosinases, TGG1 and TGG2 redundantly function in ABA and MeJA signaling in Arabidopsis guard cells. Plant Cell Physiol 50:1171–1175

Ito T, Nakata M, Fukazawa J, Ishida S, Takahashi Y (2010) Alteration of substrate specificity: the variable N-terminal domain of tobacco Ca^{2+}-dependent protein kinase is important for substrate recognition. Plant Cell 22:1592–1604

Jammes F, Song C, Shin D, Munemasa S, Takeda K, Gu D, Lee S, Cho D, Giordo R, Sritubtim S, Leonhardt N, Ellis BE, Murata Y, Kwak JM (2009) MAP kinases MPK9 and MPK12 are preferentially expressed in guard cells and positively regulate ROS-mediated ABA signalling. Proc Natl Acad Sci U S A 106:20520–20525

Jammes F, Yang X, Xiao S, Kwak JM (2011) Two Arabidopsis guard cell-preferential MAPK genes, *MPK9* and *MPK12*, function in biotic stress response. Plant Signal Behav 6:1875–1877

Jang J-H, Lee CS, Hwang D, Ryu SH (2012) Understanding of the roles of phospholipase D and phosphatidic acid through their binding partners. Prog Lipid Res 51:71–81

Jensen MK, Hagedorn PH, de Torres-Zabala M, Grant MR, Rung JH, Collinge DB, Lyngkjaer MF (2008) Transcriptional regulation by an NAC (NAM-ATAF1,2-CUC2) transcription factor attenuates ABA signaling for efficient basal defence towards *Blumeria graminis* f. sp. *hordei* in *Arabidopsis*. Plant J 56:867–880

Jeworutzki E, Roelfsema MR, Anschütz U, Krol E, Elzenga JT, Felix G, Boller T, Hedrich R, Becker D (2010) Early signaling through the *Arabidopsis* pattern recognition receptors FLS2 and EFR involves Ca^{2+}-associated opening of plasma membrane anion channels. Plant J 62:367–378

Jiang C-J, Shimono M, Sugano S, Kojima M, Yazawa K, Yoshida R, Inoue H, Hayashi N, Sakakibara H, Takatsuji H (2010) Abscisic acid interacts antagonistically with salicylic acid signaling pathway in rice-*Magnaporthe grisea* interaction. Mol Plant Microbe Interact 23:791–798

Johnson RR, Wagner RL, Verhey SD, Walker-Simmons MK (2002) The abscisic acid-responsive kinase PKABA1 interacts with a seed-specific abscisic acid response element-binding factor, TaABF, and phosphorylates TaABF peptide sequences. Plant Physiol 130:837–846

Johnston CA, Temple BR, Chen J-G, Gao Y, Moriyama EN, Jones AM, Siderovski DP, Willard FS (2007) Comment on "A G protein-coupled receptor is a plasma membrane receptor for the plant hormone abscisic acid". Science 318:914c

Joshi-Saha A, Valon C, Leung J (2011) Abscisic acid signal off the STARTing block. Mol Plant. doi:10.1093/mp/ssr055

Kaliff M, Staal J, Myrenas M, Dixelius C (2007) ABA is required for *Leptosphaeria maculans* resistance via ABI- and ABI4-dependent signaling. Mol Plant Microbe Interact 20:335–345

Kang Z, Buchenauer H (2000) Ultrastructural and immunocytochemical investigation of pathogen development and host responses in resistant and susceptible wheat spikes infected by *Fusarium culmorum*. Physiol Mol Plant Pathol 57:255–268

Kang J, Hwang J, Lee M, Kim Y, Assmann SM, Martinola L, Lee Y (2010) PDR-type ABC transporter mediates cellular uptake of the phytohormone abscisic acid. Proc Natl Acad Sci U S A 107:2355–2360

Kaplan B, Davydov O, Knight H, Galon Y, Knight MR, Fluhr R, Fromm H (2006) Rapid transcriptome changes induced by cytosolic Ca^{2+} transients reveal ABRE-related sequences as Ca^{2+}-responsive *cis* elements in *Arabidopsis*. Plant Cell 10:2733–2748

Kidd BN, Cahill DM, Manners JM, Schenk PM, Kazan K (2011) Diverse roles of the Mediator complex in plants. Semin Cell Dev Biol 22:741–748

Kim DS, Hwang BK (2014) An important role of the pepper phenylalanine ammonia-lyase gene (PAL1) in salicylic acid-dependent signaling of the defence response to microbial pathogens. J Exp Bot. doi:10.1093/jxb/eru109

Kim SY, Thomas TL (1998) A family of novel basic leucine zipper proteins bind to seed-specification elements in the carrot Dc3 gene promoter. J Plant Physiol 152:607–613

Kim SY, Chung HJ, Thomas TL (1997) Isolation of a novel class of bZIP transcription factors that interact with ABA-responsive and embryo-specification elements in the Dc3 promoter using a modified yeast one-hybrid system. Plant J 11:1237–1251

Kim JA, Agrawal GK, Rakwal R, Han KS, Kim KN, Yun CH, Heu S, Park SY, Lee YH, Jwa NS (2003) Molecular cloning and mRNA expression analysis of a novel rice (*Oryza sativa* L.) MAPK kinase kinase, OsEDR1, an ortholog of Arabidopsis AtEDR1, reveal its role in defense/stress signalling pathways and development. Biochem Biophys Res Commun 300:858–876

Kim T-W, Guan S, Sun Y, Deng Z, Tang W, Shang JX, Sun Y, Burlingame AL, Wang ZY (2009) Brassinosteroid signal transduction from cell-surface receptor kinases to nuclear transcription factors. Nat Cell Biol 11:1254–1260

Kim T, Böhmer M, Hu H, Nishimura N, Schroeder JI (2010) Guard cell signal transduction network: advances in understanding abscisic acid, CO_2 and Ca^{2+} signaling. Annu Rev Plant Biol 61:561–591

Kim TH, Hauser F, Ha T, Xue S, Böhmer M, Nishimura N, Munemasa S, Hubbard K, Peine N, Lee BH, Lee S, Robert N, Parker JE, Schroeder JI (2011) Chemical genetics reveals negative regulation of abscisic acid signaling by a plant immune response pathway. Curr Biol 21:990–997

Kline KG, Sussman MR, Jones AM (2010) Abscisic acid receptors. Plant Physiol 154:479–482

Klüsener B, Young JJ, Murata Y, Allen GJ, Mori IC, Hogouvieux V, Schroeder JI (2002) Convergence of calcium signaling pathways of pathogenic elicitors and abscisic acid in Arabidopsis guard cells. Plant Physiol 130:2152–2163

Kobayashi Y, Yamamoto S, Minami H, Kagaya Y, Hattori T (2004) Differential activation of the rice sucrose nonfermenting 1-related protein kinase2 family by hyperosmotic stress and abscisic acid signaling. Plant Cell 16:1163–1177

Kobayashi Y, Murata M, Minami H, Yamamoto S, Kagaya Y, Hobo T, Yamamoto A, Hattori T (2005) Abscisic acid-activated SNRK2 protein kinases function in the gene-regulation pathway of ABA signal transduction by phosphorylation of ABA response element –binding factors. Plant J 44:939–949

Kobayashi M, Ohura L, Kawakita K, Yokota N, Fujiwara M, Shimamoto K, Doke N, Yoshioka H (2007) Calcium-dependent protein kinases regulate the production of reactive oxygen species by potato NADPH oxidase. Plant Cell 19:1065–1080

Koga H, Dohi K, Mori M (2004) Abscisic acid and low temperatures suppress the whole plant-specific resistance reaction of rice plants to the infection of *Magnaporthe grisea*. Physiol Mol Plant Pathol 65:3–9

Köhler B, Blatt MR (2002) Protein phosphorylation activates the guard cell Ca^{2+} channel and is a prerequisite for gating by abscisic acid. Plant J 32:185–194

Kudla J, Batistić O, Hashimoto K (2010) Calcium signals: the lead currency of plant information processing. Plant Cell 22:541–563

Kuhn JM, Boisson-Dernier A, Dizon MB, Maktabi MH, Schroeder JI (2006) The protein phosphatase AtPP2CA negatively regulates abscisic acid signal transduction in *Arabidopsis*, the effects of *abh1* on AtPP2CA mRNA. Plant Physiol 140:127–139

Kuromori T, Miyaji T, Yabuuchi H, Shimizu H, Sugimoto E, Kamiya A, Moriyama Y, Shinozaki K (2010) ABC transporter AtABCG25 is involved in abscisic acid transport and responses. Proc Natl Acad Sci U S A 107:2361–2366

Kwak JM, Moon JH, Murata Y, Kuchitsu K, Leonhardt N, Delong A, Schroeder JI (2002) Disruption of a guard cell-expressed protein phosphatase 2A regulatory subunit, RCN1, confers abscisic acid insensitivity in *Arabidopsis*. Plant Cell 14:2849–2861

Kwak JM, Mori IC, Pei ZM, Leonhardt N, Torres MA, Dangl JL, Bloom RE, Bodde S, Jones JD, Schroeder JI (2003) NADPH oxidase *AtrbohD* and *AtrbohF* genes function in ROS-dependent ABA signaling in *Arabidopsis*. EMBO J 22:2623–2633

Lackman P, González-Guzmán M, Tilleman S, Carqueijeiro I, Cuéllar-Pérez A, Moses T, Seo M, Kanno Y, Häkkinen ST, Van Montagu MCE, Thevelein JM, Maaheimo H, Oksman-Caldentey KM, Rodriguez PL, Rischer H, Goossens A (2011) Jasmonate signaling involves the abscisic acid receptor PYL4 to regulate metabolic reprogramming in *Arabidopsis* and tobacco. Proc Natl Acad Sci U S A 108:5891–5896

Leckie CP, McAinsh MR, Allen GJ, Sanders D, Hetherington AM (1998) Abscisic acid-induced stomatal closure mediated by cyclic ADP-ribose. Proc Natl Acad Sci U S A 95:15833–15842

Lecourieux D, Ranjeva R, Pugin A (2006) Calcium in plant defence-signalling pathways. New Phytol 171:249–269

Lee Y, Choi YB, Suh S, Lee J, Assmann SM, Joe CO, Kelleher JF, Crain RC (1996) Abscisic acid-induced phosphoinositide turnover in guard cell protoplasts of *Vicia faba*. Plant Physiol 110:987–996

Lee SC, Lan W, Buchanan BB, Luan S (2009) A protein kinase-phosphatase pair interacts with an ion channel to regulate ABA signaling in plant guard cells. Proc Natl Acad Sci U S A 106:21419–21424

Lemichez E, Wu Y, Sanchez J-P, Mettouchi A, Mathur J, Chua N-H (2001) Inactivation of AtRac1 by abscisic acid is essential for stomatal closure. Genes Dev 15:1808–1816

L'Haridon F, Besson-Bard A, Binda M, Serrano M, Abou-Mansour E, Balet F, Schoonbeek HJ, Hess S, Mir R, Leon J, Lamotte O, Metraux J-P (2011) A permeable cuticle is associated with the release of reactive oxygen species and induction of innate immunity. PLoS Pathog 7:e1002148

Li J, Wang XQ, Watson MB, Assman SM (2000) Regulation of abscisic acid-induced stomatal closure and anion channels by guard cell AAPK kinase. Science 287:300–303

Li J, Kinoshita T, Pandey S, Ng CK-Y, Gygi SP, Shimazaki K, Assmann SM (2002) Modulation of an RNA-binding protein by abscisic acid-activated protein kinase. Nature 418:793–797

Liu Y, Zhang S (2004) Phosphorylation of 1-aminocyclopropane-1-carboxylic acid synthase by MPK6, a stress-responsive mitogen-activated protein kinase, induces ethylene biosynthesis in *Arabidopsis*. Plant Cell 16:3386–3399

Liu X, Yue Y, Li W, Nie Y, Li W, Wu W-H, Ma L (2007a) A G protein-coupled receptor is a plasma membrane receptor for the plant hormone abscisic acid. Science 315:1712–1716

Liu X, Yue Y, Li W, Ma L (2007b) Response to comment on "A G protein-coupled receptor is a plasma membrane receptor for the plant hormone abscisic acid". Science 318:914d

Liu Q, Xu C, Wen C-K (2010) Genetic and transformation studies reveal negative regulation of ERS1 ethylene receptor signaling in *Arabidopsis*. BMC Plant Biol 10:60

Liu D, Chen X, Liu J, Ye J, Guo Z (2012) The rice ERF transcription factor *OsERF922* negatively regulates resistance to *Magnaporthe oryzae* and salt tolerance. J Exp Bot 63:3899–3911

Lorenzo O, Chico JM, Sánchez-Serrano JJ, Solano R (2004) JASMONATE-INSENSITIVE1 encodes a MYC transcription factor essential to discriminate between different jasmonate-regulated defense responses in *Arabidopsis*. Plant Cell 16:1938–1950

Lu C, Han MH, Guevara-Garcia A, Fedoroff V (2002) Mitogen-activated protein kinase signaling in postgermination arrest of development by abscisic acid. Proc Natl Acad Sci U S A 99:15812–15817

Lü D, Zhang X, Jiang J, An GY, Zhang I, Song C-P (2005) NO may function in the downstream of H_2O_2 in ABA-induced stomatal closure in *Vicia faba* L. J Plant Physiol Mol Biol 31:62–70

Luan S, Kudla J, Rodriguez-Concepcion M, Yalovsky S, Gruissem W (2002) Calmodulins and calcineurin B-like proteins: calcium sensors for specific signal response coupling in plants. Plant Cell 14(Suppl):S389–S400

Ludwig AA, Romeis T, Jones JDG (2004) CDPK-mediated signaling pathways: specificity and cross-talk. J Exp Bot 55:181–188

Ma Y, Szostkiewicz I, Korte A, Moes D, Yang Y, Christmann A, Grill E (2009) Regulators of PP2C phosphatase activity function as abscisic acid sensors. Science 324:1064–1068

Maathuis FJM, Ichida AM, Sanders D, Schroeder JI (1997) Roles of higher K^+ channels. Plant Physiol 114:1141–1149

Marin E, Nussaume L, Quesada A, Gonneau M, Sotta B, Huguency P, Frey A, Marion-Poll A (1996) Molecular identification of zeaxanthin epoxidase of *Nicotiana plumbaginifolia*, a gene involved in abscisic acid biosynthesis and corresponding to the ABA locus of *Arabidopsis thaliana*. EMBO J 15:2331–2342

Marten H, Konrad KR, Dietrich P, Roelfsema RG, Hedrich R (2007) Ca^{2+}-dependent and –independent abscisic acid activation of plasma membrane anion channels in guard cells of *Nicotiana tabacum*. Plant Physiol 143:28–37

Mauch-Mani B, Mauch F (2005) The role of abscisic acid in plant-pathogen interactions. Curr Opin Plant Biol 8:409–414

McAinsh MR, Pittman JK (2009) Shaping the calcium signature. New Phytol 181:275–294

McCourt P, Creelman R (2008) The ABA receptors – we report you decide. Curr Opin Plant Biol 11:474–478

McCudden CR, Hains MD, Kimple RJ, Siderovski DP, Willard FS (2005) G-protein signaling: back to the future. Cell Mol Life Sci 62:551–577

McDonald KL, Cahill DM (1996) Influence of abscisic acid and the abscisic acid biosynthesis inhibitor, norflurazon, on interaction between *Phytophthora sojae* and soybean (*Glycine max*). Eur J Plant Pathol 105:651–658

Melcher K, Ng LM, Zhou XE, Soon FF, Xu Y, Suino-Powell KM, Park SY, Weiner JJ, Fujii H, Chinnusamy V, Kovach A, Li Jun, Wang Y, Li Jiayang, Peterson FC, Jensen DR, Yong EL, Volkman BF, Cutler SR, Zhu JK, Xu HE (2009) A gate-latch-lock mechanism for hormone signalling by abscisic acid receptors. Nature 462:602–608

Melcher K, Xu Y, Ng L-M, Zhou E, Soon F-F, Chinnusamy V, Suino-Powell KM, Kovach A, Tham FS, Cutler SR, Li J, Yong E-L, Zhu J-K, Xu HE (2010a) Identification and mechanism of ABA receptor antagonism. Nat Struct Mol Biol 17:1102–1108

Melcher K, Zhou XE, Xu HE (2010b) Thirsty plants and beyond: structural mechanisms of abscisic acid perception and signalling. Curr Opin Struct Biol 20:722–729

Melotto M, Underwood W, Koczan J, Nomura K, He SY (2006) Plant stomata function in innate immunity against bacterial invasion. Cell 126:969–980

Melotto M, Underwood W, He SY (2008) Role of stomata in plant innate immunity and foliar bacterial diseases. Annu Rev Phytopathol 46:101–122

Mengiste T, Chen X, Salmeron J, Dietrich R (2003) The *Botrytis Susceptible 1* gene encodes an R2R3MYB transcription factor protein that is required for biotic and abiotic stress responses in Arabidopsis. Plant Cell 15:2551–2565

Merlot S, Gosti F, Guerrier D, Vavasseaur A, Giraudat J (2001) The ABI1 and ABI2 protein phosphatases 2C act in a negative feedback regulatory loop of the abscisic acid signalling pathway. Plant J 25:295–303

Milborrow BV (2001) The pathway of biosynthesis of abscisic acid in vascular plants: a review of the present state of knowledge of ABA biosynthesis. J Exp Bot 52(359):1145–1164

Mishra G, Zhang W, Deng F, Zhao J, Wang X (2006) A bifurcating pathway directs abscisic acid effects on stomatal closure and opening in *Arabidopsis*. Science 312:264–266

Miyaloudama AS, Heintz D, Debayle D, Rahier A, Camara B, Bouvier F (2009) Abscisic acid negatively regulates elicitor-induced synthesis of capsidiol in wild tobacco. Plant Physiol 150:1556–1566

Miyazono K, Miyakawa T, Sawano Y, Kubota K, Kang HJ, Asano A, Miyauchi Y, Takahashi M, Zhi Y, Fujita Y, Yoshida T, Kodaira KS, Yamaguchi-Shinozaki K, Tanokura M (2009) Structural basis of abscisic acid signalling. Nature 462:609–614

Moes D, Himmelbach A, Korte A, Haberer G, Grill E (2008) Nuclear localization of the mutant protein phosphatase abi1 is required for sensitivity towards ABA responses in *Arabidopsis*. Plant J 54:806–819

Mohr PG, Cahill DM (2003) Abscisic acid influences the susceptibility of *Arabidopsis thaliana* to *Pseudomonas syringae* pv. *tomato* and *Peronospora parasitica*. Funct Plant Biol 30:461–469

Mohr PG, Cahill DM (2004) Suppression by abscisic acid of lignin production and monolignol pathway gene expression in interactions of Arabidopsis with oomycete and bacterial pathogens. Phytopathology: abstracts of American Society of Plant Biologists annual meeting, p 74

Mohr PG, Cahill DM (2007) Suppression by ABA of salicylic acid and lignin accumulation and the expression of multiple genes, in *Arabidopsis* infected with *Pseudomonas syringae* pv. *tomato*. Funct Integr Genomics 7:181–191

Moore JW, Loake CJ, Spoel SH (2011) Transcription dynamics in plant immunity. Plant Cell 23:2809–2820

Mori IC, Murata Y, Yang Y, Munemasa S, Wang Y, Andreoli S, Tiriac H, Alonso JM, Harper JF, Ecker JR, Kwak JM, Schroeder JI (2006) CDPKs CPK6 and CPK3 function in ABA regulation of guard cell S-type anion and Ca^{2+}-permeable channels and stomatal closure. PLoS Biol 4:e327

Mou Z, Fan W, Dong X (2003) Inducers of plant systemic acquired resistance regulate NPR1 function through redox changes. Cell 113:935–944

Mucha E, Fricke I, Schaefer A, Wittinghofer A, Berken A (2011) Rho proteins of plants-functional cycle and regulation of cytoskeletal dynamics. Eur J Cell Biol 90:934–943

Muller AH, Hansson M (2009) The barley magnesium chelatase 150-kb subunit is not an abscisic acid receptor. Plant Physiol 150:157–166

Munemasa S, Oda K, Watanabe-Sugimoto M, Nakamura Y, Shimoishi Y, Murata Y (2007) The coronatine-insensitive 1 mutation reveals the hormonal signaling interaction between abscisic acid and methyl jasmonate in *Arabidopsis* guard cells. Specific impairment of ion channel activation and second messenger production. Plant Physiol 143:1398–1407

Munemasa S, Mori IC, Murata Y (2011) Methyl jasmonate signaling and signal crosstalk berween methyl jasmonate and abscisic acid in guard cells. Plant Signal Behav 6:939–941

Munemasa S, Muroyama D, Nagahashi H, Nakamura Y, Mori IC, Murata Y (2013) Regulation of reactive oxygen species –mediated abscisic acid signaling in guard cells and drought tolerance by glutathione. Front Plant Sci 4:472. doi:10.3389/fpls.2013.00472

Murata Y, Pei ZM, Mori IC, Schroeder J (2001) Abscisic acid activation of plasma membrane Ca^{2+} channels in guard cells requires cytosolic NAD(P)H and is differentially disrupted upstream and downstream of reactive oxygen species production in *abi1-1* and *abi2-1* protein phosphatase 2C mutants. Plant Cell 13:2513–2523

Mustilli A-C, Merlot S, Vavasseur A, Fenzi F, Giraudat J (2002) *Arabidopsis* OST1 protein kinase mediates the regulation of stomatal aperture by abscisic acid and acts upstream and downstream of reactive oxygen species production. Plant Cell 14:3089–3099

Nakashima K, Fujita Y, Kanamori N, Katagiri T, Umezawa T, Kidokoro S, Maruyama K, Yoshida T, Ishiyama K, Kobayashi M, Shinozaki K, Yamaguchi-Shinozaki K (2009) Three Arabidopsis SnRK2 protein kinases, SRK2D/SnRK2.2, SRK2E/SnRK2.6/OST1 and SRK21/SnRK2.3 involved in ABA signaling are essential for the control of seed development and dormancy. Plant Cell Physiol 50:1345–1363

Negi J, Matsuda O, Nagasawa T, Oba Y, Takahashi H, Kawai-Yamada M, Uchimiya H, Hashimoto M, Iba K (2008) CO2 regulator SLAC1 and its homologues are essential for anion homeostasis in plant cells. Nature 452:483–486

Neill SJ, Desikan R, Clarke A, Hurst RD, Hancock JT (2002) Hydrogen peroxide and nitric oxide as signaling molecules in plants. J Exp Bot 53:1237–1247

Nicaise V, Roux M, Zipfel C (2009) Recent advances in PAMP-triggered immunity against bacteria: pattern recognition receptors watch over and raise the alarm. Plant Physiol 150:1638–1647

Nishimura N, Hitomi K, Arvai AS, Rambo RP, Hitomi C, Cutler SR, Schroeder JI, Getzoff ED (2009) Structural mechanism of abscisic acid binding and signaling by dimeric PYR1. Science 326:1373–1379

Nishimura N, Sarkeshik A, Nito K, Park SY, Wang A, Carvalho PC, Lee S, Caddell DF, Cutler SR, Chory J, Yates JR, Schroeder JI (2010) PYR/PYL/RCAR family members are major *in-vivo* ABI1 protein phosphatase 2C-interacting proteins in *Arabidopsis*. Plant J 61:290–299

Oki K, Inaba N, Kitagawa K, Fujioka S, Kitano H, Fujisawa Y, Kato H, Iwasaki Y (2009) Function of the α subunit of rice heterotrimeric G protein in brassinosteroid signaling. Plant Cell Physiol 50:161–172

Pandey S, Chen J-G, Jones AM, Assmann SM (2006) G-protein complex mutants are hypersensitive to abscisic acid regulation of germination and postgermination development. Plant Physiol 141:243–256

Pandey S, Nelson DC, Assmann SM (2009) Two novel GPCR-type G proteins are abscisic acid receptors in *Arabidopsis*. Cell 136:136–148

Pandey S, Wang R-S, Wilson L, Li S, Zhao Z, Gookin TE, Assmann SM, Albert R (2010) Boolean modeling of transcriptome data reveals novel modes of heterotrimeric G-protein action. Mol Syst Biol 6:372

Park JM, Park CJ, Lee SB, Ham BK, Shin R, Paek KH (2001) Overexpression of the tobacco *Tsi1* gene encoding an EREBP/AP2-type transcription factor enhances resistance against pathogen attack and osmotic stress in tobacco. Plant Cell 13:1035–1046

Park SY, Fung P, Nishimura N, Jensen DR, Fujii H, Zhao Y, Lumba S, Santiago J, Rodrigues A, Chow TF, Alfred SE, Bonetta D, Finkelstein R, Provart NJ, Desveaux D, Rodriguez PL, McCourt P, Zhu JK, Schroeder JI, Volkman BF, Cutler SR (2009) Abscisic acid inhibits type 2C protein phosphatases via the PYR/PYL family of START proteins. Science 324:1068–1071

Pei Z-M, Murata Y, Benning G, Thomine S, Klüsener B, Allen GJ, Grill E, Schroeder JI (2000) Calcium channels activated by hydrogen peroxide mediate abscisic acid signaling in guard cells. Nature 406:731–734

Perfus-Barbeoch I, Jones AM, Assmann SM (2004) Plant heterotrimeric G protein function: insights from *Arabidopsis* and rice mutants. Curr Opin Plant Biol 7:719–731

Potuschak T, Lechner E, Parmentier Y, Yanagisawa S, Grava S, Koncz C, Genschik P (2003) EIN3-dependent regulation of plant ethylene hormone signaling by two *Arabidopsis* F box proteins EBF1 and EBF2. Cell 115:679–689

Raghavendra AS, Gonugunta VK, Christmann A, Grill L (2010) ABA perception and signalling. Trends Plant Sci 15:395–401

Razem FA, El-Kereamy A, Abrams SR, Hill RD (2006) The RNA-binding protein FCA is an abscisic acid receptor. Nature 439:290–294

Razem FA, El-Kereamy A, Abrams SR, Hill RD (2008) Retraction. The RNA-binding protein FCA is an abscisic acid receptor. Nature 456:824

Reddy ASN, Ali GS, Celesnik H, Day IS (2011a) Coping with stresses: roles of calcium- and calcium/calmodulin-regulated gene expression. Plant Cell 23:2010–2032

Reddy ASN, Ben-Hur A, Day IS (2011b) Experimental and computational approaches for the study of calmodulin interactions. Phytochemistry 72:1007–1019

Rezzonico E, Flury N, Meins F, Beffa R (1998) Transcriptional down-regulation by abscisic acid of pathogenesis-related β-1,3-glucanase genes in tobacco cell cultures. Plant Physiol 117:585–592

Risk JM, Macknight RC, Day CL (2008) FCA does not bind abscisic acid. Nature 456:E5–E6

Risk JM, Day CL, Macknight RC (2009) Reevaluation of abscisic acid-binding assays shows that G-Protein-Coupled Receptor 2 does not bind abscisic acid. Plant Physiol 150:6–11

Roelfsema MR, Hedrich R (2004) ABA depolarizes guard cells in intact plants, through a transient activation of R- and S-type anion channels. Plant J 37:578–588

Romeis T, Ludwig AA, Martin R, Jones JD (2001) Calcium-dependent protein kinases play an essential role in a plant defence response. EMBO J 20:5556–5567

Saez A, Apostolova N, Gonzalez-Guzman M, Gonzalez-Garcia MP, Nicolas C, Lorenzo O, Rodriguez PL (2004) Gain-of-function and loss-of-function phenotypes of the protein phosphatase 2C HAB1 reveal its role as a negative regulator of abscisic acid signaling. Plant Physiol 37:354–369

Saez A, Robert N, Maktabi MH, Schroeder JL, Serrano R, Rodriguez PL (2006) Enhancement of abscisic acid sensitivity and reduction of water consumption in Arabidopsis by combined inactivation of the protein phosphatases type 2C ABI1 and HAB. Plant Physiol 141:1389–1399

Saito N, Munemasa S, Nakamura Y, Shimoishi Y, Mori IC, Murata Y (2008) Roles of RCN1, regulatory A subunit of protein phosphatase 2A, in methyl jasmonate signaling and signal crosstalk between methyl jasmonate and abscisic acid. Plant Cell Physiol 49:1396–1401

Saito N, Nakamura Y, Mori IC, Murata Y (2009) Nitric oxide functions in both methyl jasmonate signaling and abscisic acid signaling in Arabidopsis guard cells. Plant Signal Behav 4:119–120

Sakamoto H, Matsuda O, Iba K (2008) *ITN1*, a novel gene encoding an ankyrin-repeat protein that affects the ABA-mediated production of reactive oxygen species and is involved in salt-stress tolerance in *Arabidopsis thaliana*. Plant J 56:411–422

Sánchez-Vallet A, López C, Ramos B, Delgado-Cerezo M, Riviere P, Llorente F, Fernández PV, Miedes E, Estevez JM, Grant M, Molina A (2012) Disruption of abscisic acid signaling constitutively activates Arabidopsis resistance to the necrotrophic fungus *Plectosphaerella cucumerina*. Plant Physiol 160:2109–2124

Santiago J, Dupeux F, Round A, Antoni R, Park SY, Jamin M, Cutler SR, Rodriguez PL, Marquez JA (2009a) The abscisic acid receptor PYR1 in complex with abscisic acid. Nature 462:665–668

Santiago J, Rodrigues A, Saez A, Rubio S, Antoni R, Dupeux F, Park SY, Marquez JA, Cutler SR, Rodriguez PL (2009b) Modulation of drought resistance by the abscisic acid receptor PYL5 through inhibition of clade A PP2Cs. Plant J 60:575–578

Sato A, Sato Y, Fukao Y, Fujiwara M, Umezawa T, Shinozaki K, Hibi T, Taniquchi M, Miyake H, Goto DB, Uozumi N (2009) Threonine at position 306 of the KAT1 potassium channel is essential for channel activity and is a target site for ABA-activated SnRK2/OST1/SnRK2.6 protein kinase. Biochem J 424:439–448

Schachtman DP, Goodger JG (2008) Chemical root to shoot signaling under drought. Trends Plant Sci 9:236–243

Schmidt K, Pflugmacher M, Klages S, Maser A, Mock A, Stahl DJ (2008) Accumulation of the hormone abscisic acid (ABA) at the infection site of the fungus *Cercospora beticola* supports the role of ABA as a repressor of plant defence in sugar beet. Mol Plant Pathol 9:661–673

Schroeder JI, Hagiwara S (1989) Cytosolic calcium regulates ion channels in the plasma membrane of *Vicia faba* guard cells. Nature 338:427–430

Schroeder JI, Hagiwara S (1990) Repetitive increases in cytosolic Ca^{2+} of guard cells by abscisic acid activation of nonselective Ca^{2+} permeable channels. Proc Natl Acad Sci U S A 87:9305–9309

Schwartz SH, Tan BC, McCarty DR, Welch W, Zeevaart JA (2003) Substrate specificity and kinetics for VP14, a carotenoid cleavage dioxygenase in the ABA biosynthetic pathway. Biochim Biophys Acta 1619:9–14

Schweighofer A, Hirt H, Meskiene I (2004) Plant PP2C phosphatases: emerging functions in stress signaling. Trends Plant Sci 9:236–243

Seo PJ, Park C-M (2010) MYB96-mediated abscisic acid signals induce pathogen resistance response by promoting salicylic acid biosynthesis in *Arabidopsis*. New Phytol 186:471–483

Seo M, Peeters AJ, Koiwai H, Oritani T, Marion-Poll A, Zeevaart JA, Koornneef M, Kamiya Y, Koshiba T (2000) The *Arabidopsis* aldehyde oxidase (AAO3) gene product catalyzes the final step in abscisic acid biosynthesis in leaves. Proc Natl Acad Sci U S A 97:12908–12913

Shang Y, Yan L, Liu ZQ, Cao Z, Mei C, Xin Q, Wu FQ, Wang XF, Du SY, Jiang T, Zhang XF, Zhao R, Sun HL, Liu R, Yu YT, Zhang DP (2010) The Mg-chelatase H subunit of *Arabidopsis* antagonizes a group of WRKY transcription repressors to relieve ABA-responsive genes of inhibition. Plant Cell 22:1909–1935

Sheard LB, Zheng N (2009) Plant biology signal advance for abscisic acid. Nature 462:575–576

Sheen J (1996) Ca^{2+}-dependent protein kinase and stress signal transduction in plants. Science 274:1900–1902

Sheen J (1998) Mutational analysis of protein phosphatase 2C involved in abscisic acid signal transduction in higher plants. Proc Natl Acad Sci U S A 95:975–980

Shen YY, Wang XF, Wu FQ, Du SY, Cao Z, Shang Y, Wang XL, Peng CC, Yu XC, Zhu SY, Fan RC, Xu YH, Zhang DP (2006) The Mg-chelatase H subunit is an abscisic acid receptor. Nature 443:823–826

Sirichandra C, Gu D, Hu H, Davanture M, Lee S, Djaoui M, Valot B, Zivy M, Leung J, Merlot S (2009) Phosphorylation of Arabidopsis AtrbohF NADPH oxidase by OST1 protein kinase. FEBS Lett 583:2982–2986

Sirichandra C, Davanture M, Turk BE, Zivy M, Valot B, Leung J, Merlot S (2010) Arabidopsis ABA-activated kinase OST1 phosphorylates the bZIP transcription factor ABF3 and creates a 14-3-3 binding site involved in its turnover. PLoS One 5:e13935

Snedden WA, Fromm H (2001) Calmodulin as a versatile calcium signal transducer in plants. New Phytol 151:35–66

Solano R, Stepanova A, Chao Q, Ecker JR (1998) Nuclear events in ethylene signaling: a transcriptional cascade mediated by ETHYLENE –INSENSITIVE3 and ETHYLENE-RESPONSE FACTOR1. Genes Dev 12:3703–3714

Song CP, Agarwal M, Ohta M, Guo Y, Halfter U, Wang P, Zhu JK (2005) Role of an *Arabidopsis* AP2/EREBP-type transcriptional repressor in abscisic acid and drought stress responses. Plant Cell 17:2384–2396

Soosaar JL, Murch-Smith TM, Dinesh-Kumar SP (2005) Mechanisms of plant resistance to viruses. Nat Rev Microbiol 3:789–798

Spoel SH, Dong X (2008) Making sense of hormone crosstalk during plant immune response. Cell Host Microbe 3:348–351

Stael S, Wurzinger B, Mair A, Mehlmer N, Vothknecht UC, Teige M (2012) Plant organellar calcium signaling: an emerging field. J Exp Bot 63:1525–1542

Staxén I, Pical C, Montgomery LT, Gray JE, Hetherington AM, McAinsh MR (1999) Abscisic acid induces oscillations in guard-cell cytosolic free calcium that involve phosphoinositide-specific phospholipase C. Proc Natl Acad Sci U S A 96:1779–1784

Sugiyama Y, Uraji M, Watanabe-Sugimoto M, Okuma E, Munemasa S, Shimoishi Y, Nakamura Y, Mori IC, Iwai S, Murata Y (2012) FIA functions as an early signal component of abscisic acid signal cascade in *Vicia faba* guard cells. J Exp Bot 63:1357–1365

Suharsono U, Fujisawa Y, Kawasaki T, Iwasaki Y, Satoh H, Shimamoto K (2002) The heterotrimeric G protein subunit acts upstream of the small GTPase Rac in disease resistance of rice. Proc Natl Acad Sci U S A 99:13307–13312

Szostkiewicz I, Richter K, Kepka M, Demmel S, Ma Y, Korte A, Assaad FF, Christmann A, Grill E (2010) Closely related receptor complexes in their ABA selectivity and sensitivity. Plant J 61:25–35

Tahtiharju S, Palva T (2001) Antisense inhibition of protein phosphatase 2C accelerates cold acclimation in *Arabidopsis thaliana*. Plant J 26:461–470

Takahashi F, Yoshida R, Ichimura K, Mizoguchi T, Seo S, Yonezawa M, Maruyama K, Yamaguchi-Shinozaki K, Shinozaki K (2007) The mitogen-activated protein kinase cascade MKK3-MPK6 is an important part of the jasmonate signal transduction pathway in Arabidopsis. Plant Cell 19:805–818

Tan B-C, Joseph LM, Deng W-T, Liu L, Li Q-B, Cline K, McCarty DR (2003) Molecular characterization of the *Arabidopsis* 9-*cis* epoxycarotenoid dioxygenase gene family. Plant J 35:44–56

Tesmer JJG (2010) The quest to understand heterotrimeric G protein signaling. Nat Struct Mol Biol 17:650–652

Thaler JS, Bostock RM (2004) Interactions between abscisic acid-mediated responses and plant resistance to pathogens and insects. Ecology 85:48–58

Ton J, Mauch-Mani B (2004) β-Amino-butyric acid-induced resistance is based on ABA-dependent priming for callose. Plant J 38:119–130

Ton J, Jakab G, Toquin V, Iavicoli A, Flors V, Maeder MN, Métraux J-P, Mauch-Mani B (2005) Dissecting the β-aminobutyric acid-induced priming pathways in *Arabidopsis*. Plant Cell 17:987–999

Ton J, Flors V, Mauch-Mani B (2009) The multifaceted role of ABA in disease resistance. Trends Plant Sci 14:310–317

Trouverie J, Vidal G, Zhang Z, Sirichandra C, Madiona K, Amiar Z, Prioul JL, Jeannette E, Rona JP, Brault M (2008) Anion channel activation and proton pumping inhibition involved in the plasma membrane depolarization induced by ABA in *Arabidopsis thaliana* suspension cells are both ROS dependent. Plant Cell Physiol 49:1405–1507

Trusov Y, Chakravorty D, Botella JR (2012) Diversity of heterotrimeric G-protein γ subunits in plants. BMC Res Notes 5:608

Umezawa T, Sugiyama N, Mizoguchi M, Hayashi S, Myouga F, Yamaguchi-Shinozaki K, Ishihama Y, Hirayama T, Shinozaki K (2009) Type 2C protein phosphatases directly regulate abscisic acid-activated protein kinases in *Arabidopsis*. Proc Natl Acad Sci U S A 106: 17588–17593

Umezawa T, Nakashima K, Miyakawa T, Kuromori T, Tanokura M, Shinozaki K, Yamaguchi-Shinozaki K (2010) Molecular basis of the core regulatory network in ABA responses: sensing, signaling and transport. Plant Cell Physiol 51:1821–1839

Umezawa T, Sugiyama N, Takahashi F, Anderson JC, Ishihama Y, Peck SC, Shinozaki K (2013) Genetics and phosphoproteomics reveal a protein phosphorylation network in the abscisic acid signaling pathway in *Arabidopsis thaliana*. Sci Signal 6(270):rs8. doi:10.1126/scisignal.2003509

Uno Y, Furihata T, Abe H, Yoshida R, Shinozaki K, Yamaguchi-Shinozaki K (2000) Arabidopsis basic leucine zipper transcription factors involved in abscisic acid-dependent signal transduction pathway under drought and high-salinity conditions. Proc Natl Acad Sci U S A 97:11632–11637

Uraji M, Katagiri T, Okuma E, Te W, Hossain M-A, Masuda C, Miura A, Nakamura Y, Mori I-C, Shinozaki K, Murata Y (2012) Cooperative function of PLDδ and PLDα1 in abscisic acid-induced stomatal closure in Arabidopsis. Plant Physiol 159:450–460

van Loon LC, Rep M, Pieterse C (2006) Significance of inducible defense-related proteins in infected plants. Annu Rev Phytopathol 44:71–728

Verma DP, Hong Z (2001) Plant callose synthase complexes. Plant Mol Biol 47:693–701

Vidhyasekaran P (2007) Fungal pathogenesis in plants and crops: molecular biology and host defense mechanisms, 2nd edn. CRC Press, Taylor Francis Group, Boca Raton, pp 510

Vidhyasekaran P (2014) PAMP signals in plant innate immunity: signal perception and transduction. Springer, Dordrecht, pp 442

Vlad F, Rubio S, Rodrigues A, Sirichandra C, Belin C, Robert N, Leung J, Rodriguez PL, Lauriere C, Meriot S (2009) Protein phosphatases 2C regulate the activation of the Snf1-related kinase OST1 by abscisic acid in *Arabidopsis*. Plant Cell 21:3170–3184

Vranová E, Inzé D, Van Brueusegem F (2002) Signal transduction during oxidative stress. J Exp Bot 53:1227–1236

Wang X (2005) Regulatory functions of phospholipase D and phosphatidic acid in plant growth, development, and stress responses. Plant Physiol 139:566–573

Wang P, Song C-P (2008) Guard-cell signaling for hydrogen peroxide and abscisic acid. New Phytol 178:703–718

Wang XF, Zhang DP (2008) Abscisic acid receptors: multiple signal perception sites. Ann Bot 101:311–317

Wang XQ, Ullah H, Jones AM, Assmann SM (2001) G protein regulation of ion channels and abscisic acid signaling in Arabidopsis guard cells. Science 292:2070–2072

Wang PC, Du YY, An GY, Zhou Y, Miao C, Song C-P (2006a) Analysis of global expression profiles of *Arabidopsis* genes under abscisic acid and H_2O_2 applications. J Integr Plant Biol 48:62–74

Wang W, Esch JJ, Shiu S-H, Agula H, Binder BM, Chang C, Patterson S, Bleecker AB (2006b) Identification of important regions for ethylene binding and signaling in the transmembrane domain of the ER1 ethylene receptor of *Arabidopsis*. Plant Cell 18:3429–3442

Wang W-H, Yi X-Q, Han A-D, Liu T-W, Chen J, Wu F-H, Dong X-J, He J-X, Pei Z-M, Pei Z-M, Zheng H-L (2012) Calcium-sensing receptor regulates stomatal closure through hydrogen peroxide and nitric oxide in response to extracellular calcium in *Arabidopsis*. J Exp Bot 63:177–190

Ward JM, Pei Z-M, Schroeder JI (1995) Roles of ion channels in initiation of signal transduction in higher plants. Plant Cell 7:833–844

Ward EWB, Cahill DM, Bhattacharyya M (1989) Abscisic acid suppression of phenylalanine ammonia lyase activity and mRNA, and resistance of soybeans to *Phytophthora megasperma* f sp *glycinea*. Plant Physiol 91:23–27

Wasilewska A, Vlad F, Sirichandra C, Redko Y, Jammes F, Valon C, Frel dit Frey N, Leung J (2008) An update on abscisic acid signaling in plants and more. Mol Plant 1:198–217

Weiner JJ, Peterson FC, Volkman BF, Cutler SR (2010) Structural and functional insights into core ABA signaling. Curr Opin Plant Biol 13:495–502

Whenham RJ, Fraser RSS, Brown LP, Payne JA (1986) Tobacco mosaic virus-induced increase in abscisic acid concentration in tobacco leaves – intracellular location in light and dark-green areas, and relationship to symptom development. Planta 168:592–598

Wildermuth MC, Dewdney J, Wu G, Ausubel FM (2001) Isochorismate synthesis is required to synthesize salicylic acid for plant defense. Nature 414:562–565

Wilkinson S, Davies WJ (2010) Drought, ozone, ABA and ethylene: new insights from cell to plant to community. Plant Cell Environ 33:510–525

Wu Y, Zhang D, Chu JY, Boyle P, Wang Y, Brindle ID, De Luca V, Després C (2012) The Arabidopsis NPR1 protein is a receptor for the plant defense hormone salicylic acid. Cell Rep 1:639–647

Xie Z, Zhang ZL, Zou X, Huang J, Ruas P, Thompson D, Shen QJ (2005) Annotations and functional analyses of the rice WRKY gene superfamily reveal positive and negative regulators of abscisic acid signaling in aleurone cells. Plant Physiol 137:176–189

Xiong I, Yang Y (2003) Disease resistance and abiotic stress tolerance in rice are inversely modulated by an abscisic acid-inducible mitogen-activated protein kinase. Plant Cell 15:745–759

Xu J, Audenaert K, Hofte M, De Vleesschauwer D (2013) Abscisic acid promotes susceptibility to the rice leaf blight pathogen *Xanthomonas oryzae* pv. *oryzae* suppressing salicylic acid-mediated defenses. PLoS One 8:e67413

Yalowsky S, Baluska F, Jones A (2010) Integrated G proteins signaling in plants. Springer, Heidelberg

Yamamoto S, Suzuki K, Shinshi H (1999) Elicitor-responsive, ethylene-independent activation of GCC box-mediated transcription that is regulated by both protein phosphorylation and dephosphorylation in cultured tobacco cells. Plant J 20:571–579

Yan J, Tsuichihara N, Etoh T, Iwai S (2007) Reactive oxygen species and nitric oxide are involved in ABA inhibition of stomatal opening. Plant Cell Environ 30:1320–1325

Yang Z (2002) Small GTPases: versatile signaling switches in plants. Plant Cell 14(Suppl):S375–S388

Yasuda M, Ishikawa A, Jikumaru Y, Seki M, Umezawa T, Asami T, Maruyama-Nakashita A, Kudo T, Shinozaki K, Yoshida S, Nakashita H (2008) Antagonistic interaction between systemic acquired resistance and the abscisic acid-mediated abiotic stress response in *Arabidopsis*. Plant Cell 20:1678–1692

Yazawa K, Jiang C-J, Kojima M, Sakakibara H, Takatsuji H (2012) Reduction of abscisic acid levels or inhibition of abscisic acid signaling in rice during the early phase of *Magnaporthe oryze* infection decreases its susceptibility to the fungus. Physiol Mol Plant Pathol 78:1–7

Yin P, Fan H, Hao Q, Yuan X, Wu D, Pang Y, Yan C, Li W, Wang J, Yan N (2009) Structural insights into the mechanism of abscisic acid signaling by PYL proteins. Nat Struct Mol Biol 16:1230–1236

Yong G, Tingting L, Yang L, Cabda R, Yun Z, Maolin W (2010) Isolation and characterization of gene encoding G protein α subunit protein responsive to plant hormones and abiotic stresses in *Brassica napus*. Mol Biol Rep 37:3957–3965

Yoo SD, Cho Y-H, Sheen J (2009) Emerging connections in the ethylene signaling network. Trends Plant Sci 14:270–279

Yoon JY, Hamayun M, Lee SK, Lee I-J (2009) Methyl jasmonate alleviated salinity stress in soybean. J Crop Sci Biotech 12:63–68

Yoshida R, Hobo T, Ichimura K, Mizoguchi T, Takahashi S, Takahashi F, Alonso J, Ecker JR, Shinozaki K (2002) ABA-activated SnRK2 protein kinase is required for dehydration stress signaling in *Arabidopsis*. Plant Cell Physiol 43:1473–1483

Yoshida R, Umezawa T, Mizoguchi T, Takahashi S, Takahashi F, Shinozaki K (2006) The regulatory domain of SRK2E/OST1/SnRK26 interacts with ABI1 and integrates abscisic acid (ABA) and osmotic stress signals controlling stomatal closure in *Arabidopsis*. J Biol Chem 281:5310–5318

Yoshida T, Fujita Y, Sayama H, Kidokoro S, Maruyama K, Mizoi J, Shinozaki K, Yamaguchi-Shinorzaki K (2010) AREB1, AREB2, and ABF3 are master transcription factors that cooperatively regulate ABRE-dependent ABA signaling involved in drought stress tolerance and regulate for full activation. Plant J 61:672–685

Yu X-C, Zhu S-Y, Gao G-F, Wang X-J, Zhao R, Zou K-Q, Wang X-F, Zhang X-Y, Wu F-Q, Peng C-C, Zhang D-P (2007) Expression of a grape calcium-dependent protein kinase ACPK1 in *Arabidopsis thaliana* promotes plant growth and confers abscisic acid-hypersensitivity in germination, post germination growth, and stomatal movement. Plant Mol Biol 64:531–538

Zeng W, Melotto M, He SY (2010) Plant stomata: a checkpoint of host immunity and pathogen virulence. Curr Opin Biotechnol 21:599–603

Zhang DP, Wu ZY, Li XY, Zhao ZX (2002) Purification and identification of a 42-kilodalton abscisic acid-specific-binding protein from epidermis of broad bean leaves. Plant Physiol 128:714–725

Zhang Y, Goritschnig S, Dong X, Li X (2003) A gain of function mutation in a plant disease resistance gene leads to constitutive activation of downstream signal transduction pathways in suppressor of *npr1-1*, constitutive 1. Plant Cell 15:2636–2646

Zhang H, Zhang D, Chen J, Yang Y, Huang Z, Huang D, Wang X-C, Huang R (2004a) Tomato stress-responsive factor TSRF1 interacts with ethylene responsive element GCC box and regulates pathogen resistance to *Ralstonia solanacearum*. Plant Mol Biol 55:825–834

Zhang W, Qin C, Zhao J, Wang X (2004b) Phospholipase Dα1-derived phosphatidic acid interacts with ABI1 phosphatase 2C and regulates abscisic acid signaling. Proc Natl Acad Sci U S A 101:9508–9513

Zhang WH, Yu LJ, Zhang YY, Wang XM (2005) Phospholipase D in the signaling networks of plant response to abscisic acid and reactive oxygen species. Biochim Biophys Acta 1736:1–9

Zhang AY, Jiang MY, Zhang JH, Tan M, Hu XL (2006) Mitogen-activated protein kinase is involved in abscisic acid-induced antioxidant defense and acts downstream of reactive oxygen species production in leaves of maize plants. Plant Physiol 141:475–487

Zhang Z, Yao W, Dong N, Liang H, Liu H, Huang R (2007) A novel ERF transcription activator in wheat and its induction kinetics after pathogen and hormone treatments. J Exp Bot 58:2993–3003

Zhang W, He SY, Assmann SM (2008) The plant innate immunity response in stomatal guard cells invokes G-protein-dependent ion channel regulation. Plant J 56:984–996

Zhang Y, Zhu H, Zhang Q, Li M, Yan M, Wang R, Wang L, Welti R, Zhang W, Wang X (2009) Phospholipase Dα1 and phosphatidic acid regulate NADPH oxidase activity and production of reactive oxygen species in ABA-mediated stomatal closure in *Arabidopsis*. Plant Cell 21:2357–2377

Zhang W, Jeon BW, Assmann SM (2011) Heterotrimeric G-protein regulation of ROS signalling and calcium currents in *Arabidopsis* guard cells. J Exp Bot 62:2371–2379

Zhang H, Gao Z, Zheng X, Zhang Z (2012) The role of G-proteins in plant immunity. Plant Signal Behav 7:1284–1288

Zhou J, Zhang H, Yang Y, Zhang Z, Zhang H, Hu X, Chen J, Wang X-C, Huang R (2008) Abscisic acid regulates TSRF1-mediated resistance to *Ralstonia solanacearum* by modifying the expression of GCC box-containing genes in tobacco. J Exp Bot 59:645–652

Zhu S-Y, Yu X-C, Wang X-J, Zhao R, Li Y, Fan R-C, Shang Y, Du S-Y, Wang X-F, Wu F-Q, Xu Y-H, Zhang X-Y, Zhang D-P (2007) Two calcium-dependent protein kinases, CPK4 and CPK11, regulate abscisic acid signal transduction in *Arabidopsis*. Plant Cell 19:3019–3036

Zou X, Seemann JR, Neuman D, Shen QJ (2004) A WRKY gene from Creosote Bush encodes an activator of the abscisic acid signaling pathway. J Biol Chem 279:55770–55779

Chapter 6
Auxin Signaling System in Plant Innate Immunity

Abstract Plant hormones play important role in plant immune signaling system. While salicylic acid (SA), jasmonic acid (JA), and ethylene (ET) are integral components of signal transduction systems involved in activation of plant innate immunity, auxin is involved in the modulation of various plant defense signaling pathways. Indole-3-acetic acid (IAA) is the naturally occurring auxin in plants. IAA is synthesized from tryptophan via multiple pathways including indole-3-pyruvic acid (IPA), indole-3-acetamide (IAM), tryptamine (TAM), and indole-3-acetonitrile (IAN) pathways. TIR1/AFB proteins, which are F-box subunits of the SCF ubiquitin complex, have been identified as the receptors for auxins. AUX/IAA proteins are nuclear proteins, and the only function of the AUX/IAA proteins is to repress transcription of auxin-regulated genes. Auxin response factor (ARF) proteins are the transcription factors that bind to the cis-element in auxin-responsive promoters and trigger the expression of auxin-responsive genes. The AUX/IAA protein forms a heterodimer with ARF and represses the transcriptional regulatory function of ARFs. Auxin promotes the degradation of the AUX/IAA repressor through ubiquitin–proteasome pathway and consequently activates auxin-responsive gene expression by removing the inhibition of ARF activity. The auxin receptor TIR1 F-box protein directly links auxin perception to degradation of the AUX/IAA proteins, resulting in depression/activation of auxin response genes. Auxin rapidly alters the expression of hundreds of genes within minutes by removing the inhibitory function of AUX/IAA proteins. Auxin modulates SA, JA, ET, abscisic acid (ABA), and cytokinin signaling systems in the induction or suppression of plant immune responses. Interplay between auxin signaling system and G-protein-, mitogen-activated protein kinase (MAPK)-, reactive oxygen species (ROS), nitric oxide (NO)-, ion channels, and small RNA-mediated signaling systems has been reported. Auxin negatively or positively regulates plant defense responses, resulting in disease susceptibility or disease resistance. Several plant pathogens are known to modulate plant auxin signaling system to cause disease. Auxin, auxin response factors, auxin-responsive genes have been shown to modulate plant defense responses. Every event in auxin signaling and metabolism, including auxin biosynthesis, auxin perception, auxin conjugation, auxin transport, signaling derepression, and inactivation, has been correlated with plant defense or plant disease development.

P. Vidhyasekaran, *Plant Hormone Signaling Systems in Plant Innate Immunity*, Signaling and Communication in Plants 2, DOI 10.1007/978-94-017-9285-1_6

6.1 Auxin as a Signaling Molecule

The plant hormone auxin, indole-3-acetic acid (IAA), is a simple signaling molecule involved in the modulation of different signaling pathways (Hayashi 2012; Peer et al. 2013; Sauer et al 2013; Tatsuki et al. 2013). Auxin binds to TIR1/AFB nuclear receptors, which are F-box subunits of SCF ubiquitin ligase complex (Dharmasiri et al. 2005; Mockaitis and Estelle 2008; Parry et al. 2009; Calderon-Villalobos et al. 2010). The auxin signal is then modulated by the Aux/IAA repressors and the auxin response factor (ARF) transcription factors (Hayashi 2012). The specificity of the auxin-regulated gene expression is regulated by the expression of these regulatory proteins (Hayashi 2012). While salicylic acid (SA), jasmonic acid (JA), and ethylene (ET) are clearly integrated components of signal transduction cascades leading to disease resistance, auxin shows a more complex interplay with various defense signaling systems (Bari and Jones 2009; Savatin et al. 2011). Auxin is involved in the modulation of reactive oxygen species (ROS) (Peer et al. 2013), ethylene (Stepanova et al. 2008; Tatsuki et al. 2013), salicylic acid (Wang et al. 2007), jasmonate (JA) (Tiryaki and Staswick 2002; Nagpal et al. 2005; Tabata et al. 2010), abscisic acid (Rock and Sun 2005), and cytokinin (Naseem and Dandekar 2012) signaling. It also activates K^+ channels, H^+-ATPases, and Rho GTPases (G-proteins) (Hayashi 2012; Takahashi et al. 2012). Nitric oxide (NO) influences auxin signaling through S-nitrosylation of the *Arabidopsis* TRANSPORT INHIBITOR RESPONSE 1(TIR1)/AUXIN SIGNALING F-BOX (AFB) auxin receptor (Beard et al. 2012; Terrile et al. 2012). Almost every event in auxin signaling and metabolism, including auxin perception, signaling derepression, biosynthesis, and inactivation, is intimately correlated with plant defense or plant disease development (Fu and Wang 2011; Cui et al. 2013).

6.2 Auxin Biosynthesis

Indole-3-acetic acid (IAA) is the naturally occurring auxin in plants (Hayashi 2012). Tryptophan is the main precursor for IAA in plants (Zhao 2010). IAA is biosynthesized from tryptophan via multiple pathways. The different branches of the tryptophan-dependent pathway are distinguished by branch-specific intermediates, namely, indole-3-pyruvic acid (IPA), indole-3-acetamide (IAM), tryptamine (TAM), and indole-3-acetonitrile (IAN) (Park et al. 2003; Lau et al. 2008). The IPA pathway is the common IAA biosynthetic pathway in plants. In this pathway, tryptophan is converted to IPA through the action of TRYPTOPHAN AMINOTRANSFERASE OF ARABIDOPSIS (TAA) family of amino transferases, and the IPA is converted to IAA through the action of YUCCA (YUC) family of flavin monooxygenases (Fig. 6.1; Lau et al. 2008; Stepanova et al. 2008; Won et al. 2011; Mashiguchi et al. 2011; Brumos et al. 2013; Dai et al. 2013).

IAN pathway operates only in relatively few plant species. In this pathway, tryptophan is converted to indole-3-acetaldoxime (IAOx) through the action of CYP79B family members, the cytochrome P450s (Sugawara et al. 2009). IAOx is converted to IAN through the action of nitrilases (Park et al. 2003) (Fig. 6.2).

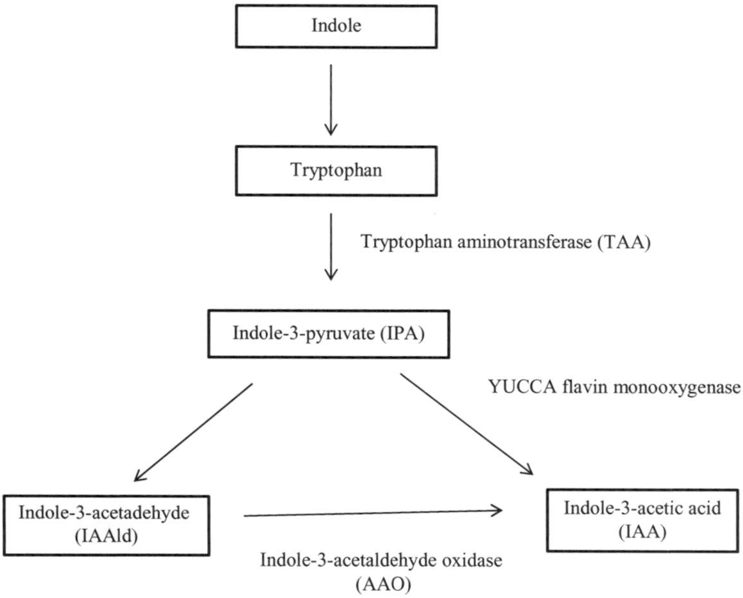

Fig. 6.1 Biosynthesis of auxin through IPA pathway in *Arabidopsis* (Adapted from Won et al. 2011; Mashiguchi et al. 2011)

Fig. 6.2 Biosynthesis of IAA through IAN pathway (Adapted from Zhao et al. 2001; Park et al. 2003; O'Donnell et al. 2003; Mashiguchi et al. 2011)

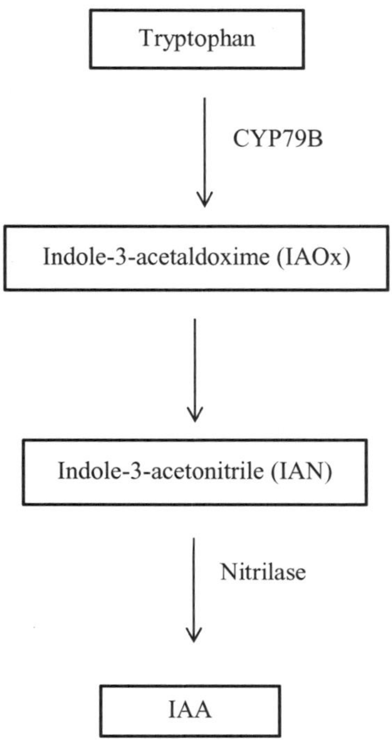

IAM pathway has been reported in a wide range of plant species. However, it is still not known how IAM is produced from tryptophan (Lehmann et al. 2010). IAM is converted to IAA by an amidase, *Arabidopsis* AMIDASE 1 (AMI1) in *Arabidopsis* (Pollmann et al. 2003). Tryptophan is converted to tryptamine (TAM), which is then converted into *N*-hydroxy-TAM (HTAM) by the YUCCA family of flavin monooxygenases (Cheng et al. 2007; Quittenden et al. 2009; Mashiguchi et al. 2011). The underlying pathway from HTAM to IAA is not yet known (Mashiguchi et al. 2011). Tryptophan-independent pathway for biosynthesis of IAA in plants has also been suggested (Strader and Bartel 2008; Quittenden et al. 2009).

6.3 Auxin Signaling Pathway

6.3.1 Auxin-Binding Proteins/Receptors

Transport Inhibitor Response 1(TIR1)/Auxin Signaling F-Box (AFB) proteins, which are F-box subunits of the SCF ubiquitin complex, have been identified as the receptors for auxins (Dharmasiri et al. 2005; Tan et al. 2007; Parry et al. 2009; Greenham et al. 2011; Hayashi 2012). Auxin binds to TIR1 (Mockaitis and Estelle 2008). IAA binds to the surface of TIR1 with hydrophobic and hydrogen bonds (Hayashi 2012). In *Arabidopsis*, TIR1 and five TIR1 homolog proteins (AFB1, AFB2, AFB3, AFB4, and AFB5) redundantly function as auxin receptors (Greenham et al. 2011). TIR1 is one of the well-characterized F-box proteins, and the F-box proteins are the substrate receptors of SCF-type ubiquitin protein ligase (E3) complexes. Through their F-box domain, F-box proteins bind to the SKP1 adaptor protein while their other diverse protein–protein interaction domains bind to the substrates (Calderon-Villalobos et al. 2010). The auxin-binding site in the TIR1 has been identified. The crystal structure of TIR1 revealed an inositol hexakisphosphate (InsP6) molecule tightly bound to the protein at a functionally important position of the auxin receptor. Surrounded by more than 10 conserved positively charged residues at the concave surface of the TIR1 leucine-rich repeat (LRR) domain, InsP6 interacts with the auxin-binding pocket from underneath and is in direct contact with the basic residue binding to the carboxyl group of the hormone. InsP6 may be a functional cofactor of TIR1 (Calderon-Villalobos et al. 2010).

6.3.2 Auxin–IAA Proteins

The Aux/IAA family of proteins plays a central role in the auxin signal transduction that regulates auxin-responsive gene expression (Mockaitis and Estelle 2008). Aux/IAA encode the rapidly turned over transcriptional repressors of auxin-inducible genes (Leyser 2006). The TIR1 F-box protein acts as an auxin receptor and directly links auxin perception to degradation of the AUX/IAA proteins. Auxin acts as a "molecular glue" to increase the strength of the interaction between TIR1 and the Aux/IAA protein, thus promoting their ubiquitination and degradation (Tan et al. 2007).

Auxin/indole-3-acetic acid (Aux/IAA) proteins are nuclear proteins that play a key role in regulating the expression of auxin response genes (Hagen and Guilfoyle 2002; Li et al. 2011). The Aux/IAA proteins play an important role in the transcriptional regulation of specific sets of genes (Hayashi 2012). Twenty-nine genes encoding Aux/IAA proteins have been detected in *Arabidopsis*, and they are induced by auxins themselves (Li et al. 2011). The size of the Aux/IAA proteins ranges from 18 to 36 kDa (Liscum and Reed 2002). All of the Aux/IAA proteins function as transcriptional repressors (Li et al. 2011). Aux/IAA proteins contain four conserved domains, I, II, II, and IV. Domain I of Aux/IAA functions as an active repression domain and interacts with the TOPLESS (TPL) transcriptional corepressor to form a complex that represses the transcriptional function of auxin response factor (ARF) (Szemenyei et al. 2008). Domain II is responsible for the stability of Aux/IAA proteins and directly binds to the TIR1/AFB auxin receptors in an auxin-dependent manner (Dreher et al. 2006; Li et al. 2011). Domains III and IV contain a protein–protein interaction domain that shares homology with the C-terminal domain (CTD) of ARF proteins. Aux/IAA proteins form a heterodimer with ARF through an interaction between domains III and IV and the CTD (Muto et al. 2006; Li et al. 2011). Aux/IAA proteins function as repressors of auxin response gene expression when auxin concentrations in a cell are low. At elevated auxin concentrations, these repressors are destroyed via the ubiquitin–proteasome pathway, resulting in derepression/activation of auxin response (Li et al. 2011).

6.3.3 Auxin Response Factor (ARF) Proteins

ARF family proteins are transcription factors that bind to the cis-element in auxin-responsive promoters. The Aux/IAA protein forms a heterodimer with ARF to repress the transcriptional regulatory function of ARF. The intracellular level of Aux/IAA protein repressors determines the ARF transcriptional activity (Mockaitis and Estelle 2008). Auxin promotes the degradation of the Aux/IAA repressor and consequently activates auxin-responsive gene expression by removing the inhibition of ARF activity. In this pathway, auxin controls the ubiquitination of Aux/IAA repressors via the $SCF^{TIR1/AFB}$-proteasome pathway (Hayashi 2012).

DNA-binding auxin response factor (ARF) transcription factors interact with AUX/IAA transcription inhibitors. Auxin response factor (ARF) protein family members have been shown to bind to the auxin response element (ARE) TGTCTC (Ulmasov et al. 1997). In *Arabidopsis thaliana* genome 22 ARF genes have been reported (Liscum and Reed 2002). They contain an amino-terminal DNA-binding domain and carboxy-terminal dimerization domains, known as domains III and IV (Tiwari et al. 2003). The composition of the middle region between the DNA-binding and dimerization domains varies in the different ARFs (Leyser 2006). Five of the 22 ARFs with glutamine-rich middle regions act as transcriptional activators.

The remaining 17, which have serine-rich, serine–glycine-rich, serine–proline-rich, or serine–proline–leucine-rich middle regions, act as repressors of transcription (Leyser 2006).

6.3.4 Auxin-Inducible Gene Expression

ARF gene families consisting of 23 loci in *Arabidopsis* encode proteins that interact with *Aux/IAA* gene families consisting of 29 loci to mediate auxin responses (Remington et al. 2004). Auxin rapidly alters the expression of hundreds of genes within minutes (Chapman and Estelle 2009). The well-known auxin-induced responses involve rapid changes in gene expression (Rademacher et al. 2011). The early auxin-inducible genes include three major gene families: the Aux/IAA (Auxin/IAA inducible genes), GH3s (Gretchen Hagen3), and the SAURs (small auxin upregulated RNAs) (Jain et al. 2006; Leyser 2006; Hayashi 2012; Yuan et al. 2013). Members of each of these families are up-regulated in response to auxin addition (Leyser 2006). The promoters of these three gene families contain specific auxin response elements (AREs). A number of putative AREs have been defined within upstream promoter regions of auxin-responsive genes, such as one or more copies of a conserved motif, TGTCTC, which confer auxin responsiveness (Jain et al. 2006). The TGTCTC element has been shown to be sufficient to mediate strong auxin-inducible expression of the genes. The DNA-binding domain of auxin-responsive factors (ARFs) binds to AREs and regulates the expression of auxin-responsive genes (Jain et al. 2006).

Auxin-signaling genes include *TRANSPORT INHIBITOR RESPONSE1* (*TIR1*), *AUXIN RESISTANT2* (*AXR2*), *AUXIN RESISTANT3* (*AXR3*), *AUXIN RESISTANT6* (*AXR6*), *SUPPRESSOR OF G2 ALLELE OF SKP1B* (*SGT1B*), and *RING-BOX1* (*RBX1*). TIR1 gene encodes an auxin receptor while AXR2 and AXR3 encode the IAA/AUX proteins IAA7 and IAA17, respectively, which act as repressors of auxin-inducible gene expression (Mockaitis and Estelle 2008). SGT1B encodes a regulatory protein involved in SCF^{TIR1}-mediated degradation of Aux/IAA proteins (Gray et al. 2003).

6.3.5 Ubiquitin–Proteasome System in Auxin Signaling Pathway

Ubiquitin- and proteasome-mediated degradation of proteins plays an important role in auxin signaling pathway (Leyser 2006; Maraschin Fdos et al. 2009; Calderon-Villalobos et al. 2010; Hayashi 2012). Proteasomes are large protein complexes located in the nucleus and the cytoplasm, and the main function of the proteasome is to degrade unneeded or damaged proteins by proteolysis. Proteins are tagged for degradation by a small protein called ubiquitin (Pickart and Eddins 2004). Ubiquitin acts as a covalent molecular tag and its attachment requires three distinct enzymatic activities. A ubiquitin-activating enzyme, E1, activates ubiquitin C-terminal carboxyl group by adenylation and then forms a thioester bond with cysteinyl sulfhydryl residue on the E1 protein itself (Lee et al. 2011). The adenylated ubiquitin is then transferred to a cysteine of a second enzyme, ubiquitin-conjugating enzyme (E2). In the last step, a member of a highly diverse class of enzymes known as ubiquitin ligases (E3) recognizes the specific protein to be ubiquitinated and catalyzes the transfer of ubiquitin from E2 to this target protein (Pickart and Eddins 2004; Smalle and Vierstra 2004; Dreher and Callis 2007). The 26S proteasome recognizes polyubiquitinated proteins and transfers them to the catalytic core. After delivery to the proteasome mediated in part by ubiquitin-binding proteins, the polyubiquitylated substrate can be deubiquitylated by the proteasome's regulatory cap or associated proteases. The deubiquitylated substrate is fed into the proteolytic core of the proteasome where it is cleaved into small peptides (Book et al. 2005; Zhu et al. 2005; Dreher and Callis 2007).

The ubiquitin ligases (E3) play a key role in auxin signaling system. The E3 ubiquitin ligases belong to the Skp, Cullin, F-box containing complex (SCF complex), which are composed of four primary subunits: SKP1 (for S-phase Kinase-associated Protein 1), a cullin family member protein (e.g., CUL1), a RING finger protein, and an F-box protein (Dreher and Callis 2007). Within this complex, the F-box protein directly binds the substrate through protein–protein interaction domains, the cullin binds the RING finger protein, together they recruit E2 ubiquitin-conjugating enzyme, and SKP1 helps to link the F-box protein and cullin (Deshaies 1999). The auxin signal is transduced by E3 ubiquitin ligase–proteasome pathway. The auxin receptor TIR1 encodes an F-box protein and F-box proteins are subunits of SCF-type ubiquitin–protein ligases. F-box proteins contain an amino-terminal F-box motif, which interacts with Skp1. At the carboxyl terminus, they have protein–protein interaction domains, which are responsible for recruiting the target protein to the SCF for ubiquitination. The TIR1 F-box protein has carboxy-terminal leucine-rich repeats (LRRs) (Leyser 2006).

The only function of the Aux/IAA proteins is to repress transcription of auxin-regulated genes. Aux/IAAs recruit the corepressor TOPLESS (TPL) to promoters through interactions with both TPL and ARF proteins (Szemenyei et al. 2008; Parry et al. 2009). The presence of auxin and the subsequent degradation of the Aux/IAA proteins release this repression and permit ARF-dependent transcription (Tiwari et al. 2003). Auxin signaling pathway involves an auxin-receptor/ubiquitin ligase, DNA-binding auxin response factor (ARF) transcription factors, and their interacting auxin/indole-3-acetic acid (Aux/IAA) transcriptional inhibitors. Auxin promotes the degradation of Aux/IAA proteins through the auxin receptor and hence releases the inhibition of ARF transcription factors (Rademacher et al. 2011).

In the auxin signaling, intracellular auxin is perceived by the TIR/AFB receptors, thereby triggering degradation of Aux/IAA proteins. These Aux/IAAs under low concentrations, form dimers with ARF transcription factors, thereby blocking the activity of the activating ARFs. Once freed from the Aux/IAAs, these ARFs regulate the expression of auxin-responsive genes (Lau et al. 2008).

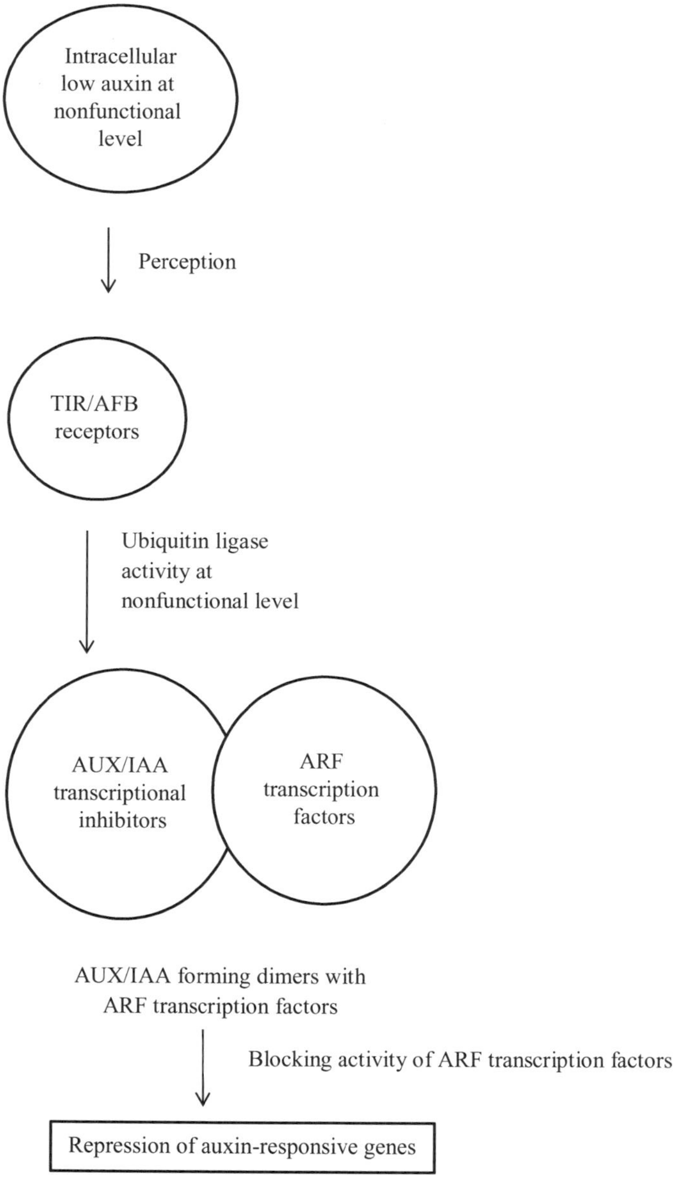

Fig. 6.3 Function of AUX/IAA proteins as repressors of auxin-responsive genes by forming dimers with ARF transcription factors (Adapted from Li et al. 2011)

Aux/IAA proteins function as repressors of auxin response gene expression when auxin concentrations in a cell are low (Fig. 6.3; Li et al. 2011). At elevated auxin concentrations, these repressors are destroyed via the ubiquitin–proteasome pathway, resulting in derepression/activation of auxin response genes (Fig. 6.4; Li et al. 2011).

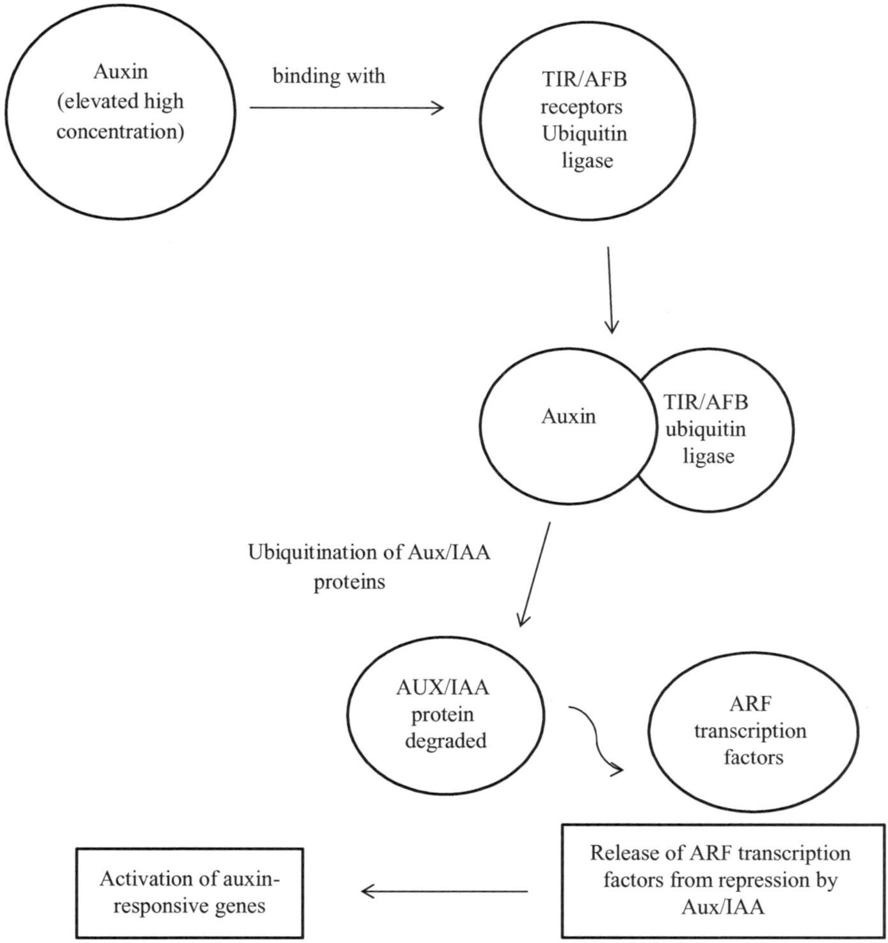

Fig. 6.4 Auxin-induced destruction of Aux/IAA proteins by TIR ubiquitin ligase to release inhibition of ARF transcription factors and activate Auxin- responsive genes (Adapted from Li et al. 2011)

6.3.6 Auxin Homeostasis

Auxin conjugation is mostly controlled by GH3 family proteins (Woodward and Bartel 2005). GH3s encode auxin-conjugating enzymes, which act to reduce free auxin levels (Leyser 2006). The GH3 proteins have acyl acid amido synthetase activity (Ding et al. 2008; Fu et al. 2011; Westfall et al. 2012). The group II GH3 proteins with amido synthetase activity use various amino acids as substrates to form IAA–amido conjugates (Westfall et al. 2010). The IAA–alanine and IAA–leucine conjugates serve to temporarily store IAA and can be hydrolyzed to release free IAA when needed. In contrast IAA–aspartic acid and IAA–glutamate sequester

IAA for degradation (Staswick et al. 2005). IAA–amino synthetase prevents free IAA accumulation in plants (Ding et al. 2008). Rice plants overexpressing GH3-8 gene showed reduced accumulation of free IAA, and in these plants, IAA–Asp concentration was about twofold higher than in the wild-type plants (Ding et al. 2008). The results suggest that GH3-8 suppresses the free IAA accumulation by increasing the synthesis of IAA–Asp. Synthesis of the IAA conjugates is important to maintain IAA homeostasis (Fu and Wang 2011).

6.3.7 Auxin Transport

Auxin is unique among plant hormones in that it is actively moved around the plant by a series of transmembrane pumps or pump components (Blakeslee et al. 2005; Leyser 2006; Petrášek and Friml 2009; Hayashi 2012; Yu et al. 2013). Auxin is transported within the plant either through phloem (known as nonpolar auxin transport) or via one cell to another (known as polar auxin transport [PAT]) (Kidd et al. 2011). The polar transport stream is a highly stable route for auxin movement in the plant. Polarized auxin transport is driven by a chemiosmotic H^+ gradient between the neutral cytoplasm and the acidic apoplast, which is generated by plasma membrane ATPases. In tissues where auxin concentrations are high, lipophilic diffusion of IAA is augmented by gradient-driven H^+ symport activity. This activity is associated with the AUX/LAX family of amino-acid permease-like proteins in *Arabidopsis* (Blakeslee et al. 2005). AUXIN RESISTANT1/LIKE AUXIN RESISTANT1 (AUX1/LAX) functions both in basipetal auxin transport and in acropetal transport in a phloem-based auxin transport stream (Swarup et al. 2001). AUX1 exhibits a basal plasma membrane localization in plant cells, where it appears to function in conjunction with apically localized efflux carriers (Swarup et al. 2001; Blakeslee et al. 2005). Auxin synthesized in leaf tissues is actively transported to other parts of the plant in a polar fashion to form and maintain auxin gradients (Zhao 2010; Yu et al. 2013). The polarly localized efflux complexes are characterized by the PIN-FORMED (PIN) family of proteins (Friml and Palme 2002). PIN proteins are members of the major efflux facilitator family of integral membrane proteins that are essential for polarized auxin transport (Friml 2003). PAT is controlled by the localization of auxin influx and efflux carriers to the plasma membrane (Titapiwatanakun and Murphy 2009). AUX1 controls auxin influx and it acts as auxin influx facilitator. Localization of AUX1 is regulated by AUXIN RESISTANT4 (AXR4) (Swarup et al. 2001; Dharmasiri et al. 2006). Auxin efflux is regulated by the PIN proteins together with multidrug resistance/phosphoglycoprotein (MDR/PGP) ATP-binding cassette transporters (Titapiwatanakun and Murphy 2009). The gene called *BIG* (also known as *ASA1* [ATTENUATED SHADE AVOIDANCE], *TIR3* [TRANSPORT INHIBITOR RESPONSE3]) encodes a calossin-like protein required for PAT (Gil et al. 2001). Ubiquitination is also required in auxin transport. It has been shown that the auxin-efflux facilitator PIN2/EIR2 is ubiquitinated during root gravitropism (Abas et al. 2006).

6.4 Pathogen Infection Elevates Auxin Biosynthesis in Plants

Elevated levels of endogenous plant IAA have been observed during pathogen infection. IAA levels were found to increase in response to *Xanthomonas campestris* pv. *campestris* infection in *Arabidopsis* (Fig. 6.5; O'Donnell et al. 2003). When the plants were inoculated with another bacterial pathogen, *Pseudomonas syringae* pv. *tomato*, similar increase in IAA accumulation was observed (O'Donnell et al. 2003). Increased IAA concentration was observed in *A. thaliana* plants when they were inoculated with the bacterial pathogen *Pseudomonas syringae* pv. *tomato* DC3000 (Schmelz et al. 2004). The IAA content increased 100-fold in tobacco plants infected with *Ralstonia solanacearum* (Sequeira and Kelman 1962). A five-fold increase in free IAA levels has been reported in *F. oxysporum*-infected roots of cotton (Dowd et al. 2004).

The rice bacterial blight pathogen *Xanthomonas oryzae* pv. *oryzae* infection significantly increased the content of free IAA in the infected tissues (Ding et al. 2008). Susceptible rice plants accumulated more IAA than resistant plants during infection with the bacterial blight pathogen (Ding et al. 2008; Fu et al. 2011). The increase in IAA levels may be due to the increased biosynthesis of IAA in pathogen-inoculated plants. Genes that increased after *X. campestris* pv. *campestris* infection in *A. thaliana* included various auxin-regulated genes and *NIT2*, a gene-encoding nitrilase which is involved in IAN auxin biosynthesis pathway (O'Donnell et al. 2003). Nitrilases convert indole-3-acetonitrile (IAN) to IAA in this pathway (Fig. 6.2; Vorwerk et al. 2001). The results suggest that the bacterial infection triggers IAA biosynthesis by activating the key enzyme nitrilase in the IAA biosynthesis pathway.

The rice bacterial blight pathogen *Xanthomonas oryzae* pv. *oryzae* induced accumulation of IAA in rice plants. The bacterial infection significantly induced the expression of indole-3-acetaldehyde oxidase (AAO) genes, *AAO1* (2.6-fold), *AAO2*

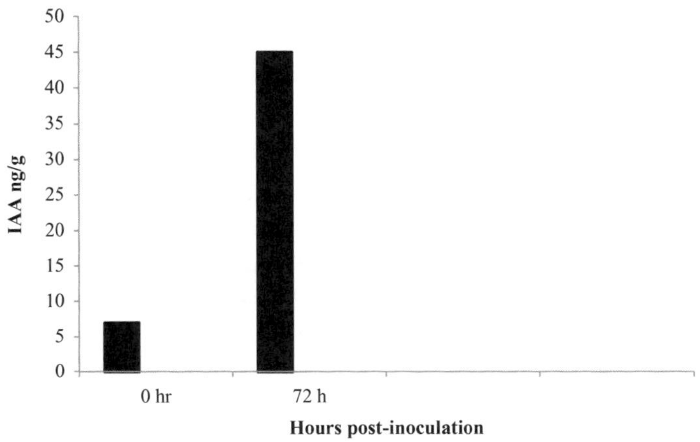

Fig. 6.5 *Xanthomonas campestris*-induced accumulation of IAA in *Arabidopsis thaliana* (Adapted from O'Donnell et al. 2003)

(11.7-fold), *AAO3* (17.5-fold), and nitrilase gene *NIT1* (2.1-fold) (Fu et al. 2011). Indole-3-acetaldehyde oxidase plays an important role in IPA pathway, while nitrilase plays key role in the IAN pathway in auxin biosynthesis (Figs. 6.1 and 6.2; O'Donnell et al. 2003; Park et al. 2003; Woodward and Bartel 2005; Won et al. 2011; Mashiguchi et al. 2011). The results suggest that the pathogen-induced accumulation of IAA is synthesized endogenously in rice.

The expression of *AAO3* and *NIT1* in rice plants was also induced by IAA treatment. Treatment of rice with IAA also rapidly induced GH3-2 (Fu et al. 2011). These results suggest pathogens secreting IAA may induce rice to synthesize its own IAA, which further helps the infection by pathogens in susceptible rice.

Expression of *AAO3* and *NIT1* genes, which were putatively involved in IAA biosynthesis, was markedly induced in a susceptible cultivar after *X. oryzae* pv. *oryzae* infection (Ding et al. 2008). *X. oryzae* pv. *oryzae*, *X. oryzae* pv. *oryzicola*, and *Magnaporthe grisea* infection all induced AAO3 and NIT1 in both susceptible and resistant rice lines. However, the transcript levels of *AAO3* and *NIT1* were significantly higher in susceptible lines than in resistant lines after *X. oryzae* pv. *oryzae*, *X. oryzae* pv. *oryzicola*, or *M. grisea* infection (Fu et al. 2011). These results suggest that the susceptible reaction may be associated with rapidly increased endogenous biosynthesis of IAA.

Infection with *P. syringae* pv. *tomato* activates genes related to IAA biosynthesis and represses Aux/IAA family and auxin transporter genes, suggesting that the bacterial pathogen impacts auxin signaling through activating IAA production, altering IAA movement, and derepressing the auxin pathway (Thilmony et al. 2006). The repressed genes due to *P. syringae* pv. *tomato* infection in *Arabidopsis* include four *AUX/IAA* and 18 *SAUR* (small auxin up-regulated) genes, six auxin transporters (*PIN3*, *PIN4*, *PIN7*, *AUX1*, *LAX1*, *LAX3*), and an auxin response factor gene (*ARF18*) (Thilmony et al. 2006). The genes induced due to the bacterial pathogen infection include three IAA–amino acid hydrolases (ILR1, ILL5, ILL6), three IAA–amido synthases (DFL2, GH3.3, GH3.12), IAA8, and NIT3 (Thilmony et al. 2006; Fig. 6.6). AUX/IAA genes are rapidly induced by auxin (Thilmony et al. 2006), and they may act in a feedback loop repressing the auxin response signaling pathway (Tiwari et al. 2004). IAA–amino acid hydrolases and IAA–amido synthases convert IAA from an amino acid conjugate to the free form (Staswick et al. 2005). NIT3 is involved in IAA biosynthesis (Thilmony et al. 2006).

6.5 Antagonism Between Auxin Signaling and PAMPs/Elicitors-Triggered Signaling Systems

The immune responses induced by the oomycete *Phytophthora megasperma* PAMP glucan were inhibited by auxin in soybean cells (Leguay and Jouanneau 1987). Auxin and the oomycete PAMP worked in an opposite manner in the inducibility of enzymes and pathogenesis-related proteins that play a role in the induced defense responses in soybean (Leguay and Jouanneau 1987). The bacterial PAMP flg22

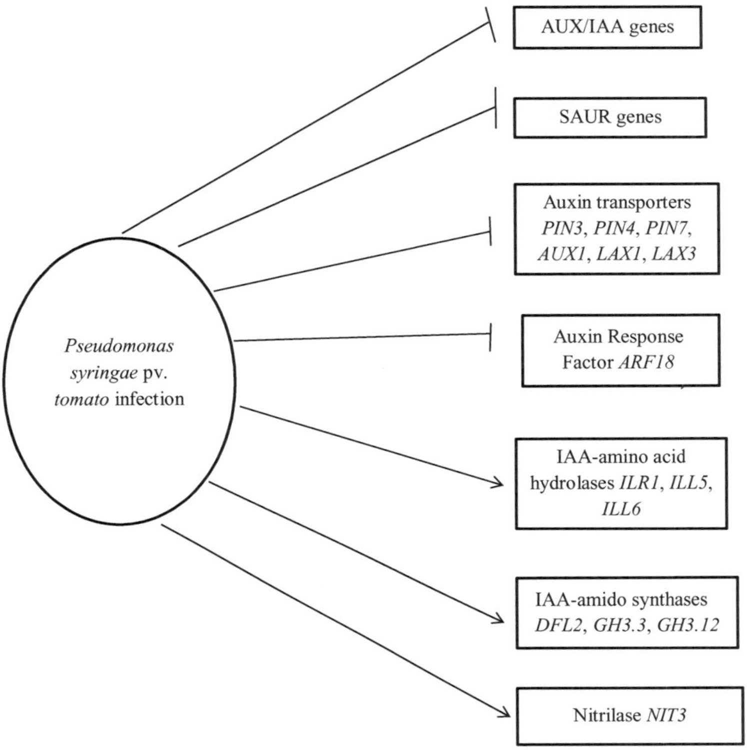

Fig. 6.6 *Pseudomonas syringae* pv. *tomato* infection-triggered repression or induction of genes involved in auxin signaling pathway in *Arabidopsis* (Adapted from Thilmony et al. 2006)

possesses auxin-antagonistic activity (Navarro et al. 2006). Flg22 treatment triggered biogenesis and accumulation of several small RNAs in *Arabidopsis* and the majority of the small RNAs have been identified as microRNAs (miRNAs) (Li et al. 2010). Flg22 induced a two-fold increase in microRNA (miR393) accumulation in *Arabidopsis* seedlings. Flg22 elicited a subset of mRNAs, including TIR1 (Transport Inhibitor Response 1) and two of its three functional paralogs, AFB2 and AFB3 (auxin signaling F-box proteins 2 and 3). The F-box proteins TIR1, AFB2, and AFB3 were specifically cleaved by miR393. A two- to threefold reduction in the levels of TIR1, AFB1, AFB2, and AFB3 was observed 30 min after flg22 elicitation (Navarro et al. 2006). TIR1 is part of the ubiquitin ligase complex SCF^{TIR1} that interacts with Aux/IAA proteins to promote their degradation. Aux/IAA proteins repress auxin signaling through heterodimerization with auxin response factors (ARFs). These transcription factors ARFs bind to auxin-responsive elements (AuxREs) in promoters of primary auxin response genes and activate (or repress) transcription. The PAMP flg22 triggered events that contributed to rapid down-regulation of the primary auxin response genes *GH3-like*, *BDL/IAA12*, and *AXR3/IAA17* (Navarro et al. 2006). Transgenic *Arabidopsis* plants overexpressing *miR393a* showed resistance to the pathogen (Navarro et al. 2006). Augmenting auxin signaling through overexpressing a TIR1 paralog that is partially refractory to miR393 enhanced susceptibility to the bacterial pathogen, and conversely, repressing auxin signaling through *miR393* overexpression increased bacterial resistance. These results suggest that down-regulation of auxin signaling, resulting in ARF inactivation, is part of a PAMP-induced immune response (Navarro et al. 2006). It is known that auxin promotes susceptibility to bacterial diseases (O'Donnell et al. 2003). It negatively regulates the expression of genes encoding the auxin receptors TIR1, AFB2, and AFB3 (Navarro et al. 2006). The bacterial PAMP flg22 has been shown to induce the accumulation of miR393, which contributes to plant resistance against bacterial pathogens by negatively regulating the mRNA level of F-box auxin receptors TIR1, AFB2, and AFB3 (Navarro et al. 2006).

Li et al. (2010) showed that the PAMP flg22-triggered accumulation of another miRNA, the miR160a, besides miR393 already reported in *Arabidopsis*. Flg22 induces *miR160a* accumulation and represses its target genes *ARF16* and *ARF17* (Li et al. 2010). Flg22 treatment up-regulated the miR160a, which targets *ARF10*, *ARF16*, and *ARF17* (Mallory et al. 2005). ARF proteins bind auxin-responsive elements to activate or repress transcription of primary auxin response genes (Hagen and Guilfoyle 2002).

Flg22 regulates the expression of another microRNA miR167. Flg22-regulated miR167 targets auxin response factors *ARF6* and *ARF8* (Rhoades et al. 2002; Jones-Rhoades and Bartel 2004). Flg22 treatment repressed the expression of *ARF10*, *ARF16*, and *ARF17*, but did not significantly alter *ARF6* and *ARF8* (Li et al. 2010).

6.6 Antagonism Between Auxin Signaling and HAMP/ Endogenous Elicitor-Triggered Signaling Systems

The endogenous elicitor oligogalacturonides (OGs) showed auxin-antagonistic activity in *Arabidopsis*. OGs inhibit the expression of several auxin up-regulated genes. Auxin was able to counteract the OG-induced protection against the fungal pathogen *Botrytis cinerea* (Savatin et al. 2011). Besides elicitors of pathogen origin (PAMPs), several elicitors of host origin (endogenous elicitors) have also been reported to trigger immune responses. These endogenous elicitors are called host-associated molecular patterns (HAMPs). Fragments of pectic polysaccharide homogalacturonan (HGA) called oligogalacturonides (OGs) function as danger signals activating plant innate immune responses, and these OGs are called endogenous elicitors/host-associated molecular patterns (HAMPs) (Ryan et al. 2007; Caffali and Mohnen 2009; Galletti et al. 2009; Zipfel 2009; Huffaker et al. 2011). Oligogalacturonides are generated by the action of endopolygalacturonases (PGs) secreted by pathogens during the infection process. The PGs generate OGs by hydrolyzing pectin. The activity of OGs depends on the degree of polymerization (DP). The OG oligomers with a degree of polymerization (DP) between 9 and 15 have been shown to be most potent inducers of defense responses in several plants

(Poinssot et al. 2003; Aziz et al. 2004; Moscatiello et al. 2006). The formation of the most active OG oligomers is favored by when microbial PGs interact with PG-inhibiting proteins (PGIPs), specific Leu-rich repeat recognition proteins located in the plant apoplast (De Lorenzo and Ferrari 2002; Casasoli et al. 2009; Savatin et al. 2011).

PAMPs and HAMPs have been reported to induce almost similar early signaling systems (Galletti et al. 2009). The transcriptome analysis revealed that both of them induced expression of almost same genes involved in Ca^{2+} signaling, ROS signaling, NO signaling, MAPK signaling, SA signaling, JA signaling, ET signaling, and ABA signaling systems in *Arabidopsis* (Denoux et al. 2008). Both OGs and the PAMP flg22 activate defense responses against various pathogens (Zipfel et al. 2004; Ferrari et al. 2007). Auxin was shown to abolish the protection effect of OGs against the fungal pathogen *Botrytis cinerea* in *Arabidopsis* (Savatin et al. 2011). The auxin alone did not significantly affect susceptibility against *B. cinerea* but abolished the protection by OGs (Savatin et al. 2011).

OGs have shown antagonistic action against auxins in the induction of plant immune responses. The OG–auxin (indole-3-acetic acid, IAA) antagonism has been widely reported (Branca et al. 1988; Savatin et al. 2011). OGs inhibited the expression of several endogenous genes that are early up-regulated by IAA in *Arabidopsis*. These genes include Aux/IAA genes (*IAA5*, *IAA19*, *IAA20*, and *IAA22*), *SAUR-AC1*, and *GH3.3*. OGs inhibited the accumulation of transcripts of all these genes. Endogenous levels of free IAA were unaffected in *Arabidopsis* seedlings and adult leaves treated with OGs. The results suggest that antagonism does not occur at the auxin biosynthesis level and it may occur only at the perception or transduction level (Savatin et al. 2011). TIR1 and five TIR1 homolog proteins (AFB1, AFB2, AFB3, AFB4, and AFB5) are the auxin receptors detected in *Arabidopsis* (Greenham et al. 2011). Savatin et al. (2011) analyzed the expression of *TIR1*, *AFB1*, *AFB2*, and *AFB3* genes in *Arabidopsis* seedlings treated with OGs and found no significant variation compared to the water treatment. The studies suggest that OG–auxin antagonism does not occur at the auxin signal perception level.

Aux/IAA proteins play an important role in the auxin signal transduction system and they regulate the auxin-responsive gene expression (Mockaitis and Estelle 2008; Li et al. 2011). Aux/IAA proteins are repressors of auxin response gene expression. Auxin promotes the degradation of the Aux/IAA repressor, and at elevated auxin concentration induced by elicitors, the auxin destroys the repressor, resulting in activation of auxin-responsive gene expression (Hayashi 2012). Degradation of Aux/IAA proteins induced by exogenous auxin was not affected by OGs in *Arabidopsis* (Savatin et al. 2011). It suggests that OGs do not interfere with the Aux/IAA–auxin interaction stage.

The early auxin-inducible genes include *AXR* genes (*IAA* genes) encoding Aux/IAA proteins, *SAUR* genes, and *GH3s* (Rademacher et al. 2011; Yuan et al. 2013). Treatment with OGs reduced IAA-induced accumulation of *IAA5*, *SAUR16*, and *SAUR-AC1* transcripts in *Arabidopsis*. Inhibition of the expression of these genes by OGs was observed within 30 min after treatment. It suggests that the OG-triggered cascade that affects auxin signaling occurs rapidly (Savatin et al. 2011).

6.7 Interplay Between Auxin Signaling and Mitogen-Activated Protein Kinase Mediated Signaling Systems

Mitogen-activated protein kinase (MAPK) cascades are major pathways downstream of PAMP/PRR signaling complex (Hettenhausen et al. 2012; Zhang et al. 2012). A typical MAPK signaling module consists of three interconnected protein kinases: a MAP kinase kinase kinase (MAPKKK), a MAP kinase kinase (MAPKK), and a MAP kinase (MAPK) (Mészáros et al. 2006). MAP kinase cascade involves sequence of phosphorylation events. MAPKs function at the bottom of the three-kinase cascade and are activated by MAPKKs through phosphorylation on the Thr and Tyr residues in their activation motif between the kinase subdomains VII and VIII. The activity of MAPKKs is, in turn, regulated by MAPKKKs via phosphorylation of two Ser/Thr residues in the activation loop of MAPKKs. MAPKKKs receive signals from upstream receptors and activate the MAPK signaling cascade (Kovtun et al. 1998; Ichimura et al. 2002; Li et al. 2012).

GH3 genes are involved in auxin signaling system, and GH3 proteins with amido synthetase activity use various amino acids as substrates to form IAA–amido conjugates from free IAA (Westfall et al. 2010, 2012 Fu et al. 2011). Synthesis of the IAA conjugates is important to maintain IAA homeostasis (Fu and Wang 2011). NPK1 is a specific plant MAPKKK, which activates a MAPK cascade that leads to the suppression of early auxin response gene transcription (Kovtun et al. 1998). NPK1 has been shown to suppress the early auxin response *GH3* gene in maize (Kovtun et al. 1998). ANP1 is a specific *Arabidopsis* MAPKKK, which initiates the MAPK cascade, but blocks the action of auxin (Kovtun et al. 2000). In *Arabidopsis* protoplasts, IAA treatment dramatically increased GH3 promoter activity. Constitutively active ANP1 effectively suppressed the GH3 promoter induction by auxin (Fig. 6.7; Kovtun et al. 2000).

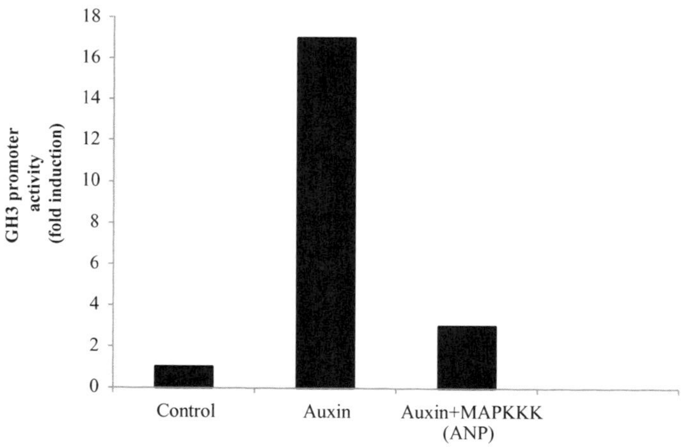

Fig. 6.7 Repression of the auxin-inducible GH3 promoter by the MAPKK ANP1 (Adapted from Kovtun et al. 2000)

6.8 Nitric Oxide Modulates Auxin Signaling

Nitric oxide (NO) is a gaseous readily diffusible free radical identified as a signaling molecule in plants (Besson-Bard et al. 2008). It is a key mediator for rapid induction of plant immune responses (Bellin et al. 2013). Involvement of NO in defense signaling has been well demonstrated (Perchepied et al. 2010; Chun et al. 2012). Interaction between auxin and NO has been reported. IAA is known to trigger NO accumulation in plants (Hu et al. 2005; Correa-Aragunde et al. 2006). NO activates auxin signaling. Evidence has been presented to show that NO acts as a positive modulator of the TIR1/AFB auxin signaling pathway (Terrile et al. 2012). The addition of an NO moiety to a cysteine (Cys) thiol to form an S-nitrosothiol (SNO) is termed S-nitrosylation (Malik et al. 2011). S-nitrosylation is considered as post-translational modification which is similar to phosphorylation (Lindermayr et al. 2005; Astier and Lindermeyr 2012). NO rapidly induces reversible *S*-nitrosylation of proteins involved in signal transduction (Terrile et al. 2012). NO modulates auxin signaling through S-nitrosylation of the TIR1 auxin receptor. TIR1 S-nitrosylation enhanced TIR1-Aux/IAA interaction, facilitating Aux/IAA degradation and subsequently promoting activation of gene expression (Terrile et al. 2012).

6.9 Interaction Between Auxin and Salicylic Acid (SA) Signaling Systems

6.9.1 Repression of Auxin Signaling Pathway by Salicylic Acid

Many pathogens produce IAA in planta (Glickmann et al. 1998; Chung et al. 2003; Maor et al. 2004; Vandeputte et al. 2005) or manipulate IAA biosynthesis pathway in plants to interfere with host's defense systems to cause disease. SA is the key signal triggering plant immune responses and systemic acquired resistance (SAR). It has been demonstrated that SA induces disease resistance through repression of auxin signaling pathway (Wang et al. 2007). Role of SA in plant innate immunity was studied by using an SA analog, benzothiadiazole S-methyl ester (BTH) in *Arabidopsis thaliana.* BTH treatment repressed the expression of several hundreds of genes. Among the 1,133 BTH-repressed genes, 21 genes were found to be related to auxin signal transduction. These repressed genes included genes for the auxin receptors TIR1 and AFB1, *IAA/AUX* family genes, and auxin-inducible *SAUR* genes and *AUX1* and *PIN7* genes encoding an auxin importer and exporter, respectively (Wang et al. 2007). The two genes up-regulated by BTH treatment included genes encoding auxin-conjugating enzymes, IAA–amido synthases that conjugate amino acids to auxin (Fig. 6.8; Wang et al. 2007). These results suggest that BTH treatment might also affect auxin homeostasis by lowering the levels of free auxin. Collectively these studies suggest that auxin signaling is involved in conferring susceptibility and SA induces disease resistance by repressing the auxin signaling. SA might

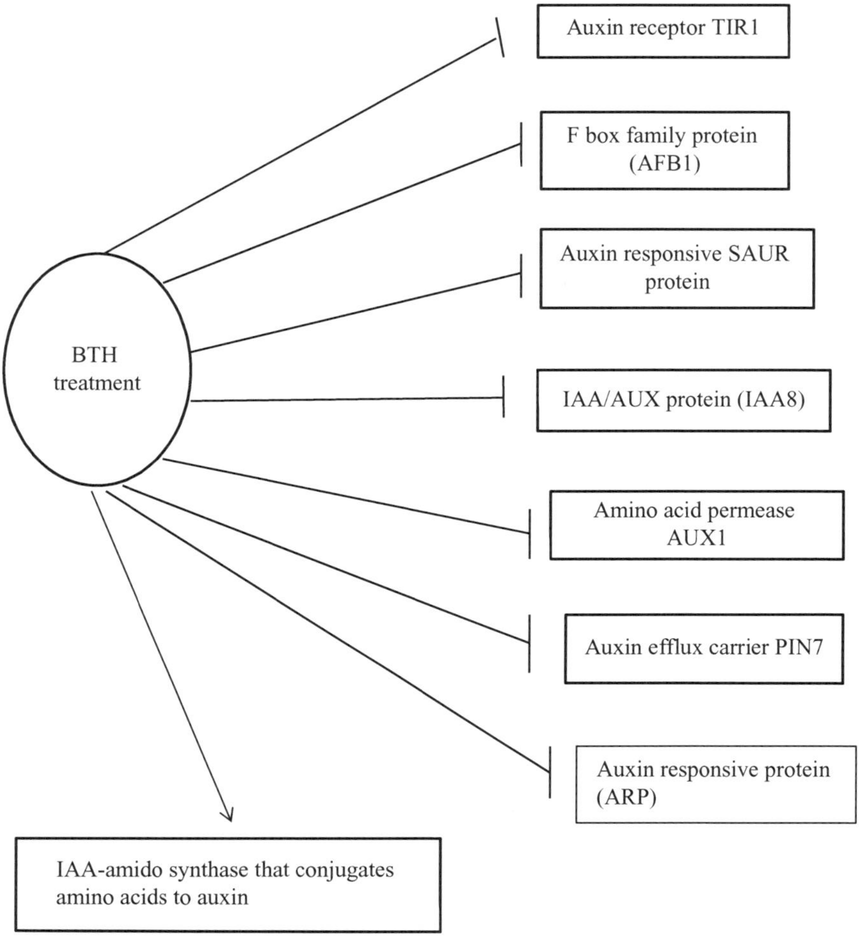

Fig. 6.8 Salicylic acid analog (BTH)-triggered repression/activation of auxin signaling components (Adapted from Wang et al. 2007)

inhibit auxin signaling through transcriptional repression of the auxin receptor genes. Reduction in TIR1 and ABF1 levels may lead to reduced degradation of Aux/IAA repressors resulting in inhibition of auxin responses (Wang et al. 2007).

6.9.2 Auxin Signaling Compromises the Induction of SA Signaling

Auxin may suppress the induction of SA signaling. In plants overexpressing the auxin receptor *AFB1*, the accumulation of SA or induction of SA response pathway were impaired after *P. syringae* pv. *tomato* DC3000 challenge inoculation in

Arabidopsis (Robert-Seilaniantz et al. 2011). Treatment of the transgenic plants expressing the GUS reporter gene under the control of the PR1 promoter with the PAMP flg22 triggered a cascade of events that included SA production and PR1 expression. When the plants were pretreated with the synthetic auxin NAA, induction of PR1 production by the PAMP flg22 was prevented (Robert-Seilaniantz et al. 2011). The results suggest that the auxin has a negative effect on SA signaling.

6.9.3 Auxin Response Gene (GH3) Modulates SA Signaling

SA plays a key role in plant disease resistance, whereas auxin is mostly involved in plant disease susceptibility (Zhang et al. 2007). One of the three major early auxin-inducible gene families includes *GH3*s (*Gretchen Hagen3*) (Jain et al. 2006; Leyser 2006; Ding et al. 2008; Hayashi 2012; Yuan et al. 2013). The *GH3* genes are represented as a multigene family in the *Arabidopsis* genome comprising of 20 members (Hagen and Guilfoyle 2002). Auxin treatment enhances the transcription of *GH3* genes (Jain et al. 2006). *GH3*s encode auxin-conjugating enzymes, which act to reduce free auxin levels (Staswick et al. 2005).

GH3.5, a member of the GH3 family of early auxin-responsive genes in *Arabidopsis*, encodes a protein possessing adenylation activity on both IAA and SA (Zhang et al. 2007). It has been shown that GH3.5 acts as a bifunctional modulator in both SA and auxin signaling during *Pseudomonas syringae* infection (Zhang et al. 2007). GH3.5 adenylated both IAA and SA in vitro (Staswick et al. 2002). Overexpression of the *GH3.5* gene led to elevated accumulation of SA and increased expression of PR-1 in local and systemic tissues in response to avirulent pathogens. In contrast, T-DNA insertional mutant of *GH3.5* accumulated high levels of free IAA after pathogen infection and impaired different resistance-gene-mediated resistance (Zhang et al. 2007). It was shown that SA and auxin pathways were simultaneously augmented in *gh3.5* mutant after infection with an avirulent pathogen. The SA pathway was amplified by *GH3.5* through inducing SA-responsive genes, whereas the auxin pathway was derepressed through up-regulating IAA biosynthesis and down-regulating auxin-repressor genes (Zhang et al. 2007). These studies suggest that *GH3.5* positively regulates the SA signaling pathway in plant defense and modulates the auxin pathway to enhance host susceptibility. It is also suggested that *GH3.5* both positively and negatively affects various aspects of plant defense by modulating both SA and auxin signaling systems.

The interaction between auxin signaling and SA signaling systems may vary depending on compatible and incompatible plant–pathogen interactions. During virulent pathogen infection, GH3.5 modulates the auxin pathway to enhance disease susceptibility by increasing IAA production and derepressing the auxin pathway. During avirulent pathogen infection, SA accumulation is rapidly induced whereas IAA does not accumulate; thus, GH3.5 might mainly modulate the SA pathway to enhance disease resistance (Zhang et al. 2007).

6.10 Role of Auxin Signaling in Systemic Acquired Resistance (SAR)

Systemic acquired resistance (SAR) is an induced/heightened defense response against wide range of pathogens that is activated throughout a plant following local infection (Liu et al. 2011a). After localized foliar infections by diverse pathogens or by treatment with synthetic compounds, plants develop whole–plant immunity, called SAR (Conrath et al. 2006; Beckers and Conrath 2007; Conrath 2009; Jung et al. 2009; Mukherjee et al. 2010; Luna et al. 2012; Camañes et al. 2012; Slaughter et al. 2012). Induction of SAR requires accumulation of the endogenous signaling molecule salicylic acid (SA), which mediates activation of a large set of PR genes (Durrant and Dong 2004). In this process, distal (systemic) leaves become primed to activate stronger defense response upon secondary infection (Beckers et al. 2009; Luna et al. 2012; Slaughter et al. 2012). SAR is associated with priming of defense (Kohler et al. 2002; Jung et al. 2009; Luna et al. 2012) and the priming results in a faster and stronger induction of defense mechanisms after pathogen attack (Conrath 2011). The potentiated induction of defense responses becomes apparent only after further pathogen challenge and has often been overseen when studying SAR (Kohler et al. 2002; Conrath et al. 2006; Ahmad et al. 2010). The induced resistance is detected in the uninoculated systemic tissue of a pathogen-infected plant (Liu et al. 2011b).

The establishment of SAR may require translocation of a signal from the pathogen-inoculated leaf to the distal organs, where defenses are activated (Chaturvedi et al. 2008). Some mobile signal(s) produced at the site of primary infection may have to carry the SAR message and travel throughout the plant. Several attempts have been made to identify the systemic signal (s) (Park et al. 2007; Truman et al. 2007; Chaturvedi et al. 2008; Jung et al. 2009; Park et al. 2009; Shah 2009; Manosalva et al. 2010; Chanda et al. 2011; Liu et al. 2011a, b). Liu et al. (2011b) suggested that methyl salicylate may be the systemic signal involved in triggering SAR. A plastid glycerolipid may be the mobile signal molecule triggering the SAR (Chaturvedi et al. 2008). Chanda et al. (2011) reported that the cooperative interaction of a lipid transfer protein (DIR1) and a lipid-derived molecule (glycerol-3 phosphate) orchestrates the induction of SAR in plants. Azelaic acid is another lipid mobile signal involved in signaling SAR (Jung et al. 2009; Shah 2009).

Recent studies suggest that auxin is highly mobile and may be involved in SA-triggered SAR. No SAR effect was observed in auxin F-box receptor mutant *tir1 Arabidopsis* plants challenged with virulent *Pseudomonas syringae* pv. *maculicola* in secondary leaves following an initial inoculation with *P. syringae* pv. *tomato* (Truman et al. 2010). It has also been shown that SAR response was completely abolished in the auxin polar transport mutants *aux1* and *axr4*. The *axr4* acts in the same pathway as *aux1*, actively disrupting trafficking of AUX1 to the plasma membranes of epidermal cells (Dharmasiri et al. 2006). SAR was also abolished in *cyp79b* double mutant. The *cyp79b* double mutant is blocked in the synthesis of indole-3-acetaldoxime, which is involved in the biosynthesis of IAA through IAN pathway (Truman et al. 2010). These results suggest that auxin signaling is involved in SAR response.

Auxin may not have direct role in inducing SAR. Although SA, JA, and ET are clearly integrated components of signal transduction cascades inducing disease resistance, auxin acts only by modulating various defense signaling systems (Bari and Jones 2009; Truman et al. 2010; Savatin et al. 2011). SAR has been shown to depend on SA signaling. SA or its functional analog benzo(1,2,3)thiadiazole-7-carbothioic acid S-methyl ester (BTH) and a SA biosynthesis inducing chemical probenazole are known to induce SAR (Hukkanen et al. 2007; Mandal et al. 2008; Ziadi et al. 2008; Umemura et al. 2009). By contrast, abscisic acid (ABA) has been shown to suppress chemically induced SAR (Yasuda et al. 2008; de Torres-Zabala et al. 2009). JA is known to promote susceptibility to the bacterial pathogen *Pseudomonas syringae* pv. *tomato* (Koornneef and Pieterse 2008). The antagonism between JA and SA has been widely reported (Koornneef et al. 2008; El Oirdi et al. 2011; Pieterse et al. 2012; Zheng et al. 2012). SA, JA, and ABA levels in the systemic leaves of *aux1* mutant *Arabidopsis* plants inoculated with *P. syringae* pv. *tomato* were measured (Truman et al. 2010). Levels of free SA and conjugated SA were significantly reduced in *aux1* mutant responding to *P. syringae* pv. *tomato* inoculation. By contrast, both JA and ABA, which are known to increase susceptibility to the bacterial pathogen, accumulated to higher levels in *aux1* mutant plants responding to the bacterial pathogen compared with the wild-type control plants (Truman et al. 2010). The results suggest that auxin signaling increases SA levels, which trigger SAR, and suppresses ABA and JA signaling, which induces susceptibility to the bacterial pathogen. Auxin has been shown to act antagonistically to ABA (Rock and Sun 2005). Enhanced ABA levels in *aux1* may act to suppress both SA biosynthesis and responses to SA (Truman et al. 2010).

6.11 Interactions Between Auxin and Jasmonate Signaling Systems

Both JA and auxin play an important role in plant immune signal transduction. These plant signals have been shown to interact with each other in complementary ways in their signaling roles. Both the signaling systems involve an ubiquitin–proteasome pathway that includes an SCF-type E3 ubiquitin ligase complex (Gray and Estelle 2000). COI1 (CORONATINE INSENSITVE1) is a key player in the jasmonate perception and signal transduction pathway. COI1 has been identified as a receptor for jasmonates (Gfeller et al. 2010). It is involved in the direct perception of the jasmonate signal (Yan et al. 2009). The gene *CoI1* is required for all the JA-mediated responses (Devoto et al. 2005; Balbi and Devoto 2008; Kazan and Manners 2008). The COI1 locus encodes an F-box protein that is part of an E3 ubiquitin ligase complex involved in ubiquitin-mediated protein degradation (Balbi and Devoto 2008). COI1 protein associates with CUL1 (CULLIN1), RBX1

(RING-BOX PROTEIN1), and the Skp1-like proteins ASK1 (ARABIDOPSIS S-PHASE KINASE-ASSOCIATED PROTEIN1) and ASK2 to form the SCF^{COI1} (SKP1, CDC53p/CUL1, F-box^{COI1}) ubiquitin ligase complex in *A. thaliana* (Devoto et al. 2002; Xu et al. 2002; Pauwels and Goossens 2011).

In case of auxin signaling, the F-box protein is TIR1, which is closely related to the jasmonate response factor encoded by COI1. TIR1 encodes an F-box protein, which is a subunit of SCF-type ubiquitin–protein ligases. The F-box protein interacts with Skp1. At the carboxyl terminus, the F-box protein has protein–protein interaction domains, which are responsible for recruiting the target protein to the SCF for ubiquitination (Leyser 2006). COI1 in the JA signaling system is related to the TIR1 component of the auxin signaling SCF complex. This suggests that jasmonate and auxin use a similar ubiquitin–proteasome signaling mechanism (Tiryaki and Staswick 2002). However, *coi* is not altered in its response to auxin (Feys et al. 1994). These results suggest that although the JA and auxin may act through a common signaling intermediate, the auxin and JA signaling pathways may be separate signaling pathways.

AXR1 is an auxin response gene playing important role in auxin signaling. AXR1 protein is required for posttranslation modification of CUL1, a component of the SCF complex involved in the degradation of AUX/IAA proteins (Chapman and Estelle 2009). In auxin signaling, removal of AUX/IAA family of transcription factor repressors through auxin-stimulated binding of SCF^{TIR1} complexes and targeting to the ubiquitin-proteasome pathway is the important component in auxin signaling-mediated immune responses (Llorente et al. 2008). The degradation of AUX and IAA proteins is compromised in the *axr1* mutant (Kidd et al. 2011).

AXR1 activates the ubiquitin-like RUB protein and the activated RUB protein is then conjugated to cullins (del Pozo et al. 1998, 2002). Along with cullin, the SCF complex consists of SKP1, RBX1, and an F-box protein. The F-box protein involved in auxin signaling is TIR1 and in JA signaling COI1 shows similar function (Xie et al. 1998). It has been demonstrated that the auxin and JA proteasome pathways are directly connected through AXR1, which acts upstream of TIR1/COI1 (Tiryaki and Staswick 2002). AXR1 has been shown to be necessary for defense against the opportunistic pathogen *Pythium irregulare* in *Arabidopsis* (Tiryaki and Staswick 2002). It suggests that both auxin and JA signaling systems may be involved in the defense response.

JAR1 (Jasmonate resistant1) is an important component in the JA signaling system. *JAR1* encodes the enzyme JA–amino synthetase, which activates conjugation of JA to an amino acid (Staswick and Tiryaki 2004). This enzyme may be involved in JA-Ile biosynthesis. In addition to Ile, the JAR family of related GH3 enzymes has the potential to conjugate other amino acids, such as Trp, Val, and Leu. The JA-Trp, JA-Val, and JA-Leu also participate in JA signaling pathway (Wang et al. 2008). To assess the interplay of auxin and JA signaling systems, the relative defense responses of the *Arabidopsis* mutants *axr1*, *jar1* and *jar1/axr1* against *Pytium irregulare* were employed. The mutants showed increased disease incidence. The results showed that disease symptoms appeared earlier in the double mutant jar1/axr1 than in *jar1* and *axr1* plants. Maximum disease incidence was observed in the double

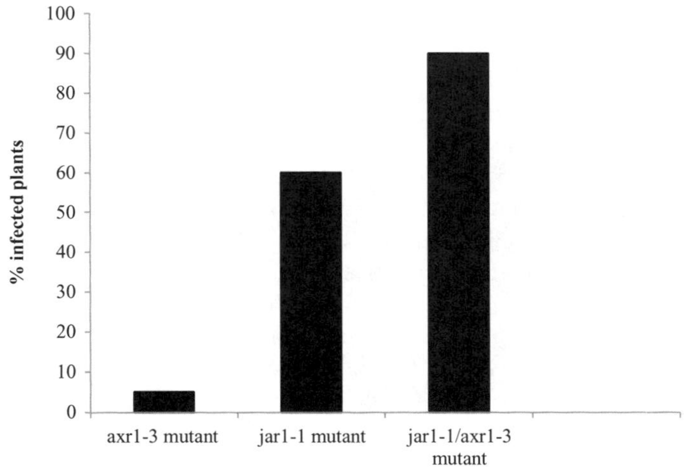

Fig. 6.9 Additive double mutant effect of auxin and jasmonate signaling systems in inducing resistance against *Pythium irregulare* in *Arabidopsis* (Adapted from Tiryaki and Staswick 2002)

mutant, followed by *jar1* and *axr1* mutant plants (Fig. 6.9; Tiryaki and Staswick 2002). The results suggest a link between auxin and JA-mediated defense response and an additive effect of *JAR1* and *AXR1* in triggering disease resistance.

IAA treatment induced expression of *LOX2* and *AOS* genes (Tiryaki and Staswick 2002), which are involved in JA biosynthesis (Delker et al. 2007; Vidhyasekaran 2007; Balbi and Devoto 2008; Bertoni 2012). Expression of the transcripts of the JA-responsive *AtVSP* was also induced by IAA treatment. *LOX2* and *AOS* induction was also suppressed in *axr1* mutant plants (Tiryaki and Staswick 2002). These results indicate another link between JA and auxin signaling.

GH3-8 encodes IAA–amido synthetase, which functions in maintenance of auxin homeostasis by conjugating IAA to amino acids in auxin signaling system (Staswick et al. 2005). GH3-8-overexpressing plants showed suppression of free IAA accumulation (Ding et al. 2008). Overexpression of *GH3.8*, which suppressed IAA accumulation, also suppressed the accumulation of endogenous JA. Expression of both *LOX* and *AOS* genes was suppressed in *GH3.8* overexpressing plants (Ding et al. 2008).

6.12 Interaction Between Auxin and Ethylene Signaling Systems

Arabidopsis plants infected with *Agrobacterium tumefaciens* showed elevated levels of the ethylene precursor 1-aminocyclopropane-1-carboxylate (ACC). The overproduction of ACC in infected *Arabidopsis* tissues correlated with the induction of genes like *ACS6* and *ACS8*, encoding ACC synthase (ACS) involved in ACC biosynthesis, by the T-DNA-harboring strain C58 exclusively (Lee et al. 2009). The C58 strain produced more auxin than avirulent strain. Auxin is known to induce the transcription of *ACS8* and the conversion of Met to ethylene (Yu and Yang 1979). Arabidopsis-derived auxin together with the auxin released by agrobacteria may stimulate ACC biosynthesis (Lee et al. 2009). Auxin increases ACS transcription, thus stimulating ethylene biosynthesis (Tsuchisaka and Theologis 2004). The oncogenic strain C58, which produces IAA triggered the transcription of genes encoding the ET receptor ETR2 and an ET response factor (Lee et al. 2009).

Auxin has been shown to trigger biosynthesis of ethylene in peaches (Tatsuki et al. 2013). Increased IAA concentration in the peach fruits showed increased ethylene production. Application of 1-naphthalene acetic acid, a synthetic auxin, to peaches induced a high level of *ACS1* expression and a large amount of ethylene production. Application of an anti-auxin, a (phenylethyl-2-one)-IAA reduced the levels of *ACS1* expression and ethylene production in peaches (Tatsuki et al. 2013). Similarly, ethylene application promotes the expression of IAA biosynthetic genes (Stepanova et al. 2005, 2008), increases IAA biosynthesis (Swarup et al. 2007), and increases IAA levels (Růžička et al. 2007). These results suggest that IAA accumulation in plant tissues may trigger ethylene biosynthesis.

Application of ethylene biosynthetic precursor ACC triggered an increase in the rate of IAA biosynthesis. Conversely application of the ACC synthase inhibitor 1-aminoethoxyvinyl-glycine (AVG) resulted in a significant decrease in the rate of IAA biosynthesis (Fig. 6.10; Swarup et al. 2007). Ethylene treatment has been shown to up-regulate the expression of auxin-responsive genes *IAA1* and *IAA2* (Abel and Theologis 1995; Swarup et al. 2007). ACC may also modulate IAA transport system in the auxin signaling pathway. It has been shown that ACC treatments significantly increased IAA transport in *Arabidopsis* (Negi et al. 2008). Collectively, these results demonstrate the importance of ethylene in auxin biosynthesis and signal transduction system.

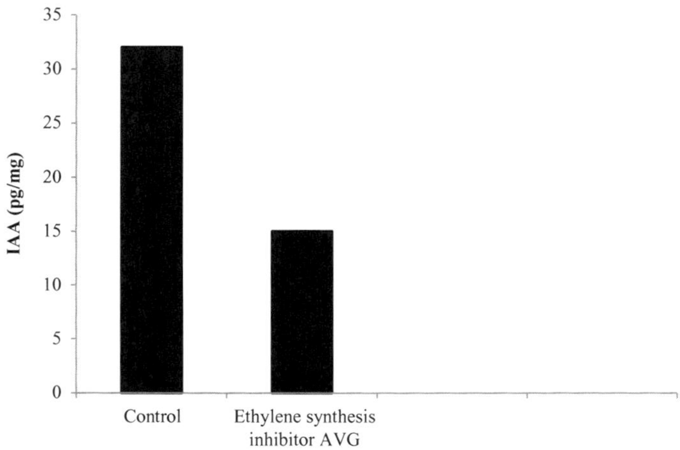

Fig. 6.10 Role of ethylene in IAA biosynthesis (Adapted from Swarup et al. 2007)

6.13 Interaction Between Small RNAs and Auxin Signaling Systems

Small RNAs (sRNAs) are nonprotein-coding RNAs of 20 to 30-nucleotide length (Havecker et al. 2010; Kulcheski et al. 2011; Wang et al. 2011). These small RNAs modulate gene expression at both the transcriptional and posttranscriptional levels (Zhao et al. 2012a, b). They act as sequence-specific repressors of target gene expression, either at the transcriptional level through DNA and/or histone methylation or at the posttranscriptional level through transcriptional cleavage or translational inhibition (Ramachandran and Chen 2008). Small RNA molecules act as mobile signals that direct mRNA cleavage and DNA methylation in recipient cells (Ciomperlik et al. 2011; Molnár et al. 2011). They are big contributors to plant innate immunity (Chellappan et al. 2009; Dunoyer et al. 2010; Mosher et al. 2010; Molnár et al. 2011).

PAMPs have been reported to trigger the expression of several small RNAs, and the majority of the small RNAs have been identified as microRNAs (miRNAs) (Li et al. 2010). MicroRNA393 (miRNA393) detected in *Arabidopsis* negatively regulates the expression of genes encoding the auxin receptors TIR1/AFB2 and AFB3. It results in stabilization of the Aux/IAA repressors of auxin response factors (Navarro et al. 2006). Targeting of the auxin receptor mRNAs results in direct suppression of auxin signaling (Robert-Seilaniantz et al. 2011). The microRNA miR393 makes the plant more resistant to auxin by targeting the auxin receptors (Parry et al. 2009). The miR393 indirectly prevents the release of auxin response factors (ARFs) from their negative regulators. As a consequence, ARF proteins, such as ARF1 and ARF9, cannot induce or repress their targeted genes (Robert-Seilaniantz et al. 2011). It has been shown that auxin signaling repressed SA levels and signaling in *Arabidopsis* (Robert-Seilaniantz et al. 2011). The microRNA miR393 has been shown to repress auxin signaling, preventing it from antagonizing SA signaling. The action of miR393 on auxin signaling prevents the negative effect of auxin on SA signaling (Robert-Seilaniantz et al. 2011).

The bacterial PAMP flg22 triggers the accumulation of another microRNA called miR160a. The miR160a accumulation results in repression of the auxin response factor genes *ARF16* and *ARF17* (Li et al. 2010). ARF proteins bind auxin-responsive elements to activate or repress transcription of primary auxin response genes (Li et al. 2010). The expression of ARF16 and ARF17 was reduced in the miR160a overexpression plants, whereas ARF10 was not significantly altered (Li et al. 2010). The miRNA miR160 has been shown to regulate expression of auxin response transcription factor, ARF17 (Mallory et al. 2005). The transgenic plants expressing *miR167* down-regulate *ARF6* and *ARF8* mRNA (Wu et al. 2006).

Down-regulation of auxin signaling by microRNAs results in ARF inactivation (Navarro et al. 2006). MicroRNA represses auxin signaling and increases antibacterial immune responses. MicroRNA repression of auxin signaling increased bacterial resistance (Navarro et al. 2006). Overexpression of miR393 rendered plants more resistant to biotroph pathogens and more susceptible to necrotroph pathogens.

In contrast, overexpression of the auxin receptor AFB1, whose mRNA is partially resistant to miR393 degradation, rendered the plant more susceptible to biotrophic pathogens (Navarro et al. 2006; Robert-Seilaniantz et al. 2011). Augmenting auxin signaling through overexpressing a TIR1 paralog that is partially refractory to miR393 enhanced susceptibility to the bacterial pathogen, and conversely, repressing auxin signaling through miR393 overexpression increased bacterial resistance. These studies suggest that small RNA signaling suppresses auxin signaling and modulates plant immune responses.

6.14 Auxin Signaling May Promote Susceptibility

6.14.1 Enhanced Auxin Levels Promote Susceptibility

Auxin signaling system has been shown to be involved in promoting susceptibility to pathogens and inducing disease development. Auxin signaling appears to be mostly involved in disease susceptibility rather than in disease resistance (O'Donnell et al. 2003; Kidd et al. 2011). Plant pathogens have been reported to produce auxin during their infection process. Auxin produced by pathogens is an important pathogenicity determinant in certain plant–pathogen interactions. The rice blast pathogen *Magnaporthe oryzae* produced the auxin indole-3-acetic acid during its biotrophic phase of infection. The invaded rice tissues responded to the exogenous fungal auxin, and the increased auxin responses were observed around the apex of the infection hyphae in the infected tissues. The fungus-produced auxin may play an important role in pathogenesis in early biotrophic stages of infection (Tanaka et al. 2011).

Agrobacterium tumefaciens produces auxin and the auxin stimulates gall formation in plants infected with the pathogen (Lee et al. 2009). *A. tumefaciens* strain C58 with T-DNA-encoded oncogenes is a virulent strain which promotes tumor induction in *Arabidopsis*, while the disarmed avirulent strain GV3101 is without T-DNA oncogenes (Lee et al. 2009). The virulent strain C58 produced twice as much auxin as the strain GV3101 without oncogenes (Fig. 6.11; Lee et al. 2009). *Pseudomonas syringae* strains are known to produce IAA and cause diseases (Glickmann et al. 1998; O'Donnell et al. 2003).

The exogenous auxin also triggered accumulation of endogenous auxin which promotes disease susceptibility (Lee et al. 2009). Exogenous application of the auxin analog 2,4-dichlorophenoxyacetic acid enhanced *P. syringae* pv. *tomato* disease symptoms in *Arabidopsis* (Navarro et al. 2006). Application of the auxin analog 1-naphthalene-acetic acid promoted disease symptom development in *Pst*-infected plants, suggesting that elevated auxin levels within host tissue promote *Pst* virulence (Chen et al. 2007). Treating rice lines resistant to the bacterial blight pathogen *Xanthomonas oryzae* pv. *oryzae* (*Xoo*) with either IAA or 2,4-D significantly promoted disease symptoms. The lesion area of these treated rice lines increased about sevenfold after *Xoo* infection compared with that of control plants (Ding et al. 2008).

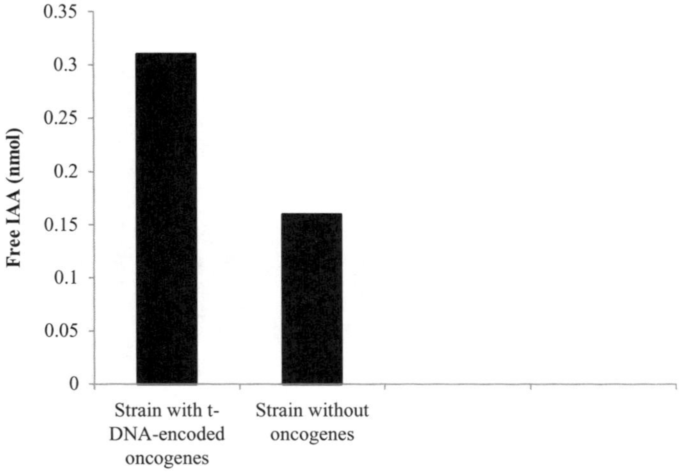

Fig. 6.11 IAA production in *Agrobacterium tumefaciens* bacterial cells in vitro (Adapted from Lee et al. 2009)

The pathogen infection may also elevate endogenous auxin levels and modulate auxin signaling system to induce disease development. *Pseudomonas syringae* pv. *tomato* (*Pst*) strain 3000 infection elevated free auxin levels in *A. thaliana* (Chen et al. 2007). *Xanthomonas oryzae pv. oryzae* infection also induced the accumulation of free IAA in rice plants. It triggered the expression of *AAO1*, *AAO2*, *AAO3*, and *NIT1* genes functioning in IAA biosynthesis pathways and induced accumulation of free IAA. The results suggest that the bacterial infection enhances endogenous synthesis of IAA which may promote disease development (Ding et al. 2008). Indole-3-acetic acid (IAA), the primary plant auxin, increased in plant tissues infected by *Pseudomonas syringae* pv. *tomato* DC3000 (O'Donnell et al. 2003). Moreover, the type III effector AvrRpt2 modulates host IAA levels to promote pathogen virulence and disease development in *Arabidopsis* (Chen et al. 2007; Cui et al. 2013). Infection with *P. syringae* pv. *tomato* DC3000 activates genes related to IAA biosynthesis and represses Aux/IAA family and auxin transporter genes, suggesting that the bacterial pathogen impacts auxin signaling probably through activating IAA production, altering IAA movement, and derepressing the auxin pathway (Thilmony et al. 2006).

6.14.2 Role of Auxin Receptors in Promoting Disease Susceptibility

It has been reported that auxin signaling promotes susceptibility to *Pseudomonas syringae* pv. *tomato* in *Arabidopsis* (Navarro et al. 2006). TIR1 and its paralogs AFB1 (Auxin signaling F-box protein 1), AFB2, and AFB3 are receptors for auxin

(Dharmasiri et al. 2005; Kepinski and Leyser 2005). When auxin signaling was augmented through overexpressing a *TIR1* paralog (*AFB1*) in *Arabidopsis*, the transformed plants showed enhanced susceptibility to *P. syringae* pv. *tomato* (Navarro et al. 2006). The results suggest that the auxin signaling promotes susceptibility.

6.14.3 Role of Aux/IAA Proteins in Promoting Susceptibility

Several plant pathogens are known to modulate plant auxin signaling system to cause disease (Wang et al. 2007; Cui et al. 2013). The *Pseudomonas syringae* type III effector AvRpt2 promotes auxin response by stimulating the turnover of Aux/IAA proteins (Cui et al. 2013). Aux/IAA proteins are transcription repressors. Auxin binds to its receptor F-box protein TIR1, a subunit of the ubiquitin ligase complex, and stabilizes the interaction with Aux/IAA proteins. This promotes SCF^{TIR1} ubiquitin complex catalyzed degradation of AUX/IAA proteins. This process relieves AUX-/IAA-mediated repression of auxin response factors (ARFs) and results in the transcription of auxin-responsive genes (Li et al. 2011; Rademacher et al. 2011; Hayashi 2012; Cui et al. 2013). The bacterial effector AvrRpt2 specifically reduces the protein stability of the negative regulators AXR2 and AXR3 in auxin signaling. AvrRpt2 encodes a Cys protease and Cys protease activity is required for AvrRpt2-mediated AXR2 degradation (Cui et al. 2013).

Auxin signaling is required for inducing susceptibility to the bacterial pathogen *Pseudomonas syringae* and the biotrophic oomycete pathogen *Hyaloperonospora parasitica* in Arabidopsis (Wang et al. 2007). The *sgt1b* mutant showed increased resistance to the necrotrophic fungal pathogen *F. culmorum* (Cuzick et al. 2009). The *sgt1b* mutant showed increased resistance to *Pseudomonas syringae* pv. *tomato* (Uppalapati et al. 2011). The Arabidopsis *sgt1b* mutant showed increased susceptibility to the biotrophic pathogen *Hyalop eronospora paraitica* (Tör et al. 2002).

The increased disease susceptibility observed in auxin-signaling mutants such as *axr1* and *axr2* against necrotrophic fungal pathogens may be due to the effects of compromised protein ubiquitination pathways (Llorente et al. 2008). The treatment of wild-type plants with a proteasome inhibitor increased susceptibility to the fungal necrotrophs whereas this treatment did not have any further effect on the susceptibility phenotypes of *axr1* and *axr2* (Llorente et al. 2008). The results suggest that the increased susceptibility observed in the auxin mutants is associated with a compromised protein ubiquitination.

Auxin signaling system has been shown to promote susceptibility to the wilt pathogen *Fusarium oxysporum* in *Arabidopsis*. Several auxin-signaling mutants, *axr1*, *axr2*, *axr3*, and *sgt1b*, showed increased resistance against *F. oxysporum* in *Arabidopsis. AXR2* and *AXR3* encode the IAA/AUX proteins IAA7 and IAA17, respectively (Kidd et al. 2011). AXR1 protein is required for post-translational modification of CUL1, a component of the SCF complex involved in the degradation of AUX/IAA proteins (Chapman and Estelle 2009). The degradation of AUX and IAA proteins is compromised in the *axr1* mutant (Kidd et al. 2011). SGT1B encodes a regulatory protein involved in SCF^{TIR1}-mediated degradation of Aux/IAA proteins (Gray et al. 2003). *SGT1B* is required for SCF^{TIR1}-mediated protein degradation and the *sgt1b* mutant is compromised in protein ubiquitination (Gray et al. 2003). These results suggest that Aux/IAA proteins, downstream of the auxin receptor in the auxin signaling system, contribute to the susceptibility of *Arabidopsis* to *Fusarium oxysporum*.

Viral infection also modulates auxin signaling to promote disease development. The helicase domain of the *Tobacco mosaic virus* (TMV) 126- and/or 183-kDa replicase protein(s) was found to interact with the *Arabidopsis* Aux/IAA protein IAA26 (also named PAP1), a regulator of auxin response genes (Padmanabhan et al. 2005). The viral replicase protein was found to disrupt the localization and stability of interacting Aux/IAA proteins in *Arabidopsis*. During TMV infection, interaction with the viral replicase may promote the destabilization and/or inappropriate sequestration of IAA26, thus interfering with its function as transcriptional regulator of auxin response genes. About 30 % of the *Arabidopsis* genes displaying transcriptional alterations in response to TMV contained multiple auxin response promoter elements (Padmanabhan et al. 2005). The results suggest that the TMV replicase protein interferes with the plant's auxin response system to induce disease development (Padmanabhan et al. 2005). TMV-induced disease symptoms can be attributed, in part, to the ability of the viral replicase protein to disrupt the localization and subsequent function of interacting AUX/IAA proteins (Padmanabhan et al. 2006). Similar replicase–Aux/IAA interaction affecting virus disease development has been reported in tomato (Padmanabhan et al. 2008). The interacting Aux/IAA proteins have been shown to regulate cellular factors that affect virus infection (Padmanabhan et al. 2008). The ability to affect AUX/IAA function may provide an important advantage to TMV in establishing an infection. The ability of TMV to interact with the specific Aux/IAAs corresponded with an enhanced ability of the virus to spread in the leaf tissues. TMV replicase–Aux/IAA interactions selectively enhance virus pathogenicity in tomato tissues where AUX/IAA proteins accumulate (Padmanabhan et al. 2008). It is suggested that TMV disrupts interacting AUX/IAA functions as a means of manipulating its cellular environment and promoting its own replication and spread (Padmanabhan et al. 2008).

6.14.4 Role of the Auxin-Responsive GH3 Genes in Promoting Disease Susceptibility

GH3 is an auxin-responsive gene and several *GH3* genes have been identified in plants (Staswick et al. 2005; Jain et al. 2006). The *GH3-8* gene has been detected in rice line showing resistance to the bacterial pathogen *X. oryzae* pv. *oryzae*. When the gene was overexpressed in a susceptible rice cultivar, the transformed susceptible cultivar showed enhanced resistance against the bacterial pathogen.

The enhanced resistance was clearly associated with overexpression of *GH3-8* (Ding et al. 2008). GH3-8 encodes IAA–amido synthetase, which functions in maintenance of auxin homeostasis by conjugating IAA to amino acids (Staswick et al. 2005). GH3-8-overexpressing plants showed suppression of free IAA accumulation (Ding et al. 2008). Up-regulation of GH3-8 and disease resistance was shown to suppress auxin signaling, besides the suppression of auxin accumulation. IAA induced the expression of auxin signaling-related genes belonging to Aux/IAA and ARF families, including *IAA4*, *IAA24*, *ARF6a*, and *ARF8*, in wild-type plants, but not in GH3-8-overexpressing plants (Ding et al. 2008). Collectively, these results suggest that auxin is involved in disease susceptibility, and suppression of auxin accumulation leads to activation of defense responses.

The auxin-responsive GH3-8 has been shown to down-regulate both SA and JA signaling systems, which are known to play important role in plant innate immune signaling (Fig. 6.12; Ding et al. 2008). The SA levels in *GH3-8*-overexpressing rice plants were lower than those of wild-type plants. The transcript levels of *PAD4*, a key component in SA signaling, were found to be lower in GH3-8-overexpressing

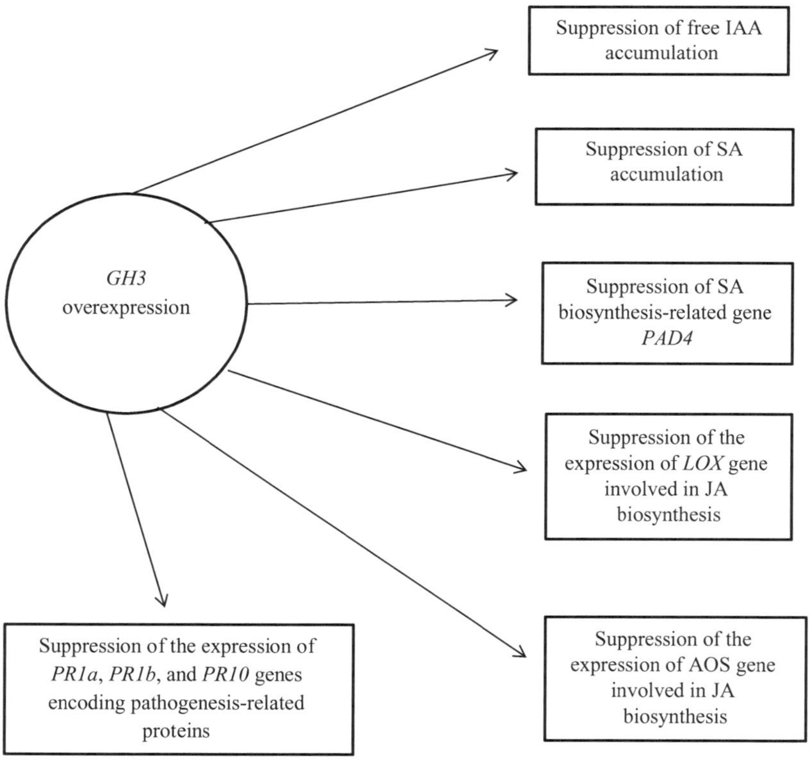

Fig. 6.12 Suppression of plant immune signaling systems by overexpression of *GH3* in rice plants (Adapted from Ding et al. 2008)

plants (Ding et al. 2008). PAD4 is involved in the production of SA (Rustérucci et al. 2001; Mateo et al. 2004; Lippok et al. 2007). It has been suggested that PAD4 is required for amplification of weak signals to a level sufficient for activation of SA signaling (Jirage et al. 1999). Overexpression of *GH3.8* significantly suppressed the accumulation of endogenous JA. Lipoxygenase (LOX) and allene oxide synthase (AOS) are important enzymes in JA biosynthesis pathway (Balbi and Devoto 2008; Vidhyasekaran 2007; Bertoni 2012), and expression of both *LOX* and *AOS* genes was suppressed in *GH3.8*-overexpressing plants (Ding et al. 2008). The transcript levels of the defense-responsive pathogenesis-related (PR) genes *PR1*and *PR10* were also lower in the *GH3.8*-overexpressing plants (Ding et al. 2008). These results suggest that auxin signaling may suppress defense responses by down-regulating the plant immune responses.

6.14.5 Conjugated Auxin Promotes Plant Disease Susceptibility

Conjugation of free auxin has been proposed to function in permanent inactivation and temporary storage of auxin as well as in detoxifying excess IAA (LeClere et al. 2002). One of the responses induced by the activation of auxin signaling is the formation of the conjugated forms of auxin through the action of GH3 proteins (Fu et al. 2011; González-Lamothe et al. 2012). Fungal and bacterial pathogens hijack the host auxin metabolism in *Arabidopsis thaliana*, leading to the accumulation of a conjugated form of the hormone, indole-3-acetic acid (IAA)–Asp, to promote disease development. IAA–Asp increases pathogen progression in the plant by regulating the transcription of virulence genes (González-Lamothe et al. 2012).

It has been shown that *Arabidopsis* accumulates the conjugated form of IAA, IAA–Asp, in response to the necrotrophic fungal pathogen *Botrytis cinerea* and the hemibiotroph bacterial pathogen *Pseudomonas syringae* pv. *tomato* strain DC3000 (González-Lamothe et al. 2012). IAA–Asp treatment enhanced both the fungal and bacterial disease development in *Arabidopsis*. IAA–Asp treatment promoted *P. syringae* pv. *tomato* bacterial growth in both tomato and *Nicotiana benthamiana*, and these plants had around 8–10 times more bacteria than control plants (González-Lamothe et al. 2012). It is suggested that the effect of IAA–Asp on disease susceptibility might be exerted on the pathogen, by contributing to the regulation of the transcription of virulence genes. The expression of the effector genes *AvrPto* and *HopAO1* of *P. syringae* pv. *tomato* and *BcatrB* gene of *Botrytis cinerea* was significantly reduced in *gh3.2* mutant plants (González-Lamothe et al. 2012). These results suggest that the GH3.2 is involved in activating expression of the virulence genes in both the pathogens and this action is mediated by IAA–Asp. Manipulation of auxin signaling may be one of the strategies used by pathogens to overcome plant immune pathways to cause the disease. The fungal and bacterial pathogens may manipulate auxin metabolism to produce a conjugated auxin and perceive it as a signal to activate its virulence.

6.14.6 Role of Auxin Transport System in Promoting Disease Susceptibility

Auxin transport also may have a role in auxin-induced susceptibility. The gene called *BIG* was responsive to infection by the fungal pathogen *Fusarium oxysporum* in *Arabidopsis*. It encodes a calossin-like protein required for polar auxin transport (PAT). The *big* mutant was found to be more resistant to *F. oxysporum,* suggesting that polar auxin transport is involved in triggering susceptibility (Kidd et al. 2011). PAT is controlled by localization of auxin influx and efflux carriers to the plasma membrane. AUXIN RESISTANT1 (AUX1) controls auxin efflux and its localization is regulated by AUXIN RESISTANT4 (AXR4) (Dharmasiri et al. 2006; Titapiwatanakun and Murphy 2009). Both *aux1* and *axr4* mutants showed increased resistance to *F. oxysporum* (Kidd et al. 2011). Auxin efflux is regulated by the pin-formed (PIN) proteins. The *pin2* mutant plants were found to be more resistant to *F. oxysporum* (Kidd et al. 2011). The *Arabidopsis aux1* mutant was compromised in SAR response (Truman et al. 2010). The *pin2* mutants were found to be compromised in ISR (INDUCED SYSTEMIC RESISTANCE) induced by *Pseudomonas fluorescens* (Iavicoli et al. 2003). Chemical inhibition of PAT by 2,3,5-triiodobenzoic acid (TIBA) treatment also reduced the wilt disease development caused by *F. oxysporum* in *Arabidopsis*. Collectively the studies with auxin transport mutants show that the auxin transport system is involved in plant disease susceptibility.

6.15 Auxin Signaling May Promote Plant Disease Resistance

6.15.1 Overexpression of Auxin-Responsive Genes Promote Disease Resistance

GH3 is an auxin-responsive gene and several *GH3* genes have been identified in plants (Staswick et al. 2005; Jain et al. 2006). The GH3 gene family consists of several members and the rice GH3 gene consists of at least 12 members (Jain et al. 2006). The *GH3-8* gene has been detected in rice line showing resistance to the bacterial pathogen *X. oryzae* pv. *oryzae*. *GH3-8* gene encodes an IAA–amido synthetase that maintains auxin homeostasis by conjugating excess IAA to amino acids. It is capable of catalyzing the synthesis of IAA–Asp (Ding et al. 2008). When the gene was overexpressed in a susceptible rice cultivar, the transformed susceptible cultivar showed enhanced resistance against the bacterial pathogen. The enhanced resistance was clearly associated with overexpression of *GH3-8* (Ding et al. 2008).

Rice plants overexpressing *GH3-8* showed higher accumulation of IAA–Asp and enhanced resistance against the rice bacterial blight pathogen *Xanthomonas oryzae* pv. *oryzae* (Ding et al. 2008). Accumulation of GH3-8 was found to accompany suppression of the expression of expansin genes. Expansins are wall-loosening proteins of plant cell walls (McQueen-Mason et al. 1992). The loosening cell walls may make the cell vulnerable to invaders and make the plants susceptible to pathogens. The suppression of expansin genes may be important in conferring resistance against pathogens. The transgenic rice plants overexpressing expansin genes showed significantly increased susceptibility to *X. oryzae* pv. *oryzae* (Ding et al. 2008). Overexpression of *GH3-8* in rice plants resulted in enhanced disease resistance, which was accompanied by the suppression of expansin genes in rice (Ding et al. 2008). The results suggest that GH3-8 is involved in enhancing disease resistance by suppressing the expression of expansin genes.

Transgenic rice plants overexpressing *GH3-2* gene, which is involved in auxin signaling, showed enhanced resistance against the bacterial blight pathogen X. *oryzae* pv. *oryzae*, the bacterial leaf streak pathogen *X. oryzae* pv. *oryzicola*, and the rice blast fungal pathogen *Magnaporthe oryzae* (previously called *M. grisea*; Choi et al. 2013) (Fu et al. 2011). The rice bacterial and fungal pathogens secrete IAA in planta and also induce rice plant to synthesize its own IAA at the infection site. IAA induces the production of expansins, the cell wall-loosening proteins, and makes rice vulnerable to pathogens. Activation of *GH3-2* inactivates IAA by catalyzing the formation of an IAA–amino acid conjugate, which results in the suppression of expansin genes (Fu et al. 2011). The results suggest that the auxin signaling component GH3-2 mediates basal resistance by suppressing pathogen-induced IAA accumulation.

Overexpression of OsGH3.1 in rice significantly reduced free auxin content and activated a number of defense-related genes. Rice plants overexpressing *OsGH3.1* showed reduction in auxin content and cell wall loosening and enhanced resistance against the rice blast pathogen *M. oryzae* (Domingo et al. 2009).

Auxin signaling has been shown to be required for resistance to necrotrophic pathogens such as *Botrytis cinerea* and *Plectosphaerella cucumerina* (Llorente et al. 2008). *Arabidopsis* auxin signaling mutants *axr1*, *axr2*, and *axr6* that have defects in the auxin-stimulated SCF (Skp1-Cullin-F-box) ubiquitination pathway exhibit increased susceptibility to *B. cinerea* and *P. cucumerina*. Pharmacological inhibition of auxin transport or proteasome function each compromise the fungal disease resistance in plants (Llorente et al. 2008). These results suggest that auxin signaling positively regulates disease resistance against the necrotrophic fungal pathogens in *Arabidopsis*.

6.15.2 Auxin Response Factors Modulate Plant Defense Responses

The ARFs are transcription factors that control the induction or repression of genes after auxin treatment. Mutations in *arf1* and *arf9* affected resistance to the biotrophic oomycete pathogen *Hyaloperonospora parasitica* and the necrotrophic fungal pathogen *Alternaria brassicicola*. The *arf9* mutant was more resistant to *H. parasitica* and more susceptible to *A. brassicicola*. The *arf1* mutant was more resistant to *H. parasitica*, but its resistance to *A. brassicicola* was unaltered (Robert-Seilaniantz et al. 2011). The results suggest that ARF1 and ARF9 suppress biotroph resistance while ARF9 positively enhances necrotroph resistance.

Ubiquitin-mediated protein degradation has been shown to be an important component of auxin signaling-mediated plant resistance to the necrotrophic fungal pathogens. An effective *Arabidopsis* auxin response relies on the removal of AUX/IAA family of transcription factor repressors through auxin-stimulated binding of SCF^{TIR1} complexes and targeting to the ubiquitin–proteasome pathway (Llorente et al. 2008). The *axr2* mutant that has a stabilized AXR2 repressor protein and the *axr1* and *axr6* mutants that are impaired in two components (AXR1/RUB1 and AXR6/Cullin1) of the SCF E3 ubiquitin ligase complexes and defective in IAA/AUX degradation showed enhanced susceptibility to *B. cinerea* and *P. cucumerina* (Llorente et al. 2008). The results suggest that ubiquitin-mediated protein degradation is an important component in auxin signaling-mediated resistance to the fungal pathogens.

6.15.3 Exogenous Application of Auxin Induces Plant Disease Resistance

Macrophomina phaseolina causes charcoal rot disease in a wide range of plant species. No genetic resistance against the pathogen has been identified in most field crops. Auxin has been suggested to play an important role in resistance against *M. phaseolina* (Mah et al. 2012). Genes involved in auxin homeostasis, polar auxin transport, and auxin signaling were regulated in *Medicago truncatula* infected by *M. phaseolina*. The differential expression patterns of these auxin-related genes suggested that the host susceptibility may be partly due to the suppression of auxin response in the host by *M. phaseolina*. This conclusion was strengthened by the finding that exogenous application of auxin enhanced the charcoal rot disease resistance in *Medicago truncatula* (Mah et al. 2012). These results suggest that auxin signaling may play an important role in the activation of defense responses.

Auxins have been shown to induce resistance against common scab disease of potato caused by *Streptomyces scabiei* (Tegg et al. 2008). An *Arabidopsis axr1* mutant showed hypersusceptibility to *Sclerotinia sclerotiorum* (Stotz et al. 2011). It suggests that AXR1 plays an important role in triggering defense responses against *Sclerotinia sclerotiorum*.

References

Abas L, Benjamins R, Malencia N, Paciorek T, Wirniewska J, Moulinier-Anzola JC, Sieberer T, Friml J, Luschnig C (2006) Intracellular trafficking and proteolysis of the *Arabidopsis* auxin-efflux facilitator PIN2 are involved in root gravitropism. Nat Cell Biol 8:249–256

Abel S, Theologis A (1995) A polymorphic bipartite motif signals nuclear targeting of early auxin-inducible proteins related to PS-IAA4 from pea (*Pisum sativum*). Plant J 8:87–96

Ahmad S, Gordon-Weeks R, Pickett J, Ton J (2010) Natural variation in priming of basal resistance: from evolutionary origin to agricultural exploitation. Mol Plant Pathol 11:817–827

Astier J, Lindermeyr C (2012) Nitric oxide-dependent posttranslational modification in plants: an update. Int J Mol Sci 13:15193–151208

Aziz A, Heyraud A, Lambert B (2004) Oligogalacturonide signal transduction, induction of defence-related responses and protection of grapevine against *Botrytis cinerea*. Planta 218:767–774

Balbi V, Devoto A (2008) Jasmonate signalling network in *Arabidopsis thaliana*: crucial regulatory nodes and new physiological scenarios. New Phytol 177:301–318

Bari R, Jones JDG (2009) Role of plant hormones in plant defence responses. Plant Mol Biol 69:473–488

Beard RA, Anderson DJ, Bufford JL, Tallman G (2012) Heat reduces nitric oxide production required for auxin-mediated gene expression and fate determination in tree tobacco guard cell protoplasts. Plant Physiol 159:1608–1623

Beckers GJM, Conrath U (2007) Priming for stress resistance: from the lab to the field. Curr Opin Plant Biol 10:425–431

Beckers GJM, Jaskiewicz M, Liu Y, Underwood WR, He SY, Zhang S, Conrath U (2009) Mitogen-activated protein kinases 3 and 6 are required for full priming of stress responses in *Arabidopsis thaliana*. Plant Cell 21:944–953

Bellin D, Asai S, Delledonne M, Yoshioka H (2013) Nitric oxide as a mediator for defense responses. Mol Plant Microbe Interact 26:271–277

Bennett MJ, Marchant A, Green HG, May ST, Ward SP, Millner PA, Walker AR, Schulz B, Feldmann KA (1996) *Arabidopsis AUX1* gene: a permease-like regulator of root gravitropism. Science 273:948–950

Bertoni G (2012) Oxylipins and plant palatability. Plant Cell 24:1305

Besson-Bard A, Courtois C, Gauthier A, Dahan J, Dobrowolska G, Jeandroz S, Pugin A, Wendehenne D (2008) Nitric oxide in plants: production and cross-talk with Ca^{2+} signaling. Mol Plant 1:218–228

Blakeslee JJ, Peer WA, Murphy AS (2005) Auxin transport. Curr Opin Plant Biol 8:494–500

Book AJ, Yang P, Scalf M, Smith LM, Viersta RD (2005) Tripeptidyl peptidase II. An oligomeric protease complex from *Arabidopsis*. Plant Physiol 138:1046–1057

Branca C, De Lorenzo G, Cervone F (1988) Competitive inhibition of the auxin-induced elongation of by α-D-oligogalacturonides in pea stem segments. Physiol Plant 72:499–504

Brumos A, Alonso JM, Stepanova AN (2013) Genetic aspects of auxin biosynthesis and its regulation. Physiol Plant. doi:10.1111/ppl.12098

Brutus A, Sicilia F, Macone A, Cervone F, De Lorenzo G (2010) A domain swap approach reveals a role of the plant wall-associated kinase 1 (WAK1) as a receptor of oligogalacturonides. Proc Natl Acad Sci U S A 107:9452–9457

Caffali KH, Mohnen D (2009) The structure, function, and biosynthesis of plant cell wall pectic polysaccharides. Carbohydr Res 344:1879–1900

Calderon-Villalobos LI, Tan X, Zheng N, Estelle M (2010) Auxin perception – structural insights. Cold Spring Harb Perspect Biol 2010:a005546

Camañes G, Pastor V, Cerezo M, Garcia-Andrade J, Vicedo B, Garcia-Agustin P, Flors V (2012) A deletion in *NRT2.1* attenuates *Pseudomonas syringae*-induced hormonal perturbation, resulting in primed plant defenses. Plant Physiol 158:1054–1066

Carrier DJ, Bakar NT, Swarup R, Callaghan R, Napier RM, Bennett MJ, Kerr ID (2008) The binding of auxin to the *Arabidopsis* auxin influx transporter AUX1. Plant Physiol 148:529–535

Casasoli M, Federici L, Spinelli F, Di Matteo A, Vella N, Scaloni F, Fernandez-Recio J, Cervone F, De Lorenzo G (2009) Integration of evolutionary and desolvation energy analysis identifies functional sites in a plant immunity protein. Proc Natl Acad Sci U S A 106:7666–7671

Celenza JL, Quirl JA, Smolen GA, Merrikh H, Silvestro AR, NormanlyJ Bender J (2005) The *Arabidopsis* ATR1 Myb transcription factor controls indolic glucosinolate homeostasis. Plant Physiol 137:253–262

Cernadas RA, Benedetti CE (2009) Roles of auxin and gibberellin in citrus canker development and in the transcriptional control of cell-wall remodeling genes modulated by *Xanthomonas axonopodis* pv. *citri*. Plant Sci 177:190–195

Chanda B, Xia Y, Mandal MK, Yu K, Sekine K-T, Gao Q-M, Selote D, Hu Y, Stromberg A, Navarre D, Kachroo A, Kachroo P (2011) Glycerol-3-phosphate is a critical mobile inducer of systemic immunity in plants. Nat Genet 43:421–427

Chapman EJ, Estelle M (2009) Mechanism of auxin-regulated gene expression in plants. Annu Rev Genet 43:265–285

Chaturvedi R, Krothapalli K, Makandar R, Nandi A, Sparks AA, Roth MR, Welti R, Shah J (2008) Plastid omega3-fatty acid desaturase-dependent accumulation of a systemic acquired resistance inducing activity in petiole exudates of *Arabidopsis thaliana* is independent of jasmonic acid. Plant J 54:106–117

Chellappan P, Zhang X, Jin H (2009) Host small RNAs are big contributors to plant innate immunity. Curr Opin Plant Biol 12:465–472

Chen Z, Agnew JL, Cohen JD, He P, Shan I, Sheen J, Kunkel BN (2007) *Pseudomonas syringae* type III effector AvrRpt2 alters *Arabidopsis thaliana* auxin physiology. Proc Natl Acad Sci U S A 104:20131–20136

Chen QF, Westfall CS, Hicks LM, Wang SP, Jez JM (2010) Kinetic basis for the conjugation of auxin by a GH3 family indole-acetic acid-amido synthetase. J Biol Chem 285:29780–29786

Cheng Y, Dai X, Zhao Y (2007) Auxin synthesized by the YUCCA flavin monooxygenases is essential for embryogenesis and leaf formation in *Arabidopsis*. Plant Cell 19:2430–2439

Choi J, Park S-Y, Kim B-R, Roh J-H, Oh IS, Han SS, Lee YH (2013) Comparative analyses of pathogenicity and phylogenetic relationship in *Magnaporthe grisea* species complex. PLoS One 8(20):e57196

Chun HJ, Park HC, Koo SC, Lee JH, Park CY, Choi MS, Kang CH, Baek D, Cheong YH, Yun DJ, Chung WS, Cho MJ, Kim MC (2012) Constitutive expression of mammalian nitric oxide synthase in tobacco plants triggers disease resistance to pathogens. Mol Cells 34:463–471

Chung KR, Shilts T, Erturk U, Timmer LW, Ueng PP (2003) Indole derivatives produced by the fungus *Colletotrichum acutatum* causing lime anthracnose and postbloom fruit drop of citrus. FEMS Microbiol Lett 70:1852–1854

Ciomperlik J, Omarov RT, Scholthof HB (2011) An antiviral RISC isolated from *Tobacco rattle virus*-infected plants. Virology 412:117–124

Conrath U (2009) Priming of induced plant defense responses. Adv Bot Res 51:361–395

Conrath U (2011) Molecular aspects of defence priming. Trends Plant Sci 16:524–531

Conrath U, Beckers GJM, Flors V, Garcia-Agustin P, Jakab G, Mauch F, Newman M-A, Pieterse CMJ, Poinssot B, Pozo MJ, Pugin A, Schaffrath U, Ton J, Wendehenne D, Zimmerli L, Mauch-Mani B (2006) Priming: getting ready for battle. Mol Plant Microbe Interact 19:1062–1071

Correa-Aragunde N, Graziano M, Chevalier C, Lamattina L (2006) Nitric oxide modulates the expression of cell cycle regulatory genes during lateral root formation in tomato. J Exp Bot 57:581–588

Cui F, Wu S, Sun W, Coaker G, Kunkel B, He P, Shan L (2013) The *Pseudomonas syringae* type III effector AvrRpt2 promotes pathogen virulence via stimulating Arabidopsis auxin/indole acetic acid protein turnover. Plant Physiol 162:1018–1029

Cuzick A, Maguire K, Hammond-Kosack KE (2009) Lack of the plant signalling component SGT1b enhances disease resistance to *Fusarium culmorum* in Arabidopsis buds and flowers. New Phytol 181:901–912

Dai X, Mashiguchi K, Chen Q, Kasahara H, Kamiya Y, Ojha S, DuBois J, Ballou D, Zhao Y (2013) The biochemical mechanism of auxin biosynthesis by an arabidopsis YUCCA flavin-containing monooxygenase. J Biol Chem 288:1448–1457

Davies RT, Goetz DH, Lasswell J, Anderson MN, Bartel B (1999) IAR3 encodes an auxin conjugate hydrolase from *Arabidopsis*. Plant Cell 11:365–382

De Lorenzo G, Ferrari S (2002) Polygalacturonase-inhibiting proteins in defense against phytopathogenic fungi. Curr Opin Plant Biol 5:295–299

de Torres-Zabala M, Bennett MH, Truman WH, Grant MR (2009) Antagonism between salicylic and abscisic acid reflects early host-pathogen conflict and moulds plant defense responses. Plant J 59:375–386 del Pozo JC, Timpte C, Tan S, Callis J, Estelle M (1998) The ubiquitin-related protein RUB1 and auxin response in *Arabidopsis*. Science 280:1760–1763

del Pozo JC, Dharmasiri S, Hellmann H, Walker L, Gray WM, Estelle M (2002) AXR1-ECR1-dependent conjugation of RUB1 to the *Arabidopsis* cullin AtCUL1 is required for auxin response. Plant Cell 14:421–433

Delker C, Zolman BK, Miersch O, Wasternack C (2007) Jasmonate biosynthesis in *Arabidopsis thaliana* requires peroxisomal β-oxidation enzymes – Additional proof by properties of *pex6* and *aim1*. Phytochemistry 68:1542–1650

Denancé N, Sánchez-Vallet A, Goffner D, Molina A (2013) Disease resistance or growth: the role of plant hormones in balancing immune responses and fitness costs. Front Plant Sci 4:Article 155. pp 1–12

Denoux C, Galletti R, Mammarella N, Gopalan S, Werck D, De Lorenzo G, Ferrari S, Ausubel FM, Dewdney J (2008) Activation of defense response pathways by OGs and Flg22 elicitors in Arabidopsis seedlings. Mol Plant 1:423–445

Deshaies RJ (1999) SCF and cullin/ring H2-based ubiquitin ligases. Annu Rev Cell Dev Biol 15:435–467

Devoto A, Nieto-Rostro M, Xie D, Ellis C, Harmston R, Patrick E, Davis J, Sherratt L, Coleman M, Turner JG (2002) COI1 links jasmonate signaling and fertility to the SCF ubiquitin-ligase complex in *Arabidopsis*. Plant J 32:457–466

Devoto A, Ellis C, Magusin A, Chang HS, Chilcott C, Zhu T, Turner JG (2005) Expression profiling reveals COI1 to be a key regulator of genes involved in wound- and methyl jasmonate-induced secondary metabolism, defence, and hormone interactions. Plant Mol Biol 58:497–513

Dharmasiri N, Dharmasiri S, Estelle M (2005) The F-box protein TIR1 is an auxin receptor. Nature 435:441–445

Dharmasiri S, Swarup R, Mockaitis K, Dharmasiri N, Singh SK, Kowalchyk M, Marchant A, Mills S, Sandberg G, Bennett MJ, Estelle M (2006) AXR4 is required for localization of the auxin influx facilitator AUX1. Science 312:1218–1220

Ding X, Cao Y, Huang L, Zhao C, Li X, Wang S (2008) Activation of the indole-3-acetic acid-amido synthetase GH3-8 suppresses expansin expression and promotes salicylate- and jasmonate-independent basal immunity in rice. Plant Cell 20:228–240

Domingo C, Andrés F, Tharreau D, Iglesias DJ, Talón M (2009) Constitutive expression of *OsGH3.1* reduces auxin content and enhances defense response and resistance to a fungal pathogen in rice. Mol Plant Microbe Interact 22:201–210

Dowd C, Wilson IW, McFadden H (2004) Gene expression profile changes in cotton root and hypocotyl tissues in response to infection with *Fusarium oxysporum* f.sp. *vasinfectum*. Mol Plant Microbe Interact 17:654–667

Dreher K, Callis J (2007) Ubiquitin, hormones and biotic stress in plants. Ann Bot 99:787–822

Dreher KA, Brown J, Saw RF, Callis J (2006) The *Arabidopsis* Aux/IAA protein family has diversified in degradation and auxin responsiveness. Plant Cell 18:699–714

Dunoyer P, Schott G, Himber C, Meyer D, Takeda A, Carrington JC, Voinnet O (2010) Small RNA duplexes function as mobile silencing signals between plant cells. Science 328:912–916

Durrant WE, Dong X (2004) Systemic acquired resistance. Annu Rev Phytopathol 42:185–209

El Oirdi M, El Rahman TA, Rigano L, El Hadrami A, Rodriguez MC, Daay F, Vojnov A, Bouarab K (2011) *Botrytis cinerea* manipulates the antagonistic effects between immune pathways to promote disease development in tomato. Plant Cell 23:2405–2421

Ferrari S, Galletti R, Denoux C, De Lorenzo G, Ausubel FM, Dewdney J (2007) Resistance to *Botrytis cinerea* induced in Arabidopsis by elicitors is independent of salicylic acid, ethylene, or jasmonate signaling but requires PHYTOALEXIN DEFICIENT3. Plant Physiol 144:367–379

Ferrari S, Galletti R, Pontiggia D, Manfredni D, Lionetti V, Bellincampi D, Cervone F, De Lorenzo G (2008) Transgenic expression of a fungal endo-polygalacturonase increases plant resistance to pathogens and reduces auxin sensitivity. Plant Physiol 146:669–681

Feys BJF, Benedetti CE, Penfold CN, Turner JG (1994) Arabidopsis mutants selected for resistance to the phytotoxin coronatine are male sterile, insensitive to methyl jasmonate, and resistance to a bacterial pathogen. Plant Cell 6:751–759

Friml J (2003) Auxin transport –shaping the plant. Curr Opin Plant Biol 6:7–12

Friml J, Palme K (2002) Polar auxin transport – old questions and new concepts? Plant Mol Biol 49:273–284

Fu J, Wang S (2011) Insights into auxin signaling in plant-pathogen interactions. Frontiers Plant Sci 2:Article 74. p 7

Fu J, Liu H, Li Y, Yu H, Li X, Xiao J, Wang S (2011) Manipulating broad-spectrum disease resistance by suppressing pathogen-induced auxin accumulation in rice. Plant Physiol 155:589–602

Galletti R, De Lorenzo G, Ferrari S (2009) Host-derived signals activate plant innate immunity. Plant Signal Behav 4:33–34

Galon Y, Aloni R, Nachmias D, Snir O, Feldmesser E, Scrase-Field S, Boyce JM, Bouché N, Knight MR, Fromm H (2010) Calmodulin-binding transcription activator 1 mediates auxin signaling and responds to stresses in Arabidopsis. Planta 232:165–178

Geisler M, Blakeslee J, Bouchard R, Lee O, Vincenzetti V, Bandyopadhyay A, Titapiwatanakun B, Peer WA, Bailly A, Richards EL, Ejendal KF, Smith AP, Baroux C, Grossniklaus U, Müller A, Hrycyna CA, Dudler R, Murphy AS, Martinoia E (2005) Cellular efflux of auxin catalyzed by the *Arabidopsis* MDR/PGP transporter AtPGP1. Plant J 44:179–194

Gfeller A, Liechti R, Farmer EE (2010) Arabidopsis jasmonate signaling pathway. Sci Signal 3(109):cm4

Gil P, Dewey E, Friml J, Zhao Y, Snowden KC, Putterill J, Palme K, Estelle M, Chory J (2001) BIG: a calossin-like protein required for polar auxin transport in *Arabidopsis*. Genes Dev 15:1985–1997

Glickmann E, Garden L, Jacquet S, Hussain S, Elasri M, Petit A, Dessaux Y (1998) Auxin production is a common feature of most pathovars of *Pseudomonas syringae*. Mol Plant Microbe Interact 11:156–162

González-Lamothe R, El Oirdi M, Brisson N, Bouarab K (2012) The conjugated auxin indole-3-acetic acid-aspartic acid promotes plant disease development. Plant Cell 24:762–777

Grant MR, Jones JDG (2009) Hormone (Dis)harmony moulds plant health and disease. Science 324:750–752

Gray WM, Estelle M (2000) Function of the ubiquitin-proteasome pathway in auxin-response. Trends Biochem Sci 25:133–138

Gray WM, Muskett PR, Chuang HW, Parker JE (2003) *Arabidopsis* SGT1b is required for SCF^{TIR1}-mediated auxin response. Plant Cell 15:1310–1319

Greenham K, Santner A, Castillejo C, Mooney S, Sairanen I, Ljung K, Estelle M (2011) The AFB4 auxin receptor is a negative regulator of auxin signaling in seedlings. Curr Biol 21:520–525

Guilfoyle TJ, Hagen G (2007) Auxin response factors. Curr Opin Plant Biol 10:453–460

Hagen G, Guilfoyle T (2002) Auxin-responsive gene expression: genes, promoters and regulatory factors. Plant Mol Biol 49:373–385

Havecker ER, Wallbridge LM, Hardcastle TJ, Bush MS, Kelly KA, Dunn RM, Schwach F, Doonan JH, Baulcombe DC (2010) The *Arabidopsis* RNA-directed DNA methylation argonautes functionally diverge based on their expression and interaction with target loci. Plant Cell 22:321–334

Hayashi K-I (2012) The interaction and integration of auxin signaling components. Plant Cell Physiol 53:965–975

Hayashi K, Tan X, Zheng N, Hatate T, Kimura Y, Kepinski S, Nozaki H (2008) Small-molecule agonists and antagonists of F-box protein-substrate interactions in auxin perception and signaling. Proc Natl Acad Sci U S A 105:5632–5637

Hettenhausen C, Baldwin IT, Wu J (2012) Silencing *MPK4* in *Nicotiana attenuata* enhances photosynthesis and seed production but compromises abscisic acid-induced stomatal closure and guard cell-mediated resistance to *Pseudomonas syringae* pv. *tomato* DC3000. Plant Physiol 158:759–776

Hu X, Neill SJ, Tang Z, Cai W (2005) Nitric oxide mediates gravitropic bending in soybean roots. Plant Physiol 137:663–670

Huffaker A, Dafoe NJ, Schmelz EA (2011) ZmPep1 an ortholog of *Arabidopsis* elicitor peptide 1 regulates maize innate immunity and enhances disease resistance. Plant Physiol 155:1325–1338

Hukkanen AT, Kokko HI, Buchala AJ, McDougall GJ, Stewart D, Kárenlampi SO, Karjalainen RO (2007) Benzothiadiazole induces the accumulation of phenolics and improves resistance to powdery mildew in strawberries. J Agric Food Chem 55:1862–1870

Iavicoli A, Boutet E, Buchala A, Metraux JP (2003) Induced systemic resistance in *Arabidopsis thaliana* in response to root inoculation with *Pseudomonas fluorescens* CHAO. Mol Plant Microbe Interact 16:851–858

Ichimura K, Shinozaki K, Tena G, Sheen J, Henry Y (2002) Mitogen-activated protein kinase cascades in plants: a new nomenclature. Trends Plant Sci 7:301–308

Jain M, Kaur N, Tyagi AK, Khurana JP (2006) The auxin-responsive *GH3* gene family in rice (*Oryza sativa*). Funct Integr Genomics 6:36–46

Jirage D, Tootle TL, Reuber TL, Frost LN, Feys BJ, Parker JE (1999) *Arabidopsis thaliana* PAD4 encodes a lipase-like gene that is important for salicylic acid signaling. Proc Natl Acad Sci U S A 96:13583–13588

Jones-Rhoades MW, Bartel DP (2004) Computational identification of plant microRNAs and their targets, including a stress-induced miRNA. Mol Cell 14:787–799

Jung HW, Tschaplinski TJ, Wang L, Glazebrook J, Greenberg JT (2009) Priming in systemic plant immunity. Science 324:89–91

Kazan K, Manners JM (2008) Jasmonate signaling: toward an integrated view. Plant Physiol 146:1459–1468

Kazan K, Manners JM (2009) Linking development to defense: auxin in plant-pathogen interactions. Trends Plant Sci 14:373–382

Kepinski S (2007) The anatomy of auxin perception. Bioessays 29:953–956

Kepinski S, Leyser O (2005) The *Arabidopsis* F-box protein TIR1 is an auxin receptor. Nature 435:446–451

Kidd BN, Kadoo NY, Dombrecht B, Tekeoglu M, Gardiner DM, Thatcher LF, Aitken EA, Schenk PM, Manners JM, Kazan K (2011) Auxin signaling and transport promote susceptibility to the root-infecting fungal pathogen *Fusarium oxysporum* in Arabidopsis. Mol Plant Microbe Interact 24:733–748

Kohler A, Schwindling S, Conrath U (2002) Benzothiadiazole-induced priming for potentiated responses to pathogen infection, wounding, and infiltration of water into leaves requires the *NPR1/NIM1* gene in Arabidopsis. Plant Physiol 128:1046–1056

Koornneef A, Pieterse CMJ (2008) Cross talk in defense signaling. Plant Physiol 146:839–844

Koornneef A, Leon-Reyes A, Ritsema T, Verhage A, Den Otter FC, Van Loon LC, Pieterse CMJ (2008) Kinetics of salicylate-mediated suppression of jasmonate signaling reveal a role for redox modulation. Plant Physiol 147:1358–1368

Kovtun Y, Chiu WL, Zeng W, Sheen J (1998) Suppression of auxin signal transduction by a MAPK cascade in higher plants. Nature 395:716–720

Kovtun Y, Chiu WL, Tena G, Sheen J (2000) Functional analysis of oxidative stress-activated-mitogen-activated protein kinase cascade in plants. Proc Natl Acad Sci U S A 2940–2945

Kulcheski F, de Oliveira LFV, Molina LG, Almerão MP, Rodrigues FA, Marcolino J, Barbosa JF, Stolf-Moreira R, Nepomuceno AL, Marcelino-Guimarães FC, Abdelnoor RV, Nascimento LC, Carazzolle MF, Pereira GAG, Margis R (2011) Identification of novel soybean microRNAs involved in abiotic and biotic stresses. BMC Genomics 12:307

Lau S, Jürgens G, De Smet I (2008) The evolving complexity of the auxin pathway. Plant Cell 20:1738–1746

LeClere S, Tellez R, Rampey RA, Matsuda SPT, Bartel B (2002) Characterization of a family of IAA-amino acid conjugate hydrolases from Arabidopsis. J Biol Chem 277:20446–20452

Lee C-W, Efetova M, Engelmann JC, Kramell R, Wasternack C, Ludwig-Müller J, Hedrich R, Deeken R (2009) *Agrobacterium tumefaciens* promotes tumor induction by modulating pathogen defense in *Arabidopsis thaliana*. Plant Cell 21:2948–2962

Lee DH, Choi HW, Hwang BK (2011) The pepper E3 ubiquitin ligase *RING1* gene, *CaRING1*, is required for cell death and the salicylic acid-dependent defense response. Plant Physiol 156:2011–2025

Leguay JJ, Jouanneau JP (1987) Auxin (2,4-dichlorophenoxyacetic acid) starvation and treatment with glucan elicitor isolated from *Phytophthora megasperma* induces similar responses in soybean-cultured cell suspensions. Dev Genet 8:351–364

Lehmann T, Hoffmann M, Hentrich M, Pollmann S (2010) Indole-3-acetamide-dependent auxin biosynthesis: a widely distributed way of indole-3-acetic acid production? Eur J Cell Biol 89:895–905

Leyser O (2006) Dynamic integration of auxin transport and signaling. Curr Biol 16:R424–R433

Li Y, Zhang Q, Zhang J, Wu L, Qi Y, Zhou J-M (2010) Identification of microRNAs involved in pathogen-associated molecular pattern-triggered plant innate immunity. Plant Physiol 152:2222–2231

Li H, Tiwari SB, Hagen G, Guilfoyle TJ (2011) Identical amino acid substitutions in the repression domain of auxin/indole-3-acetic acid proteins have contrasting effects on auxin signaling. Plant Physiol 155:1252–1263

Li G, Meng X, Wang R, Mao G, Han L, Liu Y, Zhang S (2012) Dual-level regulation of ACC synthase activity by MPK3/MPK6 cascade and its downstream WRKY transcription factor during ethylene induction in Arabidopsis. PLoS Genet 8(6):e1002767. doi:10.1371/journal.pgen.1002767

Lindermayr C, Saalbach G, Durner J (2005) Proteomic identification of *S*-nitrosylated proteins in *Arabidopsis*. Plant Physiol 137:921–930

Lippok B, Birkenbihl RP, Rivory G, Brummer J, Schmelzer E, Logemann E, Somssich IE (2007) Expression of *AtWRKY33* encoding a pathogen- or PAMP-responsive WRKY transcription factor is regulated by a composite DNA motif containing W box elements. Mol Plant Microbe Interact 20:420–429

Liscum E, Reed JW (2002) Genetics of Aux/IAA and ARF action in plant growth and development. Plant Mol Biol 49:387–400

Liu PP, Montgomery TA, Fahlgren N, Kasschau KD, Nonogaki H, Carrington JC (2007) Repression of AUXIN RESPONSE FACTOR10 by microRNA160 is critical for seed germination and post-germination stages. Plant J 52:133–146

Liu P-P, von Dahl CC, Klessig DF (2011a) The extent to which methyl salicylate is required for signaling systemic acquired resistance is dependent on exposure to light after infection. Plant Physiol 157:2216–2226

Liu P-P, von Dahl CC, Park S-W, Klessig DF (2011b) Interconnection between methyl salicylate and lipid-based long-distance signaling during the development of systemic acquired resistance in Arabidopsis and tobacco. Plant Physiol 155:1762–1768

Llorente E, Muskett P, Sáuchez Vallet A, Lopez G, Ramos B, Sánchez Rodriguez C, Jordá L, Parker J, Molina A (2008) Repression of the auxin response pathway increases *Arabidopsis* susceptibility to necrotrophic fungi. Mol Plant 1:496–509

Luna E, Bruce TJA, Roberts MR, Flors V, Ton J (2012) Next generation systemic acquired resistance. Plant Physiol 158:844–853

Mah KM, Uppalapati SR, Tang Y, Allen S, Shuai B (2012) Gene expression profiling of *Macrophomina phaseolina* infected *Medicago truncatula* roots reveals a role for auxin in plant tolerance against the charcoal rot pathogen. Physiol Mol Plant Pathol 79:21–30

Malik SI, Hussain A, Yun BW, Spoel SH, Loake GJ (2011) GSNOR-mediated de-nitrosylation in the plant defence response. Plant Sci 181:540–544

Mallory AC, Bartel DP, Bartel B (2005) MicroRNA-directed regulation of *Arabidopsis* AUXIN RESPONSE FACTOR7 is essential for proper development and modulates expression of early auxin response genes. Plant Cell 17:1360–1375

Mandal B, Mandal S, Csinos AS, Martinez N, Culbreath AK, Pappu HR (2008) Biological and molecular analyses of the acibenzolar-S-methyl-induced systemic acquired resistance in flue-cured tobacco against *Tomato spotted wilt virus*. Phytopathology 98:196–204

Manosalva PM, Park S-W, Forouhar F, Tong L, Fry WE, Klessig DF (2010) *Methyl Esterase 1 (StMES1)* is required for systemic acquired resistance in potato. Mol Plant Microbe Interact 23:1151–1163

Maor R, Haskin S, Levi-Kadmi H, Sharon A (2004) In planta production of indole-3-acetic acid by *Colletotrichum gloeosporioides* f. sp. *aeschynomene*. Appl Environ Microbiol 70:1852–1854

Maraschin Fdos S, Memelink J, Offringa R (2009) Auxin induced SCF^{TIR1}-mediated polyubiquitination marks AUX/IAA proteins for degradation. Plant J 59:100–109

Mashiguchi K, Tanaka K, Sakai T, Sugawara S, Kawaide H, Natsume M, Hanada A, Yaeno T, Shirasu K, Yao H, McSteen P, Zhao Y, Hayashi K-I, Kamiya Y, Kasahara H (2011) The main auxin biosynthesis pathway in *Arabidopsis*. Proc Natl Acad Sci U S A 108:18512–18517

Mateo A, Muhlenbock P, Rustérucci C, Chang CC, Miszalski Z, Karpinska B, Parker JE, Mullineaux PM, Karpinski S (2004) *LESION SIMULATING DISEASE 1* is required for acclimation to conditions that promote excess excitation energy. Plant Physiol 136:2818–2830

McQueen-Mason S, Durachko DM, Cosgrove DJ (1992) Two endogenous proteins that induce cell wall extension in plants. Plant Cell 4:1425–1433

Mészáros T, Helfer A, Hatzimasoura E, Magyar Z, Serazetdinova L, Rios G, Bardóczy V, Teige M, Koncz C, Peck S, Bögre L (2006) The Arabidopsis MAP kinase kinase MKK1 participates in defence responses to the bacterial elicitor flagellin. Plant J 48:485–495

Mockaitis K, Estelle M (2008) Auxin receptors and plant development: a new signaling paradigm. Annu Rev Cell Dev Biol 24:55–80

Molnár A, Melnyk C, Baulcombe DC (2011) Silencing signals in plants: a long journey for small RNAs. Genome Biol 12:215

Moscatiello R, Mariani P, Sanders D, Maathuis FJ (2006) Transcriptional analysis of calcium-dependent and calcium-independent signalling pathways induced by oligogalacturonides. J Exp Bot 57:2847–2865

Mosher RA, Lewsey MG, Shivaprasad PV (2010) RNA silencing in plants: flash report. Silence 1:13

Mukherjee M, Larrimore KE, Ahmed NJ, Bedick TS, Barghouthi NT, Traw MB, Barth C (2010) Ascorbic acid deficiency in *Arabidopsis* induces constitutive priming that is dependent on hydrogen peroxide, salicylic acid, and the *NPR1* gene. Mol Plant Microbe Interact 23:340–351

Muto H, Nagao I, Demura T, Fukuda H, Kinjo M, Yamamoto KT (2006) Fluorescence cross-correlation analyses of the molecular interaction between an Aux/protein, MSG2/IAA19, and protein-protein interaction domains of auxin response factors of *Arabidopsis* expressed in HeLa cells. Plant Cell Physiol 47:1095–1101

Muto H, Watahiki MK, Nakamoto D, Kinjo M, Yamamoto KT (2007) Specificity and similarity of functions of the Aux/IAA genes in auxin signaling of arabidopsis revealed by promoter-exchange experiments among MSG2/IAA19, AXR2/IAA7, and SLR/IAA 141. Plant Physiol 144:187–196

Nagpal P, Ellis CM, Weber H, Ploense SE, Barkawi LS, Guilfoyle LS, Hagen G, Alonso JM, Cohen JD, Farmer EE, Ecker JR, Reed JW (2005) Auxin response factors ARF6 and ARF8 promote jasmonic acid production and flower maturation. Development 132:4107–4718

Naseem M, Dandekar T (2012) The role of auxin-cytokinin antagonism in plant-pathogen interactions. PLoS Pathog 8(11):e1003026. doi:10.1371/journal ppat 1003026

Navarro L, Dunoyer P, Jay F, Arnold B, Dhamasiri N, Estelle M, Voinnet O, Jones JDG (2006) A plant miRNA contributes to antibacterial resistance by repressing auxin signaling. Science 312:436–439

Negi S, Ivanchenko MG, Muday GK (2008) Ethylene regulates lateral root formation and auxin transport in *Arabidopsis thaliana*. Plant J 55:175–187

O'Donnell PJ, Schmelz EA, Moussatche P, Lund ST, Jones JB, Klee HJ (2003) Susceptible to intolerance – a range of hormonal actions in a susceptible *Arabidopsis* pathogen response. Plant J 33:245–257

Okrent RA, Brooks MD, Wildermuth MC (2009) *Arabidopsis* GH3.12 (PBS3) conjugates amino acids to 4-substituted benzoates and is inhibited by salicylate. J Biol Chem 284:9742–9754

Okushima Y, Overvoorde PJ, Arima K, Alonso JM, Chan A, Chang C, Ecker JR, Hughes B, Lui A, Nguyen D, Onodera C, Quach H, Smith A, Yu G, Theologis A (2005) Functional genomic analysis of the *AUXIN RESPONSE FACTOR* gene family members in *Arabidopsis thaliana*: unique and overlapping functions of *ARF7* and *ARF19*. Plant Cell 17:444–463

Padmanabhan MS, Goregaoker SP, Golem S, Shiferaw H, Culver JN (2005) Interaction of the *Tobacco mosaic virus* replicase protein with the Aux/IAA protein PAP1/IAA26 is associated with disease development. J Virol 79:2549–2558

Padmanabhan MS, Shiferaw H, Culver JN (2006) The *Tobacco mosaic virus* replicase protein disrupts the localization and function of interacting Aux/IAA proteins. Mol Plant Microbe Interact 19:864–873

Padmanabhan MS, Kramer SR, Wang X, Culver JN (2008) *Tobacco mosaic virus* replicase-auxin/indole acetic acid protein interactions: reprogramming the auxin response pathway to enhance virus infection. J Virol 82:2477–2485

Park WJ, Kriechbaumer V, Müller A, Piotrowski M, Meeley RB, Gierl A, Glawischnig E (2003) The nitrilase ZmNIT2 converts indole-3-acetonitrile to indole-3-acetic acid. Plant Physiol 133:794–802

Park JE, Park JY, Kim YS, Staswick PE, Jeon J, Yun J, Kim SY, Kim J, Lee YH, Park CM (2007) GH3-mediated auxin homeostasis links growth regulation with stress adaptation response in *Arabidopsis*. J Biol Chem 282:10036–10046

Park SW, Liu PP, Forouhar F, Vlot AC, Tong L, Tietjen K, Klessig DF (2009) Use of synthetic salicylic acid analog to investigate the roles of methyl salicylate and its esterases in plant disease resistance. J Biol Chem 284:7307–7317

Parry G, Calderon-Villalobos LI, Prigge M, Peret B, Dharmasiri S, Itoh H, Lechner E, Gray WM, Bennett M, Estelle M (2009) Complex regulation of the TIR1/AFB family of auxin receptors. Proc Natl Acad Sci U S A 106:22540–22545

Pauwels L, Goossens A (2011) The JAZ proteins: a crucial interface in the jasmonate signaling cascade. Plant Cell 23:3089–3100

Peer WA, Cheng Y, Murphy AS (2013) Evidence of oxidative attenuation of auxin signaling. J Exp Bot 64:2629–2639

Perchepied L, Balagué C, Riou C, Claudel-Renard C, Riviére N, Grezes-Besset B, Roby D (2010) Nitric oxide participates in the complex interplay of defense-related signaling pathways controlling disease resistance to *Sclerotinia sclerotiorum* in *Arabidopsis thaliana*. Mol Plant Microbe Interact 23:846–860

Petrášek J, Friml J (2009) Auxin transport routes in plant development. Development 136:2175–2688

Pickart CM, Eddins MJ (2004) Ubiquitin: structures, functions, mechanisms. Biochim Biophys Acta 1695:55–72

Pickett FB, Wilson AK, Estelle M (1990) The *aux1* mutation of *Arabidopsis* confers both auxin and ethylene resistance. Plant Physiol 94:1462–1466

Pieterse CMJ, Leon-Reyes A, van der Does D, Verhage A, Koornneef A, van Pelt JA, van Wees SCM (2012) Networking by small-molecule hormones in plant immunity. Induced resistance against insects and diseases. IOBC-WPRS Bull 83:77–80

Poinssot B, Vandelle E, Bentéjac M, Adrian M, Levis C, Brygoo Y, Garin J, Sicilia F, Coutos-Thévenot P, Pugin A (2003) The endopolygalacturonase I from *Botrytis cinerea* activates grapevine defense reactions unrelated to its enzymatic activity. Mol Plant Microbe Interact 16:553–564

Pollmann S, Neu D, Weiler EW (2003) Molecular cloning and characterization of an amidase from *Arabidopsis thaliana* capable of converting indole-3-acetamide into the plant growth hormone, indole-3-acetic acid. Phytochemistry 62:293–300

Polverari A, Molesini B, Pezzotti M, Buonaurio R, Marte M, Delledonne M (2003) Nitric oxide-mediated transcriptional changes in *Arabidopsis thaliana*. Mol Plant Microbe Interact 16:1094–1105

Quittenden LJ, Davies NW, Smith JA, Molesworth PP, Tivendale ND, Ross JJ (2009) Auxin biosynthesis in pea: characterization of the tryptamine pathway. Plant Physiol 151:1130–1138

Rademacher EH, Moller B, Lokerse AS, Llavata-Peris CI, van den Berg W, Weijers D (2011) A cellular expression map of the *Arabidopsis* AUXIN RESPONSE FACTOR gene family. Plant J 68:597–606

Ramachandran V, Chen X (2008) Small RNA metabolism in *Arabidopsis*. Trends Plant Sci 13:368–374

Remington DL, Vision TJ, Guilfoyle TJ, Reed JW (2004) Contrasting modes of diversification of the *Aux/IAA* and *ARF* gene families. Plant Physiol 135:1738–1752

Rhoades MW, Reinhart BJ, Lim LP, Burge CB, Bartel B, Bartel DP (2002) Prediction of plant microRNA targets. Cell 110:513–520

Robert-Seilaniantz A, Navarro L, Bari R, Jones JDG (2007) Pathological hormone imbalances. Curr Opin Plant Biol 10:372–379

Robert-Seilaniantz A, MacLean D, Jikumaru Y, Hill L, Yamaguchi S, Kamiya Y, Jones JDG (2011) The microRNA mi393 re-directs secondary metabolite biosynthesis away from camalexin and towards glucosinolates. Plant J 67:218–231

Rock CD, Sun X (2005) Crosstalk between ABA and auxin signaling pathways in roots of *Arabidopsis thaliana* (l.) Heynh. Planta 222:98–106

Rustérucci C, Aviv DH, Holt BF, Dangl JL, Parker JE (2001) The disease resistance signaling components *EDS1* and *PAD4* are essential regulators of the cell death pathway controlled by *LSD1* in Arabidopsis. Plant Cell 13:2211–2224

Rustérucci C, Espunya MC, Diaz M, Chabannes M, Martinez MC (2007) *S*-Nitrosoglutathione reductase affords protection against pathogens in Arabidopsis, both locally and systemically. Plant Physiol 143:1282–1292

Růžička K, Ljung K, Vanneste S, Podhorska R, Beeckman T, Friml J, Benkova E (2007) Ethylene regulates root growth effects on auxin biosynthesis and transport dependent auxin distribution. Plant Cell 19:2197–2212

Ryan CA, Huffaker A, Yamaguchi Y (2007) New insights into innate immunity in Arabidopsis. Cell Microbiol 9:1902–1908

Sauer M, Robert S, Kleine-Vehn J (2013) Auxin: simply complicated. J Exp Bot 64:2565–2577

Savatin DV, Ferrari S, Sicilia E, De Lorenzo G (2011) Oligogalacturonide auxin antagonism does not require post-transcriptional gene silencing or stabilization of auxin response repressors in *Arabidopsis thaliana*. Plant Physiol 157:1163–1174

Schmelz EA, Engelberth J, Turulinson JH, Block A, Alborn HT (2004) The use of vapor phase extraction in metabolic profiling of phytohormones and other metabolites. Plant J 39:790–808

Sequeira L (1965) Origin of indoleacetic acid in plants infected by *Pseudomonas solanacearum*. Phytopathology 55:1232–1236

Sequeira L, Kelman A (1962) The accumulation of growth substances in plants infected by *Pseudomonas solanacearum*. Phytopathology 52:439–448

Shah J (2009) Plants under attack: systemic signals in defence. Curr Opin Plant Biol 12:459–464

Slaughter A, Daniel X, Flors V, Luna E, Hohn B, Mauch-Mani B (2012) Descendants of primed *Arabidopsis* plants exhibit resistance to biotic stress. Plant Physiol 158:835–843

Smalle J, Vierstra RD (2004) The ubiquitin 26S proteasome proteolytic pathway. Annu Rev Plant Biol 55:555–590

Spaepen S, Vanderleyden J, Remans R (2007) Indole-3-acetic acid in microbial and microorganism-plant signaling. FEMS Microbiol Rev 31:425–448

Staswick PE (2009) The tryptophan conjugates of jasmonic and indole-3-acetic acids are endogenous auxin inhibitors. Plant Physiol 150:1310–1321

Staswick PE, Tiryaki I (2004) The oxylipin signal jasmonic acid is activated by an enzyme that conjugates it to isoleucine in *Arabidopsis*. Plant Cell 16:2117–2127

Staswick PE, Tiryaki I, Rowe ML (2002) Jasmonate response locus JAR1 and several related *Arabidopsis* genes encode enzymes of the firefly luciferase superfamily that show activity on jasmonic, salicylic, and indole-3-acetic acids in an assay for adenylation. Plant Cell 14:1405–1415

Staswick PE, Serban B, Rowe M, Tiryaki I, Maldonado MT, Maldonado MC, Suza W (2005) Characterization of an *Arabidopsis* enzyme family that conjugates amino acids to indole-3-acetic acid. Plant Cell 17:616–627

Stepanova AN, Hoyt JM, Hamilton AA, Alonso JM (2005) A link between ethylene and auxin uncovered by the characterization of two root-specific ethylene –insensitive mutants in *Arabidopsis*. Plant Cell 17:2230–2242

Stepanova AN, Yun J, Likhacheva AV, Alonso JM (2007) Multilevel interactions between ethylene and auxin in Arabidopsis roots. Plant Cell 19:2169–2185

Stepanova AN, Robertson-Hoyt J, Yun J, Benavente LM, Xie DY, DoleZal K, Schlereth A, Jürgens G, Alonso JM (2008) TAA1-mediated auxin biosynthesis is essential for hormone crosstalk and plant development. Cell 133:177–191

Stotz HU, Jikumaru Y, Shimada Y, Sasaki E, Stingl N, Mueller MJ, Kamiya Y (2011) Jasmonate-dependent and COI1-independent defense responses against *Sclerotinia sclerotiorum* in *Arabidopsis thaliana*: auxin is part of COI1-independent defense signaling. Plant Cell Physiol 52:1941–1956

Strader LC, Bartel B (2008) A new path to auxin. Nat Chem Biol 4:337–339

Strader LC, Chen GL, Bartel B (2010) Ethylene directs auxin to control root cell expansion. Plant J 64:874–884

Sugawara S, Hishiyama S, Jikumaru Y, Hanada A, Nishimura T, Koshiba T, Zhao Y, Kamiya Y, Kasahara H (2009) Biochemical analyses of indole-3-acetaldoxime-dependent auxin biosynthesis in *Arabidopsis*. Proc Natl Acad Sci U S A 106:5430–5435

Swarup R, Friml J, Marchant A, Ljung K, Sandberg G, Palme K, Bennett M (2001) Localization of the auxin permease AUX1 suggests two functionally distinct hormone transport pathways operate in the *Arabidopsis* root apex. Genes Dev 15:2648–2653

Swarup R, Kargul J, Marchant A, Zadik D, Rahman A, Mills R, Yemm A, May S, Williams L, Millner P, Tsurumi S, Moore I, Napier R, Kerr ID, Bennett MJ (2004) Structure-function analysis of the presumptive *Arabidopsis* auxin permease AUX1. Plant Cell 16:3069–3083

Swarup R, Perry P, Hagenbeek D, Van Der Straeten D, Beemster GT, Sandberg G, Bhalerao R, Ljung K, Bennett MJ (2007) Ethylene upregulates auxin biosynthesis in Arabidopsis seedlings to enhance inhibition of root cell elongation. Plant Cell 19:2186–2196

Swarup K, Benková E, Swarup R, Casimiro I, Péret B, Yang Y, Parry G, Nielsen E, De Smet I, Vanneste S, Levesque MP, Carrier D, James N, Calvo V, Ljung K, Kramer E, Roberts R, Graham N, Marillonnet S, Patel K, Jones JD, Taylor CG, Schachtman DP, May S, Sandberg G, Friml J, Kerr I, Beeckman T, Laplaze L, Bennett MJ (2008) The auxin influx carrier LAX3 promotes lateral root emergence. Nat Cell Biol 10:946–954

Szemenyei H, Hannon M, Long JA (2008) TOPLESS mediates auxin-dependent transcriptional repression during *Arabidopsis* embryogenesis. Science 319:1384–1386

Tabata R, Ikezaki M, Fujibe T, Aida M, Tian CE, Ueno Y, Yamamoto KT, Machida Y, Nakamura K, Ishiguro S (2010) *Arabidopsis* auxin response factor6 and 8 regulate jasmonic acid biosynthesis and floral organ development via repression of class 1 KNOX genes. Plant Cell Physiol 51:164–175

Takahashi K, Hayashi K-I, Kinoshita T (2012) Auxin activates the plasma membrane H^+-ATPase by phosphorylation during hypocotyl elongation in Arabidopsis. Plant Physiol 159:632–641

Tan X, Calderon-Villalobos LI, Sharon M, Zheng C, Robinson CV, Estelle M, Zheng M (2007) Mechanism of auxin perception by the TIR1 ubiquitin ligase. Nature 446:640–645

Tanaka E, Koga H, Mori M, Masashi M (2011) Auxin production by the rice blast fungus and its localization in host tissue. J Phytopathol 159:522–530

Tao Y, Ferrer JL, Ljung K, Pojer F, Hong FX, Long JA, Li L, Moreno JF, Bowman ME, Ivans LJ, Cheng Y, Lim J, Zhao Y, Ballaré CL, Sandberg G, Noel JP, Chory J (2008) Rapid synthesis of auxin via a new tryptophan-dependent pathway is required for shade avoidance in plants. Cell 133:164–176

Tatsuki M, Nakajima N, Fujii H, Shimada T, Nakano M, Hayashi K-I, Hayama H, Yoshioka H, Nakamura Y (2013) Increased levels of IAA are required for system 2 ethylene synthesis causing fruit softening in peach (*Prunus persica* L. Batsch). J Exp Bot 64:1049–1059

Tegg RS, Gill WM, Thompson HK, Davies NW, Ross JJ, Wilson CR (2008) Auxin-induced resistance to common scab disease of potato linked to inhibition of thaxtomin A toxicity. Plant Dis 92:1321–1328

Terrile MC, Paris R, Calderón-Villalobos LI, Iglesias MJ, Lamattina L, Estelle M, Casalongué CA (2012) Nitric oxide influences auxin signaling through S-nitrosylation of the *Arabidopsis* TRANSPORT INHIBITOR RESPONSE 1 auxin receptor. Plant J 70:492–500

Thatcher LF, Manners JM, Kazan K (2009) *Fusarium oxysporum* hijacks COI1-mediated jasmonate signaling to promote disease development in *Arabidopsis*. Plant J 58:927–939

Thilmony R, Underwood W, He SY (2006) Genome-wide transcriptional analysis of the *Arabidopsis thaliana* interaction with the plant pathogen *Pseudomonas syringae* pv. *tomato* DC3000 and the human pathogen *Escherichia coli* O157:H7. Plant J 46:34–53

Timpte C (2001) Auxin binding protein: curiousor and curiouser. Trends Plant Sci 6:586–590

Tiryaki I, Staswick PE (2002) An Arabidopsis mutant defective in jasmonate response is allelic to the auxin signaling mutant *axr1*. Plant Physiol 130:887–894

Titapiwatanakun B, Murphy AS (2009) Post-transcriptional regulation of auxin transport proteins: cellular trafficking, protein phosphorylation, protein maturation, ubiquitination, and membrane composition. J Exp Bot 60:1093–1107

Tiwari SB, Wang XJ, Hagen G, Guilfoyle TJ (2001) AUX/IAA proteins are active repressors and their stability and activity are modulated by auxin. Plant Cell 13:2809–2822

Tiwari SB, Hagen G, Guilfoyle TJ (2003) The roles of auxin response factor domains in auxin-responsive transcription. Plant Cell 15:533–543

Tiwari SB, Hagen G, Guilfoyle TJ (2004) Aux/IAA proteins contain a potent transcriptional repression domain. Plant Cell 16:533–543

Tör M, Gordon P, Cuzick A, Eulgem T, Sinapidou E, Mert-Turk F, Can C, Dangl JL, Holub EB (2002) Arabidopsis SGT1b is required for defense signaling conferred by several downy mildew resistance genes. Plant Cell 14:993–1003

Tromas A, Braun N, Muller P, Khodus T, Paponov IA, Palme K, Ljung K, Lee K, Lee J-Y, Benfey P, Murray JAH, Scheres B, Perrot-Rechenmann C (2009) The AUXIN BINDING PROTEIN 1 is required for differential auxin responses mediating root growth. PLoS One 4:e6648

Truman W, Bennett MH, Kubigsteltig I, Turnbull C, Grant M (2007) *Arabidopsis* systemic immunity uses conserved defense signaling pathways and is mediated by jasmonates. Proc Natl Acad Sci U S A 104:1075–1080

Truman WM, Bennett MH, Turnbull CGN, Grant MR (2010) *Arabidopsis* auxin mutants are compromised in systemic acquired resistance and exhibit aberrant accumulation of various indolic compounds. Plant Physiol 152:1562–1573

Tsuchisaka A, Theologis A (2004) Unique and overlapping expression patterns among the Arabidopsis 1-amino-cyclopropane-1-carboxylate synthase gene family members. Plant Physiol 136:2982–3000

Ulmasov T, Hagen G, Guilfoyle TJ (1997) ARF1, a transcription factor that binds to auxin response elements. Science 276:1865–1868

Umemura K, Satou J, Iwata M, Uozumi N, Koga J, Kawano T, Koshiba T, Anzai H, Mitomi M (2009) Contribution of salicylic acid glucosyltransferase, OsSGT1, to chemically induced disease resistance in rice plants. Plant J 57:463–472

Uppalapati SR, Ishiga Y, Ryu CM, Ishiga T, Wang K, Noel LD, Parker JE, Mysore KS (2011) SGT1 contributes to coronatine signaling and *Pseudomonas syringae* pv. *tomato* disease symptom development in tomato and *Arabidopsis*. New Phytol 189:83–93

Van der Does D, Zamioudis C, Leon-reyes A, van Wees SCM (2012) Hormonal modulation of plant immunity. Annu Rev Cell Dev Biol 28:489–521

Vandeputte O, Oden S, Mol A, Vereecke D, Goethals K, El Jaziri M, Prinsen E (2005) Biosynthesis of auxin by the gram-positive phytopathogen *Rhodococcus fascians* is controlled by compounds specific to infected plant tissues. Appl Environ Microbiol 71:1169–1177

Vidhyasekaran P (2007) Fungal pathogenesis in plants and crops: molecular biology and host defense mechanisms, IIth edn. CRC Press, Taylor Francis Group, Boca Raton, p 510

Vorwerk S, Biemacki S, Hillebrand H, Janzik I, Müller A, Weiler EW, Piotrowski M (2001) Enzymatic characterization of the recombinant *Arabidopsis thaliana* nitrilase subfamily encoded by the NI2/NIT1/NIT3-gene cluster. Planta 212:508–516

Wang D, Pajerowska-Mukhtar K, Culler AH, Dong X (2007) Salicylic acid inhibits pathogen growth in plants through repression of the auxin signaling pathway. Curr Biol 17:1784–1790

Wang Z, Cao G, Wang X, Miao J, Liu X, Chen Z, Qu L-J, Gu H (2008) Identification and characterization of COI1-dependent transcription factor genes involved in JA-mediated response to wounding in *Arabidopsis* plants. Plant Cell Rep 27:125–135

Wang J, Liu C-Y, Zhang L-W, Wang J-L, Hu G-H, Ding J-J, Chen Q-S (2011) MicroRNAs involved in the pathogenesis of Phytophthora root rot of soybean (*Glycine max*). Agric Sci China 10:1150–1167

Wen N, Chu Z, Wang S (2003) Three types of defense-responsible genes are involved in resistance to bacterial blight and fungal blast diseases in rice. Mol Genet Genomics 269:331–339

Westfall CS, Herrmann J, Chen Q, Wang S, Jez JM (2010) Modulating plant hormones by enzyme action: the GH3 family of acyl acid amido synthetases. Plant Signal Behav 61:752–766

Westfall CS, Zubieta C, Herrmann J, Kapp U, Nanao MH, Jez JM (2012) Structural basis for prereceptor modulation of plant hormones by GH3 proteins. Science 336:1708–1711

Won C, Shen XL, Mashiguchi K, Zheng ZY, Dai XH, Cheng YF, Kasahara H, Kamiya Y, Chory J, Zhao Y (2011) Conversion of tryptophan to indole-3-acetic acid by TRYPTOPHAN AMINOTRANSFERASES OF ARABIDOPSIS and YUCCAs in *Arabidopsis*. Proc Natl Acad Sci U S A 108:18518–18523

Woodward AW, Bartel B (2005) Auxin: regulation, action, and interaction. Ann Bot (Lond) 95:707–735

Wu MF, Tian Q, Reed JW (2006) Arabidopsis *microRNA167* controls patterns of *ARF6* and *ARF8* expression, and regulates both female and male reproduction. Development 133:4211–4218

Xie DX, Feys BF, James S, Nieto-Rostro M, Turner JG (1998) *COI1*: an *Arabidopsis* gene required for jasmonate-regulated defense and fertility. Science 280:1091–1094

Xu L, Liu F, Lechner E, Genschik P, Crosby WL, Ma H, Peng W, Huang D, Xie D (2002) The SCF(COI1) ubiquitin-ligase complexes are required for jasmonate response in *Arabidopsis*. Plant Cell 14:1919–1935

Yan J, Zhang C, Gu M, Bai Z, Zhang W, Qi T, Cheng Z, Peng W, Luo H, Nam F, Wang Z, Xie D (2009) The Arabidopsis CORONATINE INSENSITIVE1 protein is a jasmonate receptor. Plant Cell 21:2220–2236

Yang S, Zhang Q, Guo J, Charkowski A, Glick BR, Ibekwe AM, Cooksey DA, Yang CH (2007) Global effect of indole-3-acetic acid biosynthesis on multiple virulence factors of *Erwinia chrysanthemi* 3937. Appl Environ Microbiol 73:1079–1088

Yasuda M, Ishikawa A, Jikumaru Y, Seki M, Umezawa T, Asami T, Maruyama-Nakashita A, Kudo T, Shinozaki K, Yoshida S, Nakashita H (2008) Antagonistic interaction between systemic acquired resistance and the abscisic acid-mediated abiotic stress response in *Arabidopsis*. Plant Cell 20:1678–1692

Yu Y-B, Yang SF (1979) Auxin-induced ethylene production and its inhibition by aminoethoxyvinylglycine and cobalt ion. Plant Physiol 64:1074–1077

Yu H, Karampelias M, Robert S, Peer WA, Swarup R, Ye S, Ge L, Cohen J, Murphy A, Friml J, Estelle M (2013) ROOT ULTRAVIOLET B-SENSITIVE1/WEAK AUXIN RESPONSE3 is essential for polar auxin transport in Arabidopsis. Plant Physiol 162:965–976

Yuan H, Zhao K, Lei H, Shen X, Liu Y, Liao X, Li T (2013) Genome-wide analysis of the GH3 family in apple (*Malus* X *domestica*). BMC Genomics 14:297

Zhang Z, Li Q, Li Z, Staswick PE, Wang M, Zhu Y, He Z (2007) Dual regulation role of *GH3.5* in salicylic acid and auxin signaling during Arabidopsis-*Pseudomonas syringae* interaction. Plant Physiol 145:450–464

Zhang H, Li D, Wang M, Liu J, Teng W, Cheng B, Huang Q, Wang M, Song W, Dong S, Zheng X, Zhang Z (2012) The *Nicotiana benthamiana* mitogen-activated protein kinase cascade and WRKY transcription factor participate in $Nep1_{Mo}$-triggered plant responses. Mol Plant Microbe Interact 25:1639–1653

Zhao Y (2010) Auxin biosynthesis and its role in plant development. Annu Rev Plant Biol 61:49–64

Zhao Y, Christensen SK, Fankhauser C, Cashman JR, Cohen JD, Weigel D, Chory J (2001) A role for flavin monooxygenase-like enzymes in auxin biosynthesis. Science 291:306–309

Zhao Y, Hull AK, Gupta NR, Goss KA, Alonso J, Ecker JR, Normanly J, Chory J, Celenza JL (2002) Trp-dependent auxin biosynthesis in *Arabidopsis*: involvement of cytochrome P450s CYP79B2 and CYP79B3. Genes Dev 16:3100–3112

Zhao Y, Yu Y, Zhai J, Ramachandran V, Dinh TT, Meyers BC, Mo B, Chen X (2012a) The Arabidopsis nucleotidyltransferase HESO1 uridylates unmethylated small RNAs to trigger their degradation. Curr Biol 22:689–694

Zhao Y-T, Wang M, Fu S-X, Yang W-C, Qi C-K, Wang X-J (2012b) Small RNA profiling in two *Brassica napus* cultivars identifies microRNAs with oil production- and development-correlated expression and new small RNA classes. Plant Physiol 158:813–823

Zheng X-y, Spivey NW, Zeng W, Liu P-P, Fu ZQ, Klessig DF, He SY, Dong X (2012) Coronatine promotes *Pseudomonas syringae* virulence by activating a signaling cascade that inhibits salicylic acid accumulation. Cell Host Microbe 11:587–596

Zhu Q, Wani G, Wang QE, El-mahdy M, Snapka RM, Wani AA (2005) Deubiquitination by proteasome is coordinated with substrate translocation for proteolysis in vivo. Exp Cell Res 307:436–451

Ziadi S, Barbedette S, Godard JF, Monot C, Le Corre D, Silue D (2008) Production of pathogenesis-related proteins in the cauliflower (*Brassica oleracea* var *botrytis*)-downy mildew (*Peronospora parasitica*) pathosystem treated with acibenzolar-S-methyl. Plant Pathol 50:579–586

Zipfel C (2009) Early molecular events in PAMP-triggered immunity. Curr Opin Plant Biol 12:414–420

Zipfel C, Robatzek S, Navarro L, Oakeley EJ, Jones JDG, Felix G, Boller T (2004) Bacterial disease resistance in *Arabidopsis* through flagellin perception. Nature 428:764–767

Chapter 7
Cytokinin Signaling System in Plant Immunity

Abstract Cytokinins play an important role in the modulation of plant innate immunity. Cytokinins regulate the host defense responses either positively or negatively depending on the concentrations of cytokinins available at the infection site. Cytokinin perception and signaling involve two-component phosphorelay system. Binding of cytokinin to sensor histidine kinase (AHKs) receptors initiates a phosphotransfer cascade that culminates in the phosphorylation of response regulator proteins (ARRs) which are responsible for the regulation of cytokinin signaling system. The two types of ARR transcription activators differentially modulate the defense responses. Type-A ARRs negatively regulate cytokinin signaling, and transgenic lines overexpressing type-A ARRs display decreased defense responses, while the type-B ARR transcription factor triggers enhancement of defense responses. Pathogen-associated molecular patterns modulate the expression of both type-A and type-B *ARR* genes. The ARR transcription activators regulate the expression of various transcription factors belonging to MYB and AP2/ERF families, which play an important role in plant immune responses. Cytokinin homeostasis in the cytokinin signaling pathway regulates plant immune responses. Cytokinin augments PAMP-triggered immunity by enhancing callose deposition. Cytokinin may trigger the phytoalexin accumulation and confer resistance against pathogens. Cytokinins have also been shown to prime plant cells for the potentiated activation of defense-related genes. Cytokinins systemically induce resistance against pathogen infection, and this resistance is orchestrated by endogenous cytokinin and salicylic acid signaling. Cytokinins enhance resistance against pathogens via a direct induction of defense-related genes, in cooperation with SA, by modulating the NPR1/TGA3-mediated signaling. Cytokinin signaling in concert with SA signaling activates defense responses. Type-A ARRs negatively regulate SA-dependent basal immunity. Cytokinin signaling system may also function independent of SA signaling system. ABA modulates the expression of genes encoding proteins involved in cytokinin biosynthesis and degradation. Synergism and antagonism between auxin and cytokinin signaling have been reported. Auxin positively or negatively regulates the expression of the genes involved in cytokinin homeostasis. Auxin also causes repression of cytokinin responses by activating type-A ARRs, which negatively regulate cytokinin signaling.

P. Vidhyasekaran, *Plant Hormone Signaling Systems in Plant Innate Immunity*, Signaling and Communication in Plants 2, DOI 10.1007/978-94-017-9285-1_7

7.1 Cytokinin Signaling in Plant Immune System

Cytokinins are phytohormones that are involved in various regulatory processes throughout plant development (Frébort et al. 2011; Großkinsky et al. 2011). They execute their action via the molecular machinery of signal perception and transduction (Frébort et al. 2011). Cytokinins act as both long-range and local signals (Hwang and Sakakibara 2006). They may also play important role in modulation of plant innate immunity (Choi et al. 2010; Großkinsky et al. 2011; Naseem et al. 2012; Pieterse et al. 2012). Cytokinin signaling system may regulate positively or negatively the plant defense responses (Choi et al. 2010; Argueso et al. 2012). Treatment with low concentrations of exogenous cytokinin led to enhanced susceptibility of *Arabidopsis* plants to the oomycete pathogen *Hyaloperonospora arabidopsidis*. In contrast, treatment with higher levels of cytokinin led to decreased susceptibility to the pathogen (Argueso et al. 2012). Low concentrations of cytokinin increased susceptibility against the wheat powdery mildew pathogen *Blumeria graminis*, while higher concentrations induced enhanced resistance against the pathogen (Babosha 2009). These results suggest that cytokinins may regulate the host defense responses either positively or negatively depending on the concentrations of cytokinins available at the infection site.

In the cytokinin signaling pathway, two types of ARR transcription activators have been reported. Type-A ARRs are negative regulators of cytokinin signaling (Hwang and Sheen 2001; Kiba et al. 2003). Transgenic lines overexpressing type-A ARRs display decreased defense responses (Argueso et al. 2012), while a type-B ARR transcription factor has been shown to trigger enhancement of defense responses (Choi et al. 2010). The results suggest that different components in the cytokinin signaling system may differently regulate defense responses. PAMP treatment altered the expression of both type-A and type-B *ARR* genes, but the expression of type-A *ARR* genes was most affected (Argueso et al. 2012).

7.2 Cytokinin Biosynthesis

Several cytokinin compounds have been reported in plants (Hutchison and Kieber 2002). *Trans*-zeatin and isopentenyladenine-type cytokinins are the predominant plant cytokinins, with the *cis*-isomer being present only in minor quantities with low or no activity (Haberer and Kieber 2002; Frébort et al. 2011). Increase in biosynthesis of cytokinins has been reported in many plant–pathogen interactions (Sziraki et al. 1980; Whenham 1989; Großkinsky et al. 2011). Cytokinin biosynthesis pathway is highly complex, and the simplified major pathways are presented in Fig. 7.1. The first step in the biosynthesis of cytokinins is catalyzed by adenosine phosphate-isopentenyltransferase (IPT) using adenosine 5′-phosphates (ATP, ADP, or AMP) and dimethylallyl diphosphate (DMAPP) (Hwang and Sakakibara 2006; Sakakibara 2006). IPTs from plants preferentially utilize ADP or ATP over AMP as a prenyl

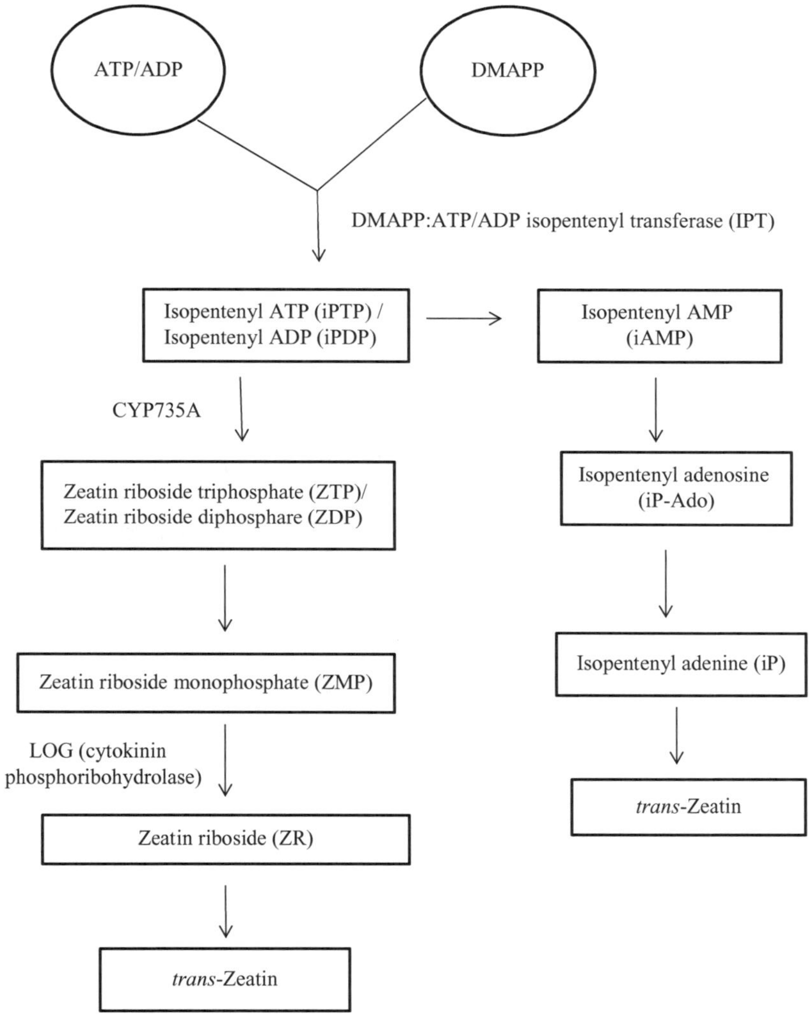

Fig. 7.1 Cytokinin biosynthesis pathway in plants (Adapted from Kakimoto 2003; Hwang and Sakakibara 2006)

acceptor and DMAPP as a prenyl donor. DMAPP is produced via the mevalonate pathway (Hwang and Sakakibara 2006). Isopentenyltransferase (IPT) is the key enzyme involved in cytokinin biosynthesis. Nine *ipt* genes have been detected in *Arabidopsis* genome (Takei et al. 2001; Kakimoto 2001). The IPT utilizes ATP and ADP as a substrate. The products of the plant enzyme are isopentenyladenosine-5′-triphosphate (iPTP) and isopentenyladenosine-5′-diphosphate (iPDP), which can be subsequently interconverted to zeatin. The product of iPTP/iPDP is isopentenyladenosine-5′-monophosphate (iPMP), which is converted to isopentenyladenosine (iPA) and finally to *trans*-zeatin (Fig. 7.1; Kakimoto 2003; Hwang and Sakakibara 2006).

It has been suggested that tRNA prenyl modification also contributes to cytokinin production. Some of the tRNA species with anticodons complementary to codons beginning with uridine carry a prenylated adenosine adjacent to the anticodon. When the tRNA is degraded, the adenosine is released as a cytokinin (Hwang and Sakakibara 2006). The prenylation is catalyzed by tRNA-isopentenyltransferase (tRNA-IPT). tRNA degradation has been suggested to be a source of *cis*-zeatin type of cytokinins (Hwang and Sakakibara 2006).

7.3 Cytokinin Degradation

Cytokinin biosynthesis and degradation processes contribute to cytokinin homeostasis (Wang et al. 2011). Cytokinin oxidase/dehydrogenase catalyzes the irreversible degradation of cytokinins and is responsible for the majority of metabolic cytokinin deactivation in plants (Wang et al. 2011). A series of genes involved in cytokinin biosynthesis and metabolism involved in the cytokinin homeostasis include those encode IPT, CYP735A, adenosine nucleosidase, adenine phosphoribosyltransferase, cytokinin oxidase/dehydrogenase (CKX), and cytokinin glycosyltransferases. The steady-state level of cytokinin nucleobases, which is responsible for cytokinin activity, appears to be tightly controlled by the expression of key enzymes for cytokinin biosynthesis and metabolism (Hwang and Sakakibara 2006).

The genes involved in cytokinin homeostasis (cytokinin synthases and cytokinin oxidases/dehydrogenases) were strongly down-regulated in *Arabidopsis* plants infected with the obligate biotrophic protist *Plasmodiophora brassicae* (Siemens et al. 2006). Stabilization of cytokinin enhances disease resistance against *Verticillium longisporum* in *Arabidopsis* (Reusche et al. 2013). Levels of *trans-zeatin* decrease in *V. longisporum*-infected plants. Potentially, induction of cytokinin oxidase/dehydrogenase expression by *Verticillium* infection contributed to the observed decreases in cytokinin levels. Stabilization of *Arabidopsis* cytokinin levels by both pharmacological and genetic approaches inhibited *Verticillium* proliferation and coincided with reduced disease symptom development (Reusche et al. 2013). These results suggest that cytokinin homeostasis in the cytokinin signaling pathway is important in plant defense responses.

7.4 Cytokinin Signal Perception and Transduction

7.4.1 Cytokinin Receptors

CRE1/WOL/AHK4, a histidine kinase, has been identified as a cytokinin receptor in the cytokinin signaling pathway in *Arabidopsis* (Inoue et al. 2001; Suzuki et al. 2001; Aoyama and Oka 2003; Hwang and Sakakibara 2006; Yamada et al. 2001). Two closely related His kinases (AHK2 and AHK3) have also been suggested to act as cytokinin receptors (Hwang and Sheen 2001; Haberer and Kieber 2002). Both His kinases are highly homologous to the entire CRE1 sequence, including putative ligand-binding domain, and also share with CRE1 the presence of two response regulator domains in their C terminus (Haberer and Kieber 2002). The N-terminus of these cytokinin receptors consists of two to three transmembrane domains and an extracellular domain, followed by a histidine kinase domain, a receiver-like domain, and a receiver domain. The cyclase/histidine-kinase-associated sensing extracellular (CHASE) domain, which presumably acts as the cytokinin-sensing domain, has been found in the extracellular domains of the cytokinin receptors (Anantharaman and Arvind 2001; Mougel and Zhulin 2001). All the three cytokinin receptors (AHK2, AHK3, AHK4) are hybrid His kinases that contain a fused receiver domain in addition to an input (a cytokinin-binding CHASE domain) and a His kinase domain (Inoue et al. 2001; Suzuki et al. 2001; Jennifer et al. 2007).

7.4.2 Cytokinin Phosphorelay Signaling System

There are three major steps in the cytokinin phosphorelay signaling system: AHK sensing and signaling, AHP nuclear translocation, and ARR-dependent transcriptional activation. In response to cytokinin binding, the cytokinin receptors autophosphorylate on a conserved His residue and relay this phosphoryl group to *Arabidopsis* response regulators (ARRs) via an intermediate set of histidine phosphotransfer (hpt) proteins called the *Arabidopsis* Hpt proteins (AHPs) (Jennifer et al. 2007). Upon perception of cytokinins, the histidine kinase cytokinin receptor is dimerized and autophosphorylated. The phosphorylated receptor initiates the His–Asp phosphorelay via AHP proteins to ARR proteins. AHP proteins serve as phosphorelay carriers between the cytokinin receptors and nuclear cytokinin responses (Hwang and Sakakibara 2006; Choi et al. 2010; Frébort et al. 2011). Thus, the cytokinin signaling system involves a multistep histidine kinase (AHK) → histidine-containing phosphotransmitter (AHP) → response regulator (ARR) phosphorelay signaling circuitry, which is initiated by the cytokinin receptor histidine protein kinases in *Arabidopsis* (Fig. 7.2; Kiba et al. 2005; Jennifer et al. 2007; Choi et al. 2010; Frébort et al. 2011).

Cytokinin signaling is mediated by a multistep two-component circuitry through histidine (His) and aspartate phosphorelay. This phosphorelay system consists of two proteins: the histidine kinase (HK) and the response regulator (RR). HKs contain two functional domains, a variable ligand sensor and a conserved His-containing catalytic protein kinase. The RRs contain a conserved Asp-containing receiver domain and a variable output domain. Upon signal perception, phosphoryl groups are transferred from His to Asp. In the cytokinin pathway, they are transferred from His to Asp to His to Asp. The His to Asp phosphotransfer signal transduction mechanism involves three common signaling domains: the transmitter (or His kinase), the receiver, and the histidine-containing phosphotransfer (Hpt) domain. A sensor kinase has a His kinase domain, and a response regulator has a receiver domain containing a phosphoaccepting

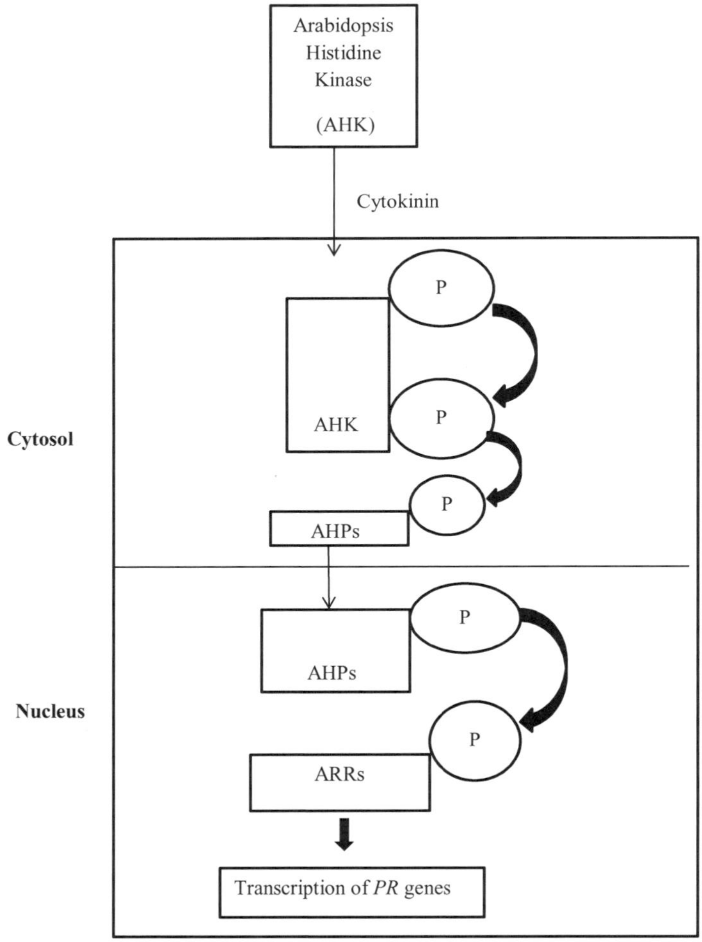

Fig. 7.2 Phosphorelay system in cytokinin signaling system (Adapted from Choi et al. 2010)

aspartate, whereas a histidine-containing phosphotransfer domain serves as a mediator of the histidine-to-aspartate phosphotransfer (Imamura et al. 1998). This multistep phosphorelay results from the addition of a receiver domain to the HK to form a hybrid HK and a phosphotransfer protein, which consists of a conserved His-containing domain (Müller and Sheen 2007a, b).

Cytokinins bind to a CHASE (cyclase- and HK-associated sensing extracellular) domain of transmembrane receptors with a hybrid HK (Anantharaman and Aravind 2001). The cytokinin binding triggers phosphorylation at the conserved His residues within the receptors. In the cytokinin signaling pathway, the phosphoryl group is first transferred from His to a conserved Asp residue in receptor's receiver domain.

It is then transferred to a His phosphotransfer protein (HPT) that translocates to the nucleus, where the phosphoryl group gets passed over to a conserved Asp residue in the receiver domain of an RR protein (Müller and Sheen 2007a, b). There are two types of cytokinin RRs, type-A and type-B RRs. Both have a conserved receiver domain at the N-terminus. Type-B RRs contain extra C-terminal domains for DNA binding and transcriptional activation. The type-A RRs have a short C-terminal domain, and they form a negative feedback loop in the pathway (Müller and Sheen 2007a, b; Argueso et al. 2012). Type-A ARRs act as negative regulators of cytokinin responses (Hwang and Sheen 2001; Kiba et al. 2003; To et al. 2004; Lee et al. 2007), while type-B ARRs positively regulate cytokinin signaling (Hwang and Sheen 2001; Sakai et al. 2001; Mason et al. 2005; Argyros et al. 2008). In *Arabidopsis*, the final targets of the canonical phosphorelay system are a set of *Arabidopsis* response regulator (ARR) proteins containing a receiver domain with a conserved Asp phosphorylation site. The type-A ARRs and type-B ARRs play an important role in cytokinin signaling (Fig. 7.3; Jennifer et al. 2007; Argueso et al. 2012). The type-A ARRs are primary transcriptional targets of cytokinin signaling, and they act as negative regulators of cytokinin signaling (Kiba et al. 2003; To et al. 2004; Jennifer et al. 2007). The type-B ARRs are positive regulators of cytokinin signaling that control the transcription of a subset of cytokinin-regulated targets, including the type-A ARRs (Hwang and Sheen 2001; Mason et al. 2005).

The translocation of AHP proteins into the nucleus is proposed to transfer a phosphoryl group to type-B ARRs. The phosphotransfer modulates the transcriptional activity of ARR proteins, which in turn activate the expression of primary cytokinin response genes including type-A *ARR*s (Hwang and Sakakibara 2006;

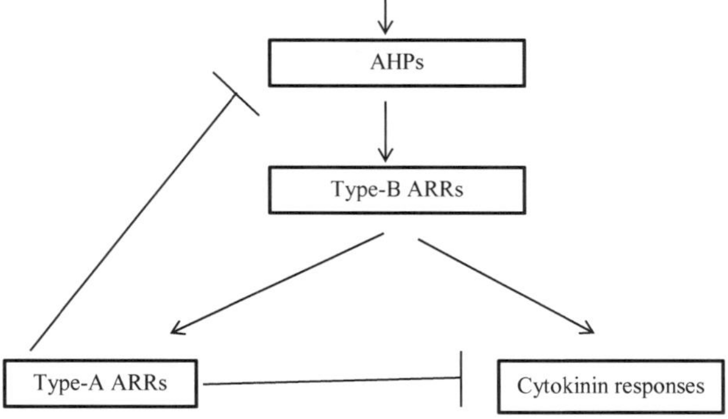

Fig. 7.3 Cytokinin signaling pathway (Adapted from Argueso et al. 2012)

Argueso et al. 2012). Type-A ARRs are negative regulators in cytokinin signaling and provide a negative feedback mechanism by which the primary cytokinin signal response is down-regulated (Hwang and Sheen 2001; To et al. 2004). The type-A ARRs act as negative regulators of cytokinin signaling (Kiba et al. 2003; Lee et al. 2007). Overexpression of a type-A ARR gene (*ARR7*) showed a distinctly repressive impact on various groups of the cytokinin-regulated genes. In particular, the induction of all A-type ARRs except for one ARR (ARR22), AHK1, and AHK4 was suppressed by ARR7. Cytokinin-induced expression of most of the 12 expansin genes was repressed by ARR7. Up-regulation of five cytokinin oxidase genes by cytokinin was negatively affected by ARR7 (Lee et al. 2007). These results suggest that type A acts as a transcriptional repressor of a variety of early cytokinin-regulated genes.

7.5 Cytokinin-Responsive Genes

Cytokinin realizes at least part of its signaling output through the regulation of gene expression (Brenner et al. 2012). Several genes have been found to be up-regulated or down-regulated in response to cytokinin signaling (Rashotte et al. 2003; Brenner et al. 2012; Tsai et al. 2012). The expression of at least 30 genes was up-regulated, while the expression of at least 40 genes was down-regulated at multiple time points after application of cytokinin in *Arabidopsis*. The cytokinin up-regulated genes included the type-A *Arabidopsis* response regulators (ARRs); cytokinin oxidase, which encodes an enzyme that degrades cytokinins; and several transcription factors. Cytokinin down-regulated genes include several peroxidases and kinases and an E3 ubiquitin ligase. Cytokinin receptor mutants (*cre1*) revealed down-regulation of the type-A ARRs in *Arabidopsis* (Rashotte et al. 2003). The transcription factors identified as being cytokinin responsive include those belonging to the AP2/EREBP family, MYB family, GATA family, or bHLH family (Kiba et al. 2005). The transcriptional regulation of transcription factors of the MYB family is mostly mediated by type-B response regulators. MYB and AP2/EREBP family transcription factors play an important role in the modulation of plant innate immune responses (Grennan 2008; Van der Ent et al. 2008; Sun et al. 2010). In *Arabidopsis*, only a single cytokinin response element, to which type-B response regulators bind, has been identified, and it has 5′-GAT (T/C)-3′ as a core sequence (Brenner et al. 2012). Genes encoding type-A response regulators displayed the most stable response to cytokinin (Brenner et al. 2012). Transcription of a number of cytokinin metabolism genes (CKX4, CKX5, CYP735A2, UGT76C2) is also modulated by cytokinins, indicating homeostatic mechanisms operating at the transcriptional level (Brenner et al. 2012). All type-A OsRRs and OsHK4 were up-regulated in response to cytokinin in rice. Multiple cytokinin oxidase (*OsCKX*) genes were also up-regulated by cytokinin (Tsai et al. 2012).

7.6 Cytokinins May Be Involved in Triggering Defense Responses

7.6.1 Cytokinins Confer Resistance Against Pathogens

Cytokinins have been shown to confer resistance against several fungal, bacterial, and viral pathogens (Pogány et al. 2004; Choi et al. 2010; Großkinsky et al. 2011; Argueso et al. 2012). Cytokinin has been shown to protect *Arabidopsis* against the necrotrophic fungal pathogen *Alternaria brassicicola* (Choi et al. 2010). Elevated cytokinin levels have been shown to confer resistance against different viruses (Sano et al. 1996; Pogány et al. 2004). Application of cytokinin resulted in the induction of defense gene expression and decreased replication of *White clover mosaic virus* in bean (Clarke et al. 1998). Increased cytokinin levels enhanced plant resistance to *Tobacco necrosis virus* (TNV) in *Phaseolus vulgaris* (Fayza and Sabrey 2006). Higher levels of cytokinin increased resistance to *Pseudomonas syringae* pv. *tomato* in *Arabidopsis* (Choi et al. 2010) and against *P. syringae* pv. *tabaci* in tobacco (Großkinsky et al. 2011).

Isopentenyltransferase is a key enzyme in cytokinin biosynthesis, and it catalyzes the rate-limiting step in cytokinin biosynthesis (Kakimoto 2003; Hwang and Sakakibara 2006; Großkinsky et al. 2011). Transgenic tobacco plants expressing the bacterial *ipt* gene from *Agrobacterium tumefaciens* T-DNA encoding isopentenyltransferase were developed. The *ipt*-expressing regions in the transgenic leaf inoculated with the pathogen *Pseudomonas syringae* pv. *tabaci* showed high accumulation of cytokinins. In *ipt*-expressing regions challenged with *P. syringae* pv. *tabaci*, pathogen resistance was markedly enhanced (Großkinsky et al. 2011). Tobacco plants transformed with an *ipt* gene were found to overproduce cytokinin. The cytokinin overproducing transgenic tobacco lines showed resistance to *Tobacco necrosis virus* (TNV). The number of virus-induced necrotic lesions as well as the coat protein content was significantly lower in the transgenic tobacco, compared with the control (Pogány et al. 2004). These studies suggest that the induction of *ipt* expression, resulting in cytokinin accumulation at the site of infection, is highly effective in activating plant immune responses.

Application of kinetin, a naturally occurring adenine-derived cytokinin, enhanced resistance against *P. syringae* pv. *tabaci* in tobacco leaves. Exogenously supplied cytokinin induced resistance before as well as after the bacterial infection. A single kinetin pulse was sufficient to confer at least resistance for 7 days (Großkinsky et al. 2011). The efficacy of kinetin treatment in inducing defense responses against various pathogens differed widely. Although the cytokinin treatment was effective in controlling *P. syringae* pv. *tabaci*, it had no effect on *P. syringae* pv. *phaseolicola-induced* hypersensitive response in tobacco (Großkinsky et al. 2011).

7.6.2 Cytokinin Augments Plant Immune Responses by Enhancing Callose Deposition

Cytokinin augments PAMP-triggered immunity in *Arabidopsis* by enhancing callose deposition upon the PAMP flg22 treatment (Choi et al. 2010). When plants were treated with flg22, the transgenic plants expressing the cytokinin biosynthetic gene *IPT3* accumulated higher number of callose deposits in leaves than the wild-type plants. The transgenic plants expressing the gene (*CKX*) encoding cytokinin-degrading enzyme cytokinin oxidase/dehydrogenase showed significantly lower levels of callose deposition (Choi et al. 2010). The results suggest that cytokinin triggers disease resistance by augmenting callose deposition induced by the PAMP.

7.6.3 Cytokinin May Trigger Accumulation of Antimicrobial Phytoalexins to Confer Disease Resistance

Cytokinins have been shown to induce resistance in tobacco against *Pseudomonas syringae* pv. *tomato* in *Arabidopsis*, which is predominantly mediated by increased synthesis of the two major antimicrobial phytoalexins, scopoletin and capsidiol, in the early phase of infection, efficiently restricting pathogen growth (Großkinsky et al. 2011). The cytokinin-mediated resistance against *P. syringae* pv. *tabaci* through increased accumulation of phytoalexins in tobacco was independent of SA signaling (Großkinsky et al. 2011).

Cytokinin-treated *Arabidopsis* plants accumulated significantly more camalexin than did control plants. Enhanced camalexin accumulation upon cytokinin treatment occurred in late time points (72 h after cytokinin application). The difference in SA accumulation between treatments of *P. syringae* pv. *tomato* with and without cytokinin was not significant. The results showed that cytokinin does not cause higher SA accumulation to protect *Arabidopsis* against the infection with *P. syringae* pv. *tomato* and that cytokinin treatment has a positive effect on camalexin accumulation (Naseem et al. 2012). The results suggest that cytokinin may trigger the phytoalexin accumulation and confer resistance against pathogens.

7.6.4 Cytokinins Induce Priming of Plant Cells for Activation of Defense-Related Genes

In addition to the direct activation of defense-related genes, cytokinins have also been shown to prime plant cells for the potentiated activation of defense-related genes. Pretreatment with cytokinin conferred hyperactivation of the marker defense gene *Pr1* transcription after the pathogen inoculation (Choi et al. 2010). Priming is a unique physiological state in which a faster and stronger induction of defense mechanisms occurs only after pathogen attack (Conrath 2011; Camañes et al. 2012; Slaughter et al. 2012). Priming is involved in "systemic acquired resistance" (Kohler et al. 2002; Jung et al. 2009; Conrath 2011; Luna et al. 2012). Priming can be inherited epigenetically from disease-exposed plants, and descendants of primed plants exhibit next-generation systemic acquired resistance (Luna et al. 2012; Slaughter et al. 2012). Priming is also involved in "induced systemic resistance" (Akram et al. 2008; Kravchuk et al. 2011; Mathys et al. 2012; Martinez-Medina et al. 2013). Cytokinins act as both long-range and local signals (Hwang and Sakakibara 2006). Endogenous cytokinins systemically induce resistance against pathogen infection, and this resistance is orchestrated by endogenous cytokinin and salicylic acid signaling (Choi et al. 2011).

7.6.5 Cytokinin May Modulate SA Signaling System to Trigger Immune Responses

Cytokinins have been shown to induce resistance against *Pseudomonas syringae* pv. *tomato* DC3000 by activating NPR1-dependent salicylate signaling system in *Arabidopsis thaliana* (Choi et al. 2010). Treatment with high concentrations of cytokinin leads to decreased susceptibility through a process that requires SA accumulation and activation of SA-dependent defense gene expression (Argueso et al. 2012). Cytokinins acting in concert with SA signaling systemically induce resistance against pathogen infection (Choi et al. 2011). Cytokinins have been shown to induce resistance against *Pseudomonas syringae* pv. *tomato* DC3000 by activating TGA3/NPR1 (for TGACG [TGA] MOTIF-BINDING PROTEIN3/NONEXPRESSOR OF PR PROTEINS1)-dependent salicylate signaling system in *Arabidopsis thaliana* (Choi et al. 2010).

7.6.6 Cytokinins May Induce Resistance Independently of SA Signaling System

There are also reports that cytokinin signaling system may function independent of SA signaling system. Increase in endogenous cytokinins has been shown to mediate enhanced resistance against *P. syringae* pv. *tabaci* in tobacco. The cytokinin-mediated activation of plant innate immunity was shown to be independent of SA signaling in tobacco (Großkinsky et al. 2011). Transgenic plants expressing *nahG* gene encoding the SA-degrading salicylate hydroxylase from *Pseudomonas putida* are more susceptible to infection by *P. syringae* pv. *tomato* (Delaney et al. 1994). However, the cytokinin treatment was still able to confer resistance against the bacterial pathogen (Großkinsky et al. 2011), suggesting that cytokinin functions independently of SA. The cytokinin-mediated resistance in tobacco was not only independent of salicylic acid system but also independent of jasmonate and ROS signaling systems (Großkinsky et al. 2011). The results suggest that cytokinins may act in a signaling pathway in other than the well-known defense signaling pathways.

7.6.7 Cytokinins May Modulate Redox Signaling to Trigger Immune Responses

Signaling through the redox active molecule H_2O_2 is important in inducing plant innate immune responses (Desikan et al. 2005). Along with increase in H_2O_2, the transcript levels of genes encoding antioxidant proteins also increase (Joo et al. 2005). The *NPR1* gene is a master regulator of the SA-mediated induction of defense genes (Cao et al. 1997). Nuclear localization of NPR1 protein is essential for its function (Kinkema et al. 2000; Meur et al. 2006). Without induction, NPR1 protein forms an oligomer and is excluded from the nucleus. Redox changes cause monomeric NPR1 to emerge and accumulate in the nucleus and activate *PR* gene expression (Kinkema et al. 2000; Mou et al. 2003). The NPR1 oligomer-to-monomer reaction is catalyzed by antioxidants (Tada et al. 2008). The buildup of antioxidants that enhance the general cellular reducing capacity would have reduced the NPR1 complex resulting in activation of NPR1 (Cumming et al. 2004).

Cytokinin signaling appears to trigger the activity of antioxidant enzymes including glutathione S-transferase, catalase, and ascorbate oxidase and also increase the levels of nonenzymatic antioxidants (Pogány et al. 2004). A cytokinin-overproducing tobacco line transformed with an *ipt* gene encoding isopentenyltransferase involved in cytokinin biosynthesis has been developed (Pogány et al. 2004). The transgenic plants showed enhanced resistance against *Tobacco necrosis virus* (TNV). The number of virus-induced necrotic lesions and the viral coat protein content were lower in the transgenic tobacco plants expressing the *ipt* gene (Pogány et al. 2004). The activity of three antioxidant enzymes (ascorbate peroxidase, glutathione S-transferase, and catalase) and amount of the antioxidant ascorbic acid were higher in the transgenic tobacco plant overproducing cytokinin. Diminished H_2O_2 accumulation was also observed in the transgenic tobacco plants (Pogány et al. 2004). These results suggest that cytokinin may modulate redox signaling system and trigger defense responses.

7.7 Cytokinins May Induce Susceptibility

Cytokinins are phytohormones that are involved in various regulatory processes throughout plant development (Großkinsky et al. 2011). They are also produced by pathogens (Akiyoshi et al. 1987; Morris et al. 1991). The cytokinin production has been shown to be essential for infection by several pathogens (Crespi et al. 1994; Barash and Manulis-Sasson 2007; Pertry et al. 2009, 2010). *Rhodococcus fascians* provokes the formation of differentiated leafy galls in *Arabidopsis*. The pathogen produced three classical cytokinins (isopentenyladenine, *trans*-zeatin, and *cis*-zeatin) and their 2-methylthio (2MeS) derivatives. The bacterial cytokinins were perceived by the cytokinin receptor AHK3 and in parallel strongly activated *AHK4* expression. These cytokinins were responsible for the tissue proliferation and symptom development (Pertry et al. 2009, 2010).

Cytokinin has been shown to induce susceptibility against some pathogens. *Botrytis cinerea* infection was strongly enhanced in cytokinin-pretreated tobacco leaves (Großkinsky et al. 2011). Cytokinins have been shown to play a key role in the development of clubroot disease caused by the obligate protist *Plasmodiophora brassicae* in *Arabidopsis*. Genes involved in cytokinin homeostasis (cytokinin synthases and cytokinin oxidases/dehydrogenases) were down-regulated during the disease development. Cytokinin oxidase/dehydrogenase (the enzymes which degrade cytokinins) overexpressing lines were disease resistant, indicating the importance of cytokinin in clubroot disease development (Siemens et al. 2006). The cytokinin may alter the host physiology, which would facilitate maximum pathogen access early in hemibiotrophic/biotrophic host plant–pathogen interactions (Walters et al. 2008).

Infection of barley plants by *Pyrenophora teres* and maize plants by *Drechslera maydis* was characterized by "green island" formation and high accumulation of cytokinins at infection sites. By virtue of accumulation of cytokinins, infection sites may be acting as metabolic sinks helping proliferation of the pathogens (Angra-Sharma and Sharma 1999).

7.8 Interplay Between Cytokinin and SA Signaling Pathways in Plant Immune System

7.8.1 Cytokinin May Enhance SA Biosynthesis

Isochorismate synthase (ICS) is the key enzyme in the isochorismate pathway involved in SA biosynthesis (Wildermuth et al. 2001; Gaille et al. 2002, 2003). Two ICS genes, *ICS1* and *ICS2*, have been identified in *Arabidopsis* (Garcion et al. 2008). Cytokinin treatment induced the expression of *ICS1* within 30 min which led to elevated SA production in *Arabidopsis* (Choi et al. 2010). Within 4 h of the PAMP flg22 application, the expression of *ICS1* gene was hyperactivated in the presence of the cytokinin *t*-zeatin (Fig. 7.4; Choi et al. 2010). *ICS1* gene required for SA biosynthesis is mutated in the *eds16* mutant, and the *eds16* mutant plants displayed a substantially reduced response to cytokinin as compared to wild-type *Arabidopsis* plants (Argueso et al. 2012). Collectively these results suggest that cytokinin response is largely dependent on SA biosynthesis. Elevated endogenous cytokinin levels in wounded tobacco leaves also led to the induction of salicylic acid (Sano et al. 1994, 1996).

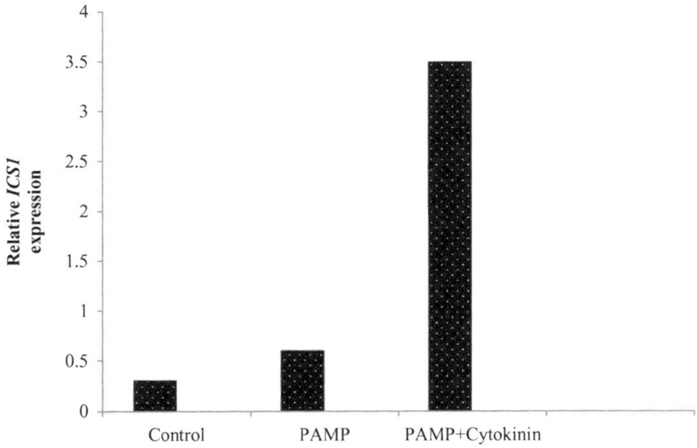

Fig. 7.4 Cytokinin-induced enhanced expression of the SA biosynthetic gene *ICS1* in *Arabidopsis* (Adapted from Choi et al. 2010)

7.8.2 Type-B ARR Interacts with TGA3 of SA Signaling Pathway to Trigger Immune Responses

NPR1 is an important regulator of responses downstream of SA (Fu et al. 2012; Moreau et al. 2012; Spoel and Dong 2012; Wu et al. 2012). NPR1 functions as a transcriptional co-activator (Rochon et al. 2006). Several TGA family members of bZIP transcription factors interact with NPR1 (Chern et al. 2001; Jakoby et al. 2002; Fitzgerald et al. 2005). TGA2, TGA3, TGA5, TGA6, and TGA7 interact with NPR1 in planta when transiently expressed (Després et al. 2000; Kim and Delaney 2002). NPR1 as a cofactor of TGA transcription factor enhances binding of TGA transcription factors to the promoter of *PR1* gene to activate transcription of *PR1* gene (Fan and Dong 2002; Johnson et al. 2008). In the cytokinin signaling pathway, type-A and type-B response regulators (ARRs) are known to activate transcription of various cytokinin-responsive genes. Among 11 type-A and 12 type-B ARRs in *Arabidopsis*, only the type-B ARRs function as transcriptional activators (Hwang and Sheen 2001; Hwang et al. 2002; Mason et al. 2005). Among the type-B ARRs, ARR2 was found to be involved in triggering defense responses (Hwang and Sheen 2001; Choi et al. 2010). ARR2 of the cytokinin signaling has been shown to interact with TGA3 of SA signaling pathway and triggers SA-mediated defense responses (Fig. 7.5; Choi et al. 2010).

TGA3 recruits ARR2 to the defense gene *Pr1* promoter when SA signaling is activated. The binding of ARR2 to the *Pr1* promoter was increased by 2.7-fold when SA was applied. The transcriptional activation of *Pr1* by ARR2 required TGA binding to the LS7 *cis*-acting element of the *Pr1* promoter (Choi et al. 2010). Elevated levels of endogenous cytokinins in the cytokinin biosynthetic gene *Ipt*

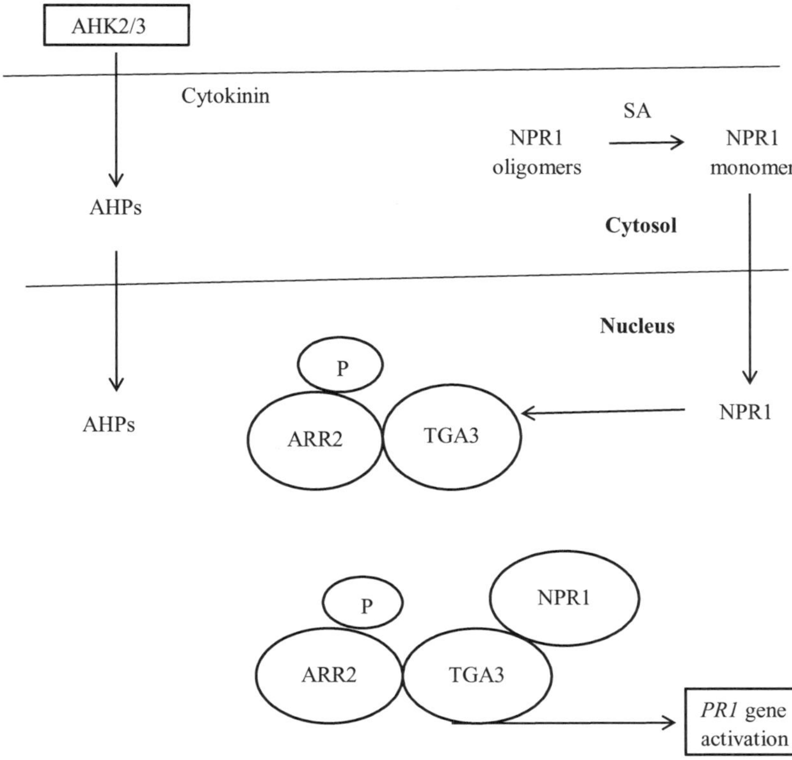

Fig. 7.5 Cytokinin activates SA signaling to trigger defense responses (Adapted from Choi et al. 2010)

overexpression plants as perceived by the AHK2/AHK3 receptors exerted a defense response, which resulted in enhanced resistance against *Pseudomonas syringae* pv. *tomato* DC3000 (*Pto*) in *Arabidopsis*. ARR2 induced expression of *Pr1*, a marker gene for plant immunity, and this induction required NPR1 and TGA3 (Choi et al. 2010). These results suggest that cytokinins enhance resistance against pathogens via a direct induction of defense-related genes, in cooperation with SA, by modulating the NPR1/TGA3-mediated signaling.

7.8.3 Type-A ARRs Negatively Regulate SA-Dependent Immune Responses

There are 10 genes encoding type-A ARRs in *Arabidopsis*. The type-A *arr* multiple mutants showed decreased susceptibility to the oomycete pathogen *Hyaloperonospora arabidopsidis* (Argueso et al. 2012). These results suggest that the type-A ARRs have a negative role in plant immunity. Transgenic *Arabidopsis* lines overexpressing type-A *ARR* genes showed enhanced susceptibility. Phosphorylation of type-A ARRs on a conserved residue (Asp87) in the receiver domain is required for type-A ARR activation and function in cytokinin signaling. Type-A ARRs were found to negatively regulate SA-dependent basal immunity. The type-A ARRs suppress defense gene expression downstream of SA accumulation (Argueso et al. 2012).

7.8.4 Cytokinin Synergistically Acts with SA to Trigger Immune Responses

An overlap between cytokinin- and SA-regulated genes involved in plant immune responses has been reported. The overlapping genes include those encoding chitinase and the PR protein PR-1 and several disease resistance genes (Naseem et al. 2012). Individually, activation of SA in comparison to cytokinin strongly activated PR-1 during infection by *Pseudomonas syringae* pv. *tomato* (*Pto*), while combined activation of SA, cytokinin, and *Pto* further augmented the PR-1signal. Deletion of SA from the network brought the PR-1 signal back to the level that was observed when cytokinin plus *Pto* alone was applied (Naseem et al. 2012). The results suggest that cytokinin synergistically acts with SA to trigger defense responses.

Cytokinin signaling in concert with SA signaling activates defense responses against *Hyaloperonospora arabidopsidis* in *Arabidopsis* (Argueso et al. 2012). Argueso et al. (2012) showed that cytokinin up-regulates plant immunity via an elevation of SA-dependent defense responses and SA in turn feedback inhibits cytokinin signaling. The crosstalk between cytokinin and SA signaling networks may help plants fine-tune defense responses against pathogens (Argueso et al. 2012).

7.9 Interaction Between Cytokinin and Abscisic Acid Signaling Systems

ABA treatment decreased the expression of several genes involved in cytokinin biosynthesis and degradation in rice (Fig. 7.6; Tsai et al. 2012). The expression of *OsIPT2*, *OsIPT4*, and *OsIPT7* genes encoding isopentenyltransferase was down-regulated in response to ABA. ABA treatment also down-regulated *CYP735A* genes encoding cytokinin hydroxylases, which are also involved in the cytokinin biosynthesis pathway (Takei et al. 2004b; Tsai et al. 2012). Expression of several *OsCKX* genes encoding cytokinin oxidase/dehydrogenase (CKX) was also down-regulated. CKX is involved in degradation of cytokinin. Multiple *LOGL* genes encoding cytokinin phosphoribohydrolase ("Lonely Guy") were altered both positively and negatively in response to ABA treatment (Tsai et al. 2012). ABA had

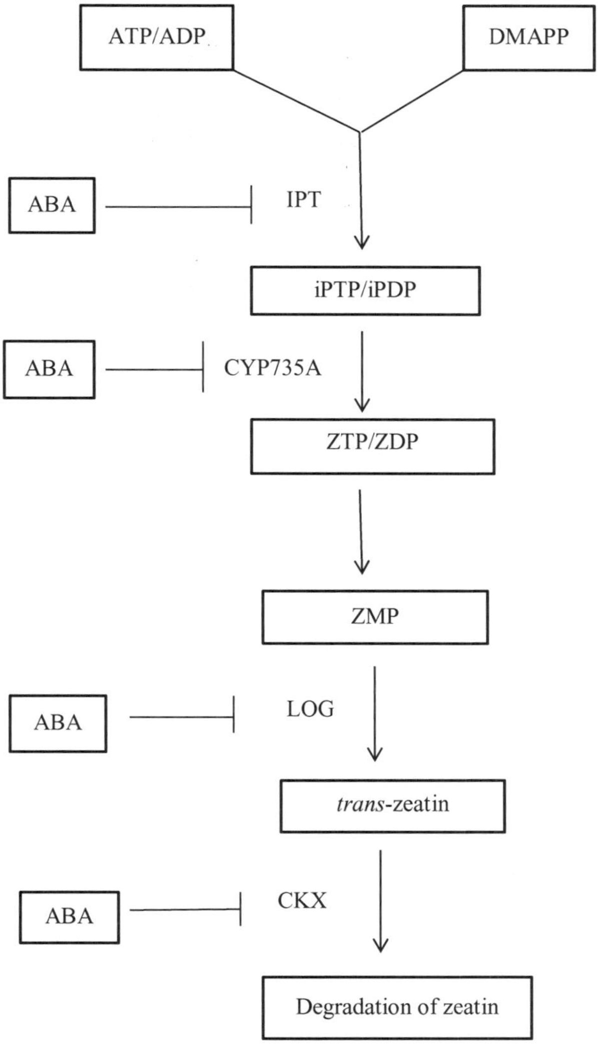

Fig. 7.6 Interplay between cytokinin and ABA signaling in rice plants (Adapted from Tsai et al. 2012)

only a modest effect on the level of transcripts of genes encoding cytokinin signaling elements. In the signaling pathway, *OsHK3* and *OsRR4* (a type-A response regulator) were slightly elevated in response to ABA (Tsai et al. 2012). ABA treatment down-regulated multiple *IPT*, *CKX*, and *CYP735A* genes in *Arabidopsis* (Takei et al. 2004a, b; Nishiyama et al. 2011). In maize leaf disks, *CKX* genes were up-regulated by ABA (Brugière et al. 2003). Collectively, these results suggest that ABA modulates the expression of genes encoding proteins involved in cytokinin biosynthesis and degradation.

7.10 Interplay Between Cytokinin and Auxin Signaling Systems in Plant Immunity

Auxin has been shown to positively or negatively regulate the expression of the genes involved in cytokinin homeostasis (Müller and Sheen 2008; Zhao et al. 2010; Tsai et al. 2012). Auxin affects both cytokinin biosynthesis and responsiveness. Auxin triggered a rapid down-regulation of zeatin biosynthesis in *Arabidopsis* (Nordström et al. 2004). Auxin treatment resulted in up-regulation of *IPT* genes in *Arabidopsis* (Miyawaki et al. 2004), while it repressed *IPT* expression in pea (Tanaka et al. 2006). Auxin down-regulated the expression of *CYP735a* genes in *Arabidopsis* (Takei et al. 2004b). Auxin up-regulated the expression of *LOG/LOGL* genes in rice (Tsai et al. 2012). Auxin up-regulated the expression of *CKX* genes encoding the cytokinin-degrading enzyme cytokinin oxidase in rice (Tsai et al. 2012) and pea (Tanaka et al. 2006; Shimizu-Sato et al. 2009). These results suggest that auxin differentially modulates cytokinin biosynthesis and signaling. Reciprocally, cytokinin antagonistically impacts the signaling of auxin (Stepanova and Alonso 2011). Synergism between auxin and cytokinin signaling has also been reported (Hwang et al. 2012).

Cytokinin–auxin antagonism has been reported in the *Arabidopsis–Pseudomonas syringae* pv. *tomato* (*Pto*) interaction. Auxin has been shown to enhance susceptibility to the bacterial pathogen, and elevated plant auxin levels repressed the defense-related *PR1* gene expression (Kazan and Manners 2009). In contrast, elevated cytokinin levels mediate resistance and induction of *PR1* (Choi et al. 2011; Naseem et al. 2012). During the course of infection, *Pto* enhances de novo auxin accumulation, whereas the level of cytokinin decreases during infection (Robert-Seilaniantz et al. 2011; Naseem et al. 2012). *Pto* infection resulted in decrease in *t*-zeatin levels in *Arabidopsis*, and the suppression of the cytokinin accumulation favors disease development (Naseem et al. 2012). Overexpression of *CKX4* gene, which encodes the cytokinin-degrading enzyme, favors fast bacterial growth (Choi et al. 2010; Naseem and Dandekar 2012). Auxin biosynthesis genes were up-regulated during infection with *Pto*. In contrast, the genes encoding enzymes of cytokinin biosynthesis were found to be down-regulated, while those of cytokinin catabolism were induced (Thilmony et al. 2006). These studies suggest that auxin and cytokinin play opposing role in plant immune responses (Naseem and Dandekar 2012). Molecular basis of the cytokinin–auxin antagonism is not yet fully understood. Cytokinin has been reported to inhibit the auxin transport (Moubayidin et al. 2009; Stepanova and Alonso 2011). Cytokinin induces the activation of type-B ARRs (Choi et al. 2011). The primary cytokinin response transcription factor, ARR1, has been shown to activate the gene *SHY2/IAA3*, a repressor of auxin signaling (Dello Ioio et al. 2008). The results suggest that ARRs in the cytokinin signaling may suppress auxin signaling (Naseem and Dandekar 2012). Auxin may also cause repression of cytokinin responses by activating type-A ARRs, which negatively regulate cytokinin signaling (Moubayidin et al. 2009; Argueso et al. 2012).

References

Akiyoshi DE, Regier DA, Gordon MP (1987) Cytokinin production by *Agrobacterium* and *Pseudomonas* spp. J Bacteriol 169:4242–4248

Akram A, Ongena M, Duby F, Dommes J, Thonart P (2008) Systemic resistance and lipoxygenase-related defence response induced in tomato by *Pseudomonas putida* strain BTP1. BMC Plant Biol 8:113

Anantharaman V, Aravind I (2001) The CHASE domain: a predicted ligand-binding module in plant cytokinin receptors and other eukaryotic and bacterial receptors. Trends Biochem Sci 26:579–582

Angra-Sharma R, Sharma DK (1999) Cytokinins in pathogenesis and disease resistance of *Pyrenophora teres*-barley and *Drechslera maydis*-maize interactions during early stages of infection. Mycopathogia 148:87–95

Aoyama T, Oka A (2003) Cytokinin signal transduction in plant cells. J Plant Res 116:221–231

Argueso CT, Ferreira FJ, Epple P, To JPC, Hutchison CE, Schaller GE, Dangl JL, Kieber JJ (2012) Two-component elements mediate interactions between cytokinin and salicylic acid in plant immunity. PLoS Genet 8(1):e1002448

Argyros R, Mathews D, Chiang Y, Palmer C, Thibault D, Etheridge N, Argyros D, Mason M, Kieber J, Schaller G (2008) Type B response regulators of Arabidopsis play key roles in cytokinin signaling and plant development. Plant Cell 20:2102–2116

Babosha AV (2009) Regulation of resistance and susceptibility in wheat-powdery mildew pathosystem with exogenous cytokinins. J Plant Physiol 166:1892–1903

Barash I, Manulis-Sasson S (2007) Virulence mechanisms and host specificity of gall-forming *Pantoea agglomerans*. Trends Microbiol 15:538–545

Brenner WG, Ramireddy E, Heyl A, Schmulling T (2012) Gene regulation by cytokinin in *Arabidopsis*. Front Plant Sci 3:8

Brugiěre N, Jiao S, Hantke S, Zinselmeier C, Roessler JA, Niu X, Jones RJ, Habben JE (2003) Cytokinin oxidase gene expression in maize is localized to the vasculature, and is induced by cytokinins, abscisic acid, and abiotic stress. Plant Physiol 132:1228–1240

Camañes G, Pastor V, Cerezo M, Garcia-Andrade J, Vicedo B, Garcia-Agustin P, Flors V (2012) A deletion in *NRT2.1* attenuates *Pseudomonas syringae*-induced hormonal perturbation, resulting in primed plant defenses. Plant Physiol 158:1054–1066

Cao H, Glazebrook J, Clarke J, Volko S, Dong X (1997) The Arabidopsis *npr1* gene that controls systemic acquired resistance encodes a novel protein containing ankyrin repeats. Cell 88:57–63

Chern M-S, Fitzgerald H, Yadav RC, Canlas PE, Dong X, Ronald PC (2001) Evidence for a disease-resistance pathway in rice similar to the *NPR1*-mediated signaling pathway in Arabidopsis. Plant J 27:101–113

Choi J, Huh SU, Kojima M, Sakakibara H, Paek K-H, Hwang I (2010) The cytokinin-activated transcription factor ARR2 promotes plant immunity via TGA3/NPR1-dependent salicylic acid signaling in *Arabidopsis*. Dev Cell 19:284–295

Choi J, Choi D, Lee R, Ryu CM, Hwang I (2011) Cytokinins and plant immunity: old foes or new friends. Trends Plant Sci 16:388–394

Clarke SF, Burritt DJ, Guy PL (1998) Influence of plant hormones on virus replication and pathogenesis-related proteins in *Phaseolus vulgaris* L infected with white clover mosaic potexvirus. Physiol Mol Plant Pathol 53:195–207

Conrath U (2011) Molecular aspects of defence priming. Trends Plant Sci 16:524–531

Crespi M, Vereecke D, Temmerman W, Van Montagu M, Desomer J (1994) The *fas* operon of *Rhodococcus fascians* encodes new genes required for efficient fasciation of host plants. J Bacteriol 176:2492–2501

Cumming RC, Andon NL, Haynes PA, Park M, Fischer WH, Schubert D (2004) Protein disulfide bond formation in the cytoplasm during oxidative stress. J Biol Chem 279:21749–21758

Delaney TP, Uknes S, Vernooij B, Friedrich L, Weymann K, Negrotto D, Gaffney T, Gut-Rella M, Kessmann H, Ward E, Ryals J (1994) A central role of salicylic acid in plant disease resistance. Science 266:1247–1250

Dello Ioio R, Nakamura K, Moubayidin L, Perilli S, Taniguchi M, Morita MT, Aoyama T, Costantino P, Sabatini S (2008) A genetic framework for the control of cell division and differentiation in the root meristem. Science 322:1380–1384

Desikan R, Hancock JT, Bright J, Harrison J, Weir I, Hooley R, Neill SJ (2005) A role for ETR1 in hydrogen peroxide signaling in stomatal guard cells. Plant Physiol 137:831–834

Després C, DeLong C, Glaze S, Liu E, Fobert PR (2000) The Arabidopsis NPR1/NIM1 protein enhances the DNA binding activity of a subgroup of the TGA family of bZIP transcription factors. Plant Cell 12:279–290

Fan W, Dong X (2002) In vivo interaction between NPR1 and transcription factor TGA2 leads to salicylic acid-mediated gene activation in *Arabidopsis*. Plant Cell 14:1377–1389

Fayza AF, Sabrey YMM (2006) Induction of resistance in *Phaseolus vulgaris* against TNV by salicylic acid and kinetin. Int J Agric Biol 8:47–51

Fitzgerald HA, Canlas PE, Chern M-S, Ronald PC (2005) Alteration of TGA factor activity in rice results in enhanced tolerance to *Xanthomonas oryzae* pv. *oryzae*. Plant J 43:335–347

Frébort I, Kowalska M, Hluska T, Frébortova J, Galuszka P (2011) Evolution of cytokinin biosynthesis and degradation. J Exp Bot 62:2431–2452

Fu ZQ, Yang S, Saleh A, Wang W, Ruble J, Oka N, Mohan R, Spoel SH, Tada Y, Zheng N, Dong X (2012) NPR3 and NPR4 are receptors for the immune signal salicylic acid in plants. Nature 486:228–232

Gaille C, Kast P, Hass D (2002) Salicylate biosynthesis in *Pseudomonas aeruginosa*. Purification and characterization of PchB, a novel bifunctional enzyme displaying isochorismate pyruvate-lyase and chorismate mutase activities. J Biol Chem 277:21768–21775

Gaille C, Reimmann C, Haas D (2003) Isochorismate synthase (PchA), the first and rate-limiting enzyme in salicylate biosynthesis of *Pseudomonas aeruginosa*. J Biol Chem 278:16893–16898

Garcion C, Lohmann A, Lamodiere E, Catinot J, Buchala A, Doermann P, Métraux J-P (2008) Characterization and biological function of the *ISOCHORISMATE SYNTHASE2* gene of the Arabidopsis. Plant Physiol 147:1279–1287

Grennan AK (2008) Ethylene response factors in jasmonate signaling and defense response. Plant Physiol 146:1457–1458

Großkinsky DK, Naseem M, Abdelmohsen UR, Plickert N, Engelke T, Griebel T, Zeier J, Novák O, Strnad M, Pfeifhofer H, Graaff EVD, Simon U, Roitsch T (2011) Cytokinins mediate resistance against *Pseudomonas syringae* in tobacco through increased antimicrobial phytoalexin synthesis independent of salicylic acid signaling. Plant Physiol 157:815–830

Haberer G, Kieber J (2002) Cytokinins. New insights into a classic phytohormone. Plant Physiol 128:354–362

Hutchison CE, Kieber JJ (2002) Cytokinin signaling in *Arabidopsis*. Plant Cell 14:547–559

Hwang I, Sakakibara H (2006) Cytokinin biosynthesis and perception. Physiol Plant 126:528–538

Hwang I, Sheen J (2001) Two-component circuitry in *Arabidopsis* cytokinin signal transduction. Nature 413:383–389

Hwang I, Chen H-C, Sheen J (2002) Two-component signal transduction pathways in Arabidopsis. Plant Physiol 129:500–515

Hwang I, Sheen J, Muller B (2012) Cytokinin signaling networks. Annu Rev Plant Biol 63:353–380

Imamura A, Hanaki N, Umeda H, Nakamura A, Suzuki T, Ueguchi C, Mizuno T (1998) Response regulators implicated in His-to-Asp phosphotransfer signaling in *Arabidopsis*. Proc Natl Acad Sci U S A 95:2691–2696

Inoue T, Higuchi M, Hashimoto Y, Seki M, Kobayashi M, Kato T, Tabata S, Shinozaki K, Kakimoto T (2001) Identification of CRE1 as a cytokinin receptor from *Arabidopsis*. Nature 409:1060–1063

Jakoby M, Weisshaar B, Droge-Laser W, Vicente-Carbajosa J, Tiedemann J, Kroj T, Parcy F (2002) bZIP transcription factors in *Arabidopsis*. Trends Plant Sci 7:106–111

Jennifer PC, Deruére J, Maxwell BB, Morris VF, Hutchison CE, Ferreira FJ, Eric Schaller G, Kieber JJ (2007) Cytokinin regulates type-A *Arabidopsis* response regulator activity and protein stability via two-component phosphorelay. Plant Cell 19:3901–3914

Johnson C, Mhatre A, Arias J (2008) NPR1 preferentially binds to the DNA-inactive form of *Arabidopsis* TGA2. Biochim Biophys Acta 1779:583–589

Joo JH, Wang S, Chen JG, Jones AM, Fedoroff NV (2005) Different signaling and cell death roles of heterotrimeric G protein α and β subunits in the *Arabidopsis* oxidative stress response to ozone. Plant Cell 17:957–970

Jung HW, Tschaplinski TJ, Wang L, Glazebrook J, Greenberg JT (2009) Priming in systemic plant immunity. Science 324:89–91

Kakimoto T (2001) Identification of plant cytokinin biosynthetic enzymes as dimethylallyl diphosphate:ATP/ADP isopentenyltransferases. Plant Cell Physiol 42:677–685

Kakimoto T (2003) Perception and signal transduction of cytokinins. Annu Rev Plant Biol 54:605–627

Kazan K, Manners JM (2009) Linking development to defense: auxin in plant-pathogen interactions. Trends Plant Sci 14:373–382

Kiba T, Yamada H, Sato S, Kato T, Tabata S, Yamashino T, Mizuno T (2003) The type-A response regulator, ARR15, acts as a negative regulator in the cytokinin-mediated signal transduction in *Arabidopsis thaliana*. Plant Cell Physiol 44:868–874

Kiba T, Naitou T, Koizumi N, Yamashino T, Sakakibara H, Mizuno T (2005) Combinatorial microarray analysis revealing arabidopsis genes implicated in cytokinin responses through the His→Asp phosphorelay circuitry. Plant Cell Physiol 46:339–355

Kim HS, Delaney TP (2002) Overexpression of TGA5, which encodes a bZIP transcription factor that interacts with NIM1/NPR1, confers SAR-independent resistance in *Arabidopsis thaliana* to *Peronospora parasitica*. Plant J 32:151–163

Kinkema M, Fan W, Dong X (2000) Nuclear localization of NPR1 is required for activation of PR gene expression. Plant Cell 12:2339–2350

Kohler A, Schwindling S, Conrath U (2002) Benzothiadiazole-induced priming for potentiated responses to pathogen infection, wounding, and infiltration of water into leaves requires the *NPR1/NIM1* gene in *Arabidopsis*. Plant Physiol 128:1046–1056

Kravchuk Z, Vicedo B, Flors V, Camañes G, González-Bosch C, Garcia-Agustin P (2011) Priming for JA-dependent defenses using hexanoic acid is an effective mechanism to protect Arabidopsis against *B. cinerea*. J Plant Physiol 186:359–366

Lee DJ, Park JY, Ku SJ, Ha YM, Kim S, Kim MD, Oh MH, Kim J (2007) Genome-wide expression of *Arabidopsis* response regulator (ARR7) overexpression in cytokinin response. Mol Genet Genomics 277:115–137

Luna E, Bruce TJA, Roberts MR, Flors V, Ton J (2012) Next generation systemic acquired resistance. Plant Physiol 158:844–853

Martinez-Medina A, Fernández I, Sánchez-Guzmán MJ, Jung SC, Pascual JA, Pozo MJ (2013) Deciphering the hormonal signalling network behind the systemic resistance induced by *Trichoderma harzianum* in tomato. Front Plant Sci 4:206

Mason MG, Mathews DE, Argyros DA, Maxwell BB, Kieber JJ, Alonso JM, Ecker JRS, Shaller GE (2005) Multiple type-B response regulators mediate cytokinin signal transduction in Arabidopsis. Plant Cell 17:3007–3018

Mathys J, De Cremer K, Timmermans P, Van Kerckhove S, Lievens B, Vanhaecke M, Cammue BP, De Coninck B (2012) Genome-wide characterization of ISR induced in *Arabidopsis thaliana* by *Trichoderma hamatum* T382 against *Botrytis cinerea* infection. Front Plant Sci 3:108. doi:10.3389/fpls.2012.00108

Meur G, Budatha M, Gupta AD, Prakash S, Kirti PB (2006) Differential induction of NPR1 during defense responses in *Brassica juncea*. Physiol Mol Plant Pathol 68:128–137

Miyawaki K, Matsumoto-Kitano M, Kakimoto T (2004) Expression of cytokinin biosynthetic isopentenyltransferase genes in Arabidopsis: tissue specificity and regulation by auxin, cytokinin, and nitrate. Plant J 37:128–138

Moreau M, Tian M, Klessig DF (2012) Salicylic acid binds NPR3 and NPR4 to regulate NPR1-dependent defense responses. Cell Res. doi:10.1038/cr 2012.100

Morris RO, Jameson PE, Laloue M, Morris JW (1991) Rapid identification of cytokinins by an immunological method. Plant Physiol 95:1156–1161

Mou Z, Fan W, Dong X (2003) Inducers of plant systemic acquired resistance regulate NPR1 function through redox changes. Cell 113:935–944

Moubayidin L, Di Mambro R, Sabatini S (2009) Cytokinin-auxin crosstalk. Trends Plant Sci 14:557–562

Mougel C, Zhulin IB (2001) CHASE: an extracellular sensing domain common to transmembrane receptors from prokaryotes, lower eukaryotes and plants. Trends Biochem Sci 26:582–584

Müller B, Sheen J (2007a) Advances in cytokinin signaling. Science 318:68–69

Müller B, Sheen J (2007b) Arabidopsis cytokinin signaling pathway. Sci STKE 2007:cm5

Müller B, Sheen J (2008) Cytokinin and auxin interaction in root stem-cell specification during early embryogenesis. Nature 453:1094–1097

Naseem M, Dandekar T (2012) The role of auxin-cytokinin antagonism in plant-pathogen interactions. PLoS Pathog 8(11):e1003026. doi:10.1371/journal ppat 1003026

Naseem M, Philippi N, Hussain A, Wangorsch G, Ahmed N, Dandekar T (2012) Integrated systems view on networking by hormones in *Arabidopsis* immunity reveals multiple crosstalk for cytokinin. Plant Cell 24:1793–1814

Nishiyama R, Watanabe Y, Fujita Y, Le DT, Kojima M, Werner T, Vankova R, Yamaguchi-Shinozaki K, Kakimota T, Sakakibara H, Schmülling T, Tran LS (2011) Analysis of cytokinin mutants and regulation of cytokinin metabolic genes reveals important regulatory roles of cytokinins in drought, salt and abscisic acid biosynthesis. Plant Cell 23:2169–2183

Nordström A, Tarkowski P, Tarkowska D, Norbaek R, Åstot C, Dolezal K, Sandberg G (2004) Auxin regulation of cytokinin biosynthesis in *Arabidopsis thaliana*: a factor of potential importance for auxin-cytokinin-regulated development. Proc Natl Acad Sci U S A 101:8039–8044

Pertry I, Václaviková K, Depuydt S, Galuszka P, Spichal L, Temmerman W, Stes E, Schmülling T, Kakimoto T, Van Montagu MCE, Strnad M, Holsters M, Tarkowski P, Vereecke D (2009) Identification of *Rhodococcus fascians* cytokinins and their modus operandi to reshape the plant. Proc Natl Acad Sci U S A 106:929–934

Pertry I, Václaviková K, Gemrotova M, Spichal L, Galuszka P, Depuydt S, Temmerman W, Stes E, De Keyser A, Riefler M, Biondi S, Novák O, Schmülling T, Strnad M, Tarkowski P, Holsters M, Vereecke D (2010) *Rhodococcus fascians* impacts plant development through the dynamic Fas-mediated production of a cytokinin mix. Mol Plant-Microbe Interact 23:1164–1174

Pieterse CM, van der Does D, Zamioudis C, Leon-Reyes A, van Wees SC (2012) Hormonal modulation of plant immunity. Annu Rev Cell Biol 28:489–521

Pogány M, Koehl J, Heiser I, Elstner E, Barna B (2004) Juvenility of tobacco induced by cytokinin gene introduction decreases susceptibility to *Tobacco necrosis virus* and confers tolerance to oxidative stress. Physiol Mol Plant Pathol 65:39–47

Rashotte AM, Carson SD, To JP, Kieber JJ (2003) Expression profiling of cytokinin action in *Arabidopsis*. Plant Physiol 132:1998–2011

Reusche M, Kláskova J, Thole K, Truskina J, Novák O, Janz D, Strnad M, Spichal L, Lipka V, Teichmann T (2013) Stabilization of cytokinin levels enhances *Arabidopsis* resistance against *Verticillium longisporum*. Mol Plant-Microbe Interact 26:850–860

Robert-Seilaniantz A, Grant M, Jones JDG (2011) Hormone crosstalk in plant disease and defense: more than just jasmonate-salicylate antagonism. Annu Rev Phytopathol 49:317–343

Rochon A, Boyle P, Wignes T, Fobert PR, Després C (2006) The coactivator function of *Arabidopsis* NPR1 requires the core of its BTB/POZ domain and the oxidation of C-terminal cysteines. Plant Cell 18:3670–3685

Sakai H, Honma T, Aoyama T, Sato S, Kato T, Tabata S, Oka A (2001) ARR1, a transcription factor for genes immediately responsive to cytokinins. Science 294:1519–1521

Sakakibara H (2006) Cytokinins: activity, biosynthesis, and translocation. Annu Rev Plant Biol 57:431–449

Sano H, Seo S, Orudgev S, Youssefian K, Ishizuka K, Ohashi Y (1994) Expression of the gene for a small GTP binding protein in transgenic tobacco elevates endogenous cytokinin levels, abnormally induces salicylic acid in response to wounding, and increases resistance to tobacco mosaic virus infection. Proc Natl Acad Sci U S A 91:10556–10560

Sano H, Seo S, Koizumi N, Niki T, Iwamura H, Ohashi Y (1996) Regulation by cytokinins of endogenous levels of jasmonic and salicylic acids in mechanically wounded tobacco plants. Plant Cell Physiol 37:762–769

Shimizu-Sato S, Tanaka M, Mori H (2009) Auxin-cytokinin interactions in the control of shoot branching. Plant Mol Biol 69:429–4435

Siemens J, Keller I, Sarx J, Kunz S, Schuller A, Nagel W, Schmülling T, Parniske M, Ludwig-Müller J (2006) Transcriptome analysis of *Arabidopsis* clubroots indicate a key role for cytokinins in disease development. Mol Plant Microbe Interact 19:480–494

Slaughter A, Daniel X, Flors V, Luna E, Hohn B, Mauch-Mani B (2012) Descendants of primed *Arabidopsis* plants exhibit resistance to biotic stress. Plant Physiol 158:835–843

Spoel SH, Dong X (2012) How do plants achieve immunity? Defence without specialized immune cells. Nat Rev Immunol 12:89–100

Stepanova AN, Alonso JM (2011) Bypassing transcription: a shortcut in cytokinin-auxin interactions. Dev Cell 4:606–610

Sun F, Liu P, Xu J, Dong H (2010) Mutation in RAP2.6L, a transactivator of the ERF transcription factor family, enhances *Arabidopsis* resistance to *Pseudomonas syringae*. Physiol Mol Plant Pathol 74:295–302

Suzuki T, Miwa K, Ishikawa K, Yamada H, Aiba H, Mizuno T (2001) The *Arabidopsis* sensor kinase, AHK4, can respond to cytokinin. Plant Cell Physiol 42:107–113

Sziráki I, Balázs E, Király Z (1980) Role of different stresses in inducing systemic acquired resistance to TMV and increasing cytokinin level in tobacco. Physiol Plant Pathol 16:277–280

Tada Y, Spoel SH, Pajerowska-Mukhtar K, Mou Z, Song J, Dong X (2008) Plant immunity requires conformational changes of NPR1 via S-nitrosylation and thioredoxin. Science 321:952–956

Takei K, Sakakibara H, Sugiyama T (2001) Identification of genes encoding adenylate isopentenyltransferase, a cytokinin biosynthesis enzyme, in *Arabidopsis thaliana*. J Biol Chem 276:26405–26410

Takei K, Ueda N, Aoki K, Kuromori T, Hirayama T, Shinozaki K, Yamaya T, Sakakibara H (2004a) *AtIPT3*, an *Arabidopsis* isopentenyltransferase gene, is a key determinant of macronutrient-responsive cytokinin biosynthesis. Plant Cell Physiol 45:1053–1062

Takei K, Yamaya T, Sakakibara H (2004b) *Arabidopsis CYP735A1* and *CYP735A2* encode cytokinin hydroxylases that catalyze the biosynthesis of *trans*-zeatin. J Biol Chem 279:41866–41872

Tanaka M, Takei K, Kojima M, Sakakibara H, Mori H (2006) Auxin controls local cytokinin biosynthesis in the nodal stem in apical dominance. Plant J 45:1028–1036

Thilmony R, Underwood W, He SY (2006) Genome-wide transcriptional analysis of the *Arabidopsis thaliana* interaction with the plant pathogen *Pseudomonas syringae* pv *tomato* DC3000 and the human pathogen *Escheria coli* 0157:H7. Plant J 46:34–53

To JP, Haberer G, Ferreira FJ, Deruere J, Mason MG, Schaller GE, Alonso JM, Ecker JR, Kieber JJ (2004) Type-A *Arabidopsis* response regulators are partially redundant negative regulators of cytokinin signaling. Plant Cell 16:658–671

Tsai YC, Weir NR, Hill K, Zhang W, Kim HJ, Shiu S-H, Schaller GE, Kieber JJ (2012) Characterization of genes involved in cytokinin signaling and metabolism from rice. Plant Physiol 158:1666–1684

Van der Ent S, Verhagen BWM, Van Doorn R, Bakker D, Verlaan MG, Pel MJC, Joosten RG, Proveniers MCG, Van Loon LC, Ton J, Pieterse CMJ (2008) *MYB72* is required in early signaling steps of rhizobacteria-induced systemic resistance in *Arabidopsis*. Plant Physiol 146:1293–1304

Walters DR, McRoberts N, Fitt BD (2008) Are green islands red herrings? Significance of green islands in plant interactions with pathogens and pests. Biol Rev Camb Philos Soc 83:79–102

Wang J, Ma X-M, Kojima M, Sakakibara H, Hou B-K (2011) N-Glucosyltransferase UGT76C2 is involved in cytokinin homeostasis and cytokinin response in *Arabidopsis thaliana*. Plant Cell Physiol 52:2200–2213

Whenham RJ (1989) Effect of systemic tobacco mosaic virus infection on endogenous cytokinin concentration in tobacco (*Nicotiana tabacum* L.) leaves: consequences for the control of resistance and symptom development. Physiol Mol Plant Pathol 35:85–95

Wildermuth MC, Dewdney J, Wu G, Ausubel FM (2001) Isochorismate synthesis is required to synthesize salicylic acid for plant defense. Nature 414:562–565

Wu Y, Zhang D, Chu JY, Boyle P, Wang Y, Brindle ID, De Luca V, Desprès C (2012) The Arabidopsis NPR1 protein is a receptor to the plant defense hormone salicylic acid. Cell Rep 1:639–647

Yamada H, Suzuki T, Terada K, Takei K, Ishikawa K, Miwa K, Mizyuno T (2001) The *Arabidopsis* AHK4 histidine kinase is a cytokinin-binding receptor that transduces cytokinin signals across the membrane. Plant Cell Physiol 42:1017–1023

Zhao Z, Andersen SU, Ljung K, Dolezal K, Miotk A, Schultheiss SJ, Lohmann JU (2010) Hormonal control of the shoot stem-cell niche. Nature 465:1089–1092

Chapter 8
Gibberellin Signaling in Plant Innate Immunity

Abstract Gibberellins (GA) are important plant growth hormones involved in plant innate immunity. GA modulates plant defense responses, mostly by regulating SA–JA–ET signaling systems, which form the backbone of the immune system. Key components in the GA signaling pathway include the GA receptor GID1, the DELLA proteins, and the F-box proteins. Upon GA binding, the GA-dependent conformational change causes the GID1 N-terminal helical lid domain to behave like "molecular glue" to form the GA–GID1–DELLA complex. DELLAs repress GA-dependent defense responses, whereas GA relieves their repressive activity. GA lifts DELLA repression by targeting DELLA for destruction via the ubiquitin–proteasome pathway. GA modulates plant disease resistance or susceptibility by inducing the degradation of DELLA proteins. GA interacts antagonistically with JA signaling to modulate plant immunity. GA attenuates the expression of several JA-responsive genes. Brassinosteroids (BR) have been shown to negatively regulate innate immune responses induced by GAs. BR and GA cause -cross-inhibitory effects on the reciprocal hormone biosynthesis pathways to interact in a mutually antagonistic manner. GA is also involved in triggering systemic acquired resistance (SAR). Pathogen triggers overexpression of DELLA proteins to suppress GA-regulated defense responses.

8.1 Role of Gibberellins in Plant Immune Responses

Plant immune responses are regulated by an array of cross-communicating plant hormone signaling systems. Salicylic acid (SA), jasmonic acid (JA), and ethylene (ET) are the archetypal defense hormones, and their importance in the hard wiring of the plant immune system is well established (Robert-Seilaniantz et al. 2011; De Vleesschauwer et al. 2012). SA–JA–ET signaling systems form the central backbone of the immunity (Pieterse et al. 2012). Gibberellins (GA) and other plant growth hormones (auxins, cytokinins, abscisic acid, and brassinosteroids) may be involved in modulating plant immune responses by regulating host defense responses triggered by the SA–JA–ET signaling systems (De Vleesschauwer et al. 2012; Pieterse et al. 2012; Naseem et al. 2012; Naseem and Dandekar 2012).

P. Vidhyasekaran, *Plant Hormone Signaling Systems in Plant Innate Immunity*, Signaling and Communication in Plants 2, DOI 10.1007/978-94-017-9285-1_8

Gibberellins have been shown to regulate plant immune responses by modulating JA and SA signaling systems (Navarro et al. 2006; Yang et al. 2008; Qi et al. 2014). GA regulates the plant innate immune responses either positively or negatively. It induces susceptibility or resistance against different fungal and bacterial pathogens (Yang et al. 2008; De Vleesschauwer et al. 2012; Qin et al. 2013). GA has been shown to enhance systemic acquired resistance against pathogens (Xia et al. 2010).

8.2 Biosynthesis of Gibberellins

Gibberellins, a large family of tetracyclic, diterpenoid plant hormones, play an important role in regulating diverse processes throughout plant development (Itoh et al. 2008; Yang et al. 2008; Davière and Achard 2013; Colebrook et al. 2014). Although 126 GAs have been reported in plants, fungi, and bacteria, only a very few GAs (GA1, GA3, GA4, GA7) have been reported to be biologically active (Yamaguchi 2008). GA4 and GA1 have been reported to be highly bioactive in plants (Olszewski et al. 2002). The simplified biosynthesis pathway of these two bioactive GAs is presented in Fig. 8.1

Geranylgeranyl diphosphate, the common precursor for diterpenes, is converted to *ent*-copalyl diphosphate (CDP) catalyzed by *ent*-copalyl diphosphate synthase (CPS). CDP is then converted to *ent*-kaurene by the action of *ent*-kaurene synthase (KS). The P450 monooxygenase *ent*-kaurenoic acid oxidase (KO) converts *ent-kaurene* to *ent*-kaurenoic acid, which is converted to GA_{12} aldehyde by the action of *ent*-kaurenoic acid oxidase (KAO). GA13 oxidase converts GA_{12} aldehyde to GA_{12}, which is then converted to GA_{53} by 13-hydroxylation. Both GA_{12} and GA_{53} are substrates for the multifunctional enzyme GA20-oxidase in parallel pathways. GA_{12} and GA_{53} are converted to various GA intermediates and bioactive GAs, including GA_1 and GA_4, by a series of oxidation steps catalyzed by 2-oxoglutarate-dependent dioxygenases, GA20-oxidases and GA3-oxidases (Fig. 8.1; Xu et al. 1999; Olszewski et al. 2002).

8.3 GA Signaling Pathway

8.3.1 GA Signal Receptors

Key components in GA signaling pathway include the GA receptor GID1 (Gibberellin Insensitive Dwarf1), the growth inhibitors DELLA (Asp–Glu–Leu–Leu–Ala) proteins and the F-box proteins SLY1 (SLEEPY1) and SNZ (SNEEZY) in *Arabidopsis*, and GID2 (Gibberellin Insensitive Dwarf2) in rice (Davière and Achard 2013). GA signaling is controlled through direct regulation of hormone accumulation mediated by changes in GA20ox and GA3ox expression in response to various stimuli (Hauvermale et al. 2012). GA signal is perceived by a soluble receptor protein GID1 in *Arabidopsis* and rice. *GID1* gene was first cloned from

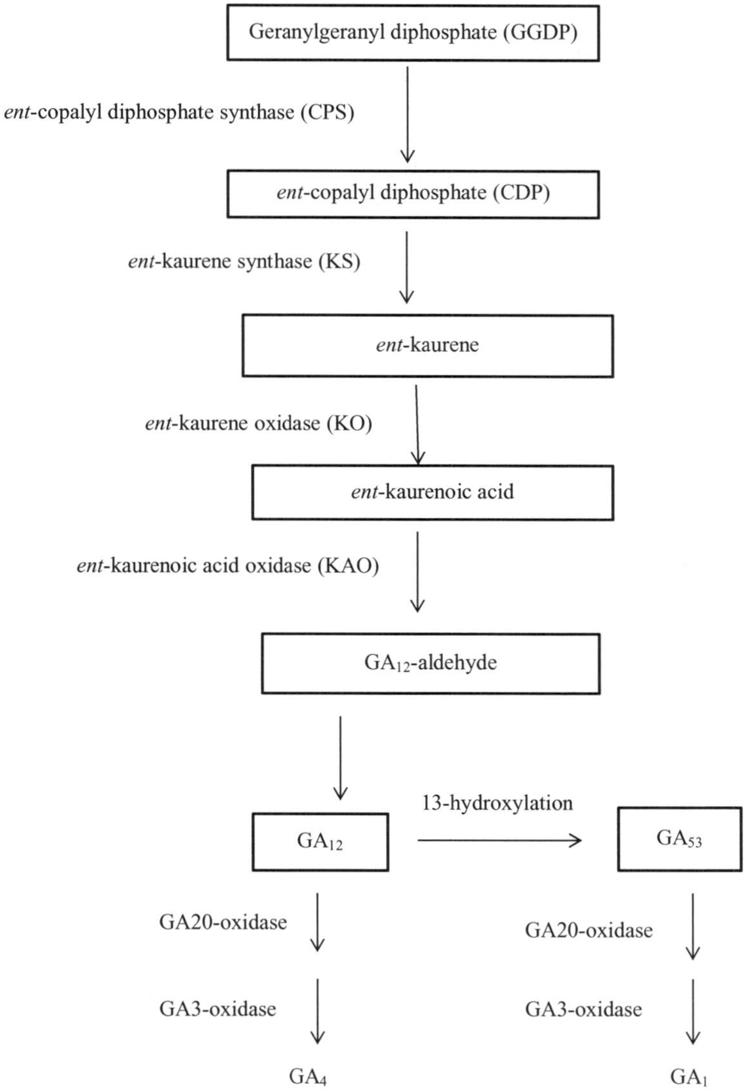

Fig. 8.1 Biosynthesis of bioactive GA4 and GA1 gibberellins (Adapted from Olszewski et al. 2002)

rice, and only a single copy of the gene was detected in rice (Ueguchi-Tanaka et al. 2005; Hartweck 2006). *Arabidopsis* has three functional GID1 orthologs, *GID1a*, *GID1b*, and *GIDc*. The loss of all three *GID1* genes is required for a complete loss of GA response (Griffiths et al. 2006; Iuchi et al. 2007; Willige et al. 2007; Hauvermale et al. 2012). In barley a single copy of the *GID1* homolog *GSE1* has been reported (Chandler et al. 2008).

GID1 encodes a hormone-sensitive lipase, and the lipase catalytic domain binds GAs. The N-terminal "lid" domain of GID1 interacts hydrophobically with the γ-lactone ring of GA_4 and, upon GA binding, folds over the GA binding pocket (Hauvermale et al. 2012). Upon the binding of bioactive GA, the hydroxyl group of the GA molecule becomes hydrogen bound to the Tyr31 residue of GID1, inducing the conformational change in the N-terminal extension to cover the GA pocket (Murase et al. 2008; Shimada et al. 2008). This GA-dependent conformational change causes the GID1 N-terminal helical lid domain to behave like "molecular glue" to interact with DELLA proteins (Murase et al. 2008). The upper surface of the lid binds with the DELLA and TVHYNP regions of DELLAs to form the GA–GID1–DELLA complex (Willige et al. 2007; Davière and Achard 2013).

8.3.2 DELLA Proteins, Repressors of GA Signaling

DELLA proteins are intracellular repressors of GA signaling. Gain-of-function mutations in DELLA genes lead to reduced GA signaling, whereas loss of function leads to increased GA signaling (Hauvermale et al. 2012). DELLAs repress almost all known GA-dependent processes, whereas GA relieves their repressive activity (Achard and Genschik 2009). The DELLA repressors have an N-terminal DELLA regulatory domain containing the conserved amino acid sequence Asp Glu Leu Leu Ala (DELLA) and the TVHYNP domain (Davière and Achard 2013). DELLAs have a C-terminal GRAS (for GA1, RGA, and SCARECROW) functional domain (Zhang et al. 2011; Hauvermale et al. 2012; Sun et al. 2012). The N-terminal DELLA regulatory domain is an intrinsically disordered domain that folds and becomes structured upon GID1 protein folding (Sun et al. 2010). Mutations in the DELLA and TVHYNP regions of the regulatory domain interfere with the ability to bind the GID1 receptor (Silverstone et al. 2007). A Ser/Thr/Val-rich domain is another DELLA regulatory domain which is involved in the regulation of DELLA function by phosphorylation (Silverstone et al. 2007; Dai and Xue 2010).

The *Arabidopsis* genome encodes five DELLAs (GA-INSENSITIVE [GAI], REPRESSOR OF GAI-3 [RGA], RGA-LIKE1 [RGL1], RGL2, and RGL3) that play distinct but also overlapping functions in repressing GA responses (Wen and Chang 2002; Wild et al. 2012; Davière and Achard 2013). The DELLA gene of rice is called *SLENDER RICE1* (*SLR1*) (Itoh et al. 2002; Gomi et al. 2004; Shimada et al. 2006; Hirano et al. 2012), and the DELLA gene of barley is called *SLENDER1* (*SLN1*) (Chandler et al. 2002).

8.3.3 Suppression of the Repressive Activity of DELLAs by Proteasome-Dependent Degradation of DELLAs

GA binding to GID1 has been shown to stimulate the formation of GA–GID1–DELLA complex. The GA–GID1–DELLA complex binds GA_4 with a higher affinity than GID1 alone, indicating that DELLA binding stabilizes the receptor–hormone interaction (Nakajima et al. 2006; Ueguchi-Tanaka et al. 2007). DELLAs are known to repress GA responses (Thomas and Sun 2004; Hong et al. 2012). DELLAs inhibit the activity of several transcriptional regulators (Hou et al. 2010; Zhang et al. 2011; Hong et al. 2012). DELLAs block the DNA-binding capacity of transcription factors (de Lucas et al. 2008; Feng et al. 2008). DELLAs repress GA responses by also directly activating the transcription of downstream genes (Hirano et al. 2012).

In the absence of GA, DELLAs accumulate and repress GA responses. The formation of the GA–GID1–DELLA complex stimulates the degradation of the DELLAs. The formation of the GA–GID1–DELLA complex induces conformational changes in the GRAS domain of DELLA that enhance recognition between the VHIID and LHRII motifs of DELLA and the F-box protein SLEEPY1 (SLY1) in *Arabidopsis*/GID2 in rice (McGinnis et al. 2003; Hirano et al. 2010). F-box proteins are components of the SCF (SKP1, CULLIN, F-BOX) E3 ubiquitin ligase complexes, which catalyze the attachment of polyubiquitin chains to target proteins for their subsequent degradation by the 26S proteasome (Lechner et al. 2006; Vidhyasekaran 2014). The $SCF^{SLY1/GID2}$ complex promotes the ubiquitylation and subsequent destruction of DELLAs by the 26S proteasome (McGinnis et al. 2003; Dill et al. 2001, 2004). These studies suggest that GA lifts DELLA repression of GA responses by targeting DELLA for destruction via the ubiquitin–proteasome pathway (Fig. 8.2; Hauvermale et al. 2012; Davière and Achard 2013).

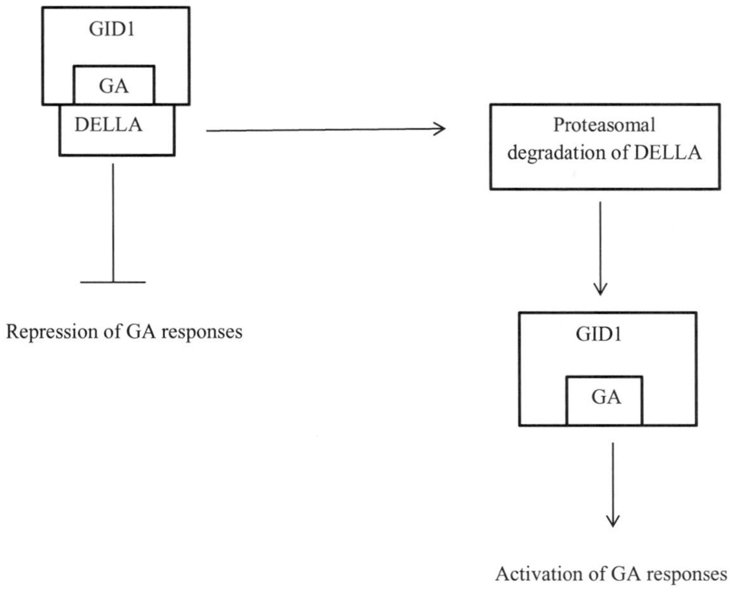

Fig. 8.2 GA lifts DELLA repression of GA responses by targeting DELLA for destruction via ubiquitin-proteasome pathway (Adapted from Hauvermale et al. 2012; Davière and Achard 2013)

8.4 GA Triggers Susceptibility or Resistance Against Different Pathogens

8.4.1 GA Triggers Resistance Against Pathogens

GA induces resistance or susceptibility against different pathogens. GA has been shown to modulate plant disease resistance by inducing the degradation of DELLA proteins, a class of nuclear growth-repressing proteins that act as central suppressors of GA signaling (Navarro et al. 2008). Rice has a DELLA gene, designated *SLENDER RICE1* (*SLR1*). Several GA-deficient and/or GA-insensitive rice mutants have been identified, and all of these mutants were shown to overaccumulate *SLR1* (Ueguchi-Tanaka et al. 2008). All mutants, either impaired in GA biosynthesis, insensitive to GA, or both, showed increased susceptibility to *Pythium graminicola* compared with wild-type rice plants. Enhanced susceptibility to *P. graminicola* was also observed in two independent *SLR1* gain-of-function mutants. Further, *slr1-1*, a loss-of-function allele displaying a constitutive GA response phenotype, showed increased resistance against the oomycete pathogen (De Vleesschauwer et al. 2012). The mutations of the DELLA proteins in *Arabidopsis*, which result in activation of GA signaling, make the plants more resistant to the bacterial pathogen *Pseudomonas syringae* pv. *tomato* (Robert-Seilaniantz et al. 2007). GA promotes resistance against *Pseudomonas syringae* pv. *tomato* by degrading DELLA proteins (Navarro et al. 2008). These results suggest that GA induces resistance by negatively regulating the DELLA protein. Treatment of rice plants with increasing concentrations of GA enhanced resistance to *P. graminicola* in a concentration-dependent manner. Conversely, depletion of endogenous GA levels using the GA biosynthesis inhibitor uniconazole promoted disease susceptibility (De Vleesschauwer et al. 2012).

8.4.2 GAs May Negatively Regulate Plant Defense Responses and Induce Susceptibility

GAs may induce susceptibility against some pathogens. GAs play negative roles in resistance to both the bacterial pathogen *Xanthomonas oryzae* pv. *oryzae* and the fungal pathogen *Magnaporthe oryzae* in rice (Yang et al. 2008). *OsGA20ox3* gene encoding GA20-oxidase is involved in the biosynthesis of bioactive gibberellins. GA20ox catalyzes consecutive steps of oxidation in the late part of the GA biosynthetic pathway (Olszewski et al. 2002; Qin et al. 2013). *OsGA20ox3*-overexpressing rice plants produced more GA, and the GA-overproducing plants were more susceptible to *M. oryzae* and *X. oryzae* pv. *oryzae* compared with the wild-type plants. The susceptibility of wild-type plants to *X. oryzae* pv. *oryzae* was increased by exogenous application of GA3 (Qin et al. 2013). Conversely, *OsGA20ox3-suppressing* RNAi lines showed enhanced disease resistance against the rice pathogens *M. oryzae* and *X. oryzae* pv. *oryzae* and increased expression of defense-related genes. These results suggest that GA suppresses plant defense responses and induces susceptibility against rice pathogens (Qin et al. 2013).

EUI (ELONGATED UPPERMOST INTERNODE) is a P450 monooxygenase, which deactivates the biologically active GAs (Zhu et al. 2006). The loss-of-function mutants of *EUI* accumulate large amounts of bioactive GAs (Xu et al. 2004; Zhu et al. 2006; Yang et al. 2008). The GA-overproducing *eui* mutant rice plants have been found to be more susceptible to bacterial blight pathogen *X. oryzae* pv. *oryzae* and to the blast pathogen *M. oryzae* (Yang et al. 2008). GA_3 application suppressed disease resistance of the *Eui* overexpressors, while the GA biosynthesis inhibitor uniconazole restored basal resistance in the RNAi lines. Exogenous application of GA_3 increases susceptibility to *X. oryzae* pv. *oryzae*, while inhibitor of GA synthesis (uniconazole) induces disease resistance. Similarly, uniconazole restored disease resistance of the *eui* mutant, and GA_3 decreased disease resistance of the *Eui* overexpressors against *X. oryzae* pv. *oryzae* (Yang et al. 2008). These results suggest that GAs negatively regulate rice disease resistance and GAs play an important role in inducing susceptibility against both bacterial and fungal diseases (Yang et al. 2008). The knockout of the *Eui* gene encoding a GA-deactivating enzyme compromises disease resistance, while overexpression of the *Eui* gene increases disease resistance against rice bacterial blight caused by *X. oryzae* pv. *oryzae* and the rice blast caused by *M. oryzae* (Yang et al. 2008). Transgenic rice plants overexpressing *eui* gene show enhanced resistance against both *X. oryzae* pv. *oryzae* and *M. oryzae* (Yang et al. 2008).

GA level was found to be reduced during virus infection in rice (Zhu et al. 2005). The outer capsid protein P2 of *Rice dwarf virus* (RDV) has been shown to interact with the host GA biosynthesis enzyme *ent*-kaurene oxidase (KAO) in vivo, leading to reduced biosynthesis of gibberellins. The reduced KAO expression in the infected plants led to reduced rice dwarf symptom development (Zhu et al. 2005). It suggests that increased KAO expression induces susceptibility and triggers severe disease symptom development. The interaction between the viral capsid protein P2 and rice *ent*-kaurene oxidase-like proteins may decrease phytoalexin biosynthesis and make plants highly susceptible to pathogens (Zhu et al. 2006). DELLA proteins are the repressors of GA signaling. The mutations of the DELLA proteins in *Arabidopsis*, which result in activation of GA signaling, make the plants more susceptible to the necrotrophic fungal pathogen *Alternaria brassicicola* (Robert-Seilaniantz et al. 2007). These results suggest that GA signaling induces susceptibility by suppressing the DELLA proteins.

8.5 Interplay of GA Signaling System with SA Signaling System in Modulating Plant Immune System

Interplay between GA and SA signaling systems in plant innate immunity has been reported. GA expedites SA accumulation (Navarro et al. 2008; Alonso-Ramirez et al. 2009) and promotes resistance against *Pseudomonas syringae* pv. *tomato* by degrading DELLA proteins (Navarro et al. 2008). Loss-of-function mutants in DELLAs, the suppressors of GA signaling, up-regulate the SA-mediated defense and down-regulate JA/ET-mediated defense in *Arabidopsis* (Robert-Seilaniantz et al. 2007). Transgenic poplar plants overproducing mutant DELLA protein *gai1* or *rgi1* accumulated less SA (Busov et al. 2006). RGL3 is a DELLA protein detected in *Arabidopsis*. It represses GA-dependent responses in the GA signaling system (Hauvermale et al. 2012; Davière and Achard 2013). SA-dependent *PR-1* transcripts displayed stronger induction in *P. syringae*-infected leaves of *rgl3-5* mutant plants compared to the wild-type plants. The results suggest that the DELLA protein RGL3 represses the SA pathway (Wild et al. 2012).

Eui is a P450 monooxygenase, which deactivates the biologically active GAs (Zhu et al. 2006). Transgenic plants overexpressing *eui* accumulate low levels of GA, while *eui* mutant plants accumulate high levels of bioactive GAs. *Eui* overexpression results in suppression of SA accumulation (Yang et al. 2008), whereas *eui* mutant plants showed increased accumulation of SA (Fig. 8.3; Yang et al. 2008). These results suggest that low levels of GA result in reduction in SA biosynthesis, and, in contrast, increased GA levels result in increased SA biosynthesis. SA-dependent *PR-1* transcripts displayed stronger induction in *P. syringae*-infected leaves of *rgl3-5* mutant plants compared to the wild-type plants. The results suggest that the DELLA protein RGL3 represses the SA pathway (Wild et al. 2012). Salicylate (SA) level was decreased in the rice plants overexpressing *Eui*, which show enhanced disease resistance against the bacterial blight and blast pathogens, while the JA level was found to be reduced in *eui* mutant plants, which show enhanced susceptibility to the pathogens (Yang et al. 2008).

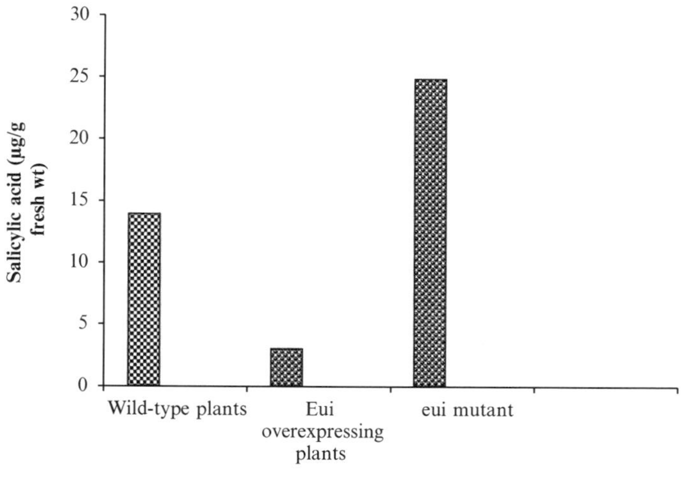

Fig. 8.3 Role of *eui* in modulating SA levels in mutant and transgenic rice plants (Adapted from Yang et al. 2008)

8.6 Interplay of GA and JA Signaling Systems in Modulating Plant Immune System

8.6.1 Antagonistic Interaction Between GA and JA Signaling Systems

GA interacts antagonistically with JA signaling in rice immunity (Yang et al. 2013), while GA attenuates the JA-induced expression of a number of JA-responsive genes (Cao et al. 2006; Hou et al. 2008, 2010). Both JA and GA signaling systems modulate plant immune responses. Binding of bioactive JA or GA ligands to cognate receptors leads to proteasome-dependent degradation of specific transcriptional repressors (the JAZ or DELLA family of proteins), which, at the resting state, represses cognate transcription factors involved in defense (Yang et al. 2012). JA interferes with gibberellin signaling cascade. JA delays GA-mediated DELLA protein degradation. The JASMONATE ZIM-DOMAIN (JAZ) protein JAZ9 inhibits RGA (a DELLA protein) interaction with the transcription factor PIF3 (phytochrome-interacting factor 3) (Yang et al. 2012). Conversely, GA interferes with JA signaling system. GA attenuates the JA-induced expression of a number of JA-responsive genes (Cao et al. 2006; Hou et al. 2008, 2010). DELLAs contribute to plant defense by interacting with JAZ proteins (Hou et al. 2010; Yang et al. 2012; Wild et al. 2012).

8.6.2 JA Induces Enhanced Expression of DELLA Genes Involved in GA Signaling

JA signaling has been shown to activate expression of DELLA genes involved in GA signaling pathway. JA-induced *RGL3* expression works via the COI1/MYC2-dependent signaling pathway. JA-mediated induction of RGL3 expression was abolished in the *coi1-1* mutant, indicating that *RGL3* is downstream of COI1 (Wild et al. 2012). JA enhances *RGL3* transcripts and protein levels. Binding of GA to its receptor GID1 promotes the GID1–DELLA interaction which in turn stimulates the interaction between DELLAs and the specific E3 ubiquitin ligase $SCF^{SLY1/GID2}$ complex, leading to subsequent degradation of DELLAs by the 26S proteasome (Wild et al. 2012).

P. syringae pv. *tomato* produces a JA-mimicking phytotoxin, coronatine (COR), which can induce a set of JA-responsive genes to promote susceptibility (Laurie-Berry et al. 2006). *P. syringae pv. tomato* induced *RGL3* expression in leaves 3 days after inoculation with the pathogen in *Arabidopsis*. Conversely, *RGL3* expression was not induced by a *P. syringae* strain defective in COR synthesis (Wild et al. 2012). The results suggest that *P. syringae* uses the COR-dependent pathway to modulate *RGL3* expression.

8.6.3 DELLAs Modulate JA Responses by Degrading JAZ Proteins and/or Sequestering JAZs into Inactive Complexes

DELLA proteins are the repressors of expression of GA-responsive genes. A quadruple-DELLA mutant, which lacks four out of the five *Arabidopsis* DELLA proteins, was partially insensitive to gene induction by methyl jasmonate (Navarro et al. 2008). The constitutively active dominant DELLA mutant *gai* was sensitized for JA-responsive gene induction. The results suggest that DELLAs are involved in JA signaling (Navarro et al. 2008). The elevated resistance of DELLA mutant *gai* to the necrotrophic fungal pathogen *Alternaria brassicicola* and susceptibility to the hemibiotroph *Pseudomonas syringae* pv. *tomato* strain DC3000 were attenuated in the JA-insensitive *coi1-16* mutant (Navarro et al. 2008). The results suggest that the interplay between GA and JA signaling systems plays an important role in modulating plant immune responses.

The DELLA protein RGA-LIKE3 (RGL3) protein has been shown to be essential to enhance the JA-mediated responses in *Arabidopsis* (Wild et al. 2012). JA rapidly induced RGL3 expression in COI1–MYC2-dependent manner. MYC2 has been shown to bind directly to RGL3 promoter. MYC2 induced RGL3 accumulation, and the activated RGL3 binds with JAZ proteins, key repressors of JA signaling (Wild et al. 2012). These results suggest that JA/MYC2-dependent accumulation of RGL3 represses JAZ activity, which in turn enhances the expression of JA-responsive genes (Fig. 8.4). It has been shown that induction of primary JA-responsive genes is reduced in the *rgl3-5* mutant and enhanced in transgenic lines overexpressing *RGL3*. The induction of JA-dependent gene expression, including *PDF1.2*, *TAT1*, and *LOX2*, was severely compromised in *rgl3-5*-infected leaves in comparison with wild-type-infected leaves (Wild et al. 2012). *RGL3* positively regulates JA-mediated resistance to the necrotrophic fungal pathogen *Botrytis cinerea* and susceptibility to the hemibiotrophic bacterial pathogen *Pseudomonas syringae* (Wild et al. 2012).

The JA receptor JAZ proteins have been identified as suppressors of jasmonate signaling. JAZ proteins suppress the expression of the JA-responsive genes (Chini et al. 2007; Chung and Howe 2009; Howe 2010; Pauwels and Goossens 2011; Sheard et al. 2010; Wager and Browse 2012; Wasternack and Hause 2013). DELLA proteins RGA, GAI, RGL1, and RGL2 directly bind to and thereby inhibit the JAZ proteins JAZ1, JAZ3, and JAZ9 (Hou et al. 2010). DELLAs modulate JA responses by repressing JAZ proteins (Schwechheimer 2012). RGL3 interacts with JAZs and modulates MYC2-dependent gene expression (Wild et al. 2012). The interaction of the DELLA proteins with JAZ proteins has been reported (Hou et al. 2010; Wild et al. 2012; Yang et al. 2012). The JAZ proteins JAZ1 and JAZ8 directly interact with MYC2. RGL3 has been shown to be essential to fully enhance JA-induced MYC2 gene expression (Wild et al. 2012). JA enhances MYC2 transcriptional activity by two different mechanisms. JA stimulates the SCF^{COI1}-dependent degradation of the JAZs, thus releasing MYC2 transcriptional activity. JA may

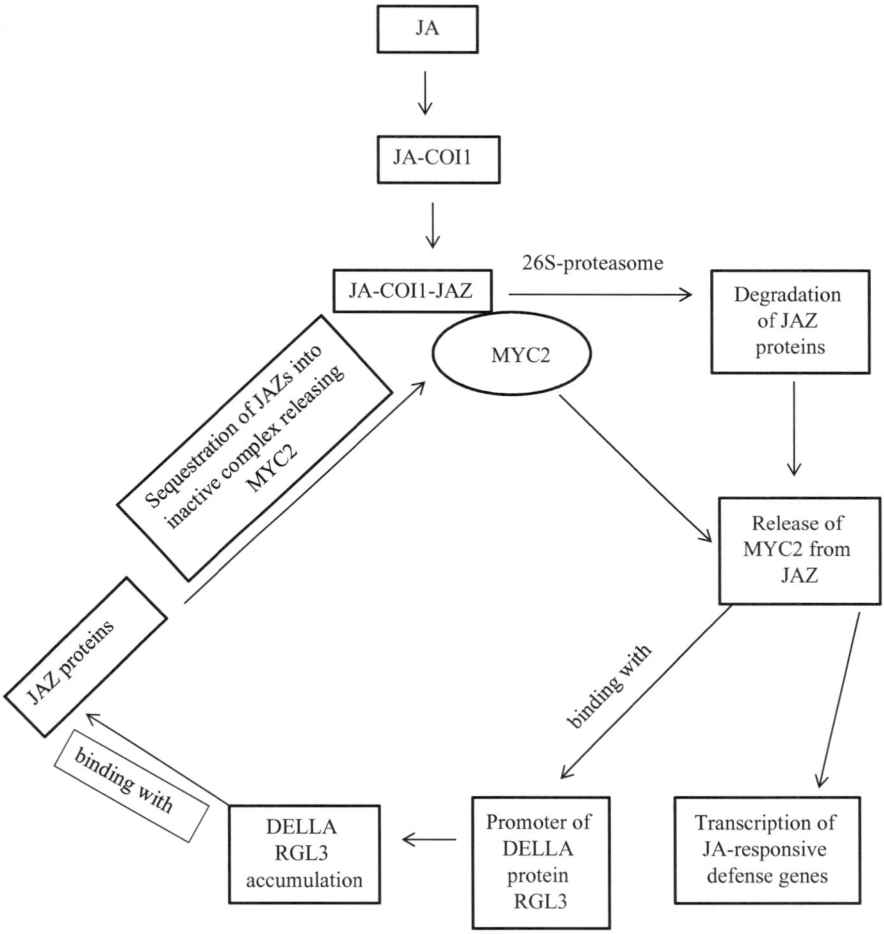

Fig. 8.4 Interplay between GA and JA signaling systems in induction of JA-responsive defense gene expression (Adapted from Wild et al. 2012)

also enhance the accumulation of RGL3, which in turn may sequester JAZs into inactive complexes, unable to restrain MYC2 activity (Fig. 8.4; Hou et al. 2010; Wild et al. 2012).

8.6.4 GA Attenuates the Expression of JA-Responsive Genes

GA attenuates the JA-induced expression of a number of JA-responsive genes (Cao et al. 2006; Hou et al. 2008, 2010). The expression of these genes is dependent on the transcription activators MYC2, MYC3, and MYC4 and their negative regulators, JAZ proteins (Cheng et al. 2011; Fernández-Calvo et al. 2011; Hong et al. 2012; Niu et al. 2011; Kazan and Manners 2013; Wasternack and Hause 2013). MYC2-regulated genes include the JA-responsive defense genes *PDF1.2*, *CHIB/PR3*, and *HEL/PR4*, and they are negatively regulated by MYC2 (Anderson et al. 2004; Lorenzo et al. 2004; Dombrecht et al. 2007). MYC2 positively regulates the JA-responsive genes *VSP* and *LOX* (Lorenzo et al. 2004). MYC2/MYC3/MYC4 proteins act redundantly in the activation of RGL3 in response to JA (Wild et al. 2012).

8.7 Interplay Between GA and Brassinosteroids Signaling Systems in Plant Immune Responses

Brassinosteroids (BR) have been shown to negatively regulate innate immune responses induced by GAs in rice. GA induces resistance against *Pythium graminicola* in rice. BR suppresses GA signaling. Supplying the brassinosteroid brassinolide (BL) along with GA induced disease susceptibility. The results suggest that GA-mediated defense mechanisms are repressed by BR (De Vleesschauwer et al. 2012). Endogenous GA levels have been shown as an important reservoir needed to mount a full innate immune response in rice plants against the root pathogen *P. graminicola*. BR inhibits GA responses in the roots by interfering with GA metabolism. BR-induced susceptibility develops coincidently with decreases in endogenous GA content (De Vleesschauwer et al. 2012). The rice DELLA protein SLR1 is transiently up-regulated in response to pathogen inoculation, suggesting that pathogen triggers overexpression of *SLR1* to suppress GA-regulated defense responses (De Vleesschauwer et al. 2012).

In rice plants, BR signaling negatively regulates innate immune responses against *Pythium graminicola*. In contrast, GA positively regulates the plant immunity against the pathogen (De Vleesschauwer et al. 2012). GA modulates plant disease resistance by inducing the degradation of DELLA proteins, a class of nuclear growth-repressing proteins that act as central suppressors of GA signaling (Navarro et al. 2008).

Pythium graminicola strongly induced the GA-degrading enzyme *GIBBERELLIN 2-β-DIOXYGENASE3* (*OsGA2ox3*) in roots of rice plants grown in the presence or absence of the most active brassinosteroid, brassinolide (BL). In contrast, the GA biosynthesis enzyme *GIBBERELLIN 20-OXIDASE3* (*OsGA20ox3*) was severely down-regulated following *P. graminicola* inoculation (De Vleesschauwer et al. 2012). *P. graminicola* recruits the BR pathway to actively suppress GA biosynthesis in rice. Exogenously administered BR inhibited *GA20ox3* and induced *GA2ox3* within 8 h of inoculation with *P. graminicola*. When rice roots were treated with GA, a fast and strong down-regulation of the BR response gene *OsBLE2* and the BR biosynthesis gene *OsDWARF2* was observed (De Vleesschauwer et al. 2012). Collectively these results suggest that BR and GA cause cross-inhibitory effects on the reciprocal hormone biosynthesis pathways to interact in a mutually antagonistic manner.

BR also may inhibit GA action through transcriptional activation of GA repressor genes. Several proteins with an inhibitory function in GA signaling have been characterized in rice plants. These include the *O*-linked GlcNAc transferase SPINDLY (Shimada et al. 2006) and the casein kinase EARLY FLOWERING1 (EL1) (Dai and Xue 2010). Expression of SPINDLY was up-regulated by BR treatment alone, and the most dramatic changes in SPINDLY kinetics were observed following the pathogen *P. graminicola* inoculation. Similar changes in *EL1* transcript accumulation were observed in rice plants inoculated with *P. graminicola* (De Vleesschauwer et al. 2012). These results suggest that *P. graminicola* triggers the expression of GA repressor proteins in a BR-dependent manner. Brassinolides may not intervene with GA-mediated SLR1 protein turnover per se but may indirectly promote SLR1 stabilization by inactivating GA through the combined repression of GA synthesis and transcriptional activation of GA responses (De Vleesschauwer et al. 2012).

8.8 Interplay Between GA and Auxin Signaling Systems

Interplay between GA and auxin signaling systems in plant immune system has not been studied in detail. During infection process of the citrus canker pathogen *Xanthomonas axonopodis* pv. *citri*, changes in the expression of genes involved in gibberellin biosynthesis and in auxin signaling and mobilization were observed. Auxin was found to promote canker development, whereas an inhibitor of auxin transport reduced pustule formation. Conversely, an inhibitor of gibberellin biosynthesis significantly inhibited the auxin-induced transcription of genes and also the appearance of canker tissues (Cernadas and Benedetti 2009). The results suggest that a crosstalk between gibberellin and auxin controls the transcription of genes involved in the bacterial disease development.

Auxin and auxin transport have been shown to control DELLA protein abundance (Fu and Harberd 2003). GA may control auxin transport and PIN (PIN-FORMED) protein abundance (Willige et al. 2011). GA promotes the degradation of the auxin efflux carriers PIN1 and PIN2 (Willige et al. 2011). PIN proteins are members of the major efflux facilitator family of integral membrane proteins that are essential for polarized auxin transport (Friml 2003). The repressed genes due to *Pseudomonas syringae* pv. *tomato* infection in *Arabidopsis* include six auxin transporters (*PIN3*, *PIN4*, *PIN7*, *AUX1*, *LAX1*, *LAX3*) (Thilmony et al. 2006). The role of these auxin transporters in plant immunity has to be studied in detail.

8.9 GA May Be Involved in Triggering Systemic Acquired Resistance (SAR)

GA has been shown to induce systemic acquired resistance (SAR) against *Pseudomonas syringae* in *Arabidopsis* (Xia et al. 2010). SAR is a heightened state of defense against a broad spectrum of pathogens activated throughout a plant following a local infection (Liu et al. 2011a). The induced resistance is detected in the uninoculated systemic tissue of a pathogen-infected plant (Liu et al. 2011b; Návarová et al. 2012; Shah and Zeier 2013). Xia et al. (2010) showed that an abnormal cuticle, which is the result of a nonallelic mutation in the *GLABRA1* (*GL1*) gene, is associated with SAR. Besides *GL1*, other mutations affecting trichome formation also compromised cuticle and thereby SAR. Exogenous application of GA on *gl1* plants compensated the effect of *gl1* and induced SAR. The results suggest that GA might participate in inducing systemic resistance to bacterial pathogens by restoring cuticle formation (Xia et al. 2010).

8.10 Pathogen May Suppress GA Signaling Pathway to Cause Disease

GA positively regulates the plant immunity against the pathogen *Pythium graminicola* in rice (De Vleesschauwer et al. 2012). GA modulates plant disease resistance by inducing the degradation of DELLA proteins, a class of nuclear growth-repressing proteins that act as central suppressors of GA signaling (Navarro et al. 2008). The rice DELLA gene, *SLENDER RICE1* (*SLR1*), plays an important role in GA signaling. Several GA-deficient and/or GA-insensitive rice mutants have been identified, and all of these mutants were shown to overaccumulate *SLR1* (Ueguchi-Tanaka et al. 2008). *SLR1* is transiently up-regulated in response to pathogen inoculation, suggesting that pathogen triggers overexpression of *SLR1* to suppress GA-regulated defense responses (De Vleesschauwer et al. 2012).

References

Achard P, Genschik P (2009) Releasing the brakes of plant growth: how GAs shut down DELLA proteins. J Exp Bot 60:1085–1092

Alonso-Ramirez A, Rodriguez D, Reyes D, Jiménez JA, Nicolás G, López-Climent M, Gómez-Cadenas A, Nicolás C (2009) Evidence for a role of gibberellins in salicylic acid-modulated early plant responses to abiotic stress in *Arabidopsis* seeds. Plant Physiol 150:1335–1344

Anderson JP, Badruzsaufari E, Schenk PM, Manners JM, Desmond OJ, Ehlert C, Maclean DJ, Ebert PR, Kazan K (2004) Antagonistic interaction between abscisic acid and jasmonate-ethylene signaling pathways modulates defense gene expression and disease resistance in *Arabidopsis*. Plant Cell 16:3460–3479

Busov V, Meilan R, Pearce DW, Rood SB, Ma C, Tschaplinski TJ, Strauss SH (2006) Transgenic modification of *gai* and *rgl1* causes dwarfing and alters gibberellins, root growth, and metabolite profiles on *Populus*. Planta 224:288–299

Cao D, Cheng H, Wu W, Soo HM, Peng J (2006) Gibberellin mobilizes distinct DELLA dependent transcriptomes to regulate seed germination and floral development in *Arabidopsis*. Plant Physiol 142:509–525

Cernadas RA, Benedetti CE (2009) Role of auxin and gibberellin in citrus canker development and in the transcriptional control of cell-wall remodeling genes modulated by *Xanthomonas axonopodis* pv. *citri*. Plant Sci 177:190–195

Chandler PM, Marion-Poll A, Ellis M, Gubler F (2002) Mutants at the *Slender1* locus of barley cv Himalaya. Molecular and physiological characterization. Plant Physiol 129:181–190

Chandler PM, Harding CA, Ashton AE, Mulcair MD, Dixon NE, Mander LN (2008) Characterization of gibberellin receptor mutants of barley (*Hordeum vulgare* L.). Mol Plant 1:285–294

Cheng Z, Sun L, Qi T, Zhang B, Peng W, Liu Y, Xie D (2011) The bHLH transcription factor MYC3 interacts with the jasmonate ZIM-domain proteins to mediate jasmonate response in *Arabidopsis*. Mol Plant 4:279–288

Chini A, Fonseca S, Fernández G, Adie B, Chico JM, Lorenzo O, Garcia-Casado G, López-Vidriero I, Lozano FM, Ponce MR, Micol JL, Solano R (2007) The JAZ family of repressors is the missing link in jasmonate signaling. Nature 448:666–671

Chung HS, Howe GA (2009) A critical role for the TIFY motif in repression of jasmonate signaling by a stabilized splice variant of the JASMONATE ZIM-domain protein JAZ10 in Arabidopsis. Plant Cell 21:131–145

Colebrook EH, Thomas SG, Phillips AL, Hedden P (2014) The role of gibberellin signaling in plant responses to abiotic stress. J Exp Biol 217:67–75

Dai C, Xue H-W (2010) Rice early flowering1, a CK1, phosphorylates DELLA protein SLR1 to negatively regulate gibberellin signaling. EMBO J 29:1916–1927

Davière J-M, Achard P (2013) Development at a glance. Gibberellin signaling in plants. Development 140:1147–1151

de Lucas M, Davière JM, Rodriguez-Falcón M, Pontin M, Iglesias-Pedraz JM, Lorrain S, Blázquez MA, Titarenko E, Prat S (2008) A molecular framework for light and gibberellin control of cell elongation. Nature 451:480–484

De Vleesschauwer D, Van Buyten E, Satoh K, Balidion J, Mauleon R, Choi I-R, Vera-Cruz C, Kikuchi S, Höfte M (2012) Brassinosteroids antagonize gibberellin- and salicylate-mediated root immunity in rice. Plant Physiol 158:1833–1846

Dill A, Jung HS, Sun TP (2001) The DELLA motif is essential for gibberellin-induced degradation of RGA. Proc Natl Acad Sci U S A 98:14162–14167

Dill A, Thomas SG, Hu J, Steber CM, Sun TP (2004) The Arabidopsis F-box protein SLEEPY1 targets gibberellin signaling repressors for gibberellin-induced degradation. Plant Cell 16:1392–1405

Dombrecht B, Xue GP, Sprague SJ, Kirkegaard JA, Ross JJ, Reid JB, Fitt GP, Sewelam N, Schenk PM, Manners JM, Karzan K (2007) MYC2 differentially modulates diverse jasmonate-dependent functions in *Arabidopsis*. Plant Cell 19:2225–2245

Feng S, Martinez C, Gusmaroli G, Wang Y, Zhou J, Wang F, Chen L, Yu L, Iglesias-Pedraz JM, Kircher S, Schafer E, Fu X, Fan LM, Deng XW (2008) Coordinated regulation of *Arabidopsis thaliana* development by light and gibberellins. Nature 451:475–479

Fernández-Calvo P, Chini A, Fernández-Barbero G, Chico JM, Gimenez-Ibanez S, Geerinck J, Eeckhout D, Schweizer F, Godoy M, Franco-Zorrilla JM, Pauwels L, Witters E, Puga MI, Paz-Ares J, Goossens A, Reymond P, De Jaeger G, Solano R (2011) The Arabidopsis bHLH transcription factors MYC3 and MYC4 are targets of JAZ repressors and act additively with MYC2 in the activation of jasmonate responses. Plant Cell 23:701–715

Friml J (2003) Auxin transport – shaping the plant. Curr Opin Plant Biol 6:7–12

Fu X, Harberd NP (2003) Auxin promotes *Arabidopsis* root growth by modulating gibberellin response. Nature 421:740–743

Gomi K, Sasaki A, Itoh H, Ueguchi-Tanaka M, Ashikari M, Kitano H, Matsuoka M (2004) GID2, an F-box subunit of the SCF E3 complex, specifically interacts with phosphorylated SLR1 protein and regulates the gibberellin-dependent degradation of SLR1 in rice. Plant J 37:826–834

Griffiths J, Murase K, Rieu I, Zentella R, Zhang ZL, Powers SJ, Gong F, Phillips AL, Hedden P, Sun TP, Thomas SG (2006) Genetic characterization and functional analysis of the GID1 gibberellin receptors in *Arabidopsis*. Plant Cell 18:3399–3414

Hartweck LM (2006) Rice GIBBERELLIN INSENSITIVE DWARF1 is a gibberellin receptor that illuminates and raises questions about GA signaling. Plant Cell 18:278–282

Hauvermale AL, Ariizumi T, Steber CM (2012) Gibberellin signaling: a theme and variations on DELLA repression. Plant Physiol 160:83–92

Hirano K, Asano K, Tsuji H, Kawamura M, Mori H, Kitano H, Ueguchi-Tanaka M, Matsuoka M (2010) Characterization of the molecular mechanism underlying gibberellin perception complex formation in rice. Plant Cell 22:2680–2696

Hirano K, Kouketu E, Katoh H, Aya K, Ueguchi-Tanaka M, Matsuoka M (2012) The suppressive function of the rice DELLA protein SLR1 is dependent on its transcriptional activation activity. Plant J 71:443–453

Hong GJ, Xue XY, Mao YB, Wang LJ, Chen XY (2012) Arabidopsis MYC2 interacts with DELLA proteins in regulating sesquiterpene synthase gene expression. Plant Cell 24:2635–2648

Hou X, Hu WW, Shen L, Lee IY, Tao Z, Han JH, Yu H (2008) Global identification of DELLA target genes during *Arabidopsis* flower development. Plant Physiol 147:1126–1142

Hou X, Lee LYC, Xia K, Yan Y, Yu H (2010) DELLAs modulate jasmonate signaling via competitive binding to JAZs. Dev Cell 19:884–894

Howe GA (2010) Ubiquitin ligase-coupled receptors extend their reach to jasmonate. Plant Physiol 154:471–474

Itoh H, Ueguchi-Tanaka M, Sato Y, Ashikari M, Matsuoka M (2002) The gibberellin signaling pathway is regulated by the appearance and disappearance of SLENDER RICE1 in nuclei. Plant Cell 14:57–70

Itoh H, Ueguchi-Tanaka M, Matsuoka M (2008) Molecular biology of gibberellins in higher plants. Int Rev Cell Mol Biol 268:191–221

Iuchi S, Suzuki H, Kim Y-C, Iuchi A, Kuromori T, Ueguchi-Tanaka M, Asami T, Yamaguchi I, Matsuoka M, Kobayashi M, Nakajima M (2007) Multiple loss of function of Arabidopsis gibberellin receptor AtGID1s completely shuts down a gibberellin signal. Plant J 50:958–966

Kazan K, Manners JM (2013) MYC2: the master in action. Mol Plant. doi:10.1093/mp/sss128

Laurie-Berry N, Joardar V, Street JH, Kunkel BN (2006) The *Arabidopsis thaliana JASMONATE INSENSITIVE 1* gene is required for suppression of salicylic acid-dependent defenses during infection by *Pseudomonas syringae*. Mol Plant Microbe Interact 19:789–800

Lechner E, Achard P, Vansiri A, Potuschak T, Genschik P (2006) F-box proteins everywhere. Curr Opin Plant Biol 9:631–638

Liu P-P, von Dahl CC, Klessig DF (2011a) The extent to which methyl salicylate is required for signaling systemic acquired resistance is dependent on exposure to light after infection. Plant Physiol 157:2216–2226

Liu P-P, von Dahl CC, Park S-W, Klessig DF (2011b) Interconnection between methyl salicylate and lipid-based long-distance signaling during the development of systemic acquired resistance in Arabidopsis and tobacco. Plant Physiol 155:1762–1768

Lorenzo O, Chico JM, Sánchez-Serrano JJ, Solano R (2004) JASMONATE-INSENSITIVE1 encodes a MYC transcription factor essential to discriminate between different jasmonate-regulated defense responses in Arabidopsis. Plant Cell 16:1938–1950

McGinnis KM, Thomas SG, Soule JD, Strader LC, Zale JM, Sun TP, Steber C (2003) The Arabidopsis SLEEPY1 gene encodes a putative F-box subunit of an SCF E3 ubiquitin ligase. Plant Cell 15:1120–1130

Murase K, Hirano Y, Sun T, Hakoshima T (2008) Gibberellin-induced DELLA recognition by gibberellin receptor GID1. Nature 456:459–463

Nakajima M, Shimada A, Takashi Y, Kim YC, Park SH, Ueguchi-Tanaka M, Suzuki H, Katoh E, Iuchi S, Kobayashi M, Maeda T, Matsuoka M, Yamaguchi I (2006) Identification and characterization of *Arabidopsis* gibberellin receptors. Plant J 46:880–889

Naseem M, Dandekar T (2012) The role of auxin-cytokinin antagonism in plant-pathogen interactions. PLoS Pathog 8(11):e1003026. doi:10.1371/journal ppat 1003026

Naseem M, Philippi N, Hussain A, Wangorsch G, Ahmed N, Dandekar T (2012) Integrated systems view on networking by hormones in *Arabidopsis* immunity reveals multiple crosstalk by cytokinin. Plant Cell 24:1793–1814

Návarová H, Bernsdorff F, Döring A-C, Zeier J (2012) Pipecolic acid, an endogenous mediator of defense amplification and priming, is a critical regulator of inducible plant immunity. Plant Cell 24:5123–5141

Navarro L, Dunoyer P, Jay F, Arnold B, Dharmasiri N, Estelle M, Voinnet O, Jones JDJ (2006) A plant miRNA contributes to antibacterial resistance by repressing auxin signaling. Science 312:436–439

Navarro L, Bari R, Achard P, Lison P, Nemri A, Harberd NP, Jones JDG (2008) DELLAs control plant immune responses by modulating the balance of jasmonic acid and salicylic acid signaling. Curr Biol 18:650–655

Niu Y, Figueroa P, Browse J (2011) Characterization of JAZ-interacting bHLH transcription factors that regulate jasmonate responses in Arabidopsis. J Exp Bot 62:2143–2154

Olszewski N, Sun T-P, Gubler F (2002) Gibberellin signaling: biosynthesis, catabolism, and response pathways. Plant Cell 14:S61–S80

Pauwels L, Goossens A (2011) The JAZ proteins: a crucial interface in the jasmonate signaling cascade. Plant Cell 23:3080–3100

Pieterse CMJ, van der Does D, Zamioudis C, Leon-Reyes A, van Wees SCM (2012) Hormonal modulation of plant immunity. Annu Rev Cell Dev Biol 28:489–521

Qi T, Huang H, Wu D, Qi Y, Song S, Xie D (2014) *Arabidopsis* DELLA and JAZ proteins bind the WD-repeat/bHLH/MYB complex to modulate gibberellin and jasmonate signaling synergy. Plant Cell 26:1118–1133

Qin X, Liu JH, Zhao WS, Chen XJ, Guo ZJ, Peng YL (2013) Gibberellin 20-oxidase gene *OsGA20ox3* regulates plant stature and disease development in rice. Mol Plant Microbe Interact 26:227–239

Robert-Seilaniantz A, Navarro L, Bari R, Jones JDG (2007) Pathological hormone imbalances. Curr Opin Plant Biol 10:372–379

Robert-Seilaniantz A, Grant M, Jones JDG (2011) Hormone crosstalk in plant disease and defense: more than just jasmonate-salicylate antagonism. Annu Rev Phytopathol 49:317–343

Schwechheimer C (2012) Gibberellin signaling in plants – the extended version. Front Plant Sci 2:Article 2. doi:10.3389/ftpls.2011.00107

Shah J, Zeier J (2013) Long-distance communication and signal amplification in systemic acquired resistance. Front Plant Sci 4:Article 30. doi:10.3389/fpls.2013.00030

Sheard LB, Tan X, Mao H, Withers J, Ben-Nissan G, Hinds TR, Kobayashi Y, Hsu FF, Sharon M, Browse J, He SY, Rizo J, Howe GA, Zheng N (2010) Jasmonate perception by inositol phosphate-potentiated COI1-JAZ co-receptor. Nature 468:400–405

Shimada A, Ueguchi-Tanaka M, Sakamoto T, Fujioka S, Takatsuto S, Yoshida S, Sazuka T, Ashikari M, Matsuoka M (2006) The rice SPINDLY gene functions as a negative regulator of gibberellin signaling by controlling the suppressive function of the DELLA protein, SLR1, and modulating brassinosteroid synthesis. Plant J 48:390–402

Shimada A, Ueguchi-Tanaka M, Nakatsu T, Nakajima M, Naoe Y, Ohmiya H, Kato H, Matsuoka M (2008) Structural basis for gibberellin recognition by its receptor GID1. Nature 456:520–523

Silverstone AL, Tseng TS, Swain SM, Dill A, Jeong SY, Olszewski NE, Sun TP (2007) Functional analysis of SPINDLY in gibberellin signaling in *Arabidopsis*. Plant Physiol 143:987–1000

Sun X, Jones WT, Harvey D, Edwards PJ, Pascal SM, Kirk C, Considine T, Sheerin DJ, Rakonjac J, Oldfield CJ, Xue B, Dunker AK, Uversky VN (2010) N-terminal domains of DELLA proteins are intrinsically unstructured in the absence of interaction with GID1/gibberellic acid receptors. J Biol Chem 285:11557–115571

Sun X, Jones WT, Rikkerink EH (2012) GRAS proteins: the versatile roles of intrinsically disordered proteins in plant signaling. Biochem J 442:1–12

Thilmony R, Underwood W, He SY (2006) Genome-wide transcriptional analysis of the *Arabidopsis thaliana* interaction with the plant pathogen *Pseudomonas syringae* pv. *tomato* DC3000 and the human pathogen *Escherichia coli* O157:H7. Plant J 46(1):34–53

Thomas SG, Sun T-P (2004) Update on gibberellin signaling. A tale of the tall and the short. Plant Physiol 135:668–676

Ueguchi-Tanaka M, Ashikari M, Nakajima M, Itoh H, Katoh E, Kobayashi M, Chow TY, Hsing YJ, Kitano H, Yamaguchi I, Matsuoka M (2005) *GIBBERELLIN INSENSITIVE DWARF1* encodes a soluble receptor for gibberellin. Nature 437:693–698

Ueguchi-Tanaka M, Nakajima M, Katoh E, Ohmiya H, Asano K, Saji S, Hongyu X, Ashikari M, Kitano H, Yamaguchi I, Matsuoka M (2007) Molecular interactions of a soluble gibberellin receptor, GID1, with a rice DELLA protein, SLR1, and gibberellin. Plant Cell 19: 2140–2155

Ueguchi-Tanaka M, Hirano K, Hasegawa Y, Kitano H, Matsuoka M (2008) Release of the repressive activity of rice DELLA protein SLR1 by gibberellin does not require SLR1 degradation in the *gid2* mutant. Plant Cell 20:2437–2446

Vidhyasekaran P (2014) PAMP signals in plant innate immunity: signal perception and transduction. Springer, Dordrecht, pp 442

Wager A, Browse J (2012) Social network: JAZ protein interactions expand our knowledge of jasmonate signaling. Front Plant Sci 3:Article 41

Wasternack C, Hause B (2013) Jasmonates: biosynthesis, perception, signal transduction and action in plant stress response, growth and development. An update to the 2007 review in *Annals of Botany*. Ann Bot 111:1021–1058

Wen CK, Chang C (2002) Arabidopsis RGL1 encodes a negative regulator of gibberellin responses. Plant Cell 14:67–100

Wild M, Davière JM, Cheminant S, Regnault T, Baumberger N, Heintz D, Baltz R, Genschik P, Achard P (2012) The Arabidopsis DELLA RGA-LIKE3 is a direct target of MYC2 and modulates jasmonate signaling responses. Plant Cell 24:3307–3319

Willige BC, Ghosh S, Nill C, Zourelidou M, Dohmann EM, Maier A, Schwechheimer C (2007) The DELLA domain of *GA INSENSITIVE* mediates the interaction with the *GA INSENSITIVE DWARF1A* gibberellin receptor of *Arabidopsis*. Plant Cell 19:1209–1220

Willige BC, Isono E, Richter R, Zourelidon M, Schwechheimer C (2011) Gibberellin regulates PIN-FORMED abundance and is required for auxin transport-dependent growth and development in *Arabidopsis thaliana*. Plant Cell 23:2184–2195

Xia Y, Navarre D, Seebold K, Kachroo A, Kachroo P (2010) The *glabra1* mutation affects cuticle formation and plant responses to microbes. Plant Physiol 154:833–846

Xu Y-L, Li L, Gage DA, Zeevaart JAD (1999) Feedback regulation of *GA5* expression and metabolic engineering of gibberellin levels in Arabidopsis. Plant Cell 11:927–935

Xu YH, Zhu YY, Zhou HC, Li Q, Sun ZX, Liu YG, Lin HX, He ZH (2004) Identification of a 98-kb DNA segment containing the rice *Eui* gene controlling uppermost internode elongation, and construction of a TAC transgene sublibrary. Mol Genet Genomics 272:149–155

Yamaguchi S (2008) Gibberellin metabolism and its regulation. Annu Rev Plant Biol 59:225–251

Yang DL, Li Q, Deng Y-W, Lou Y-G, Wang M-Y, Zhou G-X, Zhang Y-Y, He Z-H (2008) Altered disease development in the *eui* mutants and *Eui* overexpressors indicates that gibberellins negatively regulate rice basal disease resistance. Mol Plant 1:528–537

Yang DL, Yao J, Mei CS, Tong XH, Zeng LJ, Li Q, Xiao LT, Sun TP, Li J, Deng W, Lee CM, Thomashow MF, Yang Y, He Z, He SY (2012) Plant hormone jasmonate prioritizes defense over growth by interfering with gibberellin signaling cascade. Proc Natl Acad Sci U S A 109:E1192–E1200

Yang D-L, Yang Y, He Z (2013) Roles of plant hormones and their interplay in rice immunity. Mol Plant 6:675–683

Zhang ZL, Ogawa M, Fleet CM, Zentella R, Hu J, Heo JO, Lim J, Kamiya J, Yamaguchi S, Sun TP (2011) Scarecrow-like3 promotes gibberellin signaling by antagonizing master growth repressor DELLA in *Arabidopsis*. Proc Natl Acad Sci U S A 108:2160–2165

Zhu S, Gao F, Cao X, Chen M, Ye G, Wei C, Li Y (2005) The rice dwarf virus P2 protein interacts with *ent*-kaurene oxidases in vivo, leading to reduced biosynthesis of gibberellins and rice dwarf symptoms. Plant Physiol 139:1935–1945

Zhu Y, Nomura T, Xu Y, Zhang Y, Peng Y, Mao B, Hanada A, Zhou H, Wang R, Li P, Zhu X, Mander LN, Kamiya Y, Yamaguchi S, He Z (2006) ELONGATED UPPERMOST INTERNODE encodes a cytochrome P450 monooxygenase that epoxidizes gibberellins in a novel deactivation reaction in rice. Plant Cell 18:442–456

Chapter 9
Brassinosteroid Signaling in Plant Immune System

Abstract Brassinosteroids (BRs) are growth-promoting steroidal hormones in plants; they are also involved in plant innate immunity. BR signals are perceived by the plasma membrane receptors BRI1 and co-receptor BAK1. Several positive (BSK1, BSU1, PP2A, CDG1) and negative (BKI1, BIN2, MSBP1, and 14-3-3) regulators of BR signaling control the activities of BZR1 and BES1 family of transcription factors, which regulate the expression of hundreds to thousands of genes for various BR responses. BRs either positively or negatively regulate plant innate immunity. Pathogen infection results in elevation of BR signal processing. BR triggers disease resistance against some pathogens. BAK1 is a key component in BR signaling pathway. It may be involved in triggering plant disease resistance by modulating JA signaling system. BR may also induce susceptibility. BR negates disease conferred by the SA synthetic analog benzothiadiazole, suggesting negative crosstalk between BR and SA signaling pathways. BR-mediated suppression of SA defenses occurs downstream of SA biosynthesis, but upstream of NPR1 and OsWRKY45 in the SA signaling pathway. BR triggers the expression of GA repressor proteins and suppresses GA-induced defense responses. BR signaling may also negatively regulate PAMP-triggered immunity. Enhanced BR signaling resulting from either a gain-of-function mutation in BAK1, ectopic expression of BRI1, or application of BR impedes immunity triggered by the PAMP. Crosstalk between PAMP–PRR signaling and BR synthesis pathway has been reported. BRI1–BAK signaling modulates PAMP–PRR signaling pathway. Increased BR signaling triggered by BRI1 overexpression antagonized the activities of the PAMPs/MAMPs flg22, elf18, and PGN that increased BR signaling. The bHLH transcription factor HBI1 is a positive regulator of BR-triggered resistance. HBI1 has been found to be a negative regulator of PAMP-triggered immunity (PTI). *HBI1* overexpression led to reduced PAMP-triggered responses. This inhibition was found to be correlated with reduced steady-state expression of immune marker genes, leading to increased susceptibility to *Pseudomonas syringae* in *Arabidopsis thaliana*. Activation of the BRI1 pathway leads to inhibition of PAMP-triggered immunity (PTI). Pathogens may exploit BRs as virulence factors and hijack the plant BR machinery to cause disease. An oomycete pathogen has been shown to co-opt the plant BR machinery as a decoy strategy to tap into the immune signaling circuitry and interfere with effectual SA- and GA-controlled defenses.

P. Vidhyasekaran, *Plant Hormone Signaling Systems in Plant Innate Immunity*, Signaling and Communication in Plants 2, DOI 10.1007/978-94-017-9285-1_9

9.1 Brassinosteroids Modulate Plant Immune Responses

Brassinosteroids (BRs) are a group of plant steroidal hormones that regulate various aspects of plant growth and development (Divi and Krishna 2009, 2010; Divi et al. 2010; Kim and Wang 2010; Sun et al. 2010; Vriet et al. 2012), and they are also involved in plant innate immunity (Nakashita et al. 2003; Bajguz and Hayat 2009; Jaillais et al. 2011b; Albrecht et al. 2012; Owens et al. 2012; Wang 2012). BRs are growth-promoting steroidal hormones in plants (Kim and Wang 2010). Brassinosteroids are C_{27}, C_{28}, and C_{29} steroids depending on their C-24 alkyl substituents. BRs carry an oxygen moiety at the C-3 position in combination with others at the C-2, C-6, C-22, or C-23 positions (Bishop and Koncz 2002). Brassinolide (BL) is the first brassinosteroid isolated from *Brassica napus* (Grove et al. 1979) and since then more than 50 natural analogs, collectively called brassinosteroids (BRs), have been characterized from various plants (Fujioka 1999; Shimada et al. 2001, 2006). BL is the most biologically active C_{28} and, together with its C_{28} congenes, is widely distributed in the plant kingdom (Clouse and Sasse 1998). Brassinosteroids regulate several processes through a receptor-kinase signaling pathway that controls nuclear BZR transcription factors (Sun et al. 2010).

Brassinosteroids either positively (Khripach et al. 2000; Nakashita et al. 2003) or negatively (Albrecht et al. 2012; Belkhadir et al. 2012; De Vleeschauwer et al. 2012; Nahar et al. 2013) regulate plant innate immunity. Brassinolide (BL) treatment triggers defense responses against viral (*Tobacco mosaic virus*), bacterial (*Pseudomonas syringae* pv. *tabaci*), and the fungal (*Oidium* sp.) pathogens in tobacco (Nakashita et al. 2003). BL treatment induced resistance against the rice blast pathogen *Magnaporthe oryzae* and the bacterial blight pathogen *Xanthomonas oryzae* pv. *oryzae* (Nakashita et al. 2003). Brassinosteroids activate the expression of several immunity-related genes (Li et al. 2009; Divi et al. 2010). However, there are also reports that BR signaling negatively regulates disease resistance (Belkhadir et al. 2012; De Vleeschauwer et al. 2012). BR induces susceptibility against *Pythium graminicola* in rice. *P. graminicola* exploits BRs as virulence factors and hijacks the rice BR machinery to cause disease (De Vleeschauwer et al. 2012). Brassinosteroids have been shown to modulate plant immunity at multiple levels (Albrecht et al. 2012; Wang 2012).

9.2 Biosynthesis of Brassinosteroids

9.2.1 *Pathogen Triggers Brassinosteroid (BR) Biosynthesis*

Pathogen infection induces BR biosynthesis and BR signaling pathways in disease resistant or susceptible plants. The levels of castasterone and deoxocastasterone were found to increase in tobacco leaves inoculated with *Tobacco mosaic virus* (Nakashita et al. 2003; Fig. 9.1), and castasterone and deoxocastasterone are the important intermediates in BR biosynthesis pathway.

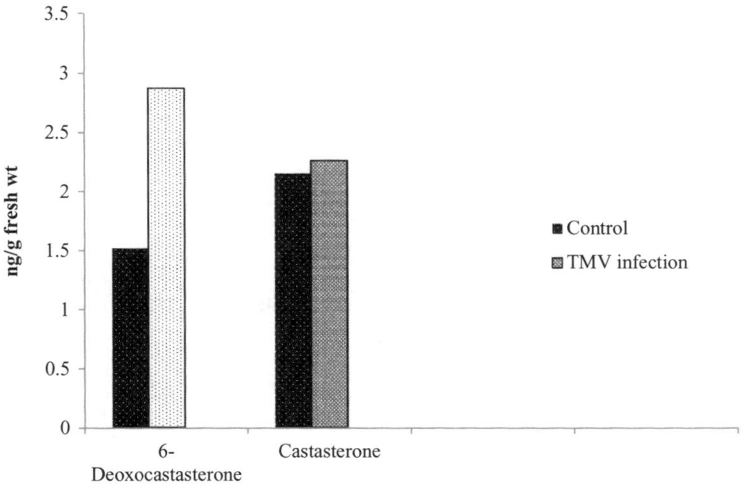

Fig. 9.1 Accumulation of BR biosynthetic pathway intermediates in tobacco leaves inoculated with *Tobacco mosaic virus* (Adapted from Nakashita et al. 2003)

Pythium graminicola infection induced a strong up-regulation of the BR biosynthetic genes *OsDWARF2* and *CONSTITUTIVE PHOTOMORPHOGENESIS AND DWARFISM1* (*OsCPD1*) with mRNA levels peaking at 7–10 times the levels found in noninoculated controls (De Vleeschauwer et al. 2012). Expression of the BR-inducible gene *BRASSINOLIDE ENHANCER2* (*OsBLE2*) responded strongly to *Pythium graminicola* infection and peaked at approximately 55 times the levels found in noninoculated controls. The pathogen-induced transcription of *OSBLE2* was even more pronounced in brassinolide-pretreated plants (De Vleeschauwer et al. 2012). These studies suggest that the oomycete *Pythium* infection triggered BR biosynthesis, BR homeostasis, and BR signaling. *Pythium* infection also showed negative feedback regulation of BR biosynthesis (BR homeostasis) (De Vleeschauwer et al. 2012).

9.2.2 BR Biosynthesis via Mevalonate Pathway

Brassinosteroids are synthesized via mevalonate pathway (Fig. 9.1). After glycolysis, pyruvic acid is oxidatively converted into acetic acid. The acetic acid then combines with coenzyme A to produce 2-acetyl-CoA. Acetoacetyl-CoA thiolase (AACT) catalyzes the condensation of two acetyl-CoA to form acetoacetyl-CoA. (*S*)-3-Hydroxy-3-methylglutaryl-CoA is synthesized by the action of hydroxy-3-methylglutaryl coenzyme A synthase (HMGS) from acetoacetyl-CoA. Mevalonate is synthesized from (*S*)-3-hydroxy-3-methylglutaryl-CoA through the action of 3-hydroxy-3-methylglutaryl coenzyme A reductase (HMGR). Mevalonate is converted to mevalonate-5-phosphate by mevalonate kinase (MVK) and then converted to mevalonate diphosphate by phosphomevalonate kinase. Another enzyme, mevalonate diphosphate decarboxylase, is involved in production of isopentenyl pyrophosphate (IPP) from mevalonate diphosphate. Dimethylallyl diphosphate is synthesized from isopentenyl pyrophosphate by the action of isopentenyl pyrophosphate isomerase. Dimethylallyl pyrophosphate (DMPP) is involved in the formation of farnesyl pyrophosphate, which is the source for synthesis of campesterol (Fig. 9.2; Bach et al. 1999; Bochar et al. 1999).

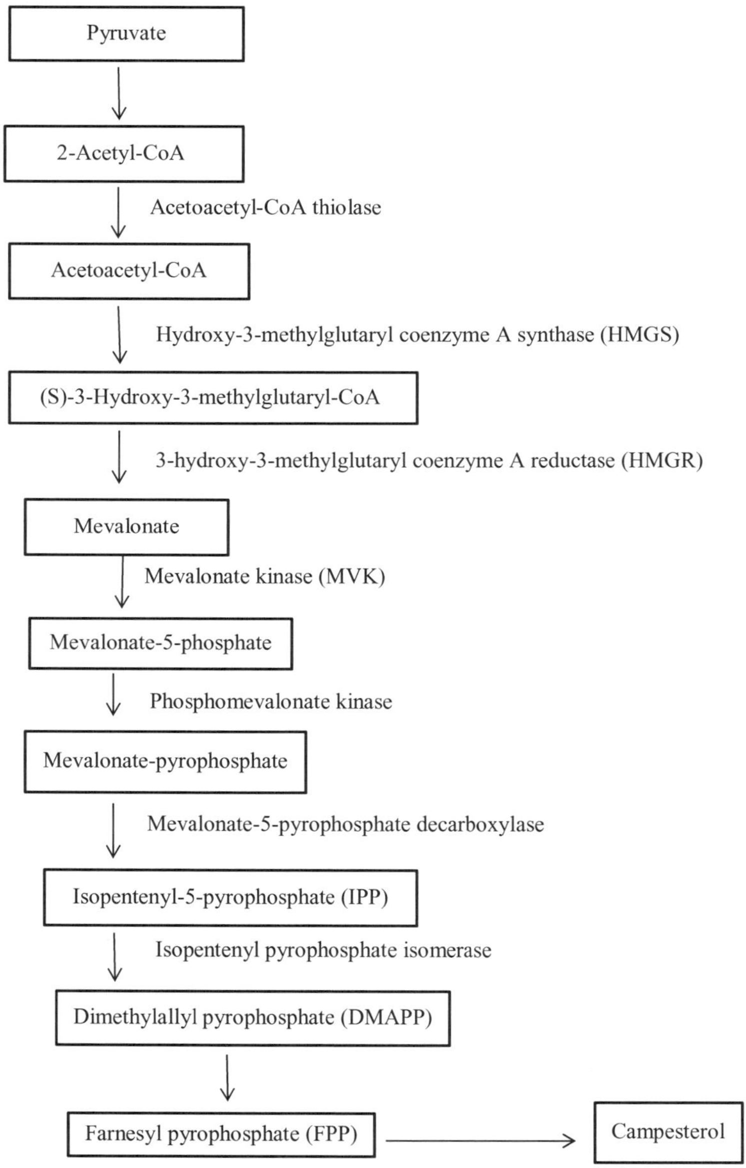

Fig. 9.2 Biosynthesis of brassinosteroids: Mevalonate pathway

9.2.3 *Early and Late C-6 Oxidation Pathways Involved in BR Biosynthesis*

Two alternative brassinosteroid biosynthetic pathways from campesterol have been reported in plants (Fig. 9.3). One is the early C-6 oxidation pathway, in which oxidation of C-6 occurs before the introduction of vicinal hydroxyls at C22 and C23 of the side chain. The other is the late C-6 oxidation pathway in which C-6 is oxidized after the introduction of hydroxyls on the side chain and the A ring (Shimada et al. 2001). The first step of the brassinosteroid biosynthesis through the C-6 oxidation pathway is the conversion of campesterol to campestanol. The biosynthetic sequence in the conversion of campesterol to campestanol has been reported to be campesterol → (24*R*)-ergost-4-en-3β-ol (4-en-3β-ol) → (24*R*)-ergost-4-en-3-one

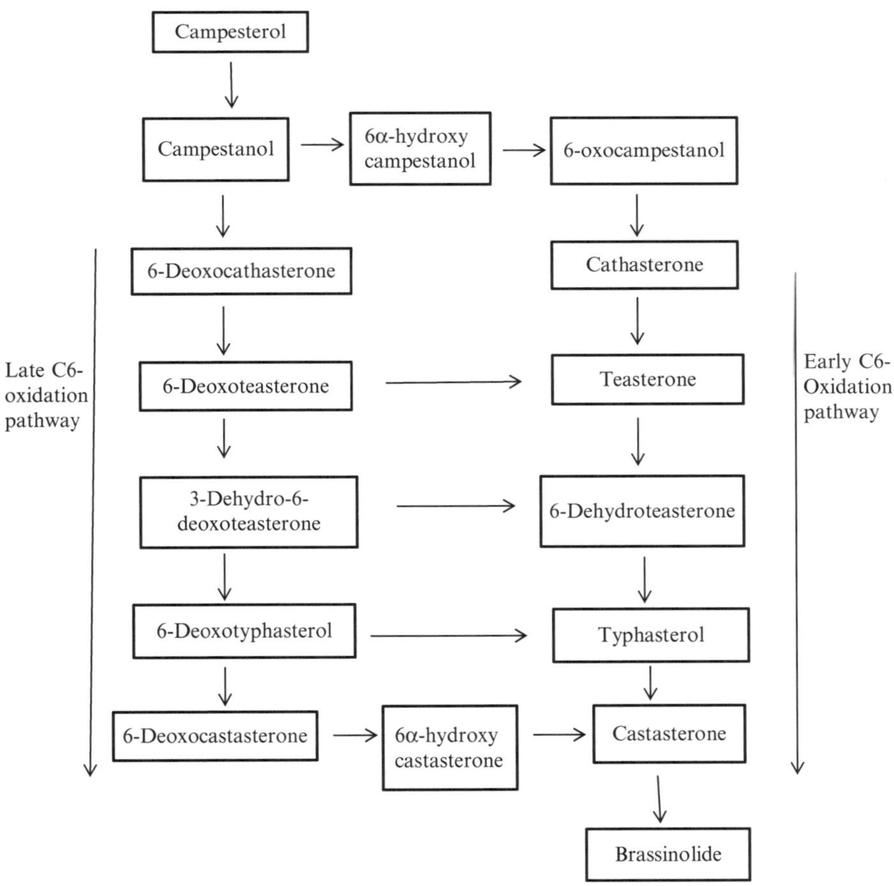

Fig. 9.3 Brassinosteroid biosynthesis: early and late C-6 oxidation pathways (Adapted from Shimada et al. 2001; Poppenberger et al. 2011)

(4-en-3-one) → (24*R*)-5α-ergostan-3-one (3-one) → campestanol (Shimada et al. 2001; Fujioka and Sakurai 1997; Fujioka et al. 2002).

Brassinosteroids synthesized in the early C-6 oxidation pathway in *Arabidopsis* include cathasterone (CT), teasterone (TE), typhasterol (Ty), castasterone (CS), and brassinolide (BL). 6-Deoxocathasterone (6-DeoxoCT), 6-deoxoteasterone (6-DeoxoTE), 6-deoxotyphasterol (6-DeoxoTY), and 6-deoxocastasterone (6-DeoxoCS) have been detected in the late C-6 oxidation pathway. Brassinolide is produced through the network of both the pathways (Noguchi et al. 2000; Shimada et al. 2001; Bishop and Koncz 2002; Poppenberger et al. 2011; Fig. 9.3). The early and late C-6 pathways are two parallel-branched brassinolide biosynthesis pathways. There is a crosstalk between these parallel pathways, implying that they are not totally autonomous and are highly networked. Both the early and late C-6 oxidation pathways are common in plant kingdom and have been reported in *Arabidopsis*, tomato, pea, and *Catharanthus roseus* (Yokota et al. 1997; Bishop et al. 1999; Noguchi et al. 1999; Nomura et al. 2001; Shimada et al. 2001; Fujioka and Sakurai 1997; Fujioka and Yokota 2003; Bishop 2007).

9.2.4 C-22 Oxidation Branch in Brassinosteroid Biosynthetic Pathway

The operation of an early C-22 oxidation branch of the brassinosteroid biosynthetic pathway has also been demonstrated. In this subpathway (22S)-22-hydroxycampesterol (22-OHCR) is converted to 6-deoxoCT via intermediates such as (22*S*,24*R*)-22-hydoxyergost-4-en-3-one (22-OH-4-en-3-one) and (22*S*,24*R*)-22-hydroxy-5α-ergostan-3-one (22-OH-3-one) (Fig. 9.4; Fujioka et al. 2002; Bak et al. 2011). Through the action of different cytochrome P450s such as CYP90A1 (CPD), CYP90B1, CYP90C1, CYP90D1, CYP85A1, and CYP85A2, different brassinosteroid intermediates such as 6-deoxocathasterone, 3-hydro-6-deoxoteasterone, 6-deoxotyphasterol, 6-deoxocastasterone, and castasterone and the most common brassinosteroid brassinolide are produced (Kim et al. 2005; Bak et al. 2011). CYP90A1 (CPD) catalyzes C-3 oxidation of the early BR intermediates (22*S*)-22-hydroxycampesterol and (22*R*,23R)-22,23-dihydroxycampesterol, as well as of 6-deoxycathasterone and 6-deoxoteasterone (Ohnishi et al. 2012).

9.2.5 Homeostasis of Brassinosteroids

Homeostasis of brassinosteroids is essential for normal growth and development in higher plants (Tanaka et al. 2005; Poppenberger et al. 2011). Biomass production and grain yield can be enhanced by controlling the level of endogenous BR (Sakamoto et al. 2006; Vriet et al. 2012). Brassinosteroid homeostasis is also a critical step in the establishment of plant immunity (Belkhadir et al. 2012).

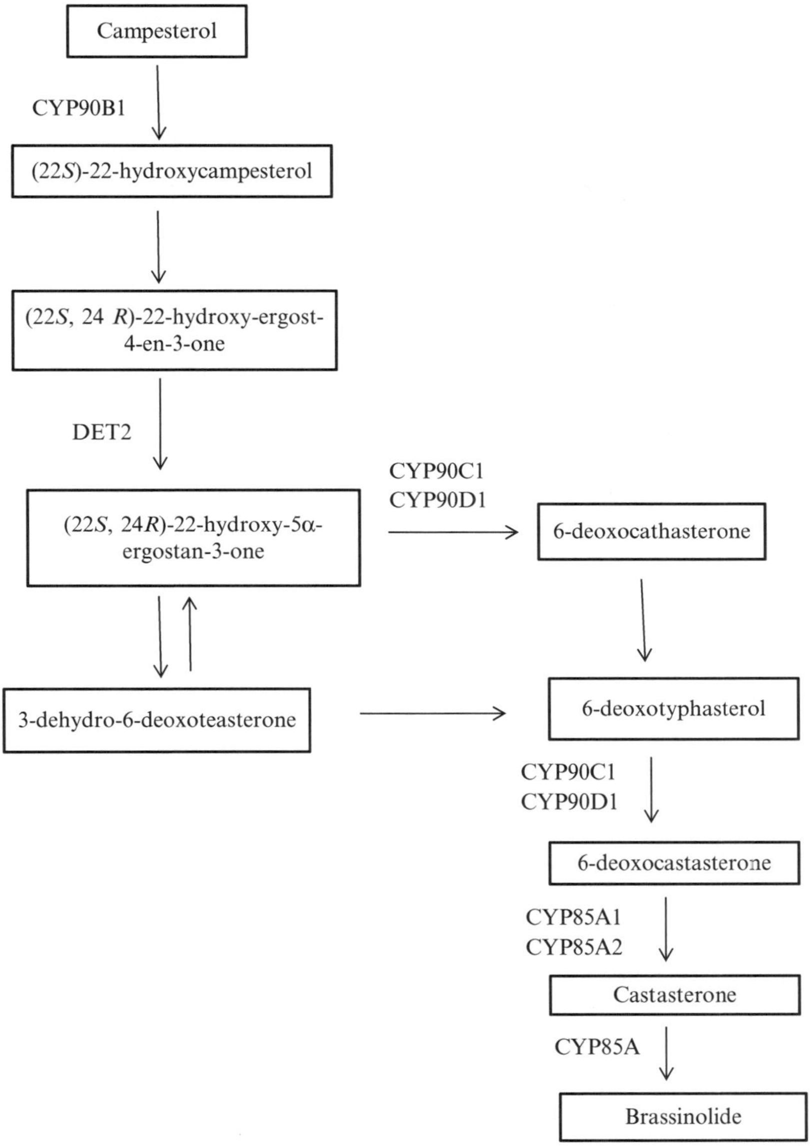

Fig. 9.4 Brassinosteroid biosynthetic pathway: C -22 oxidation branch (Adapted from Fujioka et al. 2002; Bak et al. 2011)

BR homeostasis is maintained through feedback expressions of multiple genes, each of which is involved in BR-specific biosynthesis and inactivation (Zhou et al. 2004; Tanaka et al. 2005). BR biosynthesis has to be adjusted to an optimum level for effective signal transduction and plant development. To attain BR cellular homeostasis, BR signaling is utilized to create a feedback regulatory loop that allows adjusting the expression of genes that participate in BR biosynthesis (Je et al. 2010;

Poppenberger et al. 2011). BR homeostasis is ensured by feedback expressions of multiple genes involved in its metabolism (Tanaka et al. 2005).

Five BR-specific biosynthesis genes *DEETIOLATED* (*DET2*), *DWARF4* (*DWF4*), *CONSTITUTIVE PHYTOMORPHOGENESIS AND DWARFISM* (*CPD*), *BR-6-oxidase1* (*BR6ox1*), and *ROTUNDFOLIA3* (ROT3) were up-regulated in BR-depleted wild-type *Arabidopsis* plants treated with the BR biosynthesis inhibitor brassinazole. Expressions of *FACKEL* (*FK*) and *DWARF5* (*DWF5*) genes were also induced remarkably by brassinazole, and both *FK* and *DWF5* function in sterol biosynthesis to provide precursor compounds of BR production (Tanaka et al. 2005). In contrast, in BR-excessive wild-type plants that were fed with the most active BR brassinolide, BR synthesis genes (*CPD*, *DWF4*, *BR6ox1*, *ROT3*, and *DWF7*) were down-regulated and a BR inactivation gene (*BAS1*) was up-regulated (Tanaka et al. 2005). These results strongly suggest that BR homeostasis is maintained through feedback expressions of genes involved in BR biosynthesis and inactivation. These feedback expressions of BR metabolic genes required brassinosteroid-insensitive 1 (BRI1) function (Noguchi et al. 2000; Tanaka et al. 2005).

Expressions of some metabolic genes are modulated at mRNA levels for BR homeostasis (Tanaka et al. 2005). The mRNAs of BR biosynthesis genes such as *DWF4* and *CPD* increase in response to lower amounts of endogenous BRs in BR-deficient mutants as well as in wild-type *Arabidopsis* treated with a BR biosynthesis inhibitor, brassinazole (Noguchi et al. 2000). It has been reported that BL-induced down-regulation of CPD expression is annulled in BR-insensitive *bri1* mutants in which BR reception is defective (Tanaka et al. 2005), suggesting that feedback expressions of BR metabolic genes require BR-insensitive 1 (BRI1) function. To attain BR cellular homeostasis, BR signaling is utilized to create a feedback regulatory loop that allows adjusting the expression of genes that participate in BR biosynthesis (Shimada et al. 2003; Lisso et al. 2005; Tanaka et al. 2005).

The inactivation of BRs by degradation and conjugation mechanisms contributes to the homeostasis of BRs (Nakamura et al. 2005; Je et al. 2010; Poppenberger et al. 2011). Optimal homeostasis of BRs is achieved by the balance of its signaling and biosynthesis. Signaling is initiated by binding of BR to its receptor kinase, BRI1, which eventually leads to the repression of BR biosynthetic genes. BRs activate the BRI1 receptor by stimulating its autophosphorylation (Wang et al. 2005a). The active BRI1 transmits the BR signal to the downstream genes *Bri1 Suppressor 1* (*BSU1*) and *BR INSENSITIVE2* (*BIN2*). *BSU1* encodes a phosphatase that acts as a positive regulator (Mora-Garcia et al. 2004). By contrast, *BIN2* encodes a GSK3/shaggy-like kinase and acts as a negative regulator of transcription factors (Li and Nam 2002). Bri1 Ems suppressor 1 (BES1) and brassinazole-resistant 1 (BZR1) are substrates of BIN2. BSU1 inactivates BIN2 (Kim et al. 2009). BES1 interacts with a basic helix-loop-helix transcription factor BES1-interacting Myc-like 1 (BIM1). The BES1/BIM1 complex synergistically binds to the E-box element CANNIG and activates BR-induced gene promoters (Yin et al. 2005). *BZR1* plays a dual role in feedback regulation of BR synthesis and downstream BR response (Wang et al. 2002; He et al. 2005). It enhances the expression of genes involved in cellular growth and suppresses the expression of genes involved in BR synthesis, thus producing the negative feedback suppression of BR biosynthesis.

Rice *RAVL1* gene participates in the maintenance of BR homeostasis by enhancing the expression of *BRI1* and several BR biosynthetic genes. The *RAVL1*-mediated mechanism plays a significant role ensuring the basal activities of the signaling and biosynthesis processes (Je et al. 2010). *RAVL1* is required for the expression of the BR biosynthetic genes *D2*, *D11*, and *BRD1* that are subject to BR-negative feedback. Activation of *RAVL1* was coordinated via E-box *cis*-elements in the promoters of the receptor and biosynthetic genes. RAVL1 is necessary for the response of these genes to changes in cellular BR homeostasis (Je et al. 2010).

CES, the bHLH transcription factor, is a positive regulator of BR biosynthetic gene expression. CES function is necessary for maintaining BR homeostasis. CES loss-of-function plants showed reduction in *DWF4* and *ROT3* mRNA levels. Decreased *DWF4* and *ROT3* transcript levels and an attenuated reduction of *DWF4*, *CPD*, and *ROT3* mRNA levels were observed in response to BR treatment (Poppenberger et al. 2011). CES interacts with BEE1 in vivo. CES and BEE1 act as heterodimers to regulate BR-responsive gene expression. CES is a substrate of BIN2. CES action may be controlled by BIN2 (Poppenberger et al. 2011). CES acts as a transcription factor that can bind to G-box motifs in the promoters of the cytochrome P450s CPD and CYP718 *in planta* (Poppenberger et al. 2011). *CES* localizes to the nucleus. *CES* is phosphorylated by BIN2 (Poppenberger et al. 2011). Heterodimerization of CES and BEE1 might constitute a regulatory module essential for the positive regulation of BR biosynthetic genes and other BR responses. The activity of this complex might be affected further by BR-dependent transcriptional control of *BEE1* expression as well as by posttranslational regulation via BIN2-mediated phosphorylation of CES and possibly BEE1. These regulatory switches could be essential for a fine-tuning of CES-BEE1 complex activity (Poppenberger et al. 2011).

9.3 Brassinosteroid Signaling System

9.3.1 BRI1 as a Brassinosteroid Receptor

A transmembrane receptor kinase, brassinosteroid insensitive1 (BRI1), has been implicated in brassinosteroid responses (Wang et al. 2001; Caño-Delgado et al. 2004). BRI1 functions as a receptor of brassinolide (BL), the most active brassinosteroid. The number of brassinolide-binding sites and the degree of response to brassinolide depend on the level of BRI1 protein. The BL-binding activity requires a functional BRI1 extracellular domain. Treatment of Arabidopsis seedlings with BL induces autophosphorylation of BRI1 (Wang et al. 2001). BRI1 is a receptor kinase that transduces BR signals across the plasma membrane (Wang et al. 2001).

BRI1, a leucine-rich repeat (LRR) receptor kinase localized to the plasma membrane, is a critical component of a receptor complex for brassinosteroids (Kinoshita et al. 2005). BRs directly bind to BRI1. The minimal binding domain for BRs consists of a 70-amino acid island domain (ID) located between LRR21 and LRR22 in the extracellular domain of BRI1, together with the carboxy-terminal flanking LRR (ID-LRR22). The brassinosteroids bind directly to the 94 amino acids comprising ID-LRR22 in the extracellular domain of BRI1 (Kinoshita et al. 2005).

She et al. (2011) compared the crystal structures of BRI1 in free and brassinolide-bound forms. BR was found to bind to a hydrophobicity-dominating surface groove on BRI1 (She et al. 2011). It has been suggested that brassinolide recognition by BRI1 is through an induced-fit mechanism involving stabilization of two interdomain loops that creates a nonpolar surface groove for the hormone binding (She et al. 2011).

BR1 is 1 of the 230 members of the leucine-rich repeat receptor-like kinases (LRR-RLK) in *Arabidopsis* (Li and Chory 1997). BRI1 perceives BR through its extracellular domain and transduces the signal through its cytoplasmic kinase domain (Kinoshita et al. 2005). In the absence of BR, BRI1 kinase is inactive, while BIN2, a GSK3-like kinase, phosphorylates and inactivates two homologous nuclear transcription factors called BZR1 and BZR2 (also named BES1) to repress BR-responsive gene expression (He et al. 2002; Vert and Chory 2006; Gampala et al. 2007).

Arabidopsis BRI1 receptor kinase is deactivated by autophosphorylation within glycine-rich loop (Oh et al. 2012). Serine-891 has been identified as a potential site of autophosphorylation. Autophosphorylation of serine-891 is one of the deactivation mechanisms that inhibit BRI1 activity and BR signaling in vivo. Transgenic plants expressing the BRI1 (S891A)-Flag-directed mutant accumulate higher levels of the unphosphorylated form of the BES1 transcription factor in response to exogenous BL (brassinolide) (Oh et al. 2012).

9.3.2 BRL1 as an Additional BR Receptor

Besides BRI1, another leucine-rich repeat receptor-like kinase has been reported to be involved in BR signaling in *Arabidopsis thaliana* (Zhou et al. 2004). The BRI1-like receptor kinase (named as BRL1) is a BR signal transduction regulator. BRL1 encodes a leucine-rich repeat receptor-like protein kinase (LRR-RLK). Sequence alignment revealed that BRL1 is closely related to BRI1, and it is also involved in BR perception. However, both the expression level and in vitro kinase autophosphorylation activity of BRL1 are significantly lower than those of BRI. It suggests that BRL1 is functionally redundant with BRI1 in regulating brassinosteroid signaling in *Arabidopsis* (Zhou et al. 2004).

9.3.3 *BAK1 Acts as a Co-receptor in BR Signal Reception*

BAK1 (BRI1-associated receptor kinase 1) is a serine/threonine protein kinase (Li et al. 2002). BAK1 is also called SERK3 (for Somatic EMBRYO RECEPTOR KINASE3) and it belongs to the LRR-receptor-like kinase (RLK). BAK1 belongs to the LRR type II SERK subfamily (Chinchilla et al. 2007, 2009; Heese et al. 2007; Zipfel 2008). BAK1 is a component of diverse processes, including BR signaling, light responses, cell death, and plant innate immunity (Chinchilla et al. 2009).

BRs activate BRI1- and BR-dependent activation of BRI1 precedes association with BAK1 *in planta* (Wang et al. 2008). BRI1 positively regulates BAK1 phosphorylation levels in vivo. BRI1 transphosphorylates BAK1 on specific kinase-domain residues critical for BAK1 function. BAK1 also transphorylates BRI1, thereby quantitatively increasing BRI1 kinase activity toward a specific substrate. Wang et al. (2008) proposed a sequential transphosphorylation model in which BRI1 controls signaling specificity by direct BR binding followed by substrate phosphorylation. The co-receptor BAK1 is then activated by BRI1-dependent transphosphorylation and subsequently enhances signaling output through reciprocal BRI1 transphosphorylation. The sequential transphosphorylation of the BRI1/BAK1 receptor kinase complex modulates early events in brassinosteroid signaling. Schulze et al. (2010) showed that brassinolide induces BAK1 phosphorylation within seconds. It is suggested that several LRR-RKs form tight complexes with BAK1 almost instantaneously after ligand binding and the subsequent phosphorylation events are key initial steps in signal transduction (Schulze et al. 2010). Perception of brassinosteroids requires an active complex of BRI1 and BAK1 (BR1-ASSOCIATED KINASE 1). Recognized by the extracellular LRR domain of BRI1, brassinosteroids induce a phosphorylation-mediated cascade to regulate gene expression (She et al. 2011).

BRI1/BAK1 is a receptor kinase pair mediating BR signaling (Nam and Li 2002). BAK1 and BRI1 share similar gene expression and subcellular localization patterns and physically associate with each other in plants. Overexpression of the BAK1 gene leads to a phenotype reminiscent of BRI1-overexpression transgenic plants and rescues a weak *bri1* mutation (Nam and Li 2002). These results suggest that BAK1 and BRI1 function together to mediate plant BR signaling.

BRI1 and BAK1 are BR-dependently phosphorylated. Phosphorylated forms of the two proteins persist for different lengths of time. Mutations of either protein abolished phosphorylation of the counterpart protein, implying transphosphorylation of the receptor kinases (Yun et al. 2009). Some specific amino acids have been found to be critical for formation of the receptor complex and activation of BAK1 kinase activity (Yun et al. 2009). L32E (Leucine32Glutamic acid) and L46E (Leucine46Glutamic acid) substitutions in BAK1 resulted in a loss of binding of BAK1 to BRI1, suggesting that leucine is essential for binding of BAK1 to BRI1. Similarly, threonine T455A (Threonine455Alanine) was found to be essential for the kinase activity of BAK1 (Yun et al. 2009). BAK1 has been shown to autophosphorylate on numerous serine/threonine sites as well as to transphosphorylate associated receptor kinases. Oh et al. (2010) identified Tyr-610 in the carboxy-terminal domain of BAK1 as a major site of autophosphorylation that is brassinolide-induced in vivo and requires a kinase-active BAK1.

9.3.4 Autophosphorylation of BRI1 and BAK1

The two plant receptor kinases involved in BR signaling – BRI1 and BAK1 – can autophosphorylate on tyrosines (Oh et al. 2009, 2010). BAK1 is a plant receptor kinase. Autophosphorylation of receptor kinases is recognized to be an important process in receptor kinase signaling. BAK1 was shown to autophosphorylate on Tyr residues in addition to numerous Ser/Thr residues (Oh et al. 2009, 2010). Oh et al. (2010) showed that Tyr-610 in the carboxy-terminal domain of BAK1 as a major site of autophosphorylation and phosphorylation of this residue is essential for some functions of BAK1 in vivo. In particular, the function of BAK1 as co-receptor with BRI1 in BR signaling is impaired in transgenic plants expressing the BAK1(Y610F)-Flag directed mutant. Recombinant cytoplasmic domains of BRI1 and BAK1 interact and transphorylate each other in vitro in a manner that mimics their interaction in vivo. While BAK1(Y610F) binds normally to BRI1, its ability to transphosphorylate and activate the kinase domain of BRI1 is severely compromised (Oh et al. 2010, 2011). The interaction between BAK1 and BRI1 in vitro is Mg^{2+} dependent, suggesting that cytosolic Mg^{2+} may play some role in receptor signaling in vivo (Oh et al. 2011).

BRI1 is a long-lived protein that cycles between the plasma membrane and endosomes (Geldner et al. 2007). The kinase is kept in its basal state by the C-terminal tail, which plays an autoinhibitory role, as well as by interactions of BRI1's kinase domain with an inhibitory protein, BRI1 KINASE INHIBITOR 1 (BKI1) (Wang and Chory 2006). Binding of the most active BR brassinolide (BL) in the extracellular domain causes a conformational change in the receptor that leads to autophosphorylation in several domains, including the C-terminal tail (Wang et al. 2005a, b, 2008). BRI1's kinase activity is also necessary for plasma membrane release of the inhibitory protein BKI1 (Wang and Chory 2006).

BKI1 acts through two evolutionally conserved motifs: a 20-residue conserved segment that binds the BRI1 kinase domain and a lysine–arginine-rich motif that targets BKI1 into the cytosol following ligand perception by BRI1, relieving kinase inhibition and allowing recruitment of BRI1's co-receptor, BAK1 (Jaillais et al. 2011a). A key step in BRI1 activation is the dissociation of BKI1 from the plasma membrane (PM). BRI1 is not required for BKI1 association with the PM, but BRI1 is required to release BKI1 from the PM (Wang and Chory 2006). Upon perception, BRI1 phosphorylates BKI1 on Y211 (Tyr211) to displace the kinase inhibitor into the cytosol, where it is inactive (Jaillais et al. 2011b). BKI1 has to be PM-localized to regulate BR signaling (Jaillais et al. 2011b).

9.3.5 BKI1, a Negative Regulator of BRI1 Signaling

BRs regulate dissociation of BKI1 (for BRI1 KINASE INHIBITOR 1), a negative regulator of BRI1 signaling. BRs are perceived at the plasma membrane by BRI1. A BRI1-interacting protein, BKI1, has been identified as a negative regulator of BR signaling. BRs cause the rapid dissociation of BKI1 protein from the plasma membrane in a process that is dependent on BRI1 kinase. BKI1 is a substrate of BRI1 kinase and limits the interaction of BRI1 with its co-receptor, BAK1 (Wang and Chory 2006). The results suggest that BKI1 prevents the activation of BRI1. Brassinosteroids regulate dissociation of BKI1, a negative regulator of BRI1 signaling, from the plasma membrane (Wang and Chory 2006).

BKI1 is tyrosine phosphorylated in response to brassinosteroid perception. Phosphorylation occurs within a reiterated [KR][KR] membrane targeting motif, releasing BKI1 into the cytosol and enabling formation of an active signaling complex (Jaillais et al. 2011b). BKI1 has to be PM-localized to regulate BR signaling. Release of BKI1 from the PM is triggered by tyrosine phosphorylation. BKI1 interaction with BRI, but not its inhibitory activity, is required for BKI1 localization changes and tyrosine phosphorylation (Jaillais et al. 2011b). BKI1 binding to the distal surface of the C-lobe down-regulates BRI1 activity by inhibiting its interaction with its co-receptor, BAK1. BKI1 C-terminal tail binding to BRI1 inhibits interaction with the co-receptor, BAK1 (Jaillais et al. 2011b).

9.3.6 BR Signaling Events Downstream of BR Signal Perception

BR signal transduction cascade includes BR perception by the BRI1 receptor kinase at the cell surface, activation of BRI1/BAK1 kinase complex by transphosphorylation, phosphorylation of the BSK kinases, activation of the BSU1 phosphatase, dephosphorylation and inactivation of BIN2 kinase, and accumulation of BZR transcription factors in the nucleus (Fig. 9.5; Kim and Wang 2010). The BR signaling pathway includes two receptor-like kinases (BRI1, BAK1), a plasma membrane-associated (BSK1), two phosphatases (BSU1 and PP2A), a GSK3-like kinase (BIN2), and two homologous transcription factors (BZR1 and BZR2/BES1).

Brassinolide (BL) is perceived by a cell surface receptor complex containing the receptor kinases BRI1 and BAK1, which initiates BR signaling by promoting an interaction of BSK1 with BSU1, a serine/threonine phosphatase (Kim et al. 2009). BSU1 mediates dephosphorylation and thereby inactivation of the shaggy-like kinase BIN2 that acts to suppress transcription factors of the BES1/BZR1 family, which in turn regulate the expression of BR target genes (Kim et al. 2009; Li et al. 2010). To date, BES1, BZR1, and their homologues are the only known substrates of BIN2 in plants (He et al. 2005; Yin et al. 2005; Rozhon et al. 2010). BRs bind to a receptor kinase, BRI1, at the cell surface to activate the BR response transcription factors BZR1 (Brassinozole-Resistant 1)/and BZR2 (also known as

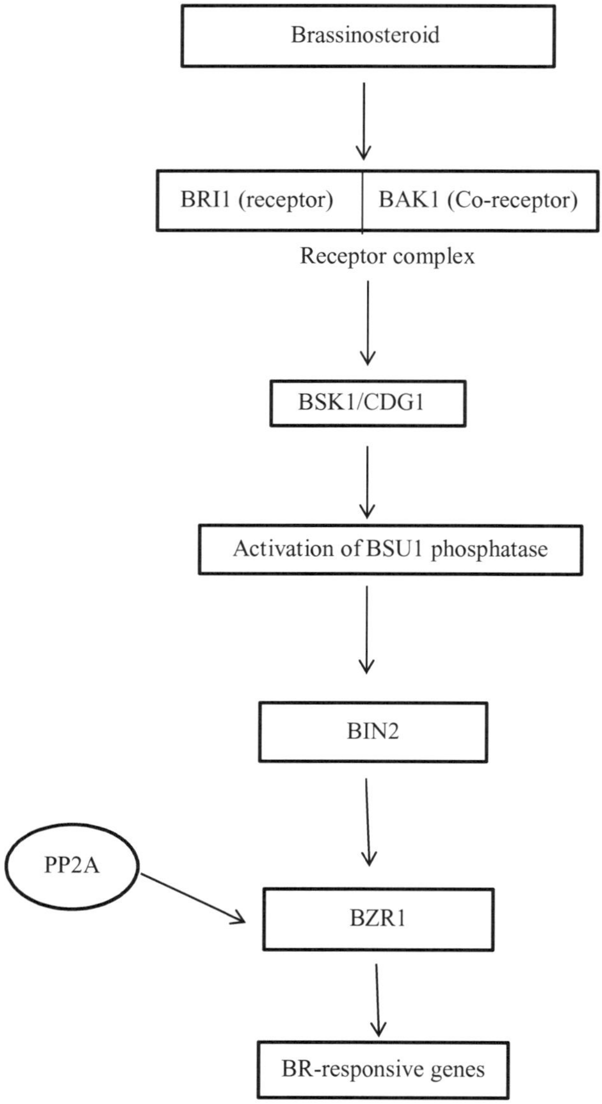

Fig. 9.5 Brassinosteroid signal transduction cascade (Adapted from Kim and Wang 2010)

BES1 [BRI1-EMS-SUPPRESSOR1]) through a signal transduction pathway. The upstream BR signaling components at the plasma membrane include BRI1 (Wang et al. 2001) and BAK1 (Li et al. 2002) receptor kinases, BIK1 that inhibits BRI1, and the plasma membrane-associated BR-signaling kinases (BSKs) (Tang et al. 2008).

Sequential phosphorylation/dephosphorylation events are involved in BR signal transduction pathway (Fig. 9.6; Sun et al. 2010; Kim et al. 2011; Tang et al. 2011; Yu et al. 2011). In plants, BR signaling relies on a phosphorylation-dependent signal transduction cascade that is initiated upon BR binding to a plasma membrane-localized receptor complex and results in the nuclear acquisition of transcription

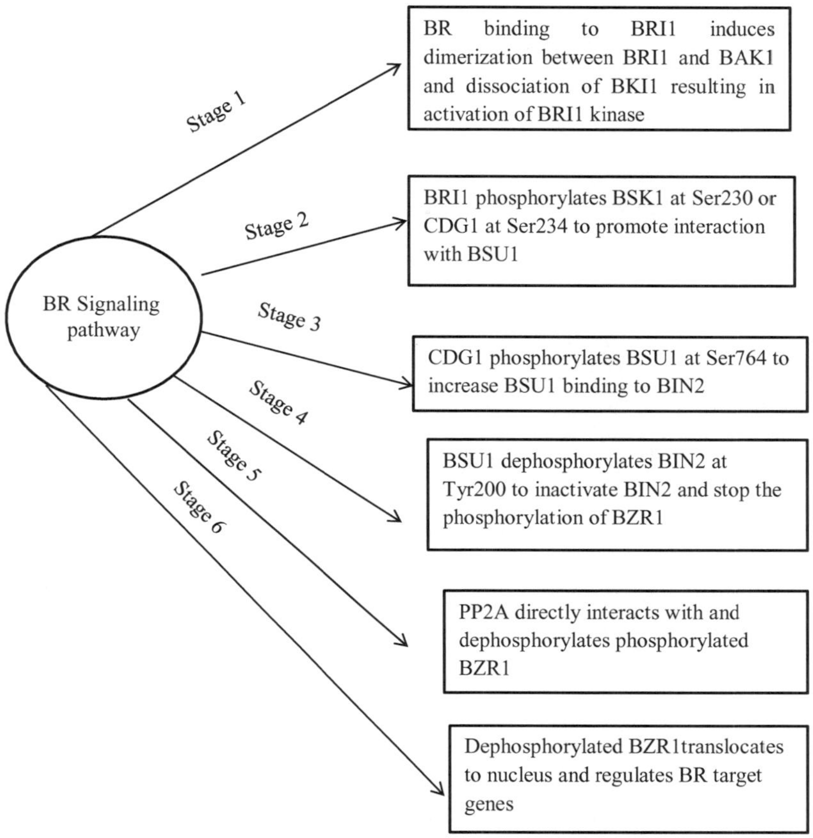

Fig. 9.6 Phosphorylation/dephosphorylation events in BR signaling pathway

factors to regulate genomic responses (Wang and He 2004; Belkhadir and Chory 2006; Poppenberger et al. 2011).

BR binding to BRI1's extracellular domain activates its kinase activity by inducing association and transphosphorylation with the co-receptor BAK1, another LRR-RLK (Li et al. 2002; Nam and Li 2002; Wang and Chory 2006; Wang et al. 2008). Activation of BRI1 also involves dissociation of BRI1 KINASE INHIBITOR 1 (BKI1) caused by tyrosine phosphorylation (Wang and Chory 2006; Jaillais et al. 2011b). BRI1 phosphorylates Ser234 to activate CDG1 kinase, and CDG1 in turn phosphorylates Ser764 to activate BSU1, which inactivates BIN2 by dephosphorylating Tyr200 of BIN2 (Kim et al. 2011). Activated BRI1 kinase initiates a signaling cascade that leads to inactivation of BIN2 and dephosphorylation of BZR1 and BZR2/BES1 by the protein phosphatase 2A (PP2A) (Tang et al. 2011). Dephosphorylated BZR1 and BZR2 accumulate in the nucleus and regulate the expression of BR target genes (He et al. 2005; Yin et al. 2005; Sun et al. 2010; Yu et al. 2011).

BRs induce dephosphorylation and accumulation of BZR1 protein. Experiments using a proteasome inhibitor, MG132, suggest that phosphorylation of BZR1 increases its degradation by the proteasome machinery (He et al. 2002). BIN2 directly interacts with BZR1 in yeast two-hybrid assays, phosphorylates BZR1 in vitro, and negatively regulates BZR1 protein accumulation in vivo. These results suggest that BIN2 phosphorylates BZR1 and targets it for degradation and that BR signaling causes BZR1 dephosphorylation and accumulation by inhibiting BIN2 activity (He et al. 2002).

BR signaling controls a wide range of target genes (Goda et al. 2002; Müssig et al. 2002). A set of basic helix-loop-helix (bHLH) transcription factors termed *BRASSINOSTEROID ENHANCED EXPRESSION* (*BEE*) is the most important set of transcription factors targeted by BR signaling. These transcription factors are rapidly up-regulated by the most active brassinosteroid, brassinolide in *Arabidopsis thaliana* (Friedrichsen et al. 2002).

Another class of BR-regulated genes includes the genes involved in feedback regulation of BR biosynthesis (Mathur et al. 1998; Bancos et al. 2002; Tanaka et al. 2005). These genes encode enzymes such as the cytochrome P450s CPD/CYP90A1, DWF4/CYP90B1, and ROT3/CYP90C1. The expression of these genes is tightly feedback controlled from the BR signaling pathway. Their transcription was strongly repressed following BR application and significantly up-regulated in response to the inhibition of BR biosynthesis (Shimada et al. 2003; Lisso et al. 2005; Tanaka et al. 2005).

9.3.7 Brassinosteroid-Signaling Kinases (BSKs)

Brassinosteroid-signaling kinases (BSKs) mediate signal transduction from the receptor kinase BRI1 in *Arabidopsis* (Tang et al. 2008). Three homologous BR-signaling kinases (BSK1, BSK2 and BSK3) have been identified in *Arabidopsis*. BSK1 is a member of the receptor-like cytoplasmic kinase (RLCK) XII family (Tang et al. 2008). RLCKs are homologous to transmembrane RLKs but contain no extracellular or transmembrane domains (Shiu et al. 2004). BSK1 is localized at the plasma membrane and is phosphorylated at Ser230 by BRI1 upon BR treatment (Tang et al. 2008). The BSKs are phosphorylated by BRI1 in vitro and interact with BRI1 in vivo.

BSKs are the substrates of BRI1 kinase that activate downstream BR signal transduction (Tang et al. 2008). In the absence of BR, BSKs are associated with BRI1. Upon BR activation of BRI1, BSKs are phosphorylated and then dissociate from the receptor complex, to activate downstream signaling (Tang et al. 2008). Such ligand-induced dissociation from a pre-existing receptor complex potentially provides faster signaling than ligand-induced recruitment of a free component into the receptor complex. Both BSKs and BRI1 are substrates of the BRI1 kinase, but they play distinct roles in BR signaling (Tang et al. 2008). First, BR induces BRI1–BAK1 interaction but reduces BRI1–BSK1 and BRI1–BSK3 interactions. Second, overexpression of

BSK3 suppresses the *bri1-116* allele, whereas overexpression of BAK1 only suppresses weak alleles but not a strong allele of *bri1* nor a double mutant containing the weak *bri-5* allele and the BR biosynthetic mutation *det2-1* (Li et al. 2002), suggesting that BSK3 functions downstream of BRI1 whereas BAK's action on downstream BR response requires a functional BRI1 (Tang et al. 2008). BAK1 most likely mediates activation of BRI1 kinase rather than signal transduction to specific downstream components in the BR signaling pathway. In contrast, the BSKs directly mediate signal transduction from BRI1 to downstream BR response (Tang et al. 2008).

Signal transduction from BRI1 to BIN2 involves the BR SIGNALING KINASE 1 (BSK1) (Tang et al. 2008) and the BRI1 SUPPRESSOR 1 (BSU1) phosphatase (Kim et al. 2009). Ser230 phosphorylation increases BSK1 binding to BSU1 (Tang et al. 2008; Kim et al. 2009). BSU1 inactivates BIN2 by dephosphorylating its Tyr200 residue and BR treatment enhances BSU1 dephosphorylation of BIN2 (Kim et al. 2009).

9.3.8 BRI1 SUPPRESSOR 1 (BSU1) Phosphatase

Perception of brassinolide (BL), the most active brassinolide by the membrane-associated receptor kinase BRI1, triggers the dephosphorylation and accumulation in the nucleus of the transcriptional modulators BES1 and BZR1. BSU1 encodes a nuclear-localized serine–threonine protein phosphatase with an N-terminal Kelch-repeat domain. BSU1 is able to modulate the phosphorylation state of BES, counteracting the action of the glycogen synthase kinase 3 BIN2, and leading to increased steady-state levels of dephosphorylated BES1. BES1 is subject to antagonistic phosphorylation and dephosphorylation reactions in the nucleus, which fine-tune the amplitude of the response to BL (Mora-Garcia et al. 2004).

In contrast, the BSU1 phosphatase is a positive regulator of BR signaling (Mora-Garcia et al. 2004). Overexpression of BSU1 increases the dephosphorylated BZR2/BES1 and activates BR responses (Mora-Garcia et al. 2004; Belkhadir et al. 2006). However, BSU1 does not interact with or effectively dephosphorylate BZR2/BES1 in vitro (Mora-Garcia et al. 2004; Belkhadir et al. 2006). It is suggested that BR induces rapid dephosphorylation of BZR1 and BZR2 by inhibiting BIN2 and/or activating BSU1.

During BR signaling in *Arabidopsis*, BSU1 phosphatase and BIN2 kinase regulate the signal intensity by determining the phosphorylation status of the transcription factors BZR1 and BES1. BIN2 and BSU1 are nucleocytoplasmic regulators that modulate the subcellular localization of BES1, with differential activities between the nucleus and the cytosol (Ryu et al. 2010b). The nuclear BIN2 induced phosphorylation and nuclear export of BES1 more efficiently than cytosolic BIN2. The cytoplasmic BSU1 mediated the dephosphorylation and nuclear translocation of BES1 more efficiently than the nuclear one. The predominance of cytosolic BSU1 and nuclear BIN2 may be required for the efficient subcellular localization of BES1 in BR signaling (Ryu et al. 2010b).

9.3.9 MSBP1 Negatively Regulates Brassinosteroid Signaling

A membrane steroid-binding protein 1 (MSBP1) has been shown to be a negative regulator of brassinosteroid signaling (Song et al. 2009). It can bind to the BR and suppress BR signaling. MSBP1 specifically interacts with the extracellular domain of BAK1 in a BR-independent manner. Suppressed BR responses by increased expression of MSBP1 could be recovered by overexpressing BAK1 or its intracellular kinase domain, suggesting that MSBP1 may suppress BR signaling through interacting with BAK. Both MSBP1 and BAK1 are localized to plasma membrane and endocytic vesicles. MSBP1 accelerates BAK1 endocytosis, which results in suppressed BR signaling by shifting the equilibrium of BAK1 toward endosomes. Enhanced MSBP1 expression reduces the interaction between BRI1 and BAK. These results suggest that MSBP1 interferes with interaction between BRI1 and BAK and not at BR–BRI1 binding stage (Song et al. 2009).

9.3.10 CDG1 in BR Signal Transduction

CONSTITUTIVE DIFFERENTIAL GROWTH1 (CDG1), a member of the RLCK VIIc family, is involved in activation of BR signaling (Muto et al. 2004). CDG1 has been shown to positively regulate BR signaling (Kim et al. 2011). CDG1 and its homolog CDL1 (for CDG1-like1) mediate signal transduction from BRI1 to BSU1 (a phosphatase) in the BR signaling system (Kim et al. 2011). BRI1 phosphorylation sites have been identified in CDG1 while CDG1 phosphorylation sites have been detected in BSU1. Mutations of these phosphorylation sites compromised the BR signaling functions (Kim et al. 2011). CDG1 phosphorylates the Ser764 residues in BSU1's phosphatase domain to increase BSU1's ability to dephosphorylate Tyr200 residue of BIN2. CDG1 is phosphorylated by BRI1 at Ser44, Ser47, and Ser234, and mutations of these phosphorylation sites compromised CDG's ability to activate BR signaling in vivo. Kim et al. (2011) demonstrated that BRI1 phosphorylates Ser234 to activate CDG1 kinase, and CDG1 in turn phosphorylates Ser764 to activate BSU1, which inactivates BIN2 (a GSK3-like kinase) by dephosphorylating Tyr200 of BIN2. In vitro assays demonstrate that BSU1 and either CDG1 or BSK1 are the minimum components required for signal transduction from BRI1 to BIN2 (Kim et al. 2011). Overexpression of CDG1 has been shown to activate the BR signal transduction pathway in *Arabidopsis* (Kim et al. 2011). CDG1 is anchored to the plasma membrane by palmitoylation and the plasma membrane localization is required for CDG1 to function in BR signal transduction. CDG1 was found to act upstream of BIN2 in the BR signaling pathway (Kim et al. 2011).

9.3.11 BIN2 Negatively Regulates BR Signaling

Glycogen synthase kinase 3 (GSK3) is a highly conserved serine/threonine protein kinase (Peng et al. 2010). BIN2 (*B*RASSINOSTEROID-*IN*SENSITIVE 2) is an *Arabidopsis* GSK3-like kinase that negatively regulates BR signaling by phosphorylating BES1 (*B*RI1 *E*MS *S*UPPRESSOR 1). BIN2 is inhibited in response to BR perception at the cell surface to relieve its inhibitory effects on downstream targets. BIN2 is regulated by proteasome-mediated protein degradation. Exogenous application of a BR synthesis inhibitor increased the amount of BIN2 proteins, while application of an active BR decreased the amount of BIN2 proteins. Treatment of a proteasome inhibitor, MG132, prevented the BR-induced BIN2 depletion and also nullified effect of BR on the BIN2 kinase activity (Peng et al. 2008). The results suggest that proteasome-mediated protein degradation constitutes an important regulatory mechanism for restricting the BIN2 activity.

BIN2 negatively regulates BR signaling by phosphorylating BES1 and BZR1 (BRASSINAZOLE-RESISTANT 1), two highly similar transcription factors critical for bringing about characteristic BR responses (Peng et al. 2010). BIN2 interacts directly with BZR1 through a 12-amino acid BIN2-docking motif adjacent to the C-terminus of BZR1. Deletion of this motif inhibits the phosphorylation and subsequent degradation of BZR1. It suggests that BIN2 utilizes a direct kinase–substrate docking mechanism to phosphorylate BZR1 and regulate its protein stability (Peng et al. 2010).

As a negative regulator of BR signaling, BIN2 phosphorylates BZR1 and BZR2 at numerous sites to inhibit their activities through multiple mechanisms (Gendron and Wang 2007). These include accelerating proteasome-mediated degradation (He et al. 2002), promoting nuclear export and cytoplasmic retention by the 14-3-3 proteins (Gampala et al. 2007; Ryu et al. 2007), and inhibiting DNA binding and transcriptional activity (Vert and Chory 2006; Gendron and Wang 2007; Gampala et al. 2007).

Brassinolide (the most active BR) treatment induces dephosphorylation of BIN2. BSU1 and its closest homolog BSL1 show manganese-dependent phosphatase activities. BSU1 inhibits BIN2 kinase activity but does not dephosphorylate pre-phosphorylated BZR1 in vitro. BSU1 directly inhibits BIN2 phosphorylation of BZR1 (Kim et al. 2009). BIN2 phosphorylation of BZR1 promotes BZR1 cytoplasmic retention by the 14-3-3 proteins while unphosphorylated BZR1 accumulates in the nucleus (Gampala et al. 2007). Co-expression of BIN2 with BZR1 increased phosphorylation and cytoplasmic retention of BZR1 (Kim et al. 2009). BR treatment induces proteasome-mediated degradation of BIN2 (Peng et al. 2008). During BR signaling in *Arabidopsis*, BSU1 phosphatase and BIN2 kinase regulate the signal intensity by determining the phosphorylation factors BZR1 and BES1 (Ryu et al. 2010b).

9.3.12 Protein Phosphatase 2A (PP2A)

A protein phosphatase 2A (PP2A) activates BR-responsive gene expression by dephosphorylating BZR1 (Tang et al. 2011). When BR levels are low, the GSK3-like kinase BIN2 phosphorylates and inactivates the BZR1 transcription factor. PP2A has been identified as a BZR1-interacting protein. Genetic analyses demonstrated a positive role for PP2A in brassinosteroid signaling and BZR1 dephosphorylation. Members of the B′ regulatory subunits of PP2A directly interact with BZR1's putative PEST domain containing the site of the *bzr1-1D* mutation. Interaction with and dephosphorylation by PP2A are enhanced by the *bzr1-1D* mutation, reduced by two intragenic *bzr1-1d* suppressor mutations, and abolished by deletion of the PEST domain. The studies reveal the crucial function of PP2A in dephosphorylating and activating BZR1 (Tang et al. 2011).

9.3.13 BZR1 and BES1 (BZR2) Transcription Factors

BR signaling triggers the activities of the BR response transcription factors BZR1 (Brassinozole-Resistant 1) and BZR2 (also known as BES1 [BRI1-EMS-SUPPRESSOR1]), which regulate large number of target genes (Yu et al. 2011). Yu et al. (2011) have identified 1,609 putative BES1 target genes. BES1 targets contribute to BR responses and interactions with other hormonal signaling pathways. BZR1 and BES1/BZR2 bind to specific promoter sequences to mediate BR-responsive gene expression (He et al. 2005; Yin et al. 2005). BIN2 and BSU1 control the phosphorylation status of the two homologous transcription factors: BZR1 and BES1/BZR2. BZR1 and BES1/BZR2 are inhibited by BIN2-mediated phosphorylation and activated by BR-induced dephosphorylation (Gampala et al. 2007).

BZR1 and BES1/BZR2 are key effectors of BR action. The dominant *bzr1-1D* and *bes1-D* mutations suppress the phenotypes of BR-deficient or insensitive mutants. BZR1 and BES1/BZR2 have been shown to directly bind to promoters of BR-responsive genes, but with different binding site sequences and transcriptional activities (Gampala et al. 2007). BZR1 has an optimal binding site of CGTG(T/C/)G sequence, which was called the BR response element (BRRE). The BRRE is conserved in the promoter of BR-repressed genes such as *CPD*, *DWF4*, *ROT3*, and *BR6OX* (He et al. 2005). BZR1 functions as a transcriptional repressor in vivo to mediate feedback inhibition of BR biosynthetic genes. In contrast, BES1/BZR2 is a transcription factor that binds to and activates BR target gene promoters. It interacts with a bHLH-type transcription factor, BIM1, to synergistically bind to E box (CANNTG) sequences present in many BR-induced promoters to regulate BR-induced genes (Yin et al. 2005).

BR acts by modulating the phosphorylation status of the two key transcription factors, BZR1 and BES1 through the action of BRASSINOSTEROID INSENSITIVE1

(BIN1)/BRI1 ASSOCIATED RECEPTOR KINASE1 (BAK1) receptors and a GSK3 kinase, BIN2 (Ryu et al. 2007). BZR1 functions as a nucleocytoplasmic shuttling protein and GSK3-like kinases induce the nuclear export of BZR1 by modulating BZR1 interaction with the 14-3-3 proteins. BR-activated phosphatase mediates rapid nuclear localization of BZR1. Besides the phosphorylation domain for 14-3-3 binding, another phosphorylation domain in BZR1 is required for the BIN2-induced nuclear export of BZR1. Mutations of putative phosphorylation and 14-3-3 recognition sites identified in BZR1 led to the constitutive nuclear localization of BZR1 and a strong BR response. The results suggest that the phosphorylation status of BZR1 is important for its subcellular distribution, which modulates BR signaling between the nucleus and cytosol (Ryu et al. 2007).

Ryu et al. (2007) identified two domains carrying putative phosphorylation sites that modulated the subcellular localization of BZR1. Positioned in the first domain, Ser-130 and Ser-134 residues are critical for the nuclear export of BZR1. The second putative phosphorylation domain of BZR1 has a short stretch of amino acid sequence that is similar to the conserved 14-3-3 recognition motif. BZR1 appears to interact with 14-3-3 $_{\kappa}$ and 14-3-3ε proteins, and Ser-173 and Thr-177 residues in the putative 14-3-3 binding domain are essential for this interaction. These results suggest that Ser-173 or Thr-177 of BZR1 is directly targeted by phosphorylation for binding to the 14-3-3 proteins (Ryu et al. 2007).

BES1 functions as a nucleocytoplasmic signal transmitter, and its subcellular localization modulates the output intensity of the BR signal (Ryu et al. 2010a). BIN2 and other group II glycogen synthase kinase 3 (GSK3)-like kinases phosphorylate BES1 and induce its nuclear export by regulating its binding affinity with 14-3-3 proteins. Twelve putative phosphorylation residues have been identified in BES1. Two of these residues, Ser171 and Thr175, have been found to be critical for interaction with 14-3-3 proteins. The other putative phosphorylation sites in the N-terminal region are required for the BIN2-mediated nuclear export of BES1. Mutations of these motifs resulted in increased nuclear accumulation and enhanced BR responses in transgenic plants (Ryu et al. 2010a). These results suggest that the spatial redistribution of BES1 is important for the regulation of the BR signaling output.

BZR1 is a nuclear component of the BR signal transduction pathway (Wang et al. 2002). A dominant mutation *bzr1-1D* suppresses BR-deficient and BR-insensitive (bri1) phenotypes and enhances feedback inhibition of BR biosynthesis. BZR1 protein accumulates in the nucleus. BZR1 is a positive regulator of the BR signaling pathway that mediates both downstream BR responses and feedback regulation of BR biosynthesis (Wang et al. 2002). BZR1 and BZR2 directly bind to DNA and regulate BR-responsive gene expression (Goda et al. 2002, 2004; Wang et al. 2002; Yin et al. 2002; He et al. 2005; Yin et al. 2005). BES1 recruits histone demethylases ELF6 and REF6 as well as a transcription elongation factor IWS1 to regulate target gene expression. Identification of BES1 and BZR1 target genes established a transcriptional network for BR response (Ye et al. 2011).

9.3.14 Function of 14-3-3 Proteins in Regulation of Activities of BZR Transcription Factors

14-3-3 proteins are phosphopeptide-binding proteins. They participate in various signal transduction and regulatory processes by interacting with diverse target proteins in a sequence-specific and phosphorylation-dependent manner (Bridges and Moorhead 2005; Jaspert et al. 2011). The proteins are also involved in the regulation of the activities of transcription factors (Sehnke et al. 2002). The 14-3-3 proteins interact specifically with BZR1 protein that has been phosphorylated by BIN2. Mutations of a BIN2 phosphorylation site in BZR1 that abolish its binding to 14-3-3 proteins result in constitutive BR-response phenotypes similar to *bzr1-1D*. Such mutant BZR1 proteins are constitutively localized in the nucleus but are not affected in either protein accumulation or DNA binding (Gampala et al. 2007). BR regulates nuclear localization of BZR1 and BZR2 proteins (Gampala et al. 2007). BR induces rapid nuclear localization of BZR1 (Gampala et al. 2007). 14-3-3 binding increases cytoplasmic localization of BZR1 and mutation of the 14-3-3 binding site increases the nuclear localization of BZR1. Phosphorylated BZR1 is retained in the cytoplasm by binding to 14-3-3 proteins, and BR-induced dephosphorylation abolishes 14-3-3 binding to allow nuclear localization and transcriptional regulation of BZR1 (Gampala et al. 2007). It has been demonstrated that 14-3-3 proteins function in BR signal transduction by regulating the subcellular localization and activity of both BZR1 and BZR2/BES1 (Gampala et al. 2007).

14-3-3 proteins play an important role in BR signal transduction. BIN2-catalyzed phosphorylation inhibits the function of the BZR1 and BZR2/BES1 proteins not only by inhibiting DNA binding but also by promoting binding to the 14-3-3 proteins, which increases cytoplasmic retention of BZR1 and BZR2/BES1 (Gampala et al. 2007). Phosphorylation can inhibit BZR1 and BZR2/BES1 through four mechanisms: increasing proteolysis, inhibiting DNA binding, promoting 14-3-3 binding, and cytoplasmic retention (Gampala et al. 2007). BIN2 phosphorylation promotes cytoplasmic retention of BZR1 by promoting 14-3-3 binding, and BR-induced dephosphorylation increases nuclear localization of BZR1 by abolishing 14-3-3 binding (Gampala et al. 2007).

The important role of OsBZR1, the rice homolog of *Arabidopsis* BZR1, and 14-3-3 proteins in brassinosteroid signal transduction in rice has been demonstrated (Bai et al. 2007). Suppressing OsBZR1 expression by RNAi resulted in reduced sensitivity and altered BR-responsive gene expression in transgenic rice plants. Mutation of a putative 14-3-3-binding site of OsBZR1 abolished its interaction with the 14-3-3 proteins. Such mutant OsBZR1 proteins showed an increased nuclear distribution compared with the wild-type protein, suggesting that 14-3-3 proteins directly inhibit OsBZR1 function by reducing its nuclear localization (Bai et al. 2007). Collectively these results demonstrate a conserved function of OsBZR1 and an important role of 14-3-3 proteins in brassinosteroid signal transduction in rice.

9.4 Pathogen Modulates Brassinosteroid Signaling System in Infected Plants

Brassinosteroids either positively (Khripach et al. 2000; Nakashita et al. 2003) or negatively (Albrecht et al. 2012; Belkhadir et al. 2012; De Vleeschauwer et al. 2012; Nahar et al. 2013; Malinovsky et al. 2014) regulate plant innate immunity. Pathogen infection induces BR biosynthesis and BR signaling pathways in disease-resistant or susceptible plants. *Pythium graminicola* infection in rice plants induced a strong up-regulation of BR biosynthetic, BR-responsive, and BR signal processing genes and also genes involved in BR homeostasis (Fig. 9.7; De Vleeschauwer et al. 2012). *P. graminicola* infection in rice plants induced a strong up-regulation of the BR biosynthetic genes *OsDWARF2* and *CONSTITUTIVE PHOTOMORPHOGENESIS AND*

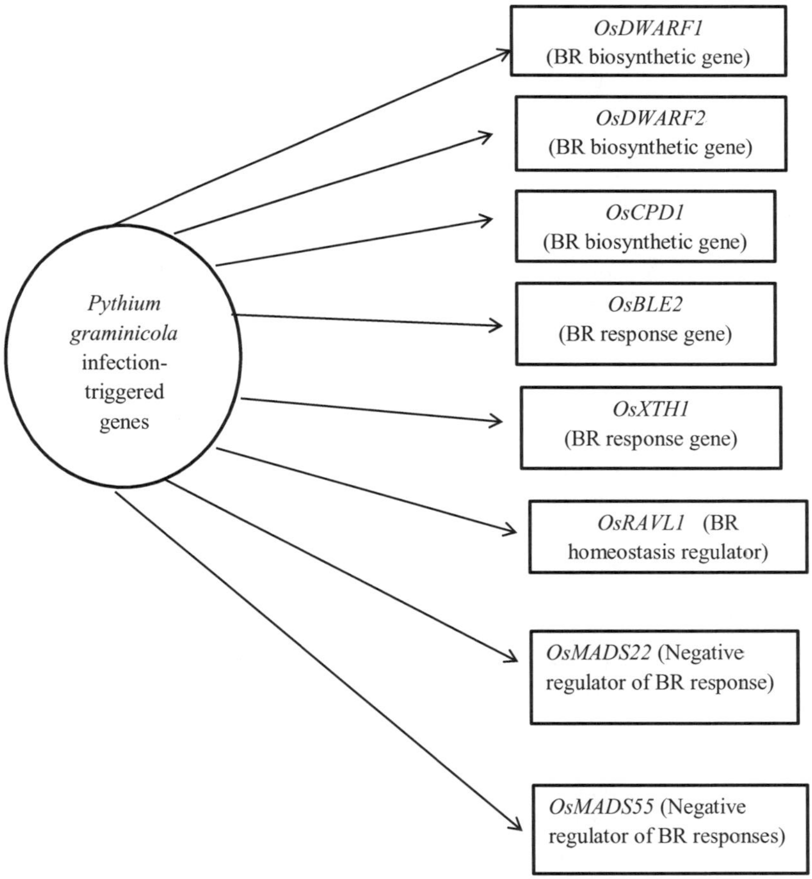

Fig. 9.7 Pathogen-triggered enhanced expression of genes involved in BR-signaling machinery in rice plants (Adapted from De Vleesschauwer et al. 2012)

DWARFISM1 (*OsCPD1*) with mRNA levels peaking at 7–10 times the levels found in noninoculated controls (De Vleeschauwer et al. 2012). Expression of the BR-inducible gene *BRASSINOLIDE ENHANCER2* (*OsBLE2*) responded strongly to *Pythium graminicola* infection and peaked at approximately 55 times the levels found in noninoculated controls. The pathogen-induced transcription of *OSBLE2* was even more pronounced in brassinolide-pretreated plants (De Vleeschauwer et al. 2012). Another BR-response gene encoding endoglucan transglucosylase OsXTH1 has been found to be up-regulated in *P. graminicola*-infected plants (De Vleeschauwer et al. 2012).

Pythium infection also showed negative feedback regulation of BR biosynthesis (BR homeostasis) (De Vleeschauwer et al. 2012). *RAVL1* has been shown to be necessary for the negative feedback regulation of BR biosynthesis (BR homeostasis) in rice (Je et al. 2010). *OsRAVL1* plays an important role in basal activity of both BR signaling and biosynthesis in rice. *OsRAVL1* mRNA levels responded strongly to *Pythium graminicolum* inoculation in rice plants, showing an approximately 15-fold induction over noninoculated controls. However, in the oomycete-inoculated brassinolide-(BL) treated plants, O*sRAVL1* expression was less pronounced, presumably due to the already strong activation of BR signaling in the pathogen-inoculated plants (De Vleeschauwer et al. 2012). These studies suggest that the oomycete *Pythium* infection triggered BR biosynthesis, BR homeostasis, and BR signaling (De Vleeschauwer et al. 2012).

Pathogen infection results in elevation of BR signal processing (De Vleeschauwer et al. 2012). The MADS-box gene family encodes transcription factors with a conserved DNA-binding domain, called the MADs box (Lee et al. 2008). Rice short vegetative phase (SVP)-group MADS-box proteins, OsMADS22 and OsMADS55, have been reported to be negative regulators of brassinosteroid responses (Duan et al. 2006; Lee et al. 2008). Rice plants pretreated with the most active brassinosteroid, brassinolide (BL), showed a strong down-regulation of the BR-negative regulator *MADS-box PROTEIN55* (*OsMADS55*), when the plants were inoculated with *P. graminicola* (De Vleeschauwer et al. 2012). The results suggest that the oomycete pathogen elevates BR signal processing. Infection of rice plants by *P. graminicola* was associated with major transcriptional reprogramming of various BR-associated genes (De Vleeschauwer et al. 2012). The genes altered in infected plants included *OsDWARF* and *OSDWARF2*, which are involved in BR biosynthesis. The expression of putative BR receptor gene *BRI1-LIKE 3* (*BRL3*) and the negative BR signaling regulator gene *MADS-box PROTEIN1* was also altered in infected plants. Various *cis*-elements known to be responsive to BRs were up-regulated in the infected plants. The *cis*-elements up-regulated due to infection included two E-box elements (CANNTG), various motifs containing the BR-response element (BRRE), and the G-box CACGTG (De Vleeschauwer et al. 2012). The levels of castasterone and deoxocastasterone, the important intermediates in BR biosynthesis pathway, were found to increase in tobacco leaves inoculated with *Tobacco mosaic virus* (Nakashita et al. 2003). Global analysis of tomato gene expression during *Potato spindle tuber viroid* (PSTVd) infection revealed a complex array of changes affecting hormone signaling. A marked up-regulation of genes involved in response to stress stimuli was observed only when exogenous brassinosteroid was applied to PSTVd-infected plants, suggesting that brassinosteroid-mediated signaling is involved in viroid disease induction (Owens et al. 2012).

9.5 BR Signaling Triggers Plant Disease Resistance

Brassinosteroid signaling system has been shown to positively or negatively induce plant defense responses. Nakashita et al. (2003) demonstrated that brassinolide (BL) application enhanced resistance against the rice blast pathogen *Magnaporthe oryzae* and the rice bacterial blight pathogen *Xanthomonas oryzae* pv. *oryzae*. The tobacco plants treated with BL showed enhanced resistance against *Tobacco mosaic virus* (TMV) (Nakashita et al. 2003). BL treatment also induced resistance to the bacterial pathogen *Pseudomonas syringae* pv. *tabaci* and the powdery mildew fungal pathogen *Oidium* sp. BL-induced resistance does not seem to require SA signaling system (Nakashita et al. 2003).

9.6 BAK1 in the BR Signaling Pathway Triggers Plant Disease Resistance

BAK1 is a key component in BR signaling pathway (Schwessinger et al. 2011). It may be involved in triggering plant disease resistance by modulating JA signaling system. BAK1 may be a signal "amplifier" rather than an integral component of downstream signaling pathways (Nicaise et al. 2009). BRI1–BAK1 interaction leads to the transphorylation of their respective kinase domains and the subsequent enhancement of BRI1 signaling output (Wang et al. 2008), suggesting that BAK is a signal amplifier rather than an integral component of downstream signaling pathways (Nicaise et al. 2009). Schulze et al. (2010) showed that brassinolide induces BAK1 phosphorylation within seconds. Several LRR-RKs form tight complexes with BAK1 almost instantaneously after ligand binding and the subsequent phosphorylation events are key initial steps in signal transduction (Schulze et al. 2010).

BAK1 appears to be a key component in plant innate immune system and suppression of *BAK1* gene expression results in suppression of the plant innate immunity. *BAK1*-silenced plants were found to be more susceptible to adapted and nonadapted *Pseudomonas* and to the oomycete *Hyaloperonospora parasitica* (Heese et al. 2007). Arabidopsis *bak1* mutants are also extremely susceptible to necrotrophic fungi *Botrytis cinerea* and *Alternaria brassicicola* (Kemmerling et al. 2007). BAK1 has been shown to be required for basal resistance against the late blight pathogen *Phytophthora infestans* in *Nicotiana benthamiana* (Chaparro-Garcia et al. 2011). Transgenic *Arabidopsis* expressing *AvrPto*, a bacterial effector that targets the BR co-receptor BAK1, not only displays compromised immunity but also exhibits phenotypes similar to these of BR-insensitive mutants (Shan et al. 2008). *AvrPto* expression does not affect the plant response to other plant hormones, raising the possibility that AvrPto specifically binds BAK1 to modify BR signaling and attenuate host defenses (Shan et al. 2008). These results demonstrate the importance of BAK1 in immune signaling. BAK1 positively modulates wound-induced accumulation of jasmonic acid and JA-isoleucine (JA-Ile). The results suggest that brassinosteroid-induced BAK1 may activate JA signaling system triggering immune responses (Yang et al. 2011a, b). JA was found to induce resistance against the rice blast pathogen *Magnaporthe oryzae* (Schweizer et al. 1998), which was similar to the effect of the brassinosteroid BL treatment (Nakashita et al. 2003).

9.7 BR Signaling Machinery Negatively Regulates Plant Immune Responses and Induces Susceptibility

Brassinolide (BL) markedly promoted susceptibility to *Pythium graminicola*, whereas treatment with brassinazole (BRZ), a highly specific inhibitor of BR biosynthesis, induced substantial levels of resistance in rice plants. The extensive transcriptional reprogramming of various BR biosynthetic, signaling, and response genes in pathogen-inoculated roots and the overrepresentation of BR-responsive *cis*-elements in the promoters of *P. graminicola*-dependent rice genes strongly suggest that *P. graminicola* hijacks the rice BR biosynthesis and signaling machinery to cause diseases (De Vleeschauwer et al. 2012).

9.8 Brassinosteroid Signaling Negatively Regulates Salicylate-Mediated Immunity

BR signaling induced susceptibility to the rice root pathogen *Pythium graminicola* by suppressing the plant innate immunity. BR treatment induced susceptibility, while an inhibitor of BR signaling, brassinazole (BRZ), induced disease resistance activating plant immunity. These results suggest that BRs are negative regulators of innate immunity in rice plants. In contrast, SA signaling triggered immune responses against the pathogen in rice plants. Application of the brassinosteroid (BR) brassinolide (BL) negated resistance conferred by the SA synthetic analog benzothiadiazole (BTH), suggesting negative crosstalk between BR and SA signaling pathways. Exogenous application of BL decreased the pathogen-induced strong up-regulation of the transcription of the SA regulatory genes *OsNPR1* and *OsWRKY45*. The temporal expression of both the SA regulatory genes was inversely correlated with that of the BR marker genes *OsBLE2*, *OsCPD1*, and *OsRAVL1* (De Vleeschauwer et al. 2012).

These studies suggest that BR-mediated suppression of SA defenses occurs downstream of SA biosynthesis, but upstream of NPR1 and OsWRKY45 in the SA signaling pathway (De Vleeschauwer et al. 2012).

9.9 BR Signaling Negatively Regulates Gibberellic Acid (GA)-Mediated Plant Immune Responses

GA has been shown to induce resistance against some pathogens. GA enhanced resistance to *Pythium graminicola* in rice (De Vleeschauwer et al. 2012). In contrast, BR negatively regulates innate immune responses in rice. BR suppresses GA signaling. Supplying the brassinosteroid brassinolide (BL) along with GA induced disease susceptibility. The results suggest that GA-mediated defense mechanisms are repressed by BR (De Vleeschauwer et al. 2012). BR inhibits GA responses in the roots by interfering with GA metabolism. BR-induced susceptibility develops coincidently with decreases in endogenous GA content (De Vleeschauwer et al. 2012). An interaction between BZR1 and DELLAs mediates direct signaling cross-talk between brassinosteroids and gibberellins in *Arabidopsis* (Li et al. 2012).

P. graminicola recruits the BR pathway to actively suppress GA biosynthesis in rice. Exogenously administered BR inhibited the BR biosynthetic gene *GA20ox3* and induced the BR-degradation pathway gene *GA2ox3* within 8 h of inoculation with *P. graminicola* (De Vleeschauwer et al. 2012). The *O*-linked GlcNAc transferase SPINDLY (Shimada et al. 2006) and the casein kinase EARLY FLOWERING1 (EL1) (Dai and Xue 2010) in rice plants show inhibitory function in GA signaling. Expression of SPINDLY was up-regulated by BR treatment alone and the most dramatic changes in SPINDLY kinetics were observed following the pathogen *P. graminicola* inoculation. Similar changes in *EL1* transcript accumulation were observed in rice plants inoculated with *P. graminicola* (De Vleeschauwer et al. 2012). These results suggest that BR triggers the expression of GA repressor proteins and suppresses GA-induced defense responses.

The most active brassinosteroid brassinolide-fed rice plants disrupted endogenous GA content and induced susceptibility. The brassinosteroid biosynthesis inhibitor brassinazole (BRZ) treatment triggered resistance against the pathogen. BL treatment positively regulated the abundance of the DELLA proteins, which are intracellular repressors of GA signaling (Li et al. 2012; Qin et al. 2013). Gain-of-function mutations in DELLA genes leads to reduced GA signaling, whereas loss of function leads to increased GA signaling (Hauvermale et al. 2012). The brassinosteroid BL treatment suppressed GA-mediated defense responses (Qin et al. 2013). These results suggest that BR triggers susceptibility to *P. graminicola* by inducing indirect stabilization of the DELLA protein.

9.10 Interplay Between BR and PAMP–PRR Signaling Systems

9.10.1 PAMP–PRR Signaling Complex

PAMPs are the molecules derived from pathogens triggering innate immunity. PAMPs bind to plant pattern recognition receptors (PRRs) and induce defense responses (Nicaise et al. 2009; Tsuda and Katagiri 2010). PAMPs are perceived as alarm/danger signals by cognate PRRs, and the PAMP–PRR complex activates the plant immune system (He et al. 2007; Wan et al. 2008; Iriti and Faoro 2009). PAMPs trigger increased transcription of PRR genes and accumulation of PRR proteins (Qutob et al. 2006; Lohmann et al. 2010). Most of the PRRs are receptor kinases and the PAMPs induce autophosphorylation of the kinase domain of PRRs (Kanzaki et al. 2008; Xiang et al. 2008). The plant immune system uses several second messengers to encode information generated by the PAMP/PRR signaling complex and deliver the information downstream of PRRs to proteins which decode/interpret signals and initiate defense gene expression (van Verk et al. 2008; Mersmann et al. 2010; Boudsocq et al. 2010; Hwang and Hwang 2011).

PRRs interact with additional transmembrane proteins that act as signaling adapters or amplifiers to achieve full functionality (Zipfel 2009). These transmembrane proteins include BAK1 (for *B*RI1 (BRASSINOSTEROID INSENSITIVE1)-*A*SSOCIATED *K*INASE1), BIK1 (*B*OTRYTIS-*I*NDUCED *K*INASE*1*), and BIR1 (*B*RANCHING *I*NHIBITING *R*ECEPTOR*1*) (Postel et al. 2010; Zhang and Zhou 2010; Zhang et al. 2010). BAK1 is also a component of brassinosteroid (BR) signaling pathway (Chinchilla et al. 2009). BAK1 was originally identified as a BRI1-associated receptor kinase mediating brassinosteroid signaling (Li et al. 2002; Nam and Li 2002). BRs are perceived by BRI1, which is structurally similar to the PRR FLS2 (Belkhadir et al. 2006). BAK1 is a positive regulator of PAMP-triggered plant immunity, and it acts as an adapter of multiple LRR-RKs that act in defense signaling, including the PRRs FLS2, EFR, PEPR1, and PEPR2 (Chinchilla et al. 2007; Ryan et al. 2007; Gao et al. 2009; Postel et al. 2010; Schulze et al. 2010). It also acts as an adapter of the receptor kinases BIR1, which seem to act as part of a presumed PRR complex(es) and/or at a downstream step in the signaling cascade (Saijo 2010).

BAK1 forms a complex with the RLK BIR1 to negatively regulate defense responses (Gao et al. 2009). The PAMP flg22 interacts with BAK1 to phosphorylate BIK1, which seems to act as positive regulator of the PAMP signaling pathway (Lu et al. 2010; Zhang et al. 2010). BIK1 is an essential component in PAMP signal transduction, which links the PAMP receptor complex to downstream intracellular signaling (Lu et al. 2010). PRRs form tight complexes with the receptor kinase BAK1 instantaneously after ligand binding (Schulze et al. 2010). FLS2–BAK1 heteromerization occurs almost instantaneously after perception of the PAMP flg22. Flg22 can induce formation of a stable FLS2–BAK1 complex in microsomal membrane preparations in vitro. Schulze et al. (2010) detected de novo phosphorylation of both FLS2 and BAK1 within 15 s of stimulation with flg22. Similarly, the active brassinosteroid brassinolide induces BAK1 phosphorylation within seconds.

9.10.2 Crosstalk Between BR Biosynthesis Pathway and PAMP–PRR Signaling

Crosstalk between PAMP–PRR signaling and BR synthesis pathway has been reported. Transgenic *Arabidopsis* plants ectopically overexpressing *DWF4* (*35S::DWF4*) were used to investigate whether PAMP–PRR signaling is altered under conditions of excess BR biosynthesis. *DWF4* encodes a C-22 hydroxylase, which is a rate-limiting step of BL biosynthesis. The *35S::DWF4* plants show reduced responses to flg22 (Belkhadir et al. 2012). Increasing the endogenous pool of bioactive BR antagonizes flg22-induced responses (Belkhadir et al. 2012). Brassinazole (BRZ), a triazole compound, reversibly and specifically blocks DWF4 activity. In the presence of BRZ, the response of a cytochrome P450 gene, *CYP71A12*, to the PAMP flg22 was greatly diminished. The cytochrome P450 gene is involved in BR biosynthesis. It suggests that transient depletion of endogenous BR pools strongly suppresses flg22 signaling in root cells. The sustained activation of the FLS2 pathway could not block BKI1 displacement from the plasma membrane, even at low doses of BL. It suggests that sustained FLS2 signaling does not modulate early BR signaling outputs. Collectively these results suggest the existence of one-way crosstalk between the balanced activities of BR synthesis and FLS2 signaling (Belkhadir et al. 2012).

9.10.3 Overexpression of BRI1 Antagonizes BAK1-Mediated PAMP–PRR Signaling

Enhanced BR signaling as observed in gain-of-function mutations in *BAK1* (Jaillais et al. 2011b) and *BRI1* (Belkhadir et al. 2012) both impair responses to PAMP-triggered immunity. The results suggest that BRI1–BAK signaling modulates PAMP–PRR signaling pathway. BAK1 is a common co-receptor for the pattern recognition receptors activated by various PAMPs including flg22, elf18, and peptidoglycan. Similarly BAK1 is a co-receptor for the BR receptor, BRI1. Signaling downstream of BAK1 differs between BRI1 and FLS2 (PRR for flg22) pathways. BIK1 is bound to FLS2 and dissociates in a BAK1-dependent manner upon flg22 binding. BIK1 and paralogs positively regulate PAMP-triggered immunity responses downstream of FLS2 (Lu et al. 2010; Zhang et al. 2010). FLS2 is ubiquitinated by the BAK1-associated ubiquitin ligases PUB12 and PUB13 and degraded (Lu et al. 2011). FLS2 activation leads to rapid bursts of calcium and ROS, activation of MAP kinases and calcium-dependent protein kinases, ultimately leading to PAMP-triggered immunity (Segonzac and Zipfel 2011).

Plants overexpressing *BRI1* under the control of the strong CaMV35S promoter (*35S::BRI1*) failed to respond to the PAMPs flg22, elf18, and PGN treatments (Belkhadir et al. 2012). Increased BR signaling triggered by BRI1 overexpression antagonized the activities of the PAMPs/MAMPs flg22, elf18, and PGN. In contrast, overexpression of BR signaling did not alter the expression of chitin-triggered immune responses. Chitin is a PAMP that triggers BAK1-independent immune responses, while the activities of the PAMPs flg22, elf18, and PGN (peptidoglycan) are dependent on BAK1 (Belkhadir et al. 2012). Only BAK1-mediated pathogen-triggered immunity (PTI) is impaired in *35S::BRI1* plants. The lack of BAK1-mediated PTI response in *35S::BRI1* plants may be the consequence of a limiting pool of BAK1 that is unavailable for PAMP signaling. To prove this hypothesis, BAK1 dosage was increased by introducing a *BAK1-HA* transgene into *35S::BRI1* plants and tested flg22 responsiveness. The presence of extra copies of BAK1 restored flg22 sensitivity in *35S::BRI1* plants (Belkhadir et al. 2012). Thus there was a direct link between restoration of FLS2 function and BAK1 dosage in the specific context of heightened BRI1 levels. These results suggest that increased BR signaling triggered by BRI1 overexpression can antagonize the activities of the BAK1-dependent PAMP signaling pathways (Belkhadir et al. 2012). The antagonistic effects on PAMP signaling observed in *35S::BRI1* plants are likely the consequence of increased levels of BRI1 and not of increased intrinsic BRI1 signaling activity (Belkhadir et al. 2012).

Overexpression of BRI1 antagonizes BAK1-mediated programmed (spontaneously developed) cell death response. Increased BR signaling mediated by BRI1 overexpression antagonizes the programmed (inappropriate) cell death signaling triggered by increased BAK1 dosage (Belkhadir et al. 2012). Programmed cell death signaling is an important component in plant innate immunity (Vidhyasekaran 2007).

Arabidopsis hypermorphic BRI1^{sud1}allele plants phenocopied every BR-dependent morphological aspect of *35S::BRI1* plants, but displayed dramatically increased responses to flg22. Flg22 infiltrated into *Arabidopsis* leaves induces PAMP-triggered immunity that inhibit *Pseudomonas syringae* pv. *tomato* (Pto) DC3000 growth (Belkhadir et al. 2012). When *Pto DC3000hrc* mutant, which is deficient in its type III secretion system (T3SS) (T3SS allows bacterial effectors delivery to host cells for maximal virulence), was infiltrated together with flg22, bacterial growth was inhibited approximately 10-fold on *BRI1* plants and approximately 100-fold on *BRI1^{sud1}* plants. These studies suggest that the FLS2 signaling in BRI^{sud1} plants can be further stimulated and acts effectively to limit bacterial growth. In contrast, there was no flg22-dependent inhibition of *Pto DC3000hrc* growth on *35S::BRI1* or *bak1* plants. It has been suggested that loss of signaling in *35S::BRI1* plants may be due to BAK1's inability to contribute to flg22 signaling. Increased BR signaling via BRI1^{sud1} primes BR and flg22 signaling through BAK1. When BRI1 levels are increased, the stimulatory effects of BR on flg22 signaling are negated, presumably by increased recruitment of BAK1 to BRI1 (Belkhadir et al. 2012). It has been reported that enhanced BR signaling resulting from either a gain-of-function mutation in BAK1 (Jaillais et al. 2011b), ectopic expression of BRI1 (Belkhadir et al. 2012), or application of BL (Albrecht et al. 2012) impedes immunity of *P. syringae* conditioned by the conserved microbial signature flg22.

9.10.4 *BR-Responsive Transcription Regulator BZR1 May Suppress PAMP–PRR Signaling System*

BR may also antagonize plant defense signaling via its transcriptional regulator BZR1. BZR1 is a nuclear component of the BR signal transduction pathway (Wang et al. 2002). BR acts by modulating the phosphorylation status of the transcription factors BZR1 through the action of BAK1 receptors and a GSK3 kinase, BIN2 (Ryu et al. 2007). Signal transduction from BRI1 to BIN2 involves the BSK1 (Tang et al. 2008) and the BSU1 phosphatase (Kim et al. 2009). BIN2 negatively regulates BR signaling by phosphorylating the BR-responsive transcription factor BZR1 critical for bringing about characteristic BR responses (Peng et al. 2010). BZR1 directly binds to DNA and regulates BR-responsive gene expression (Wang et al. 2002; Yin et al. 2002, 2005; He et al. 2005). Upon activation by BR, BZR1 binds and represses the promoters of gene encoding the flagellin receptor FLS2. Suppression of the expression of the PRR by BR suppresses the PAMP–PRR signaling system-triggered immune responses (Fig. 9.8; Sun et al. 2010).

9.10.5 *Antagonistic Regulation of PAMP–Triggered Immunity by the bHLH transcription Factor HBI1*

The basic helix-loop-helix (bHLH) transcription factor HOMOLOG OF BRASSINOSTEROID ENHANCED EXPRESSION2 INTERACTING WITH IBH1 (HBI1) is a positive regulator of BR-triggered responses (Malinovsky et al. 2014). *HBI1* expression is enhanced by BR treatment. *HBI1*-overexpressing plants are more resistant to the BR biosynthetic inhibitor brassinazole. HBI1 has been found to be a negative regulator of PAMP-triggered immunity (PTI) in *Arabidopsis thaliana* (Malinovsky et al. 2014). *HBI1* expression was down-regulated in response to different PAMPs. *HBI1* overexpression led to reduced PAMP-triggered responses. This inhibition was found to be correlated with reduced steady-state expression of immune marker genes, leading to increased susceptibility to *Pseudomonas syringae* in *Arabidopsis thaliana* (Malinovsky et al. 2014). These results reveal the existence of antagonism between BR and PTI signaling systems.

9.10.6 *Activation of BRI1 Pathway Leads to Inhibition of PAMP-Triggered Immunity*

Studies conducted using exogenous application of biologically active 24-epibrassinolide (epiBL) in *Arabidopsis thaliana* showed that activation of the BRI1 pathway leads to inhibition of PAMP-triggered immunity (PTI) (Albrecht

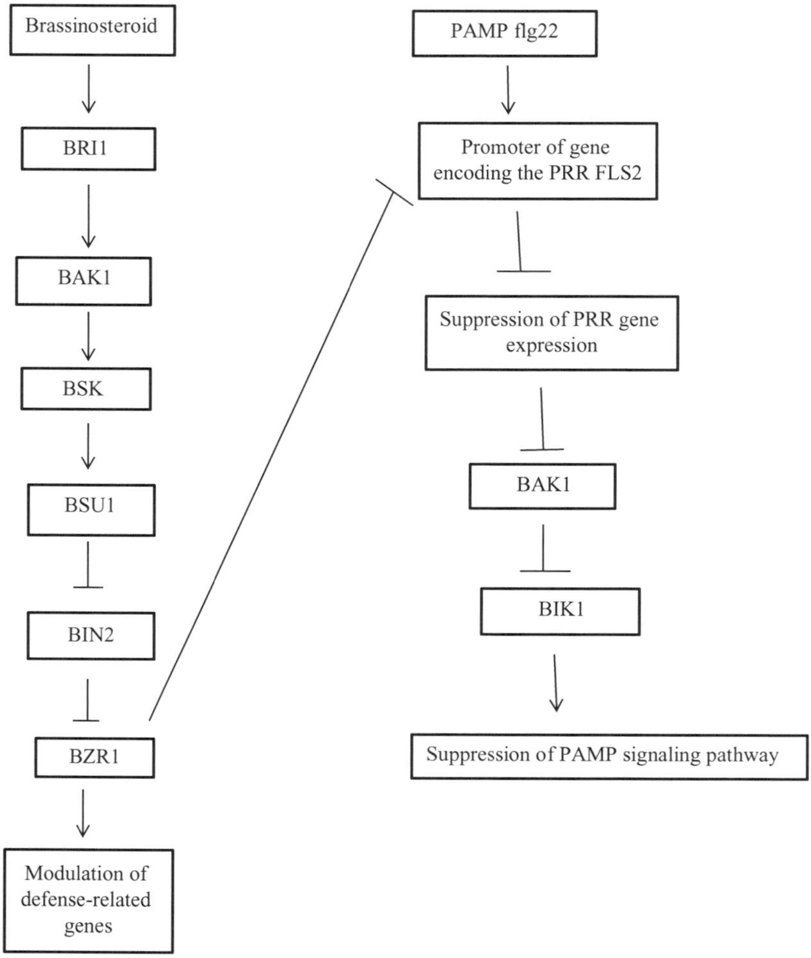

Fig. 9.8 Suppression of PAMP-PRR signaling pathway by brassinosteroid-responsive transcription factor BZR1 (Adapted from Sun et al. 2010)

et al. 2012). EpiBL treatment inhibited the chitin-induced ROS burst. It also inhibited ROS burst induced by the PAMPs flg22 and elf18. The epiBL-mediated inhibition of the flg22- and chitin-induced ROS bursts also occurred in the null *bak1-4* mutant. While flg22-FLS2 and elf18-EFR PAMP–PRR signaling systems are BAK1-dependent, chitin-CERK1 system is BAK1 independent (Albrecht et al. 2012). However, both the BAK1-dependent and BAK1-independent PAMP–PRR signaling systems are found to be inhibited by epiBL treatment. Albrecht et al. (2012) suggested that epiBL inhibition of FLS2 may be independent or downstream of BAK1–BIK1 in the signaling systems.

9.10.7 *Inhibition of PRR-Mediated Immune Signaling by BR Perception*

Albrecht et al. (2012) showed a unidirectional inhibition of the pattern recognition receptor FLS2-mediated immune signaling by BR perception. This effect occurred downstream or independently of complex formation with BAK1 and associated downstream phosphorylation. BRs also inhibited signaling triggered by the BAK1-independent recognition of the fungal PAMP chitin (Albrecht et al. 2012). Studies conducted using exogenous application of biologically active 24-epibrassinolide (epiBL) in *Arabidopsis thaliana* showed that activation of the BRI1 pathway leads to inhibition of PAMP-triggered immunity (PTI) (Albrecht et al. 2012). EpiBL treatment inhibited the chitin-induced ROS burst. It also inhibited ROS burst induced by the PAMPs flg22 and elf18. The epiBL-mediated inhibition of the flg22- and chitin-induced ROS bursts also occurred in the null *bak1-4* mutant. While flg22-FLS2 and elf18-EFR PAMP–PRR signaling systems are BAK1 dependent, chitin-CERK1 system is BAK1 independent (Albrecht et al. 2012). However, both the BAK1-dependent and BAK1-independent PAMP–PRR signaling systems are found to be inhibited by epiBL treatment.

9.11 Pathogen Hijacks Brassinosteroid Signaling Machinery to Cause Disease

Pythium graminicola exploits BRs as virulence factors and hijacks the rice BR machinery to cause disease (De Vleeschauwer et al. 2012). BRs suppress root immunity to *Pythium graminicola*, resulting in increased pathogen proliferation and substantially enhanced disease susceptibility. *P. graminicola* co-opts the plant BR machinery as a decoy strategy to tap into the immune signaling circuitry and interfere with effectual SA- and GA-controlled defenses (De Vleeschauwer et al. 2012).

9.12 Crosstalk Between BR and Other Hormone Signaling Systems

Brassinosteroids may interact with other plant hormones in modulating plant immune responses. BRs interfere with SA signaling system by suppressing the transcription of the SA regulatory genes *OsNPR1* and *OsWRKY45* in rice plants (De Vleeschauwer et al. 2012). In both *Arabidopsis* and cucumber, BL treatment induces the expression of the classical SA marker gene *PR1* (Xia et al. 2009; Divi et al. 2010).

BR induces the expression of *OPR3*, encoding a 12-oxo-phytodienoic acid reductase, the enzyme involved in JA biosynthesis (Müssig et al. 2000). The BR co-receptor BAK1 has been shown to trigger the accumulation of JA and JA-isoleucine in *Nicotiana attenuata* (Yang et al. 2011a). Application of low levels of BL suppressed the expression of *OsAOS2* and *OsJAmyb*, two important genes of the JA signaling pathway in rice plants, and this was accompanied by an enhanced susceptibility to the nematode *Meloidogyne graminicola* infection (Nahar et al. 2013). The endogenous levels of BR and the JA-intermediate OPDA were inversely correlated in rice plants (Nahar et al. 2013).

BRs regulate the biosynthesis of ethylene via stabilizing the ACC synthase, the key enzyme involved in ethylene biosynthesis (Hansen et al. 2009). Ethylene and BR may promote the biosynthesis of each other (Shi et al. 2006; Zhang et al. 2009b). BRs counteract the effect of the ethylene biosynthesis inhibitor AVG (Shi et al. 2006). Ethylene could overcome the inhibitory effect of the BR biosynthesis inhibitor brassinazole (BRZ) (Shi et al. 2006).

BRs and auxins synergistically increase the expression of transcription of auxin-induced genes (Vert et al. 2008). A direct connection has been reported between the BR-regulated BIN2 kinase and ARF2, a member of the auxin response factor family of transcriptional regulators (Vert et al. 2008). Phosphorylation by BIN2 results in loss of ARF2 DNA binding and repression activities. BIN2 increases expression of auxin-induced genes by directly inactivating repressor ARFs, leading to synergistic increases in transcription (Vert et al. 2008).

Abscisic acid (ABA) may regulate BR signaling (Zhang et al. 2009b; Divi et al. 2010). Early responsive genes of ABA, *RD29A* and *RD22*, are also regulated by BRs in *Arabidopsis* (Kagale et al. 2007). Exogenously applied ABA rapidly inhibits BR signaling through phosphorylation of *BES1* in *Arabidopsis* (Zhang et al. 2009a). BR negatively regulates innate immune responses in rice by suppressing GA signaling by BR (De Vleeschauwer et al. 2012). These studies suggest that BR may activate or suppress SA, JA, ET, ABA, and auxin signaling systems; however, their role in plant innate immunity has not yet been studied in detail.

References

Albrecht C, Boutrot F, Segonzac C, Schwessinger B, Gimenez-Ibanez S, Chinchilla D, Rathjen JP, de Vries SC, Zipfel C (2012) Brassinosteroids inhibit pathogen-associated molecular pattern-triggered immune signaling independent of the receptor kinase BAK1. Proc Natl Acad Sci U S A 109:303–308

Bach TJ, Boronat A, Campos N, Ferrer A, Vollack KU (1999) Mevalonate biosynthesis in plants. Crit Rev Biochem Mol Biol 34:107–122

Bai MY, Zhang LY, Gampala SS, Zhu SW, Song WY, Chong K, Wang ZY (2007) Functions of OsBZR1 and 14-3-3 proteins in brassinosteroid signaling in rice. Proc Natl Acad Sci U S A 104:13839–13844

Bajguz A, Hayat S (2009) Effects of brassinosteroids on the plant responses to environmental stresses. Plant Physiol Biochem 47:1–8

Bak S, Beisson F, Bishop G, Hamberger B, Höfer R, Paquette S, Werck-Reichhart D (2011) Cytochrome p450. Arabidopsis Book 9:e0144. doi:10.1199/tab.0144

Bancos S, Nomura T, Sato T, Molnar G, Bishop GJ, Koncz C, Yokota T, Nagy F, Szekeres M (2002) Regulation of transcript levels of the *Arabidopsis* cytochrome p450 genes involved in brassinosteroid biosynthesis. Plant Physiol 130:504–513

Belkhadir Y, Chory J (2006) Brassinosteroid signaling: a paradigm for steroid hormone signaling from the cell surface. Science 314:1410–1411

Belkhadir Y, Wang X, Chory J (2006) Arabidopsis brassinosteroid signaling pathway. Sci STKE. 2006:cm5

Belkhadir Y, Jaillais Y, Epple P, Balsemäo-Pires E, Dangl JL, Chory J (2012) Brassinosteroids modulate the efficiency of plant immune responses to microbe-associated molecular patterns. Proc Natl Acad Sci U S A 109:297–302

Bishop GJ (2007) Refining the plant steroid hormone biosynthesis pathway. Trends Plant Sci 12:377–380

Bishop GJ, Koncz C (2002) Brassinosteroids and plant steroid hormone signaling. Plant Cell 14(Suppl):S97–S110

Bishop G, Nomura T, Yokota T, Harrison K, Noguchi T, Fujioka S, Takatsuto S, Jones JD, Kamiya Y (1999) The tomato DWARF enzyme catalyzes C-6 oxidation in brassinosteroid biosynthesis. Proc Natl Acad Sci U S A 96:1761–1766

Bochar DA, Freisen JA, Stauffacher CV, Rodwell VW (1999) Biosynthesis of mevalonic acid from acetyl-CoA. In: Barton D, Nakanishi K (eds) Comprehensive natural products chemistry, vol 2. Elsevier Science Ltd, Amsterdam, pp 15–44

Boudsocq M, Willmann MR, McCormack M, Lee H, Shan L, He P, Bush J, Cheng SH, Sheen J (2010) Differential innate immune signalling via Ca^{2+} sensor protein kinases. Nature 464:418–422

Bridges D, Moorhead BG (2005) 14-3-3 proteins: a number of functions for a numbered protein. Sci STKE 2005:re10

Caño-Delgado A, Yin Y, Yu C, Vafeados D, Mora-Garcia S, Cheng JC, Nam KH, Li J, Chory J (2004) BRL1 and BRL3 are novel brassinosteroid receptors that function in vascular differentiation in Arabidopsis. Development 131:5341–5351

Chaparro-Garcia A, Wilkinson RC, Gimenez-Ibanez S, Findlay K, Coffey MD, Zipfel C, Rathjen JP, Kamoun S, Schornack S (2011) The receptor-like kinase SERK3/BAK1 is required for basal resistance against the late blight pathogen *Phytophthora infestans* in *Nicotiana benthamiana*. PLoS One 6(1):e16608

Chinchilla D, Zipfel C, Robatzek S, Kemmerling B, Nürnberger T, Jones JDG, Felix G, Boller T (2007) A flagellin-induced complex of the receptor FLS2 and BAK1 initiates plant defence. Nature 448:497–500

Chinchilla D, Shan L, He P, de Vries S, Kemmerling B (2009) One for all: the receptor-associated kinase BAK1. Trends Plant Sci 14:535–541

Clouse SD, Sasse JM (1998) Brassinosteroids: essential regulators of plant growth and development. Annu Rev Plant Physiol Plant Mol Biol 49:427–451

Dai C, Xue H-W (2010) Rice early flowering1, a CK1, phosphorylates DELLA protein SLR1 to negatively regulate gibberellin signaling. EMBO J 29:1916–1927

De Vleeschauwer D, Van Buyten E, Satoh K, Balidion J, Mauleon R, Choi I-R, Vera-Cruz C, Kikuchi S, Höfte M (2012) Brassinosteroids antagonize gibberellin- and salicylate-mediated root immunity in rice. Plant Physiol 158:1833–1846

Divi UK, Krishna P (2009) Brassinosteroid: a biotechnological target for enhancing crop yield and stress tolerance. N Biotechnol 26:131–136

Divi UK, Krishna P (2010) Overexpression of the brassinosteroid biosynthetic gene *AtDWF4* in *Arabidopsis* seeds overcomes abscisic acid-induced inhibition of germination and increases cold tolerance in transgenic seedlings. J Plant Growth Regul 29:385–393

Divi UK, Rahman T, Krishna P (2010) Brassinosteroid-mediated stress tolerance in *Arabidopsis* shows interactions with abscisic acid, ethylene and salicylic acid pathways. BMC Plant Biol 10:151

Duan K, Li L, Hu P, Xu S-P, Xu Z-H, Xue H-W (2006) A brassinolide-suppressed rice MADS-box transcription factor, OsMDP1, has a negative regulatory role in BR signaling. Plant J 47:519–531

Friedrichsen DM, Nemhauser J, Muramitsu T, Maloof JN, Alonso J, Ecker JR, Furuya M, Chory J (2002) Three redundant brassinosteroid early response genes encode putative bHLH transcription factors required for normal growth. Genetics 162:1445–1456
Fujioka S (1999) Natural occurrence of brassinosteroids in the plant kingdom. In: Sakurai A, Yokota T, Clouse SD (eds) Brassinosteroids: steroidal plant hormones. Springer, Tokyo, pp 21–45
Fujioka S, Sakurai A (1997) Biosynthesis and metabolism of brassinosteroids. Physiol Plant 100:710–715
Fujioka S, Yokota T (2003) Biosynthesis and metabolism of brassinosteroids. Annu Rev Plant Biol 54:137–164
Fujioka S, Takatsuto S, Yoshida S (2002) An early C-22 oxidation branch in the brassinosteroid biosynthetic pathway. Plant Physiol 130:930–939
Gampala SS, Kim TW, He JX, Tang W, Deng Z, Bai MY, Guan S, Lalonde S, Sun Y, Gendron JM, Chen H, Shibagaki N, Ferl RJ, Ehrhardt D, Chong K, Burlingame AL, Wang ZY (2007) An essential role for 14-3-3 proteins in brassinosteroid signal transduction in *Arabidopsis*. Dev Cell 13:177–189
Gao M, Wang X, Wang D, Xu F, Ding X, Zhang Z, Bi D, Cheng YT, Chen S, Li X, Zhang Y (2009) Regulation of cell death and innate immunity by two receptor-like kinases in *Arabidopsis*. Cell Host Microbe 23:34–44
Geldner N, Hyman DL, Wang X, Schumacher K, Chory J (2007) Endosomal signaling of plant steroid receptor kinase BRI1. Genes Dev 21:1598–1602
Gendron JM, Wang ZY (2007) Multiple mechanisms modulate brassinosteroid signaling. Curr Opin Plant Biol 10:436–441
Goda H, Shimada Y, Asami T, Fujioka S, Yoshida S (2002) Microarray analysis of brassinosteroid-regulated genes in *Arabidopsis*. Plant Physiol 130:1319–1334
Goda H, Sawa S, Asami T, Fujioka S, Shimada Y, Yoshida S (2004) Comprehensive comparison of auxin-regulated and brassinosteroid-regulated genes in *Arabidopsis*. Plant Physiol 134:1555–1573
Grove MD, Spencer GF, Rohwedder WK, Mandava N, Worley JF, Warthen JD, Steffens GL, Flippenanderson JL, Cook JC (1979) Brassinolide, a plant growth-promoting steroid isolated from *Brassica napus* pollen. Nature 281:216–217
Hansen M, Chae H, Kieber J (2009) Regulation of ACS protein stability by cytokinin and brassinosteroid. Plant J 57:606–614
Hauvermale AL, Ariizumi T, Steber CM (2012) Gibberellin signaling: a theme and variations on DELLA repression. Plant Physiol 160:83–92
He JX, Gendron JM, Yang Y, Li J, Wang ZY (2002) The GSK3-like kinase BIN2 phosphorylates and destabilizes BZR1, a positive regulator of the brassinosteroid signaling pathway in *Arabidopsis*. Proc Natl Acad Sci U S A 99:10185–10190
He J-X, Gendron MJ, Sun Y, Gampala SSL, Gendron N, Sun CQ, Wang Z-Y (2005) BZR1 is a transcriptional repressor with dual roles in brassinosteroid biosynthesis and growth responses. Science 307:1634–1638
He P, Shan L, Sheen J (2007) Elicitation and suppression of microbe-associated molecular pattern-triggered immunity in plant-microbe interactions. Cell Microbiol 9:1385–1396
Heese A, Hann DR, Gimenez-Ibanez S, Jones AME, He K, Li J, Schroeder JI, Peck SC, Rathjen JP (2007) The receptor-like kinase SERK3/BAK1 is a central regulator of innate immunity in plants. Proc Natl Acad Sci U S A 104:12217–12222
Hwang IS, Hwang BK (2011) The pepper mannose-binding lectin gene *CaMBL1* is required to regulate cell death and defense responses to microbial pathogens. Plant Physiol 155:447–463
Iriti M, Faoro F (2009) Chitosan as a MAMP, searching for a PRR. Plant Signal Behav 4:66–68
Jaillais Y, Belkhadir Y, Balsemäo-Pires E, Dangl JL, Chory J (2011a) Extracellular leucine-rich repeats as a platform for receptor/co-receptor complex formation. Proc Natl Acad Sci U S A 108:8503–8507
Jaillais Y, Hothorn M, Belkhadir Y, Dabi T, Nimchuk ZL, Meyerowitz EM, Chory J (2011b) Tyrosine phosphorylation controls brassinosteroid receptor activation by triggering membrane release of its kinase inhibitor. Genes Dev 25:232–237

Jaspert N, Throm C, Oecking C (2011) Arabidopsis 14-3-3 proteins: fascinating and less fascinating aspects. Front Plant Sci 2:96. doi:10.3389/fpls.00096

Je BI, Piao HL, Park SJ, Park SH, Kim CM, Xuan YH, Park SH, Huang J, Choi YD, An G, Wong HL, Fujioka S, Kim MC, Shimamoto K, Han C-D (2010) *RAV1-Like1* maintains brassinosteroid homeostasis via the coordinated activation of *BRI1* and biosynthetic genes in rice. Plant Cell 22:1777–1791

Kagale S, Divi UK, Krochko JE, Keller WA, Krishna P (2007) Brassinosteroid confers tolerance in *Arabidopsis thaliana* and *Brassica napus* to a range of abiotic stresses. Planta 225:353–364

Kanzaki H, Saitoh H, Takahashi Y, Berberich T, Ito A, Kamoun S, Terauchi R (2008) NbLRK1, a lectin-like receptor kinase protein of *Nicotiana benthamiana*, interacts with *Phytophthora infestans* INF1 elicitin and mediates INF1-induced cell death. Planta 228:977–987

Kemmerling B, Schwedt A, Rodriguez P, Mazzotta S, Frank M, Qamar SA, Mengiste T, Betsuyaku S, Parker JE, Müssig C, Thomma BPHJ, Albrecht C, de Vries SC, Hirt H, Nürnberger T (2007) The BRI1-associated kinase 1, BAK1, has a brassinolide-independent role in plant cell-death control. Curr Biol 17:1116–1122

Khripach V, Zhabinskii V, Groot AD (2000) Twenty years of brassinosteroids: steroidal plant hormones warrant better crops for XX1 century. Ann Bot 86:441–447

Kim T-W, Wang Z-Y (2010) Brassinosteroid signal transduction from receptor kinases to transcription factors. Annu Rev Plant Biol 61:681–704

Kim G-T, Fujioka S, Kozuka T, Tax FE, Takatsuto S, Yoshida S, Tsukaya H (2005) CYP90C1 and CYP90D1 are involved in different steps in the brassinosteroid biosynthesis pathway in *Arabidopsis thaliana*. Plant J 41:710–721

Kim T-W, Guan S, Sun Y, Deng Z, Tang W, Shang JX, Sun Y, Burlingame AL, Wang ZY (2009) Brassinosteroid signal transduction from cell-surface receptor kinases to nuclear transcription factors. Nat Cell Biol 11:1254–1260

Kim T-W, Guan S, Burlingame AL, Wang Z-Y (2011) The CDG1 kinase mediates brassinosteroid signal transduction from BRI1 receptor kinase to BSU1 phosphatase and GSK3-like kinase BIN2. Mol Cell 19(43):561–571

Kinoshita T, Caño-Delgado A, Seto H, Hiranuma S, Fujioka S, Yoshida S, Chory J (2005) Binding of brassinosteroids to the extracellular domain of plant receptor kinase BRI1. Nature 433:167–171

Lee S, Choi SC, An G (2008) Rice SVP-group MADS-box proteins, OsMAD22 and OsMAD55, are negative regulators of brassinosteroid responses. Plant J 54:93–105

Li J, Chory JA (1997) A putative leucine-rich repeat receptor kinase involved in brassinosteroid signal transduction. Cell 90:929–938

Li J, Nam KH (2002) Regulation of brassinosteroid signaling by a GSK3/SHAGGY-like kinase. Science 295:1299–1301

Li J, Wen J, Lease KA, Doke JT, Tax FE, Walker JC (2002) BAK1, an Arabidopsis LRR receptor-like protein kinase, interacts with BRI1 and modulates brassinosteroid signaling. Cell 110:213–222

Li J, Yu X, Thompson A, Guo M, Yoshida S, Asami T, Chory J, Yin Y (2009) *Arabidopsis* MYB30 is a direct target of BES1 and cooperates with BES1 to regulate brassinosteroid-induced gene. Plant J 58:275–286

Li L, Ye H, Guo H, Yin Y (2010) *Arabidopsis* IWS1 interacts with transcription factor BES1 and is involved in plant steroid hormone brassinosteroid regulated gene expression. Proc Natl Acad Sci U S A 107:3918–3923

Li Q-F, Wang C, Jiang L, Li S, Sun SS, He J-X (2012) An interaction between BZR1 and DELLAs mediates direct signaling crosstalk between brassinosteroids and gibberellins in *Arabidopsis*. Sci Signal 5:ra72

Lisso J, Steinhauser D, Altmann T, Kopka J, Mussig C (2005) Identification of brassinosteroid-related genes by means of transcript co-response analyses. Nucleic Acids Res 33:2685–2696

Lohmann GV, Shimoda Y, Nielsen W, Jørgensen FG, Grossmann C, Sandal N, Sørensen K, Thirup S, Madsen LH, Tabata S, Sato S, Stougaard J, Radutoiu S (2010) Evolution and regulation of the *Lotus japonica* LysM receptor gene family. Mol Plant Microbe Interact 23:510–521

Lu D, Wu S, Gao X, Zhang Y, Shan L, He P (2010) A receptor-like cytoplasmic kinase, BIK1, associates with a flagellin receptor complex to initiate plant innate immunity. Proc Natl Acad Sci U S A 107:495–501

Lu D, Lin W, Gao X, Wu S, Cheng C, Avita J, Heese A, Devarenne TP, He P, Shan L (2011) Direct ubiquitination of pattern recognition receptor FLS2 attenuates plant innate immunity. Science 17:1439–1442

Malinovsky FG, Batoux M, Schwessinger B, Youn JH, Stransfeld L, Win J, Kim SK, Zipfel C (2014) Antagonistic regulation of growth and immunity by the Arabidopsis basic helix-loop-helix transcription factor homolog of brassinosteroid enhanced expression2 interacting with increased leaf inclination1 binding bHLH1. Plant Physiol 164:1443–1455

Mathur J, Molnar G, Fujioka S, Takatsuto S, Sakurai A, Yokota T, Adam G, Voigt B, Nagy F, Maas C, Schell J, Koncz C, Szekeres M (1998) Transcription of the *Arabidopsis CPD* gene, encoding a steroidogenic cytochrome P450, is negatively controlled by brassinosteroids. Plant J 14:593–602

Mersmann S, Bourdais G, Rietz S, Robatzek S (2010) Ethylene signaling regulates accumulation of the FLS2 receptor and is required for the oxidative burst contributing to plant immunity. Plant Physiol 154:391–400

Mora-Garcia S, Vert G, Yin Y, Cano-Delgado A, Cheong H, Chory J (2004) Nuclear protein phosphatases with Kelch-repeat domains modulate the response to brassinosteroids in *Arabidopsis*. Genes Dev 18:448–460

Müssig C, Biesgen C, Lisso J, Uwer U, Weiler EW, Altmann T (2000) A novel stress-inducible 12-oxophytodienoate reductase from *Arabidopsis thaliana* provides a potential link between brassinosteroid-action and jasmonic acid synthesis. J Plant Physiol 157:143–152

Müssig C, Fischer S, Altmann T (2002) Brassinosteroid-regulated gene expression. Plant Physiol 129:1241–1251

Muto H, Yabe N, Asami T, Hasunuma K, Yamamoto KT (2004) Overexpression of constitutive differential growth 1 gene, which encodes a RLCK VII-subfamily protein kinase, causes abnormal differential and elongation growth after organ differentiation in *Arabidopsis*. Plant Physiol 136:3124–3133

Nahar K, Kyndt T, Hause B, Höfte M, Gheysen G (2013) Brassinosteroids suppress rice defense against root-knot nematodes through antagonism with jasmonate pathway. Mol Plant Microbe Interact 26:106–115

Nakamura M, Satoh T, Tanaka S, Mochizuki N, Yokota T, Nagatani A (2005) Activation of the cytochrome P450 gene, CYP72C1, reduces the levels of active brassinosteroids in vivo. J Exp Bot 56:833–840

Nakashita H, Yasuda M, Nitta T, Asami T, Fujioka S, Arai Y, Sekimata K, Takatsuto S, Yamaguchi I, Yoshida S (2003) Brassinosteroid functions in a broad range of disease resistance in tobacco and rice. Plant J 33:887–898

Nam KH, Li J (2002) BRI1/BAK1, a receptor kinase pair mediating brassinosteroid signaling. Cell 110:203–212

Nicaise V, Roux M, Zipfel C (2009) Recent advances in PAMP-triggered immunity against bacteria: pattern recognition receptors watch over and raise the alarm. Plant Physiol 150:1638–1647

Noguchi T, Fujioka S, Choe S, Takatsuto S, Yoshida S, Yuan H, Feldmann KA, Tax FE (1999) Brassinosteroid-insensitive dwarf mutants of *Arabidopsis* accumulate brassinosteroids. Plant Physiol 121:743–752

Noguchi T, Fujioka S, Choe S, Takatsuto S, Tax FE, Feldmann KA (2000) Biosynthetic pathways of brassinolide in Arabidopsis. Plant Physiol 124:201–209

Nomura T, Sato T, Bishop GJ, Kamiya Y, Takatsuto S, Yokota T (2001) Accumulation of 6-deoxocathasterone and 6-deoxocastasterone in Arabidopsis, pea and tomato is suggestive of common rate-limiting steps in brassinosteroid biosynthesis. Phytochemistry 57:171–178

Oh MH, Wang X, Kota U, Goshe MB, Clouse SD, Huber SC (2009) Tyrosine phosphorylation of the BRI1 receptor kinase emerges as a component of brassinosteroid signaling in *Arabidopsis*. Proc Natl Acad Sci U S A 107:17827–17832

Oh MH, Wang X, Wu X, Zhao Y, Clouse SD, Huber SC (2010) Autophosphorylation of Tyr-610 in the receptor kinase BAK1 plays a role in brassinosteroid signaling and basal defense gene expression. Proc Natl Acad Sci U S A 107:17827–17832

Oh MH, Wu X, Clouse SD, Huber SC (2011) Functional importance of BAK1 tyrosine phosphorylation in vivo. Plant Signal Behav 6:400–405

Oh MH, Wang X, Clouse SD, Huber SC (2012) Deactivation of the Arabidopsis BRASSINOSTEROID INSENSITIVE 1 (BRI1) receptor kinase by autophosphorylation within the glycine-rich loop. Proc Natl Acad Sci U S A 109:327–332

Ohnishi T, Godza B, Watanabe B, Fujioka S, Hategan L, Ide K, Shibata K, Yokota T, Szekeres M, Mizutani M (2012) CYP90A1/CPD, a brassinosteroid biosynthetic cytochrome P450 of *Arabidopsis*, catalyzes C-3 oxidation. J Biol Chem 287:31551–31560

Owens RA, Tech KB, Shao JY, Sano T, Baker CJ (2012) Global analysis of tomato gene expression during *Potato spindle tuber viroid* infection reveals a complex array of changes affecting hormone signaling. Mol Plant Microbe Interact 25:582–598

Peng P, Yan Z, Zhu Y, Li J (2008) Regulation of the Arabidopsis GSK3-like kinase BRASSINOSTEROID-INSENSITIVE 2 through proteasome-mediated protein degradation. Mol Plant 1:338–346

Peng P, Zhao J, Zhu Y, Asami T, Li J (2010) A direct docking mechanism for a plant GSK3-like kinase to phosphorylate its substrates. J Biol Chem 285:24646–24653

Poppenberger B, Rozhon W, Khan M, Husar S, Adam G, Luschnig C, Fujioka S, Sieberer T (2011) CESTA, a positive regulator of brassinosteroid synthesis. EMBO J 30:1149–1161

Postel S, Kufner I, Beueter C, Mazzotta S, Schwedt A, Borlotti A, Halter T, Kemmerling B, Nürnberger T (2010) The multifunctional leucine-rich repeat receptor kinase BAK1 is implicated in *Arabidopsis* development and immunity. Eur J Cell Biol 89:169–174

Qin X, Liu JH, Zhao WS, Chen XJ, Guo ZJ, Peng YL (2013) Gibberellin 20-oxidase gene *OsGA20ox3* regulates plant stature and disease development in rice. Mol Plant Microbe Interact 26:227–239

Qutob D, Kemmerling B, Brunner F, Küfner I, Engelhardt S, Gust AA, Luberacki B, Seitz HU, Stahl D, Rauhut T, Glawischnig E, Schween G, Lacombe B, Watanabe N, Lam E, Schlichting R, Scheel D, Nau K, Dodt G, Hubert D, Gijzen M, Nürnberger T (2006) Phytotoxicity and innate immune responses induced by NEP1-like proteins. Plant Cell 18:3721–3744

Rozhon W, Mayerhofer J, Petuschnig E, Fujioka S, Jonak C (2010) *Arabidopsis* GSK3, functions in the brassinosteroid signalling pathway. Plant J 62:215–223

Ryan CA, Huffaker A, Yamaguchi Y (2007) New insights into innate immunity in Arabidopsis. Cell Microbiol 9:1902–1908

Ryu H, Kim K, Cho H, Park J, Choe S, Hwang I (2007) Nucleocytoplasmic shuttling of BZR1 mediated by phosphorylation is essential in *Arabidopsis* brassinosteroid signaling. Plant Cell 19:2749–2762

Ryu H, Cho H, Kim K, Hwang I (2010a) Phosphorylation dependent nucleocytoplasmic shuttling of BES1 is a key regulatory event in brassinosteroid signaling. Mol Cell 29:283–290

Ryu H, Kim K, Cho H, Hwang I (2010b) Predominant actions of cytosolic BSU1 and nuclear BIN2 regulate subcellular localization of BES1 in brassinosteroid signaling. Mol Cells 29:291–296

Saijo Y (2010) ER quality control of immune receptors and regulators in plants. Cell Microbiol 12:716–724

Sakamoto T, Morinka Y, Ohnishi T, Sunohara H, Fujioka S, Ueguchi-Tanaka M, Mizutani M, Sakata K, Takatsuto S, Yoshida S, Tanaka H, Kitano H, Matsuoka M (2006) Erect leaves caused by brassinosteroid deficiency increase biomass production and grain yield in rice. Nat Biotechnol 24:105–109

Schulze B, Mentzel T, Jehle A, Mueller K, Beeler S, Boller T, Felix G, Chinchilla D (2010) Rapid heteromerization and phosphorylation of ligand-activated plant transmembrane receptors and their associated kinase BAK1. J Biol Chem 285:9444–9451

Schweizer P, Buchala A, Dudler RA, Metraux JP (1998) Induced systemic resistance in wounded rice plants. Plant J 14:475–481

Schwessinger B, Roux M, Kadota Y, Ntoukakis V, Sklenar J, Jones A, Zipfel C (2011) Phosphorylation-dependent differential regulation of plant growth, cell death, and innate immunity by the regulatory receptor-like kinase BAK1. PLoS Genet 7(4):e1002046

Segonzac C, Zipfel C (2011) Activation of plant pattern-recognition receptors by bacteria. Curr Opin Microbiol 14:54–61

Sehnke PC, DeLille JM, Ferl RJ (2002) Consummating signal transduction: the role of 14-3-3 proteins in the completion of signal-induced transitions in protein activity. Plant Cell 14(Suppl):S339–S354

Shan L, He P, Li J, Heese A, Peck SC, Nümberger T, Martin GB, Sheen J (2008) Bacterial effectors target the common signaling partner BAK1 to disrupt multiple MAMP receptor-signaling complexes and impede plant immunity. Cell Host Microbe 4:17–27

She J, Han Z, Kim TW, Wang J, Cheng W, Chang J, Shi S, Wang J, Yang M, Wang ZY, Chai J (2011) Structural insight into brassinosteroid perception by BRI1. Nature 474:472–476

Shi Y, Zhu S, Mao X, Feng J, Qin Y, Zhang L, Cheng J, Wei LP, Wang ZY, Zhu YX (2006) Transcriptome profiling, molecular biological and physiological studies reveal a major role for ethylene in cotton fiber cell elongation. Plant Cell 18:651–664

Shimada Y, Fujioka S, Miyauchi N, Kushiro M, Takatsuto S, Nomura T, Yokota T, Kamiya Y, Bishop GJ, Yoshida S (2001) Brassinosteroid-6-oxidases from *Arabidopsis* and tomato catalyze multiple C-6 oxidations in brassinosteroid biosynthesis. Plant Physiol 126:770–779

Shimada Y, Goda H, Nakamura A, Takatsuto S, Fujioka S, Yoshida S (2003) Organ-specific expression of brassinosteroid-biosynthetic genes and distribution of endogenous brassinosteroids in *Arabidopsis*. Plant Physiol 131:837–842

Shimada A, Ueguchi-Tanaka M, Sakamoto T, Fujioka S, Takatsuto S, Yoshida S, Sazuka T, Ashikari M, Matsuoka M (2006) The rice SPINDLY gene functions as a negative regulator of gibberellin signaling by controlling the suppressive function of the DELLA protein, SLR1, and modulating brassinosteroid synthesis. Plant J 48:390–402

Shiu SH, Karlowski WM, Pan R, Tzeng YH, Mayer KE, Li WH (2004) Comparative analysis of the receptor-like kinase family in Arabidopsis and rice. Plant Cell 16:1220–1234

Song L, Shi QM, Yang XH, Xu ZH, Xue HW (2009) Membrane steroid-binding protein 1 (MSBP1) negatively regulates brassinosteroid signaling by enhancing the endocytosis of BAK1. Cell Res 19:864–876

Sun Y, Fan X-Y, Cao D-M, He K, Tang W, Zhu J-Y, He J-X, Bai M-Y, Zhu S, Oh E, Patil S, Kim T-W, Ji H, Wong WH, Rhee SY, Wang Z-Y (2010) Integration of brassinosteroid signal transduction with the transcription network for plant growth regulation in *Arabidopsis*. Dev Cell 19:765–777

Tanaka K, Asami T, Yoshida S, Nakamura Y, Matsuo T, Okamoto S (2005) Brassinosteroid homeostasis in Arabidopsis is ensured by feedback expressions of multiple genes involved in its metabolism. Plant Physiol 138:1117–1125

Tang W, Kim TW, Oses-Prieto JA, Sun Y, Deng Z, Zhu S, Wang R, Burlingame AL, Wang Z-Y (2008) Brassinosteroid-signaling kinases (BSKs) mediate signal transduction from the receptor kinase BRI1 in *Arabidopsis*. Science 321:557–560

Tang W, Yuan M, Wang R, Yang Y, Wang C, Oses-Prieto JA, Kim TW, Zhou HW, Deng Z, Gampala SS, Gendron JM, Jonassen EM, Lillo C, DeLong A, Burlingame AL, Sun Y, Wang ZY (2011) PP2A activates brassinosteroid-responsive gene expression and plant growth by dephosphorylating BZR1. Nat Cell Biol 13:124–131

Tsuda K, Katagiri P (2010) Comparing signaling mechanisms engaged in pattern-triggered and effector-triggered immunity. Curr Opin Plant Biol 13:459–465

van Verk MC, Pappaaioannou D, Neeleman L, Bol JF, Linthorst HJM (2008) A novel WRKY transcription factor is required for induction of *PR-1a* gene expression by salicylic acid and bacterial elicitors. Plant Physiol 140:1983–1995

Vert G, Chory J (2006) Downstream nuclear events in brassinosteroid signaling. Nature 441:96–100

Vert G, Walcher CL, Chory J, Nemhauser JL (2008) Integration of auxin and brassinosteroid pathways by Auxin Response Factor 2. Proc Natl Acad Sci U S A 105:9829–9834

Vidhyasekaran P (2007) Fungal pathogenesis in plants and crops: molecular biology and host defense mechanisms, 2nd edn. CRC Press, Taylor Francis Group, Boca Raton, pp 510

Vriet C, Russinova E, Reuzeau C (2012) Boosting crop yields with plant steroids. Plant Cell 24:842–857

Wan J, Zhang X-C, Stacey G (2008) Chitin signaling and plant disease resistance. Plant Signal Behav 3:831–833

Wang Z-Y (2012) Brassinosteroids modulate plant innate immunity at multiple levels. Proc Natl Acad Sci U S A 109:7–8

Wang X, Chory J (2006) Brassinosteroids regulate dissociation of BKI1, a negative regulator of BRI1, from the plasma membrane. Science 313:1118–1122

Wang Z-Y, He J-X (2004) Brassinosteroid signal transduction – choices of signals and receptors. Trends Plant Sci 9:91–96

Wang ZY, Seto H, Fujioka S, Yoshida S, Chory J (2001) BRI1 is a critical component of a plasma-membrane receptor for plant steroids. Nature 410:380–383

Wang ZY, Nakano T, Gendron J, He J, Chen M, Vafeados D, Yang Y, Fujioka S, Yoshida S, Asami T, Chory J (2002) Nuclear –localized BZR1 mediates brassinosteroid-induced growth and feedback suppression of brassinosteroid biosynthesis. Dev Cell 2:505–513

Wang X, Goshe MB, Soderblom EJ, Phinney BS, Kuchar JA, Li J, Asami T, Yoshida S, Huber SC, Clouse SD (2005a) Identification and functional analysis of in vivo phosphorylation sites of the *Arabidopsis* BRASSINOSTEROID-INSENSITIVE 1 receptor kinase. Plant Cell 17:1685–1703

Wang X, Li X, Meisenhelder J, Hunter T, Yoshida S, Asami T, Chory J (2005b) Autoregulation and homodimerization are involved in the activation of the plant steroid receptor BRI1. Dev Cell 8:855–865

Wang X, Kota U, He K, Blackburn K, Li J, Goshe MB, Clouse SD (2008) Sequential transphosphorylation of the BRI1/BAK1 receptor kinase complex impacts early events in brassinosteroid signaling. Dev Cell 15:220–235

Xia X-J, Wang Y-J, Zhou Y-H, Tao Y, Mao W-H, Shi K, Asami T, Chen Z, Yu J-Q (2009) Reactive oxygen species are involved in brassinosteroid-induced stress tolerance in cucumber. Plant Physiol 150:801–804

Xiang T, Zong N, Zou Y, Wu Y, Zhang J, Xing W, Li Y, Tang X, Zhu L, Chai J, Zhou J-M (2008) *Pseudomonas syringae* effector AvrPto blocks innate immunity by targeting receptor kinases. Curr Biol 18:74–80

Yang DH, Hettenhausen C, Baldwin IT, Wu J (2011a) BAK1 regulates the accumulation of jasmonic acid and the levels of trypsin proteinase inhibitors in *Nicotiana attenuata*'s responses to herbivory. J Exp Bot 62:641–652

Yang DH, Hettenhausen C, Baldwin IT, Wu J (2011b) The multifaceted function of BAK1/SERK3: plant immunity to pathogens and responses to insect herbivores. Plant Signal Behav 6:1322–1324

Ye H, Li L, Yin Y (2011) Recent advances in the regulation of brassinosteroid signaling and biosynthesis pathways. J Integr Plant Biol 53:455–468

Yin Y, Wang ZY, Mora-Garcia S, Li J, Yoshida S, Asami T, Chory J (2002) BES1 accumulates in the nucleus in response to brassinosteroids to regulate gene expression and promote stem elongation. Cell 109:181–191

Yin Y, Vafeados D, Tao Y, Yoshida S, Asami T, Chory J (2005) A new class of transcription factors mediates brassinosteroid-regulated gene expression in *Arabidopsis*. Cell 120:249–259

Yokota T, Nomura T, Nakayama M (1997) Identification of brassinosteroids that appear to be derived from campesterol and cholesterol in tomato shoots. Plant Cell Physiol 38:1291–1294

Yu X, Zola J, Aluru M, Ye H, Foudree A, Guo H, Anderson S, Aluru S, Liu P, Rodermel S, Yin Y (2011) A brassinosteroid transcriptional network revealed by genome-wide identification of BES1 target genes in *Arabidopsis thaliana*. Plant J 65:634–646

Yun HS, Bae YH, Lee YJ, Chang SC, Kim SK, Li J, Nam KH (2009) Analysis of phosphorylation of the BRI1/BAK1 complex in Arabidopsis reveals amino acid residues critical for receptor formation and activation of BR signaling. Mol Cells 27:183–190

Zhang J, Zhou J-M (2010) Plant immunity triggered by microbial molecular signatures. Mol Plant 3:783–793

Zhang S, Cai Z, Wang X (2009a) The primary signaling outputs of brassinosteroids are regulated by abscisic acid signaling. Proc Natl Acad Sci U S A 106:4543–4548

Zhang S, Wei Y, Lu Y, Wang X (2009b) Mechanisms of brassinosteroids interacting with multiple hormones. Plant Signal Behav 4:1117–1120

Zhang J, Li W, Xiang T, Liu Z, Laluk K, Ding X, Zou Y, Gao M, Zhang X, Chen S, Mengiste T, Zhang Y, Zhou JM (2010) Receptor-like cytoplasmic kinases integrate signaling from multiple plant immune receptors and are targeted by a *Pseudomonas syringae* effector. Cell Host Microbe 7:290–301

Zhou A, Wang H, Walker JC, Li J (2004) BRL1, a leucine-rich repeat receptor-like protein kinase, is functionally redundant with BR1 in regulating *Arabidopsis* brassinosteroid signaling. Plant J 40:399–409

Zipfel C (2008) Pattern-recognition receptors in plant innate immunity. Curr Opin Immunol 20:10–16

Zipfel C (2009) Early molecular events in PAMP-triggered immunity. Curr Opin Plant Biol 12:414–420

Index

A

AACT in brassinosteroid (BR) biosynthesis, 405
ABA. *See* Abscisic acid (ABA)
ABA1, 276, 280, 281
ABA2 (encoding alcohol dehydrogenase), 250, 252–254, 275, 276, 281
ABA3 (encoding molybdenum cofactor sulfurase), 252, 254, 273
ABA-activated Ca^{2+}-permeable channels, 271
ABA-activated G-proteins, 267
ABA-activated MAPK signaling system, 271
ABA-activated protein kinase (AAPK), 263
ABA-activated R-type (rapid) anion channel, 267
ABA-activated S-type (slow) anion channel, 267
ABA-induced activation of CDPKs, 270
ABA-induced activation of JA biosynthesis, 278, 288
ABA-induced activation of MAPK, 272
ABA-induced Ca^{2+} release into cytosol, 269
ABA-induced cytosolic Ca^{2+} increases, 271
ABA-induced disease resistance, 246, 251, 288
ABA-induced NO in stomatal closure, 271, 287
ABA-induced ROS production, 263, 267, 270–272
ABA-induced stomatal closure immune responses, 271, 287, 292
ABA-induced susceptibility, 247, 250
ABA-induced virus disease resistance, 287
ABA insensitive (ABI) transcription factors
  ABI1, 259, 260, 264, 265, 281, 285
  ABI2, 259, 260
  ABI3, 248, 280
  ABI4, 248
  ABI5, 280
ABA-mediated Ca^{2+} transients, 268
ABA-mediated stomatal closure, 169, 263, 267
ABA-modulated JA, SA, and ET signaling pathways, 290
ABAR/CHLH as ABA receptor, 256
ABA-regulated genes, 276
ABA response element-binding factors (ABFs), 6, 252, 265, 270, 272
ABA-responsive1 (ABR1), 247
ABA-responsive element (ABRE), 6, 245, 270, 272
ABA-responsive genes, 250, 259, 262
ABA-responsive transcription factors, 264
ABA-stimulated calcium-dependent protein kinase1 (ACPK1), 270
ABA-suppressed lignin deposition, 291
ABA-suppressed PAL activity, 278
ABA-suppressed SA biosynthesis, 278
ABA-suppressed SA signaling, 290
ABA-triggered membrane depolarization, 268
ABRE-binding factors. *See* ABA response element-binding factors (ABFs)
ABREBs, 6, 265, 272, 273. *See also* ABA response element-binding factors (ABFs)

P. Vidhyasekaran, *Plant Hormone Signaling Systems in Plant Innate Immunity*,
Signaling and Communication in Plants 2, DOI 10.1007/978-94-017-9285-1

Abscisic acid (ABA)
  aldehyde, 252, 254, 284
  biosynthesis genes, 250
  in induced systemic resistance (ISR), 248
  in JA biosynthesis, 168, 275, 276
  receptors, 1, 6, 124, 168, 245, 256
  signal in cytokinin signaling, 375
  signaling system
    conferring resistance, 285
    conferring susceptibility, 289
    triggers callose deposition, 286
  transport, 274
  transporters
    AtABCG25 in ABA transport, 274
    AtABCG40 in intercellular transport of ABA, 274
  as virulence factor, 292
Abscisic acid oxidase3 (AAO3), 252, 254, 275, 280, 281, 321–323
Abscisic alcohol, 252
ACC oxidase (ACO), 159, 196, 198, 201, 204, 205, 207, 208, 282, 283, 285
ACC synthase (ACS), 195–198, 200–202, 204–208, 282, 333, 334, 436
ACC synthase inhibitor AVG, 334, 436
ACGTGGC motif, 245, 272
ACO gene, 199, 282
ACPK1. *See* ABA-stimulated calcium-dependent protein kinase1 (ACPK1)
ACS genes, 199, 206, 230, 283, 333
Acyl amido synthetase, 319
Acyl-CoA oxidase (ACX), 126, 127, 129, 137
Acyl-CoA synthetase, 129
Adenosine phosphate-isopentenyltransferase (IPT), 360
AFB1/2/3/4/5 auxin receptors, 323–325, 328, 338
AHK2/AHK3/AHK4 cytokinin receptors, 362, 363, 371, 373
AHK sensing and signaling, 363
AHPs. *See Arabidopsis* Hpt (histidine phosphotransfer) proteins (AHPs)
ALD1 in SAR, 69, 70, 72, 73
Alkamide, 157
Allene oxide cyclase (AOC), 126, 127, 130, 137, 138, 164
Allene oxide synthase (AOS), 126, 127, 135–137, 159, 163, 164, 172, 226, 341
Allene oxide synthase (AOS) genes, 333, 340
Alternative oxidase (AOX), 59, 60, 63, 64
Amidase, 314
1-Aminocyclopropane-1-carboxylic acid (ACC), 128, 162, 197, 201–203, 208, 282, 333
Anion channel, 271, 272
Anion channel blockers, 267
Antagonism
  between ABA and ET signaling, 250, 285
  between ABA and JA signaling, 250, 276
  between ABA and SA signaling, 250, 280
  between auxin and ABA signaling, 331
  between GA and BR signaling, 383
  between GA and JA signaling, 383
  between OGs and auxin signaling, 324, 325
  between SA and JA signaling, 81, 96, 162, 163, 331
Antagonistic regulation of PAMP-triggered immunity by bHLH, 433
AOC. *See* Allene oxide cyclase (AOC)
AOS. *See* Allene oxide synthase (AOS)
AOX. *See* Alternative oxidase (AOX)
AOX-mediated alternative respiratory pathway, 59
AOX-mediated redox signaling, 62, 63
AP2/EREBP transcription factors, 154, 219, 359, 366
Arabidopsis histidine kinase (AHK), 364, 365
Arabidopsis Hpt (histidine phosphotransfer) proteins (AHPs), 363, 364
  nuclear translocation, 363
  proteins, 365
Arabidopsis response regulators (ARRs). *See* ARR-dependent transcriptional activation
Archetypal defense hormones, 383
ARF9 enhancing resistance to necrotrophs, 344
ARF1 suppressing resistance to biotrophs, 344
ARR7 (Type-A ARR), 366
ARR-dependent transcriptional activation, 363
Asp–Glu–Leu–Leu–Leu–Ala (DELLA), 384
  GAI, RGA, RGL1, RGL2, RGL3, 386
  gene, 386, 388, 396
  modulated JA responses, 392, 393
  proteins, 1, 9–11, 383, 384, 386–391, 393, 395
  proteins degradation, 383, 387
  repressing GA-dependent defense responses, 383, 386, 387, 392
AUX1, 320, 328, 330, 342
Aux/IAA degradation, 327
AUX/IAA proteins, 1, 7, 314, 318, 322, 323, 328
AUX/IAA repressors, 1, 7, 311, 312, 328

Auxin
application inducing disease resistance, 344
biosynthesis, 323
causing repression of cytokinin responses, 359
conjugation, 319
homoeostasis, 8, 319, 320, 327
involved in modulation
of ABA signaling, 312
of cytokinin signaling, 312
of ET signaling, 312
of JA signaling, 312
of ROS signaling, 312
of SA signaling, 312
perception, 311
receptor TIR1 F-box protein, 7, 311, 312, 314
response factors modulating defense responses, 343
signaling genes, 316
signaling promoting disease resistance, 342, 343
signaling promoting disease susceptibility, 336–338
signaling repression by SA, 327
signaling suppression by small RNA signaling, 336
transport, 320, 395
transporters, 323
transport system in promoting susceptibility, 342
Auxin-activated G-proteins, 312
Auxin-activated H^+-ATPases, 312
Auxin-activated K^+ channels, 312
Auxin-induced suppression of SA signaling, 328, 329
Auxin-inducible genes
Aux/IAA genes, 316, 322
Gretchen Hagen3 (GH3s), 316, 329 (*see also* GH3 genes)
small auxin upregulated RNAs (SAURs), 316, 322
Auxin-promoted susceptibility, 324, 337
Auxin-response element (ARE) TGTCTC, 315, 316
Auxin response factor (ARF), 1, 311, 312, 315, 317–319, 322, 323, 335, 343
Auxin-responsive genes, 311, 315, 317, 323, 334
Auxin-responsive genes promoting disease resistance, 342
Auxin responsive protein (ARP), 328
AUX–LAX family of amino-acid permease-like proteins, 320
AVG (ethylene biosynthesis inhibitor), 197
AXR1, 332, 333, 338, 344
AXR4, 342
AXR6, 344
Azelaic acid, 330

B

BABA-induced disease resistance, 286
Backbone of plant immune system, 383
BAK1, 9, 10, 403, 413–416, 419, 420, 430, 434
BAK1-activated JA signaling system, 428
BAK1 in immune signaling, 428
BAK1-mediated PAMP–PRR signaling, 431
BAK1-mediated programmed cell death, 432
BAK1-triggered disease resistance, 427
BEE transcription factor, 418
Benzoic acid 2-hydroxylase, 27, 44
BES1, 410, 415, 416, 419, 421, 422, 436
β-oxidation, 126, 127, 136
β-oxidation pathway, 30, 127
β-thujaplicin, 159, 225
bHLH transcription factor HBI1, 433
BIG gene encoding calossin-like protein, 342
BIK1, 416, 430, 434
BIM1, 410
BIN2, 410, 411, 415, 417, 419, 421–423, 434, 436
Bioactive GAs, 384, 385
BKI1, 415, 417, 418
Blocking of SAR by ABA, 279
BOS1 in ABA signaling, 273, 275
Branches of tryptophan-dependent pathway, 312
Brassinazole (BRZ), 428, 436
Brassinolide (BL)
biosynthesis pathways, 407, 408
promoted susceptibility, 428
Brassinosteroids (BRs), 9, 383, 403, 404
biosynthesis, 405–409, 425
biosynthetic gene
CPD1, 405, 410, 425
DWARF2, 405, 425
homeostasis, 408–410, 425, 426
in negative regulation of GA-induced immune responses, 383, 429
receptors, 9, 403
signaling-induced susceptibility, 428
signaling pathway, 417

Brassinosteroids (BRs) (*cont.*)
signaling regulators
BSK1, 403, 415, 416
BSU1, 403, 410, 415, 420, 422
CDG1, 403, 416
PP2A, 403, 415–417
signaling-triggered disease resistance responses, 427
signal processing genes, 425
Brassinosteroids-modulated plant immune responses, 404
BRI1, 9, 10, 403, 410–413, 415–420, 430, 432, 434
BRI1-activated inhibition of PAMP-triggered immunity, 433
BRI1–BAK1 interaction, 418
BRI1–BSK1 interaction, 418
BRI1–BSK3 interaction, 418
BR-induced susceptibility, 429
BR-inducible gene OsBLE2, 405
BRL1, 412
BR-mediated negative regulation of GA signaling, 429
BR-mediated suppression of SA defenses, 429
BR-responsive genes, 416, 425
BRs. *See* Brassinosteroids (BRs)
BR signaling kinases (BSKs), 418
BR-triggered susceptibility, 429
BRZ-triggered disease resistance, 429
BSK1, 416–419, 434
BSK2, 418
BSK3, 418, 419
BSU1 phosphatase, 416, 417, 419, 421, 434
bZIP transcription factors, 245, 255, 262, 273, 411, 418
BZR transcription factors, 404, 410, 415–417, 424
BZR1, 415–419, 421–424, 433, 434
BZR1-mediated suppression of PAMP–PRR signaling system, 433
BZR2, 417, 419, 421–423

C

Calcineurin B-like (CBL) protein kinase, 270
Calcium (Ca^{2+})
dependent protein kinase, 195, 202, 263, 269, 270
influx, 123, 133–135, 201, 202, 224, 225, 260, 261, 266–268
mediated ethylene biosynthesis, 201
mediated NO signaling, 123
ion channel blockers, 201
ion channels, 123, 202, 266
ion chelator EGTA, 201, 224, 225
responsive cis elements, 270
sensors, 41, 195, 269
signaling, 41, 202, 287
signals, 202
signature, 27, 40–42, 123, 201, 202, 268
signature ûtriggered JA biosynthesis, 134
wave, 123, 134, 195, 202
Callose deposition, 7, 58, 249, 368
Calmodulin (CAM), 41, 42, 225
Calmodulin-binding protein CBP60g, 27, 41, 42
Calmodulin-binding protein EICBP, 225
Calyculin A, 198
Camalexin, 368
CAM binding domain, 225
Campestanol, 407
Campesterol, 406
CAMTA transcription factors, 42
Castasterone, 407, 409, 426
CDG1 in BR signal transduction, 420
CDL1, 420
CDP. *See* *ent*-Copalyl diphosphate (CDP)
CDR1 gene in SA biosynthesis, 35, 36
Cell wall fortification, 249, 290
Chorismate, 32
Chromatin modification-induced priming, 28, 78
Chromatin remodeling, 3, 28
CIPKs, 270
CKX4/CKX5 cytokinin metabolism genes, 366, 375, 376
Cofactor of TGA transcription factors, 56
COI1-dependent pathway, 126
COI1-independent pathway, 126
COI1 in JA signal perception, 123, 139, 331, 332
COI1-JAZ interaction, 143
COI1-JAZ receptor complex, 140, 164
COI1 protein, 142, 153, 331
Conjugated auxin, 341
Conjugated auxin-promoted susceptibility, 341
Constitutive triple response1 (CTR1), 160, 195, 209–214, 216, 218, 219, 222, 223
Cooperative function
of JA and ABA signaling systems, 276
of JA and ET signaling pathways, 158, 159
ent-Copalyl diphosphate (CDP), 384, 385
ent-Copalyl diphosphate synthase (CPS), 384, 385
Copper transporter (RAN1), 210

COP9 signalosome (CSN), 143, 213
Coronatine, 199, 292, 391
C-22 oxidation branch in BR biosynthesis pathway, 408
CPD1, 422
Crosstalk
  between BR and ABA signaling, 436
  between BR and ET signaling, 436
  between BR and GA signaling, 436
  between BR and JA signaling, 435
  between BR and SA signaling, 435
  between BR biosynthesis pathway and PAMP–PRR signaling, 431
  between JA and small RNA signaling system, 169
Cryptogein, 135, 199
CTR1–MKK9–MPK3/MPK6 cascade, 222, 223
CUL3 E3 ligase, 54
Cullin RING ligases, 208
Cuticle in induced SAR, 396
Cyclase- and HK-associated sensing extracellular (CHASE), 364
Cytochrome P450, 57, 312
Cytokinin
  acting as long-range signals, 360
  biosynthesis, 360–362
  concentrations altering immune responses, 360
  conferring resistance against pathogens, 367
  degradation, 362
  homeostasis, 359, 362
  hydroxylases, 374
  in NPR1/TGA3-mediated signaling, 373
  phosphorelay signaling system, 363–365
  phosphoribohydrolase (Lonely Guy), 374
  receptors, 362, 363, 371
  in SA biosynthesis, 36
  synthase, 362
Cytokinin–auxin antagonism, 376
Cytokinin-binding CHASE domain, 363
Cytokinin-induced activation of plant innate immunity, 369
Cytokinin-induced disease susceptibility, 370
Cytokinin-induced priming, 368, 369
Cytokinin-induced TGA3/NPR1, 369
Cytokinin-modulated redox signaling system, 370, 371
Cytokinin-modulated SA signaling system, 369
Cytokinin oxidase/dehydrogenase (CKX), 362, 366, 371, 374
Cytokinin-responsive genes, 366
Cytokinin-triggered defense responses, 367

D
DAD1 in JA biosynthesis, 125
DELLA. *See* Asp–Glu–Leu–Leu–Leu–Ala (DELLA)
DGL1 in JA biosynthesis, 125
Diacylglycerol kinase (DAG), 222
Dimethylallyl diphosphate (DMAPP), 360, 361, 375
Dimethylallyl pyrophosphate (DMPP), 406
DNA methylation in transgenerational SAR, 28, 80, 81
Dormant MAPKs in SA-triggered priming, 76

E
Early C6-oxidation pathway, 407, 408
EAR motif, 123, 141
EBF1/EBF2 mRNA decay, 217
EDR1
  in ET signaling, 207, 208
  in JA signaling, 151, 152
  in SA signaling, 43, 49
EDS1
  in SA biosynthesis, 31–34, 36, 42, 280
  in SA/JA crosstalk, 87
EDS4 in SA biosynthesis, 31, 33, 34, 36
EDS5/SID1 in SA biosynthesis, 31, 33, 34, 36
Effectors suppressing ABA-dependent defenses, 292
Effectors suppressing SA signaling system, 94
EIL1
  in JA/ET signaling synergy, 160, 161
  transcription factor, 160, 161, 195, 214, 216, 231, 284
EIN2 in ET signaling, 160, 195–197, 212–214, 230, 231, 284
EIN2-targeting protein1 (ETP1), 213, 214
EIN2-targeting protein2 (ETP2), 213, 214
EIN3-binding F-box proteins, 160
EIN3 binding F-box protein 1 (EBF1), 160, 195, 215–217
EIN3 binding F-box protein 2 (EBF2), 160, 195, 215–217
EIN3/EIL family of proteins, 215, 227
EIN3 in ET signaling, 160, 161, 195, 213–217, 221, 224, 284
EIN3 in JA/ET signaling synergy, 160, 161
EIN4, 208–211, 217
EIN5/XRN4, 217
EIN7, 217
Endogenous immune signal, 29

24-epibrassinolide (epiBL), 433–435
Epoxycarotenoid, 252, 254, 284
EREBP2/AP2 domain proteins, 203, 221, 229
ERF
  branch pathway, 165
  in modulation of JA and ET signaling pathways, 159, 160
  proteins in ET signaling, 282
  transcription factors, 51, 147–149, 159, 160, 164, 219, 221, 226, 229, 230, 282–284
ERF-regulated JA-responsive gene expression, 159
ERF1, 147–149, 154, 155, 159, 170, 219, 221, 226, 229
ERF1 in JA and ET pathways convergence, 159, 160, 170
ERS1, 208–211
ERS2, 208–210
ET-binding domain (EBD), 209
Ethephon, 197, 230
Ethylene (ET)
  biosynthesis, 11, 159, 199
  and JA signaling interdependency, 225, 226
  in PAMP-triggered immunity, 224
  pathway-induced susceptibility, 230
  receptors, 195, 208
  signaling-triggered ABA biosynthesis, 283
Ethylene enhanced response5 (EER5), 213
ET-induced calmodulin-binding protein (EICBP), 225
ET-induced disease resistance, 196, 197, 231
ET-induced JA biosynthesis, 159, 226
ET-induced susceptibility, 197
ET-induced transcription of defense genes, 228, 229
ET/JA coactions, 226
ET/JA signaling, 161, 225
ET-mediated gene expression, 159
ETR1, 197, 208–211, 213, 218, 219
ETR2, 208–210, 212, 334
ETR1 as ROS sensor, 221
ET-responsive genes, 221
E3 ubiquitin ligase, 142, 143, 153, 208, 215, 317, 387, 391
E3-ubiquitin ligase SCFSLY1/GID2 complex, 387, 391
EUI-deactivating biologically active Gas, 389, 390
Exoribonuclease4 (XRN4), 216, 217
Expansin gene expression, 342, 343

F
Farnesyl pyrophosphate, 406
Fatty acid desaturase7 (FAD7) in SAR, 69
Fatty acid hydroperoxides, 125
F-box proteins, 1, 9, 383, 384
F-box proteins SLY1and SNZ in GA signaling, 384
FMO1 gene in SA biosynthesis, 36
FMO1 in SAR, 64, 70–72
14-3-3 proteins, 423, 424

G
GA. *See* Gibberellic acid (GA)
GA12 aldehyde, 384, 385
GA-degrading enzyme gibberellin 2-β-dioxygenase, 394
GA–GID1–DELLA complex, 9, 383, 386
GA-modulated disease resistance, 383
GA-modulated disease susceptibility, 383
GA3 oxidase (GA3ox), 384, 385
GA13 oxidase, 384, 385
GA20 oxidase (GA20ox), 384, 385, 388
Gα subunit-accelerating protein (GAP), 253
Gα subunit of heterotrimeric G-protein (GPA1), 131, 257, 266
GATA family transcription factors, 366
GA-triggered disease resistance, 388
GA-triggered disease susceptibility, 388
G-box motif, 164, 411
GCC-box-mediated transcription, 5, 220
GCC-box motifs, 5, 51, 88, 124, 160, 164, 165, 195, 219, 220, 226, 229, 230, 249, 282
GCR1, 266
GCR2, 257
Geranylgeranyl diphosphate, 384
GH3 enzymes, 7, 128, 319, 326, 341
GH3 genes, 325, 326, 329, 340, 342
GH3-modulated SA signaling, 329
GH3 in suppression of free IAA accumulation, 340
GH3 in suppression of SA, 340
GH3.1 gene overexpressing plants, 343
GH3-2 gene, 343
GH3-2 gene overexpressing plants, 343
GH3.5 gene, 34, 329
GH3-8, 320, 333, 339, 340
GH3-8 mediated suppression of auxin signaling, 340
GH3-8 overexpressing plants, 340–342
Gibberellic acid (GA)
  attenuates expression of JA-responsive genes, 383

biosynthesis inhibitor uniconazole, 11
enhancing induction of SAR, 384
receptor GID1, 1, 9, 383
relieving repressive activity of DELLAs, 383, 387
repressor genes, 395
repressor protein, 395
signaling pathway, 8, 384–386
triggering SAR, 383, 395, 396
Gibberellin Insensitive Dwarf2 (GID2), 384, 387
Gibberellin 20-oxidase3, 394
GID1, 383–387, 391
GID1 genes, 384, 385
Glucosyltransferase, 165
Glutamate-gated Ca^{2+} channel, 134
Glutamate receptors, 134, 135
Glutaredoxin in SA/JA crosstalk, 85, 86, 167
Glutathione-S-transferase (GST), 57
Glyceraldehyde 3-phosphate, 251
GPCR as ABA receptor, 257
G-protein-coupled receptors (GPCRs), 253
G-proteins, 39, 123, 130, 200
in expression of NADPH oxidase, 130
involved in ABA biosynthesis, 253
involved in ABA downstream signaling, 265
switching on calcium ion channels, 123
switching on Ca^{2+} signaling system, 267
switching on ET biosynthesis pathway, 195, 200
switching on JA biosynthesis pathway, 130
switching on ROS signaling system, 267
G-proteins-triggered polyamine biosynthesis, 131
GRAS (GAI1, RGA, SCARECROW), 386
GTG1/GTG2, 257, 258

H

Heterotrimeric G-proteins, 130, 265
Hexanoic acid, 5, 172, 174
Hijacking
of ABA signaling system by pathogens, 291
of BR signaling machinery by pathogens, 435
of ET signaling, 197
His phosphotransfer protein (HPT), 365
Histidine and aspartate phosphorelay, 363
Histidine kinase receptors, 359, 362, 363
His to Asp phosphotransfer signal transduction, 363
Histone
acetylation, 153
deacetylases in JA signaling, 123, 153
memory in SAR, 28
modifications, 3, 28
modifications in gene priming, 77
replacement in priming of SA-responsive loci, 79
HMGR in BR biosynthesis, 405
HMGS in BR biosynthesis, 405
H_2O_2. *See* Hydrogen peroxide (H_2O_2)
Host-associated molecular pattern (HAMP), 153, 199, 205, 324
13-HPOT, 126, 127
Hydrogen peroxide (H_2O_2)
in Ca^{2+} signaling system, 123
in ethylene biosynthesis, 203
production, 203
in SA biosynthesis, 45
signaling, 221
N-Hydroxy-TAM (HTAM), 314
Hyperpolarization-activated Ca^{2+} channels, 268
Hypomethylated genes in priming, 28

I

IAA–alanine, 319
IAA–amido synthase, 320, 322, 323, 327
IAA–amino acid hydrolases, 322, 323
IAA–Asp, 7, 319, 320, 341, 342
IAA/Aux family genes, 327
IAA–AUX proteins. *See* AUX/IAA proteins
IAA/AUX proteins degradation, 344
IAA–glutamate, 319
IAA–leucine, 319
ICS1, 371
Indole-3-acetaldehyde (IAAld), 313, 321
Indole-3-acetaldoxime (IAOx), 312, 313, 330
Indole-3-acetamide (IAM) pathway, 311–314
Indole-3-acetic acid (IAA), 311, 313, 321
homeostasis, 320, 326
production, 329, 337
transport, 8, 334
Indole-3-acetonitrile (IAN) pathway, 311, 312, 321, 330
Indole-3-pyruvic acid (IPA) pathway, 311–313
Induced systemic immunity, 171
Induced systemic resistance (ISR), 1, 4, 124, 171, 172, 196, 342
Inositol pentakisphosphate (InsP5), 140
Interaction between auxin and ET signaling system, 333

Interaction between auxin and SA signaling system, 327
Interaction between auxin and small RNA signaling systems, 335
Interaction between cytokinin and ABA signaling systems, 374. *See also* Interplay between cytokinin and ABA signaling
Interaction between JAZ1 and COI1, 140
Interaction between JAZ proteins and DELLAs, 391
Interaction between TIR1 and Aux/IAA protein, 315
Interconnection between ABA and JA signaling pathways, 274
Interplay between ABA and ET signaling systems, 282
Interplay between ABA and SA signaling systems, 278, 280
Interplay between auxin and JA signaling system, 332
Interplay between auxin and MAPK signaling systems, 326
Interplay between BR and PAMP–PRR signaling, 430
Interplay between cytokinin and ABA signaling, 375
Interplay between cytokinin and auxin signaling, 376
Interplay between GA and auxin signaling, 395
Interplay between GA and BR signaling, 394
Interplay between GA and JA signaling, 391
Interplay between GA and SA signaling, 389, 390
Interplay between JA and ABA signaling, 168, 274
Inward-rectifying K^+ channel, 287
IPT. *See* Isopentenyl transferase (IPT)
Isochorismate, 32, 279
Isochorismate pathway, 29, 31, 279
Isochorismate synthase (ICS), 31, 33, 35–38, 42, 64, 72, 94, 279, 371, 372
Isopentenyladenine (iP), 361
Isopentenyladenine-type cytokinins, 360
Isopentenyl AMP (iAMP), 361
Isopentenyl diphosphate, 251
Isopentenyl pyrophosphate, 406
Isopentenyl transferase (IPT), 360–362, 367, 374–376
Isopetenyl adenosine (iP-Ado), 361
ISR-prime, 5, 174

J

JA-ACC, 128
JA-activated defense genes, 156, 157
JA-amino acid synthetase (JAR1), 128, 143, 332, 333
JA-amino synthetase, 332
JA/ET crosstalk, 10
JA/GA crosstalk, 10
JA-Ile as the mobile signal in ISR, 173
JA-Ile/SCFCOI1 pathway, 144
JA-induced enhanced expression of DELLA genes, 391
JA-Leu, 128
JA-mediated priming, 174
JA-responsive genes, 5, 161–163, 165, 167, 168, 174, 226
Jasmonate (JA)
  biosynthesis blocking by SA, 81
  biosynthesis pathway, 127
  and ET interaction, 159, 225
  and ET signaling system, 162
  in inhibition of SA signaling, 163
  metabolites as signal molecules, 124
  proteasome pathway, 332
  receptor complex
    COI1, 4, 123, 142, 277
    inositol pentakisphosphate, 4, 123, 140
    JAZ proteins, 4, 123, 141
  and SA interaction, 11, 12, 81–88, 124
  signaling suppression by JAZ proteins, 141
Jasmonic acid carboxyl methyltransferase (JMT), 127, 128
Jasmonoyl-L-isoleucine (JA-Ile), 4, 123–125, 128, 140, 142, 143, 149, 152, 165, 172–174, 436
JA-Trp, 128
JA-Val, 128
JAZ
  degradation, 4, 142, 144, 147, 227, 392
  degradation signal (degron), 144
  dimerization, 141
  protein JAZ9 inhibits RGA DELLA protein, 391
  proteins, 140–142, 144, 158, 161, 226, 227, 391, 394

K

ent-Kaurene, 384, 385
ent-Kaurene synthase (KS), 384, 385
ent-Kaurenoic acid, 384, 385
ent-Kaurenoic acid oxidase (KAO), 384, 385
K^+ current rectifier KAT1, 260, 263
3-Ketoacyl-CoA, 126

3-Ketoacyl-CoA thiolase (KAT), 126, 127, 137
Kinetin, 367

L
Late C6-oxidation pathway, 407, 408
LAX genes, 322
Lignin, 249, 290
Link between Ca^{2+} signaling and ABA signaling systems, 270
Link between JA and auxin signaling, 333
Linolenic acid, 125–127, 129, 136
Lipase in JA biosynthesis, 135
Lipases in synthesis of oxylipins, 125
Lipid hydroperoxidase (LHP), 173
Lipid transfer protein (DIR1), 330
Lipoxygenase (LOX), 125, 126, 129, 130, 133–136, 146, 163, 164, 170, 173, 341
Lipoxygenase pathway, 173
L-Methionine, 197
Lonely Guy, 374
Long-distance mobile signals in SAR, 65, 66. *See also* Systemic mobile signal molecules involved in SAR
13-LOX enzymes, 125
LOX gene, 333, 340
Lycopene, 251

M
MADS-box gene family, 425, 426
Manipulation of signaling network, 12
MAP kinase (MAPK), 198, 261
  cascade, 27, 43, 48, 123, 137, 138, 195, 204, 205, 207, 222, 272, 326
  cascade in ethylene biosynthesis, 195, 205
  dormant MAPKs in SA-triggered priming, 76
  in JA biosynthesis, 123, 135, 137, 138
  in JA signaling pathway, 150, 151
  MPK4 in ABA-induced stomatal closure, 288
  MPK4 in SA/JA crosstalk, 86
  in SA biosynthesis, 43, 44
  signaling module, 43
  signaling system, 39, 41, 48
MAPK phosphatases (MKPs), 138
MED8 in JA signaling, 4, 150
MED15/NRB4 in regulation of SAR, 3, 75
MED16
  in JA signaling, 4
  in NPR1 function, 3
  in regulation of SAR, 3, 74
Med16/SFR6 in JA signaling, 150
MED25
  in ABA and JA signaling interplay, 277
  in ABA signaling, 245, 277
  association with MYC2, 277
  interaction with transcription factors, 277
  in JA signaling, 4, 123, 147–149, 158, 277
  physical association with ABI5, 277
Mediator complex
  in JA signaling, 147
  in SAR, 73, 74
Mediators in SA signaling, 28
Membrane steroid-binding protein 1 (MSBP1), 420
Methionine in ET biosynthesis pathway, 282
Methionine synthase, 200
Methyl esterase1 (MES1) in SAR, 64
Methyl jasmonate (MeJA), 126, 128, 155, 157, 159, 163, 165, 168, 172, 225, 275
Mevalonate
  diphosphate decarboxylase, 406
  pathway, 405, 406
Mevalonate kinase (MVK), 405, 406
MicroRNA, 170, 335
  miR160a in repression of auxin response factor genes, 335
  miR167 in auxin signaling, 324
  miR393 in auxin signaling, 323, 324, 335, 336
  miR393 targeting auxin receptors, 335
  repression of auxin signaling, 335
MicroRNA-directed RNA silencing, 124, 170
Mitochondrial signaling processes, 59
MKK3–MPK6 cascade in JA signaling, 150, 151
Mobile signaling cue in ISR, 123
Mobile signal involved in ISR, 172
Molecular glue, 383, 386
MPK3/MPK6 in EIN3 protein stabilization, 222
MPK3/MPK6 in ethylene biosynthesis, 205–207
MPK4 in ET signaling, 223
MPK4-mediated ABA signaling, 169
Multifunctional protein (MFP), 126, 127, 137
MYB44 in JA/SA crosstalk, 167
MYB96 as a signaling link in integration of ABA and SA signals, 281
MYB96-induced *SID2* expression, 282
MYB transcription factors, 36–38, 141, 154, 155, 166, 273, 276, 277, 359, 366

MYC branch pathway, 165
MYC-type transcription factors, 164
MYC2
in ABA signaling, 168, 273, 276
in inducing ISR, 146
induction by ABA, 277
interaction with JAZ proteins, 144, 145, 161, 392, 393
as master switch in JA signaling, 146
regulates JA signaling, 277, 394
as transcription factor, 4, 95, 123, 141, 142, 144, 145, 147–152, 154, 155, 168, 227, 277, 392, 393
MYC3, 146, 147, 154, 155, 161, 393, 394
MYC4, 146, 147, 154, 155, 161, 393, 394
Myrosinases, 276

N
NAC transcription factors, 96, 154, 155, 247
NADPH
in cytosol, 271
oxidase
in ABA signaling, 271
in ROS production, 123, 130, 134, 153, 260, 263, 270
in SA biosynthesis pathway, 45
NADPH oxidase-dependent ABA signaling, 270
NCED. *See* 9-*cis*-epoxycarotenoid dioxygenase (NCED)
Negative crosstalk between BR and SA signaling systems, 428
Negative regulation of BR signaling by BIN2, 421
Neoxanthin, 251, 252
Next-generation SAR, 1, 3, 28, 80
9-*cis*-epoxycarotenoid dioxygenase (NCED), 247, 252, 253, 275, 283, 285
9-*cis*-epoxycarotenoid dioxygenase3, 168
NINJA, 4, 123, 141, 142
NINJA-TPL corepressor complex, 123, 141
Nitrate reductase in NO production, 271
Nitric oxide (NO)
in ABA signaling system, 271
in auxin signaling, 312, 327
in ethylene biosynthesis enzymes, 204
in ethylene production, 195, 204
in JA biosynthesis, 135, 136
production, 135
in SA biosynthesis pathway, 45
in SA signaling system, 46
signaling, 39, 45, 46, 48
Nitric oxide synthase (NOS)-like enzyme, 271
Nitrilase, 312, 313, 321
Nitrosoglutathione as long-distance signal, 48
NO-induced SAM synthetase, 197
Non-β-oxidation pathway, 30
Non-expresser of PR gene1 (NPR1)
binding with NPR4, 54
in chromatin modification-induced priming, 78
as a co-activator in SA signaling, 272
cytosolic NPR1, 166
degradation of NPR1, 54
interacting proteins, 55
master regulator of SA signaling, 27, 52, 74
monomeric NPR1, 52, 53
negative effect over ABA biosynthesis genes, 281
nuclear localization, 3, 27, 52
oligomerization, 48, 53
proteasomal degradation, 53
regulation by MED16, 3
regulator of SAR, 48
in SA/JA crosstalk, 83, 84, 165, 167
TGA2-NPR1 complex, 3, 52, 372
transcriptional co-activator, 372
triggering SAR, 64, 75
Nonpolar auxin transport, 320
NPR1-activated transcription factors, 55
NPR3, 28, 54, 55
NPR4, 28, 54, 55

O
Oligogalacturonide (OG) elicitor, 200, 205, 324, 325
OPC-8:0, 126, 127
OPDA reductase (OPR), 126, 127, 129–131, 135–137, 164, 172
ORA59 transcription factor, 147–149, 154, 161, 162, 164, 165, 250
OsBLE2, 427, 428
OsCPD1, 427, 428
OsEDR1, 208
OsERF922, 247, 284, 289
OsHK3, 375
OsNCED3,4 in ABA biosynthesis, 284
OsNPR1, 290, 428
OsRac1, 200
OsRAVL1, 425, 426, 428
OsRR4, 375
Oxalic acid as a virulence factor, 292
Oxidative burst, 202
12-Oxo-phytodienoic acid (OPDA), 126, 127, 174, 275, 436
Oxylipins, 125

P

PAD4
in SA biosynthesis, 31–34, 36, 41, 72, 340
in SA/JA crosstalk, 87
PAMP. *See* Pathogen-associated molecular pattern (PAMP)
PAMP–PRR signaling complex, 1, 10, 124, 137, 223, 224, 430–433, 435
PAMP-triggered ABA biosynthesis, 251
PAMP-triggered expression of small RNAs, 335
PAMP-triggered immunity (PTI), 435
Pathogen-associated molecular pattern (PAMP)
cues-activated JA biosynthesis, 123, 130
signals activated G-proteins, 123, 134
Pathogen's signature, 27
Pathogen-triggered BR biosynthesis, 404, 405
Pathogen-triggered BR signaling system, 425
PBS3 (GH3.12) in SA biosynthesis, 38, 47
Pep1 peptide, 153
PEPRs, 153
PG-inhibiting proteins (PGIP), 325
Phenylalanine ammonia lyase (PAL), 27, 37, 45, 249, 270, 278, 289, 291
Phenylalanine pathway, 29, 31, 39
Phenylpropanoid pathway, 290
Phosphatidic acid (PA), 222, 264, 265, 287
Phosphatidylcholine, 264
Phospholipase A1 (PLA1), 125, 127, 138
Phospholipase A enzyme PLA1γ1, 125, 126
Phospholipase C, 222
Phospholipase D (PLD), 264, 265, 287
Phytoalexins, 368
Phytoene, 251
PIN genes, 322, 327
PIN proteins, 320, 328, 342, 395
Polar auxin transport (PAT), 8, 320, 342
Polyamines, 131
biosynthesis, 131, 132
in ROS production, 133
spermine, 131, 133
spermine in JA biosynthesis, 133
PP2A, 9, 276, 422
PP2C. *See* Type 2C protein phosphatase (PP2C)
PP2C–SnRK2 complex, 248, 255, 265
Prenylation, 362
Priming in ISR, 173, 369
Priming in SAR, 369
Priming of defense responses, 1, 28, 29, 330, 368
Priming of JA-dependent responses, 5, 174
Probenazole, 331
Proteasomal degradation, 216, 387
Proteasomal degradation of DELLAs, 387
Proteasome, 54, 208, 316
Proteasome inhibitor MG132, 418, 421
PRR-mediated immune signaling inhibition by BR perception, 435
PYR/PYL/RCAR, 1, 6, 168, 245, 254, 258, 259, 265, 274

R

RBOH-mediated ROS signaling, 131, 249
RCN1 involved in JA and ABA signaling, 276
Reactive nitrogen species (RNS), 221
Reactive oxygen species (ROS)
in ethylene signaling system, 203
in JA biosynthesis pathway, 135, 138
production, 133, 202, 249, 260, 261, 270, 287
in SA signaling system, 39, 47
sensor, 221
signaling downstream of ABA signaling system, 270
signaling in SA biosynthesis, 44, 53
Reciprocal antagonistic interaction between ABA and SA signaling systems, 280
Regulator of G-protein signaling (RGS), 253, 254
Regulatory proteins in SA biosynthesis, 31–34
Repression of auxin-responsive genes, 318
Respiratory burst oxidase (RBOH), 130
Responsive-to-antagonist1 (RAN1), 210, 211
Reversion-to-ethylene sensitivity1 (RTE1), 218
RNA-binding protein (RBP), 34, 35
RNA-dependent RNA polymerase (RDR/RDRP), 60, 61, 63
RNA polymerase II (RNAPII), 245, 277
RNA silencing, 4, 169
RNA silencing signal, 169
ROS. *See* Reactive oxygen species (ROS)
ROS-mediated ABA signaling, 272
RUB-conjugating enzyme RCE1, 208
RUB protein, 332
RUB1/RUB2, 208

S

SA. *See* Salicylic acid (SA)
SA/ABA crosstalk, 10, 11, 89
SA/auxin crosstalk, 10, 89
SA-binding protein (SABP), 47, 66, 67
SA/BR crosstalk, 90

SA/cytokinin crosstalk, 11
S-adenosyl-L-methionine (SAM), 65, 201, 204
S-adenosyl methionine synthetase, 200, 203, 294
S-AdoMet, 197, 198
S-AdoMet (SAM) synthetase, 197, 282
SA/ET crosstalk, 10, 89
SA-induced disease resistance, 90–93, 162, 163
SA-induced transcription of defense genes, 58
SA-induced virus disease resistance, 93
SA/JA crosstalk, 10, 81–88, 165
SA–JA–ET signaling systems as central backbone of innate immunity, 383
Salicylhydroxamic acid (SHAM), 59, 60
Salicylic acid (SA)
  analog BTH-induced priming, 76, 77
  biosynthesis, 32
  biosynthesis suppression by ABA, 278, 279
  in inhibition of JA biosynthesis, 83
  and JA/ET crosstalk, 162
  signaling inhibition by JA, 88
  signaling system suppression by ABA, 279
Salicylic acid carboxyl methyltransferase (SAMT), 66, 67
Salicylic acid-induced protein kinase (SIPK), 43, 50, 138, 207
SA-mediated suppression of JA signaling, 82, 87, 88
SAM synthetase, 203, 204
SA-O-β-glucoside (SAG), 33, 38, 46, 47
SAR. *See* Systemic acquired resistance (SAR)
SAR-definition, 3
SAR-induced by GA, 383
SAR priming, 72
SA-triggered priming, 75, 76
SAUR genes, 322, 323, 325, 327
SCFCOI1-E3 ubiquitin ligase, 143, 213, 215, 217, 332
SCFCOI1-mediated ubiquitination, 4, 140, 142, 143
SCFSLY1/GID2 complex. *See* E3-ubiquitin ligase SCFSLY1/GID2 complex
SCFTIR1-mediated degradation of AUX/IAA proteins, 316, 323
SCF ubiquitin ligase, 1, 142, 215, 312, 314, 331, 343, 344
SFD1 in SAR-inducing activity, 69
SGT1B in auxin signaling, 316, 338, 339
sgt1b mutant, 338
Shikimic acid pathway, 32, 279
SID2 in SA biosynthesis, 31, 33–35, 37, 40, 41, 280, 281
SLAC1, 255, 260, 262, 263
SLEEPY1 (SLY1) in GA signaling, 384
Slow (S)-type anion channel, 260, 266, 269, 276
Small-interfering RNAs (siRNAs), 60, 61, 169
Small monomeric G-proteins, 130, 266
Small RNA-directed DNA methylation, 335
Small RNA-directed mRNA cleavage, 335
Small RNA-directed RNA silencing, 3, 60, 61, 169, 170
Small RNA-mediated suppression of auxin signaling, 336
Small RNAs, 63, 169
  in auxin signaling system, 335
  molecules as mobile signals, 335
  signaling system, 170
Small ubiquitin-like modifier (SUMO). *See* SUMO in SA signaling system
SNARE in ABA-dependent disease resistance, 286
Sneezy (SNZ) in GA signaling, 384
SNF1-related protein kinase (SnRK2), 1, 6, 245, 254, 264, 265. *See also* SnRK2
S-nitrosoglutathione (GSNO), 52, 53
S-nitrosothiol (SNO), 327
S-nitrosylation, 312, 327
SnRK2, 254, 255, 259, 260, 263–265
Spermidine synthase, 132, 133
Spermine, 133
  biosynthesis, 131, 132
  synthase, 131, 133
26S proteasome, 143, 146, 215, 387, 391
Staurosporine, 198
Stomatal closure immune response, 7, 58, 59, 169
SUMO E3 ligase, 58
SUMO in SA signaling system, 57, 58
Suppression
  of GA signaling by pathogens to cause disease, 396
  of JA signaling by SA, 163
  of PAMP-PRR signaling pathway by BZR1, 434
  of plant immune responses by ABA, 289, 290
  of SA signaling system by JA, 162
Synergism between cytokinin and SA signaling, 374
Synergism between JA and ET signaling systems, 161
Synergism between SA and JA signaling pathways, 88

Synergistic interaction between ABA and ET signaling systems, 285
Synergistic interaction between ABA and SA signaling systems, 281, 282
Systemic acquired resistance (SAR), 1–4, 8, 28, 29, 48, 64, 171, 327, 330, 331, 342, 383
Systemic immunity, 1, 4, 6, 196, 197
Systemic mobile signal molecules involved in SAR, 2
  azelaic acid, 2, 28, 65, 66, 70
  dehydroabietinal, 2, 28, 65, 70
  glycerol-3-phosphate, 2, 23, 65, 66, 68, 69
  lipid transfer protein (DIR1), 2, 28, 65, 66, 68, 69
  methyl salicylate, 2, 28, 65, 66, 68
  methylsalicylate esterase (MSE), 28, 67, 68
  pipecolic acid, 2, 71–73
Systemic movement of ABA, 273
Systemic RNA silencing, 64
Systemin, 138, 139

T
TAA family of aminotransferases, 312
TCP transcription factor, 170
TGA transcription factors, 28, 55–57, 85, 86, 154, 155, 157, 167, 168, 372, 373
TGA transcription factors in SA/JA crosstalk, 85, 86, 167
Thioredoxin (TRX), 52, 53
TIFY motif in ZIM domain, 141
TIR1, 314, 323–325, 327, 328, 332, 338
TIR1/AFB nuclear receptors, 1, 7, 311, 312, 315, 317, 318
TIR1/AFB proteins, 311, 312, 319, 325
TOPLESS (TPL), 123, 141, 142, 144, 315, 317
TOPLESS (TPL) as corepressor of JA signaling pathway, 4
Toxin-mediated suppression of ABA-dependent defenses, 292
Toxins in suppression of SA signaling system, 94, 95
Transgenerational SAR, 1, 3, 28, 80
Trans-zeatin, 360–362, 375
tRNA-isopentenyltransferase (tRNA-IPT), 362
Tryptamine (TAM) pathway, 311, 314
Tryptophan as the precursor for IAA biosynthesis, 312, 313
Two-component phosphorelay system, 359
Type-A AAR in SA-dependent immune responses, 373, 374
Type-A ARR transcription activators, 359, 360, 365, 366, 372, 374, 376
Type-B ARR transcription activators, 359, 360, 365, 367, 372, 376
Type 2C protein phosphatase (PP2C), 1, 6, 245, 246, 254, 255, 258–261, 263, 265, 274

U
Ubiquitiation, 320
Ubiquitin, 316
Ubiquitin-conjugating enzyme, 220
Ubiquitin ligase, 54, 317
Ubiquitin–proteasome in auxin signaling, 316, 318, 344
Ubiquitin–proteasome pathway in ET biosynthesis, 195, 208
Ubiquitin–proteasome pathway in GA signaling, 383
Uniconazole, 11

V
Verapamil, 201
Violaxanthin, 251, 252
Viral replicase–Aux/IAA interaction, 339
Viral suppressor of RNA silencing (VSR), 61–64

W
WIZZ transcription factor, 170
Wound-induced protein kinase (WIPK), 43, 138, 207
WRKY transcription factors, 38, 49–51, 57, 58, 77, 78, 84, 96, 150, 153–155, 163, 165, 166, 170, 206, 273, 290
WRKY18 induced by SA, 50
WRKY33
  in ABA signaling, 250
  ethylene biosynthesis, 206, 207
  in JA/SA crosstalk, 96, 163, 165
WRKY45 in SA signaling, 290
WRKY50 in JA/SA crosstalk, 166
WRKY 51 in JA/SA crosstalk, 166
WRKY62
  induced by SA, 50
  in JA/SA crosstalk, 84, 165

WRKY62 (*cont.*)
in suppression of JA-responsive gene expression, 166
WRKY70
in JA/SA crosstalk, 84, 85, 166, 167
in SA-responsive gene expression, 50, 84

X
Xanthophylls, 252, 253
Xanthoxin, 252–254, 284

Y
YUCCA family of flavin monooxygenases, 312–314

Z
Zeatin degradation, 375
Zeatin riboside (ZR), 361
Zeatin riboside diphosphate (ZDP), 361, 375
Zeatin riboside monophosphate (ZMP), 361, 375
Zeatin riboside triphosphate (ZTP), 361, 375
Zeaxanthin, 251
ZmPep1, 199

Printed by Printforce, the Netherlands